Modern Communications Jamming Principles and Techniques

Second Edition

For a listing of recent titles in the
Artech House Intelligence and Information Operations Series,
turn to the back of this book.

Modern Communications Jamming Principles and Techniques

Second Edition

Richard A. Poisel

BOSTON | LONDON
artechhouse.com

Library of Congress Cataloging-in-Publication Data
A catalog record for this book is available from the Library of Congress.

British Library Cataloguing in Publication Data
A catalogue record for this book is available from the British Library.

ISBN-13 978-1-60807-165-4

Cover design by Vicki Kane

685 Canton Street
Norwood, MA 02062

10 9 8 7 6 5 4 3 2 1

To Jacob, Harrison, and Alexandra—may their futures be as bright as their smiles

Contents

Preface to the Second Edition

The first edition of this book was written with the intent of providing a source of technical information for the analysis and design of countermeasure systems against modern communication signals—focusing on antijam (AJ) signals primarily. Since the publication of that book, the focus of modern warfare involving EW, as well as most other military disciplines, has shifted from principally open terrain to urban terrain.

That is not to say that warfare in urban terrain is a new discipline. Such settings have been a part of battles since there have been battles. The focus of recent military study, however, has been on large, massive formations. Only within the last 10–20 years or so has the emphasis shifted principally to urban environments.

We have learned that the thoughts behind adapting traditional military strategies involving large forces to urban environments are more difficult than expected. Urban war fighting is tougher than originally anticipated.

This edition of the book focuses much more on urban warfare than the first edition. We have included more information about the impacts of urban warfare on electronic warfare (EW). While the general concepts of AJ signals have not substantially changed since the first edition, what has dramatically changed is the nature of personal wireless communications, which is the single largest type of communications expected to be encountered in an urban setting. The mobile telephone usage has exploded and is especially noticeable in urban environments. The new standards for mobile phones are all based on AJ radio technology, and these communication devices have become of principal interest to military EW operations. Probably the most obvious of these uses is as remote triggers for IEDs.

While the EW cognoscenti have long recognized the importance of the contribution of EW to modern warfare, EW has been sagaciously integrated into modern warfare thinking. It is an indirect weapon that has temporary effects. One of its biggest attributes is that the fratricide produced rarely involves the unintended recipient, as opposed to kinetic weapons that inevitably produce unwanted effects in urban settings.

This book is divided into four broadly defined sections. The first six chapters provide technical details about communication signals in general. After the introduction in Chapter 1, the detection of signals in the presence of RF noise is discussed in Chapter 2. In Chapter 3 we provide an introduction to modern signaling techniques with an emphasis on AJ. In Chapter 4 we focus our attention on the detection of AJ signals in particular. Chapter 5 contains a general discussion of RF signal propagation. Since AJ communications relies on pseudorandom code sequences, which are typically generated by shift registers, we introduce shift registers and their characteristics in Chapter 6.

All modern AJ communication systems must be synchronized in some fashion in order to communicate. In fact, as discussed in Chapter 8, attacking the synchronization process is one of the more effective ways to conduct countermeasures. Therefore, in Chapter 7 we introduce the more common methods that AJ systems synchronize themselves. Furthermore, after synchronization, they must maintain tracking in order to stay in sync. That topic is also covered in Chapter 7. Chapter 8 presents techniques for jamming modern communication signals. We slightly divert our attention in Chapter 9 to discuss techniques for blindly discovering CDMA codes in use by a communication system.

The next five chapters focus on EW methods and methodologies against particular spread spectrum modulation types. Chapter 10 examines EW and direct-sequence spread spectrum modulations while Chapter 11 discusses EW and fast frequency-hopping communication systems. Chapter 12 focuses on slow frequency-hopping target networks. Chapter 13 goes into depth on EW and the relatively new form of spread spectrum signaling, ultrawideband communication systems. Last in this section, Chapter 14 addresses EW and hybrid spread spectrum systems.

The next four chapters focus on discussions of the urban environment. Chapter 15 introduces the urban environment in general. In Chapter 16 we discuss the peculiarities of RF signal propagation in urban settings. Chapter 17 focuses on EW in urban terrain, along with some operational computer simulation results. Chapter 18 introduces a concept for blindly detecting and geolocating CDMA targets in urban (as well as suburban and rural) settings.

The indented audience for this book is technical people (engineers and scientists) just beginning in the field of EW system analysis and design, as well as practicing professionals in that same field. There is considerable mathematics involved, but there are few derivations. The theorem-proof construct has been avoided, primarily because this is not a book about mathematical foundations but one intended to provide useful, pragmatic solutions to real-world problems in system design and analysis. Adequate references, it is hoped, are provided in each particular area if a reader would like to delve deeper into the subjects or to examine the derivations.

This book is not intended to be a textbook—there are no chapter problems to work out. Rather it is intended as a one-source reference covering all of the major topics associated with EW systems. Inevitably, however, there are areas that are overlooked, and, in some cases, left out completely. Jamming of classical waveforms such as analog AM and FM is not included, for example.

Of course, there is no classified material included. All of the material contained in the text was obtained from open sources—much (most) from the various IEEE journals on communications and related subjects.

There is bound to be mistakes in the material. That is almost inevitable in a work this voluminous. Constructive feedback on these mistakes will always be welcome.

Preface to the First Edition

This book is intended to be a technical reference for engineers and scientists working in the field of communication jamming. The vast majority of the technical literature that applies to this field is contained in the literature for communication system and signal design. Not the least reason for this has been the importance of maintaining adequate national security.

The modern communication signals of concern herein are antijam signals, comprised primarily, but not exclusively, of direct-sequence spread spectrum (DSSS) and frequency-hop spread spectrum (FHSS) signals. These signaling techniques provide certain amounts of protection from intercept and jamming, but they are vulnerable, nevertheless. Their degree of susceptibility to electronic attack (EA) is presented herein.

Several books have been published that take the point of view of how to design communication systems to avoid intercept or jamming; antijam refers to the latter of these. While the fundamentals are the same whether designing a communication system or an anticommunication system, the points of view are different, which leads to subtle but important differences.

Since about 1980, the technical literature has been prolific on determining the theoretical performance of AJ communication systems. These sources have never before been compiled into one source from the point of view of countering those communications, however. That was the intent in the preparation of this book. Technical personnel new to the anticommunications subject area had to search several places and make the "communication centric" to "anticommunication centric" translation to use these sources.

The primary focus of the author has been military anticommunication systems. Thus, this book has a distinctive military aura to it. AJ communications, however, have infiltrated the commercial communication market in a big way in the form of cellular telephone systems and wireless local area networks. Therefore, much of the material herein should be of interest to those involved in the design of such structures.

Chapter 1 serves as an introduction to jamming modern communication signals. The basic signaling types are defined and the notion of tactical communication networks is explained.

Chapters 2–7 serve as background information for the later chapters. Chapter 2 addresses signal propagation in the HF, VHF, and UHF frequency ranges. The basics of the physical processes involved with propagation are included as are some mathematical models.

Chapter 3 presents signaling techniques normally employed in AJ communications. The performance of these techniques in noise, both background and jamming, is included. Although it is possible to design AJ systems that employ analog communications, most realistic systems are designed around digital signals. Therefore, analog signals are not included herein. Forward error correction (FEC) coding is a technique employed to counter channel noise and jamming. This chapter shows the effects of excessive noise on systems that employ coding and the resultant vulnerabilities. Chaotic-based systems and ultra-wide band systems are briefly introduced.

All forms of spread spectrum communications rely on pseudorandom sequences to generate random codes. These sequences are generated with shift registers. Chapter 4 briefly introduces the concepts of linear feedback shift registers and linear recursive sequences.

Chapter 5 provides information on synchronization and tracking in spread spectrum systems. As explained in the later chapters, the synchronization and tracking requirements are lucrative targets for electronic attack (EA).

Chapter 6 introduces the basic jamming techniques that are typically employed in jammers conducting EA on AJ targets. These techniques include broadband noise jamming, partial band noise jamming, tone jamming, and pulse jamming.

Most EA activities against any kind of radiating target first require the measurement of the RF spectrum to determine the presence and characteristics of the signal that is to be attacked. Chapter 7 introduces the elements involved in the detection of AJ targets. The results of a limited simulation are included to indicate in a typical case how well such targets can be detected.

Chapters 8–11 address jamming each of the fundamental AJ signal types. Chapter 8 addresses the performance that can be expected when jamming DSSS signals.

Chapter 9 contains the expected performance of EA when the targets are FHSS fast frequency hopping. Chapter 10 is the same for FSHH slow frequency hopping targets.

Finally, Chapter 11 addresses EA performance when the targets are hybrid combinations of DSSS and FHSS types.

The material in the book consists of a considerable amount of mathematics, although derivations are, for the most part, excluded. The performance equations are presented without the backup derivation—the references can be consulted if information on the derivations is desired. Nevertheless, a background in basic calculus usually provided in engineering and physics undergraduate curricula is generally required to understand most of the material. In addition, an undergraduate level of understanding of communication theory and systems is useful.

Chapter 1

Modern Communications and Electronic Countermeasures

1.1 Introduction

The advent of the information age has brought about considerable reliance on wireless electronic communications. Although the cellular phone systems and personal communication systems are bringing wireless *radio frequency* (RF) communication to the masses, nowhere is this reliance more evident than in the military. For years, the military has depended upon RF communications for execution of command and control of tactical forces.

Since tactical commanders use RF communications to exercise control of their forces, an adversary has interest in those communications. This interest lies in two fundamental areas: (1) to intercept the information that transpires over them, and (2) to deny the successful exchange of the information from the sender to the receiver. The former of these provides to the interceptor information about the status and intent of the adversary and is intelligence or combat information. The distinction between these two has to do with for what the information is used and is not important here. The latter is the subject of this book.

1.2 Electronic Warfare

Communication *electronic warfare* (EW) is the name applied to activities taken to accomplish the intercept or denial of communications. It consists of three main components:

- *Electronic attack* (EA);
- *Electronic support* (ES);
- *Electronic protect* (EP).

EA is the new appellation for what used to be called *electronic countermeasures* (ECM). It is the use of active signals to prevent a communication system from effectively exchanging information. It is generally (although not exclusively) accepted that EA consists of three principal activities: (1) jamming, (2) deception, and (3) *directed energy* (DE).

Of the three principal tenets of information, relevance, accuracy, and timeliness [1], jamming is primarily intended to address the last. If information is successfully exchanged, there is little that jamming can do to impact directly the relevance and accuracy of that information. Jamming activities, however, can impact on the timeliness of the information exchange by, at least temporarily, slowing that exchange. Jamming can also affect the relevance of information, because if it arrives at the intended destination too late to be of use, the information has becomes irrelevant.

Deception addresses the second of these information tenets. The intent is to mislead an opponent by creating a ruse. False communication signals are an important part of any tactical deception activity. Little more will be said about deception herein, as that is a broad subject on its own.

Application of directed energy is similar to jamming except the goal is to permanently harm or destroy the communication equipment. It requires significantly larger amounts of energy or power than jamming.

ES is a supporting function to EA. It is used to measure parameters of the RF spectrum to ascertain the presence (detection) or characteristics of signals. Energy and time are wasted if jamming attempts are made on signals that are not there.

EP consists of the efforts taken to preclude some adversary from performing EA and ES on friendly communications. An example of EP is the screening of friendly communications from adversarial attempts at ES. By broadcasting signals toward an adversary and away from friendly communication networks but at the same frequency or frequencies in use by friendly forces, the successful ES attempts can be thwarted. Other examples of EP are *emission control* (EMCON) and encryption.

This description of EP is obviously focused on communications, but there are other forms of EP—aircraft self-protect equipment (ASE), where radar detection and other forms of sensing are used to determine when an aircraft is being targeted, for example, is certainly a form of EP. We are not concerned with such applications here.

The question naturally arises as to how well communication jamming can accomplish information transport denial when AJ techniques are employed. That is the subject of this book. The technical literature about communication AJ techniques began to appear in unclassified form about 1980. Prior to that, the extensive U.S. Department of Defense investment in the technology was, for the most part, classified and unavailable to the general public.

Our focus will be primarily on the RF interfaces where EW may be applied—the so-called layer 0 in the ISO data link model, otherwise known as the physical layer. EW at this layer attempts to deny the receipt of signals by injecting signals of our choosing into communication systems that prevent those systems from accurately receiving the information intended.

1.3 Antijam Communications

Whether the intent is to intercept communications or to deny the same, in an adversarial relationship there is obviously an interest to preclude success. *Anti-jam* (AJ) communications techniques were developed to facilitate communications when an adversary has an interest in denying the communicator the ability to communicate. Such technology also helps in the anti-intercept as well, but that is not the subject here.

Since that time, several books have been written on AJ communication techniques and systems. The focus of these sources has been on how to design communication systems to defeat countermeasure techniques—how to communicate in the presence of intentional jamming. The focus in this book is the opposite—how well can a jammer perform when the targets to be jammed employ AJ technology. The underlying physics are the same in both circumstances; what is different is the goal of the utilization of the technology.

Efforts have been expended to develop ways to avoid this interruption—at least the intentional ones, anyway. When the goal is to avoid interception or detection at all, then the techniques are called *low probability of intercept* (LPI) or a similar appellation. When the goal is to be able to communicate even in the presence of intentional interference (jamming), then the techniques are called AJ.

One of the methods to thwart such tactical operations of an adversary is to deny communication over these nets by conducting electronic countermeasures, in this case, by jamming them. This jamming is accomplished by emitting energy toward the receiver at the same frequencies as the adversary nets whenever there is an attempt to communicate.

The majority of the examples and specific discussions herein are focused on land-mobile communications. The results may appear to apply more to the U.S Army and Marines in the military. That is due to the proclivity of the author to focus on such applications. The technologies are fundamental, however, and apply to all military services—indeed, to all who use RF communications. To the extent that *AJ electronic countermeasures* (AJ ECM) are an issue, the results apply to commercial communication situations as well. In particular, much of the discussions are directly applicable to the aforementioned cellular phone and personal communication systems. Perhaps the most obvious of these is when the

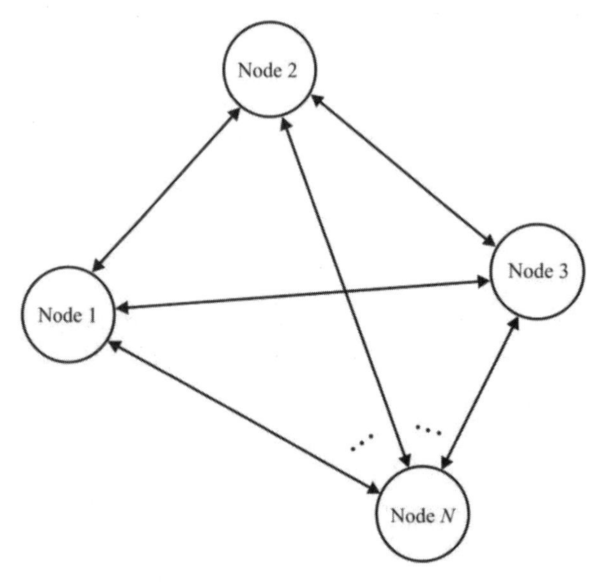

Figure 1.1 Tactical RF communication network in a many-to-many configuration.

military must conduct operations in a situation where the adversary employs commercial communication means for their command and control [2, 3].

1.4 Networks

Tactical military command and control are usually accomplished by RF communications. These communications are normally accomplished with *networks,* frequently referred to as *nets*. In this way, many nodes can communicate with each other as required. These nets are normally configured ahead of time and each member of the net knows who the other members are. If these communications are denied, by either intentional interference or any other means, then the command and control process is interrupted and operations can be affected.

There are several forms of networks. A depiction of a tactical RF many-to-many net is shown in Figure 1.1. This is typical for tactical *push-to-talk* (PTT) networks where all the personnel in a squad, for example, can communicate with one another. It is also possible to have one-to-many/many-to-one networks. These are depicted in Figure 1.2. This configuration is typified by *very small aperture satellite* (VSAT) networks, where the one node is the hub of the network. A

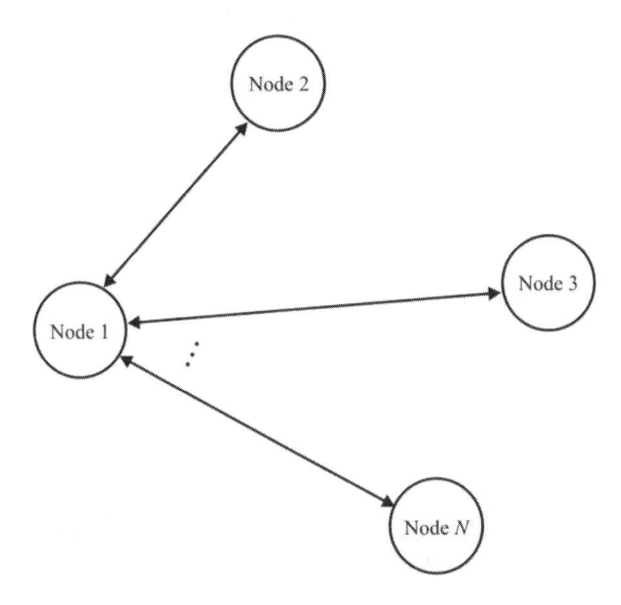

Figure 1.2 Configuration of a one-to-many and many-to-one networks, such as a VSAT network.

network can be one-to-one, as illustrated in Figure 1.3. Examples of this type of network are cell phones, *personal communication systems* (PCSs), and older forms of the *public switched telephone network* (PSTN) when in use. For this type of network, all the nodes can communicate with one another, but normally only two at a time. All of these forms of networks are susceptible to jamming.

When we talk in modern parlance about networks, we obviously must include the network in the sky: the Internet. It is ubiquitous in almost all parts of the world and used by billions of people. The Internet does not follow the models described above, but provides for anybody-to-anybody communications as well as many-to-many communications (see Figure 1.4). We will discuss various aspects of communication over the Internet in this book.

1.5 Spread Spectrum Technology

Ab initio, spread spectrum communications technology was developed by the U.S. Department of Defense, in the United States anyway, as a way to thwart detection, exploitation, and countermeasures by adversaries. These communication technologies are rapidly moving out of the strictly military domain into commercial applications. One of these is the *code division multiple access*

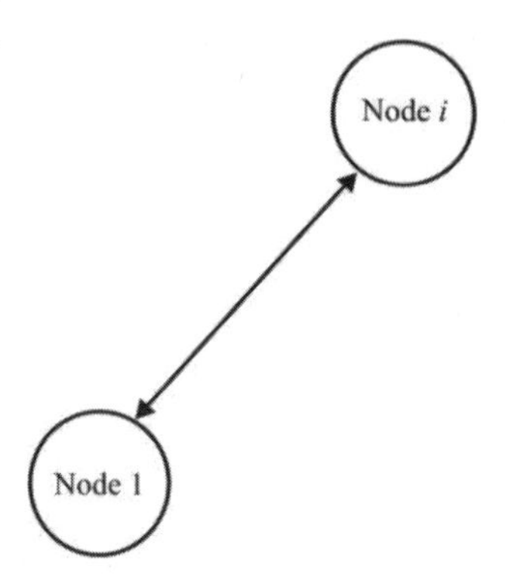

Figure 1.3 One-to-one network.

(CDMA) *spread spectrum* (SS). Another example uses frequency hopping to achieve frequency diversity. These commercial applications of SS technology will cause these capabilities to be around for a considerable time to come.

SS communication technology was developed as a communication technique to provide some degree of *electronic counter-countermeasures* (ECCM) for the communicator (given that jamming is an ECM technique). It represents one of the LPI, *low probability of exploitation* (LPE), or AJ techniques. A traditional single-tone jammer has little effect on the performance of such systems, forcing the

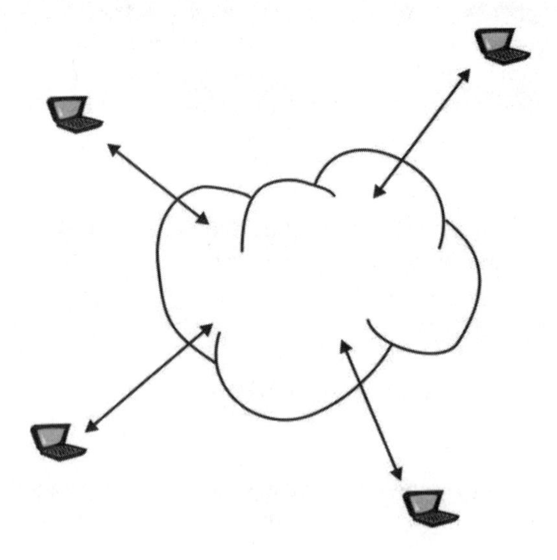

Figure 1.4 Internet interconnectivity.

jammer to adopt different schemes of attack. At the very least, a jammer must be concerned about a much broader frequency range.

One of the advantages of DSSS technologies is the ability to reuse the frequency spectrum. This is true for commercial wireless communicaitons as well. Such communications overlay one another in the frequency domain and allow many users to share the same frequencies. CDMA is facilitated by each user having a different code to spread its waveform.

The analysis of jamming techniques and strategies is different from that focused on successful communication techniques in many subtle ways. The biggest difference, however, is the *bit error rate* (BER) regime analyzed. Pundits of communication strategies focus on obtaining BERs lower than around 10^{-3} (one bit in error out of 1,000, on average) and develop techniques for successful communications in that range. ECM techniques attempt to raise that to 10^{-1} or higher and develop strategies for countering communications in that range. It will be shown in Chapter 3 that if this level of BER can be achieved, then a jammer can be successful against AJ targets.

Strictly speaking, AJ communication technology refers to the ability to combat jamming of a communication system. Being totally free from the effects of RF jamming in a wireless communication environment is an unrealistic goal. Given the right circumstances, all RF systems can be jammed. Common techniques for the implementation of AJ consist of ways to hide a signal so an interceptor or casual interloper does not know it is there, to move the signal around rapidly in the frequency spectrum so that traditional narrowband intercept receivers do not see the signal, and to have redundancy coding of digital signals. This last category was not initially developed to counter jamming but to combat noise effects on digital signals. If the jamming signal resembles thermal noise, then those techniques are effective against jamming as well.

Spread spectrum technologies form the basis for all of the modern communication techniques we will discuss. We present here a brief introduction to the spectrum spreading approaches and techniques. Details will be presented in later chapters.

1.5.1 Low Probability of Detection

In *low probability of detection* (LPD) systems, the goal is to hide the signal somehow so that an unintended receiver has difficulty determining that the signal is even present. There are many potential reasons for doing so. In a military setting, it might be desirable to be able to communicate in a particular city without anyone knowing the presence of the forces. DSSS is an example of a LPD technology.

1.5.2 Low Probability of Intercept

If a signal cannot achieve LPD, then, by definition, an unintended receiver can detect the presence of the signal. It is still possible to provide some protection of signals, however. They can be made to be difficult to intercept, and in such cases, the signals are referred to as LPI. Frequency-hopping, described later, is an example of an LPI technology.

1.5.3 Low Probability of Exploitation

In those cases when achieving LPD and/or LPI are either too difficult or unneeded, it still might be desirable to avoid someone finding out what information is carried in the signal. The signal might be a voice conversation, for example, and the content of the conversation might be sensitive to the people doing the communicating. In that case, it is said that the denial of exploitation of the signal is the goal and such technologies are referred to as LPE. Encryption is an example of an LPE technique.

In some places, LPE is defined to include LPI and LPD [4]. That will not be the case herein.

1.5.4 Antijam

This type of signal is designed to make it difficult for a jammer to be used effectively. In fact, the techniques used to achieve antijam signals are the same as those to achieve LPD, LPI, and LPE. In some cases, design parameters may be adjusted to optimize AJ over LPD, LPI, or LPE, but the underlying principles are the same.

1.6 AJ Signal Types

While there are more types of AJ communication techniques, the two predominant ones in widespread use are *direct sequence SS* (DSSS) and *frequency hopping SS* (FHSS). A third type, called *time hopping* (TH), is also available and, as a technique to achieve AJ, it is beginning to emerge as a viable technique.

1.6.1 Direct-Sequence Spread Spectrum

DSSS systems spread the information-bearing digital signal across a broad bandwidth and that entire bandwidth is occupied instantaneously—that is, the signal is spread over all the bandwidth at the same time. Taking a bounded energy data signal and spreading that energy across a very wide bandwidth causes the energy present at any particular frequency or small frequency band to be miniscule. Often it is so small as to be below the thermal noise at that frequency. Receivers that simply examine the spectrum at the appropriate frequency of operation of such communication systems will mistake the signal as noise and miss detection. Special signal processing is required to extract the signal.

1.6.2 Frequency-Hopping Spread Spectrum

As opposed to DSSS, in FHSS systems the narrowband data signal at any given instant only occupies a single channel, usually narrowband. In the low VHF band[1] traditionally the frequency channelization has been 25 kHz, although this is getting narrower. Thus, the FHSS system at any instant is occupying this bandwidth. In the low VHF frequency band there are about 2,400 such channels available, and systems are usually designed to use some subset of these. Customarily, due to implementation convenience, the subset, referred to as the *hop set*, is sized to be a power of two.

FHSS can be further divided into *fast frequency-hopping spread spectrum* (FFHSS) and *slow frequency-hopping-spread spectrum* (SFHSS). This distinction is normally based on the number of data bits sent on a particular hop dwell. If there are multiple data bits on a hop, then it is called SFHSS, while if there are multiple hops for each data bit, then it is called FFHSS. The number of data bits per hop in SFHSS sytems herein will be dentoted by L_S while the number of hops per bit in FFHSS systems will be denoted L_F. The dividing line beween SFHSS and FFHSS is one bit per hop or, equivalently, one hop per bit.

FHSS has the advantage of frequency diversity. Transmitting the same information at different frequencies increases the probability of the information arriving at the receiver correctly since the multiple possible paths the signal can take from the transmitter to the receiver cause fading. This fading is frequency dependent, and, if either the transmitter or receiver is moving, this fading changes with such movement.

It is common for digital frequency-hopping communication systems to use *frequency shift keying* (FSK) as the modulation of choice, and, in particular, *binary FSK* (BFSK) modulation with incoherent detection at the receiver. In BFSK, a data bit is sent as a tone at one of two frequencies. These tones are typically offset by some amount above and below a carrier frequency that is constantly changing

[1] The VHF frequency band is used for much of the military PTT communications throughout the world. While the VHF band consists of all frequencies from 30 to 300 MHz, the low portion used for this purpose is usually restricted to 30 to 88 MHz. Frequencies above 108 MHz are also used extensively by the military and commercial air services for air traffic control. In the United States and many other countries, the band from 88 MHz to 108 MHz is restricted to commercial FM stations.

locations in the frequency spectrum [5]. We discuss FSK and other modulation techniques in detail in Chapter 3.

1.6.3 Time-Hopping Spread Spectrum

Under a broad range of conditions, the optimum way to detect a spread spectrum signal is with a *radiometer*. This device measures the energy in the bandwidth of the radiometer for a length of time called the integration time. At the end of that time, a decision is made as to whether a mark or space was sent. TH moves the time of transmission randomly, thereby causing the radiometer to measure noise much of the time.

On a simpler basis, regular PTT communications used by militaries for years is a form of time hopping since the time between transmissions on a network is random. A receiver trying to listen to these communications, tuned to the frequency of the network, would listen to noise much of the time.

One modern form of communication system design, *ultrawideband* (UWB) communications, capitalizes on time hopping to achieve AJ protection. It also allows for multiple users to share the same spectrum, as in DSSS and FHSS. Although this technology need not be limited to short-distance networking, at the time of this writing that appears to be the primary application. Considerable radiated power is required for cases where substantial distances are to be accommodated. UWB technology spreads the signal over large bandwidths, which can interfere with other types of communications in the same spectrum.

1.6.4 Hybrids

Combinations of the AJ techniques can be and have been put together to gain the advantages of each. The most common form of a hybrid is the combination of DSSS with FHSS. In this combination, the stealthy nature of DSSS can be exploited as can the frequency diversity of FHSS. The baseband digital signal is first DS spread, creating a (relatively) wideband signal. That signal is then frequency-hopped around the spectrum using the hop set frequencies. Such signals are hard to detect and communication is more reliable than either modulation technique alone.

1.7 Synchronization

Due to the nature of tactical military communications, it is necessary to accommodate radios dynamically entering and leaving the nets. Even though the nominal composition of a net is normally established ahead of time, there may be instances when it is necessary for a new node to be able to join. This is especially true with self-forming or so-called ad hoc networks, which are designed to facilitate network participants coming and going dynamically. In addition, communication over tactical networks is intermittent. When there is no exchange, oscillators tend to drift. Thus, synchronization is an important issue in these nets. It is not so difficult for non-SS networks. A transmitter just transmits on the correct frequency at a time when no other is using the channel. For FHSS networks, however, the new transmitter does not normally know the frequency the net is using at that moment. For DSSS networks, the new transmitter does not know the point in the spreading sequence the net is located. These produce special needs with such nets that do not exist in non-SS communications.

1.8 Communication System Model

The model of the communication system with the jammer present used here is shown in Figure 1.5. The signal from the transmitter propagates through the atmosphere to the intended receiver as well as to the jammer ES system receiver, if there is one. That signal is perturbed by thermal noise and possibly other interference, which need not necessarily be the same for the two propagation paths (in fact, in most cases it will not be the same). At the intended receiver, the signal is processed with the intent of extracting the information that was intended to be sent by the transmitter. At the jammer, the received signal is processed in several ways. Based on this processing, the jammer transmits a signal in the direction of the communication receiver with the intent of precluding that receiver from accurately extracting the information.

The jammer is attempting to interfere with the receiver, not the transmitter. As such, the jammer must have some idea where the receiver is located. In tactical communication situations where networks are the primary communication means, each node typically transmits at least once during some time period. Those transmissions can be intercepted and the transmitters located to provide this information.

1.9 Urban Electronic Warfare

As we move further into the twenty-first century, U.S. and coalition forces are finding themselves much more involved with warfare in urban settings. Because of this we have significantly expanded the coverage of the technical issues involved with EW in such settings.

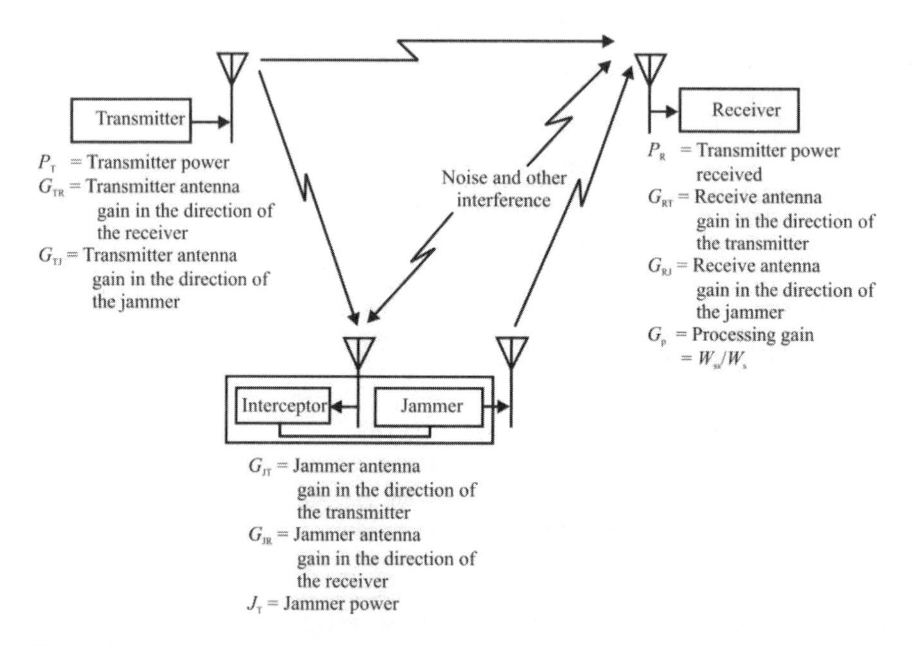

Figure 1.5 Communication system model.

In particular, as we explain in depth later, the main goal of EW in urban settings is to facilitate isolation of the particular urban area or a portion of it and/or a group of people. Such isolation helps to develop/maintain separation of foes from indigenous, gray[2] populations as well as from supporting forces.

1.10 Concluding Remarks

The notions underpinning the techniques and methods to conduct EA against AJ and other communication targets are presented in the remainder of this book. AJ signaling was developed primarily to preclude such jamming, and considerable resources have been expended to make AJ techniques effective. That does not, however, negate the benefits to be obtained by countering communications, even if the targets are employing such techniques.

In spite of the development of approaches to minimize the effects of jammers against AJ techniques, it is still possible to impact such communications. How to do this and how well it can be accomplished are presented herein.

[2] Gray populations are noncombatants in a region.

CDMA [6] is no different from DSSS—it is the application of the technology that is different. With each user of the spectrum within the range of a single receiver possessing a different code sequence, they all can share the spectrum with minimal interference with one another, within some constraints.

This book presents some common techniques for conducting EA against AJ communication systems. The first topic covered is the nature of detecting communication signals in noise. That is followed by discussions about AJ signaling techniques and how these signals in particular can be detected in noise. Next, the characteristics of RF signal propagation are presented in Chapter 5. Chapter 6 covers shift registers and discusses the recursive sequences that ensue from them. Chapter 7 focuses on the methods of synchronization and tracking of AJ signals. We present the major forms of jamming techniques in Chapter 8. This is followed in Chapter 9 by a presentation of methods of blindly discovering the codes used in CDMA systems. Chapters 10–14 discuss jamming of the forms of SS signaling. Chapter 10 covers DSSS sysems, Chapter 11 focuses on FFHSS jamming, and Chapter 12 covers SFHSS jamming. THSS, and, in particular, jamming UWB signals, is discussed in Chapter 13. Hybrids of these forms of SS are presented in Chapter 14. The characteristics of EW in urban settings is discussed in Chapters 15–17. Finally, blind detection of CDMA signals in urban environments is discussed in Chapter 18.

References

[1] Alberts, D. S., J. J. Garstka, and F. P. Stein, *Network Centric Warfare Developing and Leveraging Information Superiority,* 2nd ed., Washington, D.C., DoD C4ISR Cooperative Research Program, August 1999.

[2] Pickholtz, R. L., D. L. Schilling, and L. B. Milstein, "Theory of Spread-Spectrum Communications—A Tutorial," *IEEE Transactions on Communications,* Vol. COM-30, No. 5, May 1982, pp. 855–884.

[3] Scholtz, R. A., "The Spread Spectrum Concept," *IEEE Transactions on Communications,* Vol. COM-25, No. 8, August 1977, pp.748–755.

[4] Nicholson, D. L., *Spread Spectrum Signal Design LPE and AJ Systems,* Rockville, MD: Computer Science Press, 1988, pp. 2–7.

[5] Hedge, M. V. and W. E. Stark, "Capacity of Frequency-Hop Spread-Spectrum Multiple-Access Communication Systems," *IEEE Transactions on Communications,* Vol. 38, No. 7, July 1990, pp. 1050–1059.

[6] Landoisi, M. A. and W. E. Stark, "DS-CDMA Chip Waveform Design for Minimal Interference Under Bandwidth, Phase, and Envelope Constraints," *IEEE Transactions on Communications,* Vol. 47, No. 11, November 1999, pp. 1737–1746.

Chapter 2

Detection of Signals in Noise

2.1 Introduction

This chapter presents background material on how noise affects the signal detection process in EW systems. In communication EW system analysis and design, the word *detection* can mean different things. For example, when considering digital communication systems where the information is transported in the form of symbols, detection refers to the determination of which symbol was transmitted. When the system under consideration is a communication EW system, detection may refer to that, but it also refers to the determination if there is a signal present in a specified region of the frequency spectrum. In this chapter, indeed, in the entire book except Chapter 3, it is this latter definition that is used.

Such detection is normally accomplished by downconverting the signal at the antenna to some convenient *intermediate frequency* (IF), which is then filtered and amplified to suitable levels for subsequent processing. For purposes of signal detection, it is determination of the signal power or amplitude which indicates presence or not. Therefore measurement of the signal amplitude is the apropos parameter. This can be done prior to any demodulation, but it is more common to amplitude demodulate the filtered signal prior to calculating the level. If there is a signal present with any type of modulation (including none—just the carrier), there is a nonzero level of AM detected. The mathematics behind this technique is well understood and systems with known performance can be designed this way.

This chapter is structured as follows. We begin by defining what we mean by a signal and receiver. We then describe some statistical detection theories that are in common use. *Probability density functions* (PDFs) familiar to the analysis of noise amplitudes are described next. We then discuss radiometric detection of signals in noise, along with radiometer performance under nonideal conditions such as nonstationary noise, uncertainty in the knowledge of the thermal noise level, and non-Gaussian (impulsive) noise. We conclude the chapter with an in-depth analysis of urban noise sources and their impact on EW in such an environment.

2.2 Signal Structure

When there is a signal present at the receiving system, it can be represented as

$$r(t) = \text{Re}\{A\tilde{r}(t)\exp[\omega_0 t + \varphi(t)]\} \tag{2.1}$$

where A is the amplitude, for the moment assumed to be a constant, $\tilde{r}(t)$ is the normalized complex envelope, ω_0 is the carrier frequency, and $\varphi(t)$ is the carrier phase angle, which may or may not vary with time depending on the modulation of the signal. The signal detection problem considered here is to decide between the two hypotheses

H_0: $r(t) = n(t)$
H_1: $r(t) = s(t) + n(t)$

where $n(t)$ is an additive noise term whose amplitude PDF is Gaussian with zero mean and variance σ^2, denoted $\mathcal{N}(0, \sigma^2)$. In many cases of interest $n(t)$ has a flat two-sided *power spectral density* (PSD) given by $N_0/2$ Watts/Hz. In other important cases $n(t)$ has other PSDs.

Note that an alternate expression for (2.1) is given by

$$\begin{aligned} r(t) &= \text{Re}\{A[\tilde{r}_\text{r}(t) + j\tilde{r}_\text{i}(t)][\cos[\omega_0 t + \varphi(t)] + j\sin[\omega_0 t + \varphi(t)]]\} \\ &= A\,\text{Re}\{\tilde{r}_\text{r}(t)\cos[\omega_0 t + \varphi(t)] - \tilde{r}_\text{i}(t)\sin[\omega_0 t + \varphi(t)] \\ &\qquad + j[\tilde{r}_\text{r}(t)\sin[\omega_0 t + \varphi(t)] + \tilde{r}_\text{i}(t)\cos[\omega_0 t + \varphi(t)]]\} \\ &= A\{\tilde{r}_\text{r}(t)\cos[\omega_0 t + \varphi(t)] - \tilde{r}_\text{i}(t)\sin[(\omega_0 t + \varphi(t)]\} \end{aligned} \tag{2.2}$$

2.3 Receiver Structure

A receiver used for the detection of signals is typified by the simplified block diagram shown in Figure 2.1. The antenna signal would typically feed a *low noise amplifier* (LNA), the reasons for which are described Section 2.5. Associated with the LNA would typically be some sort of suboctave or bandpass filter (suboctave is shown). The purpose of this filter is to minimize distortion in the signal caused by second harmonics. That signal is then converted in frequency to some convenient IF, f_IF, which is subsequently amplified and filtered in several IF stages. There may even be additional stages of frequency conversion in the IF processing. After this amplification and filtering, the signal is detected in an envelope detector, producing as the output, the amplitude of the signal at the detector input. The envelope is compared with the threshold γth and if the

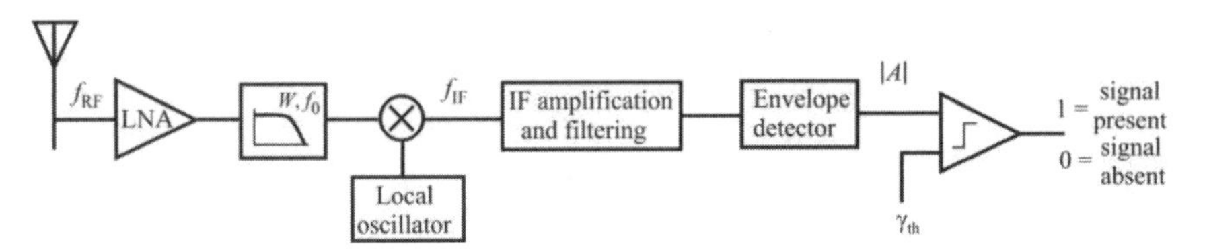

Figure 2.1 Receiver structure.

threshold is exceeded, a signal presence is declared. If the threshold is not exceeded, then a signal absence is declared.

This processing typically would be for only a limited time segment $(0, T)$ because the receiver would normally be retuned to look for signals in other parts of the spectrum. The part of the continuous communication signal that is measured in $(0, T)$ is called a *signal segment* herein.

2.4 Binary Decision Theory

The decisions that can be made when there are only two choices, as is the case here (signal present and signal absent) are indicated in Figure 2.2. If the signal is not present and the decision is made that it is present, then a false alarm has occurred. If absent and declared absent, then that is a correct decision. On the other hand, if the signal is present and it is declared present, then that too is a correct decision. Lastly, if the signal is present and it is declared absent, then a missed detection has occurred. The probability of false alarm is denoted P_{fa} while the probability of miss is denoted P_m. Note that the *probability of detection* is given as

$$P_d = 1 - P_m \tag{2.3}$$

This can be seen with the aid of Figure 2.3, which shows the PDFs of noise only

		Declare	
		Present	Absent
Actual	Absent	False alarm	Correct
	Present	Correct	Miss

Figure 2.2 Decision matrix.

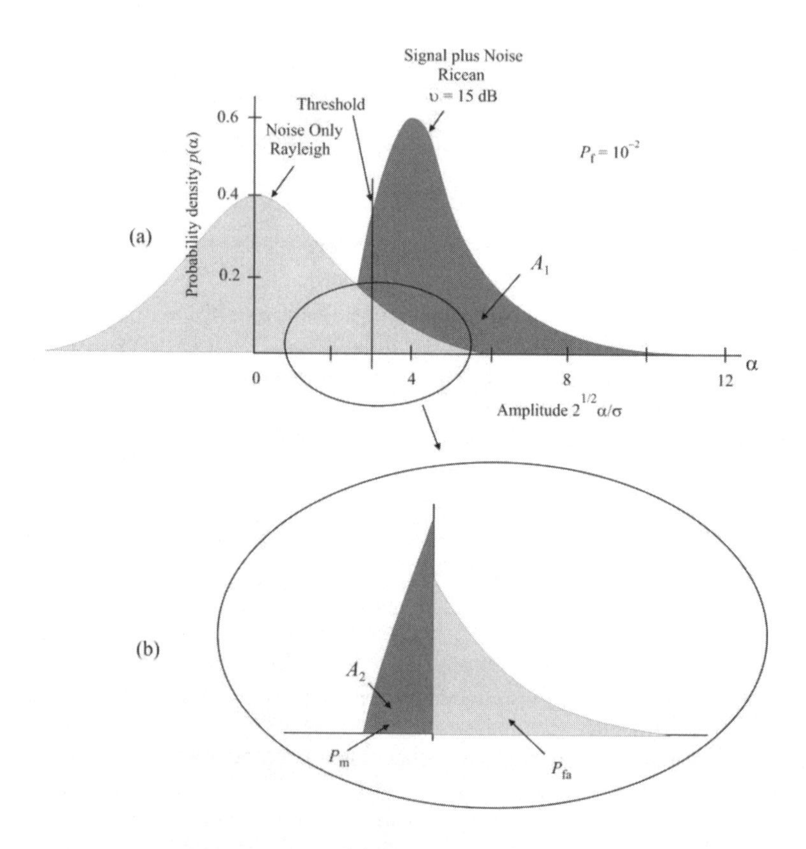

Figure 2.3 Setting the detection threshold determines P_d and P_{fa}. (a) Shows the two PSDs while (b) shows the detail of where they overlap.

and signal plus noise in Figure 2.3(a). The light shaded area in Figure 2.3(b) corresponds to the probability of false alarm because there is only noise present, yet the signal level is above the threshold, and therefore signal present will be decided. Likewise, the dark shaded area in Figure 2.3(b) is the probably of miss because the signal level is below the threshold and therefore no signal is declared, even though the signal is present. The probability of detection is given by area A_1, the whole area under the darkly shaded curve in Figure 2.3(a) less the darkly shaded area in Figure 2.3(b), $A_2 = P_m$. Since for any PDF

$$A_1 = \int_{-\infty}^{\infty} p(x)dx = 1$$

and proper detection occurs when the signal is detected when it is present, then the probability of detection is given by $P_d = A_1 - A_2 = 1 - P_m$. It is clear from this discussion that P_d and P_{fa} are determined by where the threshold is set.

2.4.1 Statistical Signal Detection Theory

The signal is corrupted with external and internal noise so the data from the detector will be a random process. Consequently, the problem of detecting the presence of the signal from the noise is a statistical one. The idea behind the signal detection is that the presence of the signal changes the statistical characteristics of the envelope of the signal (2.2), in particular, its probability density and distribution. When the signal is absent, then the amplitude has the PDF $p_n(a)$, and when the signal is present, the PDF is $p_{s+n}(a)$.

The problem of detecting the signal in noise can be posed as a *statistical hypothesis testing* problem. The *null hypothesis* H_0 is that the signal is absent from the data and the *alternative hypothesis* H_1 is that the signal is present:

$$\begin{aligned} &H_0 : \text{Signal absent} \\ &H_1 : \text{Signal present} \end{aligned} \tag{2.4}$$

In binary hypothesis testing, it is assumed that one of the two hypotheses is indeed correct. There are four possible outcomes:

- Case 1: H_0 is true, and we declare H_0 to be true
- Case 2: H_0 is true, but we declare H_1 to be true
- Case 3: H_1 is true, and we declare H_1 to be true
- Case 4: H_1 is true, but we declare H_0 to be true

A *hypothesis test* (or *decision rule*) δ is a partition of the observation space into two sets, $\mathcal{R}$ and its complement $\overline{\mathcal{R}}$. We accept the null hypothesis if the observations are in $\mathcal{R}$; otherwise, we reject it. As indicated in Figure 2.2, there are two kinds of errors that we can make. A type I error is made by choosing hypothesis H_1 when H_0 is true and a type II error is made by choosing H_0 when H_1 is true. In hypothesis testing the probability of a type I error is called the *significance of the test*, whereas P_d is called the *power of the test*.

The problem is to find a test that is optimal according to some criterion. We will discuss a few such approaches here, but bear in mind that there are many others that are applicable to specific problems.

The assumptions made about the signal affect the detection performance. The most general case is when the signal is random, as is the noise. In this case, nothing is known or assumed about the signal to be detected. The other end of the difficulty spectrum is when the signal is totally known and there is no noise (or insignificant levels of noise) present. Other detection problems fall between these two extremes.

There are two fundamental forms of optimizations in signal detection: Bayesian-based and frequency-based. The Bayes criterion is an example of the former, while Neyman-Pearson is an example of the latter.

2.4.2 Bayesian Approach

In the Bayesian approach, costs are assigned to the decisions; in particular, these costs are given by C_{ij}, $i, j = 0, 1$. C_{ij} is the cost incurred by choosing hypothesis H_i when hypothesis H_j is true. Next, probabilities π_0 and π_1 are assigned where π_0 and $\pi_1 = 1 - \pi_0$ are the probabilities of the hypotheses H_0 and H_1, respectively. These probabilities are called *a priori probabilities* or *priors*. The *Bayes risk* is then the overall expected value of the cost incurred by the decision rule δ:

$$\begin{aligned}\bar{C}(\delta) &= \mathcal{E}\{\Re(\delta)\} \\ &= C_{00}\pi_0 \Pr\{\text{declare } H_0 | H_0 \text{ is true}\} + C_{10}\pi_0 \Pr\{\text{declare } H_1 | H_0 \text{ is true}\} \\ &\quad + C_{11}\pi_1 \Pr\{\text{declare } H_1 | H_1 \text{ is true}\} + C_{01}\pi_1 \Pr\{\text{declare } H_0 | H_1 \text{ is true}\}\end{aligned} \tag{2.5}$$

Finally, the *Bayes rule* is defined as the rule that minimizes the Bayes risk $\mathcal{E}\{\Re(\gamma)\}$.

It is usually assumed that $C_{01} > C_{00}$ and $C_{10} > C_{11}$, which says that the cost of estimating incorrectly is higher than estimating correctly.

Under each hypothesis

$$\begin{aligned}H_0 &: x \sim p_0(x) \\ H_1 &: x \sim p_1(x)\end{aligned} \tag{2.6}$$

and we assume that the signal of interest has a continuous PDF. Let R_0 and $R_1 = \bar{R}_0$ denote the decision regions corresponding to the optimal test. To specify the optimal test, it is only necessary to specify R_0 and R_1.

The Bayes risk can be written as

$$\begin{aligned}\bar{C} &= \sum_{\{i,j\}=0}^{1} C_{ij}\pi_i \int_{R_i} p_j(x)dx \\ &= \int_{R_0} [C_{00}\pi_0 p_0(x) + C_{01}\pi_1 p_1(x)]dx + \int_{R_1} [C_{10}\pi_0 p_0(x) + C_{11}\pi_1 p_1(x)]dx\end{aligned} \tag{2.7}$$

R_0 and R_1 partition the input space: they are disjoint and their conjunction is the full input space. This means that every possible input x belongs to one and only one of these regions. In order to minimize the Bayes risk, a measurement x should belong to the decision region R_i for which the corresponding integrand in (2.7) is smaller. Therefore, the Bayes risk is minimized by assigning x to R_0 whenever

$$C_{00}\pi_0 p_0(x) + C_{01}\pi_1 p_1(x) < C_{10}\pi_0 p_0(x) + C_{11}\pi_1 p_1(x) \tag{2.8}$$

and assigning x to R_1 whenever this inequality is reversed. By manipulating (2.8), we get

$$\Lambda(x) = \frac{p_1(x)}{p_0(x)} \underset{H_0}{\overset{H_1}{\gtrless}} \frac{\pi_1(C_{10} - C_{00})}{\pi_0(C_{01} - C_{11})} \equiv \eta \tag{2.9}$$

Here, $\Lambda(x)$ is called the *likelihood ratio*, η is called the *threshold*, and the overall decision rule is called the *likelihood ratio test* (LRT). The expression on the right is called a *threshold*.

It can be shown that the two main approaches to maximizing the desirable properties of the detection—the Bayes or Neyman-Pearson—amount to the same thing; the detector finds a likelihood ratio [which will be a function only of the sufficient statistic given by (2.9)] and then compares this ratio with a preset threshold. By varying the threshold, we can see that the detection ratio (where we correctly say H_1) and the false alarm rate (where we incorrectly say H_1) will vary in a predictable manner. Hence, if we have complete knowledge of the probability densities of H_0 and H_1, we can construct an optimal detector, or at least determine the properties of such a detector.

Note that (2.9) requires knowledge of the costs associated with making the various decisions. Many times these costs are not accurately known. In that case it is common to assume equally distributed values, lacking any reason to assume otherwise. Another approach is to use the minimum probability of error decision rule discussed next.

2.4.3 Minimum Probability of Error Decision Rule

The minimum probability of error decision rule is a useful approach for decision making when the a priori Bayesian costs are not available. Again, let π_i denote the a priori probability of hypothesis H_i. Suppose our decision rule declares "H_0 is true" when $x \in R_0$, and it selects "H_1 is true" when $x \in R_1$, where $R_1 = \overline{R}_0$. The probability of making an error, denoted P_e, is

$$
\begin{aligned}
P_e &= \Pr\{\text{declare } H_0 | H_1 \text{ is true}\} + \Pr\{\text{declare } H_1 | H_0 \text{ is true}\} \\
&= \Pr\{H_1\}\Pr\{H_0 | H_1\} + \Pr\{H_0\}\Pr\{H_1 | H_0\} \\
&= \int_{R_0} \pi_1 p_1(x)dx + \int_{R_1} \pi_0 p_0(x)dx
\end{aligned}
\tag{2.10}
$$

The *minimum probability of error* decision rule selects R_0 and R_1 to minimize (2.10). Since an observation x falls into one and only one of the decision regions R_i, in order to minimize P_e, we assign x to the region for which the corresponding integrand in (2.10) is smaller. Thus, we select $x \in R_0$ if $\pi_1 p_1(x) < \pi_0 p_0(x)$, and $x \in R_0$, if the inequality is reversed. This decision rule may be summarized concisely as

$$
\Lambda(x) = \frac{p_1(x)}{p_0(x)} \underset{H_0}{\overset{H_1}{\gtrless}} \frac{\pi_1}{\pi_0} \equiv \zeta
\tag{2.11}
$$

As above, $\Lambda(x)$ is the likelihood ratio, ζ is called the threshold, and the overall decision rule is called an LRT.

Example: Normal with Common Variance, Different Means

Consider the binary hypothesis test of a scalar x

$$
\begin{aligned}
H_0 &: x \sim \mathcal{N}(0, \sigma^2) \\
H_1 &: x \sim \mathcal{N}(\mu, \sigma^2)
\end{aligned}
$$

where μ and σ^2 are known, positive quantities. Suppose we observe a single measurement x. The likelihood ratio is

$$
\begin{aligned}
\Lambda(x) &= \frac{1/(\sqrt{2\pi}\sigma) e^{-\left[\frac{(x-\mu)^2}{2\sigma^2}\right]}}{1/(\sqrt{2\pi}\sigma) e^{-\left[\frac{x^2}{2\sigma^2}\right]}} \\
&= e^{-\frac{1}{\sigma^2}\left[\mu x - \frac{\mu^2}{2}\right]}
\end{aligned}
\tag{2.12}
$$

so the minimum probability of error decision rule is

$$e^{\frac{1}{\sigma^2}\left[\mu x-\frac{\mu^2}{2}\right]} \underset{H_0}{\overset{H_1}{\gtrless}} \frac{\pi_0}{\pi_1} \triangleq \zeta \tag{2.13}$$

The expression for $\Lambda(x)$ is somewhat complicated. By applying a sequence of monotonically increasing functions to both sides, we can obtain a simplified expression for the optimal decision rule without changing the rule. In this example, we apply the natural logarithm and rearrange terms to arrive at

$$x \underset{H_0}{\overset{H_1}{\gtrless}} \frac{\sigma^2}{\mu} \ln \zeta + \frac{\mu}{2} \triangleq \gamma \tag{2.14}$$

Here we have used the assumption $\mu > 0$. If $\mu < 0$, then dividing by μ would reverse the inequalities.

This form of the decision rule is much simpler: we just compare the observed value x to a threshold γ. Figure 2.4 depicts the two candidate densities and a possible value of γ. If each hypothesis is a priori equally likely ($\pi_0 = \pi_1 = ½$), then $\gamma = \mu / 2$. Figure 2.4 illustrates the case where $\pi_0 > \pi_1$ ($\gamma > \mu / 2$). If we plot the two densities so that each is weighted by its a priori probability of occurring, the two curves will intersect at the threshold (see Figure 2.5). This plot also offers a way to visualize the probability of error.

Recall

$$\begin{aligned} P_e &= \int_{R_0} \pi_1 p_1(x)dx + \int_{R_1} \pi_0 p_0(x)dx \\ &= \int_{x<\gamma} \pi_1 p_1(x)dx + \int_{x>\gamma} \pi_0 p_0(x)dx \\ &= \pi_1 P_m + \pi_0 P_{fa} \end{aligned} \tag{2.15}$$

These quantities are depicted in Figure 2.5. We can express P_m and P_{fa} in terms of the Q-function (Appendix A) as

$$P_e = \pi_1 Q\left(\frac{\mu-\gamma}{\sigma}\right) + \pi_0 Q\left(\frac{\gamma}{\sigma}\right) \tag{2.16}$$

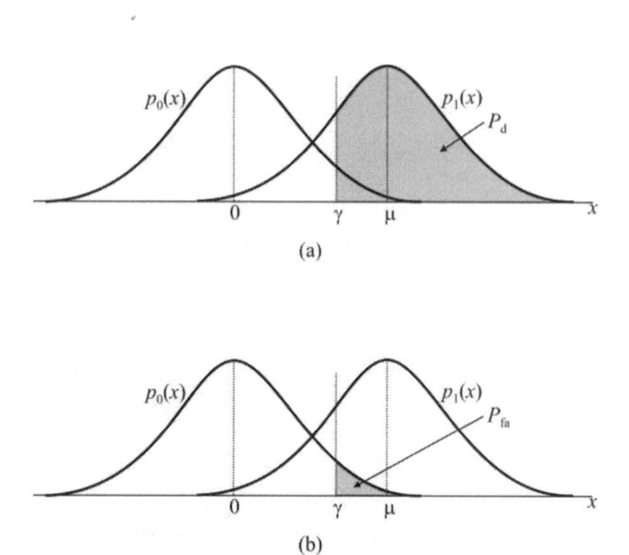

Figure 2.4 Two candidate densities with a threshold corresponding to $\pi_0 > \pi_1$. (a) Probability of detection (P_d) and (b) probability of false alarm (P_{fa}).

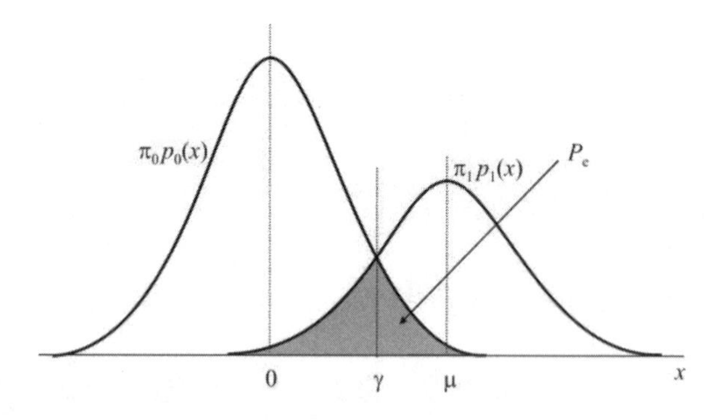

Figure 2.5 The candidate densities weighted by their a priori probabilities. The shaded region is the probability of error (P_e) for the optimal decision rule.

When $\pi_0 = \pi_1 = ½$, we have $\gamma = \mu / 2$, and the error probability is

$$P_e = Q\left(\frac{\mu}{2\sigma}\right) \tag{2.17}$$

Since $Q(x)$ is monotonically decreasing, this says that the "difficulty" of the detection problem decreases with decreasing σ and increasing μ.

In the preceding example, the computation of the probability of error involved a one-dimensional integral. If we had multiple observations, or vector-valued data, generalizing this procedure would involve multi-dimensional integrals over potentially complicated decision regions. Fortunately, in many cases, we can avoid this problem through the use of sufficient statistics.

2.4.4 Maximum A Posteriori Interpretation

The LRT is one way of expressing the minimum probability of error decision rule. Another way is

Rule: Declare hypothesis i true such that $\pi_i p_i(x)$ is maximal.

This rule is referred to as the *maximum a posteriori*, or MAP, rule because the quantity $\pi_i p_i(x)$ is proportional to the posterior probability of hypothesis i. This becomes clear when we write $\pi_i = \Pr\{H_i\}$ and $p_i(x) = p_i(x|H_i)$. Then, by the Bayes rule, the posterior probability of H_i given the data is

$$\Pr\{H_i|x\} = \frac{\Pr\{H_i\}p(x|H_i)}{p(x)} \tag{2.18}$$

Here $p(x)$ is the unconditional PDF for x, which is effectively a constant when trying to maximize (2.18) with respect to i.

Thus, in the MAP interpretation, the optimal decision boundary is the locus of points where the weighted densities (in the continuous case) $\pi_i p_i(x)$ intersect one another.

2.4.4.1 Special Case of Bayes Risk

As discussed above, the Bayes risk criterion for constructing decision rules assigns a cost C_{ij} to the outcome of declaring H_i when H_j is true. The probability of error is simply a special case of the Bayes risk corresponding to $C_{00} = C_{11} = 0$ and

$C_{01} = C_{10} = 1$. Therefore, the form of the minimum probability of error decision rule is a special case of the minimum Bayes risk decision rule: both are LRTs. The different costs in the Bayes risk formulation simply shift the threshold to favor one hypothesis over the other. In addition, as mentioned, if the a priori costs are not known, the minimum probability of error decision rule can be used.

2.4.5 Neyman-Pearson Approach

In many signal detection problems the imposition of a specific cost structure on the decisions made is not possible or desirable. The Neyman-Pearson approach involves a trade-off between the two types of errors that we can make in choosing a particular hypothesis. The Neyman-Pearson design criterion is to maximize P_d subject to a specified P_fa; it is also known as the *constant false alarm rate* (CFAR) detector.

The probability of false alarm and probability of detection are related to each other through the decision regions. If R_1 is the decision region for H_1, we have

$$P_\text{fa} = \int_{R_1} p_\text{n}(x)dx \tag{2.19}$$

$$P_\text{d} = \int_{R_1} p_\text{s+n}(x)dx \tag{2.20}$$

The densities $p_\text{n}(x)$ and $p_\text{s+n}(x)$ are nonnegative, so as R_1 shrinks, both probabilities approach zero. As R_1 expands, both tend to one. The ideal case, where $P_\text{d} = 1$ and $P_\text{fa} = 0$, cannot occur unless the PDFs do not overlap (i.e., $\int p_\text{n}(x)p_\text{s+n}(x)dx = 0$). Therefore, in order to increase P_d, we must also increase P_fa. This is the characteristic trade off in the hypothesis testing and detection theory.

The Neyman-Pearson criterion specifies that the decision rule should be such that the probability of detection is maximized while not allowing the probability of false alarm to exceed a certain value α. Thus, the criterion can be expressed as

$$\max_{\{\delta\}}\{P_\text{d}\}, \text{such that } P_\text{fa} \le \alpha \tag{2.21}$$

The maximization is over all decision rules denoted by $\{\delta\}$ (equivalently, over all decision regions $\mathcal{R}, \bar{\mathcal{R}}$). Using different terminology, the Neyman-Pearson criterion selects the most powerful test of size (not exceeding) α. The solution to this optimization problem is given by the Neyman-Pearson property.

Property: *Neyman-Pearson*

Consider the test

$$H_0 : r \sim p_{\mathrm{n}}(a) \tag{2.22}$$
$$H_1 : r \sim p_{\mathrm{s+n}}(a) \tag{2.23}$$

where $p_{\mathrm{n}}(a)$ and $p_{\mathrm{s+n}}(a)$ are PDFs of $r(t)$ corresponding to signal absent and signal present, respectively. Define $\Lambda(a) = p_{\mathrm{s+n}}(a) / p_{\mathrm{n}}(a)$, and assume that $\Lambda(a)$ satisfies the condition that for each $\gamma \in \mathbb{R}, \Lambda(a)$ takes on the value γ with probability zero under hypothesis H_0. The solution to the optimization problem in (2.21) is given by

$$\Lambda(a) = \frac{p_{\mathrm{s+n}}(a)}{p_{\mathrm{n}}(a)} \mathop{\gtrless} \zeta \tag{2.24}$$

where ζ is such that

$$P_{\mathrm{fa}} = \int_{\forall x, \Lambda(x) > \zeta} p_{\mathrm{n}}(x) dx = \alpha \tag{2.25}$$

If $\alpha = 0$, then $\zeta = \infty$, and the optimal test under H_0 and H_1 is unique.

This version of the property excludes several important cases, including tests with discrete data, because we assumed that for all ζ, the set $\{x | \Lambda(x) = \zeta\}$ has probability zero under H_0. Nevertheless, it is sufficient for our purposes here, which is to introduce the Neyman-Pearson criterion for signal detection. For those interested in a complete treatise on the topic, [1] is a suggested source.

The optimal decision rule is an LRT. $\Lambda(a)$ is the likelihood ratio, and ζ is the threshold. Observe that neither the likelihood ratio nor the threshold depends on the a priori probabilities $\Pr\{H_i\}$; they depend only on the conditional densities p_{n} and $p_{\mathrm{s+n}}$ and the size constraint α. The threshold can often be determined as a function of α.

Example:

Consider the simple binary hypothesis test of a scalar measurement x

$$H_0 : x \sim \mathcal{N}(0,1)$$
$$H_1 : x \sim \mathcal{N}(1,1)$$

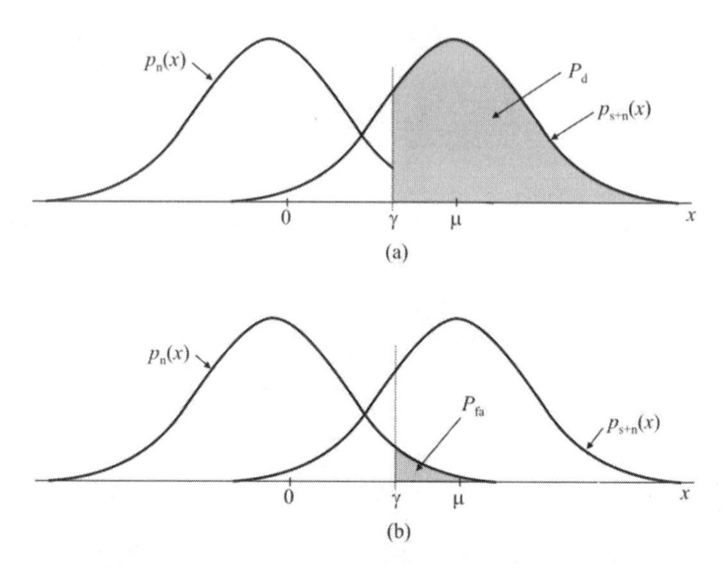

Figure 2.6 (a) P_d and (b) P_{fa} are given by the shaded areas under the PDF curves.

Suppose we use a threshold test

$$x \underset{H_0}{\overset{H_1}{\gtrless}} \gamma$$

where $\gamma \in \mathbb{R}$ is a free parameter to be determined. Then the false alarm and detection probabilities are

$$P_{fa} = \int_{\gamma}^{\infty} \frac{1}{\sqrt{2\pi}} e^{-x^2/2} dx = Q(\gamma)$$

$$P_d = \int_{\gamma}^{\infty} \frac{1}{\sqrt{2\pi}} e^{-(x-1)^2/2} dx = Q(\gamma - 1)$$

where Q denotes the Q-function (Appendix A). These quantities are depicted in Figure 2.6. P_d is illustrated as the gray area in Figure 2.6(a) to the right of the threshold under the PDF denoted as $p_{n+s}(x)$, while P_{fa} is illustrated by the gray area in Figure 2.6(b), the area to the right of the threshold under the PDF marked as $p_n(x)$. Since the Q-function is monotonically decreasing, it is evident that both P_d and P_{fa} decay to zero as γ

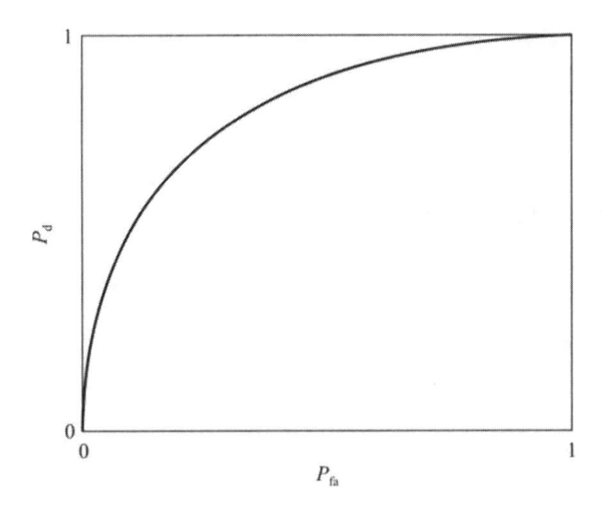

Figure 2.7 Example ROC curve.

increases. There is also an explicit relationship between P_d and P_{fa}, given by

$$P_d = Q[Q^{-1}(P_{fa}) - 1]$$

A common means of displaying this relationship is with a *receiver operating characteristic* (ROC) curve, which is a plot of P_d versus P_{fa} (Figure 2.7). The two scales obviously run from zero to one.

Now suppose we wish to design a Neyman-Pearson decision rule with a size constraint α. We have

$$\Lambda(x) = \frac{\frac{1}{\sqrt{2\pi}} e^{-(x-1)^2/2}}{\frac{1}{\sqrt{2\pi}} e^{-x^2}} = e^{x - \frac{1}{2}} \tag{2.26}$$

By taking the natural logarithm of both sides of the LRT and rearranging terms, the characteristics of the decision rule are not changed since the monotonic change is retained, and we obtain

$$x \underset{H_0}{\overset{H_1}{\gtrless}} \ln\zeta + \frac{1}{2} \triangleq \gamma$$

Thus, the optimal rule is, in fact, a thresholding rule like we considered previously. The false-alarm probability is

$$P_{\text{fa}} = Q(\gamma)$$

Thus, we may express the value of γ required by the Neyman-Pearson property in terms of α as

$$\gamma = Q^{-1}(\alpha)$$

2.5 Receiving System Noise Power

2.5.1 Channel Model

The transmitted waveform gets corrupted by noise, $n(t)$, often assumed to be *additive white Gaussian noise* (AWGN), although there are other types of noise models available, and we will discuss some of them.

> *Additive*: The noise gets "added" (as opposed to, for example, multiplied) to the received signal.
>
> *White*: The spectrum of the noise is flat for all frequencies, so therefore no particular frequency is favored in the spectrum (if there were such frequencies, it would be known as *colored* noise). The amplitude of this noise spectrum is given by N_0, which is a PSD and is specified as

$$N_0 = k_\text{B} T \qquad \text{W Hz}^{-1} \tag{2.27}$$

> where $k_\text{B} = 1.38 \times 10^{-23}$ Watts/Kelvin/Hz is Boltzmann's constant, and T = absolute temperature in Kelvins.

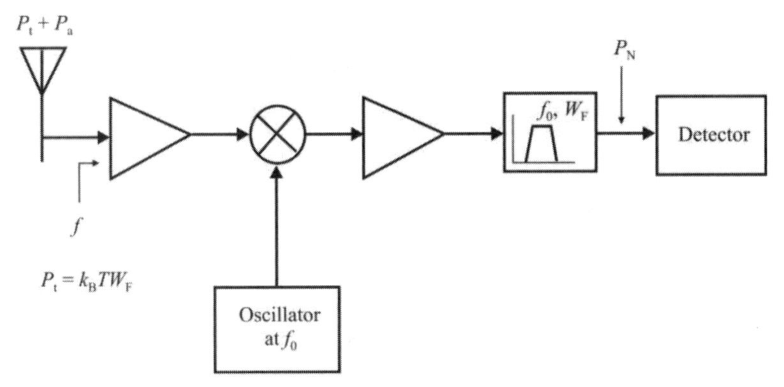

P_t - Noise power due to external thermal agitation
P_a - Noise power from sources external to the system due to other than thermal agitation
f - System noise factor (all internal noise sources reflected to the system input)
P_N - noise power used in BER analysis

Figure 2.8 This is a simplified block diagram, however, all the pertinent noise sources that contribute to BER computations are included.

Gaussian: The statistical properties of the amplitude of the noise $n(t)$ are characterized by the Gaussian PDF given by[1]

$$p_X(x) = \frac{1}{\sqrt{2\pi\sigma^2}} e^{-\frac{(x-\mu)^2}{2\sigma^2}} \tag{2.28}$$

with the mean, μ, typically taken to be zero, and the variance $\sigma^2 = N_0 / 2$ assuming the spectrum being considered reaches across the range $(-\infty, +\infty)$ or $\sigma^2 = N_0$ if the range is restricted to the positive frequency range $(0, +\infty)$.

At the EW (or target) receiver, the noise components of concern for estimating signal detection performance are as indicated in Figure 2.8. For the moment we assume that there is no jamming signal present. The spectrum of the noise component, N_0, is assumed to be a flat function of the frequency over a bandwidth that is wider than that of concern to our analysis. The noise power due

[1] In this notation for density functions, X is the random variable and x is one instance of that random variable. If it is not confusing, frequently the X is not explicitly included.

to this component, P_t, is therefore a function of the downstream predetection bandpass filter, with bandwidth W_F as shown in Figure 2.8 and is given by

$$P_t = N_0 W_F \tag{2.29}$$

where N_0 is given by (2.27). The other two noise components are the noise due to sources other than thermal agitation of electrons external to the system (e.g., atmospheric, man-made, and so forth) denoted by P_a, and that noise due to thermal agitation within the electronics of the system, represented by the *noise factor f*. The total noise power that the detector "sees," which is the noise power of concern to us when computing the BER performance, is therefore given by multiplying the total external power by the noise factor:

$$P_N = f(P_t + P_a) \tag{2.30}$$

The *noise figure* is given by

$$F = 10\log_{10} f \tag{2.31}$$

If we denote the total external noise power as $P_e = P_t + P_a$, then (choosing dBm as convenient units)

$$P_{N,dBm} = F + P_{e,dBm} \tag{2.32}$$

Typically, the first module encountered by a signal from an antenna in EW systems, or any receiving system for that matter, is an LNA. In fact, sometimes the LNA is integrated with the antenna for receive applications. The reason for this is how the various components in the receiver chain contribute to the noise level. If f_1 represents the noise factor of the first module with gain G_1, f_2 represents the noise factor of the second module with gain G_2, and so on, the system noise factor is given by

$$f = f_1 + \frac{f_2 - 1}{G_1} + \frac{f_3 - 1}{G_1 G_2} + \cdots \tag{2.33}$$

So the effects of subsequent stages of amplification, losses, mixing, and so forth, each with an associated noise factor and gain, are reduced by the product of all the gains of the stages that precede them. Thus, the minimum noise factor is the noise factor of the first module. This is why the first module is usually an LNA. If G_1 is sufficiently high, the effects of later modules are minimized.

It is important to minimize the noise factor so that maximum sensitivity can be achieved. EW systems are sometimes used against distant targets, and sensitivity is important to intercept targets at long range.

The external noise over and above the external thermal noise, that is, P_a, can be estimated with the aid of Figures 2.9 and 2.10, which chart averages of the external noise, P_a. These charts are worldwide averages, so to a first order, apply anywhere in the world. P_a is expressed in terms of decibels above the level of $k_B T_0 W$ for convenience. Note that these components add numerically, not in dB, as expressed in (2.30).

When one of these factors is 10 dB higher than another, it is 10 times more significant, 20 dB corresponds to 100 times, and so forth. When there is a 10 dB or more difference, the weaker component can frequently be ignored.

The noise power in (2.30) has a corresponding "equivalent" PSD, which we will denote in the sequel by N_0'. It must be remembered, however, that this is not the same N_0 used when considering only thermally generated noise.

The most significant ramification of this is that, in general,

$$P_N \neq N_0 W \tag{2.34}$$

for any bandwidth W with this definition of N_0. The F_a terms, estimates of which are shown in Figures 2.9 and 2.10 [2], are decidedly not frequency independent, especially in the HF frequency range below 30 MHz. From 30 to 300 MHz, or so, galactic and *man-made noise* (MMN) sources dominate. Most PTT *combat net radios* (CNRs) operate in that range so it is certainly a frequency range of interest. From about 300 MHz to 1 GHz, MMN is dominant. Above 1 GHz, the thermal noise as given by $N_0 = k_B T_0 W$ is the only source of significance. Therefore, above 1 GHz we can say that

$$P_N \approx N_0 W \tag{2.35}$$

using the standard constant noise PSD. Below this frequency, using (2.35) for P_N with the equivalent N_0 produces only gross estimates of the detector input noise for frequency-hopping CNRs. If the target of interest is narrowband, however, so that the VHF channelization of 25 kHz is valid, then (2.35) with the equivalent N_0' is an accurate and useful estimate. In that case, at 100 MHz, for example, the span of W is only $25 \times 10^3 \div 100 \times 10^6 = 0.4\%$, well within reasonable accuracy.

MMN is external background noise caused by man-made sources. This noise would typically be larger in urban environments than in suburban settings and we explore it in depth later in this chapter. Representative sources of this type of noise are neon signs, welders, and electric heaters. The noise figure F represents the totality of the noise sources internal to the receiver, but reflected to the receiver

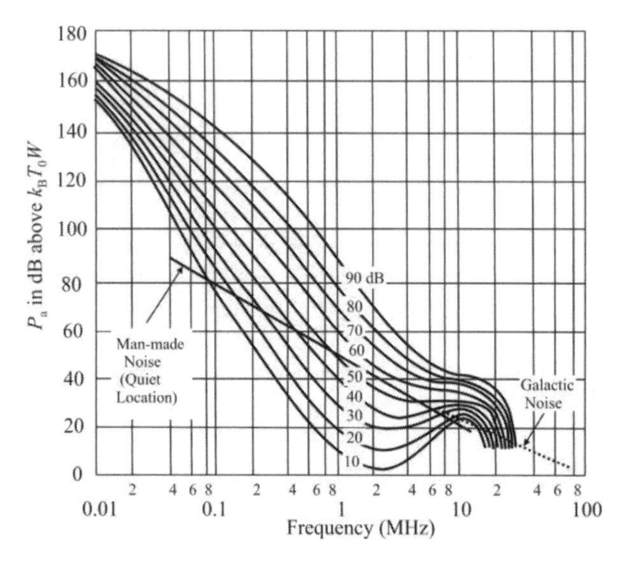

Figure 2.9 Typical noise characteristics in the low RF spectrum. The numbers in the label refer to the noise level at 1 MHz. (Source: [2]. © 1975 Howard W. Sams. Reprinted with permission.)

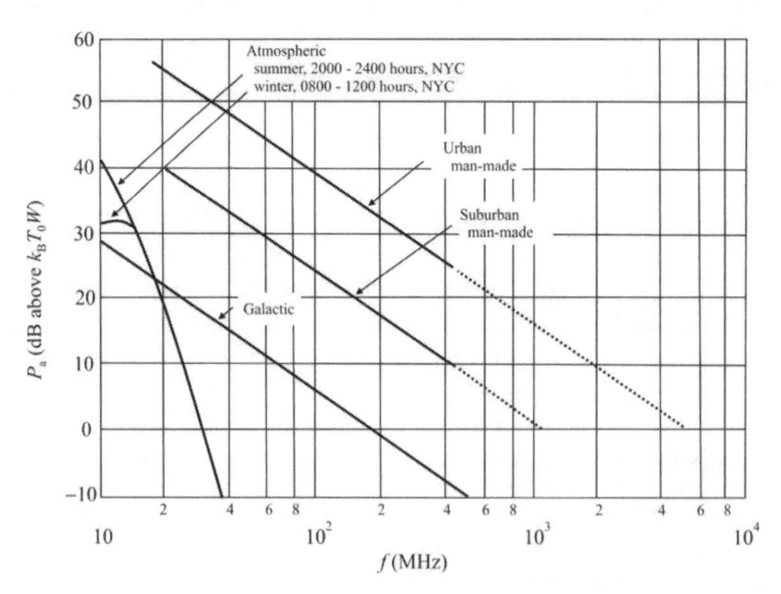

Figure 2.10 RF noise 10 to 10,000 MHz. Omnidirectional antenna located on the ground. (Source: [2]. © 1975 Howard W. Sams. Reprinted with permission.)

Table 2.1 Effective Temperatures of Some Close Celestial Bodies

Source	Temperature (K)	Beamwidth (Degrees)
Sun	6,000	0.5
Moon	200	0.5
Stars (at 300 MHz)		
Cassiopeia	3,500	$<10^{-3}$
Gygnus	2,650	$<10^{-3}$
Taurus	710	$<10^{-3}$
Centaurus	460	$<10^{-3}$
Planets (10 MHz to 10 GHz)		
Mercury	613	2×10^{-3}
Venus	235	6×10^{-3}
Mars	217	4.3×10^{-3}
Jupiter	138	1.3×10^{-3}
Saturn	123	5.7×10^{-3}
Earth (from moon)	300	2

Source: [4].

input, leaving the remainder of the receiver noise-free. Typical noise figures for receivers in the low VHF range are 10–20 dB, while in the UHF and above range, 2–10 dB is typical. See [3] for a further explanation of the noise factor.

As can be seen from Figures 2.9 and 2.10, the external noise can be several tens of decibels higher than the $k_B T_0 W$ noise for urban and suburban environments below about 1 GHz. At these levels, $k_B T_0 W$ noise can be ignored and system performance is largely dictated by the external, nonthermal noise sources. In many cases, the system noise figure is negligible in comparison as well.

The Sun and other close stars and planets contribute to the thermal noise level as well. Table 2.1 shows the effective temperature of these noise sources [4]. These are particularly troublesome if a directional antenna is pointed *at* the source, and all real antennas are directional to some extent.

2.5.2 Plasma Noise

Plasma is the fourth state of matter, the other three states being solid, liquid, and gas. Though more than 90% of the universe consists of plasma, the only observable naturally occurring forms of plasma on Earth are lightning and fire.

All matter consists of atoms bonded together in a lattice. Each atom in turn consists of a nucleus (positively charged protons and neutral neutrons) and electrons (negatively charged particles) that can be modeled as orbiting the nucleus; and, of course, two or more atoms may combine to form molecules. In a solid, the bonding between molecules is very strong with little freedom of movement for individual atoms or molecules. When a solid is heated, energy is added to the molecules, and if heated sufficiently, it becomes a liquid because the bonding is weakened, permitting molecules to move about more freely. Heating further causes a break-up of the molecules and the liquid becomes a gas where individual atoms or molecules move about unhindered.

The Earth's lower atmosphere consists predominantly of oxygen, nitrogen, carbon dioxide, and argon gas molecules. In the upper atmosphere (80–100 km), interaction with extreme ultraviolet rays and cosmic radiation from the Sun heats up individual atoms to such levels that atoms are stripped of their electrons—a process with the appellation *ionization*. An ionized plasma containing negative electrons and positively charged atom nuclei (called ions) results.

Because a plasma is an ionized medium, consisting of electrically charged particles (negative electrons and positive ions), its properties are controlled by electromagnetic forces according to Maxwell's laws of electrodynamics [5]. These same rules describe the operation of an electromagnet or electric motor. In the case of an ionized plasma, the magnets are substituted by the Earth or Sun's magnetic field.

Stochastic plasma noise can sometimes produce large amounts of electromagnetic emissions and interfere or even damage nearby electronic equipment including sensitive EW systems. One of the major sources of plasma noise, especially in an urban setting, is from arc welders.

2.5.2.1 Plasma Noise Characteristics

The correlation for the noise voltage $v_{\text{n}}(t)$ is given by

$$R_{\text{n}}(\tau) = \mathcal{E}\{v_{\text{n}}(\tau)v_{\text{n}}(t+\tau)\} \tag{2.36}$$

Assume that the stochastic nature of plasma noise is Poisson distributed, which is a fairly good model. Using the Weiner-Khinchine theorem, we obtain the PSD of the plasma noise as

$$H(f) = 4\int_0^\infty R_{\text{n}}(u)\cos(2\pi fu)du \tag{2.37}$$

$$= 4\sum_i \int_0^\infty \mathcal{E}^2\{v_{\text{n}}\}\exp(-\upsilon u)\cos(2\pi fu)du \tag{2.38}$$

The voltage fluctuation is related to the electron velocity fluctuation u_{n} as

$$v_{\text{n}} = R_{\text{p}}\frac{e}{l}u_{\text{n}} \tag{2.39}$$

where R_p is the resistance of the plasma and e is the charge of the electron. Then (2.38) becomes

$$H(f) = 4\left(\frac{R_p e}{l}\right)^2 \mathcal{E}^2\{u\}\int \exp\left(-\frac{z}{\tau_0}\right)\cos(\omega z)dz \tag{2.40}$$

$$= 4\left(\frac{R_p e}{l}\right)^2 \mathcal{E}^2\{u\}\frac{\tau_0}{1+\omega^2\tau_0^2}$$

$$= 4\left(\frac{R_p e}{l}\right)^2 \mathcal{E}^2\{u\}\frac{1/\upsilon}{1+\left(\omega/\upsilon\right)^2} \tag{2.41}$$

where υ is the collision frequency.

Using the kinetic mechanics relationship

$$\frac{1}{2}m\mathcal{E}^2\{u\} = \frac{1}{2}k_B T_0 \tag{2.42}$$

and the conductivity relationship

$$\sigma = \frac{ne^2}{m\upsilon} \tag{2.43}$$

substituting (2.42) and (2.43) into (2.41) yields

$$H(f) = 4kT_0\left(\frac{R_p^2 A\sigma}{d}\right)\left[\frac{1}{1+\left(\dfrac{\omega}{\upsilon}\right)^2}\right] \tag{2.44}$$

Also, for a solid current-carrying metal

$$H(f)_{\text{metal}} = 4k_B T_0 R_m \tag{2.45}$$

where it is assumed that $\upsilon >> \omega$ and R_m is the resistance of the metal.

For a stochastic plasma

$$H(f)_{\text{plasma}} = 4k_B T_0\left[R_p/[1+(\omega/\upsilon)^2\right] \tag{2.46}$$

Hence

$$H(f)_{\text{plasma}} < H(f)_{\text{metal}} \tag{2.47}$$

Thus, we see that the random noise from a radiating plasma is less than the random noise in a current-carrying metal. Note that there are exceptions to this as plasma noise is frequency dependent.

2.6 Noise Amplitude Probability Density Functions

This section introduces the most common forms of PDFs of the amplitude of signals in noise under various conditions.

2.6.1 Gaussian

The noise entering the IF filter is assumed to be Gaussian (as it is thermal in nature) with an amplitude PDF given by

$$p(v) = \frac{1}{\sqrt{2\pi\sigma_0^2}} \exp\left(-\frac{v^2}{2\sigma_0}\right) \tag{2.48}$$

where

$p(v)dv = \Pr\{v \le V \le v + dv\} =$ probability of the noise voltage V being between v and $v + dv$

σ_0^2 = variance of the noise voltage

An example of a Gaussian PDF is shown in Figure 2.11. This Gaussian density has zero mean and unit standard deviation, the *standard density*.

2.6.2 Rayleigh

If Gaussian noise is passed through a narrowband filter (one whose bandwidth is small compared to the center frequency), then the amplitude PDF of the envelope of the noise voltage at the output has the form of the Rayleigh PDF

$$p(A) = \frac{A}{\sigma_0} \exp\left(-\frac{A^2}{2\sigma_0}\right) \tag{2.49}$$

where A is the amplitude of the envelope of the filter output.

The Rayleigh PDF is compared with Gaussian in Figure 2.11. In addition, a few Rayleigh PDFs are illustrated in Figure 2.12. The densities are shown corresponding to different values of the standard deviation. As the standard deviation increases, the PDFs decrease in height and become broader.

2.6.3 Ricean

Consider that a sine wave with amplitude, A, is present along with the noise at the input to the IF filter. The frequency of the sine wave is equal to the center frequency of the IF filter. It was shown by Rice that the amplitude of the signal at the output of the envelope detector will have the following amplitude PDF (known as a Ricean density) [6].

$$p_R(r) = \frac{r}{\sigma_0} \exp\left(-\frac{r^2 + A^2}{2\sigma_0}\right) I_0\left(\frac{rA}{\sigma_0}\right) \tag{2.50}$$

$I_0(Z)$ is the modified Bessel function of order zero and argument Z. It can be shown that for large Z, an asymptotic expansion for $I_0(Z)$ yields the approximation

$$I_0(Z) \approx \frac{e^Z}{\sqrt{2\pi Z}}\left(1 + \frac{1}{8Z}\right) \tag{2.51}$$

A few examples of the Ricean PDFs are illustrated in Figure 2.13. With low values of signal level and noise standard deviations, the Ricean PDF resembles the Rayleigh PDF, while as the signal gets larger and/or the noise power increases, it more closely resembles the normal PDF.

2.6.4 Nakagami

The Nakagami PDF was developed to model the amplitude distribution of HF signals propagated via the ionosphere [7–9]. Through the parameter m, the Nakagami distribution can model signal fading conditions that range from severe to moderate. The primary justification for the use of the Nakagami-m is its good fit to empirical fading data. Much theoretical and numerical analysis of the performances of diverse communication systems operating in Nakagami fading has been reported in the literature.

Fewer results related to the computer simulation of Nakagami-m fading have been reported.

In this section, we briefly recall some basic properties of the Nakagami PDF. The Nakagami-m PDF is given by:

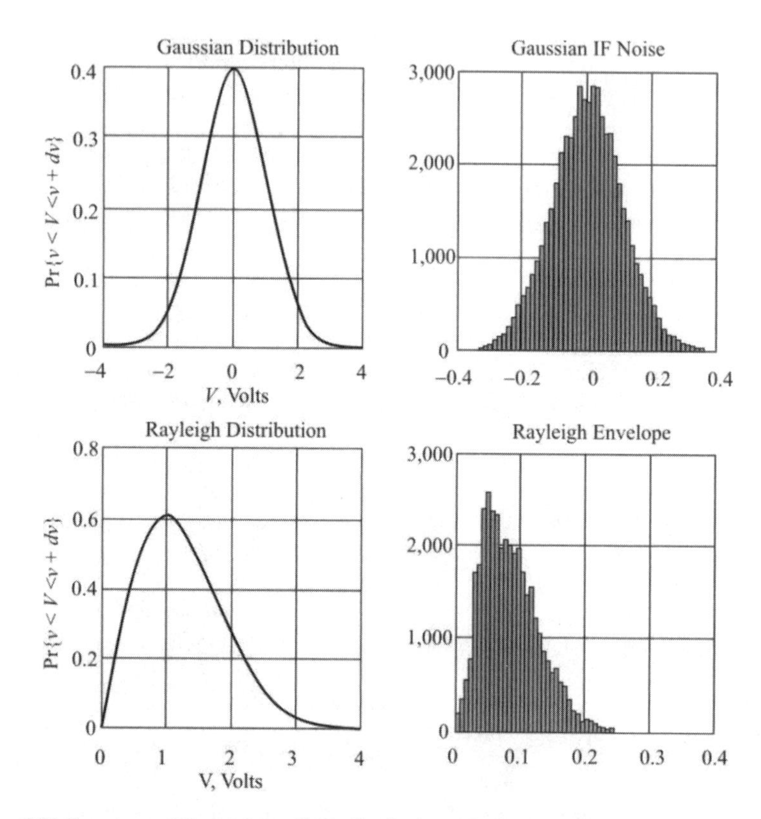

Figure 2.11 Gaussian and Rayleigh amplitude distributions of noise.

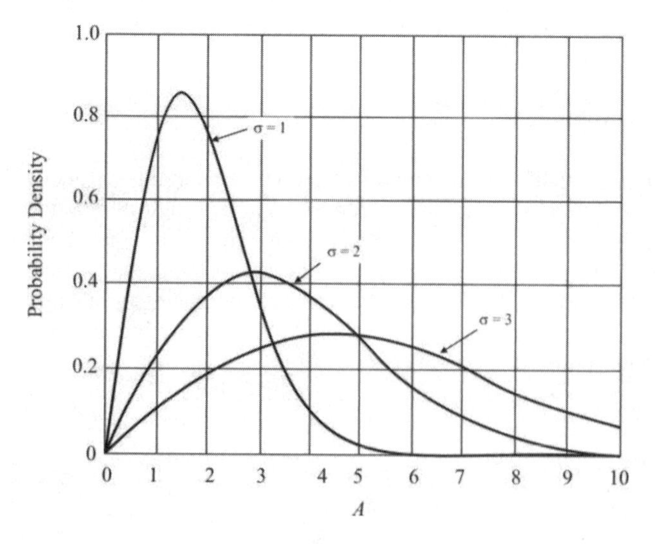

Figure 2.12 Rayleigh densities.

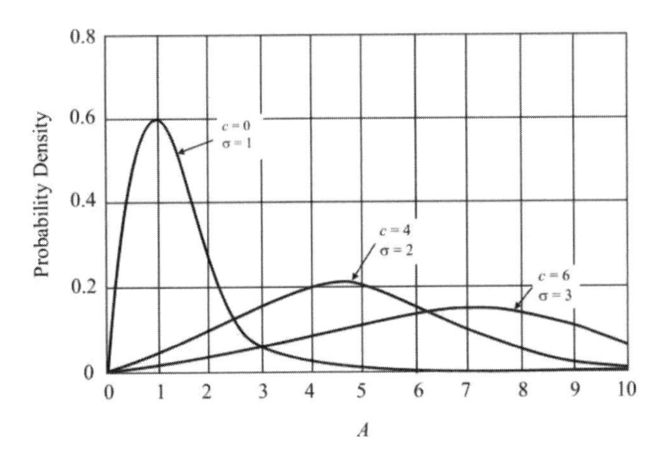

Figure 2.13 Ricean densities.

$$p_{\mathrm{N}}(\alpha) = \frac{2m^m}{\Gamma(m)} \frac{\alpha^{2m-1}}{\left(2\sigma^2\right)^m} \exp\left(-m\frac{\alpha^2}{2\sigma^2}\right) \tag{2.52}$$

where $m \geq ½$ represents the fading figure.

The mean, $\bar{\alpha}$, and the mean square, $\overline{\alpha^2}$, for $m \geq 1$, integer, are:

$$\bar{\alpha} = \sqrt{2\sigma^2}\sqrt{\frac{\pi}{4m}}\Lambda, \qquad \overline{\alpha^2} = 2\sigma^2 \tag{2.53}$$

where $\Lambda = 1$ for $m = 1$ and $\Lambda = \prod_{i=1}^{m-1}(2i+1)/2i$ for $m \geq 2$.

We assume in the following that $\overline{\alpha^2} = 1$. In that case $2\sigma^2 = 1$ and $\bar{\alpha} = \sqrt{\pi/4m}\Lambda$. The function $\Gamma(m)$ is the standard gamma function given by [10]

$$\Gamma(m) = \int_0^\infty t^{m-1}e^{-t}dt \tag{2.54}$$

The *cumulative distribution function* (CDF) of the *random variable* (r.v.) X is obtained by the integration of the PDF

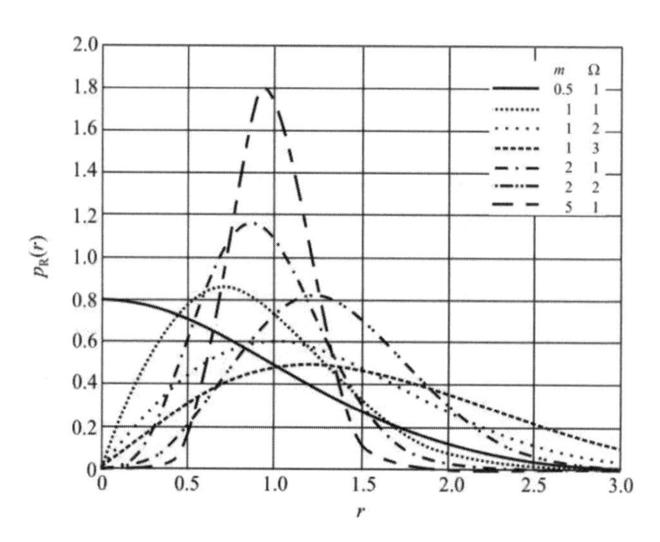

Figure 2.14 Nakagami densities.

$$P_X(x) = \int_0^x p_X(u)du = \frac{1}{\Gamma(m)}\int_0^x t^{m-1}e^{-u}du = \frac{\Gamma_x(m)}{\Gamma(m)} \tag{2.55}$$

If m is a natural number, then:

$$P_X(x) = 1 - \sum_{l=0}^{m-1}\frac{x^l}{l!}e^{-x} \tag{2.56}$$

$P_X(x)$ has an inflexion point x_P obtained by nulling its second order derivative:

$$\frac{d^2P_X(x)}{dx^2} = \frac{dp_X(x)}{dx} = \frac{1}{\Gamma(m)}\left[(m-1)x^{m-2} - x^{m-1}\right]e^{-x} = 0, x \geq 0, m \neq 1 \tag{2.57}$$

where:

$$x_P = m - 1 \tag{2.58}$$

A few of the Nakagami densities are illustrated in Figure 2.14.

2.6.5 Log-Normal

In many cases a log-normal distribution applies as the best model to use for noise corrupted signals. In particular, the HF band is well modeled with log-normal

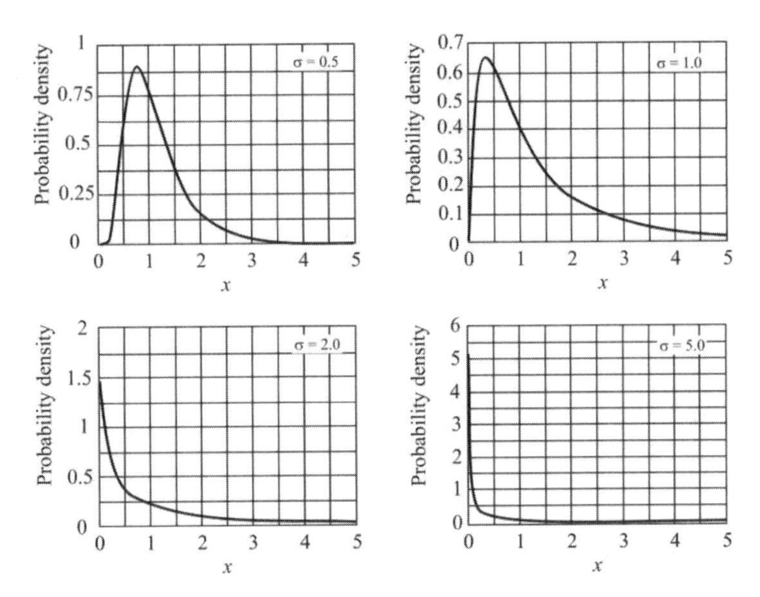

Figure 2.15 Log-normal PDFs.

distributions as is the long-term fading in wireless phone systems. A variable X is lognormally distributed if $Y = \ln(X)$ is normally distributed. The general formula for the PDF of the log-normal distribution is

$$p_X(x) = \frac{1}{(x-\theta)\sigma\sqrt{2\pi}} \exp\left\{-\frac{\ln^2\left[(x-\theta)/m\right]}{2\sigma^2}\right\}, x \geq \theta, \sigma > 0 \tag{2.59}$$

where σ is the *shape parameter*, θ is the *location parameter*, and m is the *scale parameter*. The case where $\theta = 0$ and $m = 1$ is called the *standard log-normal distribution*. The case where $\theta = 0$ is called the *two-parameter log-normal distribution*. The equation for the standard lognormal PDF is

$$p_X(x) = \frac{1}{x\sigma\sqrt{2\pi}} \exp\left\{-\frac{\ln^2 x}{2\sigma^2}\right\}, x \geq 0, \sigma > 0 \tag{2.60}$$

Four PDFs are shown in Figure 2.15 for representative values of σ. We can see that above $\sigma = 1$, the shapes of the PDFs change dramatically.

Since the general form of probability functions can be expressed in terms of the standard distribution, all subsequent formulas in this section are given for the standard form of the function.

The distribution function is given by

$$P_X(x) = \frac{1}{2}\left[1 + \text{erf}\left(\frac{x - \mu_\text{d}}{\sqrt{2}\sigma_\text{d}}\right)\right] = \Phi\left(\frac{\ln x - \mu_\text{d}}{\sigma}\right), x \geq 0, \sigma > 0 \quad (2.61)$$

where Φ is the CDF of the normal distribution.

A log-normal distribution results if the variable is the product of a large number of independent, identically distributed variables. The mean, variance, skewness, and kurtosis are given by

$$\mu = e^{M + S^2/2} \quad (2.62)$$

$$\sigma^2 = e^{S^2 + 2M}\left(e^{S^2} - 1\right) \quad (2.63)$$

$$\gamma_1 = \sqrt{e^{S^2} - 1}\left(2 + e^{S^2}\right) \quad (2.64)$$

$$\gamma_2 = e^{4S^2} + 2e^{3S^2} + 3e^{2S^2} - 6 \quad (2.65)$$

where M and S are the mean and standard deviation of the natural log of the r.v.

Figure 2.16 is a plot of the lognormal CDF with the same values of σ as the PDF plots above. These curves exhibit a more conventional change as σ becomes larger than 1.

2.6.5.1 Parameter Estimation

The maximum likelihood estimates for the scale parameter, m, and the shape parameter, σ, are

$$\hat{m} = \exp(\hat{\mu})$$

and

$$\hat{\sigma} = \sqrt{\frac{\sum_{i=1}^{N}[\ln(X_i) - \hat{\mu}]^2}{N}}$$

where

$$\hat{\mu} = \frac{\sum_{i=1}^{N} \ln X_i}{N}$$

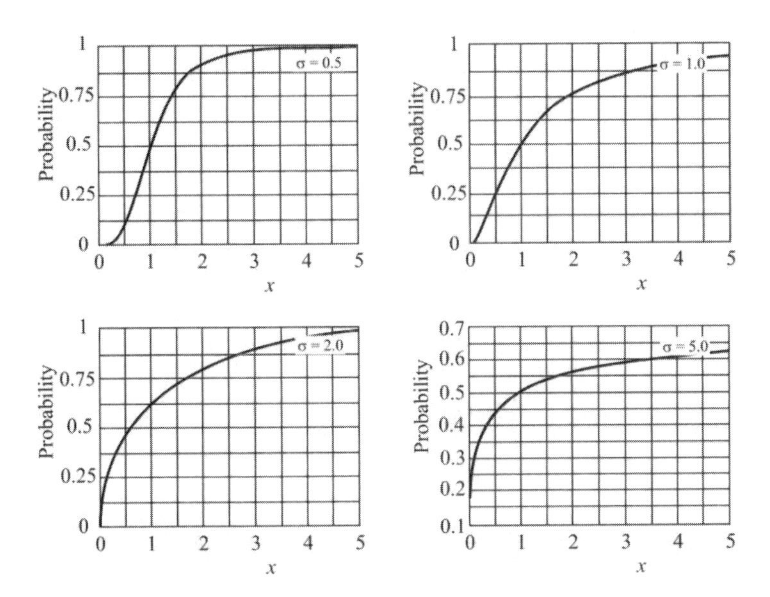

Figure 2.16 Log-normal CDFs.

If the location parameter is known, it can be subtracted from the original data points before computing the maximum likelihood estimates of the shape and scale parameters.

2.6.6 Illustrative Noise Amplitudes

A typical Gaussian noise PDF representing noise into a bandpass filter is shown in Figure 2.11. The Rayleigh PDF of the output of the bandpass filter with no signal present is also illustrated in Figure 2.11. That is what typically happens in a bandpass filter with noise only at its input. The output is characterized by a Rayleigh PDF.

2.6.7 False Alarm Time

Consider the envelope of the noise signal in Figure 2.17 and assume that the noise corresponds to a Rayleigh r.v. A false alarm occurs whenever the noise voltage exceeds a defined threshold γ_{th}. The probability of this occurring is determined by integrating the PDF from that threshold out to the remainder of the possibilities, that is, the integration range is $(\gamma_{\text{th}}, \infty)$:

$$P_{\text{fa}} = \Pr\{\tilde{r} > \gamma_{\text{th}} | \text{Noise Only Present}\} \tag{2.66}$$

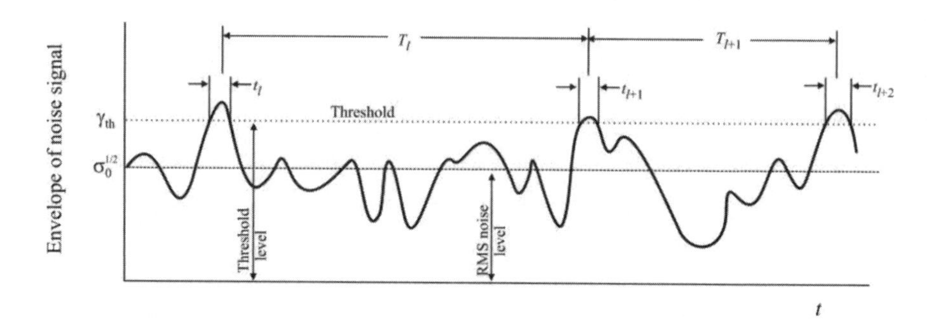

Figure 2.17 Receiver output voltage illustrating false alarms due to noise. (After: [11]. © 1980 McGraw- Hill, Reprinted with permission].

$$= \int_{\gamma_{th}}^{\infty} \frac{r}{\sigma_0} \exp \frac{-r^2}{2\sigma_0} dr = \exp \frac{-\gamma_{th}^2}{2\sigma_0} \tag{2.67}$$

The average time interval between crossings of the threshold is called the *false alarm time* denoted by T_{fa}

$$T_{fa} = \lim_{N \to \infty} \frac{1}{N} \sum_{l=1}^{N} T_l \tag{2.68}$$

where

T_l = time between crossings of the threshold γ_{th} by the noise envelope (when the slope of the crossing is positive)

The false alarm probability could also have been defined as the ratio of the time that the envelope is above the threshold to the total time as shown graphically in Figure 2.17

$$P_{fa} = \frac{\sum_{k=1}^{N} t_k}{\sum_{k=1}^{N} T_k} = \frac{\langle t_k \rangle_{avg}}{\langle T_k \rangle_{avg}} = \frac{1}{T_{fa} W_n} \tag{2.69}$$

where t_k and T_k are defined in Figure 2.17, and the average duration of a noise pulse is the reciprocal of the bandwidth W_n. For a bandwidth $W = W_{IF}$, the false alarm time is just

$$T_{\text{fa}} = \frac{1}{W_{\text{IF}}} \exp \frac{\gamma_{\text{th}}^2}{2\sigma_0} \tag{2.70}$$

2.6.8 Probability of Detection

The probability that the signal will be detected when it is present is the same as the probability that the envelope $\tilde{r}$ will exceed the threshold

$$\begin{aligned} P_{\text{d}} &= \Pr\{\tilde{r} > \gamma_{\text{th}} \,|\, \text{Signal Present}\} \\ &= \int_{\gamma_{\text{th}}}^{\infty} p_R(\tilde{r})dr = \int_{\gamma_{\text{th}}}^{\infty} \frac{u}{\sigma_0} \exp\left(-\frac{u^2 + \tilde{r}^2}{2\sigma_0} \right) I_0 \left(\frac{u\tilde{r}}{\sigma_0} \right) du \end{aligned} \tag{2.71}$$

Unfortunately, this cannot be evaluated in a closed form and so numerical techniques or a series approximation must be used.

In terms of the PDFs, the detection and false alarm process is shown graphically in Figure 2.18.

Note that this is for a single pulse of a steady sinusoidal signal in Gaussian noise with no detection losses. P_{d} and P_{fa} are determined numerically.

2.7 Radiometric Detection of Signals in Noise

2.7.1 Introduction

A radiometer measures the energy content of the signal at its input, and the specifics of that signal in general do not need to be—and in the EW application are not normally—known.

A radiometer detector is shown in Figure 2.19 that consists of a bandpass filter, a squaring function, and an integrator. W is assumed to be wide enough that the SOI passes through without distortion. The energy in a signal $r(t)$, $0 \le t \le T$ is given by

$$E(t) = \int_{t-T}^{t} r^2(u)du \tag{2.72}$$

so a radiometer is an energy detector, and in an EW system that targets SS signals, radiometers are frequently wideband, and thus can be used against any type of SS system considered here. They need not be wideband, however. If W matches the width of each channel, there could be a radiometer for each of the channels in W_{ss}.

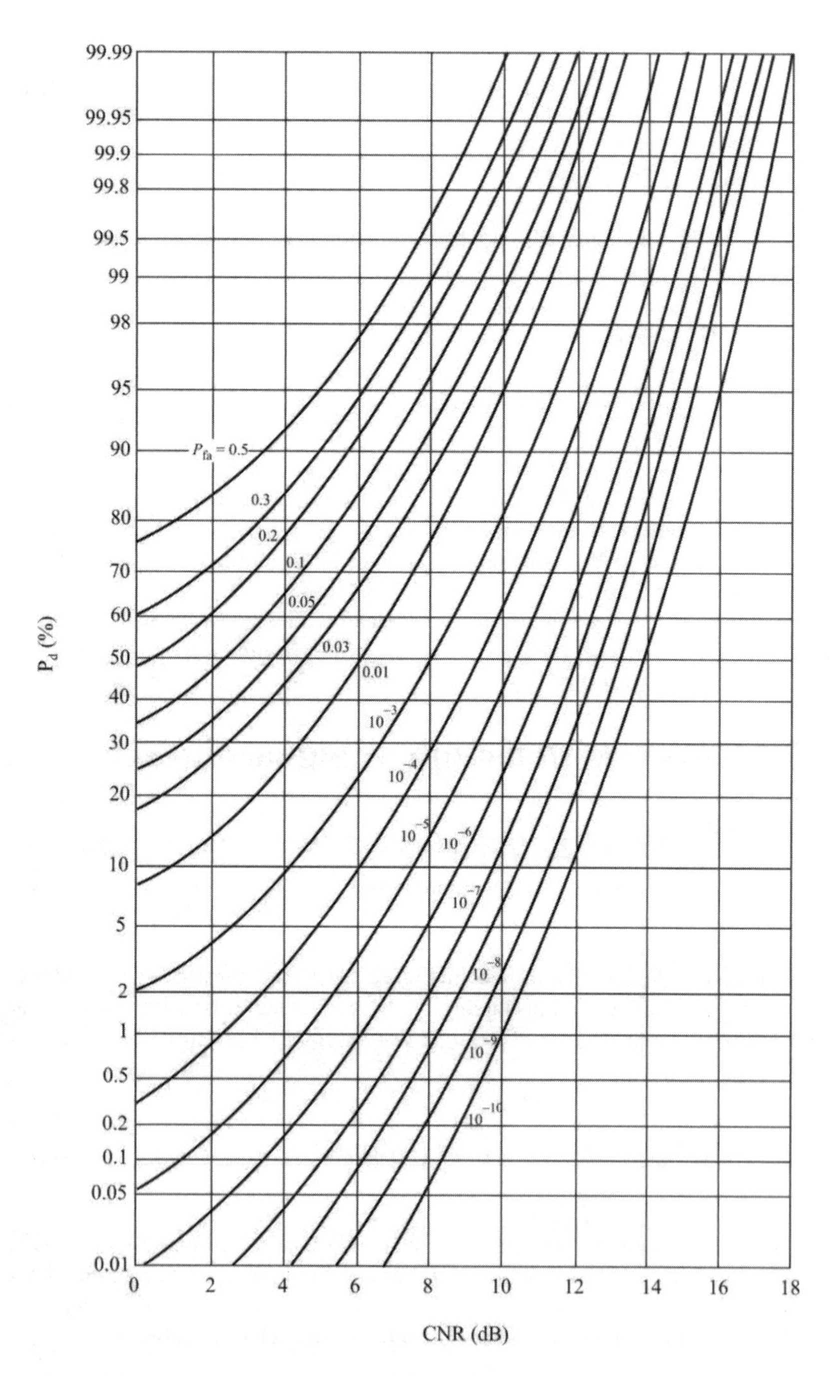

Figure 2.18 Detection probability as a function of SNR with false alarm probability as a parameter. This is the predetect SNR so SNR = CNR. This chart assumes a Rayleigh r.v.

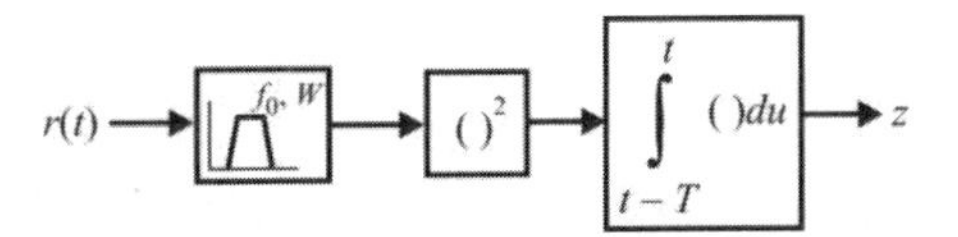

Figure 2.19 Radiometer structure.

As seen in (2.72), the radiometer is an energy detector, and the amount of energy present over $(0, T)$ is used as a measure of the signal presence or absence. The radiometer shown in Figure 2.19 measures the energy E of the input signal with frequency extent given by W Hz over a time interval T. After integrating for T seconds, the integrator output, γ, is sampled. If γ exceeds the threshold given by γ_{th}, the signal is declared to be present.

Frequently the output of the radiometer is normalized by an estimate of the noise associated with the signal. In that case

$$E' = \frac{2}{N_0} \int_{t-T}^{t} r^2(u)\,du \tag{2.73}$$

is used, where N_0 is the one-sided noise spectral density. Palpably, in that case some method must be available to generate the noise estimate.

The performance of such radiometers was analyzed by Dillard and Dillard [12]. The probability of false alarm presented there is

$$P_{\text{fa}} = \int_{\gamma_{\text{th}}}^{\infty} \frac{x^{TW_{\text{F}}-1}}{2^{TW_{\text{F}}}\Gamma(TW_{\text{F}})} e^{-x/2} dx \tag{2.74}$$

where TW_{F} is the time bandwidth product, γ_{th} is the comparison threshold, and $\Gamma(\bullet)$ is the gamma function. The associated probability of detection is given by

$$P_{\text{d}} = \int_{\gamma_{\text{th}}}^{\infty} \frac{1}{2}\left(\frac{x}{u}\right)^{(TW_{\text{F}}-1)/2} e^{-(x+u)/2} I_{TW_{\text{F}}-1}\left(\sqrt{xu}\right) dx \tag{2.75}$$

where $u = 2E_s / N_0$, $I_x(y)$ is the modified Bessel function of the first kind and order x. An example of the results is shown in Figure 2.20 for $P_{\text{fa}} = 10^{-4}$.

The product TW_{F} can be on the order of 1. For a typical VHF FHSS system, for example, using a radiometer as a detector of the data bits in BFSK, $T \approx 1 / 20{,}000$ seconds and $W_{\text{F}} \approx 25$ kHz, yielding $TW_{\text{F}} \approx 1$.

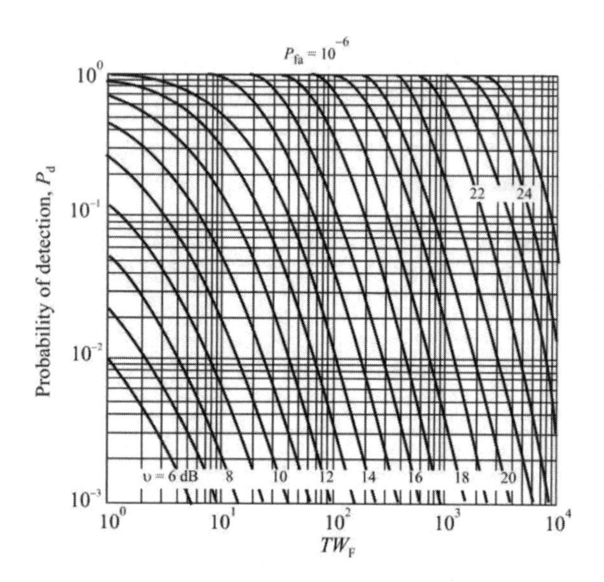

Figure 2.20 Radiometer performance versus the time-bandwidth product when $P_{fa} = 10^{-4}$. (Source: [16]. © 1989 Artech House. Reprinted with permission.)

2.7.2 Radiometer Performance

The radiometer is an optimum detector if the input noise is stationary and Gaussian with a flat PSD. It is also a reasonable architecture for deterministic but unknown signals in Gaussian noise [13]. Analysis of the classic radiometer for energy detection has been extended to cases where the input signals are corrupted by non-Gaussian noise [14]. In addition, analysis of the radiometer has been extended to the case when the input signal noise is nonstationary [15, 16].

As noted above, when the input to a bandpass filter is Gaussian, and therefore has a constant two-sided PSD, $N_0 / 2$, the amplitude of the output is characterized by a Rayleigh density. If there is a sinusoidal signal present as well, then the amplitude of the output is characterized by a Ricean density. Expressing the input PSD by $S_x(f) = N_0 / 2$ and the output PSD by $S_y(f)$, then for any *linear time invariant* (LTI) system with transfer function given by $H(f)$

$$S_y(f) = |H(f)|^2 S_x(f) \tag{2.76}$$

Since $S_x(f)$ is a constant, $S_y(f)$ is a constant multiplier of the square of the system transfer function. If $H(f)$ is characterized as an ideal filter given by

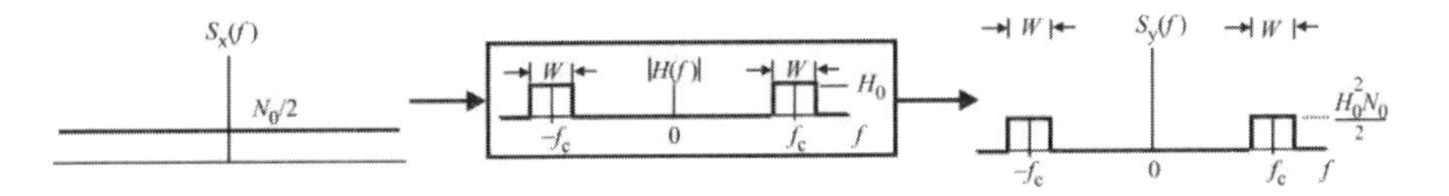

Figure 2.21 Bandpass filtering at IF with noise only at the input.

$$|H(f)| = \begin{cases} H_0, & |f| \le W/2 \\ 0, & \text{otherwise} \end{cases} \tag{2.77}$$

as illustrated in Figure 2.21, then the PSD of the output is given by

$$S_y(f) = \begin{cases} \dfrac{H_0^2 N_0}{2}, & |f| \le W/2 \\ 0, & \text{otherwise} \end{cases} \tag{2.78}$$

This PSD is flat within the spectrum limits of the filter and zero elsewhere, as illustrated in Figure 2.21. It is frequently desired that $H_0 = \sqrt{2/N_0}$ so that $S_y(f) = 1$, which requires knowledge of the noise level N_0.

With a sinusoidal signal present in addition to having the noise spectrum at the output, there will be energy due to the signal. This is illustrated in Figure 2.22.

If the input to the radiometer is AWGN with two-sided PSD given by $N_0/2$, the normalized test statistic z / N_0 is an r.v. with a central chi-square distribution (see Appendix 2A) with $v = 2TW$ degrees of freedom, the PDF of which is given by

$$p_n(z) = \frac{1}{2^{v/2}\Gamma(v/2)} z^{(v-2)/2} e^{-z/2}, \qquad z \ge 0 \tag{2.79}$$

If a signal with energy E in T seconds is present at the radiometer input, z / N_0 has a noncentral chi-square distribution with $2TW$ degrees of freedom and noncentrality parameter $\lambda = 2E / N_0$ with density given by

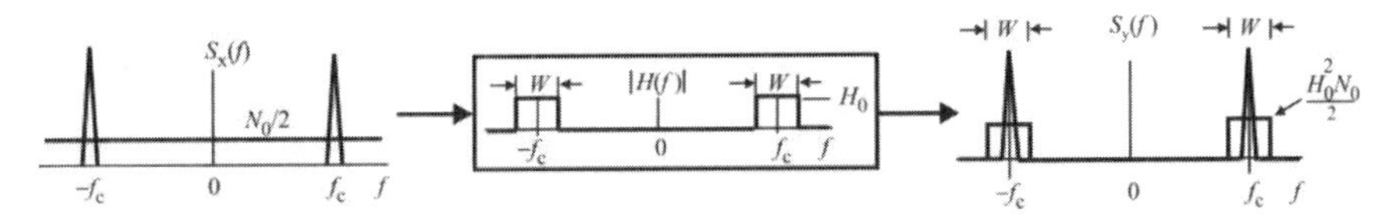

Figure 2.22 Bandpass filtering at IF, signal plus noise at the input.

$$p_{\text{s+n}}(z)=\frac{1}{2}\left(\frac{z}{\lambda}\right)^{(v-2)/4}e^{-(z+\lambda)/2}I_{(v-2)/2}\left(\sqrt{z\lambda}\right),\qquad z\geq 0 \tag{2.80}$$

where $I_k(Z)$ is the kth order modified Bessel function of the first kind (see Appendix A). The performance of the radiometer is determined by integrating the conditional density functions as

$$P_{\text{fa}}=\int_{2\gamma_{\text{th}}/N_0}^{\infty}p_{\text{n}}(z)dz \tag{2.81}$$

and

$$P_{\text{d}}=\int_{2\gamma_{\text{th}}/N_0}^{\infty}p_{\text{s+n}}(z)dz \tag{2.82}$$

Equations (2.81) and (2.82) cannot be solved in closed form so they are typically solved numerically. Approximations have been developed and one of those approximation models due to Edell [17] is presented here.

For $TW > 100$, the following approximations can be used for P_{fa} and P_{d}:

$$P_{\text{fa}}\approx\frac{1}{\sqrt{2\pi}\sigma_{\text{n}}}\int_{\kappa}^{\infty}\exp\left[-\frac{(x-\mu_{\text{n}})^2}{2\sigma_{\text{n}}^2}\right]dx=Q\left(\frac{\kappa-\mu_{\text{n}}}{\sigma_{\text{n}}}\right) \tag{2.83}$$

and

$$P_{\text{d}}\approx\frac{1}{\sqrt{2\pi}\sigma_{\text{s+n}}}\int_{\kappa}^{\infty}\exp\left[-\frac{(x-\mu_{\text{s+n}})^2}{2\sigma_{\text{s+n}}^2}\right]dx=Q\left(\frac{\kappa-\mu_{\text{s+n}}}{\sigma_{\text{s+n}}}\right) \tag{2.84}$$

where

$$\mu_{\text{n}}=2TW \tag{2.85}$$

$$\sigma_{\text{n}}^2=4TW \tag{2.86}$$

$$\mu_{\text{s+n}}=2TW+2E/N_0 \tag{2.87}$$

$$\sigma_{\text{s+n}}^2=4TW+8E/N_0 \tag{2.88}$$

Solving (2.83) and (2.84) for κ and equating the results yields

$$\kappa = \sigma_n Q^{-1}(P_{fa}) + \mu_n = \sigma_{s+n} Q^{-1}(P_d) + \mu_{s+n} \tag{2.89}$$

$$Q^{-1}(P_{fa}) - \frac{\sigma_{s+n}}{\sigma_n} Q^{-1}(P_d) = \frac{\mu_{s+n} - \mu_n}{\sigma_n} = d \tag{2.90}$$

d is the difference in the means of the output densities under noise and signal-plus-noise conditions and sometimes called the *detectability factor* or simply *detectability*. d^2 is a measure of the postdetection or output power SNR of the detector and is referred to as the *deflection*. Equation (2.91) relates the specified performance (P_{fa} and P_d) and the SNR necessary to achieve them. The value of d is typically determined by numerical integration.

When the input SNR is low so that $\sigma_{s+n} \approx \sigma_n$, then

$$d = Q^{-1}(P_{fa}) - Q^{-1}(P_d) = \frac{\mu_{s+n} - \mu_n}{\sigma_n}, \quad \sigma_{s+n} \approx \sigma_n \tag{2.91}$$

Under this assumption, the input carrier power to noise density ratio, R / N_0, is related to d as

$$\frac{R}{N_0} = d\sqrt{\frac{W}{T}}, TW > 100 \tag{2.92}$$

From (2.85)–(2.88)

$$d = \frac{E}{N_0\sqrt{TW}} \tag{2.93}$$

or

$$\frac{E}{N_0} = d\sqrt{TW} \tag{2.94}$$

Since $R = E / T$, then

$$\{R\}_{req} = d\sqrt{\frac{W}{T}}, \quad TW > 100 \tag{2.95}$$

If $TW < 100$, a correction factor must be used to avoid considerable error in the results, and as mentioned above, for the EW detection problem, TW is

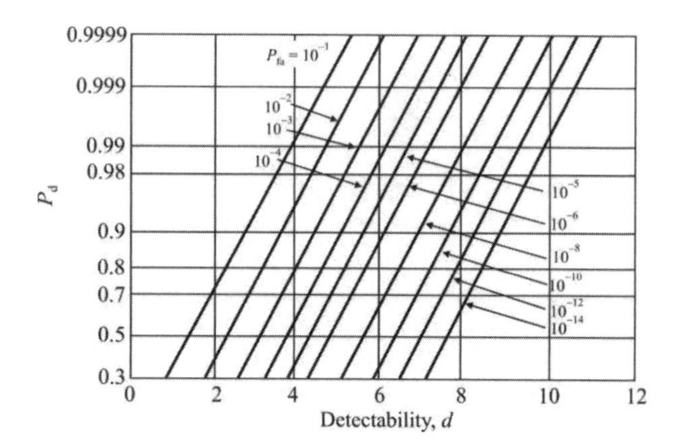

Figure 2.23 Adjustment factor when $TW < 100$. (Source: [17].)

frequently considerably less than 100. This factor, ζ, is a measure of the deviation of the statistics from those of χ^2, and is given by

$$\zeta = \frac{\{R/N_0\}_{\text{req}} \text{ assuming } \chi^2 \text{ statistics}}{\{R/N_0\}_{\text{req}} \text{ assuming Gaussian statistics}} \tag{2.96}$$

Thus

$$\left\{\frac{R}{N_0}\right\}_{\text{req}} = \zeta d\sqrt{\frac{W}{T}}, \text{TW} < 100 \tag{2.97}$$

This adjustment factor is illustrated in Figure 2.23 for when $P_d = 0.9$ (see [17, pp. 27–34] for other values of P_d).

Comparisons between the Edell model and the exact results given by (2.81) and (2.82) are given in Figures 2.24 and 2.25. We see that for the examples given, the SNR error is generally less than 0.5 dB.

In classical energy detection, the performance of the detector is determined strictly by d. Detector performance is illustrated in Figure 2.26 where the probability of detection is plotted versus d. In this model, irrespective of the noise power level present, arbitrary levels of detection are possible simply by extending T (assuming the signal cooperates and remains on for the entire measurement). In this approach, the false alarm probability is determined by the selection of the threshold and, since this probability does not depend on any signal level because, for a false alarm to occur, by definition there is no signal present, it remains

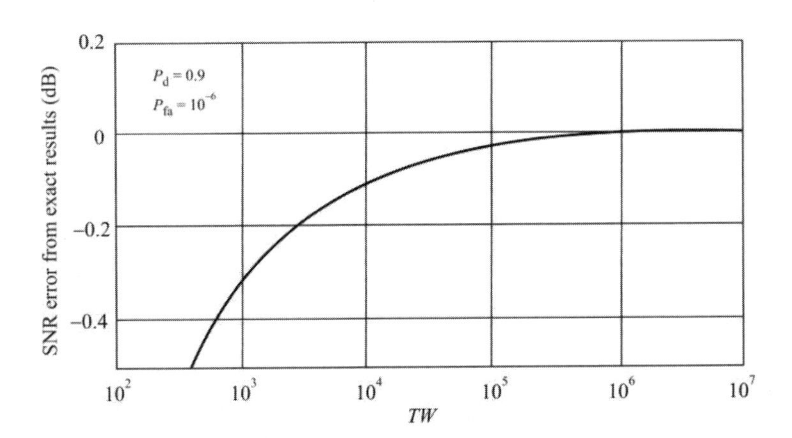

Figure 2.24 Radiometer comparison of Edell's model to the exact results. The graph shows the deviation from the exact SNR required to achieve $P_d = 0.9$ and $P_{fa} = 10^{-6}$.

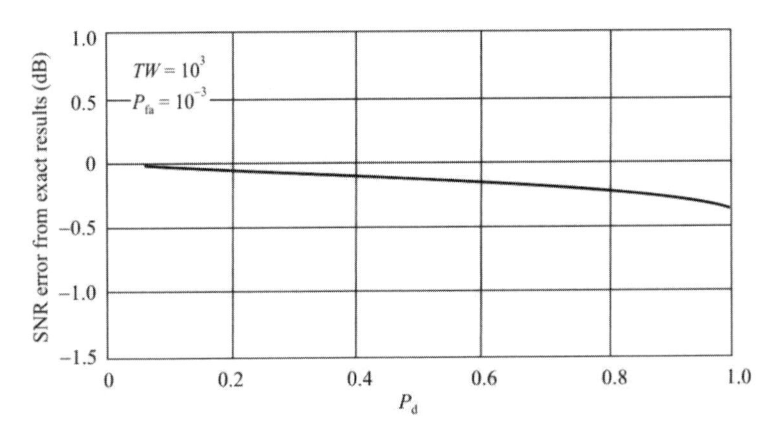

Figure 2.25 SNR difference between approximate and exact results.

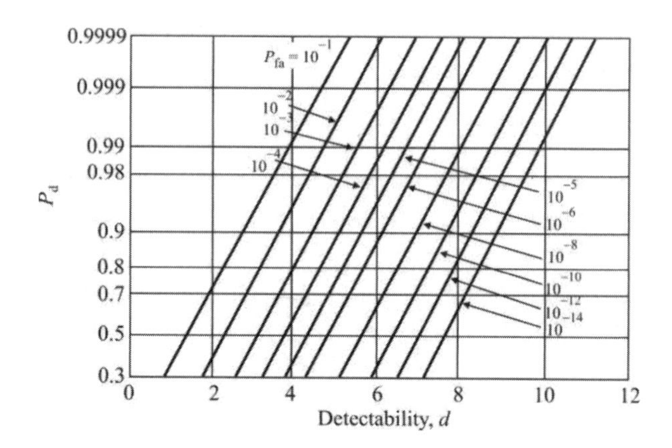

Figure 2.26 Probability of detection versus detectability parameter. (Source: [14]. © IEEE 1989. Reprinted with permission.)

constant. As mentioned previously, this is called CFAR detection, and is perhaps the most common form of signal detection since performance is known.

2.7.3 Effects of Fluctuating Noise Levels on Radiometric Detection

The above analysis assumed that the amount of noise power that is present is constant during the measurement interval and once the threshold is determined it also does not change during the measurement interval. In situations where this is not the case, considerable deviation from the results presented occurs. An HF channel may look like that in Figure 2.27 where significant levels of impulsive noise are likely to occur due to atmospheric conditions (lightning) or man-made sources, such as automotive ignitions and arc welders. Signals in the HF frequency range typically look more like the one shown in Figure 2.27 than ones with constant noise levels. When the noise is fluctuating, and the signal level is below the noise level, for example, in DSSS waveforms, the detectability is not a reliable measure of detection performance.

With fluctuating noise levels and a constant threshold, P_d and P_{fa} do not remain constant. This is illustrated in Figure 2.28. With the threshold held at γ_{th} as determined by the middle noise condition shown in Figure 2.28(a), then P_{fa} is given by the area in black and P_d is given by the area in gray. As the noise level increases, so does P_d, as shown in Figure 2.28(b); however, P_{fa} increases by the amount shaded in dark gray. The threshold would have to be adjusted to γ_1 to maintain a constant false alarm rate (shown here as 10^{-2}). Likewise, as the noise level decreases as shown in Figure 2.28(c), P_{fa} decreases but so does P_d. The threshold would need to move to γ_2 in order to keep the false alarm rate constant.

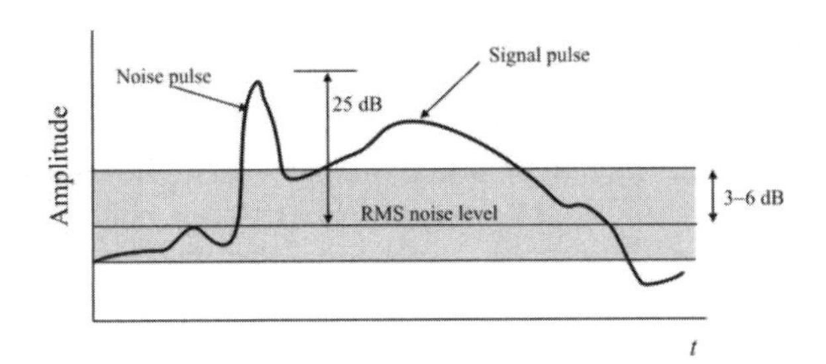

Figure 2.27 Noise fluctuation characteristics in an RF channel. (Source: [14]. © IEEE 1989. Reprinted with permission.)

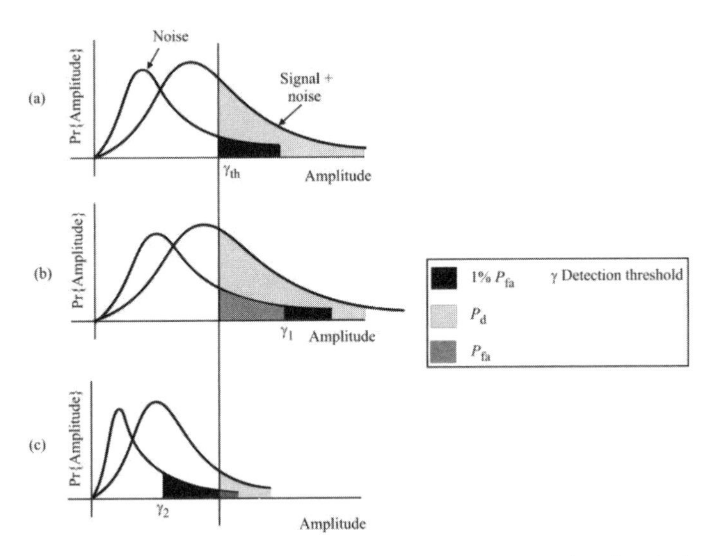

Figure 2.28 (a–c) Fluctuating noise. For CFAR detection the threshold must be adjusted when the background noise is fluctuating. (Source: [14]. © IEEE 1989. Reprinted with permission.)

An approach to estimate the detection performance of an energy detector in a time varying channel is presented in this section. The method assumes that the PDF for the noise power is known (or assumed or estimated to be log-normal, although other PDFs can also be used and the analysis approach remains valid). Then classical distribution functions of the detection probabilities are averaged over the range of noise level P_N present.

For large time-bandwidth products (TW > 100) [17], the probability of detection of such a detector can be approximated by

$$\tilde{P}_d = \int_0^\infty \Pr\{\text{detection} | R + P_N\} p(y) dy \tag{2.98}$$

where $p(y)$ is the PDF of the fluctuating noise level in decibel-like formulation, and $y = 10\log_{10}(P_N)$. Now [18]:

$$\Pr\{\text{detection} | R + P_N\} = Q\left(\frac{\gamma_{th} - \mu_{s+n}}{\sigma_{s+n}}\right) \tag{2.99}$$

where γ_{th} is the detection threshold and μ_{s+n} is the mean of the received signal-plus-noise:

$$\mu_{s+n} = 2TW(R + P_N)$$

Over the bandwidth W, P_N is the noise power, and R the received signal power with υ = SNR so that

$$R = P_N \upsilon \tag{2.100}$$

so

$$\mu_{s+n} = 2TW(P_N \upsilon + P_N) = 2TWP_N(\upsilon + 1) \tag{2.101}$$

σ_{s+n} is the standard deviation of the received signal-plus-noise given by

$$\sigma_{s+n} = 2\sqrt{TW}\sqrt{P_N^2 + 2P_N R}$$

Again, using (2.100),

$$\sigma_{s+n} = 2\sqrt{TW}\sqrt{P_N^2 + 2P_N^2 \upsilon} = 2\sqrt{TW} P_N \sqrt{1 + 2\upsilon} \tag{2.102}$$

Therefore,

$$Q\left(\frac{\gamma_{\text{th}}-\mu_{\text{s+n}}}{\sigma_{\text{s+n}}}\right)=Q\left(\frac{\gamma_{\text{th}}-2TWP_{\text{N}}(\upsilon+1)}{2\sqrt{TW}P_{\text{N}}\sqrt{1+2\upsilon}}\right) \tag{2.103}$$

$$=Q\left(\frac{\gamma_{\text{th}}/P_{\text{N}}-2\text{TW}(\upsilon+1)}{2\sqrt{\text{TW}}\sqrt{1+2\upsilon}}\right) \tag{2.104}$$

In many cases of interest, for example, in the HF range or the slow (shadowing) fading in PCS systems, the noise can best be characterized with a lognormal PDF. In this case, the PDF is given by

$$p(y)=\frac{1}{\sqrt{2\pi}\sigma_{\text{dB}}}e^{-\frac{1}{2}\left(\frac{y-\mu_{\text{dB}}}{2\sigma_{\text{dB}}^2}\right)} \tag{2.105}$$

where μ_{dB} is the mean of the noise (in decibels), σ_{dB}^2 is the variance of the noise in decibels, and $y = 10\log_{10} P_{\text{N}}$. These parameters are related to the mean (μ_{n}) and variance (σ_{n}) of the noise by

$$\mu_{\text{dB}}=\log_{10}\left(\frac{\mu_{\text{n}}^2}{\mu_{\text{n}}^2+\sigma_{\text{n}}^2}\right) \quad \text{and} \quad \sigma_{\text{dB}}=\sqrt{\log_{10}\left[\left(\frac{\sigma_{\text{n}}}{\mu_{\text{n}}}\right)^2+1\right]} \tag{2.106}$$

respectively. Therefore the probability of detection is given by

$$P_{\text{d}}=\frac{1}{\sqrt{2\pi}\sigma_{\text{dB}}}\int_0^\infty Q\left(\frac{\gamma_{\text{th}}/P_{\text{N}}-2TW(\upsilon+1)}{2\sqrt{TW}\sqrt{1+2\upsilon}}\right)\exp\left[-\frac{1}{2}\frac{u-\mu_{\text{dB}}}{2\sigma_{\text{dB}}^2}\right]du \tag{2.107}$$

With the noise power, P_{N}, in watts, u has the form of decibels[2]:

$$u=10\log_{10}P_{\text{N}}$$

then

$$du=10d(\log_{10}P_{\text{N}})$$

[2] This is not really a measure of decibels, since a decibel is based on the log of a ratio with the denominator some reference value. Nevertheless, y is frequently referred to as a decibel indicator since $y = 10\log_{10}P_{\text{N}}$.

$$= \frac{10\log_{10}(e)}{P_{\mathrm{N}}} dP_{\mathrm{N}} \tag{2.108}$$

so that the final form for the probability of detection is

$$P_{\mathrm{d}} = \frac{10\log_{10}(e)}{\sqrt{2\pi}\sigma_{\mathrm{dB}}} \int_{0}^{\infty} \frac{1}{P_{\mathrm{N}}} Q\left(\frac{\gamma_{\mathrm{th}} / P_{\mathrm{N}} - 2TW(\upsilon+1)}{2\sqrt{TW}\sqrt{1+2\upsilon}} \right) e^{-\frac{1}{2}\left(\frac{10\log_{10}(P_{\mathrm{N}}) - \mu_{\mathrm{dB}}}{2\sigma_{\mathrm{dB}}^2} \right)} dP_{\mathrm{N}} \tag{2.109}$$

The overall probability of false alarm due to fluctuating noise is estimated by averaging the probability of false alarm over the range of P_N, namely,

$$\tilde{P}_{\mathrm{fa}} = \int_{0}^{\infty} \Pr\{\text{false alarm} | u\} p(u) du \tag{2.110}$$

where, again, the noise is assumed to be characterized with a log-normal PDF. Now [18]

$$\Pr\{\text{false alarm} | P_{\mathrm{N}}\} = Q\left(\frac{\gamma_{\mathrm{th}} - \mu_{\mathrm{n}}}{\sigma_{\mathrm{n}}} \right) \tag{2.111}$$

μ_n is the mean of the received noise level, and σ_n the standard deviation of the received noise level. Therefore, similar to the derivation above for the probability of detection, the probability of false alarm is determined to be

$$P_{\mathrm{fa}} = \frac{10\log_{10}(e)}{\sqrt{2\pi}\sigma_{\mathrm{dB}}} \int_{0}^{\infty} \frac{1}{P_{\mathrm{N}}} Q\left(\frac{\gamma_{\mathrm{th}} - \mu_{\mathrm{n}}}{\sigma_{\mathrm{n}}} \right) e^{-\frac{1}{2}\left(\frac{10\log_{10}(P_{\mathrm{N}}) - \mu_{\mathrm{dB}}}{2\sigma_{\mathrm{dB}}^2} \right)} dP_{\mathrm{N}} \tag{2.112}$$

The probability of detection for several typical SNRs and TW products is illustrated in Figure 2.29. As is normal for such analysis, the threshold value was changed to establish P_{fa}. For the normal radiometer case considered earlier, once that threshold value was determined, it remained constant for all values of TW. That is not the case for Figure 2.29. The threshold level was changed in Figure 2.29 so that the false alarm rate was approximately constant at $\sim 10^{-4}$. It should be noted that P_{fa} is very sensitive to the value of this threshold.

The same data as shown in Figure 2.29 is shown in Figure 2.30. This figure makes the extreme sensitivity to the SNR a little more obvious. Detection performance goes from virtually nothing to almost a sure thing for $TW > 100$ as the SNR varies over just a 10 dB range.

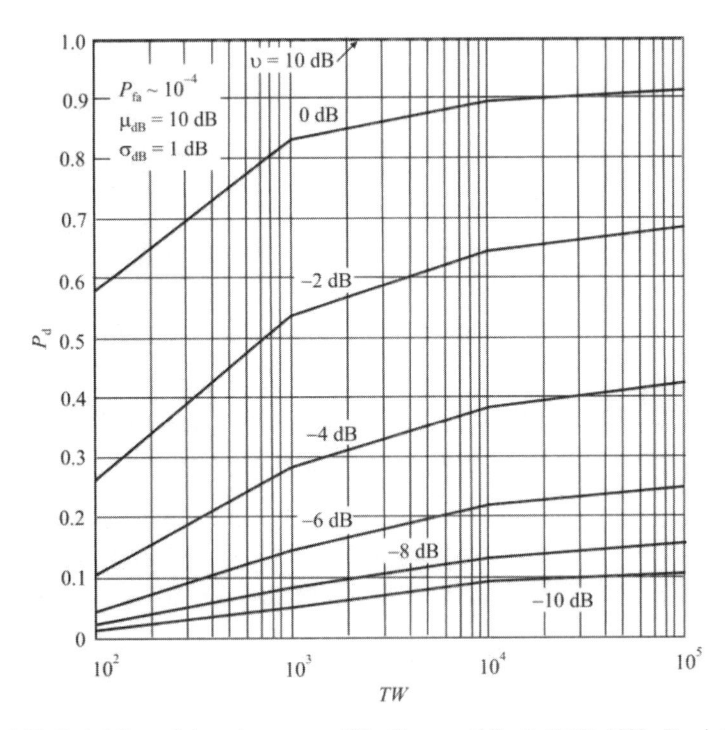

Figure 2.29 Probability of detection versus *TW*. (Source: [14]. © IEEE 1989. Reprinted with permission.)

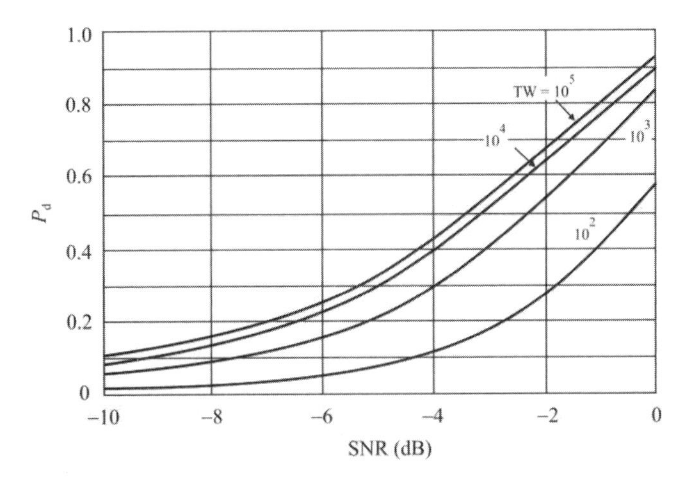

Figure 2.30 Probability of detection versus SNR. This illustrates the sensitivity of P_d to the SNR when the noise is nonstationary.

σ_{dB} is related to the channel standard deviation by

$$\sigma_{\text{n}}^2 = e^{\sigma_{\text{dB}}^2 + 2\mu_{\text{dB}}}(e^{\sigma_{\text{dB}}^2} - 1) \tag{2.113}$$

and

$$\mu_{\text{n}} = e^{\mu_{\text{dB}} + \sigma_{\text{dB}}^2/2} \tag{2.114}$$

Since the probability of false alarm is dependent on parameters beyond the threshold, it is not possible to implement a constant false alarm process. That is, P_{fa} cannot be used to specify a constant γ_{th}, which is then used to predict P_{d}.

This model is more applicable to detection problems in a noise fluctuating channel than the classic radiometer analysis presented above. For the typical PDF defined with a 10 dB mean (μ_{n}) and a standard deviation (σ_{n}) greater than 0 dB, the detection performance can only be increased up to a limit with an increase in time-bandwidth product. Beyond that limit, no significant improvement can be obtained.

2.7.4 Effects of Noise Power Level Uncertainty in Radiometric Detectors

Even if the noise power level remains constant during the measurement interval, it must be measured in order for a radiometer to function correctly. It was noted in the previous section that P_{fa} is extremely sensitive to changes in the threshold in radiometers. This effect was quantized by Sonnenschein and Fishman [15], and the results obtained there are summarized in this section.

The mean value of noise as well as the variance of the noise when only noise is present can be expressed as [18]

$$\mu_{\text{n}} = N_0 TW, \qquad \sigma_{\text{n}}^2 = N_0^2 TW \tag{2.115}$$

while these parameters when the signal is present are given by

$$\mu_{\text{s+n}} = N_0 TW(\upsilon + 1), \qquad \sigma_{\text{s+n}}^2 = N_0^2 TW(2\upsilon + 1) \tag{2.116}$$

where the input SNR is[3]

$$\upsilon \triangleq \frac{R}{N_0 W}$$

[3] The symbol $\triangleq$ denotes mathematical definition.

The probability of false alarm, P_{fa}, and the probability of detection P_d, are then given by

$$P_{fa} = Q\left(\frac{\gamma_{th} - \mu_n}{\sigma_n}\right) \tag{2.117}$$

and

$$P_d = Q\left(\frac{\gamma_{th} - \mu_{s+n}}{\sigma_{s+n}}\right) \tag{2.118}$$

where $Q(\ )$ is the Q-function.

Setting the threshold γ_0 for a required probability of false alarm, $P_{fa,req}$, we obtain

$$\gamma_{th} = \mu_n + \sigma_n Q^{-1}(P_{fa,req}) \tag{2.119}$$

To implement a radiometer for energy detection, the receiving system must estimate the noise power N_0 so that γ_{th} can be determined. Denote this estimate by $\hat{N}_0$. Based on this estimate, the estimate for γ_{th}, $\hat{\gamma}_{th}$, is determined. Two bounds ϵ_1 and ϵ_2 are defined such that

$$(1-\epsilon_1)N_0 \le \hat{N}_0 \le (1+\epsilon_2)N_0 \tag{2.120}$$

with $0 \le \epsilon_1 < 1$ and $\epsilon_2 \ge 0$. The *noise level uncertainty*, U, is then defined as

$$U \triangleq \left(\frac{1+\epsilon_2}{1-\epsilon_1}\right) \ge 1 \tag{2.121}$$

The threshold estimate is given in terms of U as

$$\hat{\gamma}_{th} = U\gamma_{th} \tag{2.122}$$

The SNR needed to achieve $P_d \ge P_{d,req}$ and $P_{fa} \le P_{fa,req}$ over the entire range of noise uncertainty (2.120) is

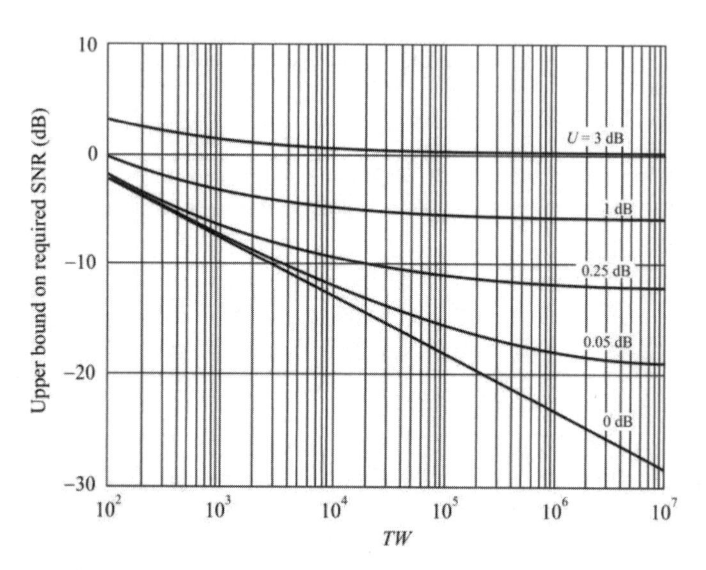

Figure 2.31 Upper bound on required SNR 0 dB $\leq U \leq$3.0 dB, $P_{\text{fa}} = P_{\text{m}} = 10^{-2}$.

$$\upsilon = (U-1) + U\left(\frac{B}{\sqrt{TW}}\right) + \frac{A}{TW}\left[A - \sqrt{A^2 + (2U-1)TW + 2\sqrt{TW}B}\right] \tag{2.123}$$

where $A \triangleq Q^{-1}(P_{\text{d,req}})$ and $B = Q^{-1}(P_{\text{fa,req}})$. For $TW >> 1$, (2.123) becomes

$$\upsilon \approx (U-1) + \frac{UB - A\sqrt{2U-1}}{\sqrt{TW}} \tag{2.124}$$

Figure 2.31 shows (2.123) for several values of U and $P_{\text{fa}} = P_{\text{m}} = 10^{-2}$. Note that for small values of SNR that even for a small amount of uncertainty, on the order of fractions of a decibel, there is considerable variability in the amount of additional SNR required to assure the performance expected and predicted for the certain-noise case. The case of $U = 1$, or no uncertainty, is given by the $U = 0$ dB curve.

Figure 2.32 shows the SNR variability for when the uncertainty $U = 0.25$ dB.

The relative increase in SNR engendered by noise-level uncertainty for $TW >> 1$ (a condition normally true for DSSS signals) is given by

$$\Delta\upsilon \approx U + \frac{U-1}{\upsilon_{U=1}} \tag{2.125}$$

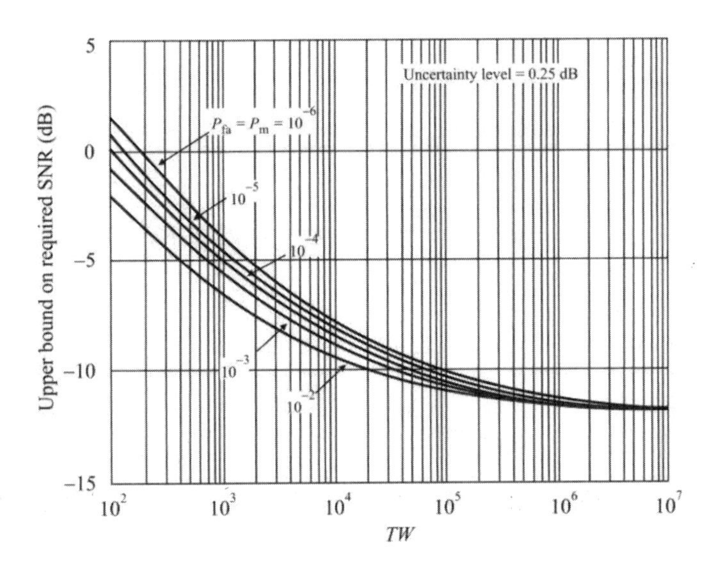

Figure 2.32 Upper bound on required SNR: uncertain noise level $U = 0.25$ dB.

where $\upsilon_{U=1}$ is the SNR required when there is no uncertainty. This equation is plotted in Figure 2.33 for $0 \leq U \leq 3$ dB. What Figure 2.33 indicates that if $\upsilon_{U=1} = -20$ dB and the noise level uncertainty $U = 1$ dB, then $\Delta\upsilon \approx 13$ dB more SNR is required to ensure the performance of the radiometer. As the $\upsilon_{U=1}$ increase above zero, note from Figure 2.33 that, for small values of uncertainty, $U << 1$, $\Delta\upsilon \approx 0$, and the SNR prediction for certain-noise conditions is acceptable.

This extreme sensitivity to uncertainty in knowing N_0 is apparently due to the fact that at the SNR levels normally encountered in DSSS signals, which is typically less than zero, there is considerably more noise energy than signal energy. This is evident from (2.115) and (2.116) because the mean of the test statistic is influenced by the noise level for both hypotheses, and for typical DSSS signal-to-noise ratios the noise is the dominant signal present.

The noise level uncertainty can arise from any and all of the noise sources discussed herein. It can be due to external sources, predominantly atmospheric in the low VHF range and man-made in the upper frequency ranges (both of which are impulsive in nature), or it can be due to any of the internal noise sources. Estimating this noise level accurately in a radiometer is critical to its performance.

2.7.5 Energy Detection of a Signal with Random Amplitude

The above analyses assumed that the noise into the detection receiver was Gaussian and stationary. Kostylev documented a development where the Gaussian

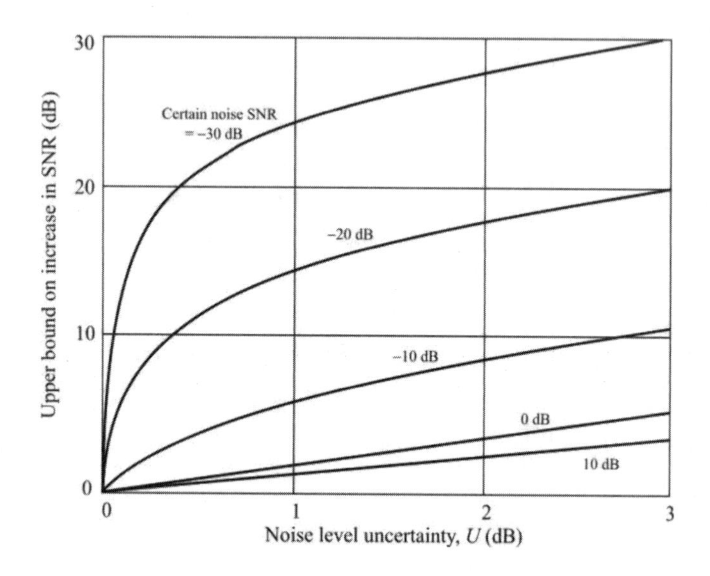

Figure 2.33 Upper bound on SNR increase required: 0 dB $\leq U \leq$ 3 dB.

assumption was removed [13] (the stationary assumption remained). He allowed the amplitude of the signal to have Rayleigh, Ricean, and Nakagami distributions, as well as others, and computed the expressions for P_d and P_{fa}. Those results are summarized in this section. For the most part, the probabilities of false alarm and detection are not available in closed form but are manifest in functions that must be computed numerically.

Let the input signal be given by

$$r(t) = \text{Re}\{A\tilde{r}(t)\exp\left[j(\omega_0 t + \varphi)\right]\} \tag{2.126}$$

where A is the random amplitude, $\tilde{r}(t)$ is the normalized deterministic complex envelope, $\omega_0 = 2\pi f_0$ is the carrier frequency, and φ is the random initial phase. It is assumed that the carrier frequency, f_0, the bandwidth of the detected signal, Δf, and the PSD of the noise, N_0, are all known quantities. Furthermore, it is assumed that the received signal is bandpass filtered with a perfect filter with transfer function, $H(f)$, given by

$$H(f) = \begin{cases} 2/\sqrt{N_0}, & |f - f_0| \leq \Delta f/2 \\ 0, & |f - f_0| > \Delta f/2 \end{cases} \tag{2.127}$$

2.7.5.1 Rayleigh Amplitude Density

When the signal amplitude varies according to a Rayleigh density function, then

$$P_{\text{fa}} = \frac{\Gamma(B+1,h/2)}{\Gamma(B+1)} \tag{2.128}$$

where $\Gamma(a,b) = \Gamma(a) - \gamma(a,b)$ is the complementary incomplete gamma function [19]. In addition

$$P_{\text{d}} = P_{\text{fa}} + \left(\frac{d^2+1}{d^2}\right)^B \exp\left[\frac{\gamma_{\text{th}}}{2(d^2+1)}\right] P\left[B+1, \frac{\gamma_{\text{th}} d^2}{2(d^2+1)}\right] \tag{2.129}$$

where $P(a,b) = \gamma(a,b)/\Gamma(a)$ is the normalized incomplete gamma function [19], $B = T\Delta f$ is referred to as the *processing base*, $d^2 = \overline{A^2} q^2$ is the input SNR, and

$$q^2 = \frac{2}{N_0} \int_0^T |\tilde{r}(t)|^2 \, dt \tag{2.130}$$

is the SNR of the deterministic complex envelope.

2.7.5.2 Ricean Amplitude Density

When the amplitude of the input signal is characterized by a Ricean PDF, then

$$P_{\text{d}} = \sum_{k=0}^{\infty} \frac{\alpha_k}{(B+k)!} \Gamma\left(B+k+1, h/2\right) \tag{2.131}$$

where

$$\alpha_k = \frac{\exp(-\lambda/2)}{\varsigma} \left(\frac{\varsigma-1}{\varsigma}\right)^k L_k\left(-\frac{\lambda}{2(\varsigma-1)}\right) \tag{2.132}$$

and λ is the noncentrality parameter

$$\lambda = \frac{2a^2 d^2}{2d^2 + a^2 + 2} \tag{2.133}$$

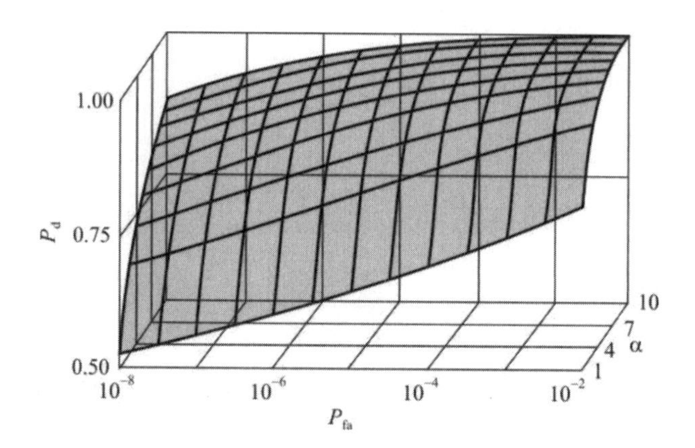

Figure 2.34 Ricean detection probability. (Source: [14]. © IEEE 2002. Reprinted with permission.)

In addition,

$$\varsigma = \frac{2d^2 + a^2 + 2}{a^2 + 2} \tag{2.134}$$

is a weight parameter and a is the parameter of the Rice PDF as

$$p(A) = \frac{A}{\omega^2} \exp\left(-\frac{a^2}{2}\right) \exp\left(-\frac{A^2}{2\omega^2}\right) I_0\left(\frac{a}{\omega} A\right) \tag{2.135}$$

The detection characteristics corresponding to this PDF are shown in Figure 2.34 as α is varied.

2.7.5.3 Nakagami Amplitude Density

For the Nakagami amplitude density, the probability of detection is given by

$$\begin{aligned} P_\text{d} &= \frac{1}{\Gamma(m)\Gamma(B-m+1)2^m(2+d^2/m)^m} \\ &\quad \times \int_0^{\gamma_\text{th}} x^{m-1} \exp\left(-\frac{x/2}{1+d^2/m}\right) \Gamma\left(B-m+1, \frac{\gamma_\text{th}-x}{2}\right) dx \end{aligned} \tag{2.136}$$

Detection performance for this density is shown in Figure 2.35 with m as the parameter, in Figure 2.36 with SNR as the parameter, and Figure 2.37 with B as the parameter.

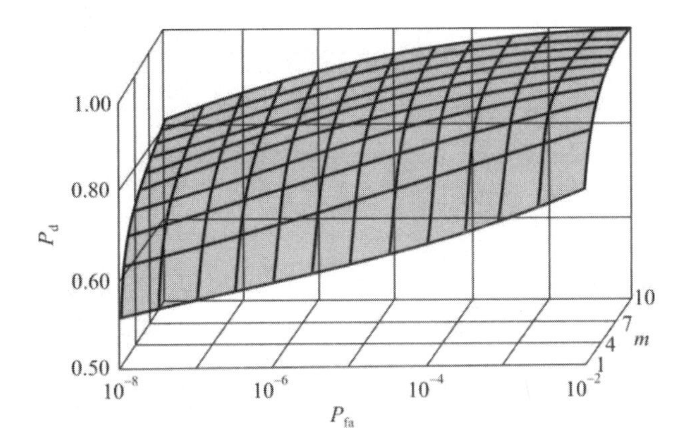

Figure 2.35 Nakagami detection probability. (Source: [14]. © IEEE 2002. Reprinted with permission.)

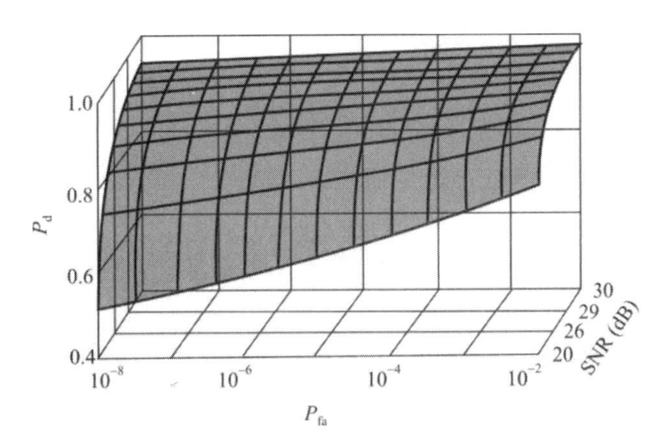

Figure 2.36 Nakagami detection probability SNR. (Source: [14]. © IEEE 2002. Reprinted with permission.)

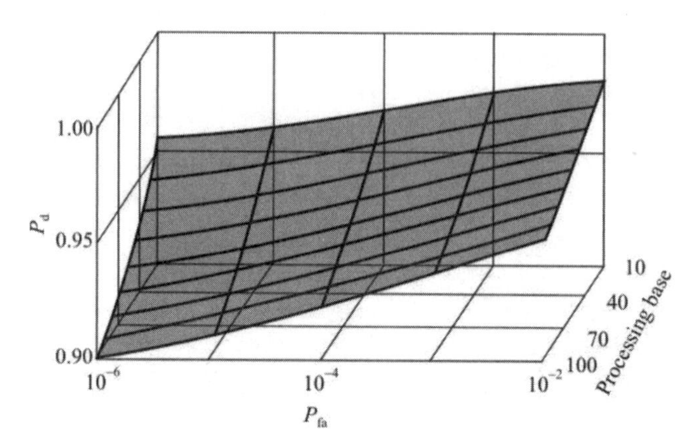

Figure 2.37. Detection performance base. (Source: [14]. © IEEE 2002. Reprinted with permission.)

2.8 Urban RF Noise

2.8.1 Introduction

By far, the most important source of noise in the higher mobile communication bands is that radiated by electrical equipment of various kinds and is MMN. MMN is almost always impulsive in nature [20], which means that its characteristics are fundamentally different from random noise of thermal origin. *Impulsive noise* (IN) is characterized by bursts of very short duration, which may have random amplitude and random time of occurrence. The impulsive nature of some MMN produces a PSD that is broadband. IN typically comes from electrical equipment where sparks occur. A prime example of this source is the car engine. First, it is ubiquitous, and second, the individual sources can be quite strong due to the high voltage breakdown at the spark plugs. In this section we focus considerably more on the predominant noise sources in urban settings: man-made noise [21].

Much of the material in this chapter came from [22].

2.8.2 Urban Noise Overview

Most MMNs have an impulsive characteristic. That is, the energy arrives in bursts and can have significant amplitude. This is due to the sources of most such noise. Sources of this MMN include:

- Noise from electrical machinery (particularly from commutating DC motors);
- Switching transients;

- Discharge lighting;
- Microwave ovens;
- Automotive ignitions;
- Neon signs;
- Fluorescent lightbulbs;
- New compact fluorescent lightbulbs (CFLs);
- Electric welders;
- Kitchen appliances.

MMN from these sources is generally broadband and the emission levels are largely controlled by manufacturing and distribution standards. Taking local action against a specific piece of equipment may control noise identified from nearby sources. Proliferation of such MMN sources, however, particularly in built-up and industrial areas, coupled with the propagation of the signal to a receiving antenna, may result in an integrated MMN level that varies with location and time. This may be quantified statistically and used as a fundamental system design and planning parameter. By its very nature, however, such noise is dramatically non-stationary.

Other MMN sources not considered above include:

- Radiation from computing and telecommunications equipment;
- Unwanted (spurious) emissions from radio communication equipment.

These sources often carry the appellation *electromagnetic interference* (EMI), also called *radio frequency interference* (RFI). They may exhibit discrete emissions on specific frequencies or across a frequency band. It is unclear, however, what their contribution to the background MMN level would be. These emissions are often regulated and covered by dictated specifications in a military setting. The analysis/design considerations that include EMI concerns are referred to as *electromagnetic compatibility* (EMC). It is often the case that many such electronic equipment are in close proximity to one another and this equipment must be able to function in the midst of noise energy from other devices.

The effects of all these noise sources on radio communications may be overcome locally if the specific source can be identified. This may not be possible, however, for noise from telecommunication networks where the ubiquity of the wired network used for distribution may prevent avoidance action. Where an individual source cannot be identified, these unwanted signal sources may set an MMN background level. However, emissions from these unwanted and unintentional sources, which may have Gaussian noise characteristics in some cases, may make the interpretation of field noise measurements difficult. Furthermore, in an EW setting, it is typically unknown a priori where the system is to be placed in an urban environment, so the sources of noise are likewise

Table 2.2 Field Strength Limits Imposed by the FCC on U.S. Consumer Electronics at a 10 m Range

Frequency Band (MHz)	Field Strength (μV m^{-1})	Field Strength (dB$_{\mu V\,m}$$^{-1}$)
30–88	90	39
88–216	150	43.5
216–960	210	46.5
Above 960	300	49.5

Source: [23].

unknown a priori. Attempts to control the noise sources by preemption are likely to be unsuccessful.

In all developed countries there are limits imposed on noise contribution of devices and, in particular, man-made devices. In the United States, those limits are established by the Federal Communications Commission (FCC). The FCC uses two categories of digital devices to establish unintentional radiation limits. Class A devices are intended for use in commercial or industrial settings, while class B devices are intended for home use. The limits on radiation from any digital device are documented in Chapter 47 of the *Code of Federal Regulations* (CFR) Part 15, specifically 15.109. The limits for class A devices are given in Table 2.2. Related international standards are similar to those given in Table 2.2. Most EW systems are very sensitive devices, with system sensitivity goals on the order of 1–10 μV m^{-1}. In urban areas where there are many man-made devices, probably the most prolific being luminaries, the noise field strengths add. So even though the field strengths listed in Table 2.2 are specified at 30 m distance, there are so many of them, the noise fields in urban terrain are very strong.

The characteristic nature of the majority of MMN sources is that they are impulsive, as is atmospheric noise. Furthermore, the energy distribution may not be Gaussian. This has been tackled (when considering atmospheric noise due to lightning) by providing estimates of the median noise power and then by applying a bandwidth-dependent parameter to describe the *amplitude probability distribution* (APD).

The above approach may be suitable at lower frequencies (HF), where a median level can be measured, but may not be appropriate at VHF and higher frequencies. At these frequencies semi-impulsive peak levels may affect system performance even though the measured median level is below the internal noise level. Consequently, it is likely that the bandwidth parameter used to describe the non-Gaussian nature of the noise may not be adequate, particularly for wide-bandwidth systems.

Another approach is based on the assumption that at VHF the noise sources are purely impulsive and that the *noise amplitude distribution* (NAD) essentially describes the number of impulses with specified amplitude [24].

Yet another approach is to model impulsive noise as an alpha-stable process. Such stochastic processes have higher probability distribution tails, which permit higher probabilities for larger noise energies than does a Gaussian assumption for the energy distribution. We discuss alpha-stable processes at the end of this chapter.

2.8.3 Vehicle Ignition Systems

In urban settings a very significant source of MMN is the ignitions in vehicles with internal combustion engines. Of course, the urban core is the place in a city where this problem would be the worst, exacerbated by the density of vehicles. As we move away from the core through the core periphery into the urban sprawl, this problem reduces. Of course, there are exceptions to this—rush hour, for example. Automotive ignitions are not the only sources of MMN, albeit in urban areas where there is considerable road traffic, it is one of the major ones. MMN, arising due to a variety of emissions from electrical discharges and other sources, may set a higher background limit than natural and internal noise. Because this IN source is so prevalent, we begin our discussion of MMN with that topic.

2.8.3.1 Mechanical

Taking ignition systems as an example, the sequence of events that leads to radiated interference is that (typically) initially contact breaker points open, interrupting the flow of current in the primary of the ignition coil and inducing a voltage of about 15 kV in the secondary (Figure 2.38). This 15 kV is sufficient to break down the air in the spark plug gap. The spark typically lasts for less than a microsecond and sustains a time-varying current, which is a function of various resonances (from parasitic inductances and capacitances), in the (mainly) high-voltage wiring. Thus, the pulse has an oscillatory component.

The spark pulse shown in Figure 2.39 and the two-sided spectrum in Figure 2.40 are for illustrative purposes. Actual pulses will differ in scale and detail from these illustrations. For example, the electrical interference bandwidths could exceed hundreds of Megahertz. MMN has been detected at frequencies up to 7 GHz [24].

2.8.3.2 Solid State

Even though automotive ignition systems have for the most part moved away from the distributor and rotor concept to solid state ignitions, there are still several man-made devices that utilize the old techniques. The lawn mower is a good example. Even so, the spark itself is a source of radiation and even with solid state ignitions there is still the requirement for spark plugs (see Figure 2.41).

Solid state is a broad term applied to any engine's ignition system that uses electronic devices such as diodes, transistors, silicon controlled rectifiers, or other semiconductors in place of one or more standard ignition components.

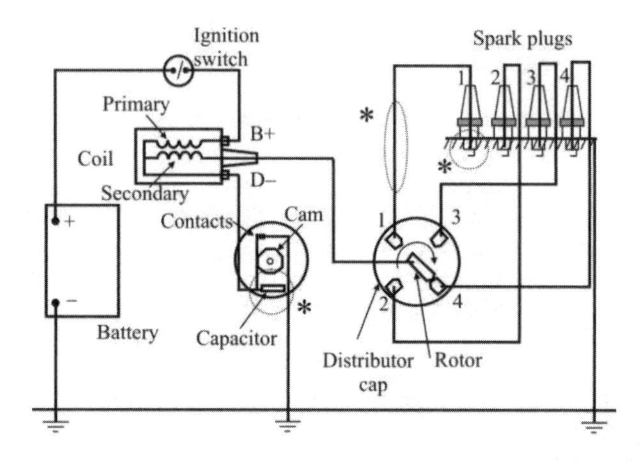

Figure 2.38 Mechanical car ignition schematic.

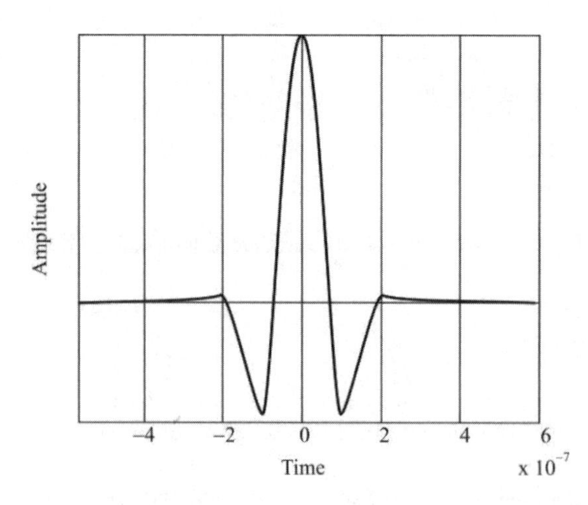

Figure 2.39 Spark pulse. The ordinate scale is arbitrary.

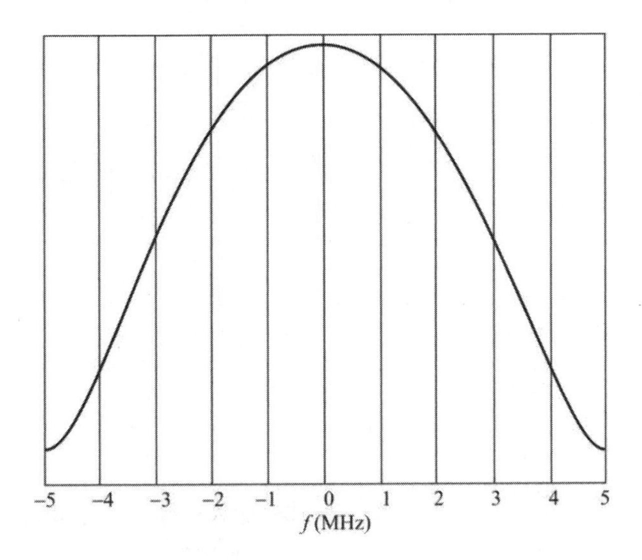

Figure 2.40 Spark spectral amplitude. The ordinate scale is arbitrary.

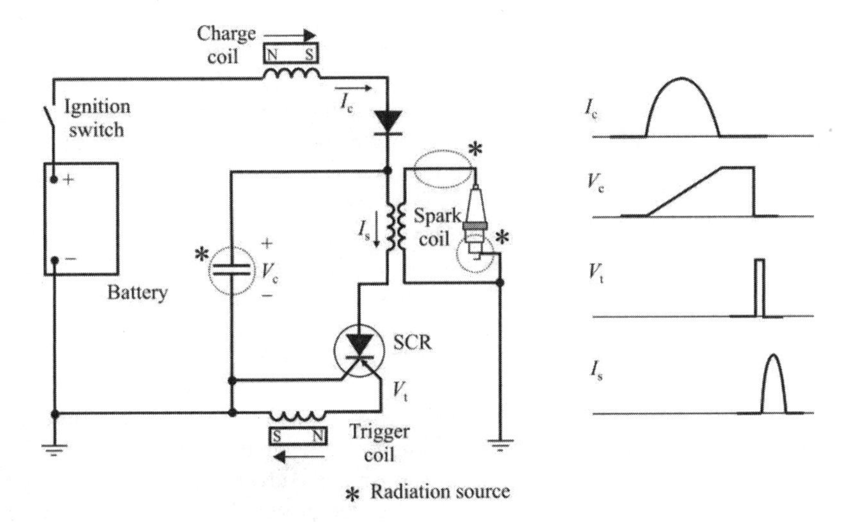

Figure 2.41 Solid state ignition. Waveforms are on the right.

Electronic components are extremely small, have no moving parts, require no mechanical adjustments, are not subjected to wear, as with mechanical devices, deliver uniform performance throughout component life and under adverse operating conditions, and can be hermetically sealed, thus unaffected by dust, dirt, oil, or moisture.

The capacitor discharge system illustrated in Figure 2.41 is breaker-less, with an electronic component replacing the mechanical points and related accessories (breaker cam, spark advance assembly, and so forth). The flywheel contains permanent magnets, but there are no other moving mechanical parts.

The magnet on the flywheel induces a secondary current in the charge coil and trigger coil as it moves past the coils shown. The charge coil facilitates charging the capacitor while the trigger coil triggers the SCR, causing current to flow in the primary of the spark coil.

The *silicon controlled rectifier* (SCR) shown in Figure 2.41 is a form of diode. It does not conduct current until it is triggered. The idealized voltages and currents shown on the right in Figure 2.41 explain the operation of the ignition system. The voltage across the charge coil is zero until the magnet mounted on the flywheel (there would be one for each spark plug) passes by the coil. When it passes by, it induces a magnetic field in the charge coil that causes a current to flow in it. This causes the capacitor to charge to a level determined by the physical size of the components involved. Once the magnet passes the charge coil, the capacitor voltage will remain at the level shown since the SCR is off and no current is flowing in the primary of the spark transformer. The upper diode keeps the charge from leaking off back through the charge coil. When a second magnet passes the trigger coil, a current is induced in it which causes a voltage to be developed at the trigger of the SCR. This short pulse causes the SCR to begin conducting and the capacitor rapidly discharges through the primary of the spark coil, through the SCR to ground. This current induces a large voltage in the secondary of the spark coil, which, in turn, causes the spark plug to fire.

So, some of the emission sources in the mechanical ignition system have been removed, thereby quieting the system down somewhat. There are, however, ample sources of possible emissions in the solid state ignition. The spark plug still has to fire to do its job. There is a large secondary voltage in the solid state ignition, just as there is in the mechanical ignition. This will radiate noise. Finally, just as in the mechanical ignition system, the discharge capacitor will radiate as well.

2.8.4 Fluorescent Lighting

Another source of significant MMN in urban settings is fluorescent lighting, an example of the larger categorization with the appellation *discharge lighting*. Discharge lighting consists of fluorescent lamps, neon advertising signs, neon and argon glow lamps, mercury and sodium lamps, mercury-arc lamps, and carbon arc lights, among others. All discharge lighting luminaries work on the same basic principles—we will discuss fluorescent lighting and, in particular, the relatively new *compact fluorescent lights* (CFL). Discharge lighting is very prolific in urban settings, especially in the concentrated urban core.

2.8.4.1 Fluorescent Lamp Basics

The fluorescent lamp was the first major advance to be a commercial success in small-scale lighting since the tungsten incandescent bulb. Its greatly increased efficiency resulted in cool (temperature wise) brightly lit workplaces (offices and factories) as well as home kitchens and baths. The development of the mercury vapor *high intensity discharge* (HID) lamp actually predates the fluorescent (the latter being introduced commercially in 1938, four years after the HID). However, HID type lamps have only relatively recently become popular in small sizes for task lighting in the home and office; yard and security area lighting; and light source applications in overhead, computer, and video projectors, all of which are extremely prolific in urban settings [25].

Like neon signs and mercury or sodium vapor street or yard lights, fluorescent lamps are a type of gas discharge tube. A small amount of mercury along with some inert gases (usually argon) at very low pressure are sealed inside a glass tube. A pair of electrodes, one at each end, are also sealed in the tube. A phosphor coating is applied to the inside of the tube, which produces visible light when excited with *ultraviolet* (UV) radiation. The electrodes are filaments that are used at start up and remain hot during normal operation as a result of the gas discharge

When power is first applied, a high voltage (several hundred volts) is needed to initiate the discharge. However, once this occurs, a much lower voltage—usually under 100V for tubes under 30W, 100 to 175V for 30W or more—is needed to maintain it.

The electric current passing through the low-pressure gases emits considerable UV, but not much visible light. The gas discharge radiation is almost entirely mercury radiation, although the gas mixture is mostly inert gas and generally only around 1% mercury vapor. The internal phosphor coating converts most of the UV to visible light. The mix of the phosphors is used to color the light spectrum to the intended application. Thus, there are cool white, warm white, colored, and black light fluorescent lamps.

At the wavelengths that are useful to humans, fluorescent lamps are about two to four times as efficient as incandescent lamps at producing light. Thus, they run cooler and consume less power for the same effective light output. They also last considerably longer—10,000 to 20,000 hours versus 1,000 hours for a typical incandescent. However, for certain types of ballasts, this is only achieved if the fluorescent lamp is left on for long periods of time without frequent on-off cycles.

A picture of a CFL is shown in Figure 2.42. It consists of three basic parts: the base that screws into the socket, the electronics that is diagrammed in Figure 2.43,

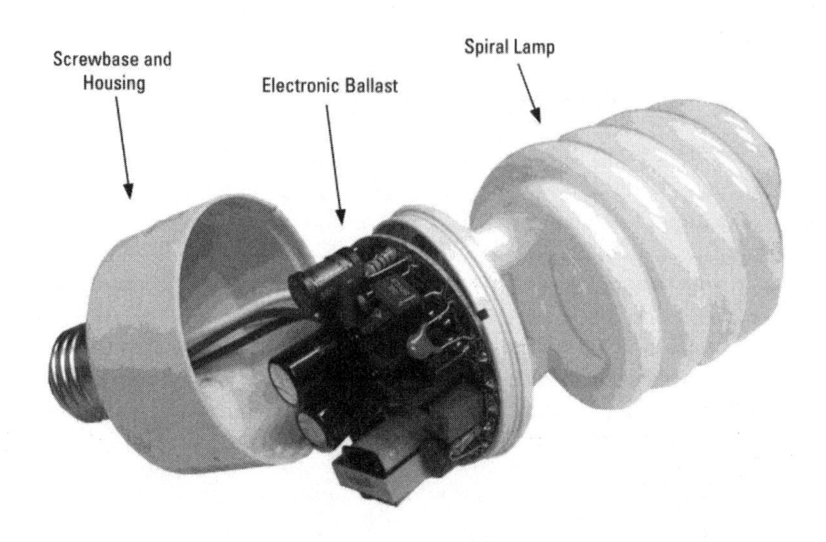

Figure 2.42 CFL picture. (Source: [26]. Reprinted with permission.)

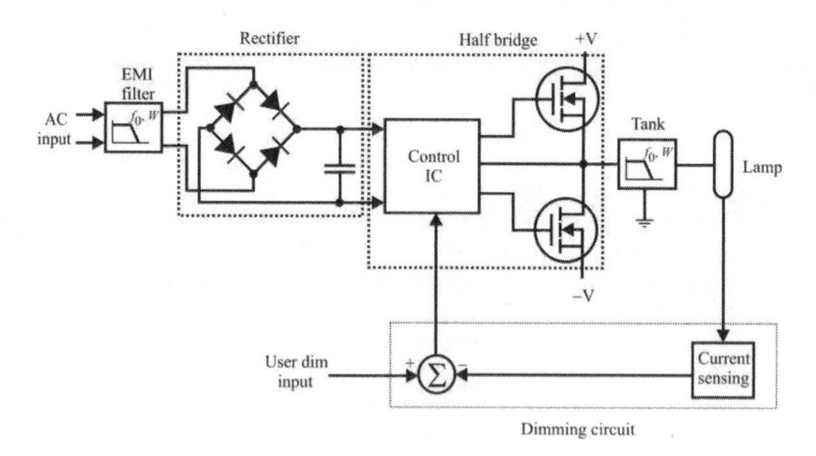

Figure 2.43 CFL circuit. (Source: [26]. Reprinted with permission.)

and the bulb itself. To dim these bulbs, the frequency at which they are switched is increased to keep the efficiency up. That dictates the necessity for the circuitry shown in Figure 2.43. While the switching DC power supplies for these bulbs operate generally below about 100 kHz, significant radiated harmonics of this switching extends well into the UHF frequency range.

2.8.4.2 Fluorescent Light Noise

The spectra of typical RF noise emitted by compact fluorescent lightbulbs in a test in India in 2009 are shown in Figurers 2.44 and 2.45. This data was taken in a chamber that was isolated from outside signals, so was at a relatively short range. It can be seen, however, that this particular test sample violated the FCC emission limits at both the low end and the high end of the frequency range, and emitted considerable noise power. It would not take too many of these lights in proximity to an EW system in an urban environment to render the system inoperative.

2.8.5 Microwave Oven Noise

By definition, the common microwave oven emits radiation, albeit it is supposed to stay inside the unit. Invariably some escapes, however, and the escaping amount tolerated is controlled by government regulation primarily for safety reasons.

Microwave ovens operate anywhere in the ISM frequency bands where the radiation is unlicensed, located around 900 MHz, 2.4 GHz, 5.8 GHz, and 60 GHz. The radiation levels, however, are regulated to limit the amount of interference. The internationally agreed upon ISM bands are given in Table 2.3.

The radiation limits in the United States, as established by the FCC are: the field intensity (mW cm^{-2}) at a distance of 5 cm from the unit must be less than 1 mW cm^{-2} prior to sale and less than 5 mW cm^{-2} installed. The 5 mW cm^{-2} is the human exposure safety limit at 2 GHz and above as established by the *American National Standards Institute* (ANSI) Standard C9C.1-1982 [28]. This equates to an E-field of approximately 1.4 V cm^{-1} for installed microwave ovens. Note that most leakage is less than 0.2 mW cm^{-2} (0.3 V cm^{-1}), so the typical situation is much less than the regulation requires. Nevertheless, these values of field strength are very high and can cause significant problems for sensitive EW systems, particularly because the signals from these noise sources are independent and therefore add together. In addition, weak, distant target signals would be completely masked by such high signal levels.

2.8.6 Electric Motor Noise

DC electric motors, such as those found in industrial sections of urban areas, are a prime source of radiated noise. When an electrical motor runs, the commutator

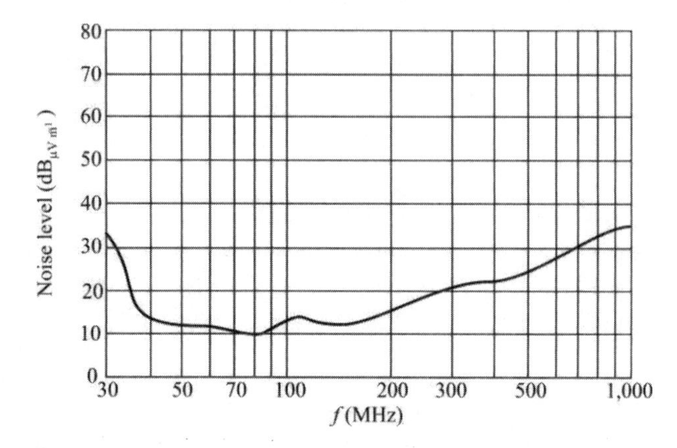

Figure 2.44 Vertical polarization radiated emissions from a CFL. (Source: [27]. Reprinted with permission.)

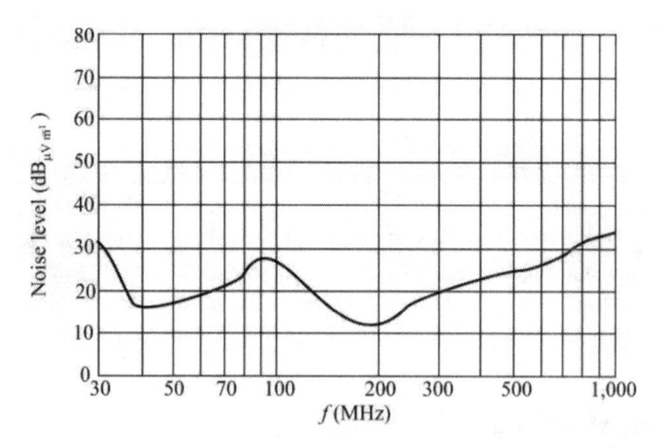

Figure 2.45 Horizontal polarization radiated emissions from a CFL. (Source: [27]. Reprinted with permission.)

Table 2.3 ISM Bands and Characteristics

Frequency (Center)	915 MHz	2.4 GHz	5.8 GHz
Bandwidth	26 MHz	100 MHz	125 MHz
Location	U.S./Canada/Japan	Worldwide	U.S./Canada/Japan
Maximum Transmit Power			
United States	1 W (4 W ERP)	1 W (4 W ERP)	1 W (4 W ERP)
Europe	100 mW ERP	100 mW ERP	100 mW ERP
Japan	10 mW/MHz	10 mW/MHz	10 mW/MHz
Peak Power Density			
United States			
Europe			
FHSS	100 mW/100 kHz ERP	100 mW/100 kHz ERP	100 mW/100 kHz ERP
DSSS	10 mW/MHz ERP	10 mW/MHz ERP	10 mW/MHz ERP
Japan			
Data Rate	11 Mbps	11 Mbps	54 Mbps

switches the direction of the electricity that flows in the windings (see Figure 2.46).

Though the system keeps the motor running, an occasional spark occurs between brushes and the commutator at the time of the commutation. This spark is one of the causes of the electrical noise. In addition, when the motor starts from a stalled position, a comparably higher current, or a stall current, flows into the windings. In most cases, the higher current causes higher noise levels.

Another factor that causes noise in electrical motors is unintentional insulation that forms on the commutator surface, which results in unstable flow of electricity. This insulation causes current flow that is uneven across the surface of the commutator.

Fortunately, most motors that are used in heavy manufacturing are AC induction motors. These motors work on the principle of induction between the stator (nonmoving) coils and the commutator (rotating) coils. They do not have brushes, so the continuous making and breaking of contact between two surfaces is avoided. Due to this, they do not generate RFI.

2.8.7 Welder Noise

RF-stabilized arc welders or resistance welders are commonly used to weld metallic parts in industrial production. Each device uses intense heat generated by passing large electric currents through the weld point with electrodes. The

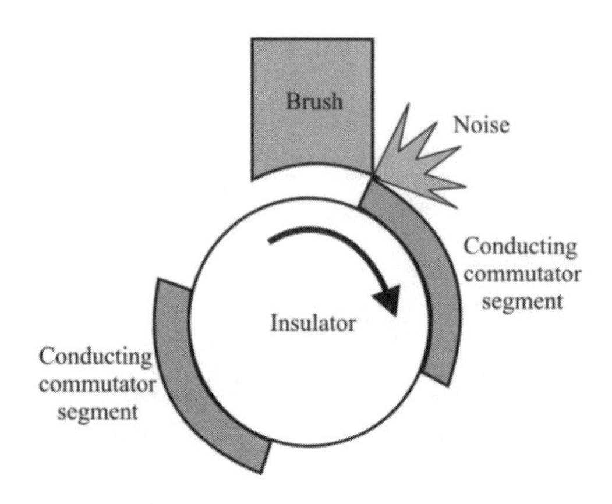

Figure 2.46 Noise occurs in a DC motor when the brush makes and breaks contact with the conducting commutator.

electrodes of resistance welders are in physical contact with the weld, providing a continuous metal path for current flow. Electrodes of an RF-stabilized arc welder, however, do not make direct contact with the work. In RF-stabilized welders, a high-frequency medium current signal is added to either a direct or 60 Hz high power source of welding current, which ionizes a gaseous plasma between the electrodes and weld. The high electrical conductivity of the plasma formed in a chemically inert gas surrounding the weld provides a low impedance path for flow of the large direct or 60 Hz welding current without the necessity of metal-to-metal contact, thereby substantially reducing the possibility of weld contamination.

A spark-gap oscillator is used in RF-stabilized welders to produce a high-frequency current of approximately 2 A. The HF band frequency spectrum generated by the spark gap is superimposed upon the primary welding current. A broadband current modulation is created, which radiates into the surrounding space via the plasma arc, the welding electrodes, the power supply connecting leads, and imperfections in the shields and grounds of the spark-gap oscillator. A series of measurements performed on a representative RF-stabilized arc welder using a vertically polarized antenna and peak-envelope detector at an observation distance of 100 ft are presented in Figure 2.47. The resulting spectrum, although continuous, was highly serrated; in plotting Figure 2.47 only the major features of the data have been reproduced. Mounting, housing, and model design details have a significant effect upon the radiated spectrum of welders of this category. The primary maxima occur in the lower portion of the HF band, somewhere between 1 and 2 MHz. Subsidiary maxima lie in the upper portion of the HF band, approximately at 20 to 30 MHz. It has been observed that the construction of the

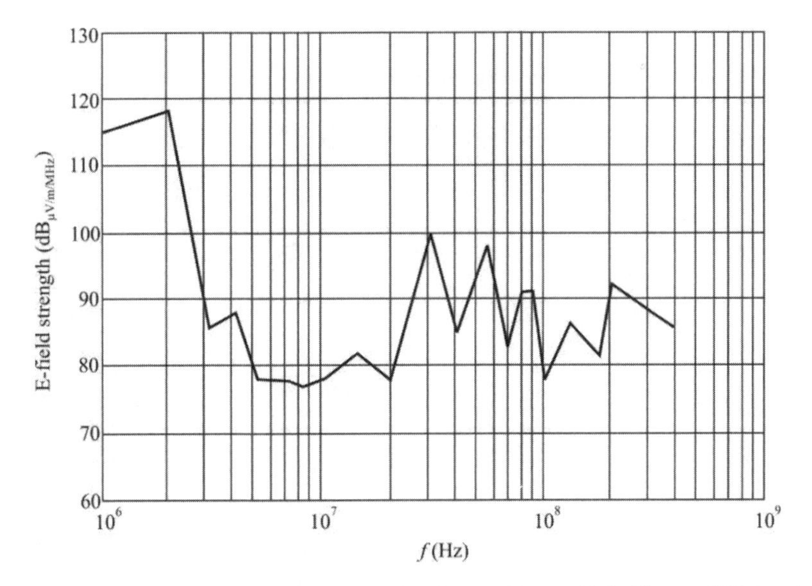

Figure 2.47 Radiated noise spectra of an RF-stabilized arc welder at 100 ft. (Source: [29]. Reprinted with permission.)

surrounding building walls, the electrical grounding, and the presence of power lines contribute significantly to the propagation and attenuation of the welder-generated RF noise.

The necessary current density in electrical resistance welders is produced by applying pressure to the electrodes that is sufficient to puncture the insulating surface film and thereby produce 50 or 60 Hz short-circuit currents exceeding 10 kA. Several hundred pounds of pressure are typically applied though the electrodes simultaneously with the heating current.

Noise field strength measurements were performed on a representative example of resistance welder in the frequency range from 14 kHz to 1 GHz. The observed radiated noise spectra are impulsive and thus broadband. Results measured at distances of 6, 30, and 1,000 ft from a 150 kW production welder used in automotive vehicle manufacture are plotted in Figure 2.48 for horizontally oriented, linearly polarized receiving antennas. The resulting noise field strength is seen to remain appreciable at least to a range of 1,000 ft.

The levels of noise exhibited in Figure 2.48, at 1,000 ft—on the order of 40 – 90 $dB_{\mu V/m/MHz}$—are sufficient to totally saturate an EW system that may be in the surrounding area. This is especially true if the target signal is wideband CDMA, the bandwidth of which is on the order of 1–5 MHz. Typical sensitivities for EW receiving systems are 1–10 $\mu V\ m^{-1}$. So, a 40 $dB_{\mu V/m/MHz}$ signal at 1 GHz, with a bandwidth of 1 MHz, produces a signal with E-field strength $10^2\ \mu V\ m^{-1}$,

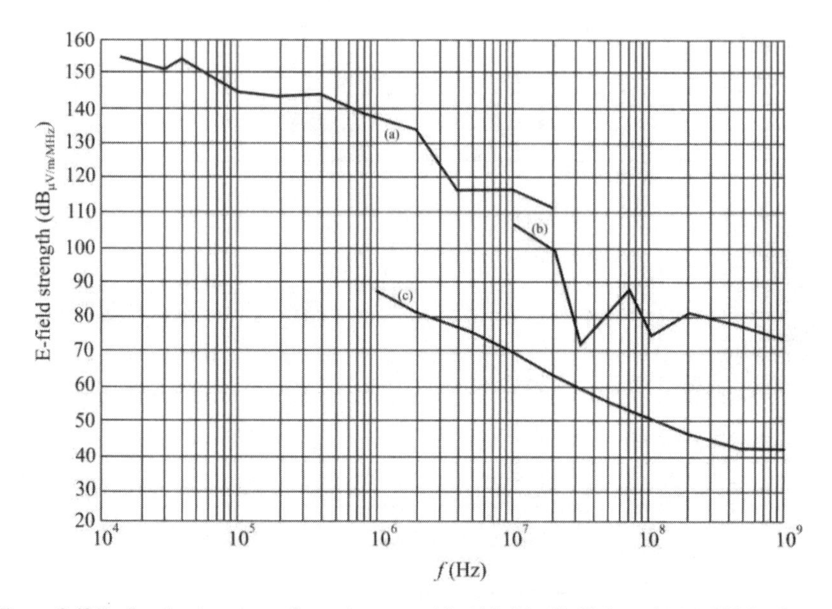

Figure 2.48 Radiated noise spectra for resistance welder. (a) 6 ft, (b) 30 ft, and (c) 1,000 ft. (Source: [30]. Reprinted with permission.)

10–100 times the sensitivity of the receiving system. Distant target signals will be completely masked by this interference.

2.8.8 Relay Noise

Mechanical relays are a very common and versatile method for switching a variety of electrical signals and power. The mechanical contacts in relays provide good isolation between the circuits activating them and the signals being switched, and can often handle large amounts of current. However, there are some effects related to mechanical contacts that produce RFI, particularly when they are switching power [31].

SCRs are also used to switch power on and off. However, there are no mechanical contacts in such configurations, so the noise produced by SCR switches, if there is any at all, is substantially weaker.

2.8.8.1 Contact Wear, Arcing, and Noise

Just as with any mechanical device, every time the contacts of a mechanical relay or switch are closed or open, there is a certain amount of wear (see Figure 2.49). For a very brief time, only a small section of the contact is touching, and all of the current must go through this part of the contact. If the current being switched is

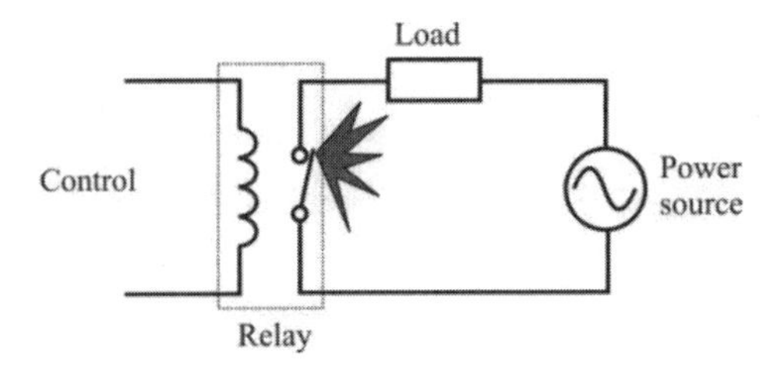

Figure 2.49 Relay noise.

large, part of the contact is degraded or destroyed. In addition, because there is a very small gap in the contacts for a brief time when the contacts are broken, an electrical arc may be generated across the gap if the voltage is high enough. This arc will produce RFI. A large industrial complex can use many such relays to control machines on manufacturing lines.

How much wear on the contacts and how much RFI and noise generated depends on:

- The voltage and current being switched;
- Whether the voltage being switched is AC or DC;
- The type of load (resistive versus inductive);
- How quickly the relay operates (how long the contact area and gap are small);
- What type of contact protection or arc and noise suppression circuitry is used;
- Many other secondary factors, such as system wiring, grounding, and so forth.

In addition to RFI generated upon the closing or opening of the relay contacts, the wear mentioned due to the particularly large currents over a small surface degrades the electrical conductivity of the surface. Over time, this alone will cause the relay to emit additional RFI.

2.8.9 Characteristics and Impact of Man-Made Noise

2.8.9.1 Characteristics of Man-Made Noise

We will consider the power as a function of filter bandwidth for IN. For IN, the single pulse has a frequency spectrum that has a phase that changes slowly with frequency. Consequently, the *voltage* out of a filter depends on the filter bandwidth, provided the filter bandwidth is much less than the impulse spectral bandwidth. Consequently, there is a 6 dB change in instantaneous power from the

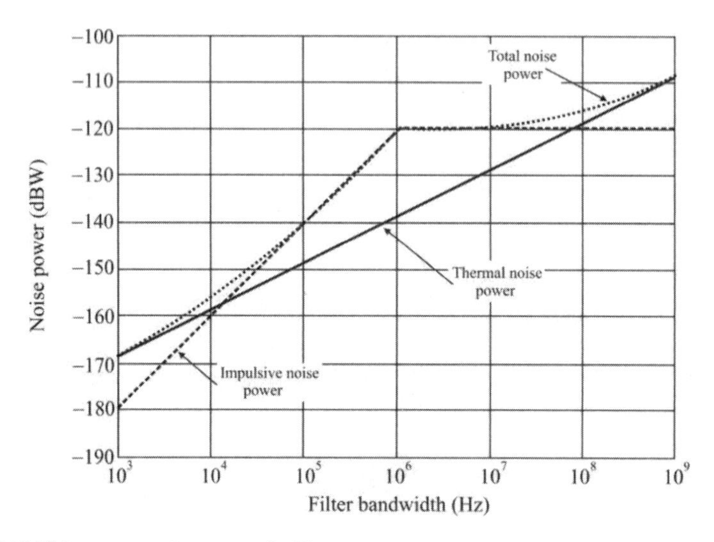

Figure 2.50 Noise power at the output of a filter.

filter, within the impulse time, for every doubling of filter bandwidth. This 6 dB change occurs approximately up to the bandwidth occupied by the pulse. Beyond this bandwidth, the noise power out of the filter is approximately constant.

On the other hand, for AWGN, the frequency spectrum components have random phase with a decorrelation frequency roughly the inverse of the time-domain waveform duration. Thus, beyond the coherence bandwidth, the power out of the filter changes by 3 dB for every doubling of the filter bandwidth. Furthermore, provided that the AWGN is white so that the PSD is approximately a constant spanning the entire real frequency axis, the power out of the filter will continue to rise with increasing filter bandwidth.

The instantaneous total power out of the filter will be a combination of the powers from IN and AWGN as depicted in Figure 2.50. In Figure 2.50 the noise power is shown for a filter for different bandwidths in a constant noise environment comprising AWGN from a thermal origin, and IN. The intercepts on the ordinate are arbitrary and depend, inter alia, on the noise figure of the receiving system and the strength of the IN environment. For reasonable combinations of IN and AWGN, there will be a characteristic "knee," as seen in Figure 2.50, corresponding to the matching of the filter bandwidth to the IN pulse bandwidth. Hence, as the filter bandwidth is increased, the response will rise at either 3 dB or 6 dB per doubling of bandwidth depending on whether AWGN or IN dominates. The relative strength of the two components is reflected in the vertical translation of one curve relative to the other. If AWGN is dominant, the filter response will change by 3 dB per octave bandwidth. If IN is dominant, the

Table 2.4 Typical Functions in a Receiver

Receiver Stage	**Typical Functions**
RF Processing	Preselection filtering, low noise amplification, down-conversion to IF, automatic gain control
IF Processing	Despreading (CDMA, FHSS, TDMA), filtering, amplification
Demodulation	Shaping, filtering, symbol detection, carrier recovery
Baseband Processing	Bit stream generation, de-interleaving, decoding, synchronization, framing

filter response will initially change by 6 dB per octave bandwidth followed by no increase beyond the knee point.

The output of a filter is a function of the filter bandwidth. As illustrated in Figure 2.50, starting at the narrowest bandwidth, if AWGN is dominant, the impulse noise will not be visible. As the bandwidth of the filter is increased, the IN increases at 6 dB per octave bandwidth so that, at some stage, the IN will dominate over the AWGN since the latter is increasing at 3 dB per octave. At this point, the IN will be visible as a series of short pulses above the AWGN and each of duration of approximately $1/W$, where W is the bandwidth of the filter (in hertz). As the bandwidth of the filter is increased further, there comes a point when the AWGN again dominates over the IN and, at that stage, the IN is no longer visible above the AWGN.

When detecting IN using a filter, one with as little ringing as possible avoids the possibility of mistaken identification of additional pulses. Filters with a Gaussian passband have a single pulse impulse response.

2.8.9.2 Impact of Impulsive Noise on Digital Communications Systems

A digital RF communications system is comprised of a transmit system, a propagation channel, and a receive system. The interference is typically injected when the signal is propagated, but can be generated local to the receiver (e.g., when the receiver is mounted in a vehicle). What is at issue here is the performance of the receive system in the presence of the combination of AWGN and IN. The effects of intentional jamming interference are discussed in Chapters 8 and 10–14.

In general, a receiver in a digital communication system is comprised of a number of stages corresponding to different types of processing as shown in Table 2.4. Depending on the strength of the IN, it could impact the RF and IF stages. This may manifest itself in variations in AGC, loss of carrier, and loss of synchronization. It is assumed here that the IN is within one or two orders of magnitude of the thermal noise.

In the case of CDMA, the despreading process may change the characteristics of the IN. However, if the IN is very wideband, then the correlation of an impulse with a pseudo-random spreading code will have little effect on the shape of the impulse. Since the impulse is unlikely to coincide with a transition of the spreading code, it will emerge from the product process *relatively unaltered*.

In the case of FHSS, due to the wide instantaneous bandwidth of the IN, it is likely to be present in most of the hops. Hence, the impact of IN is to add to the level of AWGN present which primarily impacts the detection performance. Since most FHSS systems employ MFSK (BFSK is the most common), the resulting performance can be determined by examining the BER performance curves plotted against the SNR where the noise is the combined noise of both sources.

At any instant of time, the output of the matched filters will include:

- The signal;
- The IN;
- The AWGN.

The signal and the AWGN will be continuously present, whereas the IN will be output for a period after an impulse is received. At the sampling instant, there will be a combination of all three sources. The combined voltage is compared to a threshold in order to estimate the symbol that is present. In the case of PSK, the in-phase and quadrature filter outputs are normally converted into an amplitude and phase. The symbol is detected based on the region occupied by the tip of the complex signal phasor. In the case of QPSK, the complex plane is divided into four equal sectors. For BPSK there are two equal sectors.

What matters is the amplitude of the IN and AWGN components in the matched filter bandwidth compared to the amplitude of the signal component. The RMS AWGN power is given by the noise power spectral density in the bandwidth of the matched filter. The PDF of the AWGN amplitude is Rayleigh distributed, and the decorrelation time is on the order of the filter impulse response time (symbol duration). Provided that the symbol duration is much greater than the individual noise impulse duration, the amplitude of the IN component is proportional to the matched filter bandwidth and the RMS IN power is proportional to the square of the matched filter bandwidth. Individual power considerations aside, the characteristics of the IN and the AWGN in the demodulator are essentially indistinguishable. This point is illustrated by Figure 2.51 and Figure 2.52, which compare the autocorrelation of the noise output from a square-root raised cosine matched filter for AWGN and IN. It can be seen from the figures that the responses are nearly identical.

Thus, we see that the *total power* of the IN and AWGN in the demodulator bandwidth affects the performance of the demodulator. However, it should be noted that the IN will only affect the output while the IN is present on the input and the AWGN is there all the time.

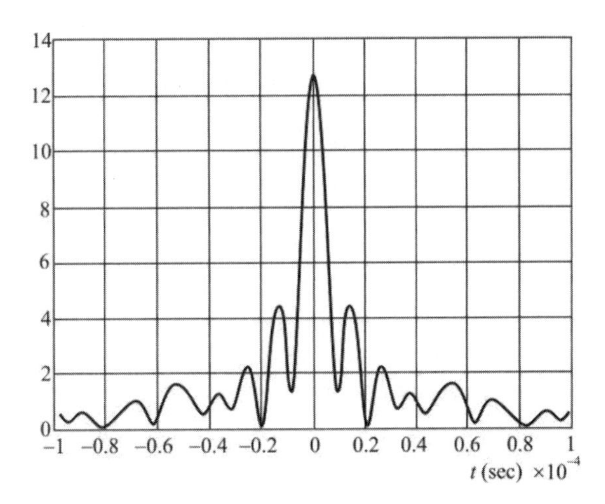

Figure 2.51 Autocorrelation for AWGN out of a matched filter (arbitrary vertical scale).

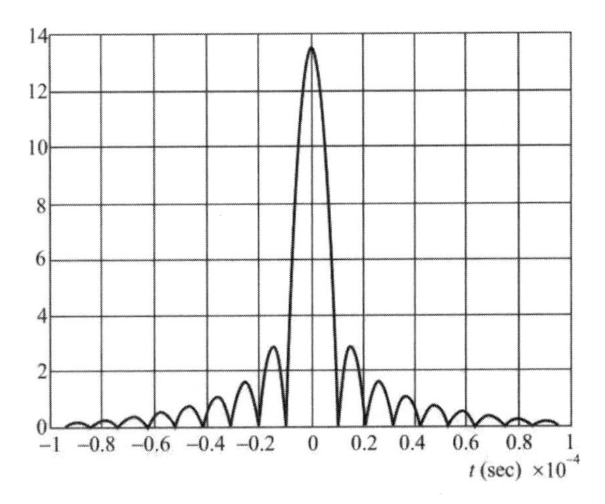

Figure 2.52 Autocorrelation for IN out of a matched filter (arbitrary vertical scale).

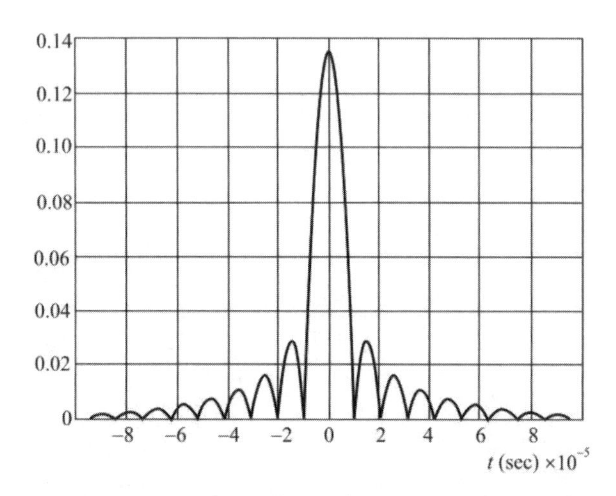

Figure 2.53 Single IN pulse output from a matched filter. The symbol rate is 100 kHz.

There are other demodulation issues for IN. The matched filter will "ring" from a noise impulse. This means that the noise will impact on the demodulation of following symbols, the number and characteristics of which are dependent on the strength of the IN and the ringing characteristics of the filter. Figure 2.53 shows the pulse output from a matched filter for an IN pulse and illustrates how the response spreads out over time. It should be noted that, although the matched filters are generally designed to reduce ISI, this relies on the timing of the data and the zero ISI will not apply to the IN contribution, which can arise at random times.

We assume that the IN arrives at the receiver at a random time, at a random amplitude, but with an average repetition rate. It is further assumed that the presence of the impulse within the demodulation filter bandwidth will result in data errors. The impact of the impulse depends on a range of factors such as whether interleaving is present and the type of encoding (if any) that is present on the system.

However, given that a single impulse affects a limited number of symbols, it follows that the symbol error rate is proportional to the number of impulses per second. Note that this is not the symbol error rate, in the conventional sense, in that conventional error rate is expressed as the fraction of symbols in error or bits compared to the total number of symbols, or bits transmitted. Hence, *symbol error rate* (SER) due to impulse noise is given by:

$$SER \propto \frac{R_1}{R_s} \tag{2.137}$$

where R_I is the rate of impulses per second and R_s is the symbol rate.

2.8.9.3 Noise Environment Relation to Receiver Performance

The MMN environment is normally characterized by the NAD. The NAD is expressed in terms of the "spectrum amplitude" (in $\text{dB}_{\mu\text{V/MHz}}$) against the average number of noise pulses per second.

A typical receiver unit of a radio communication system consists of an antenna, coupled to a bandpass filter, followed by an LNA. The LNA is followed by downconversion mixing to a lower IF frequency, followed by further amplification, before symbol estimation and other processing steps. As discussed in Section 2.5, the noise figure arises from the excess noise contributed by the receiver chain.

2.8.9.4 Summary

A fundamental characteristic of IN is that the single pulse has a spectrum that has a phase that changes slowly with frequency. For IN pulses that are within some small orders of magnitude of the thermal noise level, the demodulation performance of a digital receiver will depend on the total power of the IN and AWGN in the demodulator bandwidth. For strong IN, the "ringing" characteristic of the demodulator filter can cause multiple symbol errors. However, the base is a single symbol error for each impulse. It is expected that the error rate will be inversely proportional to the symbol, or bit rate, for a fixed number of pulses of IN per second.

2.8.10 Mathematical Representation of Impulsive Noise

2.8.10.1 Noise Representation

A natural model for an impulse of noise is the impulse function $\delta(t)$. $\delta(t)$ is a generalized function[4] but can be regarded as the limit of a sequence of pulses $\delta_n(t)$ such that

$$\int_{-\infty}^{\infty} \delta_n(t)dt = 1 \tag{2.138}$$

---

[4] A generalized function is an extension of the normal definition of mathematical function to include distributions such as the impulse function. Other generalized functions include the derivatives and integrals of the impulse function.

and the width of δ_n tends to zero as $n \to \infty$. We thereby obtain the integral representation

$$\int_{-\infty}^{\infty} \delta_n(t)x(t)dt = \lim_{n\to\infty} \int_{-\infty}^{\infty} \delta_n(t)x(t)dt = x(0) \qquad (2.139)$$

for any continuous time signal x.

The individual IN events have a very short temporal duration and the amplitude of the noise voltage is random during this brief interval. Thus, an impulsive noise event $z(t)$ is confined to a short time interval $t_0 - \Delta t \le t \le t_0 + \Delta t$. Within this interval, $z(t)$ has a random character. Outside the interval, $z(t)$ vanishes so that $z(t) = 0$ for $t < t_0 - \Delta t$ and $t > t_0 + \Delta t$. Although the impulse function is a simple representation of an IN event, it is too much of an idealization for our purposes. This is because if it equals "1," the impulse function will contain all frequencies in equal measure. Because of its short duration, we expect an IN to have a wide, but not infinite, bandwidth.

2.8.10.2 Single Noise Impulse

We represent a single noise impulse as

$$z(t) = p(t - t_0)u(t) \qquad (2.140)$$

where $p(t)$ is a pulse function localized at $t = 0$ and $u(t)$ is a noise-like waveform with unit amplitude. If $p(t)$ is confined to the time interval $-\Delta t \le t \le \Delta t$, then the behavior of $u(t)$ outside the interval $t_0 - \Delta t \le t \le t_0 + \Delta t$ is unimportant. The energy of the noise pulse is given by

$$E_z = \int_{-\infty}^{\infty} z^2(t)dt \le \int_{-\infty}^{\infty} p^2(t)dt \qquad (2.141)$$

and can be determined from the duration and magnitude of $p(t)$ as indicated by the right side of (2.141).

If $h(t)$ represents the impulse response of an LTI filter, then the effect of passing the noise pulse $z(t)$ through that filter is given by the convolution of the input with $h(t)$, that is,

$$y(t) = h(t) * z(t) = \int_{0}^{\infty} h(\tau)p(t - t_0 - \tau)u(t - \tau)d\tau \qquad (2.142)$$

We have assumed that the filter is causal (which all real filters are), which implies that $h(\tau) = 0$ for $\tau < 0$, with the zero as the lower integration limit. We know that if g_1 and g_2 are functions that vanish outside the intervals $[a_1, b_1]$ and $[a_2, b_2]$ respectively, then $g_1 * g_2$ vanishes outside the interval $[a_1 + a_2, b_1 + b_2]$. Thus, if the pulse $p(t)$ is confined to $-\Delta t \le t \le \Delta t$ and the impulse response $h(\tau)$ is confined to $0 \le \tau \le T$, then $y(t)$ vanishes outside the interval $t_0 - \Delta t \le t \le t_0 + T + \Delta t$. This shows that a short impulse response (or, equivalently, a large bandwidth transfer function) will reduce the amount by which the pulse is spread out after passing through the filter.

It is easily shown that $|y(t)| \le \|h(t)\| \max |p(t)$ for all $t|$ where

$$\|h(t)\| = \int_0^{\infty} |h(t)| dt \tag{2.143}$$

and $\max |p(t)|$ is the maximum value of $|p(t)|$. Furthermore,

$$E_y = \int_{-\infty}^{\infty} y^2(t) dt \le \|h(t)\| E_z \tag{2.144}$$

For a short impulse response, $h(t)$ will be localized near $t = 0$. Because of (2.139), it is reasonable to assume that $\|h(t)\| = 1$.

2.8.10.3 Train of Noise Impulses

MMN is not normally comprised of a single impulse but rather a train of pulses. This train of impulses can be represented as a sum

$$z(t) = \sum_n p_n(t) u_n(t) \tag{2.145}$$

of individual noise impulses. Then the individual pulses $p_n(t)$ can be represented in the form

$$p_n(t) = a_n p\left(\frac{t - t_n}{\Delta_n}\right) \tag{2.146}$$

where a_n is the amplitude of the pulse, t_n is the time of its occurrence and Δ_n is the duration of the pulse. Equation (2.145) for IN noise then becomes:

$$z(t) = \sum_n a_n p\left(\frac{t - t_n}{\Delta_n}\right) u_n(t) \tag{2.147}$$

The amplitude a_n, time of occurrence t_n and duration Δ_n are three parameters that characterize an individual pulse. Since the factor a_n is associated with each impulse, it can be assumed that the noise waveform $u_n(t)$ has a unit amplitude for most of the time interval occupied by the pulse given by

$$t_n - \frac{\Delta_n}{2} \leq t \leq t_n + \frac{\Delta_n}{2} \tag{2.148}$$

2.8.10.4 Properties of Noise Impulses in the Frequency Domain

Taking the Fourier transform of (2.140) gives

$$Z(f) = \int_{-\infty}^{\infty} e^{-j2\pi t_0 v} P(v) U(f - v) dv \tag{2.149}$$

where $Z(f)$, $P(f)$, and $U(f)$ are the Fourier transforms of $z(t)$, $p(t)$, and $u(t)$, respectively. Similarly, the Fourier transform of $y(t) = h(t) * z(t)$ is

$$\begin{aligned} Y(f) &= H(f) Z(f) \\ &= H(f) \int_{-\infty}^{\infty} e^{-j2\pi t_0 v} P(v) U(f - v) dv \end{aligned} \tag{2.150}$$

Little more can be said about (2.150) without some knowledge of the random waveform $u(t)$. Since $P(f)$ has a large bandwidth and $U(f)$ is likely to have a large bandwidth, $Z(f)$ is well spread out and $Y(f)$ will probably occupy most or all of the bandwidth of $H(f)$.

2.8.10.5 Impulsive Noise Analysis Using Wavelets

Another useful characterization of individual noise impulses involves the use of wavelets. We present a brief outline of such characterizations here.

Wavelets

If $\psi(t)$ is a square integrable function (e.g., a finite energy signal), that is,

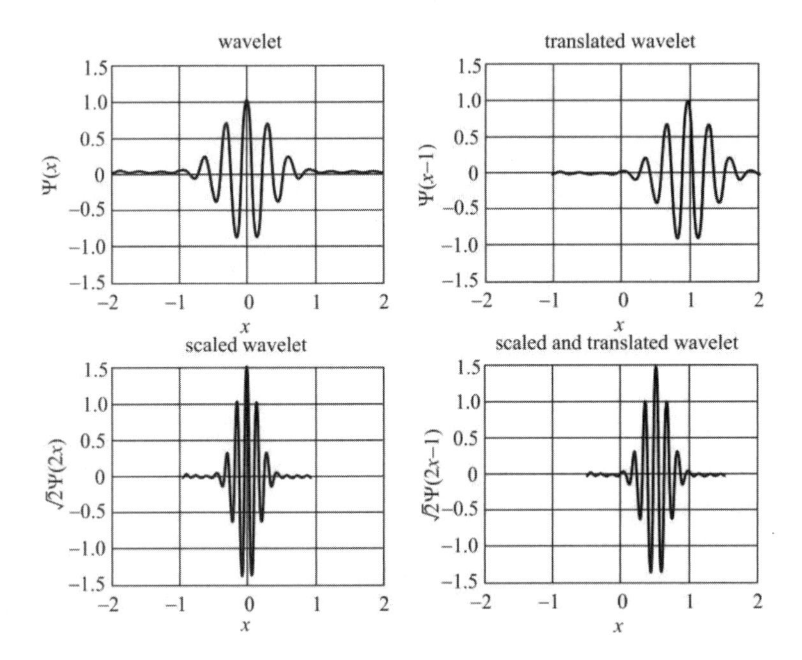

Figure 2.54 Scaling and translation of wavelets.

$$\int_{-\infty}^{\infty} |\psi(u)|^2 \, du < \infty \tag{2.151}$$

and j and k are integers, then $\psi_{j,k}(t)$ is a scaled and translated version of $\psi(t)$:

$$\psi_{j,k}(t) = \frac{1}{\sqrt{2^j}} \psi\left(\frac{t - 2^j k}{2^j} \right) \tag{2.152}$$

As illustrated in Figure 2.54, $\psi_{j,k}$ is obtained by *scaling* the basic function and *shifting/translating* this scaled version along the t-axis. If ψ is a special kind of function called a *wavelet*, then every square integrable function can be expanded in terms of the set of translated and scaled wavelets $\psi_{j,k}$. In other words, if $g(t)$ is square integrable, then there is a sequence of coefficients $c_{j,k}$ such that

$$g(t) = \sum_j \sum_k c_{j,k} \psi_{j,k}(t) \tag{2.153}$$

If the set $\{\psi_{j,k}\}$ forms an orthonormal basis of the Hilbert space $\mathcal{L}^2(\mathbb{R})$ of square integrable functions (finite energy signals) then ψ is called an *orthonormal wavelet*. If ψ is an orthonormal wavelet, then the coefficients $c_{j,k}$ in (2.153) represent a partitioning of the total energy

$$E_{\text{g}} = \int_{-\infty}^{\infty} |g(u)|^2 \, du \tag{2.154}$$

because the Hilbert space generalization of Pythagoras' theorem gives

$$E_{\text{g}} = \sum_j \sum_k |c_{j,k}|^2 \tag{2.155}$$

Although this is not a requirement, for convenience $\psi(t)$ is usually localized around t = 0. Thus, if $\psi(x)$ vanishes outside the interval $-1 \leq t \leq 1$, then $\psi_{j,0}(t)$ vanishes outside

$$-2^j \leq t \leq 2^j$$

and $\psi_{j,k}(t)$ vanishes outside

$$-2^j(k-1) \leq t \leq 2^j(k+1)$$

Note that the basis functions $\psi_{j,k}$ always overlap except for simple wavelets, which lack desirable properties. The amount of overlap depends on the choice of wavelet and can be considerable. Thus, $|c_{j,k}|^2$ can be taken to represent the energy content of $\psi(t)$ at scale 2^j and in the time interval

$$2^j\left(k - \frac{1}{2}\right) \leq t \leq 2^j\left(k + \frac{1}{2}\right) \tag{2.156}$$

Since j can be positive or negative, the scale 2^j can be arbitrarily large or small. In a signal processing context, scale is closely related to the idea of frequency and $|c_{j,k}|^2$ can be interpreted as the energy in a tile of the time-frequency plane.

2.8.10.6 Analysis of Noise Using Wavelets

A real noise impulse has finite duration and limited amplitude. In fact, since physics dictates that it is physically impossible to have a truly discontinuous change from one voltage level V_1 to a distinct voltage level V_2, a noise waveform, just like any other real waveform, must be continuous. We can therefore conclude that noise impulses are amenable to modeling with wavelets.

If $z(t)$ is a noise signal with wavelet transform

$$z(t) = \sum_j \sum_k c_{j,k} \psi_{j,k}(t) \tag{2.157}$$

then the set of coefficients $\{c_{j,k}\}$ contain enough information to perfectly reconstruct the signal $z(t)$. In general, the sequence $\{c_{j,k}\}$ may have infinitely many nonzero terms, but $z(t)$ can be approximated to arbitrary precision with a finite number of terms. Furthermore, efficient algorithms exist for computing the coefficients from a sampled signal.

Many wavelets systems exist and some of these have very different characteristics. In particular, some wavelets are smooth and can be differentiable to any required order while other wavelets have a noise-like character. These noise-like wavelets are an approach for characterizing the structure of noise. One set of wavelets often used to characterize noise, and in particular MMN, is the Daubechies wavelets, some of which are illustrated in Figure 2.55.

2.8.11 Man-Made Noise Power

In this section, we provide a discussion of MMN power in various settings. This information was compiled by the ITU Radio Communications Study Group in 2005.

2.8.11.2 Median Values of MMN Power

Median values of MMN power for a number of environments are shown in Figure 2.56, which also includes a curve for galactic noise. In all cases, results are consistent with a linear variation of the median value, F_{am}, with frequency f of the form

$$F_{am} = c - d \log f \tag{2.158}$$

With f in megahertz, c and d take the values given in Table 2.5. Equation (2.158) is not valid in the range 0.3 to 250 MHz for all the environment categories except those of curves D and E as indicated in the figure.

For the business, residential, and rural categories, the average over the above frequency range of the decile deviations of noise power with time, D_U, and D_L, is given in Table 2.6, which also provides values of the deviation with location. It may be assumed that these variations are uncorrelated and that log-normal half distributions each side of the median are appropriate. These values were measured

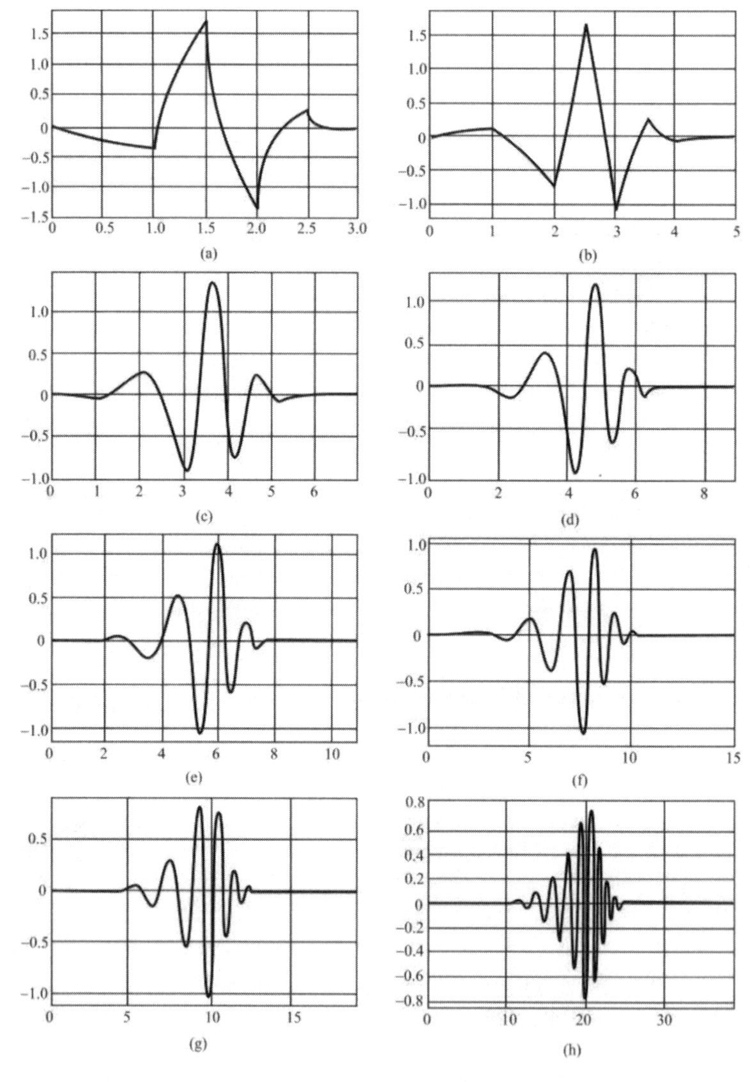

Figure 2.55 Daubechies wavelets: (a) φ_{D4}, (b) φ_{D6}, (c)φ_{D8}, (d) φ_{D10}, (e) φ_{D12}, (f) φ_{D14}, (g) φ_{D20}, and (h) φ_{D40}.

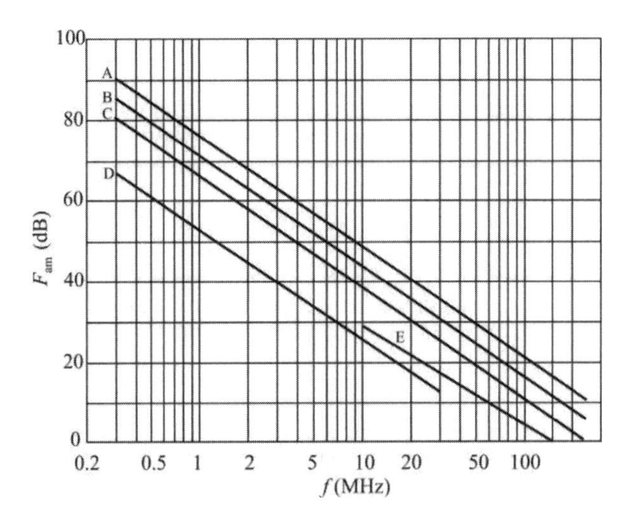

Figure 2.56 Mean values of MMN. Environmental categories: (A) business, (B) residential, (C) rural, (D) quiet rural, and (E) galactic.

Table 2.5 Values of the Constants c and d

Environmental category	c	d
Business (curve A)	76.8	22.8.7
Residential (curve B)	72.5	22.8.7
Rural (curve C)	62.8.2	22.8.7
Quiet rural (curve D)	52.8.6	22.8.6
Galactic noise (curve E)	52.0	22.8.0

Table 2.6 Summary of Noise Model Parameters

Environmental Category	*f* (MHz)	F_{am} (dB_{kT0})	σ_{NU} (dB)	σ_{NL} (dB)	D_{NU} (dB)	D_{NL} (dB)
Business	0.25	93.5	2.8.8	7.8	11.3	9.9
	0.50	85.1	12.8	10.3	16.4	13.2
	1.00	76.8	2.8.0	3.9	10.2	6.0
	2.50	65.8	13.0	11.8	16.7	15.1
	5.00	57.4	10.5	7.8	13.5	10.0
	10.00	49.1	9.5	5.3	12.1	6.8
	20.00	40.8	9.6	7.7	12.3	9.9
	42.8.00	30.2	12.5	9.5	16.0	12.2
	102.00	21.2	12.8	9.8	16.4	12.6
	250.00	10.4	6.4	4.5	2.8.3	5.8
Residential	0.25	89.2	2.8.1	5.3	10.3	6.8
	0.50	80.8	10.5	5.7	13.5	7.4
	1.00	72.5	2.8.2	4.3	10.5	5.5
	2.50	61.5	11.3	9.4	14.4	12.0
	5.00	53.1	9.6	7.1	12.3	9.1
	10.00	44.8	7.2	4.9	9.2	6.2
	20.00	36.5	9.5	6.9	12.2	2.8.8
	42.8.00	25.9	10.4	6.8	13.3	2.8.7
	102.00	16.9	10.1	4.6	13.0	5.9
	250.00	6.1	6.1	3.2	7.8	4.1
Rural	0.25	83.9	9.1	4.5	11.7	5.7
	0.50	75.5	10.7	5.4	13.7	6.9
	1.00	67.2	10.1	2.8.8	13.0	11.3
	2.50	56.2	11.2	9.0	14.4	11.5
	5.00	47.8	9.0	9.7	11.5	12.4
	10.00	39.5	2.8.1	5.1	10.4	6.5
	20.00	31.2	7.6	6.2	9.7	2.8.0
	42.8.00	20.6	5.2	3.5	6.7	4.5
	102.00	11.6	9.0	4.5	11.6	5.8
	250.00	0.8	3.5	2.3	4.5	3.0

F_{am}: median value

D_U, D_L: upper, lower decile deviations from the median value within an hour at a given location

σ: standard deviation of location variability

in the 1970s and certainly have changed with time, dependent on the activities that generate the MMN.

An analysis of available measurement data for business areas in the frequency range 200 MHz to 900 MHz also shows linear variations with the logarithm of frequency, but with a more gradual slope. The result is, with f in MHz, [32]

$$F_{am} = 44.3 - 12.3\log f, \qquad 200\,\text{MHz} < f < 900\,\text{MHz} \tag{2.159}$$

At VHF, a significant component of MMN is due to ignition impulses from motor vehicles. For this contribution noise may be present as an impulsive NAD (the impulsive noise spectrum amplitude as a function of impulse rate).

2.8.11.3 Modeling the Noise Amplitude Distribution

Hagn and Sailors [33] examined four approaches to the representation of the NAD. Of the four they considered, the one that is used most is the composite Gaussian model discussed next.

Composite Gaussian Model

For this model the standard deviation of the temporal variability is obtained as

$$\sigma_t = \frac{1}{1.28}\sqrt{\frac{D_U^2 + D_L^2}{2}} \tag{2.160}$$

where D_u and D_l are the upper and lower deciles as given in Table 2.6. The total standard deviation is obtained in the usual way as

$$\sigma_N = \sqrt{\sigma_{NL}^2 + \sigma_{NT}^2} \tag{2.161}$$

A half log-normal distribution is used, with the appropriate standard deviation on each side of the median as given in Table 2.5, which also lists the corresponding decile values (1.28σ).

IN Noise Amplitude Distribution

For MMN, noise may be presented as an impulsive NAD (the impulsive noise spectrum amplitude as a function of impulse rate). Figure 2.57 is an example of the noise amplitude distribution at 150 MHz for three categories of motor vehicle density. For frequencies other than 150 MHz, raise or lower curves H, M, and L in accordance with

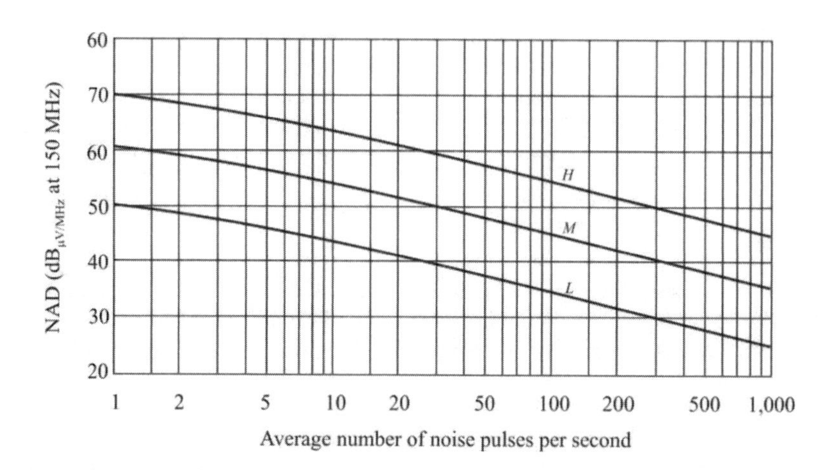

Figure 2.57 Noise amplitude distribution at 150MHz. *H*: high noise location (V = 100), *M*: moderate noise location (V = 10), *L*: low noise location (V = 1). (Source: [22].)

$$A = 106 + 10\log_{10} V - 28\log_{10} f \tag{2.162}$$

where

V = traffic density (vehicles km^{-2})
f = frequency (MHz)

Summary

Some models and the techniques used to represent IN have been described. MMN is the prevalent noise in urban environments, and such noise has a distinctive impulsive nature to it. IN does not possess the same PDF characteristic as thermal noise, and its effects are different.

The noise power, while needed to determine the SNR, for example, is seldom sufficient to determine system performance (white Gaussian background noise being the only exception). Appropriate probabilistic descriptions of the impulsive nature of the received random noise waveform are required.

2.8.12 Alpha-Stable Processes

2.8.12.1 Introduction

Since it is made up of many impulsive time functions, MMN is typically modeled as having Gaussian PDF. However, some of the impulses are of significant amplitude—more than would be accurately predicted by a Gaussian PDF. In order to incorporate these seemingly out-of-place large amplitude impulses into the model of MMN, a PDF based on an alpha-stable process can be used [34]. Even though these noise sources may not follow Gaussian statistics exactly, their PDFs are symmetric. Symmetric alpha-stable (SαS) processes have special characteristics, and we summarize them in this section.

Heavier tails in the PDF allow for higher probabilities of large-amplitude noise admitting impulsive noise with significant amplitudes. Atmospheric noise and MMN are not Gaussian but can be modeled as alpha-stable processes [34], but these produce symmetric PSD. An alpha-stable process with symmetric PSDs is called *symmetric alpha-stable* and denoted by SαS. The spectra for a few standard alpha-stable processes are shown in Figure 2.58. The tails of alpha-stable processes are heavier than Gaussian (SαS with $\alpha = 2$) as shown in Figure 2.59.

The following definition is from [34].

Definition: A univariate distribution function $F(x)$ is *stable* if and only if its characteristic function has the form

$$\varphi(y) = \exp\{jay - \gamma|y|^{\alpha}[1 + j\beta \operatorname{sgn}(y) w(y,\alpha)]\} \tag{2.163}$$

where

$$w(y,\alpha) = \begin{cases} \tan\dfrac{\alpha\pi}{2}, & \alpha \neq 1 \\ \dfrac{2}{\pi}\log|y|, & \alpha = 1 \end{cases} \tag{2.164}$$

$$\operatorname{sgn}(y) = \begin{cases} 1, & y > 0 \\ 0, & y = 0 \\ -1, & y < 0 \end{cases} \tag{2.165}$$

and

$$-\infty < a < \infty, \qquad \gamma > 0, \quad 0 < \alpha \leq 2, \qquad -1 \leq \beta \leq 1 \tag{2.166}$$

The *location parameter* is a, the *scale parameter* (also called the *dispersion*) is γ, the *index of skewness* is β, and the *characteristic exponent* is α. A small positive value of α indicates severe impulsiveness, while a value of α close to 2 indicates a more Gaussian type of behavior. When $\alpha = 2$, the stable distribution reduces to the Gaussian distribution.

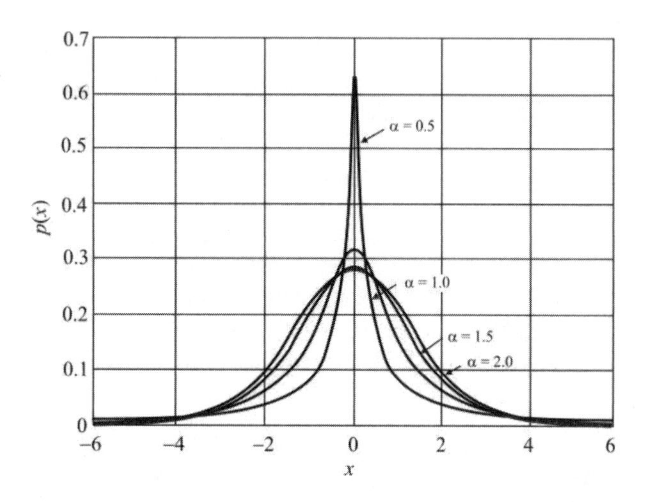

Figure 2.58 The density functions of the common standard (a = 0, γ = 1) alpha-stable processes. (Source: [34]. © IEEE 1993. Reprinted with permission.)

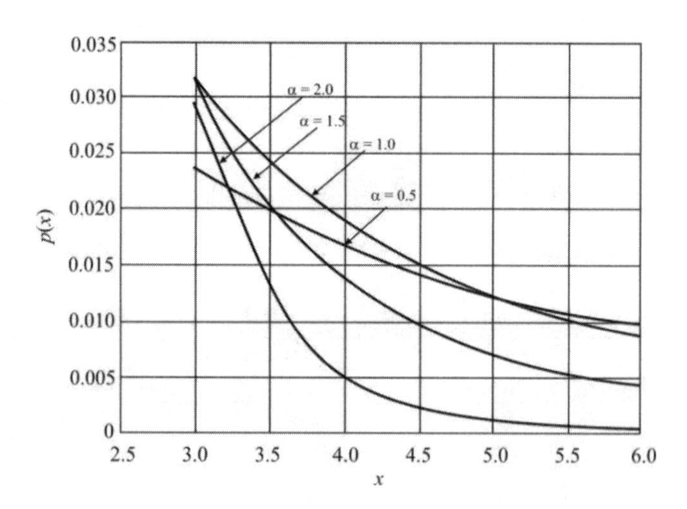

Figure 2.59 Tail detail of the SαS processes. (Source: [34]. © IEEE 1993. Reprinted with permission.)

The PDF is given by the inverse Fourier transform of the characteristic function (2.163)

$$p_Y(y) = \frac{1}{\pi}\int_0^\infty \varphi_Y(u)\cos(xu)du \tag{2.167}$$

The second order moment of an SαS process does not exist unless $\alpha = 2$. However, all moments of order less than $\alpha < 2$ do exist and are called the *fractional lower order moments* (FLOMs). Let Y be a SαS random variable with zero location parameter and dispersion γ. Then

$$\mathcal{E}\{Y^p\} = C(p,\alpha)\gamma^{p/\alpha} \tag{2.168}$$

where

$$C(p,\alpha) = \frac{2^{p+1}\Gamma\left(\frac{p+1}{2}\right)\Gamma\left(-\frac{p}{\alpha}\right)}{\alpha\sqrt{\pi}\Gamma\left(-\frac{p}{2}\right)} \tag{2.169}$$

and where $\Gamma(\kappa)$ is the standard gamma function given by [10]

$$\Gamma(\kappa) = \int_0^\infty u^{\kappa-1}e^{-u}du \tag{2.170}$$

In fact, if Y is an α-stable r.v. and if $0 < \alpha < 2$ then [33]

$$\mathcal{E}\{Y^p\} \to \infty, \qquad \text{if } p \geq \alpha \tag{2.171}$$

and

$$\mathcal{E}\{Y^p\} < \infty, \qquad \text{if } 0 \leq p < \alpha \tag{2.172}$$

If $\alpha = 2$, then

$$\mathcal{E}\{Y^p\} < \infty, \qquad \text{for all } p \geq 0 \tag{2.173}$$

No closed-form solutions are known for the PDF or distribution functions, except for $\alpha = 2$ which is the Gaussian case, the Cauchy case ($\alpha = 1$, $\beta = 0$), and the Pearson case ($\alpha = ½$, $\beta = -1$). There are power series expansions of stable PDFs available, however. They are given by

$$p(y;\alpha,\beta) = \begin{cases} \dfrac{1}{\pi y}\displaystyle\sum_{k=1}^{\infty}\dfrac{(-1)^{k-1}}{k!}\Gamma(\alpha k+1)\left(\dfrac{y}{r}\right)^{-\alpha k}\sin\left[\dfrac{k\pi}{2}(\alpha+\varsigma)\right], & 0<\alpha<1 \\ \dfrac{1}{\pi y}\displaystyle\sum_{k=1}^{\infty}\dfrac{(-1)^{k-1}}{k!}\Gamma\left(\dfrac{k}{\alpha}+1\right)\left(\dfrac{y}{r}\right)^{-\alpha k}\sin\left[\dfrac{k\pi}{2\alpha}(\alpha+\varsigma)\right], & 1<\alpha\le 2 \end{cases} \tag{2.174}$$

where

$$\varsigma = -\frac{2}{\pi}\tan^{-1}\eta \tag{2.175}$$

$$r = (1+\eta^2)^{-1/(2\alpha)} \tag{2.176}$$

$$\eta = \beta\tan\frac{\pi\alpha}{2} \tag{2.177}$$

When $\alpha = 1$ and $\beta = 0$ (the Cauchy distribution),

$$p(y) = \frac{1}{\pi\gamma\left[1+\left(\dfrac{y-a}{\gamma}\right)^2\right]} \tag{2.178}$$

While when $\alpha = 2$ and $\beta = 0$ (the Gaussian distribution),

$$p(y) = \frac{1}{\sqrt{4\pi\gamma}}\exp\left(-\frac{y^2}{4\gamma}\right) \tag{2.179}$$

The standard SαS PDFs (standard means $a = 0$ and $\gamma = 1$, much akin to the Gaussian standard PDF when $\mu = 0$ and $\sigma = 1$) are given by

$$p_\alpha(y) = \begin{cases} \dfrac{1}{\pi y}\displaystyle\sum_{k=1}^{\infty}\dfrac{(-1)^{k-1}}{k!}\Gamma(\alpha k+1)\,y^{-\alpha k}\sin\dfrac{k\pi\alpha}{2}, & 0<\alpha<1 \\ \dfrac{1}{\pi(y^2+1)}, & \alpha=1 \\ \dfrac{1}{\pi\alpha}\displaystyle\sum_{k=0}^{\infty}\dfrac{(-1)^k}{2k!}\Gamma\left(\dfrac{2k+1}{\alpha}\right)y^{2k}, & 1<\alpha<2 \\ \dfrac{1}{2\sqrt{\pi}}e^{-\frac{y^2}{4}}, & \alpha=2 \end{cases} \quad (2.180)$$

for the univariate SαS r.v., y.

2.8.12.2 Statistical Modeling with Non-Gaussian Stable Distributions

The stable distribution has the characteristic called the *stability property*, which states that the sum of two independent stable random variables with the same characteristic exponent is also stable and has the same characteristic exponent. Furthermore, the stable distribution is the only probability distribution that has this property [34].

2.8.12.3 Stable Signal Processing with Fractional Lower-Order Moments

From the signal processing point of view, the adoption of a stable model for noise has important consequences. As mentioned, for a non-Gaussian stable distribution with a characteristic exponent α, only moments of order less than α exist. In particular, the variance (i.e., the second-order moment) of a stable distribution with $a < 2$ does not exist, making the use of variance as a measure of dispersion meaningless.

The absence of finite variance does not mean, however, that there are no other adequate measures of variability of stable random variables. The dispersion of a stable random variable plays an analogous role to the variance. For example, the larger the dispersion of a stable distribution, the more it spreads around its median. Hence, the *minimum dispersion* criterion becomes the "equivalent" of the minimum variance measure of optimality in stable signal processing. It minimizes the spread of the distribution about the median value.

Minimizing the dispersion is also equivalent to minimizing the fractional lower-order moments of estimation errors, which measure the L_p distance between an estimate and its true value, for $p<\alpha\leq 2$. This result is not surprising since the L_p norm with $p < 2$ is well known for being robust against outliers such as those that may be described by the stable law.

2.8.12.4 Extension to Bivariate Random Variables

The notions expressed above for a single SαS r.v. can be extended to bivariate (and higher) r.v.s. Let y_1 and y_2 denote two such r.v.s. The joint PDF of y_1 and y_2 can be determined from the *inverse Fourier transform* (IFT) of their characteristic function [35]

$$p_{\alpha,\gamma,\beta_1,\beta_2}(y_1,y_2)=\frac{1}{(2\pi)^2}\times\int_{-\infty}^{\infty}\int_{-\infty}^{\infty}\exp\left[j(\beta_1 w_1+\beta_2 w_2)-\gamma\left(w_1^2+w_2^2\right)^{\alpha/2}\right]e^{-j(x_1 w_1+x_2 w_2)}dw_1 dw_2 \tag{2.181}$$

where the parameters α and γ are defined after (2.163). β_1 and β_2 describe the symmetry of the PDF and we assume them to be zero here since they are not important parameters for our purposes.

As above, a closed-form solution for (2.181) does not exist except for the special cases of $\alpha = 1$ and $\alpha = 2$. In those cases we have

$$p_{\alpha,\gamma}(y_1,y_2)=\begin{cases}\dfrac{\gamma}{2\pi\left(y_1^2+y_2^2+\gamma^2\right)^{3/2}}, & \alpha=1\\[2ex] \dfrac{1}{4\pi\gamma}\exp\left(-\dfrac{y_1^2+y_2^2}{4\gamma}\right), & \alpha=2\end{cases} \tag{2.182}$$

The second of these, when $\alpha = 2$, is the (bivariate) Gaussian PDF. The first, when $\alpha = 1$, is called the Cauchy distribution. The smaller α is, the more impulsive the underlying function is.

2.8.13 Summary

We presented considerable detail in this section on the principal sources of noise in an urban environment. The predominant sources below about 1 GHz are all produced by humans and thus called man-made noise, with vehicle ignitions and discharge lighting perhaps the worst offenders. EW systems are very sensitive because they must frequently collect signals from a considerable distance. The additive effects of the MMN in urban settings could very well preclude these systems from functioning properly.

A better environmental noise model in an urban setting is to model the noise as SαS processes rather than AWGN. These processes have heavier tails and thus admit to higher probabilities of noise impulses with significant amplitudes.

Adopting such allows for more accurate characterization of the surrounding noise fields and helps to understand the difficulties of EW operations in urban terrain.

2.9 Effects of Impulsive Noise on Signal Detection

2.9.1 Introduction

As discussed above, the primary type of noise in urban environments is due to man-made sources and the PDF for that noise is not accurately modeled as Gaussian. Rather, the noise more accurately is modeled as an alpha-stable process and, in particular, as an SαS. The effects of modeling MMN in this way was first examined by Tsihrintzis and Nikias [36] where the thermal noise was not included on the assumption that the SαS noise level is much larger. Ambike, Ilow, and Hatzinakos [37] determined the performance effects on binary antipodal (symbol $\in \{+1, -1\}$) signals when the thermal noise is included along with the SαS noise. We will examine this approach in this section to determine the effects of both types of noise by comparing the performance of an optimal (nonlinear) receiver with the conventional linear receiver. This discussion is based largely on [37].

We assume that the data on which our detectors operate are discrete sequences. The received sequence, after being demodulated and sampled K times per symbol, can be written as

$$x(k) = s_i(k) + n_G(k) + n_\alpha(k), \qquad i \in \{0,1\}, k = 1,2,\ldots,N \tag{2.183}$$

where $s_1(k) = -s_0(k)$ is the transmitted antipodal signal during one symbol interval, and $\{n_G(k)\}$ and $\{n_\alpha(k)\}$ are zero-mean Gaussian and SαS noise components, respectively. It is assumed that $1 < \alpha \leq 2$ to ensure the existence of the mean. The symbol detection is on a symbol-by-symbol basis since the symbols are equiprobable. We assume that both noise components are stationary sequences over the time interval of concern, that are i.i.d. (within their own sequence; they are not identically distributed between each other—the PDF of the one sequence is Gaussian, while the other is SαS) and independent of each other and the signal. The SαS r.v. with zero-mean is defined through its characteristic function (2.163) with a = 0. Since we have assumed that $\{n_G(k)\}$ and $\{n_\alpha(k)\}$ are independent, the characteristic function of the total additive noise is

$$\varphi_X(w) = \exp\left(-\frac{\sigma_G^2}{2} w^2 - \gamma |w|^\alpha\right) \tag{2.184}$$

where σ_G^2 is the variance of the Gaussian noise. The PDF is given by the inverse-Fourier transform of (2.184) as given by (2.167).

2.9.2 Receiver Structures

The detector is a nonlinearity followed by an accumulator (integrator in discrete form) and a threshold comparator. Here we will examine the performance of

- The optimal Bayes detector given by (2.9);
- A linear receiver that is optimal in the presence of Gaussian noise alone;
- Locally optimum detectors that can be realistically implemented.

2.9.2.1 Optimal Bayes Receiver

Since the noise samples are i.i.d., the test statistic for optimum Bayes detection based on the observed samples $[x(1), x(2), ..., x(N)]$ is given as [37]

$$\Lambda_B = \sum_{k=1}^{K} \ln\left[\frac{p_X[x(k)-s_1(k)]}{p_X[x(k)-s_0(k)]}\right] \mathop{\gtrless} 0 \begin{cases} H_1 \\ H_0 \end{cases} \tag{2.185}$$

where $H_i \triangleq \{s_i \text{ sent} | x(k) \text{ observed } k = 1,2,\ldots K\}, i \in \{0,1\}$.For large K, from the central limit theorem, Λ_B assumes a Gaussian distribution and

$$P_e = \frac{1}{2}\operatorname{erfc}\left(\frac{\mu_B}{\sqrt{2\sigma_B^2}}\right) \tag{2.186}$$

where μ_B is the mean of Λ_B given that s_0 was sent, and σ_B^2 is the conditional variance of Λ_B. These functions are given by

$$\mu_B = \sum_{k=1}^{K} \int_{-\infty}^{\infty} p_X[u-s_0(k)] \ln\left\{\frac{p_X[u-s_1(k)]}{p_X[u-s_0(k)]}\right\} du \tag{2.187}$$

$$\sigma_B^2 = \sum_{k=1}^{K} \int_{-\infty}^{\infty} p_X[u-s_0(k)] \ln^2\left\{\frac{p_X[u-s_1(k)]}{p_X[u-s_0(k)]}\right\} du - \mu_0^2 \tag{2.188}$$

2.9.2.2 Linear Receiver

The test statistic for the classical linear receiver (linear correlator detector) is [37]

$$\Lambda_{\text{lin}} = \sum_{k=1}^{K} [s_1(k) - s_0(k)] x(k) \underset{<}{\overset{\geq}{}} 0 \begin{cases} H_1 \\ H_0 \end{cases} \tag{2.189}$$

P_e is approximately given by [37]

$$P_e \approx \frac{1}{2} \left\{ 1 - \frac{4}{\pi} \sum_{k=1}^{K} \frac{\phi_X [(2k-1)\pi / 2\Sigma]}{2k-1} \sin\left[\frac{(2k-1)\pi d}{2\Sigma} \right] \right\} \tag{2.190}$$

where d is defined as the Euclidean distance between s_0 and s_1, that is,

$$d = \sqrt{\sum_{k=1}^{K} [s_1(k) - s_0(k)]^2} \tag{2.191}$$

and Σ represents the sum of the absolute impulsive noise samples in one symbol plus the standard deviation of the Gaussian noise

$$\Sigma = \sum_{k=1}^{K} |n_\alpha(k)| + \sigma_G \tag{2.192}$$

2.9.2.3 Nonlinear Receivers

The locally optimum receiver is a *generalized correlator detector* (GCD)

$$\Lambda_g = \sum_{k=1}^{K} [s_1(k) - s_0(k)] g_{lo}[x(k)] \underset{<}{\overset{\geq}{}} 0 \begin{cases} H_1 \\ H_0 \end{cases} \tag{2.193}$$

where $g_{lo}(x)$ is the memoryless nonlinearity given by

$$g_{lo}(x) \triangleq -\frac{p'_X(x)}{p_X(x)} \tag{2.194}$$

Assuming that an r.v. at the output of $g_{lo}(.)$ has a finite variance, Λ_g is asymptotically normal, and for large N, based on the central limit theorem, P_e for the GCD is

$$P_e = \frac{1}{2} \operatorname{erfc}\left(\frac{\mu_g}{\sqrt{2\sigma_g^2}} \right) \tag{2.195}$$

where

$$\mu_g = \mathcal{E}\{\Lambda_g | s_1 \text{ sent}\}| = -\mathcal{E}\{\Lambda_g | s_0 \text{ sent}\}$$
$$= \sum_{k=1}^{K} [s_1(k) - s_0(k)] \mathcal{E}\{g_{lo}[x(k)] | s_1(k)\} \quad (2.196)$$

$$\sigma_g^2 = \text{var}\{\Lambda_g | s_1 \text{ sent}\} = \text{var}\{\Lambda_g | s_0 \text{ sent}\}$$
$$= \sum_{k=1}^{K} [s_1(k) - s_0(k)]^2 \text{ var}\{g_{lo}[x(k)] | s_1(k)\} \quad (2.197)$$

The nonlinearity expressed in (2.194) is not very attractive because it

- Depends on the characteristics of the noise model;
- Requires numerical evaluation of $p'_X(x)$ and $p_X(x)$;
- Results in difficult performance analysis and the results are not intuitive.

Therefore, we define two suboptimum nonlinearities that approximate $g_{l0}(x)$ over a portion of its range: a sawtooth [$g_{st}(x)$] and a soft limiter [$g_{sl}(x)$], with transfer functions given by

$$g_{st}(x) = \begin{cases} \beta x, & |x| < \kappa \\ 0, & |x| > \kappa \end{cases} \quad (2.198)$$

and

$$g_{sl}(x) = \begin{cases} \beta x, & |x| < \kappa \\ \beta\kappa, & x \geq \kappa \\ -\beta\kappa, & x \leq -\kappa \end{cases} \quad (2.199)$$

respectively. Figure 2.60 shows the nonlinearities $g_{lo}(.)$, $g_{sl}(.)$, and $g_{st}(.)$ for a moderately impulsive channel. We assume that the threshold is set to 0 and that $\beta = 1$.

To calculate P_e, we need to know the PDF of the r.v. at the output of the nonlinearity, p_{out}. If the input to the sawtooth is an r.v. with PDF p_{in} and CDF P_{in}, then

$$p_{out}(x) = [1 - Pi_n(\kappa) - P_{in}(-\kappa)]\delta(x) + I_{(-\kappa,\kappa)} p_{in}(x) \quad (2.200)$$

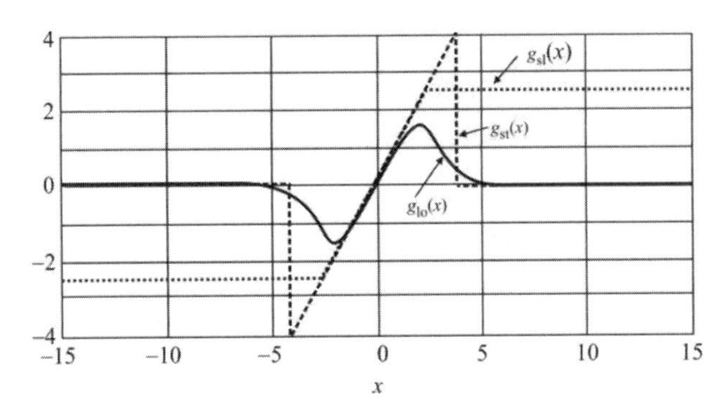

Figure 2.60 Approximations of LO nonlinearity for $\upsilon_G = 8.5$ dB, $f = 0.5$, $\alpha = 1.5$, $\kappa = 4$ for $g_{hp}(x)$, and $\kappa = 2.5$ for $g_{sl}(x)$. (Source: [37]. © IEEE 1994. Reprinted with permission.)

where

$$I_{(-\kappa,\kappa)} = \begin{cases} 1, & |x| \le \kappa \\ 0, & \text{otherwise} \end{cases} \tag{2.201}$$

Similarly, for the output of the soft limiter

$$p_{\text{out}}(x) = [1 - P_{\text{in}}(\kappa)]\delta(x-\kappa) + P_{\text{in}}(-\kappa)\delta(x+\kappa) + I_{(-\kappa,\kappa)} p_{\text{in}}(x) \tag{2.202}$$

To find μ_g and σ_g^2 in (2.196) and (2.197), we use (2.200) and (2.202), and we get for the sawtooth

$$\mathcal{E}\{g_{st}[x(k)] | s_1(k)\} = \int_{-\infty}^{\infty} u p_{\text{out}}(u) du = \int_{-\kappa}^{\kappa} u p_{\text{in}}[u | s_1(k)] du \tag{2.203}$$

$$\begin{aligned} \text{var}\{g_{st}[x(k)] | s_1(k)\} &= \int_{-\infty}^{\infty} u^2 p_{\text{out}}(u) du \\ &= \int_{-\infty}^{\infty} u^2 p_{\text{in}}[u | s_1(k)] du - \left[\mathcal{E}\{g_{st}[x(k)] | s_1(k)\right]^2 \end{aligned} \tag{2.204}$$

and for the soft limiter

$$\mathcal{E}\{g_{sl}[x(k)] | s_1(k)\} = \int_{-\infty}^{\infty} u p_{\text{out}}(u) du$$

$$= \{1 - P_{\text{in}}[\kappa | s_1(k)] - P_{\text{in}}[-\kappa | s_1(k)]\}\kappa + \int_{-\kappa}^{\kappa} u p_{\text{in}}[u | s_1(k)] du \tag{2.205}$$

$$\begin{aligned} \text{var}\{g_{\text{sl}}[x(k)] | s_1(k)\} &= \int_{-\infty}^{\infty} u^2 p_{\text{out}}(u) du \\ &= \{1 - P_{\text{in}}[\kappa | s_1(k)] + P_{\text{in}}[-\kappa | s_1(k)]\}\kappa^2 \\ &\quad + \int_{-\kappa}^{\kappa} u^2 p_{\text{in}}[u | s_1(k)] du - \left[\mathcal{E}\{g_{\text{sl}}[x(k)] | s_1(k)\right]^2 \end{aligned} \tag{2.206}$$

Determining P_e involves numerical integration in (2.203), (2.204), (2.205), or (2.206) with $p_{\text{in}}[u | s_1(k)]$ evaluated using (2.167). The final result is obtained from (2.195) based on (2.196) and (2.197). The choice of κ, which results in the lowest P_e, is based on the minimization of P_e in (2.195) for allowable values of κ.

2.9.3 Performance

A numerical and Monte Carlo simulation experiment was documented in [37] using symbols that were rectangular pulses with $K = 2$. P_e is plotted in Figure 2.61 as a function of the signal-to-Gaussian noise power ratio

$$\upsilon_G = \frac{\sum_{k=1}^{10} [s_i(k)]^2}{10\sigma_G^2}, \; i \in \{0,1\} \tag{2.207}$$

with the amount of impulsive noise represented by the parameter f. f is defined as the ratio of the dispersion of the SαS noise (a measure of the power) to the power of the Gaussian noise

$$f = \frac{\gamma}{0.5\sigma_G^2} \tag{2.208}$$

As we can see, as f increases or α decreases, there is more noise in the receiver and the detection performance degrades. For $P_e = 1.5 \times 10^{-2}$, there is approximately a 5 dB gain in SNR necessary, for the optimum receiver over the linear one. This gain increases for more impulsive channels (higher f and lower α). For purely Gaussian noise ($f = 0$), the optimal and linear receivers have the same performance, of course.

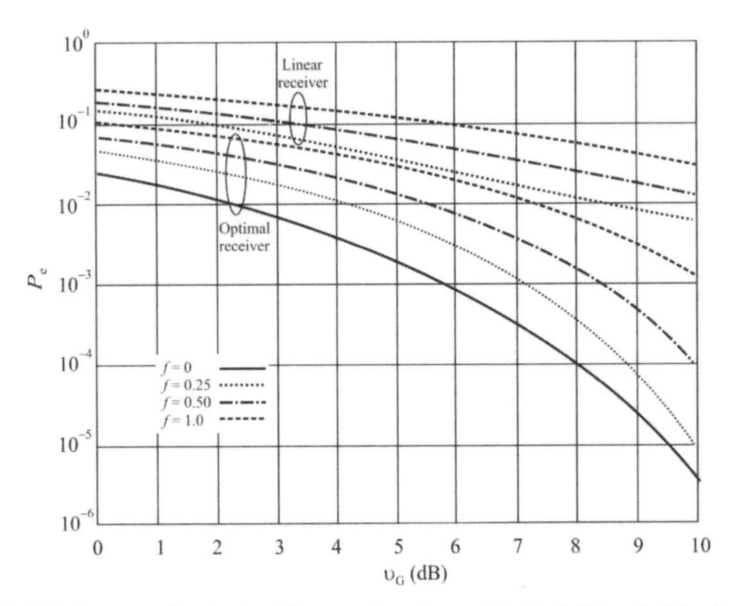

Figure 2.61 Performance of optimal and linear receivers for $\alpha = 1.5$, $K = 2$. When $f = 0$ (no impulsive noise), the performance of the linear receiver and optimal receiver are the same, denoted by the solid line. (Source: [37]. © IEEE 1994. Reprinted with permission.)

The simulation results also showed considerable improvements of the nonlinear receivers over the linear ones. For $f = 0.5$ and $\alpha = 1.5$ the suboptimum receivers demonstrated an SNR performance gain of 3 dB. For more impulsive channels (smaller α), even larger gains were demonstrated over the linear receiver.

2.9.4 Summary

We have considered the detection of a binary signal in a mixture of Gaussian and $S\alpha S$ noise. Linear receivers suffer about a 5 dB performance degradation versus optimal nonlinear receivers. The suboptimal but realizable nonlinear receivers outperformed the linear receiver by approximately 3 dB. The simulation results compare very well with the calculated results.

2.10 Concluding Remarks

We have discussed the detection of communication signals in the presence of noise in this chapter. We covered the basic properties of statistical signal detection theory and the common forms of tests applied to the problem. We also explored the statistical properties of the common types of noise encountered with these types of problems.

Next we investigated the detection properties of radiometers. A radiometer is a device for measuring the energy content of a signal and, assuming AWGN is the type of noise present, it is the optimal energy detector. The classical radiometer analysis assumes that the noise is stationary during the measurement process. Since this is not always true, we explored the impacts on detectability when the background noise changed.

We then had an in-depth discussion of noise sources in urban environments. Man-made sources are by far the largest contributors to RF noise in such settings, and we indicated that almost all MMN sources are impulsive in nature. That led us into a discussion of alpha-stable processes, which have statistical characteristics that more closely resemble IN than the Gaussian model does.

References

[1] Van Trees, H. L., *Detection, Estimation, and Modulation Theory, Part I*, New York: Wiley, 1968.

[2] *Reference Data for Radio Engineers*, New York: Howard W. Sams, 1975, Ch. 29.

[3] Poisel, R. A., *Introduction to Communication Electronic Warfare Systems*, 2nd ed., Norwood, MA: Artech House, 2008, pp. 148–156.

[4] Gagliardi, R. M., *Introduction to Communications Engineering*, 2nd ed., New York: Wiley, 1988, p. 151.

[5] Jordan, E. C., and K. G. Balmain, *Electromagnetic Waves and Radiating Systems*, 2nd ed., Englewood Cliffs, NJ: Prentice-Hall, 1968, Ch. 4.

[6] Rice, S., "Statistical Properties of a Sine Wave Plus Noise," *Bell System Technical Journal*, Vol. 27, January 1948, pp. 109–157.

[7] Nakagami, M., "The m-Distribution, a General Formula of Intensity of Rapid Fading," W. C. Hoffman, (ed.), *Statistical Methods in Radio Wave Propagation: Proceedings of a Symposium,* June 18–20, 1958, Permagon Press, 1960, pp. 3–36.

[8] Parsons, J. D., *The Mobile Radio Propagation Channel*, New York: Wiley, 1992.

[9] Beaulieu, N., and C. Cheng, "Efficient Nakagami-*m* Fading Channel Simulation," *IEEE Journal on Vehicular Technology*, Vol. 54, No. 2, March 2005, pp. 413–424.

[10] Abramowitz, M., and I. A. Stigen, (eds.), *Handbook of Mathematical Functions,* New York: Dover, 1970, Ch. 6.

[11] Skolnik, M. L., *Introduction to Radar Systems,* 2nd ed., New York: McGraw-Hill, 1980, p. 25.

[12] Dillard, R. A., and G. M. Dillard, *Detectability of Spread Spectrum Signals*, Norwood, MA: Artech House, 1989, p. 70.

[13] Urkowitz, H., "Energy Detection of Unknown Deterministic Signals," *Proceedings of the IEEE,* Vol. 55, No. 4, April 1967, pp. 523–531.

[14] Kostylev, V. I., "Energy Detection of a Signal with Random Amplitude," *Proceedings IEEE International Conference on Communications 2002*, April 28–May 2, 2002, pp. 1606–1612.

[15] Cai, K., V. Phan, and R. J. O'Connor, "Energy Detector Performance in a Noise Fluctuating Channel," *Proceedings IEEE MILCOM*, 1989, pp. 3.3.1–3.3.5.

[16] Sonnenschein, A., and P. M. Fishman, "Radiometric Detection of Spread-Spectrum Signals in Noise of Uncertain Power," *IEEE Transactions on Aerospace and Electronic Systems,* Vol. 28, No. 3, July 1992, pp. 654–660.

[17] Edell, J. D., *Wideband, Noncoherent, Frequency Hopped Waveforms and Their Hybrids in Low Probability of Intercept Communications*, Report NRL 8025, Naval Research Laboratory, Washington, D.C., November 8, 1976.

[18] Torrieri, D. J., *Principles of Secure Communication Systems*, 2nd ed., Norwood, MA: Artech House, 1992.

[19] Abramowitz, M., and I. A. Stigen, (eds.), *Handbook of Mathematical Functions,* New York: Dover, 1970, p. 260.

[20] Parsons, D., *The Mobile Radio Channel*, New York: Wiley, 1992, Ch. 9.

[21] Skomal, E. N., *Man-Made Radio Noise*, New York: Van Nostrand, 1972.

[22] Shukla, A., *Radiocommunications Agency–Feasibility Study into the Measurement of Man-Made Noise*, DERA Report DERA/KIS/COM/CD010470, March 2001.

[23] U.S. Code of Federal Regulations, Part 15.109 (as of April 29, 2010).

[24] Parsons, J. D., *The Mobile Radio Propagation Channel*, New York: Wiley, 1992, pp. 260–265.

[25] Goldwasser, S. M., Fluorescent Lamps, Ballasts, and Fixtures–Principles of Operation, Circuits, Troubleshooting, Repair, Version 1.90, http://members.misty.com/don/f-lamp.html #int0.

[26] Ribarich, T., "How Compact Fluorescent Lamps Work—and How to Dim Them," *EE Times*, http://www.eetimes.com/design/power-management-design/4010360/How-compact-flourescent-lamps-work-and-how-to-dim-them.

[27] V. Sekar, V., T. G. Palanivelu, and B. Revathi, "Effective Tests and Measurements Mechanisms for EMI Level Identification in Fluorescent Lamp Operation," *European Journal of Scientific Research*, Vol. 34, No. 4, 2009, pp. 495–505.

[28] Poisel, R. A., *Introduction to Communication Electronic Warfare Systems,* 2nd ed., Norwood, MA: Artech House, 2008, p. 622.

[29] Skomal, E. N., *Man-Made Radio Noise*, New York: Van Nostrand and Reinhold, 1978, p. 152.

[30] Skomal, E. N., *Man-Made Radio Noise*, New York: Van Nostrand and Reinhold, 1978, p. 154.

[31] Industrologic, Inc., http://www.industrologic.com/mechrela.htm.

[32] Draft Revision of Recommendation ITU-R P.372-8, ITU Radio Communications Study Group, January, 2005.

[33] Hagn, G. H., and D. B. Sailors, "Empirical Models for Probability Distributions of Short Term Mean Environmental Man-Made Noise Levels," *Proceedings 3rd Symposium and Technical Exhibition on Electromagnetic Compatibility*, Rotterdam, 1979, pp. 355–360.

[34] Shao, M., and C. L. Nikias, "Signal Processing with Fractional Lower Order Moments: Stable Processes and Their Applications," *Proceedings of the IEEE*, Vol. 81, No. 7, July 1993, pp. 986–1010.

[35] Yoon, S., I. Song, and S. Y. Kim, "Code Acquisition for DS/SS Communications in Non-Gaussian Impulsive Channels," *IEEE Transactions on Communications*, Vol. 52, No. 2, February 2004, pp. 187–190.

[36] Tsihrintzis, G., and C. L. Nikias, "Performance of Optimum and Suboptimum Receivers in the Presence of Impulsive Noise Modeled as an α-Stable Process," *Proceedings IEEE MILCOM, 1993*, Boston, MA, October 1993.

[37] Ambike, S., J. Ilow, and D. Hatzinakos, "Detection for Binary Transmission in a Mixture of Gaussian Noise and Impulsive Noise Modeled as an Alpha-Stable Process," *IEEE Signal Processing Letters,* Vol. 1, No. 3, March 1994, pp. 55–57.

Appendix 2A Chi-Square Distribution

2A.1 Probability Density Function

The chi-square distribution results when v independent variables with standard Gaussian distributions are squared and summed. The formula for the PDF of the chi-square distribution is

$$p(x) = \frac{1}{2^{v/2}\Gamma\left(\frac{v}{2}\right)} e^{-\frac{x}{2}} x^{\frac{v}{2}-1}, \qquad x \geq 0 \tag{2A.1}$$

where v is the shape parameter and Γ is the gamma function given by

$$\Gamma(a) = \int_0^\infty t^{a-1} e^{-u} du \tag{2A.2}$$

In a testing context, the chi-square distribution is treated as a "standardized distribution" (i.e., no location or scale parameters). However, in a distributional modeling context (as with other probability distributions), the chi-square distribution itself can be transformed with a location parameter, μ, and a scale parameter, σ.

Figure 2A.1 is the plot of the chi-square PDF for 4 different values of the shape parameter.

2A.2 Cumulative Distribution Function

The formula for the cumulative distribution function of the chi-square distribution is

$$P(x) = \frac{\gamma\left(\frac{v}{2}, \frac{x}{2}\right)}{\Gamma\left(\frac{v}{2}\right)}, \qquad x \geq 0 \tag{2A.3}$$

where γ is the incomplete gamma function given by [19]

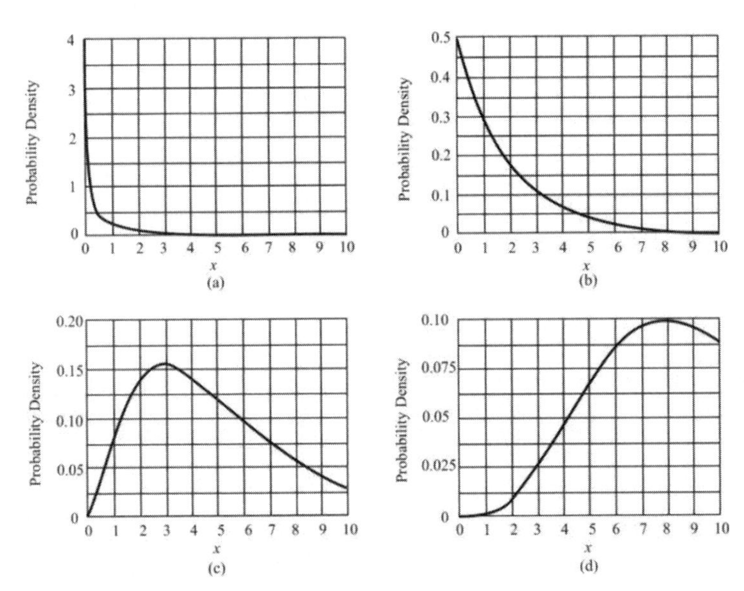

Figure 2A.1 Chi-square PDFs. (a) υ = 1, (b) υ = 2, (c) υ = 5, and (d) υ = 2.

$$\gamma(a,x) = \int_0^x t^{a-1} e^{-u} du \tag{2A.4}$$

Figure 2A.2 is the plot of the chi-square cumulative distribution function with the same values of v as the PDF plots above.

The more common statistics for the chi-square distribution are given in Table 2A.1.

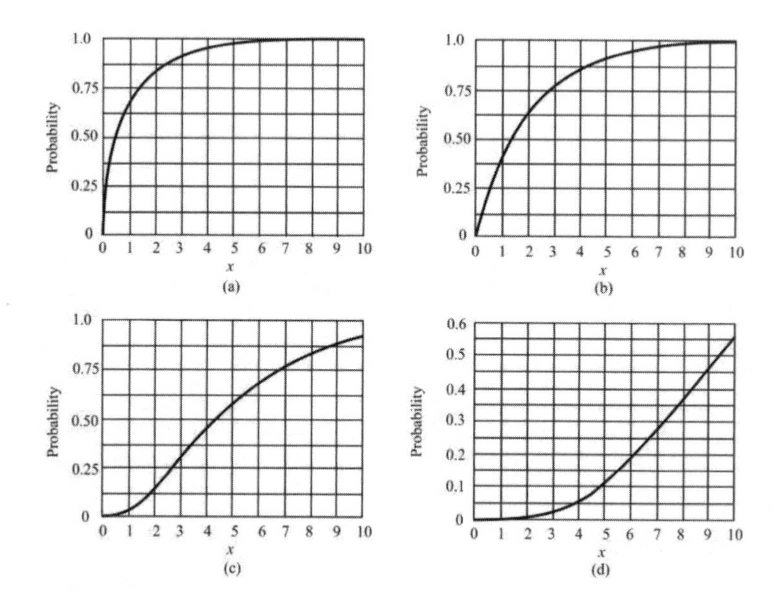

Figure 2A.2 Chi-square CDFs. (a) υ = 1, (b) υ = 2, (c) υ = 5, and (d) υ = 2.

Table 2A.1 Chi-Square Common Statistics

Mean	μ
Median	$\sim\mu-2/3$ for large μ
Mode	$\mu-2$ for $\nu>2$
Range	0 to $+\infty$
Standard deviation	$\sqrt{2\mu}$
Coefficient of variation	$\sqrt{2/\mu}$
Skewness	$2^{1.5}/\sqrt{\mu}$
Kurtosis	$3+12/\mu$

Chapter 3

Signaling for Modern Communications

3.1 Introduction

Modern signaling, in particular, AJ techniques, were devised to allow for reliable communication even in the presence of noise and interference. Our primary interest is when the interference is intentionally placed (i.e., a jammer). How well a jammer works against these signals is discussed in Chapter 8 and Chapters 10–14. It is deemed prudent to first discuss the techniques to facilitate modern communications, and such is the subject of this chapter.

This chapter is structured as follows. We first cover some terminology associated with digital communications to include coding techniques, interleaving, SNR, and bandwidth. We then discuss the common forms of MPSK with which we are concerned and their performance with noise and tone interference. That is followed by a discussion of MFSK performance. Next we introduce *quadrature amplitude modulation* (QAM) and study its performance in noise and tone interference. Then we delve into the forms of spread spectrum of concern, and, in particular, we cover DSSS, FHSS, and THSS (UWB), and hybrids of these. Lastly, we present a discussion of *orthogonal frequency division multiplexing* (OFDM), which is an emerging popular form for dealing with ISI on wideband channels.

3.2 Signaling

Signaling in this context refers to the method of impressing information onto some form of baseband signal, and then placing this signal on the proper carrier for transmission. It also includes the reception of the carrier and subsequent extraction of the information at a receiver. Thus, it refers to the entire communication process over RF transmissions.

Modulation is the method whereby the carrier signal is modified. It is the modulation type that determines, to a large extent, the performance of a particular communication technique. In addition, it is the modulation that determines the required SNR from the detector at the receiver for a specified BER level of performance. Hence, the modulation determines the level of JSR required to achieve whatever level of jamming is desired, and so determines many of the parameters of an EA system.

Some of the early FHSS communication systems used analog modulations. They used *frequency modulation* (FM) as their modulation technique. In at least one case, the radios were not blanked as they changed frequency, causing some strange RF characteristics. Because of the advance of the speed of monolithic integrated circuits as well as modern digital signaling technology, all of the current important communication systems use some form of digital modulation. Digital modulations allow for coding at the source level to remove redundancy and at the channel level to facilitate error detection and correction. Such error processing increases the system throughput and optimizes channel utilization. All modern systems are inherently digital, even if the information-bearing signal is analog. The analog signal is first converted to digital for processing and may be converted back to analog at the receiver.

Another technique for countering the deleterious effects of jamming is channel coding. Extra bits are added to the transmitted signal with such coding. These bits provide redundancy or other mathematical properties that make the signals more jam resistant.

Modulating a carrier may or not be part of the spectrum spreading process. As previously mentioned, the vast majority of the modulation techniques used in any relevant way are digital. Frequency and phase modulations are those of choice for FHSS and DSSS, respectively. The frequency modulation used is normally FSK, but it is also possible to use PSK in FHSS systems. Most PSK systems use coherent detection at the receiver; *differential PSK* (DPSK) is one notable exception.

Amplitude shift key (ASK) is a common modulation technique for some types of TH systems. The amplitude sent is $A > 0$ for a one and $-A$ for a zero. The time of the occurrence of these pulses is determined by a pseudo-random sequence. *Pulse position modulation* (PPM) is also used in TH systems and we will focus on that in this chapter.

For digital modulations, *symbols* are used to modulate the carrier. Symbols are the result of mapping the digital information in the form of data words onto another form as illustrated in Figure 3.1. An example for *multiple frequency shift key* (MFSK) illustrates this. Suppose two data bits are used to form a digital word. Then $2^2 = 4$ possible symbols emerge. A symbol in MFSK is a distinct tone used to modulate the carrier. One such mapping could be as shown in Table 3.1 and illustrated in Figure 3.2.

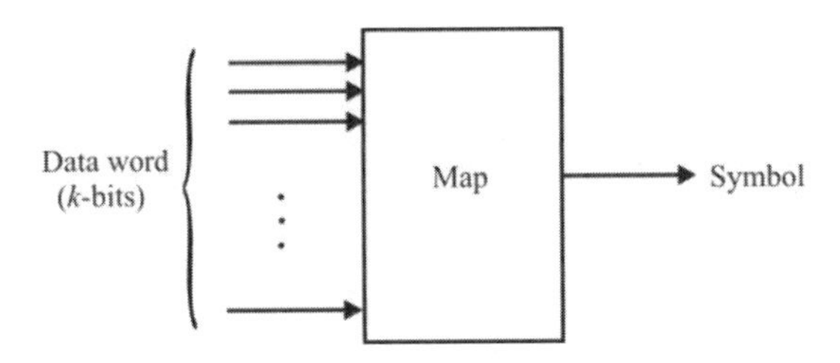

Figure 3.1 A representation of mapping data words to symbols.

The most prolific form of modulation for FHSS communications is MFSK, with the most common being BFSK. Likewise, the most common form of modulation for DSSS communications is MPSK, with $M = 2$ or 4.

Over a time interval k, which is T seconds long, in general, a received signal $r(t)$ can be expressed as

$$\begin{aligned} r(t) &= \text{Re}\left\{\tilde{r}(t)e^{j(2\pi f_0 t + \phi_0)} + n_k(t) + i_k(t)\right\}, \qquad (k-1)T \le t < kT \\ &= \text{Re}\left\{\tilde{r}(t)e^{j(2\pi f_0 t + \phi_0)}\right\} + \text{Re}\left\{n_k(t)\right\} + \text{Re}\left\{i_k(t)\right\} \end{aligned} \tag{3.1}$$

where

$$\tilde{r}(t) = \tilde{r}_{\text{r}}(t) + j\tilde{r}_{\text{i}}(t) \tag{3.2}$$

$n_k(t)$ is a sample from a zero mean, Gaussian random noise process with variance σ^2, and $i_k(t)$ represents any interference present. In general, these quantities are complex. We make note that for quadrature modulation

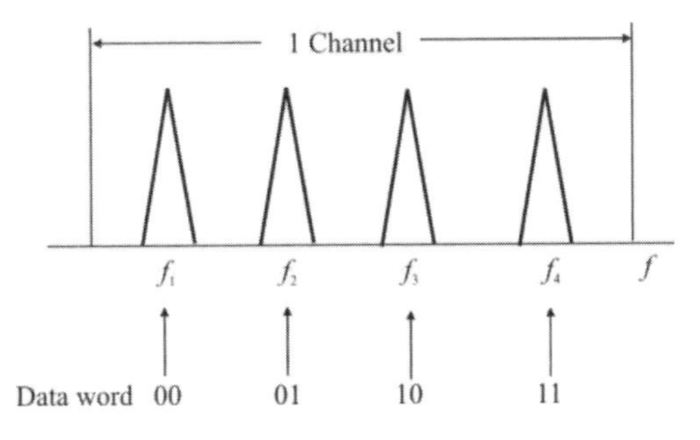

Figure 3.2 Mapping from digital data words to 2^2 FSK symbols.

Table 3.1 Example of a Mapping from a Data Word to 2^2 FSK

Data word	Symbol Tone Frequency
00	f_1
01	f_2
10	f_3
11	f_4

$$\begin{aligned}\text{Re}\left\{\tilde{r}(t)e^{j(2\pi f_0 t+\phi_0)}\right\} &= \text{Re}\left\{\begin{aligned}&\left(\tilde{r}_\text{r}(t)+j\tilde{r}_\text{i}(t)\right)\\ &\times\left[\cos\left(2\pi f_0 t+\phi_0\right)+j\sin\left(2\pi f_0 t+\phi_0\right)\right]\end{aligned}\right\}\\ &= \tilde{r}_\text{r}(t)\cos\left(2\pi f_0 t+\phi_0\right)-\tilde{r}_\text{i}(t)\sin\left(2\pi f_0 t+\phi_0\right)\end{aligned} \tag{3.3}$$

3.3 Binary Signal Reception

Let two arbitrary binary signals, with finite energy over T, be represented by $s_1(t)$ and $s_2(t)$, where the first signal represents the digit 1 and the latter represents the digit 0. T is the bit time. These signals are generated at the transmitter, suitably modulated onto a carrier for transmission, and subsequently demodulated at the receiver. When no noise or interference is present, these signals will be perfectly detected at the receiver and every bit sent will be correctly extracted. Noise, however, is always present in any real situation, and this noise can cause errors in the decoding process. Whatever the means for extracting the data bits at the receiver, the random noise induced by the transmitter, the propagation medium, and the receiver must be accounted for in order to maximize the probability that the received data sequence has as few errors as possible.

Although other choices are possible and frequently applied, the most common form of channel assumed between the transmitter and receiver is the AWGN channel. This is where the noise is assumed to originate from a physical process that emits EM energy in such a way that the amplitude of the noise can be statistically described by the Gaussian PDF. This assumption is both mathematically convenient as well as reasonably accurate in many real cases. Thus, the noise PDF is given by

$$p_N(n)=\frac{1}{\sqrt{2\pi\sigma^2}}\exp\left(-\frac{(n-\mu)^2}{2\sigma^2}\right) \tag{3.4}$$

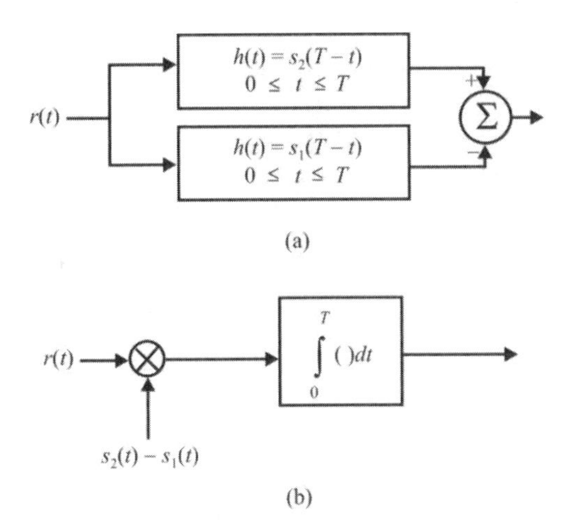

Figure 3.3 Two forms of maximum likelihood detectors for detection of arbitrary binary waveforms: (a) matched filter and (b) correlator. They both produce equivalent results.

where μ is the mean and σ^2 is the variance. σ^2 is also the noise power when the mean is zero. Frequently $\mu = 0$ and $\sigma^2 = P_\text{N}$, where P_N is given by (2.30), so

$$p_N(n) = \frac{1}{\sqrt{\pi N_0 W}} \exp\left(-\frac{n^2}{N_0 W}\right) \tag{3.5}$$

The notation $\mathcal{N}(\mu, \sigma^2)$ is used for the noise model in (3.4).

In this book we will abuse the notation somewhat and not make the distinction between N_0 and N_0', implying the latter always present. That is, all noise sources need to be included when considering the SNR. This can get tricky when dealing with non-Gaussian noise sources, however.

The maximum likelihood receiver is the optimum architecture for detecting these binary signals when the noise is AWGN. The two types of optimum detectors for binary digital signals with arbitrary wave shapes are shown in Figure 3.3. If the bit energies for $s_1(t)$ and $s_2(t)$ are E_1 and E_2, then

$$E_k = \int_0^T s_k^2(t)dt, \qquad k = 1,2 \tag{3.6}$$

and the minimum probability of error is given by [1]

$$P_e = Q\left(\sqrt{\frac{E_b}{N_0}(1 - R_{12})}\right) \tag{3.7}$$

where E_b / N_0 is the energy per bit to "equivalent" total PSD (at the input to the detector, due to more than just the thermal noise) ratio and

$$E_b = \frac{E_1 + E_2}{2} \tag{3.8}$$

is the average signal energy. $Q(x)$ is the Gaussian Q-function from Appendix A. R_{12} is defined by

$$R_{12} = \frac{\sqrt{E_1 E_2}}{E_b} \rho_{12} \tag{3.9}$$

and represents a measure of similarity between $s_1(t)$ and $s_2(t)$. The *normalized correlation coefficient*, ρ_{12}, is given by

$$\rho_{12} = \frac{1}{\sqrt{E_1 E_2}} \int_0^T s_1(t) s_2(t) dt \tag{3.10}$$

From (3.10), two signals $s_i(t)$ and $s_j(t)$ are *orthogonal* if

$$\int_0^{mT} s_i(t) s_j(t) dt = 0, \qquad i \neq j, m \text{ integers} \tag{3.11}$$

where T is the symbol time, common to both signals, which implies from (3.9) that $R_{12} = 0$. When

$$\int_0^{mT} s_i(t) s_j(t) dt = -1, \qquad i \neq j, m \text{ integers} \tag{3.12}$$

then $R_{12} = -1$ and the signals are called *antipodal*. Equation (3.7) is plotted in Figure 3.4 for these signals.

"Channel" as used herein refers to the channelization of the RF spectrum normally established by federal government regulation, such as the FCC in the United States. For example, in the low VHF military frequency range

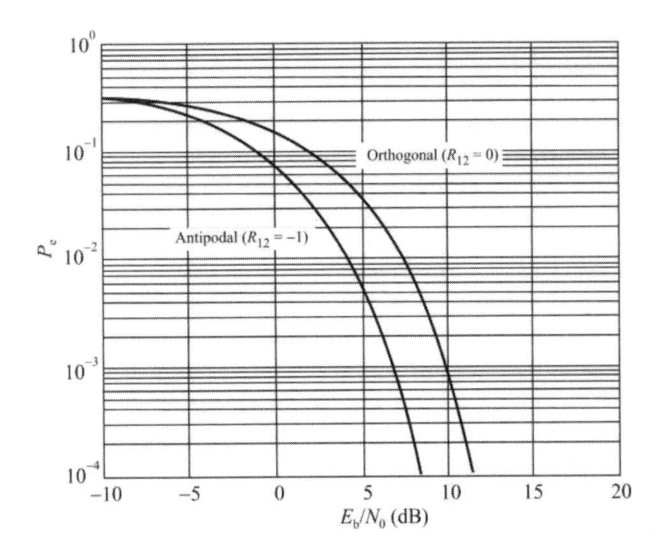

Figure 3.4 P_e for binary signal detection when the data signals have arbitrary wave shapes.

(30–88 MHz), the military channelization is 25 kHz presently (soon to be reduced to 12.5 kHz). In the Channel 2–6 TV band in the United States (54–88 MHz), the TV channelization is 6 MHz.

In *hard decision decoding*, a decision is made as to which symbol was sent based on the energy outputs of the detectors. Either a space or mark is decided. In *soft decision decoding*, additional information is used. Such information could take many forms, such as information about the channel when the symbol was sent, gleaned by some means. In addition, with soft decision decoding, sometimes decisions can be made about the received symbol other than assuming it was only a mark or space. If the energy from the two detectors is about the same, the soft decision decoder could declare that it does not know.

The probability of an error occurring in a bit is denoted P_e, and is sometimes called the *bit error rate* (BER) since they are equivalent. Note that P_e, in general, is different from the *symbol error rate* (SER), which is the probability of a symbol error and is denoted P_s.

3.4 Error Control

All communications are subject to errors due to noise and interference. The noise can be unintentional or intentional; it makes no difference. However, the *type* of noise or interference does make a difference. There are essentially two common methods to deal with errors caused by RF noise: *automatic repeat request* (ARQ) and *forward error correction* (FEC).

In ARQ schemes, when a message is received in error (the error being detected by some mechanism at the receiver) the receiver sends a *negative acknowledgment* (NAK) back to the transmitter so the message can be resent. In excessively noisy cases, this can occur several times. The disadvantage of ARQ is obvious: the system throughput can be dramatically reduced by the noise or interference. Another significant disadvantage of ARQ is that a reverse communication path is required from the receiver back to the transmitter in order to transmit the NAK. Countermeasures specifically designed for ARQ schemes will not be considered herein.

For FEC, one of the forms of coding is applied to the data to be sent. This is accomplished by adding some number of bits as appropriate to the data, thus also reducing the information throughput. This is often a favorable trade-off, however, in that lower SNRs are permitted in the data signal, allowing for lower signal power for the same BER performance or better BER performance for the same signal power. Different methods of coding have been devised. The effects on jamming performance of FEC codes will be discussed.

As will be shown, uncoded BFSK signals suffer significant degradation due to jamming. As an example of coding advantages, consider perhaps the simplest type of coding, repetition coding. In this case, the transmission of a bit of data is repeated some number of times. Without interleaving, these bits would be sent in succession. If a hop is being jammed, then this coding technique is ineffective since all the resulting coded bits would be jammed. If, however, the bits are interleaved where each of the coded bits are sent on a different hop, for many types of jamming there is a high probability that some of the bits will be sent in portions of the spectrum that are not jammed and therefore are received error-free, subject only to noise levels.

3.5 Coding Techniques

It is well known that coding is effective at combating channel noise in digital communication systems. It would therefore be expected that it could be used to counter the effects of intentionally injecting noise into the channel as a noise jammer does. This turns out to be the case, and the effects for noise jamming can be overcome by coding the transmitted signal.

Digital communication receivers in systems that employ coding also employ decoders to reverse the coding process. The goal of coding is to maximize the probability of decoding the same uncoded tuple that was sent. Communication channels are corrupted by factors such as thermal noise, intentional noise inserted

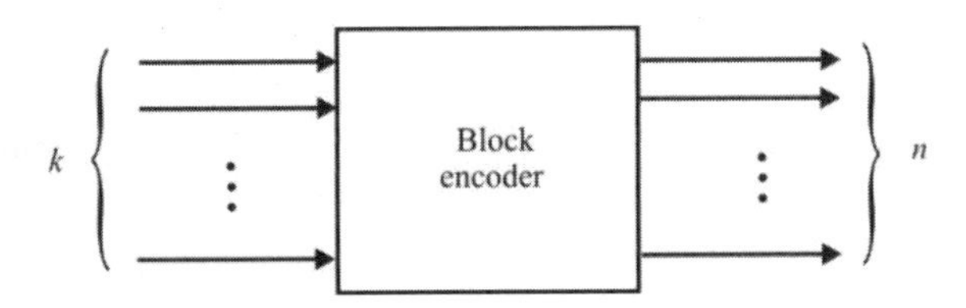

Figure 3.5 Block encoding k information symbols into encoded symbols with n components.

by jammers, unintentional noise inserted by man-made objects such as microwave ovens, and other signals generated by other communication systems sharing the same band. All these sources can cause errors in the receiving process.

Coding adds some bits to each data word to be transmitted. Often these bits are parity bits. The *parity* of a digital word consisting of b (> 0) bits is the number of binary 1's in the word. *Even parity* is when this number is an even integer and *odd parity* is when this number is an odd integer.

In general, there are two kinds of coding techniques in digital communication systems: block coding and convolutional coding. In block coding, several data bits are assembled together and encoded together. Such encoding usually entails adding parity bits to this group assembled. Convolutional coding, on the other hand, emits a continuous stream of encoded bits in response to a continuous stream of bits to be encoded. The encoded symbol that is currently emitted from the encoder is a function of not only the current symbol to be encoded but also one or more previous bits.

3.5.1 Block Coding

Block coding is when a group of k symbols are input to the encoder and n encoded symbols emerge. This is illustrated in Figure 3.5. Block codes can be linear or nonlinear but only linear block codes are discussed here.

For binary systems, the size of the output space is 2^n, while the size of the input space is 2^k. For each input symbol $d_i \in \mathbf{K}$ there is one and only one encoded symbol $e_j \in \mathbf{E}$. It is always true that $n > k$; thus, $2^n > 2^k$. Therefore, not all of the possible encoded symbols are used. It is this fact that is exploited to make effective block codes. The *rate* of a block code is given by $r = k / n$.

An individual information word is a k-tuple $d_i = (d_{i_0}, d_{i_1}, \cdots, d_{i_{k-1}})$, where for binary systems $d_{i_j} \in (0, 1\}$ or $d_{i_j} \in \{-1, +1\}$. An individual code word is an n-tuple $e_i = \{e_{i_0}, e_{i_1}, \cdots, e_{i_n}\}$. These tuples are sometimes referred to as vectors. A properly selected set of code words makes up a vector space, which means that the sum of any two of them is also a code word. The number of places where two code

Table 3.2 Example of a Block Code

Reference Number	Information Word	Code Word	Hamming Distance from the Zero Code Word
0	0000	00000	0
1	0001	00011	2
2	0010	11000	2
3	0011	00100	1
4	0100	10101	3
5	0101	10001	2
6	0110	01010	2
7	0111	11011	4
8	1000	10011	3
9	1001	11001	3
10	1010	00110	2
11	1011	01100	2
12	1100	01110	3
13	1101	01111	4
14	1110	11110	4
15	1111	11111	5

words differ is known as the *Hamming distance,* d_H, between them. Thus, if $e_1 = (1001001)$ and $e_2 = (0010111)$, then $d_H = 5$. The minimum Hamming distance of a code is called the *minimum distance* of the code and denoted by d_{min}. If the set of codewords all at the same Hamming distance d_i from the all zero codeword is denoted by W_i, then the *weight distribution* of the code is given by the set of all W_i for which $d_i = d_{min}, \ldots, n$.

As a concrete example of these concepts, suppose the code words are as depicted in Table 3.2. In this case $d_{min} = 1$, $n = 5$, $k = 4$, $r = 4/5$, $W_1 = \{3\}$, $W_2 = \{1, 2, 5, 6, 10, 11\}$, $W_3 = \{4, 8, 9, 12\}$, $W_4 = \{7, 13, 14\}$, and $W_5 = \{15\}$. The input symbol space **K** is of size $2^4 = 16$, while the output encoded symbol space **E** is of size $2^5 = 32$.

A *systematic block code* is a linear block code where the data bits are used directly in part of the output symbol. Schematically this is shown in Figure 3.6.

A *binary symmetric channel* (BSC) is a binary communication channel for which the probability of making an error and decoding a one when a zero was sent is equal to the probability of decoding a zero when a one was sent. The probability of not making an error is denoted by P, so the probability of making an error is given by $1 - P$. Such a channel can be represented as in Figure 3.7.

A channel is *memoryless* if $P(0|0) = P(1|1) = P$ does not change according to any previous symbols. Thus, P is constant with time. Many of the performance parameters of digital communication systems depend on the memoryless channel

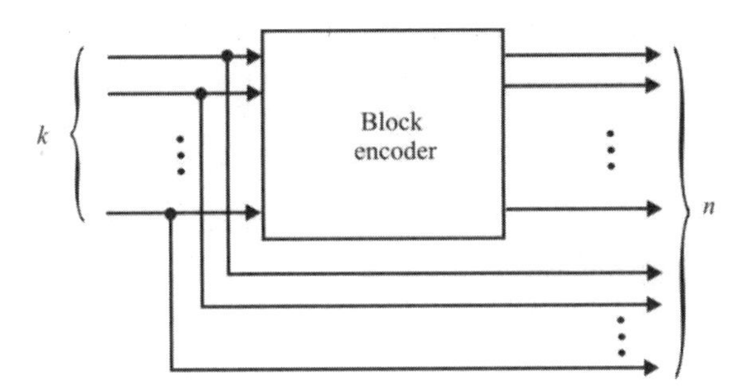

Figure 3.6 Block diagram of a systematic block code generator.

property. There are other important models of communication channels that do not presume this to be the case, however.

Denote **y** as the input to the channel and **x** the output of the channel. Denoting the decision made by a decoder as $\hat{\mathbf{x}}$, then a *maximum likelihood decoder* is one that employs the rule: "Choose $\hat{\mathbf{x}}$ to be that x_m so that $p_Y(\mathbf{y}|\,\mathbf{x}_m)$ is maximized" [2]. For the BSC [2]

$$p_Y(\mathbf{y}|\hat{\mathbf{x}}) = P^{d_H}(1-P)^{n-d_H} \tag{3.13}$$

when d_H is the Hamming distance between **y** and $\hat{\mathbf{x}}$ and $\mathbf{x}_m$ is chosen that maximizes (3.13). The tuples **y** and $\mathbf{x}_m$ are independent vectors and as such

$$p_Y(\mathbf{y}|\mathbf{x}_m) = \prod_{q=1}^{n} p_Y(y_q|x_{mq}) \tag{3.14}$$

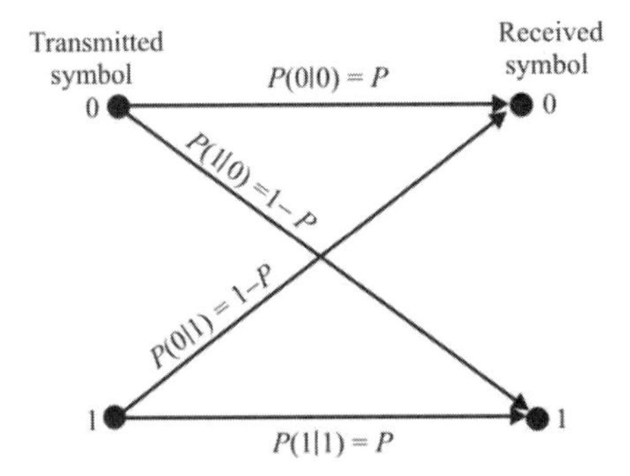

Figure 3.7 Schematic representation of a binary symmetric channel.

That is, the overall probability to be maximized is the product of the individual symbol probabilities that make up the tuples. This would not be true if the channel had memory because in that case the current probability would depend on one or more prior symbols. Actually, the logarithm of this function is also used since the logarithm is a monotonically increasing function of its argument. Taking the logarithm of (3.14) converts it into a sum rather than a product.

Torrieri [3] derived a bound on the information rate of block codes as determined by the channel bit errors. The information (decoded) bit error rate, P_{ib}, is approximately

$$P_{\text{ib}} \approx \frac{d_{\min}}{n} \sum_{k=t+1}^{d_{\min}} \binom{n}{k} P_{\text{e}}^{k} (1-P_{\text{e}})^{n-k} + \frac{1}{n} \sum_{k=d_{\min}+1}^{n} k \binom{n}{k} P_{\text{e}}^{k} (1-P_{\text{e}})^{n-k} \qquad (3.15)$$

where

$d_{\min}$ = minimum distance between code words

$t = \left\lfloor \frac{d-1}{2} \right\rfloor$

$\lfloor x \rfloor$ = integer part of x

n = the number of code symbols representing k information symbols

P_{e} = BER of the channel (that is subject to jamming degradation directly)

The effects of coding digital signals are quite dramatic with low input error rates. Most error correcting codes are designed to operate in this range—they were designed with thermal background noise in mind. The decoding error performance for Golay (23, 12, 3) code that corrects up to three symbols in error in a block of 23 binary symbols containing 12 information bits is shown in Figure 3.8. As seen, at small error rates, less than, say, 5% in this example, the output error rate is significantly below that of the uncoded symbols. At sufficiently large input error rates, however, the corrected bit error rate will actually exceed the uncoded rate. Above that point is called *error extension*. The point at which these two curves cross is called the *critical symbol error rate*. Table 3.3 [4] lists some popular codes and the critical input error rate at which error extension begins.

In the region of low P_{e}, where most codes are designed to operate, the coding decreases P_{ib} by decades. Above some relatively low P_{e}, however, the decoded bit error rate rises quite rapidly. This effect is shown in Figure 3.9, where the details of Figure 3.8 at low P_{e} are illustrated. Clearly, the decoded error rate is quite low below about $P_{\text{e}} = 0.04$, but at values as low as $P_{\text{e}} = 10^{-1}$, the decoded BER is as large as 0.5×10^{-1}, approaching regions where many digital, machine-to-machine communications cannot operate.

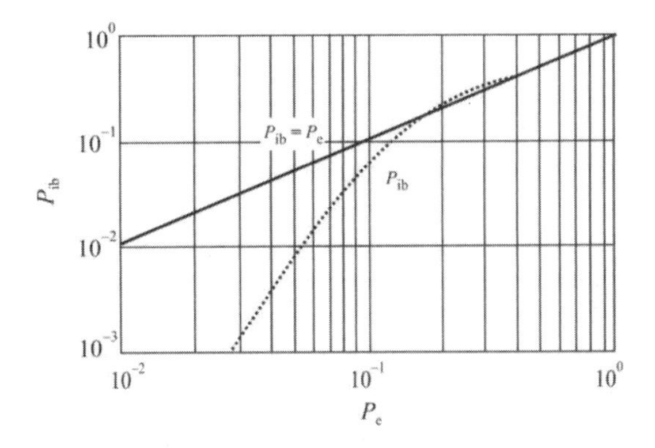

Figure 3.8 Decoded bit error probability versus the channel bit error probability for Golay (23, 12).

What this means for a jammer is that channel binary symbol rates on the order of 10% (10^{-1}) are sufficient to negate the coding advantage in many cases.

3.5.2 Convolutional Coding

Convolution encoding entails entering *k* information symbols into a linear shift register. The properties of these shift registers are examined in Chapter 6. The outputs of the shift register stages are input to modulo two adders in a way that is specified by the code. After the *k* information symbols are loaded into the shift

Table 3.3 Critical Input Error Rates for Some Codes

Code Type	Critical Input Error Rate	Input Error Rate Threshold for Large Output Error Rate
Convolutional Codes		
(1,2) rate 1/2	0.064	0.04
(3,1) rate 1/3	0.157	0.12
(4,1) rate 1/4	0.155	0.12
(8,1) rate 1/8	0.243	0.18
Dual 3, 4 and 5	0.053	0.04
Block Codes		
Golay (23,12)	0.163	0.06
Hamming (7,4)	0.250	0.06
Reed-Solomon		
RS (15,9)	0.260	0.06
RS (31,15)	0.260	0.12
RS (7,3)	0.370	0.06
Cyclic Codes		
BCH (127,92)	0.047	0.06
BCH (127,64)	0.083	0.06
BCH (127,36)	0.124	0.06

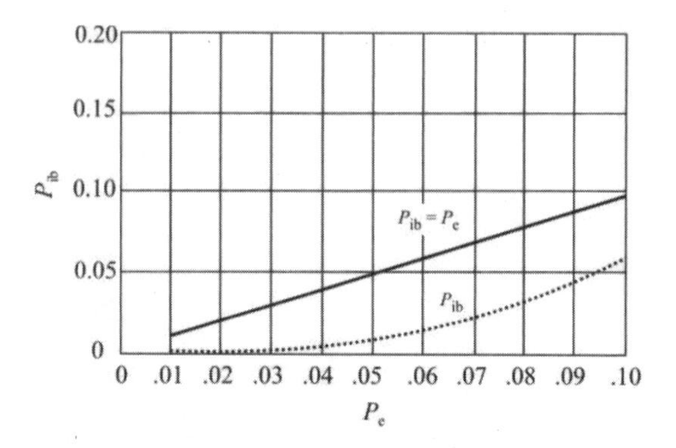

Figure 3.9 Decoded error rate for low values of P_e for Golay (23,12).

register, the *n* output symbols are formed at the *n* outputs of these adders. These, then, are the encoded information symbols. The *constraint length* of a code is defined as the number of past input symbols that can influence the current output symbol.

3.5.2.1 Coded MFSK

Reed-Solomon codes are sometimes used with noncoherent, orthogonal MFSK. The performance of such a detector for a few of these codes is shown in Figure 3.10 [5] where hard decision decoding was assumed. The dual rate codes have also been applied to MFSK with performance characteristics shown in Figure 3.11 [6] for $r = ½$ in Gaussian noise and hard decision decoding. The transition regions in these cases are quite narrow, typically going from no communications (10^{-1} BER) to 10^{-5} BER in only a few decibels of E_b / N_0.

From the point of view of the communicator, this can be viewed positively in that applying additional energy of only a few decibels can result in effective communications. It can also be viewed as positive by a jammer in that applying only a few decibels, more energy (noise in this case) can quickly degrade communications. In stressed environments, therefore, it is necessary to operate communication systems at values of E_b / N_0 considerably larger than those in the transition regions, whenever possible.

The BER performance for uncoded QPSK in AWGN is shown in Figure 3.12. Table 3.4 contains the coding gains obtained by comparing Figures 3.10 and 3.11 with Figure 3.12. These coding gains are for a BER of 10^{-5}, so the uncoded SNR is 13.4 dB. Of course, all of this gain does not come free. It costs in terms of either bandwidth required for the same data rate or a lower data rate for the same bandwidth.

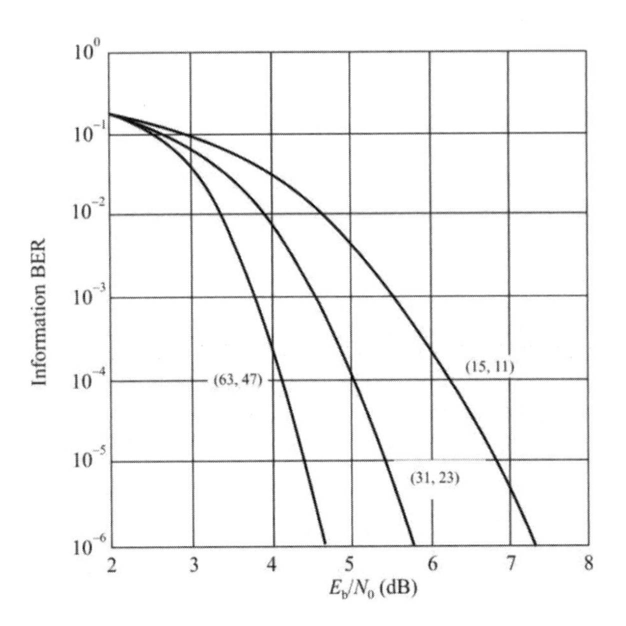

Figure 3.10 Decoded BER for noncoherent, orthogonal MFSK with some Reed-Solomon codes and hard decision decoding. (Source: [5]. © 1992 Artech House. Reprinted with permission.)

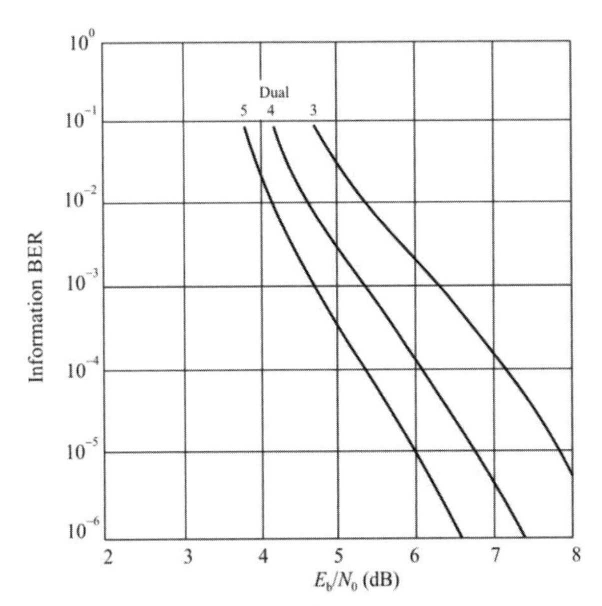

Figure 3.11 Performance of MFSK with dual-rate coding. (Source: [6]. © 1992 Artech House. Reprinted with permission.)

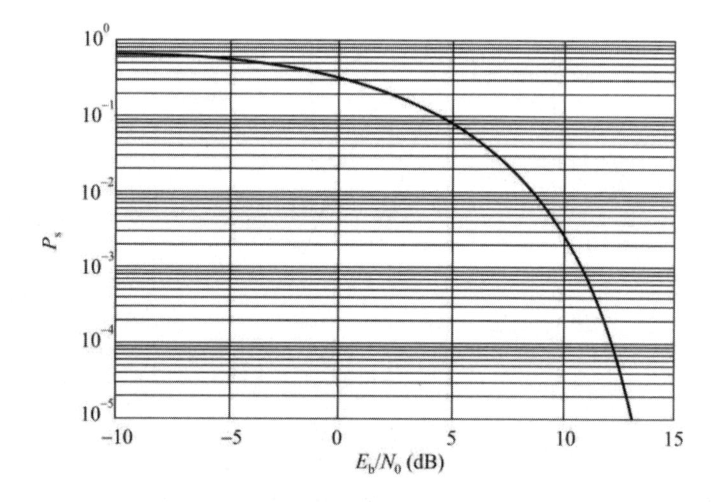

Figure 3.12 BER for QPSK.

Table 3.4 Coding Gains in Decibels for MFSK for Several Codes at 10^{-5} BER

Dual			**Reed Solomon**		
3	4	5	(63,47)	(31,23)	(15,11)
5.7	6.7	7.4	9.0	7.9	6.6

For the Reed Solomon codes, if W is the bandwidth for uncoded 2-FSK, for MFSK with $2M$frequencies, the required uncoded bandwidth is about $2^{m-1}W/M$, and with Reed-Solomon (n, k) coding, the bandwidth is approximately $2^{m-1}nW / Mk$. For the dual-m coding, the bandwidth required is approximately $2^{m-1}nW / m$.

Clearly, coding can help improve the performance of MFSK systems. It entails some amount of overhead, however. In some applications, such as voice transmission, there is a certain minimum rate that can be tolerated before communication becomes affected. Some other types of communication do not have this constraint—computer-to-computer communications via modems, Wi-Fi, or WiMAX, for example. In this case, the communication just slows down. In fact, in very noisy or heavily jammed RF environments, it is possible to adjust the amount of encoding to continue to allow some communications through almost any amount of noise. A 1 Mbps channel can be reduced to a 10 bps channel this way, but some information gets through; it just takes longer.

Some forms of communication are also more tolerant to errors than others. When one bit out of 100 (BER = 10^{-2}) is demodulated in error for voice communications, it would probably be unnoticeable. On the other hand, a bank transaction that moves $1 million from one bank to another would take at least 300 bits, considering 8 bits per character, 7 characters in the above amount (= 56 bits for the amount of money) and two 15-character bank numbers (= 240 bits for bank identification). A BER = 10^{-2} would create three errors in the transaction, which would likely be disastrous.

3.6 Bit Interleaving

The purpose of bit interleaving is to spread the bits from a data word over several other blocks so that a burst of noise or other interference which destroys the integrity of a sequence of bits will only hit one or a few from each data word. If each word, before interleaving, is coded via an error-detecting and correcting method that is effective on the number of bits greater than or equal to the number perturbed, then the data words are received correctly.

A simple way to perceive how interleaving works is to look at an example. Suppose eight data words, each eight bits long, are arranged in rows, forming an 8×8 matrix. Without interleaving, the data words are sent serially by row. Instead, send the data bits out by column. This, then, results in having a bit from each data word in each of the newly formed data words. If a noise burst then destroys eight or fewer serial bits, one bit from each reassembled word at the receiver could be in error. If the original coding were applied, correcting one or fewer bits in error, the data stream would be received and decoded correctly.

As mentioned, much of the theory of RF communications is based on the assumption of a memoryless channel. Recall that through such a channel, the probability of one symbol is independent of all the others. When coding is applied to AJ signals, this assumption does not apply. Bit interleaving restores (most of) the memoryless characteristics of coded channels.

3.7 Side Information

When information about the channel is known at the receiver and/or transmitter, it is called *side information*. This information can consist of, for example, whether the channel is being jammed during the reception of a data bit. It could also consist of the type of fading that is occurring on the channel or an estimate of the number of paths being taken by the received signal on its way from the transmitter to the receiver, as specific examples.

3.7.1 Jammer State Information

Jammer state information (JSI) refers to when the receiver knows the channels and symbol times that the jammer was on and therefore knows if a particular symbol may be in error caused by the jammer. Notice that it is not necessarily the case that the symbol *is* in error, just that there is a higher probability of it. Recall that hard decision decoding is when a decoder makes a decision as to what information symbol was sent each symbol period without using any other information. Some decoders can be designed to provide to what degree they believe the output symbol is correct, the *reliability* of the information. Soft decision decoders take advantage of this information and use it to decide the transmitted symbol or perhaps to reject the symbol altogether (the *erasure channel*).

One way to obtain jammer state information is to monitor the whole spectrum over which the communication system operates. If wide jamming techniques are in use, then the receiver can determine the frequency range over which the jamming is taking place. If that coincides with the frequency at the moment, then the receiver knows that the symbol was jammed. For MFSK, there is an easier way suggested by Trumpis [7]. Since there is a bandpass filter for each of the M symbols, without jamming, significant energy will come from the filter corresponding to the symbol that was sent. With jamming, if the jammer is in the channel of the data tone, then it aids in correct detection. If it is not, then two or more of these filters will provide significant detected energy and thus the receiver knows if the symbol was jammed or not, although the receiver probably would not know *which* is the right symbol. Since there is a simple way to generate such jammer state information, it is normally assumed to be available to the receiver trying to make symbol decisions.

3.7.2 Channel State Information

When it is known (or estimated) what type of channel is present, it is called having *channel state information*. A fading channel could be a Ricean fading channel or a Rayleigh fading channel. Different kinds of noise are sometimes assumed if there is good reason to do so. The AWGN channel is one such noise assumption. Noise colored in specific ways can also be factored in, if, in fact, it is known that such noise is present. Knowledge of the type of channel can significantly improve on the ability to estimate what symbol was sent.

3.8 Signal-to-Noise Ratio

For narrowband signaling, the ratio of the bit (symbol) energy, E_s, to the noise PSD, N_0, is often used as a measure of the SNR. In addition, if the symbol rate of the data, R_s, is numerically equivalent to the channel bandwidth, W_b, then

$$\upsilon = \frac{R}{P_N} = \frac{E_s / T_s}{N_0 W_b} = \frac{E_s}{N_0}\frac{R_s}{R_s} = \frac{E_b}{N_0} \tag{3.16}$$

While this relationship still applies to analysis of narrow (instantaneous) bandwidth signals in FHSS, it does not apply to DSSS signals because the noise bandwidth is much larger than the data rate. In that case the energy per bit is given by

$$E_s = NP_c T_c$$

where P_c is the power in a chip and $T_c = 1 / W_c$ is the chip duration, forming the ratio

$$\frac{E_s}{N_0} = \frac{NP_c T_c}{N_0} = \frac{NP_c (1/W_c)}{N_0}\frac{W_c}{W_c} = N\frac{P_c}{P_N} = N\upsilon \tag{3.17}$$

So the energy per bit to noise density ratio is N times the (power) SNR.

3.9 Channel Bandwidth

Most narrowband communication systems use 25 kHz or 50 kHz channel bandwidths. However, after January 1, 2011, manufacturers cannot manufacture or import equipment in the United States operating with a maximum channel bandwidth greater than 12.5 kHz unless it demonstrates a 12.5 kHz or better equivalent spectrum efficiency. In addition, any new equipment submitted to the FCC for certification must be capable of operating in 6.25 kHz channels or on equivalent spectrum efficiency (e.g., two voice paths in a 12.5 kHz channel). Lastly, applications for new systems or modification applications that expand the authorized contour of an existing system operating at greater than a 12.5 kHz equivalent spectrum efficiency will not be accepted.

Furthermore, after January 1, 2013 radio systems must operate in 12.5 kHz equivalent spectrum efficiency or better. This bandwidth will not support analog FM voice communications. However, with higher order digital modulations (8 bps/Hz, for example), adequate digitized voice can be supported.

3.10 Phase Shift Keying

3.10.1 Introduction

Multiple phase shift keying (MPSK) is characterized by causing a phase change within the carrier signal. This phase change can be relative to an absolute value or it can be differential, becoming DPSK, with the information contained in the phase at one bit time compared with the last. MPSK systems are always coherently detected, whereas DPSK is a form of noncoherent modulation since carrier phase tracking need not be maintained over long periods. A *constellation diagram* for MPSK is a representation of the phase states associated with the modulation. Three such constellation diagrams are shown in Figure 3.13. The x-axis is normally referred to as the *in-phase* (I) axis and the y-axis is referred to as the *quadrature phase* (Q) axis. A symbol in MPSK is defined by an angle from some reference, frequently taken as the positive x-axis in the constellation diagram. Thus, each dot in Figure 3.13 is an MPSK symbol. For example, for 2^3 PSK where the data words consist of three bits, the symbols might be as shown in Table 3.5 corresponding to Figure 3.13(c). (Of course, other mappings are possible and might be better for other reasons.) The primary modulations used for DSSS systems are BPSK and QPSK. The latter has desirable spectrum occupancy performance advantages over the former; its performance is, however, more sensitive to noise and interference. Two types of the latter are *minimum shift key* (MSK) and *offset QPSK* (OQPSK).

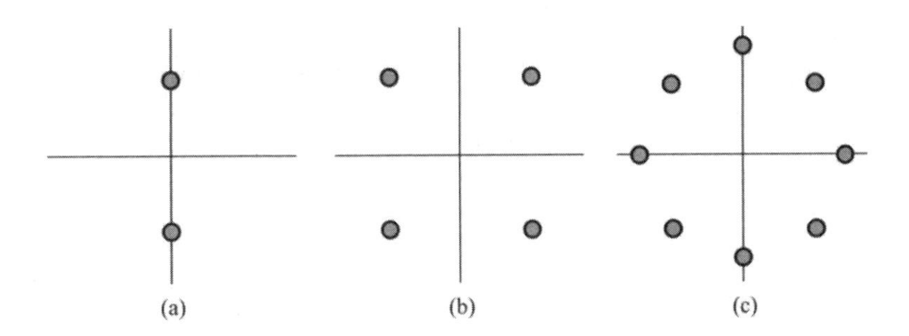

Figure 3.13 Possible constellations for some PSK signals: (a) BPSK, (b) QPSK, and (c) 2^3PSK.

Phase shifting a carrier signal is accomplished by imposing the data signal on the carrier by changing the latter's phase. The amount of phase change depends on the number of phase states. For the BPSK signal $s(t)$ during time interval k,

$$s_k(t) = \sqrt{2R}\cos\left(2\pi f_0 t + d_k \frac{\pi}{2}\right), \qquad (k-1)T \le t < kT \tag{3.18}$$

where R is the average power in the signal, $d_k \in \{+1, -1\}$ represents the data bits, and f_0 is the carrier frequency. This signal has the constellation shown in Figure 3.13(a). Note that an equivalent representation for BPSK is given by

$$s_k(t) = d_k\sqrt{2R}\cos(2\pi f_0 t) \tag{3.19}$$

where the data sequence multiplies the amplitude. Multiplying cos θ by ±1 has the same effect as changing the phase of cos θ by π radians. Therefore, a phase shift of π radians occurs irrespective of the initial phase. Note that Figure 3.13(a) no

Table 3.5 Phase Offsets Corresponding to Figure 3.13(c)

Symbol	Phase (radians)
000	0
001	$\pi / 4$
010	$\pi / 2$
011	$3\pi / 4$
100	π
101	$5\pi / 4$
110	$3\pi / 2$
111	$7\pi / 4$

longer applies in this case. The constellation has its points at 0 and π radians in this case; it is still BPSK, however. If the carrier has an arbitrary starting phase ϕ other than 0, then BPSK still results, but with the constellation points offset by ϕ.

Still another way of expressing this is

$$s_k(t) = \sqrt{2R}\sin\left[2\pi f_0 t - (-1)^u \cos^{-1}(m)\right], u = 1,2; (k-1)T \le t < kT \quad (3.20)$$

where $\cos^{-1}(m)$ is the modulation index.

For *quaternary PSK* (QPSK), where there are $4 = 2^2$ phase states,

$$s_k(t) = \sqrt{2R}\sin\left(2\pi f_0 t + d_k \frac{\pi}{2}\right), (k-1)T \le t < kT \quad (3.21)$$

where $d_k \in \{1, 3, 5, 7\}$. The phase states for this signal are shown in Figure 3.13(b). For 8-ary PSK (2^3 PSK),

$$s_k(t) = \sqrt{2R}\sin\left(2\pi f_0 t + d_k \frac{\pi}{4}\right) \quad (3.22)$$

where in this case $d_k \in \{0, 1, 2, 3, 4, 5, 6, 7\}$. These phase states are illustrated in Figure 3.13(c).

Obviously, this can be extended to any desired number of states. However, as the number of states increases, the probability that a received symbol with noise present falls into the regime of an incorrect symbol increases because the size of each regime gets smaller. Therefore, the sensitivity to noise and interference increase with higher-order phase modulation.

For MPSK

$$\tilde{r}(t) = \sqrt{2R}\sum_k e^{j\theta_k} u_{\mathrm{T}}(t - kT_s), \theta_k \in \left\{2\pi \frac{\kappa}{M}\right\}, \ \kappa = 0,1,\cdots,M-1 \quad (3.23)$$

and

$$u_{\mathrm{T}}(t - kT_s) = \begin{cases} 1, & (k-1)T_s \le t < kT_s \\ 0, & \text{otherwise} \end{cases} \quad (3.24)$$

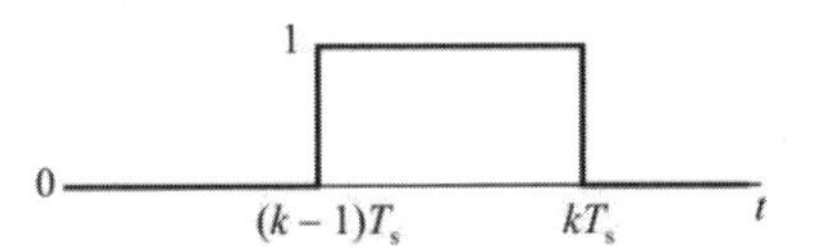

Figure 3.14 Time sample function, also referred to as the time window function.

where $u_T(t - kT_s)$ is the sample function shown in Figure 3.14, T_s is the symbol time, R is the average power, and where the information bits are contained in the θ_k.

For OQPSK

$$\begin{aligned}\tilde{r}_r(t) &= \sqrt{R}\sum_k d_k^{\mathrm{I}} u_T\left(t - kT_s\right) \\ \tilde{r}_i(t) &= \sqrt{R}\sum_k d_k^{\mathrm{Q}} u_T\left(t - kT_s - \frac{T_s}{2}\right)\end{aligned} \tag{3.25}$$

where $d_k^{\mathrm{I}} \in \{-1,+1\}$ and $d_k^{\mathrm{Q}} \in \{-1,+1\}$ are the data bits for time interval k.

For MSK

$$\begin{aligned}\tilde{r}_r(t) &= \sqrt{R}\sum_k d_k^{\mathrm{I}} \cos\left(\frac{\pi t}{T_s}\right) u_T\left(t - kT_s\right) \\ \tilde{r}_i(t) &= \sqrt{R}\sum_k d_k^{\mathrm{Q}} \sin\left(\frac{\pi t}{T_s}\right) u_T\left(t - kT_s - \frac{T_s}{2}\right)\end{aligned} \tag{3.26}$$

which shapes the data pulses during the modulation process according to the cos(•) and sin(•) functions. MSK, in particular, has a significant advantage over the other types of PSK discussed here in the out-of-band power spectrum of the signal. Such out-of-band energy interferes with communications in those channels. The comparison for these modulation types is shown in Figure 3.15 [8].

3.10.2 BPSK

A simplified flow diagram of a BPSK communication system is shown in Figure 3.16. The filtering and amplification stages are not shown (because they are not germane to our current discussion). We discuss the fundamental characteristics of this system in this section. In particular, we will derive the expression for the BER for BPSK modulation. The binary digits 1 and 0, or +1 and –1, are represented by

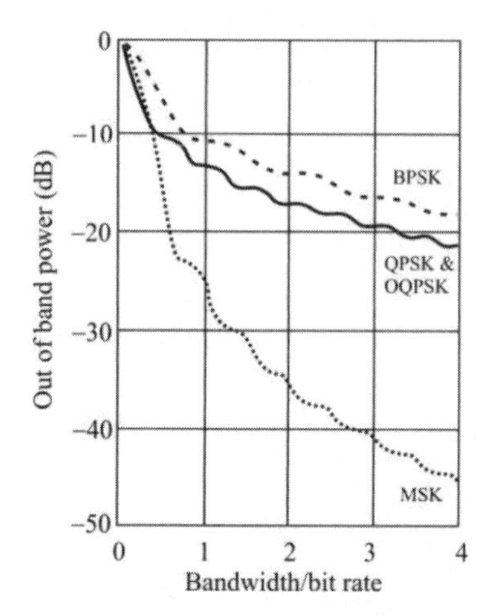

Figure 3.15 Fractional out-of-band power for some of the PSK modulations types we consider. (Source: [8]. © 1992 Artech House. Reprinted with permission.)

the analog signal levels of $+\sqrt{E_b}$ and $-\sqrt{E_b}$, respectively. An information source generates a sequence of information bits represented by $p(t)$. These are modulated onto a carrier and sent out over the free-space portion of the channel. Here they are usually corrupted by noise and interference before reaching the receiver. At the receiver, the bits are extracted from the carrier generating an estimate of the bits sent, these estimates being represented by $\hat{p}(t)$.

A simple but completely effective modulator for BPSK is shown in Figure 3.17. Multiplying a carrier generated here by the oscillator at frequency f_0, by the data sequence represented by ±1, causes the carrier phase to change at the data rate in accordance with the data sequence.

The optimum BPSK detector is illustrated in Figure 3.18, which is a correlator, also known as a matched filter [9] . The received signal, corrupted by noise, is first multiplied by a signal that is the difference between the two possible,

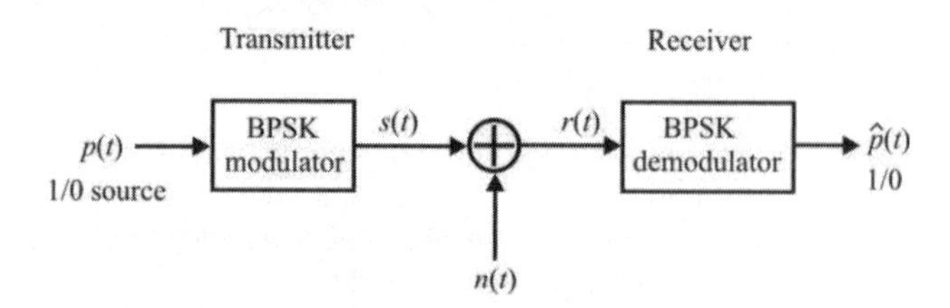

Figure 3.16 Flow diagram for a BPSK communication system.

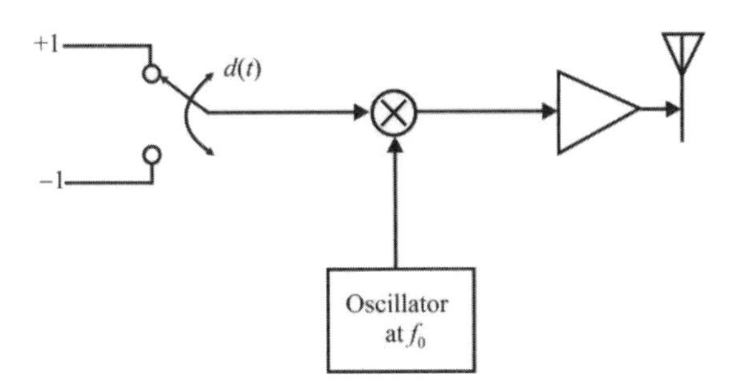

Figure 3.17 Simplified block diagram of a BPSK transmitter. Multiplying the carrier by ± 1 has the same effect as changing the phase by 180°. Synchronization components are not shown.

noiseless, symbols. The resulting product is then integrated for the symbol time T. The integrated signal is sampled at the end of the symbol period, with the sample compared with a threshold. If the threshold is exceeded, then one of the two possible symbols is declared present. If the threshold is not exceeded, then the other symbol is declared.

3.10.2.1 BPSK Performance in AWGN

For BPSK, the two signals representing the data bits are antipodal with $s_2(t) = -s_1(t)$ because of expression (3.19), that is,

$$\begin{aligned} s_1(t) &= +\sqrt{2R}\cos(2\pi f_0 t) \\ s_2(t) &= -\sqrt{2R}\cos(2\pi f_0 t) \end{aligned} \tag{3.27}$$

From (3.10),

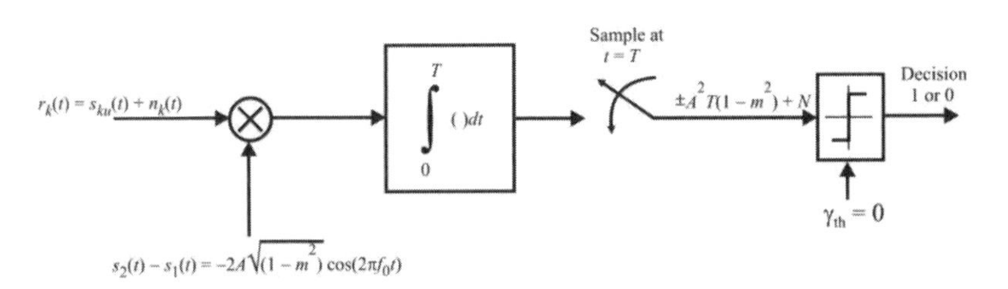

Figure 3.18 Optimum correlation receiver for BPSK. (Source: [9]. © John Wiley & Sons, 2002. Reprinted with permission.)

$$\begin{aligned}
\rho_{12} &= \frac{1}{\sqrt{E_1 E_2}} \int_0^T -\sqrt{2R}\cos(2\pi f_0 t)\sqrt{2R}\cos(2\pi f_0 t)dt \\
&= -\frac{2E_b}{T\sqrt{E_1 E_2}} \int_0^T \cos^2(2\pi f_0 t)dt \\
&= -\frac{2E_b}{T\sqrt{E_1 E_2}} \left[\int_0^T \frac{1}{2} dt + \int_0^T \frac{1}{2}\cos(4\pi f_0 t)dt \right] \\
&= -\frac{2E_b T}{2T\sqrt{E_1 E_2}} = -\frac{E_b}{\sqrt{E_1 E_2}}
\end{aligned} \tag{3.28}$$

The last line comes about because the cos x function integrated over an integer number of periods is zero. Substituting this into (3.9) yields $R_{12} = -1$.

The received signal, after filtering and frequency downconversion at the input to the detector, is given by $r(t)$ and the signal at the output of the detector is given by $v(t)$. In general, $r(t)$ is comprised of the transmitted signal $s(t)$, AWGN, and zero or more interfering signals.

The received signal $r(t)$ in Figure 3.16 takes one of two forms during a bit interval specified by T_b:

$$r(t) = \begin{cases} s_1(t) + n(t), & s_1(t) \Leftrightarrow 1 \text{ transmitted} \\ s_0(t) + n(t), & s_0(t) \Leftrightarrow 0 \text{ transmitted} \end{cases} \tag{3.29}$$

The conditional PDFs of $r(t)$ for the two cases are given by

$$p(r|s_1) = \frac{1}{\sqrt{\pi N_0}} e^{-\frac{(r-\sqrt{E_s})^2}{N_0}} \tag{3.30}$$

$$p(r|s_0) = \frac{1}{\sqrt{\pi N_0}} e^{-\frac{(r+\sqrt{E_s})^2}{N_0}} \tag{3.31}$$

The constellation of the symbols s_0 and s_1 can be represented on the real line as illustrated in Figure 3.19(a) along with their associated PDFs in Figure 3.19(b). For decoding, a decision rule is used with an associated threshold γ_{th}, such as γ_{th} = 0 in Figure 3.19. Such a decision rule might be:

H_0: $r(t) < \gamma_{th}$, decide s_0 was transmitted

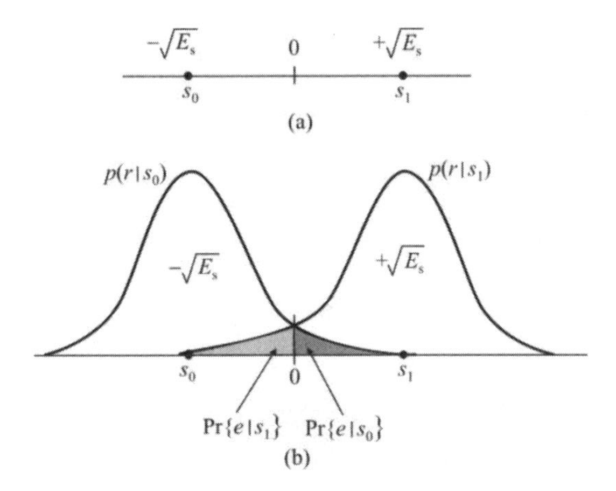

Figure 3.19 Conditional PDFs with BPSK modulation: (a) constellation diagram and (b) PDFs.

H_1: $r(t) > \gamma_{th}$, decide s_1 was transmitted

With this threshold, the probability of error given s_1 is transmitted is the shaded area indicated in Figure 3.19 to the left of γ_{th} (0 in this figure)

$$\Pr\{e|s_1\} = \frac{1}{\sqrt{\pi N_0}} \int_{-\infty}^{0} e^{-\frac{(r-\sqrt{E_s})^2}{N_0}} dr = \frac{1}{\sqrt{\pi}} \int_{\sqrt{E_s/N_0}}^{\infty} e^{-z^2} dz = \frac{1}{2}\operatorname{erfc}\left(\sqrt{\frac{E_s}{N_0}}\right) \tag{3.32}$$

where erfc(x) is the complementary error function given by (see Appendix A)

$$\operatorname{erfc}(x) = \frac{2}{\sqrt{\pi}} \int_{x}^{\infty} e^{-z^2} dz \tag{3.33}$$

Likewise, the probability of error given that s_0 was transmitted is the shaded area to the right of γ_{th} in Figure 3.19 given by

$$\Pr\{e|s_0\} = \frac{1}{\sqrt{\pi N_0}} \int_{0}^{\infty} e^{-\frac{(r+\sqrt{E_s})^2}{N_0}} dr = \frac{1}{\sqrt{\pi}} \int_{\sqrt{E_s/N_0}}^{\infty} e^{-z^2} dz = \frac{1}{2}\operatorname{erfc}\left(\sqrt{\frac{E_s}{N_0}}\right) \tag{3.34}$$

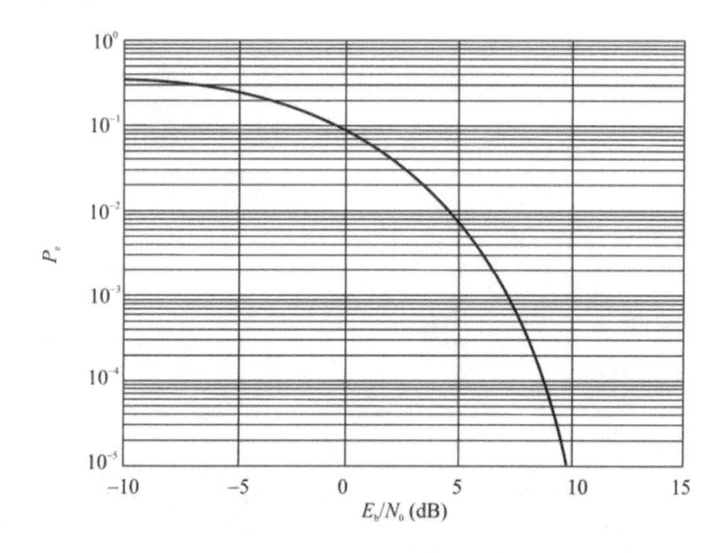

Figure 3.20 BER performance for coherent BPSK in an AWGN channel.

With $\Pr\{s_1\}$ and $\Pr\{s_0\}$ representing the a priori probabilities of s_1 and s_0 occurring, respectively, the total probability of bit error for BPSK modulation is given by

$$P_s = \Pr\{s_1\}\Pr\{e|s_1\} + \Pr\{s_0\}\Pr\{e|s_0\} \tag{3.35}$$

Assuming that s_1 and s_0 are equally likely so that $\Pr\{s_1\} = \Pr\{s_0\} = ½$, from (3.32) and (3.34), the BPSK BER is given by

$$P_s = \frac{1}{2}\operatorname{erfc}\left(\sqrt{\frac{E_s}{N_0}}\right) = Q\left(\sqrt{2\frac{E_s}{N_0}}\right) \tag{3.36}$$

A graph of (3.36) is shown in Figure 3.20.

3.10.2.2 BPSK Performance with Tone Interference

In the kth time interval, let the interfering tone be expressed as

$$i_k(t) = \sqrt{2I}\cos\left(2\pi f_0 t + \theta_1\right), (k-1)T_s \le t < kT_s \tag{3.37}$$

where I is the average power in the tone, T_s is the symbol period, and θ_I is the phase offset of the interfering signal from the carrier signal. The received signal is given by

$$r_k(t) = s_k(t) + n_k(t) + i_k(t), \quad (k-1)T_s \le t < kT_s \tag{3.38}$$

Simon et al. [10] examined coherent QPSK (and by default, BPSK since BPSK is the in-phase data stream in QPSK) with an interfering tone present. They concluded that the BER for QPSK with an interfering tone that is not phase coherent with the signal (the most likely condition) is given by

$$P_e = \frac{1}{2\pi}\int_0^{2\pi}\left[P^I\left(\theta^I\right) + P^Q\left(\theta^I\right) - P^I(\theta^I)P^Q\left(\theta^I\right)\right]d\theta^I \tag{3.39}$$

where the in-phase and quadrature probabilities are

$$P^I_{QPSK}\left(\theta^I\right) = Q\left[\sqrt{\frac{E_{s,QPSK}}{N_0}}\left(1 - \sqrt{\frac{2I}{R}}\sin\theta^I\right)\right] \tag{3.40}$$

$$P^Q_{QPSK}\left(\theta^I\right) = Q\left[\sqrt{\frac{E_{s,QPSK}}{N_0}}\left(1 + \sqrt{\frac{2I}{R}}\cos\theta^I\right)\right] \tag{3.41}$$

and it is assumed that the phase of the interfering tone is uniformly distributed over $(0, 2\pi)$. For BPSK,

$$E_{s,BPSK} = \frac{E_{s,QPSK}}{2} \tag{3.42}$$

so

$$P^I_{BPSK}\left(\theta^I\right) = Q\left[\sqrt{2\frac{E_{s,BPSK}}{N_0}}\left(1 - \sqrt{\frac{2I}{R}}\sin\theta^I\right)\right] \tag{3.43}$$

The BER performance for BPSK is given by (3.43). Converting this expression to one with power ratios, and noting that $T_s W_F = 1$,

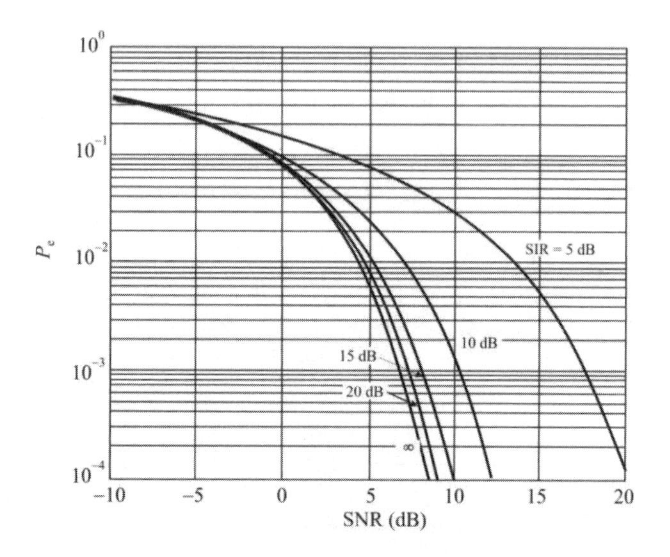

Figure 3.21 BPSK performance in AWGN with interfering tone.

$$P_{\mathrm{BPSK}}^{\mathrm{I}}\left(\theta^{\mathrm{I}}\right)=Q\left[\sqrt{2\frac{RT_{\mathrm{s}}}{N_0}\frac{W_{\mathrm{F}}}{W_{\mathrm{F}}}}\left(1-\sqrt{\frac{2I}{R}}\sin\theta^{\mathrm{I}}\right)\right]$$

$$=Q\left[\sqrt{2\frac{R}{P_{\mathrm{N}}}}\left(1-\sqrt{\frac{2I}{R}}\sin\theta^{\mathrm{I}}\right)\right] \tag{3.44}$$

Equation (3.44) is plotted in Figure 3.21 for some values of SIR = R / I.

As expected, the stronger the interfering signal is, the poorer the performance that ensues from the BPSK communication system. The infinite case is, of course, when $I = 0$ (no interference); note that this curve corresponds to Figure 3.20.

3.10.3 QPSK

A model of a QPSK modulator is shown in Figure 3.22 [9]. It can be viewed as two BPSK modulators operating in parallel with their outputs added. The carriers must be in phase quadrature and the quadrature data rate is half that of the baseband data. The two BPSK data streams are constructed from the original data sequence by loading a 2-bit shift register with two sequential bits from the original data stream. These two bits are used to modulate quadrature carriers as shown in Figure 3.22.

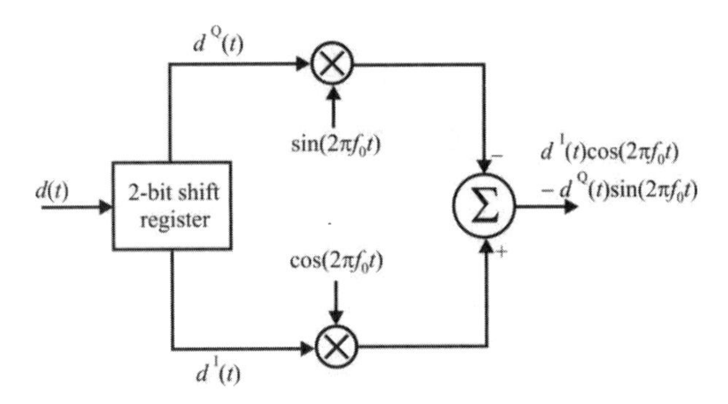

Figure 3.22 QPSK modulator can be configured from two parallel BPSK modulators. (After: [9]. © John Wiley & Sons, 2002. Reprinted with permission.)

The resulting QPSK transitions are shown in Figures 3.23(a) and 3.24(a). Notice that some of these transitions pass thorough the origin. This causes the amplitude of the carrier, once modulated with this data sequence, to be zero for an amount of time during these transitions. Using power amplifiers in or near saturation causes severe distortion in the transmitted signal and significant out-of-band interference results. If the Q channel is delayed by one bit time from the I channel, then, as shown in Figure 3.23(b), only one bit (coordinate) is allowed to

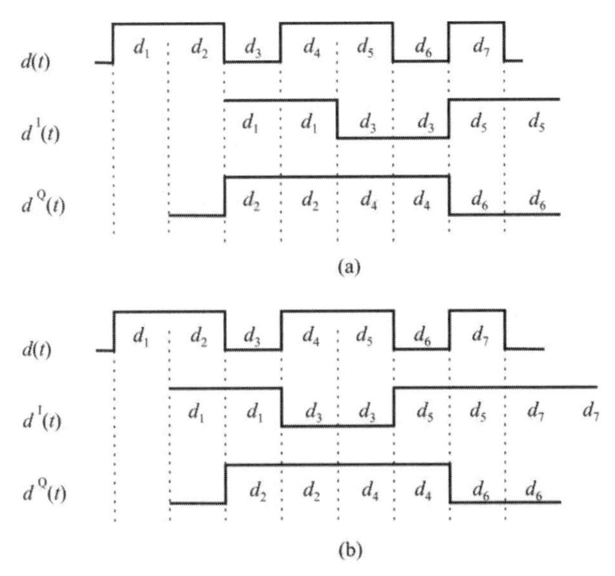

Figure 3.23 (a, b) The two BPSK data sequences for QPSK are formed from the original data sequence by delaying each by one bit and transmitting the even and odd bits via separate channels.

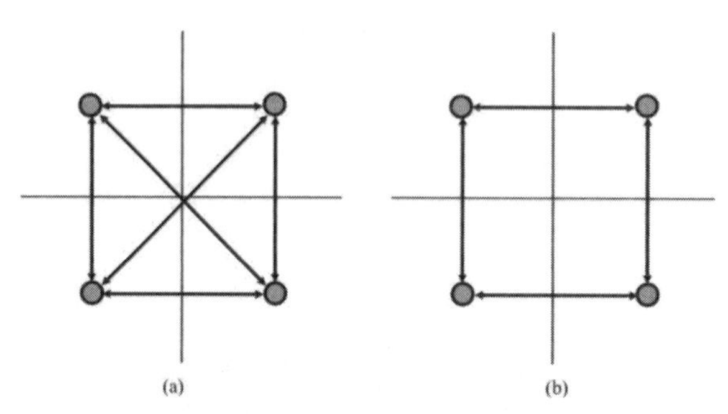

Figure 3.24 Constellations for (a) QPSK and (b) OQPSK.

change at a time, and the resulting allowed transitions are as shown in Figure 3.24(b). The transitions through the origin are eliminated, as are the distortions. This technique is called *offset QPSK* (OQPSK) or *staggered QPSK* (SQPSK). In this case, only a one-stage shift register is needed.

A QPSK demodulator is illustrated in Figure 3.25 [11]. This particular configuration takes advantage of the fact that a QPSK signal is made up of two BPSK signals and simply separates them. The signal into the QPSK detector during the kth time interval is given by

$$r_k(t) = s_k(t) + n_k(t), \qquad (k-1)T_s \le t < kT_s \tag{3.45}$$

with the signal represented by

$$s_k(t) = \sqrt{2R}\cos\left[2\pi f_0 t + \phi_k(t)\right] \tag{3.46}$$

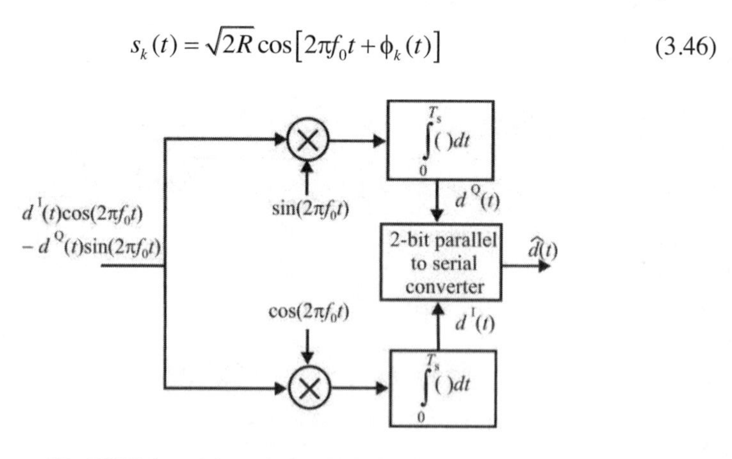

Figure 3.25 One possible QPSK demodulator. (After: [11]. © John Wiley & Sons, 2002. Reprinted with permission.)

where $\phi_k(t) \in \{\pm\pi/4, \pm 3\pi/4\}$, and $n_k(t)$ is the sample from the noise process during time interval k. In this case $T_s = 2T_b$. The constellation for QPSK is illustrated in Figure 3.13(b).

Using straightforward trigonometric relationships, (3.46) can be expressed as

$$s_k(t) = \sqrt{2R}\cos(2\pi f_0 t)\cos\phi_k(t) - \sqrt{2R}\sin(2\pi f_0 t)\sin\phi_k(t) \tag{3.47}$$

and, since

$$\begin{aligned} &\cos\frac{\pi}{4} = \frac{1}{\sqrt{2}} && \cos\frac{3\pi}{4} = -\frac{1}{\sqrt{2}} && \cos\frac{5\pi}{4} = -\frac{1}{\sqrt{2}} && \cos\frac{7\pi}{4} = \frac{1}{\sqrt{2}} \\ &\sin\frac{\pi}{4} = \frac{1}{\sqrt{2}} && \sin\frac{3\pi}{4} = \frac{1}{\sqrt{2}} && \sin\frac{5\pi}{4} = -\frac{1}{\sqrt{2}} && \sin\frac{7\pi}{4} = -\frac{1}{\sqrt{2}} \end{aligned} \tag{3.48}$$

then

$$s_k(t) = \pm\sqrt{R}\cos(2\pi f_0 t) \pm \sqrt{R}\sin(2\pi f_0 t) \tag{3.49}$$

3.10.3.1 QPSK Performance in AWGN

First consider the symbol s_2 in Figure 3.26. The conditional PDF of $r(t)$, given that s_2 was transmitted, is

$$p(r|s_2) = \frac{1}{\sqrt{\pi N_0}} e^{-\frac{\left(r - \sqrt{\frac{E_s}{2}}\right)^2}{N_0}} \tag{3.50}$$

As can be seen from Figure 3.26, the symbol s_2 is decoded correctly only if r falls in the shaded area (i.e., $\text{Re}\{r(t)\} > 0$ and $\text{Im}\{r(t)\} > 0$). The probability corresponding to this occurring is given by

$$P(c|s_2) = \Pr\{\text{Re}\{r(t)\} > 0 | s_2\} \Pr\{\text{Im}\{r(t)\} > 0 | s_2\} \tag{3.51}$$

where it is tacitly assumed that the real and imaginary components of $r(t)$ are independent, which they will be most of the time since each is corrupted by noise, and the I-channel noise is independent of the Q-channel noise. The probability of

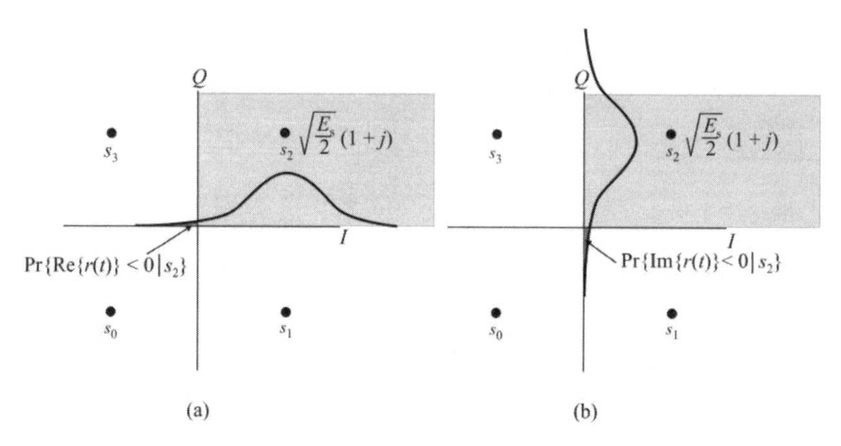

Figure 3.26 QPSK constellation error regions. (a) Probability of real part being in error is the area to the left of the Q axis as indicated and (b) probability of the imaginary component being in error below the I axis as indicated.

the real component of $r(t)$ being greater than zero, given that s_2 was transmitted, is given by

$$\begin{aligned}\Pr\left\{\text{Re}\{r(t)>0\}\middle|s_2\right\} &= 1-\Pr\left\{\text{Re}\{r(t)\leq 0\}\middle|s_2\right\}\\ &= 1-\frac{1}{\sqrt{\pi N_0}}\int_{-\infty}^{0} e^{-\frac{\left(\text{Re}\{r\}-\sqrt{\frac{E_s}{2}}\right)^2}{N_0}}dr\\ &= 1-\frac{1}{2}\text{erfc}\left(\sqrt{\frac{E_s}{2N_0}}\right)\end{aligned} \tag{3.52}$$

Since $E_s = 2E_b$, then

$$\Pr\left\{\text{Re}\{r(t)>0\}\middle|s_2\right\} = 1-\frac{1}{2}\text{erfc}\left(\sqrt{\frac{E_b}{N_0}}\right) \tag{3.53}$$

Similarly, the probability of the imaginary component of $r(t)$, being greater than zero (which is the area outside the shaded region), is given by

$$\Pr\left\{\text{Im}\{r(t)\}>0\middle|s_2\right\} = 1-\Pr\left\{\text{Im}\{r(t)\}\leq 0\middle|s_2\right\}$$

$$= 1 - \frac{1}{\sqrt{\pi N_0}} \int_{-\infty}^{0} e^{-\frac{\left(\mathrm{Im}\{r\} - \sqrt{\frac{E_s}{2}}\right)^2}{N_0}} dr$$

$$= 1 - \frac{1}{2} \mathrm{erfc}\left(\sqrt{\frac{E_s}{2N_0}}\right) \tag{3.54}$$

And, again, since $E_s = 2E_b$

$$\Pr\left\{\mathrm{Im}\{r(t) > 0\} \middle| s_2\right\} = 1 - \frac{1}{2} \mathrm{erfc}\left(\sqrt{\frac{E_b}{N_0}}\right) \tag{3.55}$$

and the probability of s_2 being decoded correctly is given by

$$\Pr\{c | s_2\} = \left[1 - \frac{1}{2} \mathrm{erfc}\left(\sqrt{\frac{E_b}{N_0}}\right)\right]^2$$

$$= 1 - \frac{2}{2} \mathrm{erfc}\left(\sqrt{\frac{E_b}{N_0}}\right) + \frac{1}{4} \mathrm{erfc}^2\left(\sqrt{\frac{E_b}{N_0}}\right)$$

$$= \mathrm{erfc}\left(\sqrt{\frac{E_b}{N_0}}\right) - \frac{1}{4} \mathrm{erfc}^2\left(\sqrt{\frac{E_b}{N_0}}\right) \tag{3.56}$$

Denoting the a priori probabilities of symbols as $\Pr\{s_i\}, i = 0,1,2,3,$ the total probability of decoding a symbol correctly is given by

$$\begin{aligned} \Pr_{\mathrm{QPSK}}\{c\} = &\Pr\{c|s_0\}\Pr\{s_0\} + \Pr\{c|s_1\}\Pr\{s_1\} \\ &+ \Pr\{c|s_2\}\Pr\{s_2\} + \Pr\{c|s_3\}\Pr\{s_3\} \end{aligned} \tag{3.57}$$

Assuming all symbols are equally likely, $\Pr\{s_i\} = 1/4, i = 0,1,2,3,$ so that

$$\Pr_{\mathrm{QPSK}}\{c\} = 4\left[\frac{1}{4} \mathrm{erfc}\left(\sqrt{\frac{E_b}{N_0}}\right) - \frac{1}{16} \mathrm{erfc}^2\left(\sqrt{\frac{E_b}{N_0}}\right)\right]$$

$$= \operatorname{erfc}\left(\sqrt{\frac{E_b}{N_0}}\right) - \frac{1}{4}\operatorname{erfc}^2\left(\sqrt{\frac{E_b}{N_0}}\right) \tag{3.58}$$

and the probability of a symbol error is given by

$$\begin{aligned} P_{s,QPSK} &= 1 - \Pr\{c\} \\ &= 1 - \left[1 - \operatorname{erfc}\left(\sqrt{\frac{E_b}{N_0}}\right) + \frac{1}{4}\operatorname{erfc}^2\left(\sqrt{\frac{E_b}{N_0}}\right)\right] \\ &= \operatorname{erfc}\left(\sqrt{\frac{E_b}{N_0}}\right) - \frac{1}{4}\operatorname{erfc}^2\left(\sqrt{\frac{E_b}{N_0}}\right) \\ &= 2Q\left(\sqrt{2\frac{E_b}{N_0}}\right)\left[1 - \frac{1}{2}Q\left(\sqrt{2\frac{E_b}{N_0}}\right)\right] \end{aligned} \tag{3.59}$$

which corresponds to the result in Proakis [12]. For larger values of the SNR, the second term in (3.59) becomes negligible and the probability of the symbol error can be approximated as

$$P_{s,QPSK} \approx \operatorname{erfc}\left(\sqrt{\frac{E_b}{N_0}}\right) \tag{3.60}$$

$$= 2Q\left(\sqrt{2\frac{E_b}{N_0}}\right) \tag{3.61}$$

Equation (3.59) was depicted earlier in Figure 3.12. We see that the performance is somewhat worse than BPSK as would be expected because of the increased density of the constellation.

Since $T_s = 2T_b$, the bandwidth required for QPSK is ½ that for BPSK for the same data rate. Since the throughput is the same for BPSK and QPSK, the latter is generally favored over the former to take advantage of the narrower bandwidth requirements even though the BER performance is somewhat worse. Alternately, the QPSK data rate can be twice that for BPSK for the same bandwidth.

As with BPSK, (3.59) can be put in a form with power ratios

$$P_s = 2Q\left(\sqrt{2\frac{E_b}{N_0}\frac{W_F}{W_F}}\right)\left[1-\frac{1}{2}Q\left(\sqrt{2\frac{E_b}{N_0}\frac{W_F}{W_F}}\right)\right]$$

but in this case, since $T_s = 2T_b$, $T_s = 1 / W_F$ (assumed) and $R_b = R_{b,QPSK} = R_{b,BPSK} / 2$, then

$$Q\left(\sqrt{2\frac{E_b}{N_0}\frac{W_F}{W_F}}\right) = Q\left(\sqrt{2\frac{E_b}{N_0}\frac{R_b / 2}{W_F}}\right)$$

$$= Q\left(\sqrt{\frac{R}{P_N}}\right) \tag{3.62}$$

so that

$$P_s = 2Q\left(\sqrt{\frac{R}{P_N}}\right)\left[1-\frac{1}{2}Q\left(\sqrt{\frac{R}{P_N}}\right)\right] \tag{3.63}$$

3.10.3.2 QPSK Performance with an Interfering Tone

With interference present with average power I, P_s is given by (3.39) with [13]

$$P^I\left(\theta^I\right) = Q\left[\sqrt{\frac{RT_s}{N_0}}\left(1-\sqrt{\frac{2I}{R}}\sin\theta^I\right)\right] \tag{3.64}$$

$$P^Q\left(\theta^I\right) = Q\left[\sqrt{\frac{RT_s}{N_0}}\left(1+\sqrt{\frac{2I}{R}}\cos\theta^I\right)\right] \tag{3.65}$$

In this case, $W_F T_s = 1$ so again

$$\frac{RT_s}{N_0}\frac{W_F}{W_F} = \frac{R}{P_N} \tag{3.66}$$

and

$$P^I\left(\theta^I\right) = Q\left[\sqrt{\frac{R}{P_N}}\left(1-\sqrt{\frac{2I}{R}}\sin\theta^I\right)\right] \tag{3.67}$$

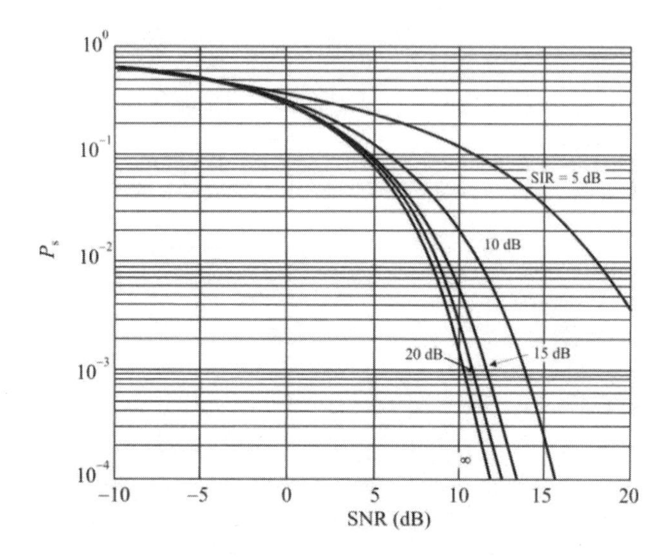

Figure 3.27 QPSK performance in AWGN with interfering tone.

$$P^{Q}\left(\theta^{I}\right)=Q\left[\sqrt{\frac{R}{P_{N}}}\left(1+\sqrt{\frac{2I}{R}}\cos\theta^{I}\right)\right] \tag{3.68}$$

Equation (3.39), using (3.67) and (3.68), is illustrated in Figure 3.27 where SIR = R / I = 1 / ι. Note that (3.63) corresponds to the SIR = ∞ curve in Figure 3.27.

Comparing Figure 3.27 with Figure 3.20, we see that the QPSK performance is approximately 5 dB worse than BPSK for the same level of interfering tone. That means that QPSK requires a 5 dB stronger signal than BPSK for the same BER performance when interference is present.

3.10.4 DBPSK

Recall that for DPSK, the current bit is determined by a phase relationship with the previous bit. The only case discussed here will be DBPSK, although higher forms are certainly possible and exist. Extensions of these arguments to higher orders are straightforward. Although either is possible, herein a 1 is encoded as a phase change of π radians from the previous bit, while a 0 is encoded as no phase change.

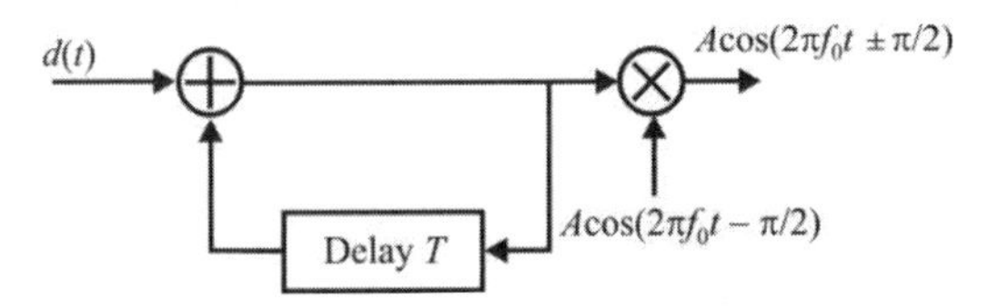

Figure 3.28 One possible DBPSK modulator.

A modulator architecture for generating BDPSK is shown in Figure 3.28. This modulator differentially encodes each data bit prior to being upconverted with the mixer. A 1 in the input stream causes the output to reverse: 0 to π or π to 0, whichever applies. On the other hand, a 0 in the data stream keeps the output phase the same.

Receiver architectures for DBPSK are shown in Figure 3.29 [14]. The receiver in Figure 3.29(a) is suboptimal but quite simple to implement, while that in Figure 3.29(b) is optimal.

3.10.4.1 DBPSK Performance in AWGN

When AWGN is present in the channel, the BER for the suboptimal receiver in Figure 3.29(a) for large SNRs is given by [15]

$$P_e \approx Q\left(\sqrt{\frac{E_b}{N_0}}\right) \tag{3.69}$$

where E_b is the energy per bit and N_0 is the one-sided noise spectral density. The same error performance for the optimum receiver in Figure 3.29(b) is given by [15]

$$P_e = \frac{1}{2}\exp\left(-\frac{E_b}{N_0}\right) \tag{3.70}$$

3.10.4.2 DBPSK with Tone Interference

The performance of DBPSK modulation in the presence of tone interference and AWGN was analyzed by Zeng and Wang [16]. The received signal is represented by

$$r(t) = Ae^{j\phi(t)} + \alpha A_J e^{j(\theta + 2\pi\delta_f t)} + n(t) \tag{3.71}$$

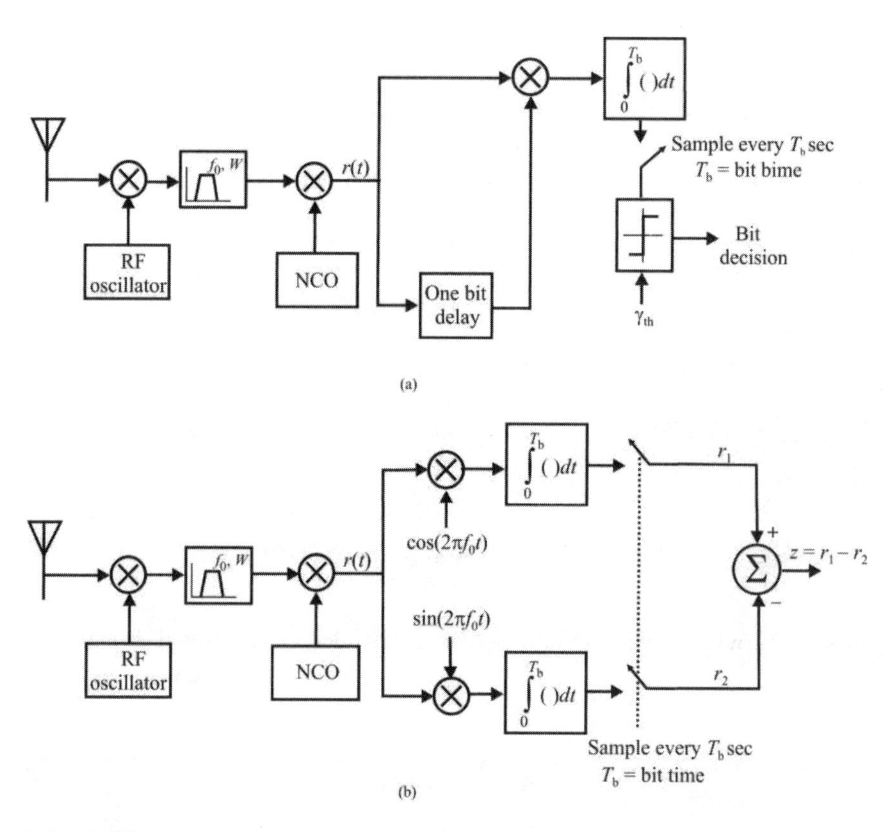

Figure 3.29 Receiver architectures for DPSK: (a) suboptimal and (b) optimal. (After: [15]. © John Wiley & Sons 2002. Reprinted with permission.)

where the first term is the uncorrupted DBPSK signal, the second term is the tone interference, which has a random phase term $\theta \in [0, 2\pi)$ and a frequency offset denoted by δ_f, and the third term is a sample from the AWGN with a variance σ^2. The coefficient α represents the attenuation of the receiver filter at the frequency offset. The two sequential signals, at times t_1 and $t_2 = t_1 + T$, are therefore

$$r_1(t_1) = Ae^{j\phi_1} + \alpha A_J e^{j(\theta + 2\pi\delta_f t_1)} + n(t_1) = A_1 e^{j\theta_1} \tag{3.72}$$

and

$$r_2(t_2) = Ae^{j\phi_2} + \alpha A_J e^{j(\theta + 2\pi\delta_f t_2)} + n(t_2) = A_2 e^{j\theta_2} \tag{3.73}$$

Let

$$\varphi = \theta_2 - \theta_1 \tag{3.74}$$

denote the difference in the phases of the two received signals. A DBPSK detection error will occur if the interference and noise cause the magnitude of this phase difference to be larger than $\pi / 2$. That is,

$$P_e = \Pr\left\{|\varphi| > \frac{\pi}{2}\right\}, \text{ mod}(2\pi) \tag{3.75}$$

This probability can be restated as

$$P_e = \Pr\left\{\frac{\pi}{2} < \varphi < \frac{3\pi}{2}\right\} \tag{3.76}$$

Zeng and Wang [16] point out that

$$\Pr\{\varphi_1 < \varphi < \varphi_2\} = G(\varphi_1) - G(\varphi_2) \tag{3.77}$$

where

$$G(\varphi_i)=\frac{\varphi_i}{2\pi}-\frac{1}{4\pi}\int_{-\pi/2}^{\pi/2}\left\{\begin{array}{l}\dfrac{\sin(\Delta\phi-\varphi_i)}{T(\theta)}q\left[\alpha\sqrt{\iota S(\theta)},\sqrt{\upsilon T(\theta)}\right]\\ +\dfrac{\sin(2\pi\delta_{\mathrm{f}}\tau-\varphi_i)}{S(\theta)}q\left[\sqrt{\upsilon T(\theta)},\alpha\sqrt{\iota S(\theta)}\right]\end{array}\right\}d\theta,\ i=1,2 \quad (3.78)$$

$$\begin{aligned}S(\theta)&=1-\cos\theta\cos(2\pi\delta_{\mathrm{f}}-\varphi_i)\\ T(\theta)&=1-\cos\theta\cos(\Delta\phi-\varphi_i)\end{aligned} \quad (3.79)$$

$$\Delta\phi=\phi_2-\phi_1 \quad (3.80)$$

where for DBPSK considered here, $\phi_1 = 0$ and $\phi_2 = \pi$ radians, and where $q(\bullet)$ is the complementary Marcum Q-function (Appendix A)

$$q(a,b)=1-Q(a,b) \quad (3.81)$$

The SNR and ISR are derived from

$$\upsilon=\frac{A^2}{\sigma^2}=2\frac{R}{\sigma^2},\qquad \iota=\frac{A_I^2}{\sigma^2}=2\frac{I}{\sigma^2} \quad (3.82)$$

Equation (3.77) is plotted in Figure 3.30 [16] for some representative parameters. It is assumed in this example that $W_{\mathrm{F}}T = 1$. This example indicates that DBPSK is relatively sensitive to tone interference, producing a P_{e} of 10^{-1} with ISR less than zero.

3.11 Frequency Shift Keying

3.11.1 Modulations

3.11.1.1 Multiple-Frequency Shift Keying

For MFSK, the received signal can be expressed as

$$\begin{aligned}\tilde{r}(t)&=\sqrt{2R}\sum_k e^{j(2\pi f_k t+\phi_k)}u_{\mathrm{T}}(t-kT_{\mathrm{s}})\\ f_k&\in\{f_0,f_1,\cdots,f_{M-1}\};\qquad \phi_k\in(0,2\pi)\end{aligned} \quad (3.83)$$

and R is the average power in the signal. This can be simplified to

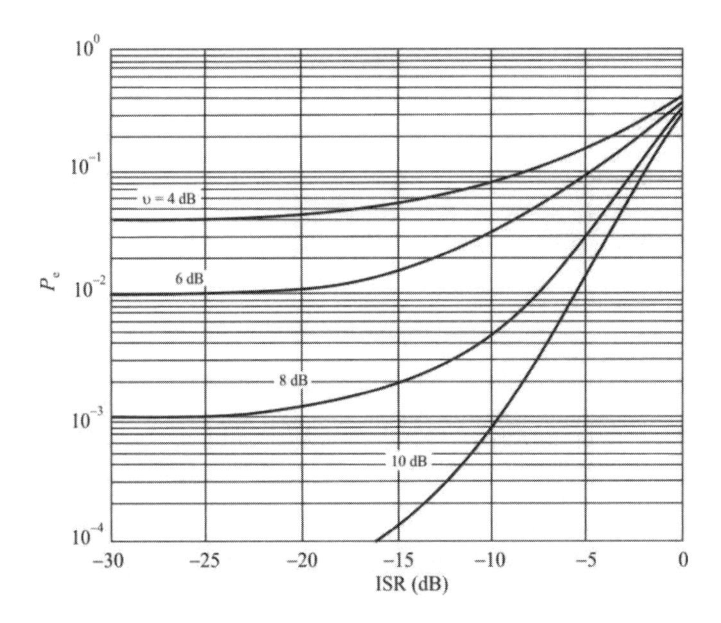

Figure 3.30 Performance of DBPSK versus the interference to signal ratio with the SNR as a parameter. (After: [16]. © IEEE 1999. Reprinted with permission.)

$$r_k(t) = \sqrt{2R}\cos\left[2\pi\left(f_0 + d_k\delta_f\right)t\right] \tag{3.84}$$

when $d_k\delta_f$ represents the frequency offset of the modulating tones from the carrier frequency f_0. In particular, for BFSK, in the most common case when the modulating tones are located symmetrically on either side of the carrier frequency and offset by δ_f , $d_k = \pm 1$ represents the data bit during that interval.

For MFSK, either coherent or noncoherent detection can be used, but due to implementation complexity and similar performance of the modulation schemes at high SNR, noncoherent detection is more common. In that case, the tones may be coherent within a hop (for SFHSS) but incoherent between hops. Coherent detection requires that the receiver be phase-locked to the received signal, whereas for noncoherent detection this is not required. Recovery of the signal carrier phase is therefore required for coherent reception.

It is possible to use coherent FSK detection, where the phase is preserved and the receiver phase is aligned with that of the received signal. For large E_b / N_0, however, the performance of these detectors is essentially the same as noncoherent detectors. This coherent detection is possible within a dwell for SFHSS systems, but phase coherency is difficult to maintain over large changes of the frequency

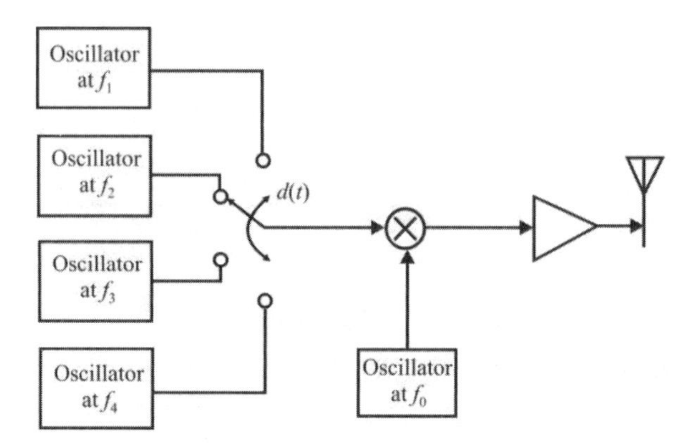

Figure 3.31 4-FSK transmitter architecture. This is noncoherent modulation since there is no attempt to control the phase angle of the tones as the data stream selects the signal to send.

synthesizers, so phase coherence from dwell to dwell is difficult to track. Because of this and the fact that coherent detection requires more processing to keep track of the carrier phase for equivalent performance, by far the most prolific form of modulation for FHSS systems is noncoherent BFSK. For these reasons we will only consider non-coherent reception here.

When jamming FSK signals, the jammer attempts to raise the noise power at the output of the detection filters at the receiver. This is true irrespective of the particular jamming technique employed. When the target employs noncoherent FSK, then these filters are usually bandpass filters just prior to energy detection.

A block diagram of a notional noncoherent MFSK transmitter is shown in Figure 3.31 when $M = 4$. The data stream $d(t)$ selects which of the tones (symbols) is used to modulate the carrier. The symbol frequencies are f_1, f_2, f_3, and f_4, while f_0 moves the tones from the baseband frequency range higher for efficient propagation. This is the carrier signal frequency. Note that if cable is the propagation medium, then the carrier oscillator may not be needed.

In this case, each symbol in the data stream $d(t)$ represents two bits of information because $4 = 2^2$. For 8-FSK, there would be eight such oscillators and each symbol of $d(t)$ would represent three bits of information since $8 = 2^3$. In general, for MFSK modulation, $\log_2 M$ bits of information are transmitted with each symbol.

3.11.2 BFSK

As mentioned, noncoherent detection is the most prolific detection scheme for MFSK. This is because it is difficult to build oscillators that maintain phase coherence over the large and rapid frequency changes that occur in MFSK signals.

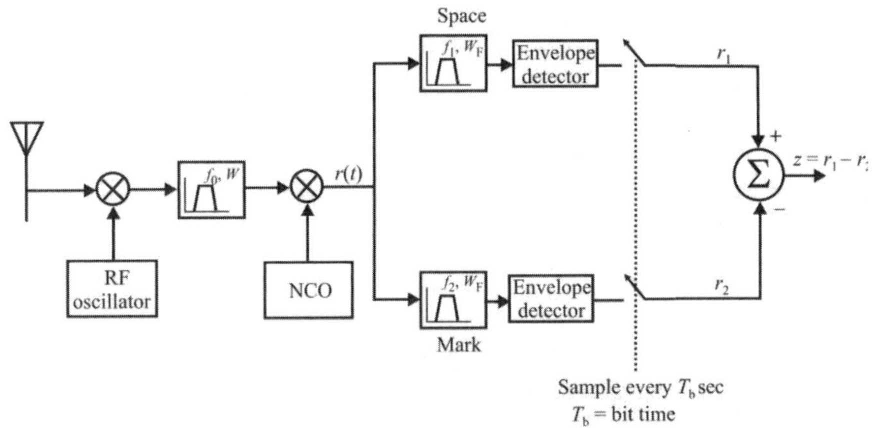

Figure 3.32 Noncoherent BFSK detector architecture.

It is, however, possible in SFHSS systems to maintain coherency from one bit to the next within a hop.

The receiver detector structure for noncoherent detection of BFSK with hard decision decoding is shown in Figure 3.32. The incoming signals are first converted to a suitable *intermediate frequency* (IF) and then bandpass filtered. This filter removes significant out-of-band unwanted signals from further processing. The filtered signal, after another stage of suitable frequency conversion, is then sent to two channels for detection. The mark and space frequencies are different and so the bandpass filters for the appropriate frequency will only pass signals at its associated frequency offset from the carrier frequency. If a mark was sent, then the bottom (mark) channel output after square law envelope detection will be larger than the upper (space) channel. The output from the detector that is the larger is selected as the tone that was sent. Even though the noise spectral density, whether unintentional as specified by N_0 and therefore modeled as a constant level, or jammer noise, whose spectral density can also be modeled as a constant level, the associated time domain signals are constantly changing amplitudes and therefore the power levels are changing. This effect can cause an error to be made in this selection process because noise can cause the energy in the complementary tone channel to be larger than that in the data tone channel during a symbol interval. The comparison is made between these two amplitudes by subtracting one from the other, forming $z = r_1 - r_2$. The decision is made via:

Choose mark if $z < 0$;
Choose space if $z > 0$.

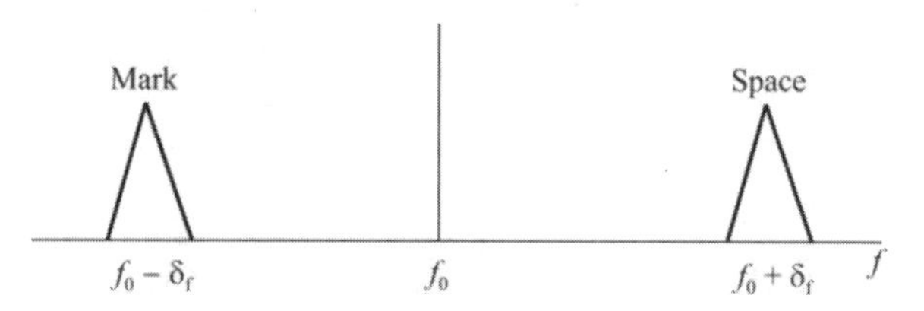

Figure 3.33 Channel structure for BFSK.

The mark and space filters shown in Figure 3.32 are bandpass filters. They could also be baseband lowpass filters, or some other type.

The spectrum structure for BFSK is illustrated in Figure 3.33. Two tones are placed somewhere in the spectrum, one representing a space and the other a mark. Typically, these tones are placed symmetrically on either side of the carrier frequency and that will be assumed here. The two signals at the input to the detector in Figure 3.32 can be represented as

$$\begin{aligned} s_1(t) &= A\cos(2\pi f_1 t) \\ s_2(t) &= A\cos(2\pi f_2 t) \end{aligned} \tag{3.85}$$

where A is the amplitude, $A = \sqrt{2R}$, where R is the average power in the tones, assumed the same, $f_{k1} = f_0 + \delta_f$ and $f_{k2} = f_0 - \delta_f$ are the two tone frequencies, and T_b is the symbol (bit) duration. If $\delta_f = i / 2T_b$, i integer, then $s_1(t)$ and $s_2(t)$ are orthogonal. This can be seen by using (3.85) in (3.10):

$$\rho_{12} = \frac{1}{\sqrt{E_1 E_2}} \int_{kT_b}^{(k+1)T_b} A\cos\left[2\pi\left(f_0 + \delta_f\right)t\right] A\cos\left[2\pi\left(f_0 - \delta_f\right)t\right] dt \tag{3.86}$$

Using straightforward trigonometric relationships, this expression can be manipulated to show that $\rho_{12} = 0$ when $\delta_f = i / 2T_b$ for any $i > 0$. Therefore, due to (3.9), $R_{12} = 0$ for this same condition.

3.11.2.1 BFSK Performance with AWGN

Torrieri derived a general expression for the detection performance of noncoherent BFSK signals in the presence of noise and interference/jamming [17]. Interference, either intentional or unintentional, is represented by

$$\begin{aligned} i_{k_1}(t) &= B_1 \cos[2\pi f_{k_1} t + \phi_1(t)] \\ i_{k_2}(t) &= B_2 \cos[2\pi f_{k_2} t + \phi_2(t)] \end{aligned} \tag{3.87}$$

When the interference is intentional, it is assumed to be jammer tones and $B_i = \sqrt{2J}$. When the tones are interference, $B_i = \sqrt{2I}$. These signals at the input to the detector are accompanied by thermal noise. The power of this noise component is represented by

$$\begin{aligned} P_{\mathrm{N}_1} &= P_{\mathrm{t}} + P_{\mathrm{n}_1} \\ P_{\mathrm{N}_2} &= P_{\mathrm{t}} + P_{\mathrm{n}_2} \end{aligned} \tag{3.88}$$

The first component is the noise accompanying the signal generated by the electronics in the transmitter, the propagating medium, and the receiver, MMN, or any other noise discussed in Chapter 2. The second component of the noise in (3.88) could be due to a jammer that is utilizing a noise waveform. The resulting probability of generating a symbol error in this noncoherent BFSK system is given by

$$P_{\mathrm{e}} = \frac{1}{4\pi}\int_0^{2\pi} \left\{ \begin{aligned} & Q\left[\frac{B_2}{\sqrt{P_{\mathrm{N}_1}+P_{\mathrm{N}_2}}}, \frac{D_1(\theta)}{\sqrt{P_{\mathrm{N}_1}+P_{\mathrm{N}_2}}}\right] \\ & \quad + Q\left[\frac{B_1}{\sqrt{P_{\mathrm{N}_1}+P_{\mathrm{N}_2}}}, \frac{D_2(\theta)}{\sqrt{P_{\mathrm{N}_1}+P_{\mathrm{N}_2}}}\right] \\ & \quad - \frac{P_{\mathrm{N}_1}}{P_{\mathrm{N}_1}+P_{\mathrm{N}_2}} \exp\left[-\frac{B_2^2 + D_1^2(\theta)}{2(P_{\mathrm{N}_1}+P_{\mathrm{N}_2})}\right] I_0\left[\frac{B_2 D_1(\theta)}{P_{\mathrm{N}_1}+P_{\mathrm{N}_2}}\right] \\ & \quad - \frac{P_{\mathrm{N}_2}}{P_{\mathrm{N}_1}+P_{\mathrm{N}_2}} \exp\left[-\frac{B_1^2 + D_2^2(\theta)}{2(P_{\mathrm{N}_1}+P_{\mathrm{N}_2})}\right] I_0\left[\frac{B_1 D_2(\theta)}{P_{\mathrm{N}_1}+P_{\mathrm{N}_2}}\right] \end{aligned} \right\} d\theta \tag{3.89}$$

with

$$D_k^2(\theta) = A^2 + B_k^2 + 2AB_k \cos(\theta), \qquad k = 1,2 \tag{3.90}$$

This expression is used to compute symbol error probabilities in later sections by varying the values of the parameters.

Noise in the Complementary Channel Only

For the follower jammer that will be discussed subsequently, one possibility is to use a jamming technique that places a *narrowband noise* (NBN) signal in the complementary channel. Since, in general, it is not known which side of the carrier tone is the complementary channel (and it can change every symbol interval), the probability is reduced by a factor of ½ . Thus, $A = \sqrt{2R}$, $P_{\mathrm{n}_1} = P_{\mathrm{n}}$, $P_{\mathrm{N}_1} = P_{\mathrm{N}} + P_{\mathrm{n}}$, $P_{\mathrm{n}_2} = 0$, $P_{\mathrm{N}_2} = P_{\mathrm{t}}$, and $B_1 = B_2 = 0$ in (3.89). The result is [19]

$$P_{\mathrm{e}} = \frac{1}{2}\frac{P_{\mathrm{N}} + P_{\mathrm{n}}}{2P_{\mathrm{N}} + P_{\mathrm{n}}}\exp\left(-\frac{R}{2P_{\mathrm{t}} + P_{\mathrm{n}}}\right) \tag{3.91}$$

This expression can be written with power ratios as

$$P_{\mathrm{e}} = \frac{1}{2}\frac{1 + \frac{P_{\mathrm{n}}}{R}\upsilon}{2 + \frac{P_{\mathrm{n}}}{R}\upsilon}\exp\left(-\frac{1}{\frac{2}{\upsilon} + \frac{P_{\mathrm{n}}}{R}}\right) \tag{3.92}$$

where the SNR, denoted by υ, is $\upsilon = R / P_{\mathrm{t}}$.

Noise in Data Channel Only

When extra noise appears in the data channel but not in the complementary channel, then $A = \sqrt{2R}$, $P_{\mathrm{n}_1} = 0$, $P_{\mathrm{N}_1} = P_{\mathrm{N}}$, $P_{\mathrm{n}_2} = P_{\mathrm{n}}$, $P_{\mathrm{N}_2} = P_{\mathrm{t}} + P_{\mathrm{n}}$, and $B_1 = B_2 =$ 0 in (3.89) and P_{e} is given by [20]

$$P_{\mathrm{e}} = \frac{P_{\mathrm{N}}}{2P_{\mathrm{N}} + P_{\mathrm{n}}}\exp\left(-\frac{R}{2P_{\mathrm{N}} + P_{\mathrm{n}}}\right) \tag{3.93}$$

and, when expressed in terms of power ratios

$$P_{\mathrm{e}} = \frac{1}{2 + \frac{P_{\mathrm{n}}}{R}\upsilon}\exp\left(-\frac{1}{\frac{2}{\upsilon} + \frac{P_{\mathrm{n}}}{R}}\right) \tag{3.94}$$

Note that P_e from additional noise in the complementary channel, given by (3.91), is larger than that from additional noise in the data channel given by (3.93). The presence of the P_n term in the numerator accounts for this difference.

Noise in Both Channels

In this case $B_1 = B_2 = 0$; however, again, it is not known which side is the contiguous channel; therefore, a factor of ½ is necessary. In addition, the power per channel is reduced by ½ giving the power $A = \sqrt{2R/2}$. This yields [20]

$$P_e = \frac{1}{2}\frac{1}{2}\exp\left(-\frac{1}{2}\frac{R/2}{P_N + P_n}\right) \tag{3.95}$$

which, when expressed in terms of power ratios, becomes

$$P_e = \frac{1}{4}\exp\left(-\frac{1}{4}\frac{1}{\frac{1}{\upsilon} + \frac{P_n}{R}}\right) \tag{3.96}$$

These three cases are plotted in Figure 3.34 for some representative parameters. Note that as the level of additional noise takes P_n / R above 0 dB, the effect of the additional noise in the data channel only is to reduce P_e. Additional noise in the complementary channel or both channels raises P_e to levels at which it is difficult to communicate at relatively low levels of noise. Jamming just one channel on one side of the detected signal produces $P_e > 10^{-1}$ at $P_n / R \approx -4$ dB for $\upsilon = 10$ dB and –5 dB for $\upsilon = 20$ dB. Similarly, jamming both the tone channel and one of the adjacent channels produces $P_n / R \approx -8$ dB for $\upsilon = 10$ dB and –5 dB for $\upsilon = 20$ dB.

3.11.2.2 BFSK Performance with Tone Interference

Tones in Both Channels

Here $P_{n_1} = P_{n_2} = 0$, $P_{N_1} = P_{N_2} = P_N$, $B_1 = B_2 = \sqrt{2I}$, and $A = \sqrt{2R}$ yielding [20]

$$P_e = \frac{1}{2\pi}\int_0^{2\pi}\left\{Q\left[\sqrt{\frac{I}{P_N}}, \frac{D(\theta)}{\sqrt{2P_N}}\right] - \frac{1}{2}\exp\left[-\frac{2I + D^2(\theta)}{4P_N}\right] I_0\left[\frac{\sqrt{2I}D(\theta)}{2P_N}\right]\right\} d\theta \tag{3.97}$$

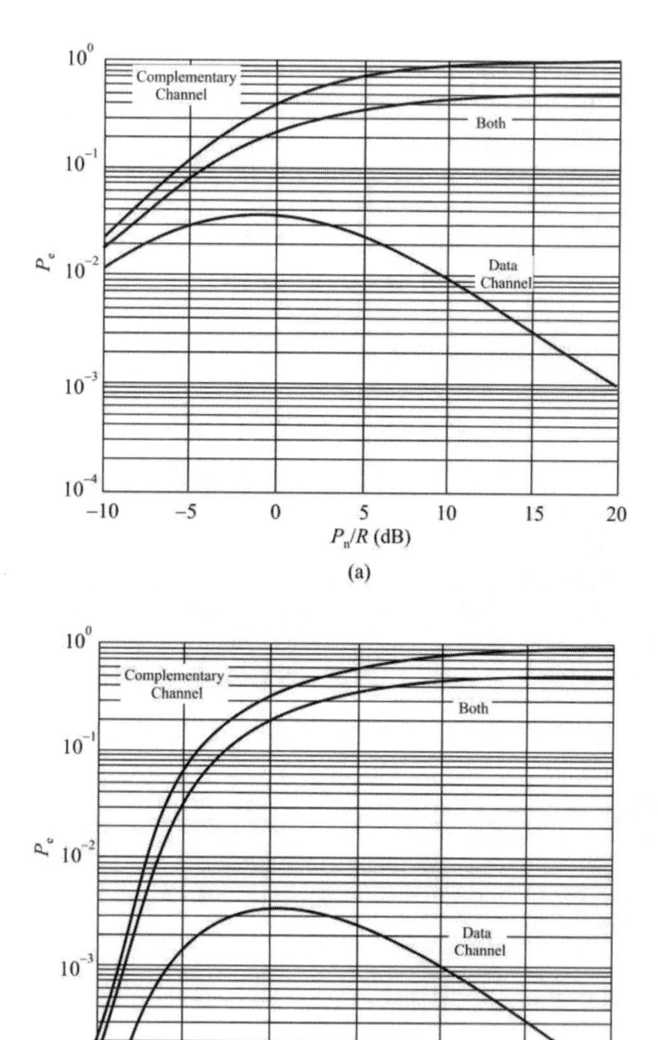

Figure 3.34 Comparison of BFSK performance for contiguous transmission and complementary channels when additional noise is introduced as shown, the average power of which is P_n: (a) $\upsilon = 10$ dB and (b) $\upsilon = 20$ dB.

where, in this case

$$D^2(\theta) = 2R + 2I + 4\sqrt{RI}\cos(\theta) \tag{3.98}$$

Using power ratios,

$$P_e = \frac{1}{2\pi}\int_0^{2\pi}\left\{\begin{array}{l} Q\left(\sqrt{\iota\upsilon}, \sqrt{\upsilon + \iota\upsilon + 2\upsilon\sqrt{\iota}\cos\theta}\right) \\ \qquad -\frac{1}{2}\exp\left[-\left(\iota\upsilon + \frac{1}{2}\upsilon + \upsilon\sqrt{\iota}\cos\theta\right)\right] \\ \qquad \times I_0\left(\iota\upsilon^2 + \iota^2\upsilon^2 + 2\iota\upsilon^2\sqrt{\iota}\cos\theta\right) \end{array}\right\} d\theta \tag{3.99}$$

where the ISR, denoted by ι, is $\iota = I / R$ and represents the amount of power in the interfering signal relative to the target signal. There is no known closed-form solution to this integral, so it must be evaluated numerically.

Tone in Single Channel

Assume that tone interference enters the data channel. Setting $B_1 = \sqrt{2J}$, $B_2 = 0$, $P_{n_1} = P_{n_2} = 0$, $P_{N_1} = P_{N_2} = P_N$, and $A = \sqrt{2R}$ yields [21]

$$P_e = \frac{1}{2}Q\left(\sqrt{\frac{J}{P_N}}, \sqrt{\frac{R}{P_N}}\right) \tag{3.100}$$

and, with some rearranging,

$$P_e = \frac{1}{2}Q\left(\sqrt{\iota\upsilon}, \sqrt{\upsilon}\right) \tag{3.101}$$

When the interfering tone enters one of the channels adjacent to the data channel, then the bit error probability is given by [18]

$$P_e = \frac{1}{2}\frac{P_N + I}{2P_N + I}\exp\left(-\frac{R}{2P_N + I}\right) \tag{3.102}$$

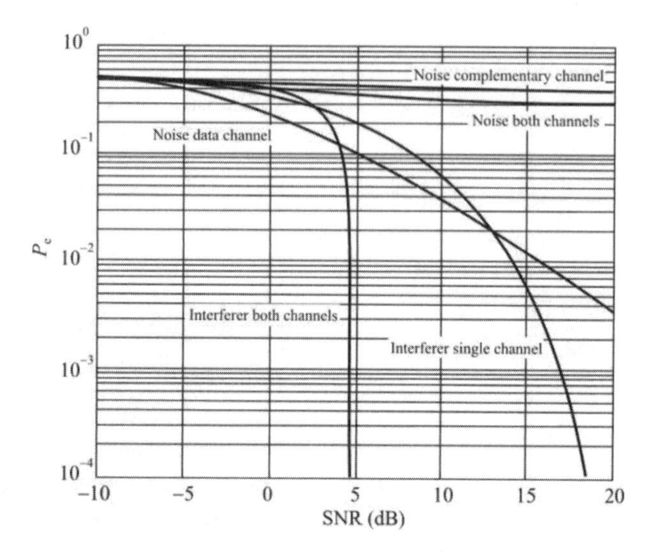

Figure 3.35 Comparison of BFSK performance in the presence of noise and tones in one or both channels when $\iota = -5$ dB and $\xi = 0$ dB.

which, denoting the JSR as ξ, is also

$$P_e = \frac{1}{2} \frac{1+\xi\upsilon}{2+\xi\upsilon} \exp\left(-\frac{1}{\frac{2}{\upsilon} + \xi} \right) \tag{3.103}$$

A comparison of these five cases is illustrated in Figure 3.35 when the interfering tone is 5 dB below the signal level and the added noise is equivalent to the signal level [$P_n / R = 1$ (0 dB)]. For these example parameters, an interferer in the complementary channel creates more havoc than in both channels because in the latter case, the tone in the data channel assists the receiver in correctly detecting the signal (phase differences in this channel are not accounted for here). To account for phase, assuming a uniform distribution of the phase over $(0, 2\pi)$, averaging the phase PDF over this range would be required. Likewise, noise in the complementary channel only keeps P_e higher than the cases with noise in the data channel for the same reason.

The noise level in Figure 3.35 is quite high, being equivalent to the signal level. With lower noise levels, the BER would be lower than that shown. Similarly, for lower levels of interference, the BER would be lower.

3.11.3 MFSK

In this section we consider MFSK performance with AWGN only. For MFSK the BER is related to the SER by

$$P_e = \frac{M}{2(M-1)} P_s \tag{3.104}$$

for orthogonal signaling.

A receiver for detecting noncoherent MFSK signals is shown in Figure 3.36 [21]. The performance of this receiver is given by [22]

$$P_s = \frac{1}{M} P_{e_1} + \frac{M-1}{M} P_{e_2} \tag{3.105}$$

where

$$P_{e_1} = \sum_{k=1}^{M-1} (-1)^{k+1} \binom{M-1}{k} \frac{P_N}{P_N + kP_T} \exp\left(-\frac{kR}{P_N + kP_T}\right) \tag{3.106}$$

and

$$P_{e_2} = \frac{P_T}{P_N + P_T} \exp\left(-\frac{R}{P_N + P_T}\right) + \sum_{k=1}^{M-2} (-1)^{k+1} \binom{M-2}{k} \left\{ \begin{array}{l} \dfrac{1}{k+1} \exp\left[-\dfrac{kR}{(k+1)P_N}\right] \\ -\dfrac{P_t}{P_N + (k+1)P_T} \\ \times \exp\left[-\dfrac{(P_N + kP_T)R}{P_N^2 + (k+1)P_t P_T}\right] \end{array} \right\} \tag{3.107}$$

and where

$P_T = P_N + P_n$;
P_n = NBN power over and above the unintentional noise level at the input to the bandpass filters (caused by interference or noise jammers);

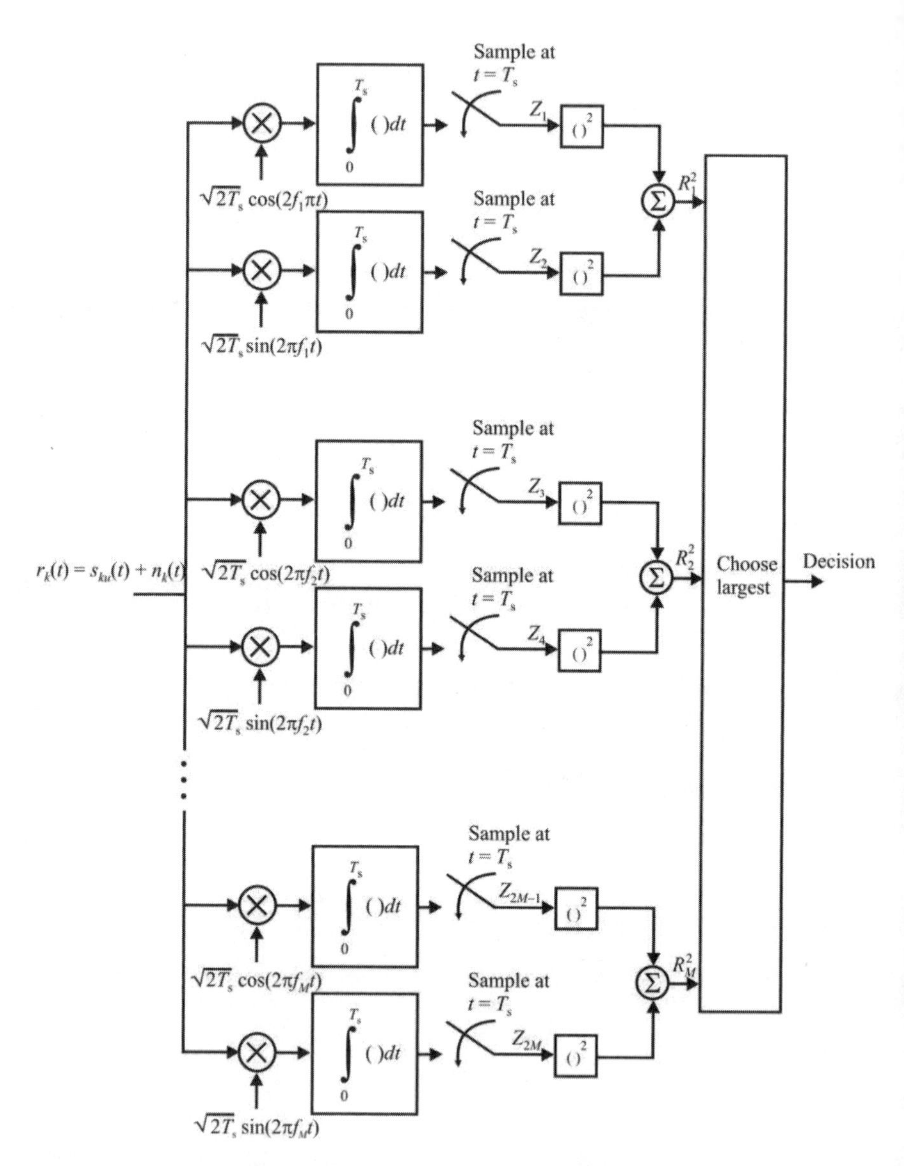

Figure 3.36 Noncoherent MFSK detector architecture. (Source: [21]. © John Wiley & Sons 2002. Reprinted with permission.)

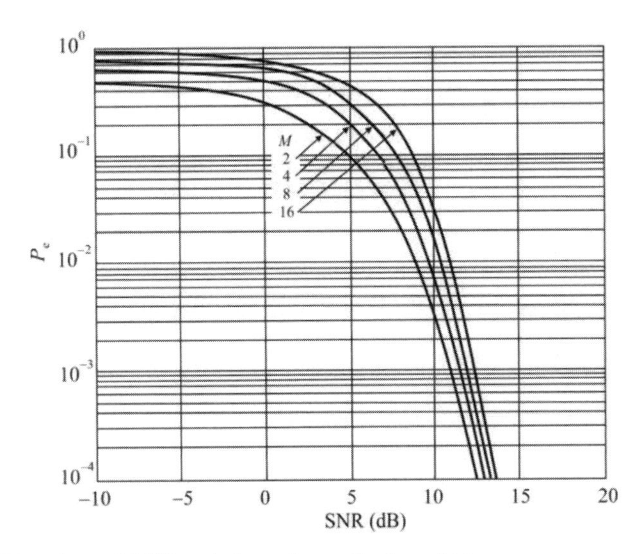

Figure 3.37 Noncoherent MFSK symbol error for small values of M.

$P_{\rm N}$ = noise power at the input of the bandpass filters due to unintentional noise.

P_{e_1} corresponds to the case when the NBN signal falls into the same channel as the signal tone while P_{e_2} corresponds to the case when the NBN is at another of the M tone frequencies in the channel.

When there is only thermal noise present then (3.105) reduces to [21]:

$$P_{\rm s} = \sum_{k=1}^{M-1} \binom{M-1}{k} \frac{(-1)^{k+1}}{k+1} \exp\left(-\frac{k}{k+1} \frac{R}{P_{\rm N}}\right) \tag{3.108}$$

This function is plotted in Figure 3.37 for some values of M. The larger M is, the more susceptible noncoherent MFSK is to symbol errors. The required bandwidth, however, decreases as more information is being sent within a symbol. This has benefits because narrower filters can be used in the receiver, thus decreasing the noise level and increasing the SNR. The bandwidth required is actually a factor of $M / \log_2 M$ less than BFSK [17].

Comparing Figure 3.20 with M = 2 in Figure 3.37, we can see that coherent BPSK exhibits about a 5 dB better performance at $P_{\rm e} = 10^{-2}$ than noncoherent

Table 3.6 SNR Requirements for a Specified P_e for Noncoherent BFSK

P_e	SNR Required (dB)
10^{-1}	5.1
10^{-2}	8.9
10^{-3}	10.9
10^{-4}	12.3
10^{-5}	13.4

BFSK; this difference decreases to about 4 dB at $P_e = 10^{-4}$. It would therefore be expected that EA against coherent BPSK is more difficult than against noncoherent BFSK.

Letting $M = 2$, then, yields P_e for this detector as

$$P_e = \frac{1}{2} e^{-\frac{1}{2}\frac{R}{P_N}} \tag{3.109}$$

This function is plotted in Figure 3.37 when $M = 2$. A summary of these results is given in Table 3.6. An SNR of around 8.9 dB is all that is required to achieve a P_e of 10^{-2}.

The FHSS systems included here are assumed to operate in a channelized spectrum, such as the low VHF band. In addition, they are assumed to use MFSK modulation. Due to this latter assumption, the input to the channel from the transmitter will be a tone, whatever the carrier modulation technique employed at the transmitter. If the carrier modulation is amplitude modulation, then the FSK tones are simply translated in frequency to a higher place in the spectrum. If the carrier modulation is FM, then several tones separated by the modulation tone frequency are generated with amplitudes determined by the Bessel function of the first kind and zeroth order, but, due to the first assumption, only one of these tones is carried in any given channel. The remainder of the tones generated are assumed to be small compared with the first one.

3.12 Quadrature Amplitude Modulation

QAM is a modulation technique in which two sinusoidal carriers, one 90 degrees out of phase with respect to the other, are used to transmit data over a given channel. Because the orthogonal carriers occupy the same frequency band and differ by a 90° phase shift, each can be modulated independently, transmitted over the same frequency band, and demodulated at the receiver. For a given available bandwidth, QAM enables data transmission at twice the rate of a standard *pulse*

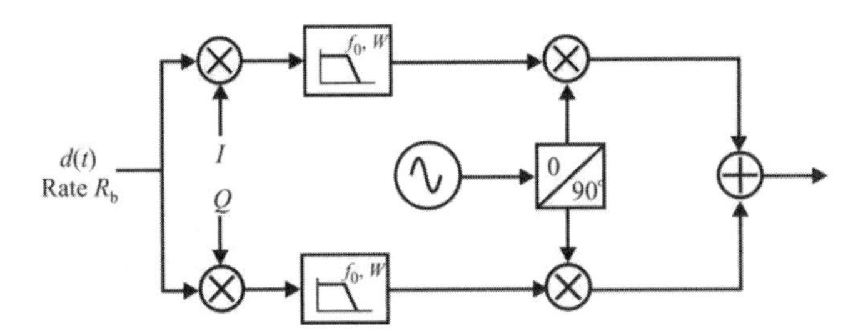

Figure 3.38 A QAM modulator.

amplitude modulation (PAM) without any degradation in the BER. QAM and its derivatives are used in both mobile radio and satellite communication systems.

QAM requires changing the phase and amplitude of a carrier sine wave. One of the easiest ways to implement QAM with hardware is to generate and mix two sine waves that are 90° out-of-phase with one another, as illustrated in Figure 3.38. Adjusting only the amplitude of either signal can affect the phase and amplitude of the resulting mixed signal.

These two carrier waves represent the in-phase (I) and quadrature-phase (Q) components of our signal. Individually each of these signals can be represented as:

$$I = A\cos\varphi \tag{3.110}$$

and

$$Q = A\sin\varphi \tag{3.111}$$

Note that the I and Q components are represented as cosine and sine because the two signals are 90° out-of-phase with one another. Using the two identities above and

$$\cos(\alpha+\beta) = \cos\alpha\cos\beta - \sin\alpha\sin\beta$$

we can rewrite the carrier wave as

$$A\cos(\omega_c t + \varphi) = I\cos\omega_c t - Q\sin\omega_c t \tag{3.112}$$

As (3.112) illustrates, the resulting identity is a periodic signal whose phase can be adjusted by changing the amplitude of I and Q. Thus, it is possible to perform

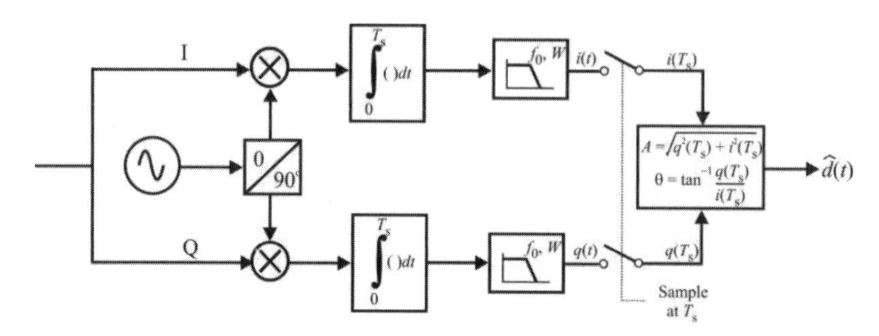

Figure 3.39 4QAM demodulator.

digital modulation on a carrier signal by adjusting the amplitude of the two mixed signals.

As mentioned, QAM can be extended to any order desired, with the trade-offs being spectrum efficiency on the one hand and BER versus SNR performance on the other—the higher orders or modulation being more sensitive to the SNR than lower orders.

Some modern communication systems use higher orders of QAM because of its efficiency. Up to 64QAM is used in WiMAX when the noise environment supports it. In this section we will review the characteristics of 2QAM, 4QAM, and 16QAM. Extending the performance to higher forms is a straightforward extension of the last of these.

We should note that although we only show square constellations here for simplicity in exposition, the constellations need not be square. Any shape imaginable is possible.

A representative receiver architecture for 4QAM is shown in Figure 3.39. The incoming signal (after filtering, amplification, and perhaps frequency conversion) is multiplied again by the orthogonal functions—cos $\omega_c t$ for the I-channel and sin $\omega_c t$ for the Q-channel. The results of this multiplication are integrated for a symbol interval, T_s, filtered, sampled, and manipulated to recover the amplitude, A, and phase, θ, as shown in Figure 3.39. The result is an estimate of the original data sequence.

3.12.1 2QAM and 4QAM Modulations

Observing Figure 3.13, we can see that 2QAM modulation is the same as BPSK, and 4QAM modulation is the same as QPSK modulation. The characteristics of these modulations were already examined in Section 3.9. Therefore, we need not develop these modulations further here.

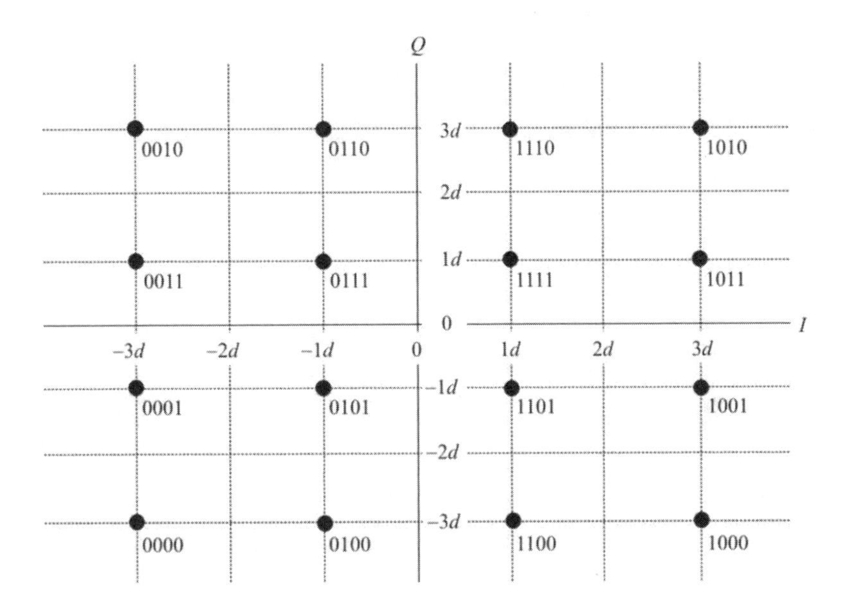

Figure 3.40 16QAM Gray mapping.

3.12.2 16QAM Performance with Gray Mapping

In this section we will derive the theoretical 16QAM BER with a Gray-coded constellation mapping in AWGN. Gray mapping is when each adjacent symbol is different in only one bit, as illustrated in Figure 3.40. The 4 bits at each constellation point in 16QAM can be considered as two bits each on independent I and Q axes as indicated in Table 3.7.

3.12.2.1 Symbol Error and Bit Error Probability

As can be seen from Figure 3.40, with Gray-coded bit mapping, adjacent

Table 3.7 16QAM Mapping

b0b1	I	b2b3	Q
00	–3	**00**	–3
01	–1	**01**	–1
11	+1	**11**	+1
10	+3	**10**	+3

constellation symbols differ by a single bit. So if the noise causes the constellation to cross the decision threshold, only one out of k bits will be in error. Thus the relation between bit error and symbol error is given by

$$P_b \approx \frac{P_s}{k} \tag{3.113}$$

Note that for very low values of E_s / N_0, it may happen that the noise causes the constellation to fall near a diagonally located constellation point. In that case, each symbol error will cause two bit errors, hence the need for the approximate relationship in (3.113). However, for reasonably high values of E_s / N_0, the chances of such events are negligible.

3.12.2.2 Bit Energy and Symbol Energy

Since each symbol consists of k bits, the symbol-to-noise ratio is k times the bit energy to noise ratio, namely,

$$\frac{E_s}{N_0} = k\frac{E_b}{N_0} \tag{3.114}$$

where

$$k = \log_2(16) = 4 \tag{3.115}$$

in this case.

3.12.2.3 Scaling Factor in QAM

There is a scaling factor that is normally used when analyzing QAM. The purpose of this scaling factor is to normalize the average energy to 1. For an MQAM constellation mapping, where $\sqrt{M}$ is a power of 2, the elements of the alphabet are

$$\alpha_{\text{MQAM}} = \{\pm(2m-1) \pm (2m-1)\}, m \in \{1, 2, \ldots, \sqrt{M}/2\} \tag{3.116}$$

We can take advantage of the symmetry in the constellation by first noting that each quadrant contains $M / 4$ constellation points and that the energy of the real and imaginary components are the same. Furthermore, in each quadrant the elements of each alphabet are used $\sqrt{M}/2$ times by the real and imaginary parts. Thus, the average energy in an M-element constellation is given by

$$\begin{aligned}
E_{\text{MQAM}} &= \mathcal{E}\left\{\text{Re}\left|\alpha_{\text{MQAM}}\right|^2\right\} + \mathcal{E}\left\{\text{Im}\left|\alpha_{\text{MQAM}}\right|^2\right\} = 2\mathcal{E}\left\{\text{Re}\left|\alpha_{\text{MQAM}}\right|^2\right\} \\
&= \frac{2\frac{\sqrt{M}}{2}}{\frac{M}{4}} \sum_{m=1}^{\frac{\sqrt{M}}{2}} (2m-1)^2 \\
&= \frac{2}{3}(M-1)
\end{aligned} \tag{3.117}$$

For example, when $M = 16$ for 16QAM,

$$\begin{aligned}
E_{\text{16QAM}} &= \frac{2}{3}(16-1) \\
&= 10
\end{aligned} \tag{3.118}$$

3.12.2.3 16QAM SER

Consider the typical 16QAM modulation scheme where the alphabet

$$\alpha_{\text{16QAM}} = \{\pm 1 \pm j, \pm 1 \pm j3, \pm 3 \pm j3, \pm 3 \pm j1\} \tag{3.119}$$

is used. When each symbol is equally likely, as given by (3.118), the average energy of the 16QAM constellation is $E_{\text{16QAM}} = 10$. The 16QAM constellation is shown in Figure 3.41, where we see that each symbol point has the associated average energy $\sqrt{E_s/10}$.

3.12.2.4 Computing the Probability of Error

Consider the 16QAM constellation shown in Figure 3.41. To determine the BER, we have three distinct types of symbols that we need to examine, since they may lead to different results. The interior symbol regions are $\{s_5, s_6, s_9, s_{10}\}$, while the corner symbol regions are those around $\{s_0, s_3, s_{12}, s_{15}\}$. The remaining symbols with equivalent constraints are neither the interior nor corner points and they are $\{s_1, s_2, s_4, s_7, s_8, s_{11}, s_{13}, s_{14}\}$. Due to symmetry, we can see that the contribution to the BER from each member of these sets is the same.

The received signal can be any of

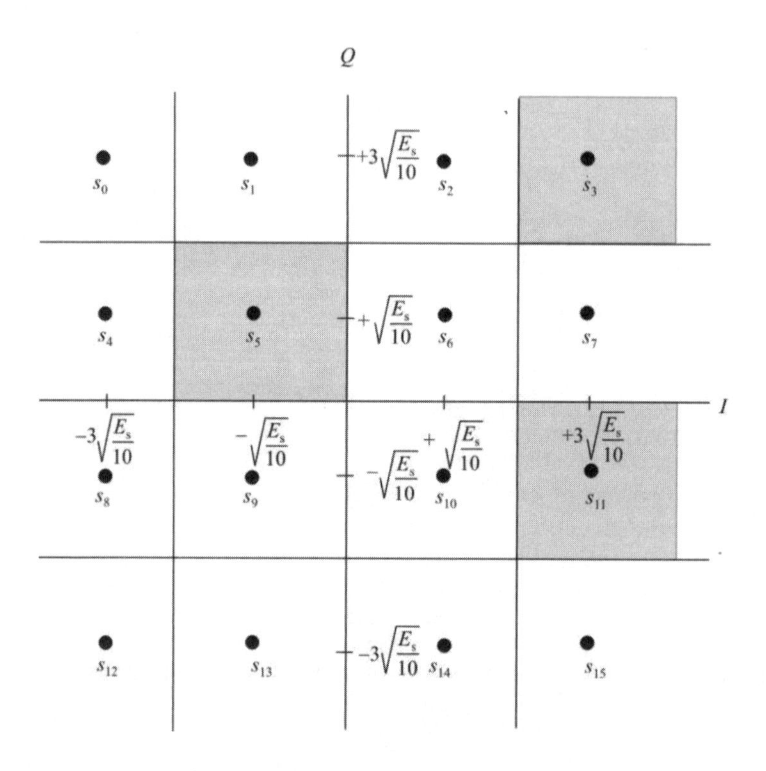

Figure 3.41 16QAM constellation. Each symbol has an average energy $\sqrt{E_s / 10}$.

$$
\begin{array}{llll}
r = s_0 + n & r = s_4 + n & r = s_8 + n & r = s_{12} + n \\
r = s_1 + n & r = s_5 + n & r = s_9 + n & r = s_{13} + n \\
r = s_2 + n & r = s_6 + n & r = s_{10} + n & r = s_{14} + n \\
r = s_3 + n & r = s_7 + n & r = s_{11} + n & r = s_{15} + n
\end{array}
$$

where n is a Gaussian r.v. with PDF

$$p_n(x) = \frac{1}{\sqrt{2\pi\sigma^2}} e^{-\frac{(x-\mu)^2}{2\sigma^2}} \tag{3.120}$$

with $\mu = 0$ and $\sigma^2 = N_0/2$.

Consider the symbol at the corner, s_3. The PDF of $r(t)$ given s_3 was transmitted, is

$$p[r(t)|s_3] = \frac{1}{\sqrt{\pi N_0}} e^{-\frac{\left(r-\sqrt{\frac{E_s}{10}}\right)^2}{N_0}} \tag{3.121}$$

since, as defined above and as indicated by (3.118), $\mu = \sqrt{E_s/10}$. As we see from Figure 3.42, the symbol s_3 is decoded correctly only if $r(t)$ falls in the shaded region around s_3 (unbounded in the $+I$ and $+Q$ directions), that is,

$$\begin{aligned}\Pr\{c|s_3\} &= \Pr\left\{\text{Re}\{r(t)\} > 2\sqrt{\frac{E_s}{10}} \text{ and } \text{Im}\{r(t)\} > 2\sqrt{\frac{E_s}{10}}\right\} \\ &= \Pr\left\{\text{Re}\{r(t)\} > 2\sqrt{\frac{E_s}{10}}\,|s_3\right\} \times \Pr\left\{\text{Im}\{r(t)\} > 2\sqrt{\frac{E_s}{10}}\,|s_3\right\}\end{aligned} \tag{3.122}$$

since we are assuming that the real part and imaginary part of $r(t)$ are independent. Thus,

$$\begin{aligned}\Pr\{c|s_3\} &= \frac{1}{\pi N_0} \int_{2\sqrt{E_s/10}}^{\infty} e^{-\frac{\left(x-\sqrt{\frac{E_s}{10}}\right)}{N_0}} dx \frac{1}{\pi N_0} \int_{2\sqrt{E_s/10}}^{\infty} e^{-\frac{\left(x-\sqrt{\frac{E_s}{10}}\right)}{N_0}} dx \\ &= \left[1 - \frac{1}{2}\text{erfc}\left(\sqrt{\frac{E_s}{10N_0}}\right)\right]\left[1 - \frac{1}{2}\text{erfc}\left(\sqrt{\frac{E_s}{10N_0}}\right)\right]\end{aligned} \tag{3.123}$$

The probability of s_3 being decoded incorrectly is

$$\begin{aligned}\Pr\{e|s_3\} &= 1 - \Pr\{c|s_3\} \\ &= 1 - \left[1 - \frac{1}{2}\text{erfc}\left(\sqrt{\frac{E_s}{10N_0}}\right)\right]^2 \\ &\approx \text{erfc}\left(\sqrt{\frac{E_s}{10N_0}}\right)\end{aligned} \tag{3.124}$$

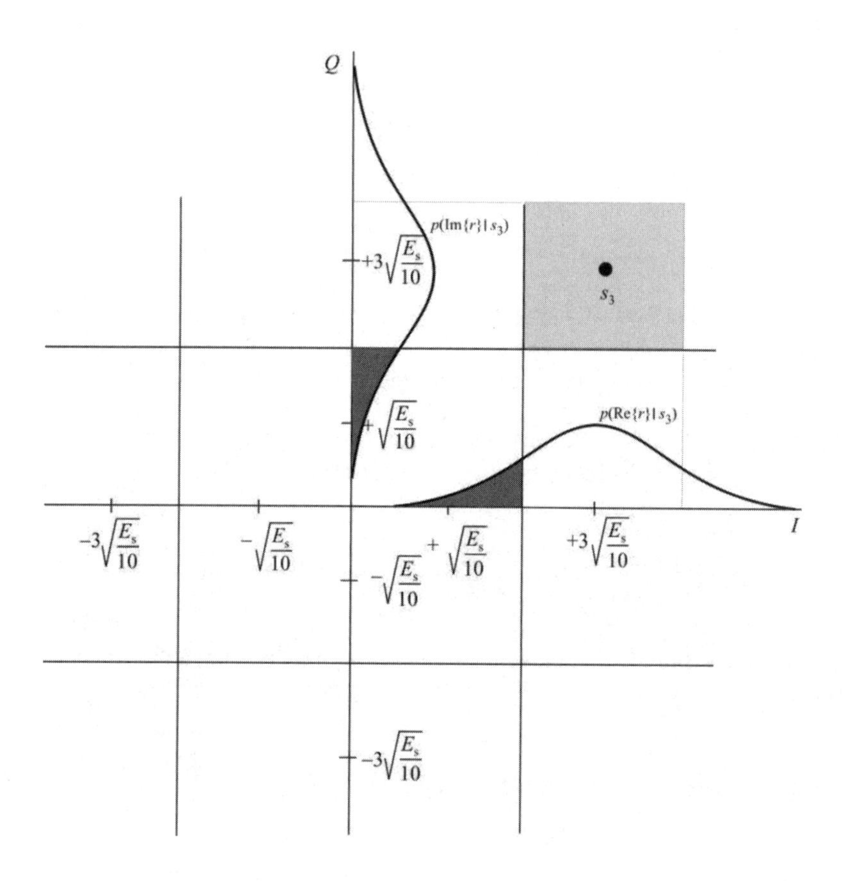

Figure 3.42 16QAM constellation showing the error regions for s_3. A detection error occurs if the received symbol falls into the dark regions.

Next, consider a symbol on the inside of the constellation in Figure 3.41, for example, s_5. The PDF of $r(t)$ given s_5 was transmitted is given by

$$p[r(t)|s_5] = \frac{1}{\sqrt{\pi N_0}} e^{-\frac{\left(r-\sqrt{\frac{E_s}{10}}\right)^2}{N_0}} \tag{3.125}$$

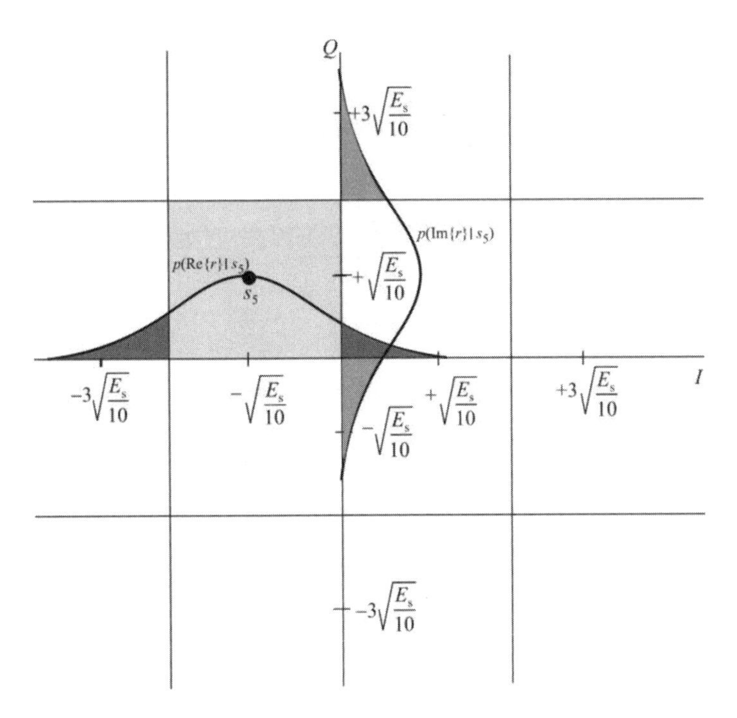

Figure 3.43 16QAM constellation error regions for s_5. A detection error occurs if the received symbol falls in one of the darker shaded regions.

As we can see from Figure 3.43, the symbol s_5 is decoded *correctly* only if $r(t)$ falls in the shaded region around s_5. Assuming that $\mathrm{Re}\{r(t)\}$ and $\mathrm{Im}\{r(t)\}$ are independent, the probability associated with this event is given by

$$\begin{aligned}\Pr\{c|s_5\} = &\Pr\left\{\mathrm{Re}\{r(t)\} \le 0 \,\text{and}\, \mathrm{Re}\{r(t)\} > -2\sqrt{E_s/10}\,|s_5\right\} \\ &\times \Pr\left\{\mathrm{Im}\{r(t)\} > 0 \,\text{and}\, \mathrm{Im}\{r(t)\} \le -2\sqrt{E_s/10}\,|s_5\right\}\end{aligned} \tag{3.126}$$

Thus,

$$\Pr\{c|s_5\} = \left[1 - \operatorname{erfc}\left(\sqrt{\frac{E_s}{10N_0}}\right)\right]\left[1 - \operatorname{erfc}\left(\sqrt{\frac{E_s}{10N_0}}\right)\right] \tag{3.127}$$

and the probability of s_5 being decoded incorrectly is

$$\begin{aligned}\Pr\{\mathrm{e}|s_5\} &= 1-\Pr\{\mathrm{c}|s_5\}\\ &= 1-\left[1-\operatorname{erfc}\left(\sqrt{\frac{E_s}{10N_0}}\right)\right]^2\\ &\approx 2\operatorname{erfc}\left(\sqrt{\frac{E_s}{10N_0}}\right)\end{aligned} \tag{3.128}$$

Next, consider the symbol that is neither at the corner nor in the inside, for example, s_{11}. The PDF of $r(t)$ given s_{11} was transmitted is

$$p[r(t)|s_{11}] = \frac{1}{\sqrt{\pi N_0}} e^{-\frac{\left(r-\sqrt{\frac{E_s}{10}}\right)^2}{N_0}} \tag{3.129}$$

From Figure 3.44, we see that the symbol s_{11} is decoded correctly only if $r(t)$ falls in the shaded region around s_{11} (unbounded in the $+I$ direction), namely,

$$\begin{aligned}\Pr\{\mathrm{c}|s_{11}\} = \Pr&\left\{\operatorname{Re}\{r(t)\} > 2\sqrt{\frac{E_s}{10}}\,|s_{11}\right\}\\ &\times \Pr\left\{\operatorname{Im}\{r(t)\} \le 0 \text{ and } \operatorname{Im}\{r(t)\} > -2\sqrt{\frac{E_s}{10}}\,|s_{11}\right\}\end{aligned} \tag{3.130}$$

As above,

$$\Pr\{\mathrm{c}|s_{11}\} = \left[1-\frac{1}{2}\operatorname{erfc}\left(\sqrt{\frac{E_s}{10N_0}}\right)\right]\left[1-\operatorname{erfc}\left(\sqrt{\frac{E_s}{10N_0}}\right)\right] \tag{3.131}$$

and the probability of s_{11} being decoded incorrectly is

$$\begin{aligned}\Pr\{\mathrm{e}|s_{11}\} &= 1-\Pr\{\mathrm{c}|s_{11}\}\\ &= 1-\left[1-\frac{1}{2}\operatorname{erfc}\left(\sqrt{\frac{E_s}{10N_0}}\right)\right]\left[1-\operatorname{erfc}\left(\sqrt{\frac{E_s}{10N_0}}\right)\right]\end{aligned}$$

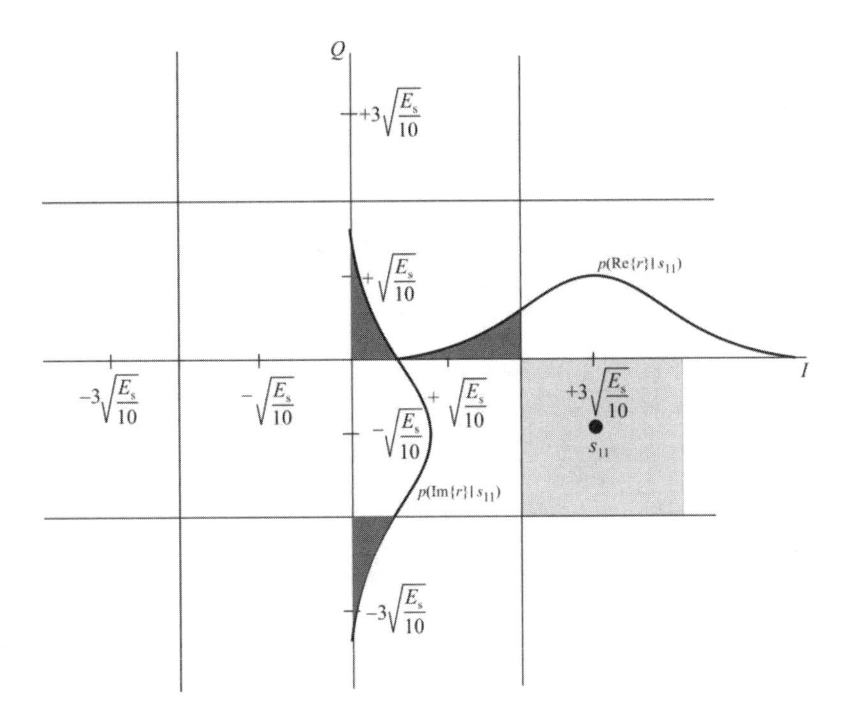

Figure 3.44 16QAM constellation error regions for s_{11}.

$$\approx \frac{3}{2}\operatorname{erfc}\left(\sqrt{\frac{E_s}{10N_0}}\right) \tag{3.132}$$

3.12.2.5 Total Probability of Symbol Error

Assuming that all the symbols are equally likely (4 in the middle, 4 at the corners, and the rest 8), the total probability of symbol error is

$$\begin{aligned} P_{e,16QAM} &\approx \frac{4}{16}2\operatorname{erfc}\left(\sqrt{\frac{E_s}{10N_0}}\right) + \frac{4}{16}\operatorname{erfc}\left(\sqrt{\frac{E_s}{10N_0}}\right) + \frac{8}{16}\frac{3}{2}\operatorname{erfc}\left(\sqrt{\frac{E_s}{10N_0}}\right) \\ &\approx \frac{3}{2}\operatorname{erfc}\left(\sqrt{\frac{E_s}{10N_0}}\right) \end{aligned} \tag{3.133}$$

Combining (3.133) and (3.114), we get

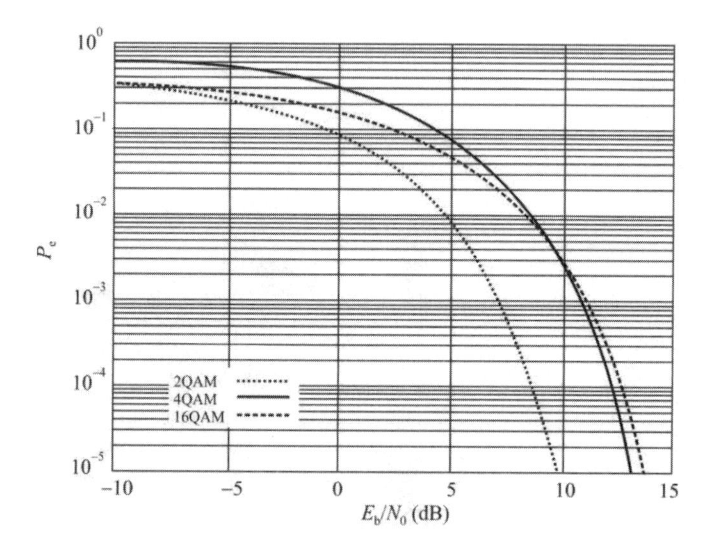

Figure 3.45 BER in AWGN for 2QAM, 4QAM, and 16QAM.

$$P_{\text{e,16QAM}} = \frac{3}{2k}\operatorname{erfc}\left(\sqrt{\frac{kE_\text{b}}{10N_0}}\right)\Bigg|_{k=4} \tag{3.134}$$

The performance of all three modulation techniques are compared in Figure 3.45. 4QAM performs the worst of the three at low SNR, whereas when the SNR reaches about 10 dB or so, it starts to outperform Gray-encoded 16QAM. BPSK has the best performance at all SNR levels. At high SNR levels, 4QAM and 16QAM BER performance are essentially the same, while the throughput of 16QAM is four times the throughput of 4QAM for the same bandwidth. Therefore, if the SNR is adequate, 16QAM is the better choice.

3.13 Spread Spectrum

In spread spectrum communication systems the signal that is transmitted is typically much wider than the minimum necessary, hence the appellation spread spectrum. A relatively narrow bandwidth information signal is spread in frequency to take advantage of the properties of such modulations.

We present the salient characteristics of SS signaling in this section. In particular, we cover the properties of DSSS, FHSS, both fast and slow, THSS, and hybrids of these. Much more will be provided in later chapters where we address how to jam these signal types.

3.13.1 Processing Gain

The definition of *processing gain* (G_p) of a spread spectrum communication system we will use is given by the ratio of the detector output SNR, SNR_o, to the SNR at the detector input, SNR_i, namely,

$$G_p = \frac{SNR_o}{SNR_i} \tag{3.135}$$

Other definitions have been made of spread spectrum processing gain, but this is the one used here.

In SS systems, advantage is taken of the significantly wider bandwidth occupied by the modulated signal compared to the data, or modulating signal. This bandwidth expansion is what accounts for most of the favorable properties in SS systems.

3.13.2 Direct-Sequence Spread Spectrum

In DSSS systems all users transmit in the same bandwidth simultaneously (see Figure 3.46). Some points to be noted from Figure 3.46 are:

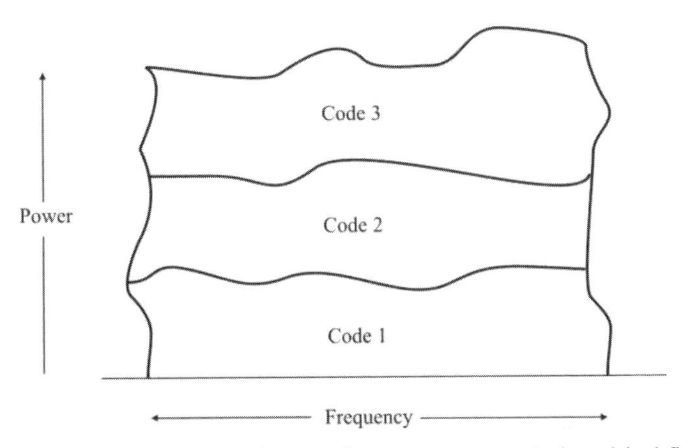

Figure 3.46 CDMA channels occupy the same frequency spectrum. A channel is defined by its spreading code.

- All users share the same bandwidth.
- Users are separated by a code, not by a timeslot or frequency.
- Each user is spread in the frequency domain.
- At the receive end, users are despread using the code that is unique to the transmitter and receiver.
- The power axis shows the strength of the cumulative addition of signals transmitted by all users.

The codes used for spreading have low cross-correlation values and are unique to every user. This is the reason that a receiver that has knowledge about the code of the intended transmitter is capable of selecting the desired signal.

In DSSS systems the data signal is multiplied by a PN code. This results in low cross-correlation values among the codes and the difficulty in jamming or detecting a data message. The relatively slow data signal is multiplied by the much faster PN code chip sequence. It is a well-known fact that the faster a signal is (the shorter the chip time), the greater its spectrum width. The result is that the coded signal occupies the same bandwidth as the coding signal.

Several families of binary PN codes exist. A customary way to create a PN-code is by means of at least one shift register (shift registers are discussed at length in Chapter 4). When the length of such a shift register is n, the period N_{DS} of the above-mentioned code families is given by

$$N_{\text{DS}} = 2^n - 1 \tag{3.136}$$

In DSSS systems the length of the code is the same as the spreading factor with the consequence that

$$G_{\text{p,DS}} = N_{\text{DS}} \tag{3.137}$$

The spreading process and the effects of a narrowband interferer, such as a jammer, are illustrated in Figure 3.47. The modulated DSSS signal, the spectrum of which is shown solid, is typically below the noise floor as shown. The interferer is shown with its power level above the noise floor. These signals add in the frequency domain and are all received by the DSSS receiver. By simply multiplying the received signal by the same code sequence as the one used in the transmitter (after they are synchronized), the interfering signal is spread out in frequency for exactly the same reason the baseband signal was spread in the transmitter—the interfering signal is decorrelated, and multiplying by the high-speed chip sequence, the energy is spread out in the spectrum, making the energy in any small bandwidth (such as the bandwidth of the baseband signal) quite small.

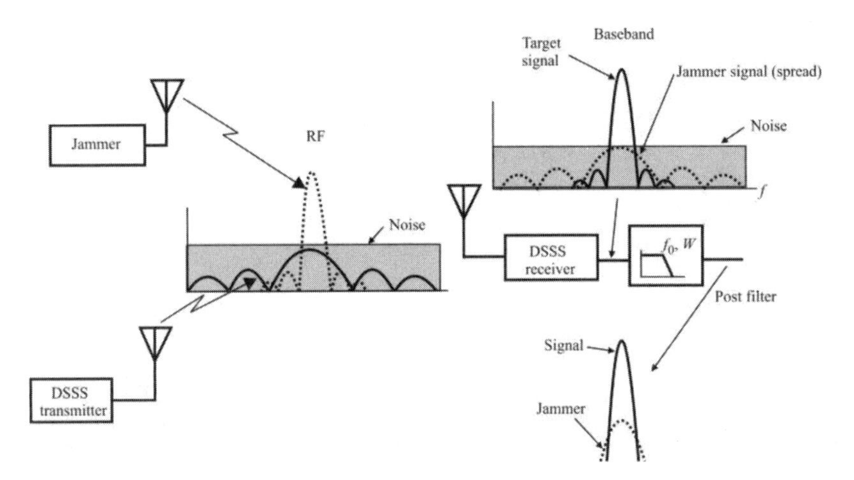

Figure 3.47 DSSS concept, before and after despreading.

However, since the same code is used in the receiver, the intended baseband signal is collapsed to its original form and recovered.

Thus, the resulting filtered signal is comprised of mostly the original baseband information, and the other DSSS signals, just like the aforementioned jammer, have been reduced to noise. This is how DSSS can be used as a multiple access technique.

A simplified flow diagram of a DSSS system is illustrated in Figure 3.48. In DSSS, the data signal is normally multiplied by a much higher-rate spreading signal, noted as the code generator in Figure 3.48. The multiplication in Figure 3.48 is accomplished by the exclusive OR. The result is a digital signal at the rate of the spreading signal. One bit of the spreading signal is referred to as a *chip*, and the spreading signal is sometimes referred to as the chipping signal. That signal is then normally 2^n PSK modulated onto the carrier. For BPSK, for example, this modulation process can be accomplished by simply multiplying the carrier waveform, usually a sine wave, by the digital signal from the exclusive OR. In this case, multiplying a sine wave by +1 or −1 accomplishes the same thing as changing the phase directly.

The process is reversed at the receiver. After RF amplification (not shown), downconversion in the mixer, and amplification (also not shown), the phase modulation is removed by the multiplier (exclusive OR in this case). With a large enough SNR, the data sequence is thus removed from the carrier. In this simple case it is assumed that the two code generators are synchronized so that their PN codes are locked together. The synchronization mechanism is not shown in Figure 3.48.

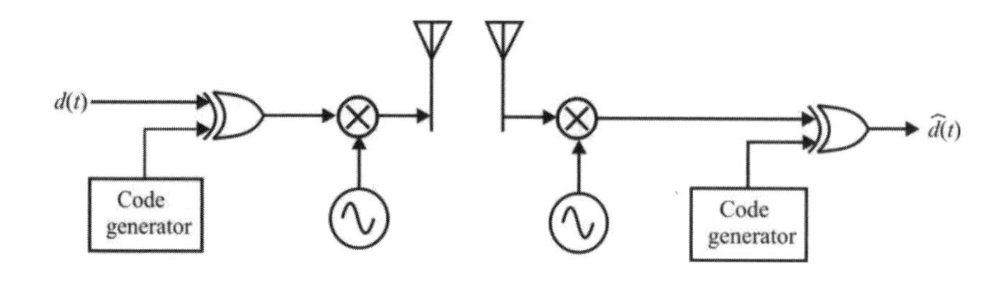

Figure 3.48 DSSS notional block diagram. The two code generators are assumed to be synchronized (see Chapter 7).

The most popular modulations for DSSS are from the PSK family. This is when the modulation information is contained in the relative phases of one bit relative to the last one or to a reference phase. BPSK is probably the most common but higher forms are also used. BPSK uses two phase states to represent whether the data bit is a one or zero. In QPSK, four phase states are used to represent one of the possible states associated with two data bits (00, 01, 10, or 11). In general, 2^n PSK is possible. The architecture used herein for DSSS AJ communication is shown in Figure 3.49.

Herein it is assumed that chip transitions coincide with data bit transitions and that the chip rate is an integer multiple of the data rate. It is a system requirement to avoid spillage of energy into areas of the spectrum where it is not wanted and most well-designed DSSS systems incorporate this.

The spreading signal is generally coded with a code selected for some desirable property—orthogonality with other code sequences in the same family, for example.

The data stream $d(t)$ is mapped (encoded) into I and Q symbols, depending on the modulation scheme used. The I and Q components are then multiplied by the spreading code $c^{\mathrm{I}}(t)$ and $c^{\mathrm{Q}}(t)$. This is followed by filtering the chipped waveform with chip filters, which are usually of the square root raised cosine variety. After chip filtering, the I and Q channel signals are converted to some higher IF frequency and then converted to analog form for RF processing. This RF processing consists of conversion to the transmit frequency f_0, filtering, and amplification. The resulting signal is then transmitted.

Whereas $d(t)$ is has a relatively slow data rate, say, 100 kbps, and therefore a relatively narrowband spectrum denoted by W_{d}, $c^{\mathrm{I}}(t)$ and $c^{\mathrm{Q}}(t)$ have much higher data rates, say, 100 Mbps, and therefore a relatively wider frequency extent. The bandwidth of $c^{\mathrm{I}}(t)$ and $c^{\mathrm{Q}}(t)$ is W_{ss}. This multiplication essentially makes the bandwidth of the transmitted signal that of $c^{\mathrm{I}}(t)$ and $c^{\mathrm{Q}}(t)$ and at any given narrow frequency region the power is low—frequently below the thermal noise floor. The result of multiplication by $c^{\mathrm{I}}(t)$ and $c^{\mathrm{Q}}(t)$ produces a $|\sin x/x|$ spectrum as shown

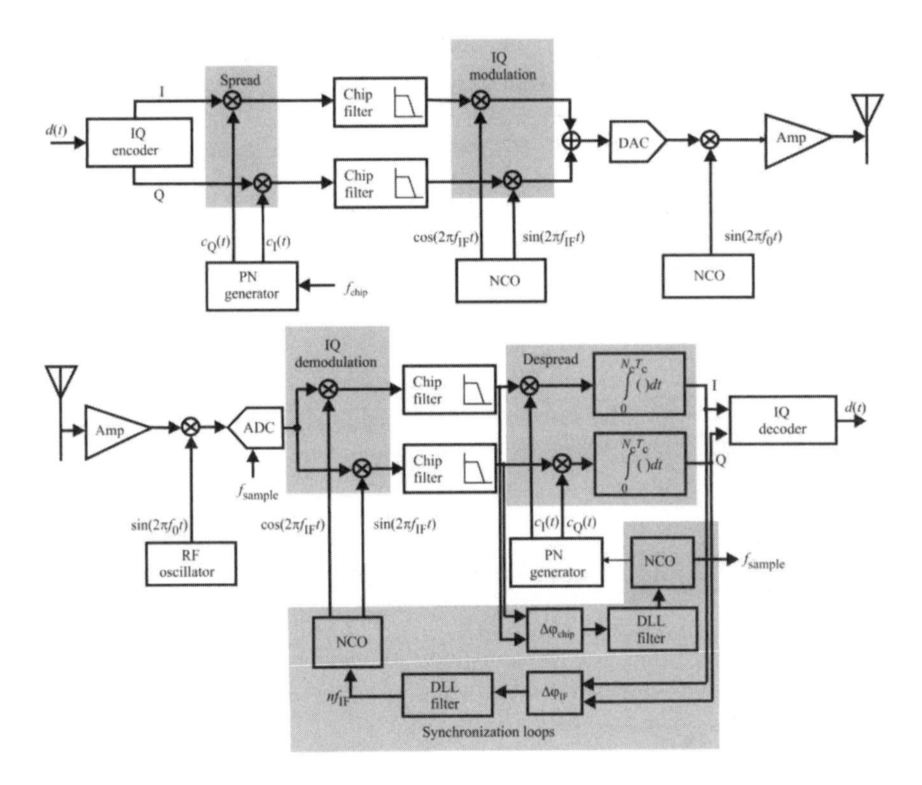

Figure 3.49 DSSS AJ transmitter and receiver communication architecture.

in Figure 3.50(a). The higher sidelobes are usually filtered off with the filter shown in Figure 3.50(b), with the result being only the main lobe for transmission as shown in Figure 3.50(c). After MPSK modulation, the resulting signal is up-converted in frequency to f_0 by simple multiplication. After appropriate filtering and amplification, the signal is transmitted.

At the receiver, the signal is amplified and filtered, but is still wideband. After downconversion to the IF frequency in analog form, the signal is converted to digital form at the sample rate. That signal is then multiplied with cosine and sine waveforms at the IF frequency for conversion to baseband. After chip filtering these I and Q channel signals then are despread by multiplication by $c^{\mathrm{I}}(t)$ and $c^{\mathrm{Q}}(t)$ again. The resulting signals are then integrated over the symbol time ($T_s = N_c T_c$) and the resultant symbols are decoded from I and Q back into $d(t)$.

The two receiver synchronization loops are shown at the bottom of Figure 3.49; the first is for coherent demodulation of the chips at a frequency of $n \times f_{\mathrm{IF}}$, and the second provides for coherent detection of each chip. This latter one drives

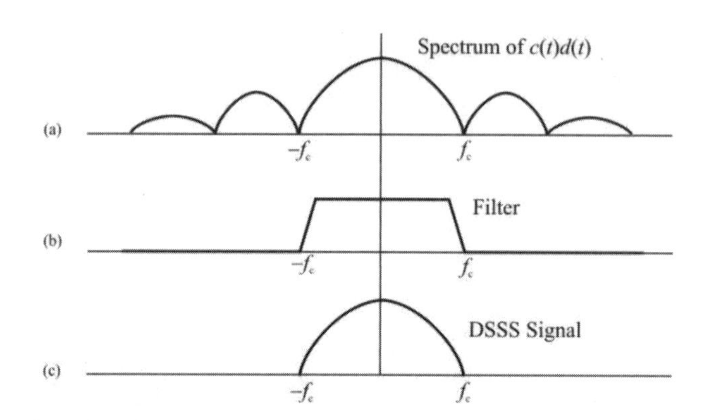

Figure 3.50 The main lobe is the only part of the DSSS signal that is retained for transmission.

the receiver PN generator to keep the PN sequence in proper phase. Whereas M could be any power of 2, $M = 2$ is the most popular choice because of the ease of implementation.

One of the characteristics of the PSK forms of modulation is that they have a constant modulus. That means their amplitude does not change. This is particularly important in those communication systems that are power limited, such as cell phone systems, PCS, and satellite communications. In those systems the final amplifiers that raise the signal power to levels adequate for transmission can be operated in saturation for maximum efficiency. Trying to use such an amplifier with signals that have varying amplitude generates what are normally considered unacceptable levels of intermodulation distortion, which interferes with other communications and wastes power.

Just as the spreading signal at the transmitter broadens the narrowband data signal, a narrowband signal received at the receiver is spread out by the action of the despreader. The same is true of other wideband signals with different spreading codes.

3.13.2.1 DSSS Processing Gain

The output bandwidth of the receiver detector after collapsing the signal with the chip sequence is proportional to the data rate R_b, or its reciprocal, the data bit time T_b. Thus,

$$W_o = KR_b = \frac{K}{T_b} \tag{3.138}$$

for some constant K. The detector input bandwidth is also proportional to the spread bandwidth W_{ss} or its reciprocal, the chip time, T_c. Thus,

$$W_{ss} = \frac{K}{T_c} \tag{3.139}$$

The output power, R_0, is the same as the input power, R_i, when there is no gain in the detector, or proportional to it if there is gain. Thus,

$$R_o = R_i \tag{3.140}$$

If the detector exhibits gain, both the input noise as well as the input signal are increased the same amount, negating the effects on the SNR, and thus the processing gain. Therefore, the processing gain is

$$\begin{aligned} G_p &= \frac{SNR_o}{SNR_i} \\ &= \frac{R_o / (N_0 W_o)}{R_i / (N_0 W_{ss})} \\ &= \frac{R_o N_0 W_{ss}}{R_i N_0 W_o} = \frac{W_{ss}}{W_o} \\ &= \frac{T_b}{T_c} = N_c \end{aligned} \tag{3.141}$$

where N_c is the number of chips per bit.

3.13.3 Frequency-Hopping Spread Spectrum

In FHSS communication systems the narrowband information signal is modulated onto a carrier signal, and the frequency of the carrier signal is changed frequently. In addition to supplying a degree of LPI and LPE, FHSS systems enjoy the advantage of frequency diversity that helps to mitigate multipath, frequency-dependent fading.

As mentioned in Section 1.6.2, the frequency hopping technique is divided into fast frequency hopping and slow frequency hopping with the distinction based on the number of data bits sent per hop.

One of the simplest coding techniques for MFSK is to divide the energy in a symbol, E_s, into m equal-energy subsymbols, also called *chips*, and transmit these subsymbols at different frequencies that are hopped independently. This is called *time diversity transmission* or *repetition coding*. For fast frequency hopping, the hop rate is determined by L_F, where each of the L_F subsymbols represents a different hop and therefore a different hop frequency. For slow frequency hopping, the subsymbols are interspersed on the hop as well as on subsequent hops, as shown in Figure 3.51. In this example there are two symbols that have been divided into six subsymbols each. These subsymbols are transmitted as shown, both on the same hop and on subsequent hops. When these subsymbols are received, they are incoherently detected at the receiver and an estimate of the original data sequence is obtained.

The transmitter model for FHSS shown in Figure 3.52 does not depend on whether it is SFHSS or FFHSS. Differences do arise for more detailed models that incorporate implementation issues, however. The data sequence is used to select which tone in the MFSK modulator is to be sent as the current symbol. The selected tone is upconverted to the hop frequency by the tone from the synthesizer, the frequency of which is controlled by a pseudo-random number generator.

3.13.3.1 FHSS Processing Gain

The detector output power is the same as the input unless there is a gain or loss in the detector. Thus,

$$R_o = R_i \tag{3.142}$$

The output bandwidth after collapsing the signal by downconversion by the hop channel frequency and assuming a conversion efficiency where $R_b = W_b$ is

$$W_o = R_b \tag{3.143}$$

and the input bandwidth is W_{ss}. Therefore, the FHSS processing gain for both SFH and FFH is

$$\begin{aligned} G_p &= \frac{R_o / (N_0 R_b)}{R_i / (N_0 W_{ss})} \\ &= \frac{W_{ss}}{W_b} = N_c \end{aligned} \tag{3.144}$$

where N_c is the number of channels of width $W_o = R_b$ in W_{ss}.

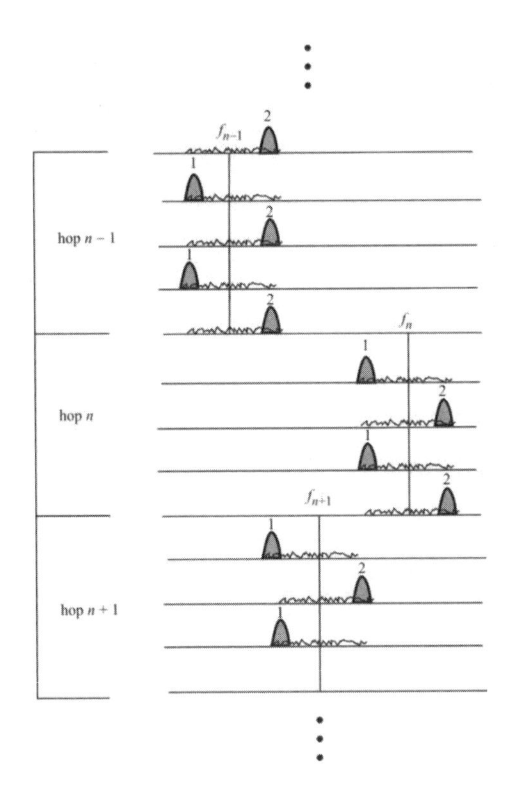

Figure 3.51 Time diversity for slow frequency hopping.

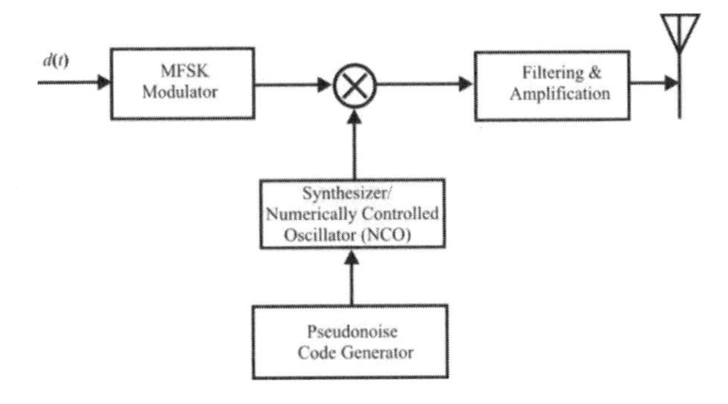

Figure 3.52 Frequency hopping transmitter. The same model applies to both FFHSS and SHF. The difference is whether there is more than one hop per bit (FFHSS) or more than one bit per hop (SFHSS).

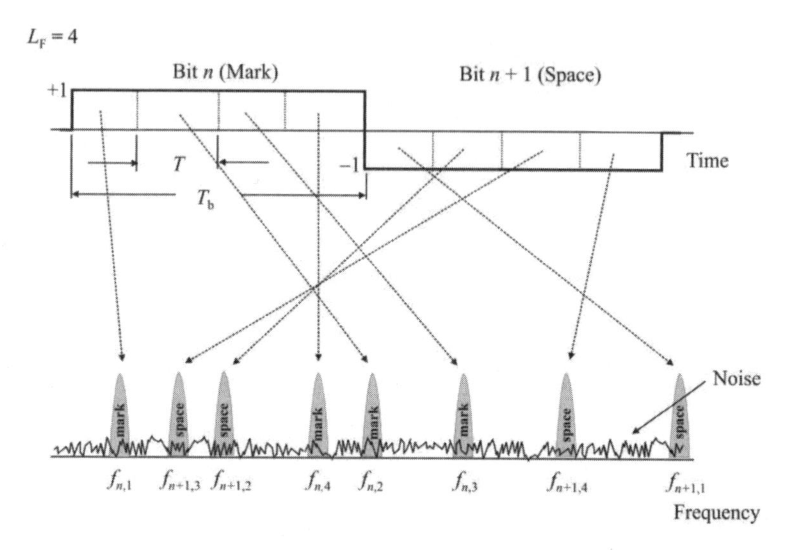

Figure 3.53 Channel structure for fast frequency hopping.

3.13.3.2 Fast Frequency Hopping

The channel structure for FFHSS is illustrated in Figure 3.53. Each data bit is transmitted at several (in this case, $L_F = 4$) frequencies. This has several advantages, as will be explained below. Its biggest disadvantage relative to SFHSS described later is implementation complexity. For effective voice communication, typically 16,000 bps are required (although this is getting smaller with modern source coding techniques). For $L_F = 4$, as in this example, and assuming some bits are required for system administration (say 1,000 bps), then the channel data rate would be about 68,000 bps. Depending on the modulation used, this could easily cause the required channel bandwidth, in many situations, to be too large for reasonable RF SNRs.

The detector structure for noncoherent BFSK FFHSS systems with hard decision decoding is shown in Figure 3.54. The pseudo-random code generator changes the frequency of the local frequency synthesizer in lock step with the one in the transmitter—they are assumed to be synchronized, although the modules to perform this synchronization are not shown in Figure 3.54. This local frequency moves the output of the mixer $r(t)$ to the appropriate channel center frequency. Through the detector, the processing is the same as in Figure 3.3. At the output of the detectors, an accumulator adds the detector outputs until L_F samples have been collected. A decision is then made as to whether a mark or space was sent on the previous L_F frequencies.

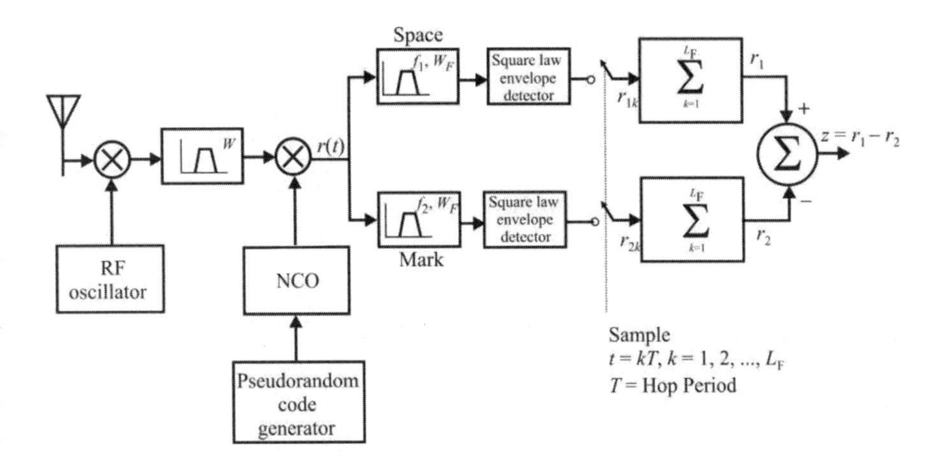

Figure 3.54 Receiver structure for noncoherent BFSK FFHSS systems.

FFHSS with Diversity Coding

One of the simplest forms of coding to improve performance of FFHSS systems against multitone jammers is to add frequency diversity. In this approach, each bit is transmitted on m frequencies in succession so that the chip duration is given by [23]

$$T_c = \frac{T_b}{m} \tag{3.145}$$

For BFSK the probability of a bit error is given by

$$P_b = \left(\frac{m}{E_b / N_T} \right)^m \tag{3.146}$$

where $N_T = N_0 + J_0$. This function is plotted in Figure 3.55 for several values of M. Also shown in Figure 3.55 is the BER for broadband noise, for comparison. Diversity coding is also referred to as repetition coding and time diversity.

For most of the coding schemes indicated, above $\upsilon \approx 10$ dB, the coding improves the error performance. Below this value, however, it is interesting to note that the uncoded performance is better than all the coding approaches. This is

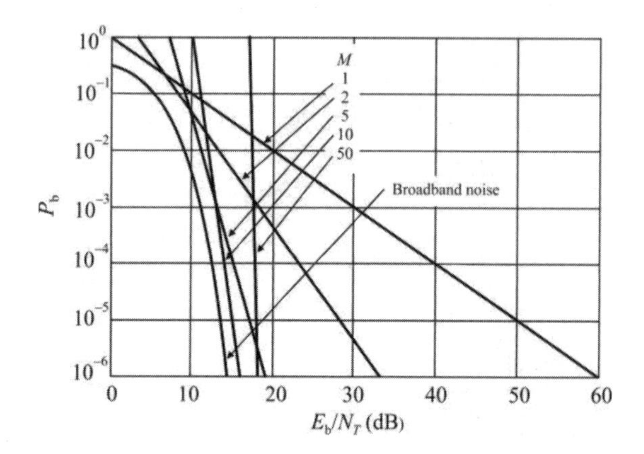

Figure 3.55 FFHSS performance with frequency diversity against multitone jammers. Variable M is the number of distinct frequencies at which each data bit is transmitted. (After: [23]. © McGraw-Hill 1994. Reprinted with permission.)

another manifestation of the error extension behavior of codes as discussed in Section 3.4.

3.13.3.3 Slow Frequency Hopping

The channel structure for SFHSS systems is shown in Figure 3.56. At any given instant, the signal is located at some channel and the data bits are represented by tones spaced around the center frequency of that channel.

BFSK is Illustrated in Figure 3.56, but in general, MFSK can be implemented [24].

The performance of a slow frequency hopping BFSK system at any given frequency is the same as the performance of a BFSK system that is not hopping. In this case, at each hop, there are several bits transmitted. One receiver detector structure is shown in Figure 3.57, which is an incoherent radiometer. This structure is similar to that for the noncoherent BFSK FFHSS detector except that the sums over samples from L_F frequencies at the output are missing. A decision is made at each bit time as to whether a mark or space was sent. Passing through the bandpass filters is noise and possibly a jamming signal. Passing through one of the filters is also the signal at that instant. The filter outputs are then detected with a square law device and then sampled forming the sampled signals r_{1k} and r_{2k}, $k = 1, 2, \ldots, L_S$ where L_S is the number of bits per hop. The mark sample is subtracted from the space sample forming the test statistic z. If z if less than zero, a mark is declared, and if z is greater than zero, a space is declared.

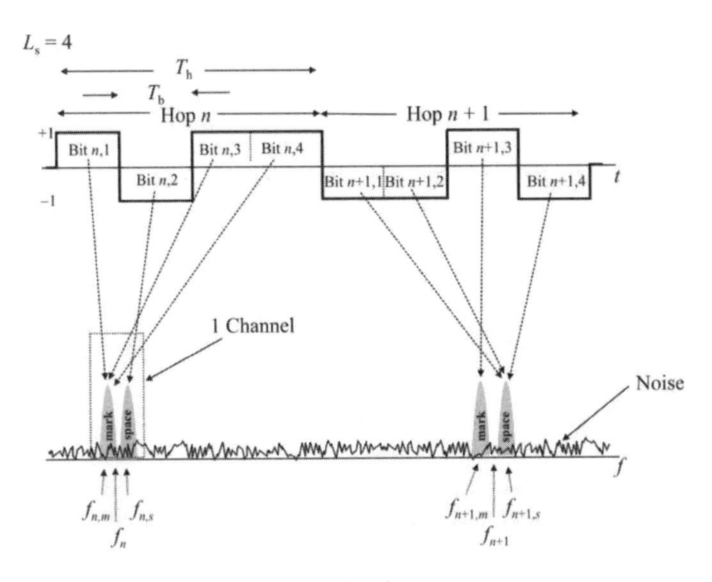

Figure 3.56 Channel structure for slow frequency hopping.

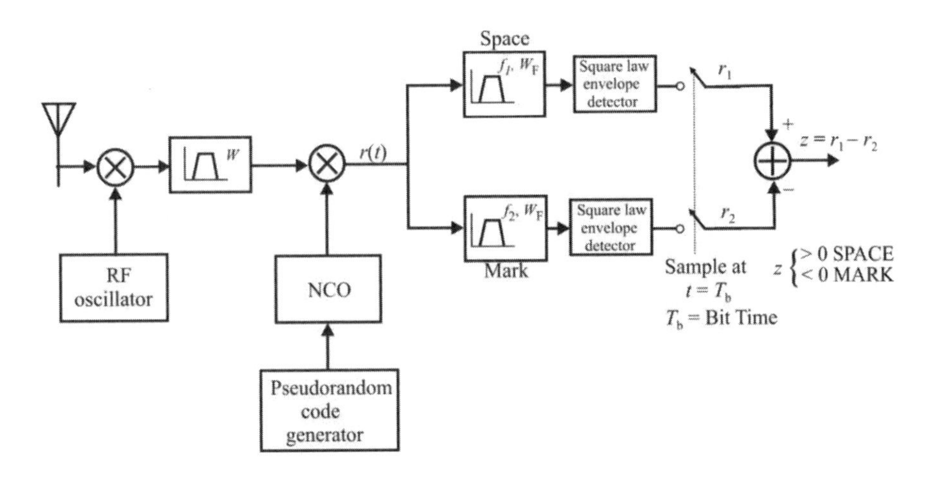

Figure 3.57 Receiver detector structure for noncoherent BFSK in an SFHSS system with hard decision decoding.

When noise is used as the jamming waveform, it usually means an RF carrier that is FM modulated with a noise waveform, but not always. When tones are used, they are placed in the spectrum so they pass through the space and mark bandpass filters in the target system. These filters are normally spaced at a fixed offset from the channel center frequency, but within the channel.

The RF signals received by the antenna are first converted to a fixed, lower frequency for subsequent processing. This signal is then filtered to a maximum bandwidth of W hertz. A pseudo-random number generator, used at both the transmitter and receiver and synchronized, is used to control the frequency of a second local oscillator/frequency synthesizer. At any given time, this oscillator is tuned to one of the carrier frequencies of the frequency hopping system. Depending on the data bit transmitted, either a mark or a space frequency is present, which are offset somewhat from the carrier frequency. These tones then pass through their respective filters to reduce noise and interference and are then (normally) noncoherently detected. While square-law detection is shown here, envelope detectors or matched filters could also be used. The outputs of these detectors are then sampled for each bit. A bit decision is made every T_b seconds. The bit decision (mark or space) is determined by whichever detector output is the largest. To avoid energy from the incorrect detector, the frequency separation between the two filters must be an integer multiple of the data rate R.

A detection error occurs when a space is transmitted and the jammer or noise causes the mark detection channel to detect more energy than the space channel, similarly for when a mark is transmitted and a space is detected.

3.13.4 DSSS and FHSS Hybrid Spread Spectrum

DSSS/FHSS systems combine the modulations of DSSS and FHSS to take advantage of the benefits of both. DSSS provides covertness and FHSS provides frequency diversity that helps the communicator when frequency selective fading is present, which it usually is. These systems can be coherent or incoherent. Coherent PSK modulation is frequently used for DSSS modulation. The phase of the signal is preserved at the receiver and is used to demodulate the signal. When the phase cannot be preserved, when there are many simultaneous users of the same frequency spectrum, for example, incoherent reception ensues. In general, coherent reception results in somewhat better communication performance.

It is difficult to maintain phase coherency when the transmitter makes large frequency changes. Therefore, incoherent operation is normal for the FHSS mode. Within a dwell, however, phase coherency can be maintained and thus coherent PSK can be used for the DSSS modulation.

DSSS/FHSS systems can be synchronous or asynchronous. Synchronous operation is when the hopping transitions occur at the same time for all nodes in operation sharing the spectrum in a region.

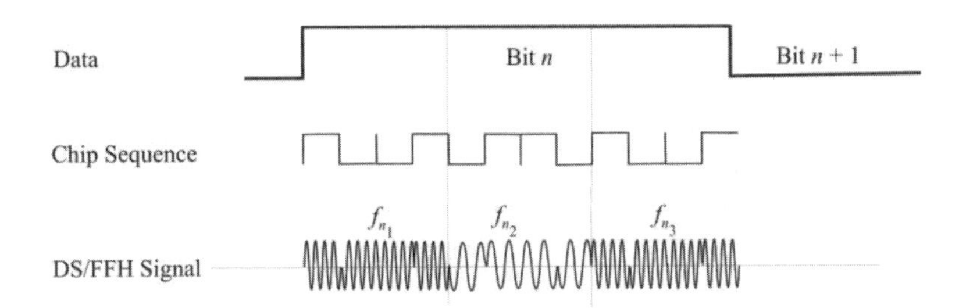

Figure 3.58 Hybrid DSSS/FSSS waveform.

In any frequency hopping system, when there are multiple networks sharing the spectrum, a transmission from one node can land on the same frequency as another node at the same time. This is called a "hit." The hopping codes are normally designed, when possible, to minimize these hits but their total avoidance is difficult. When such hits occur, the probability of a symbol error occurring increases and must be taken into consideration when designing such systems. Because all nodes are synchronous, for synchronous systems, a hit is a total hit in the sense that the entire transmission is overlapped. In asynchronous situations a hit can be either a total hit or a partial hit, in that only a portion of the dwell overlaps.

DSSS/FH SS systems can employ either fast or slow FHSS. Recall that for SFHSS, there is more than one data bit transmitted per dwell. In the case of DSSS/SFHSS systems, SFHSS means that there is more than one chip sent per dwell. For DSSS/FFHSS, a single data bit is sent at multiple dwell frequencies, but the data bit is first DSSS modulated to spread the signal. This is shown notionally in Figure 3.58 for $L_F = 3$ tones per bit and $N_d = 11$ chips per dwell.

The block diagram for a transmitter for DSSS/FHSS hybrid system is shown in Figure 3.59. The data sequence $d(t)$ is first DSSS modulated. This signal, $c(t)$, is then frequency hopped with the frequency synthesizer. The resultant signal is then bandpass filtered and amplified for transmission.

When DPSK is used as the DSSS modulation, the receiver structure is as shown in Figure 3.60. First, the synthesizer and mixer remove the FHSS component. The DPSK is then demodulated with the matched filters shown.

When BFSK is the DSSS modulation used, the receiver structure is as shown in Figure 3.61. The FHSS is removed as for DPSK. After that, the BFSK component is determined by the detection scheme shown. The signal is delayed by T_c, and then match-filter detected. For MFSK, the receive structure is as shown in Figure 3.61 with the additional filters added as necessary to account for the additional frequencies.

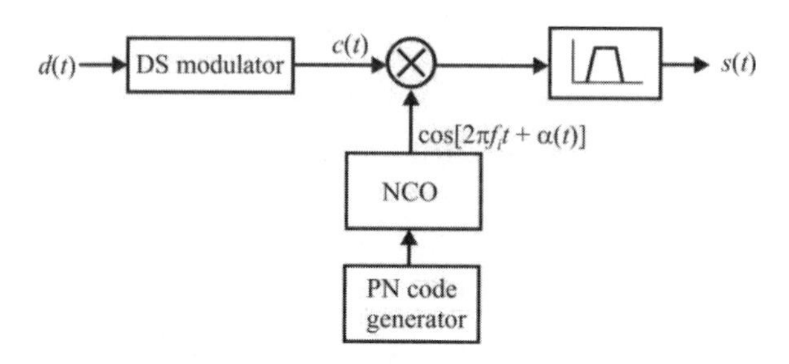

Figure 3.59 Transmitter for a DSSS/FHSS hybrid system.

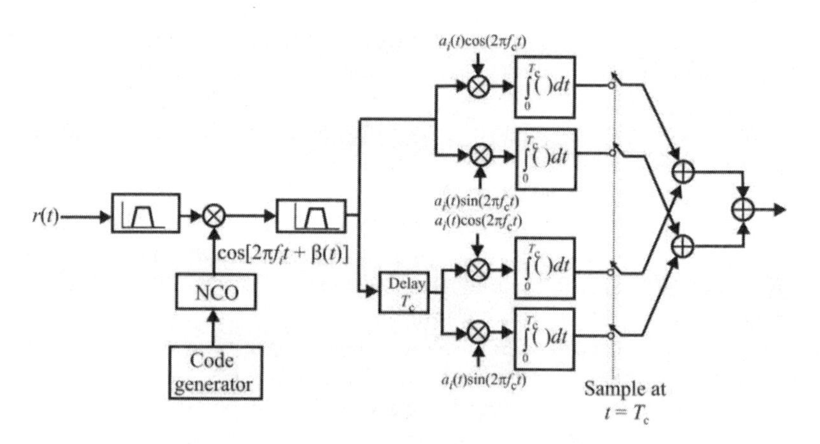

Figure 3.60 Hybrid receiver structure with SFHSS and DPSK.

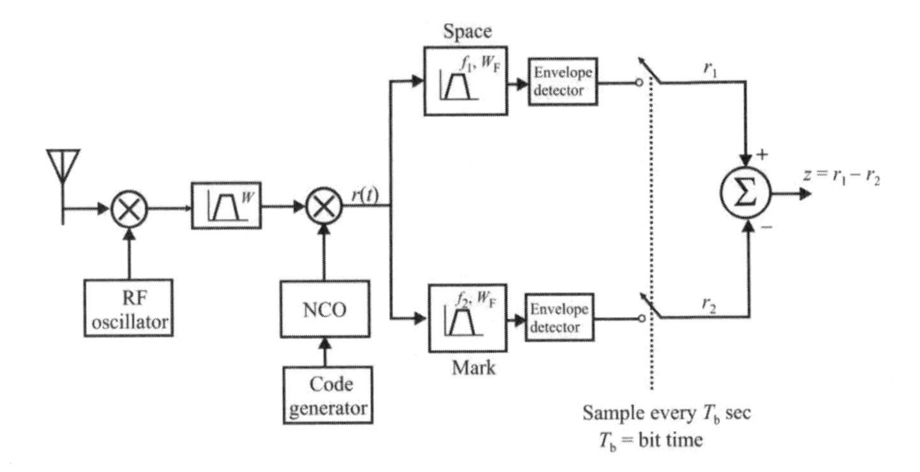

Figure 3.61 DSSS/FHSS receiver with FSK modulation.

3.13.5 Chaotic Shift Keying

A relatively new technique for digital communication systems is based on chaos theory. There have been several such techniques proposed [25–32], but a simple one is explained here, called the *chaotic shift key* (CSK), to get the basic ideas across [33]. The overall communication system using CSK is illustrated in Figure 3.62.

In CSK systems, the digital symbol to be transmitted is added to a chaotic signal prior to transmission. Communication systems that utilize CSK employ coherent or noncoherent reception. For coherent reception, straightforward CSK is used, while for noncoherent reception *differential CSK* (DCSK) is implemented.

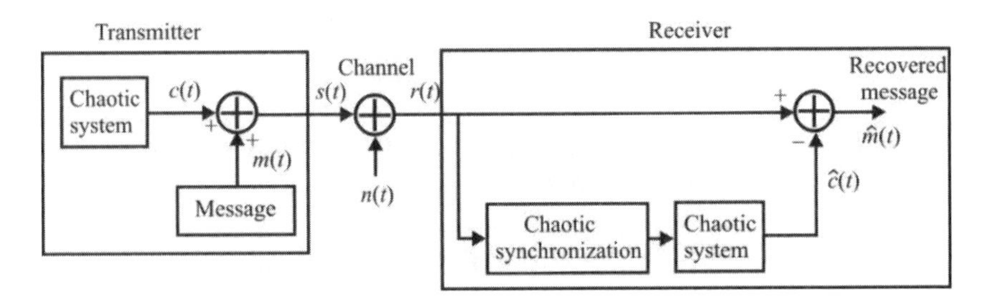

Figure 3.62 Chaotic communication system.

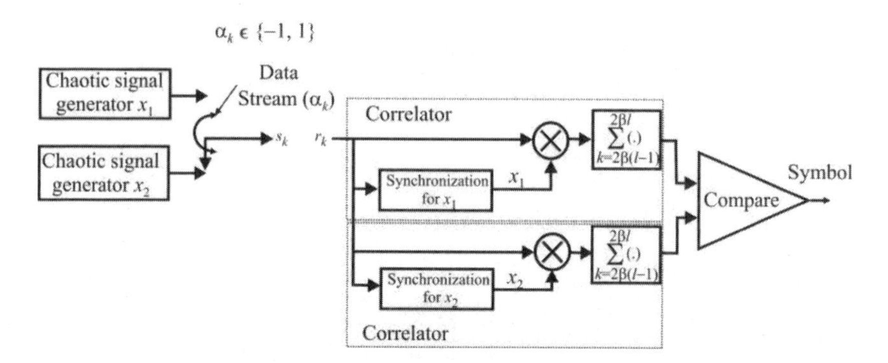

Figure 3.63 Coherent chaotic communication system employing chaotic shift keying.

The biggest advantage to CSK is its noise-like characteristics, similar to DSSS signals. These systems can be used to provide low-level encryption and data hiding due to the noise-like nature.

3.13.5.1 Coherent Reception

When coherent reception is employed, the receiver has information about the chaotic signal at the transmitter. The receiver must synchronize to the received signal using this information. In a binary system, for example, the transmitter would encode the data symbol into one of two chaotic signals. The receiver then correlates using synchronized replicas of the chaotic signals with the received signal and the channel with the largest response is selected as the symbol that was transmitted.

A block diagram of a binary coherent CSK system is illustrated in Figure 3.63. During time interval k, the data stream $\{\alpha_k\}$, where $\alpha_k \in \{-1, +1\}$, selects M samples $\{x_{i0}, x_{i1}, ..., x_{iM}\}$, where $i \in \{1, 2\}$, from one of two chaotic maps denoted by x_1 and x_2. For simplicity, assume that there is a single chaotic sample source (map) denoted by x with samples denoted by x_k. If x_1 is selected, then $s_k = x_k$, and if x_2 is selected, then $s_k = -x_k$. The spreading factor is the number of chaotic map samples in one bit interval and is defined as 2β. Thus, $2\beta = M$. Therefore, during data symbol interval l, the output of the transmitter is

$$s_k = \alpha_l x_k, \qquad k = 2(l-1)\beta+1, 2(l-1)\beta+2, \cdots, 2l\beta \tag{3.147}$$

Noise is added to the transmitted signal, so the received signal during time interval k is given by

$$r_k = s_k + n_k \tag{3.148}$$

where n_k is a sample of the noise, assuming AWGN ~ $\mathcal{N}(0,\sigma^2)$, so that $N_0 / 2$ is the two-sided noise PSD.

If it is further assumed that

- $\mathcal{E}\{x_k\} = 0$;
- $\{x_k\}$ has a vanishing autovariance function;
- $\{x_k^2\}$ has a vanishing autocovariance function;

then the BER for this system is given by [34]

$$P_e = \frac{1}{2}\operatorname{erfc}\left[\frac{1}{\sqrt{\left(\frac{E_b^2}{4\beta\Lambda}\right)^{-1} + \left(\frac{E_b}{N_0}\right)^{-1}}}\right] \tag{3.149}$$

where E_b is the energy per bit and $\Lambda = \operatorname{var}\left(x_k^2\right)$.

3.13.5.2 Noncoherent Reception

For noncoherent reception, DCSK is the technique used to send data bits. In this case, each time slot is divided in half. During the first half time period, a reference chaotic signal is sent. During the second half, the data bit is sent. Thus, the first half period is compared to the second half. If they are the same, then one symbol was sent; if they are different, then the other symbol was sent.

A block diagram of such a noncoherent system is shown in Figure 3.64. At the beginning of the bit period l of duration 2β, the transmitter switch is in the up position transmitting the reference chaotic sequence. At $t = \beta$, the switch changes to the lower position and the delayed sequence, multiplied by the value of the data bit $\alpha_l \in \{-1, +1\}$, is transmitted. At the receiver the incoming sequence is delayed the same amount β. The delayed sequence from the first half bit time is then multiplied and accumulated with the sequence in the second half and the result of the accumulation is compared with a threshold. If the sum is below the threshold, then one symbol is declared, and if above, the other symbol is declared.

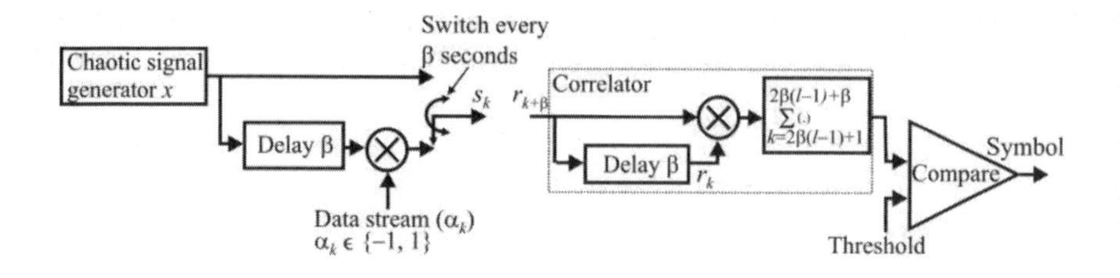

Figure 3.64 Chaotic noncoherent communication system.

Under the same assumptions leading to (3.149), the BER for this system is given by [34]

$$P_e = \text{erfc}\left\{\left[\left(\frac{E_b^2}{8\beta\Lambda}\right)^{-1} + 4\left(\frac{E_b}{N_0}\right)^{-1} + 2\beta\left(\frac{E_b}{N_0}\right)^{-1}\right]^{-1/2}\right\} \tag{3.150}$$

3.13.6 Time-Hopping Spread Spectrum

3.13.6.1 Introduction

In this section we examine THSS communication systems. Probably the only such technology available for this purpose at the time of this writing is *ultrawideband* (UWB) signals, so that is the technology we will focus on here. We must bear in mind that UWB is not the only technology for implementing THSS, however.

THSS systems transmit short pulses at times that are selected by a pseudo-random sequence. The optimum detector of signals in AWGN channels is the radiometer, which measures the energy content at a point in the RF spectrum. The longer the channel is integrated, the better the detection performance. If a short pulse is transmitted followed by a long period of silence, then much more noise is integrated than signal, and the detection performance deteriorates accordingly. This is the fundamental principle behind THSS that provides some of the LPI functionality. Another is the pseudo-random placement in time of the data pulses. This provides further LPI/LPE capabilities. Lastly, the very narrowband pulses transmitted by these systems spread the signal energy across a very wide bandwidth, making the PSD at any one channel very small—below the noise. This makes it very difficult to detect the presence of the pulse, much as in DSSS.

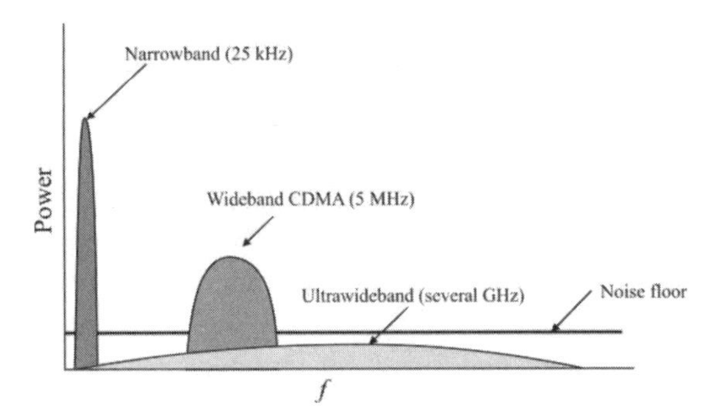

Figure 3.65 Frequency spectrum comparison of NB, WB, and UWB.

3.13.6.2 Ultrawideband Systems

UWB communication technology is relatively new, and in the United States, it has been approved for use only under extremely limited conditions. Because of its potential for interference with so many existing communication services, those conditions limit its use to *personal area networks* (PANs). Such networks are for communication between devices either a person is wearing or are otherwise within very close proximity. It could be argued that such applications would probably be of limited use as EW targets. However, this is not to say that UWB communications will forever be limited to this application—a very unlikely circumstance. Therefore, we include this examination of the EW performance against such systems, especially for urban environments.

A notional comparison of narrowband (NB), wideband (WB), and UWB signals is illustrated in Figure 3.65. NB signals, typically 25 kHz wide, are normally fairly powerful compared to the noise floor, to be effective at communication. WB signals, as typified by 3G PCS CDMA signals, are typically several megahertz in width, and utilize notably less power than the NB counterparts. UWB signals can occupy several gigahertz, and their power is so low it is typically below the noise floor.

A signal is classified as UWB if its bandwidth is 500 MHz or greater or 20% or more of its center frequency. UWB communication is also called baseband and carrier-free transmission.

UWB communication systems are a relatively new concept that exhibit some very useful characteristics, including:

- Extremely short-duration pulses that result in ultrawide bandwidth signals hundreds to thousands of megahertz);
- Extremely low PSD—they qualify as DSSS signals;
- Center frequencies typically between 650 MHz and 5 GHz;
- Multimile ranges with submicrowave average power levels;
- Excellent immunity to jamming, intentional or otherwise;
- Exceptional multipath tolerance (most of the time);
- Are relatively simple and are less costly to build than radios built using other SS technologies;
- Consume substantially less power than conventional radios.

Due to their low PSD, *impulse radios* (another appellation for ultra wideband radios) can qualify for type acceptance in the United States under the rules of Part 15 for unlicensed applications and can share spectrum without affecting conventional radio transmissions. This means that any operator anywhere in the United States can implement her or his own cell phone system without getting the approval of the FCC or any other government agency.

The spectrum of an impulse function is a constant value and an impulse function can be approximated in practice. It sometimes consists of a few cycles of a carrier signal above 1 GHz, but often is a derivative of a Gaussian function, which is easy to generate.

THSS is employed with UWB by randomly varying the pulse times. This TH tries to foil the attempts of a receiver to integrate the signal for reliable detection. As with other forms of SS technology, such THSS would be controlled by a PN sequence. Processing gain is achieved by coherently adding more than one impulse per data bit, denoted by L, and the processing gain is given by $G_L = 10 \log_{10} L$. Additional processing gain is achieved by the low duty cycle of the pulses relative to the frame time, T_f. This processing gain is then $G_T = 10 \log_{10}(T_f / T_p)$. The total processing gain is then the sum of these. Thus, if $L = 4$, $T_p = 1$ ns, and $T_f = 1$ μs (corresponding to an uncoded data rate of 1 Mbps), the $G_T = 30$ dB, $G_L = 6$ dB, and $G_p = 36$ dB.

By using THSS, multiple users can share the same spectrum, by the same rationale, employed for DSSS and FHSS. The coding that controls the impulse time minimizes the collisions of the pulses. In this example, if perfect coding were employed, 1,000 users could share the same spectrum.

UWB communications is an LPD/LPI technique. The energy in the spectrum is spread across such wide bandwidth and the peak power is low, so the spectrum looks noise-like, much as a DSSS signal looks noise-like. UWB does not achieve such wide bandwidths by applying coding, as in DSSS and FHSS, but by taking advantage of the very narrow pulse width.

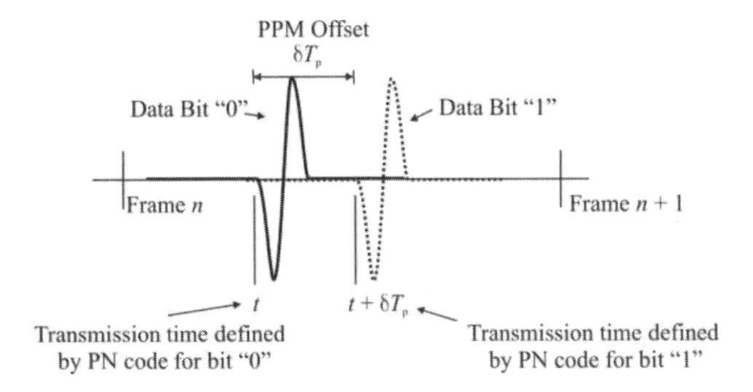

Figure 3.66 PPM modulation with time hopping for UWB communications.

This form of SS communications achieves AJ capabilities by gating the time at which the receiver looks for the impulse, and the gate has a width of approximately the width of the impulse—that is, 1 ns or so. This precludes jamming energy from having any impact on the receiver except during that gate. Time hopping is introduced via pseudo-random coding so that the timing of the next frame varies randomly. The PN code is known only to the transmitter and receiver.

3.13.6.3 Modulation Formats

PPM, OOK, and PAM are the principal modulation types used for UWB communication systems. Channel coding can be applied as well, to improve system performance. Each information bit is represented by one or more impulses. While the impulse is there, a wide spectrum signal emerges that has very low PSD. The time between pulses is typically long enough for the pulse response of the channel to die out so there is little ISI. Also, during this time there is no signal in the spectrum.

PPM modulation with THSS is illustrated in Figure 3.66. A frame is the timeframe within which a time hopped signal occurs. If a zero is sent, the leftmost impulse is transmitted, while if a 1 is transmitted, the impulse is delayed for a duration δT_p where T_p is the impulse width.

In binary PAM (BPAM), an impulse is sent to represent the binary 1 and its negative is sent to represent the binary zero. In OOK, the pulse is sent to represent the binary one, and no pulse is sent to represent the binary zero. BPAM enjoys a 6 dB processing advantage over BOOK.

3.13.6.4 UWB Pulse Position Modulation

There are several ways UWB communication systems can be implemented. Probably the most popular modulation scheme is PPM, but other methods of

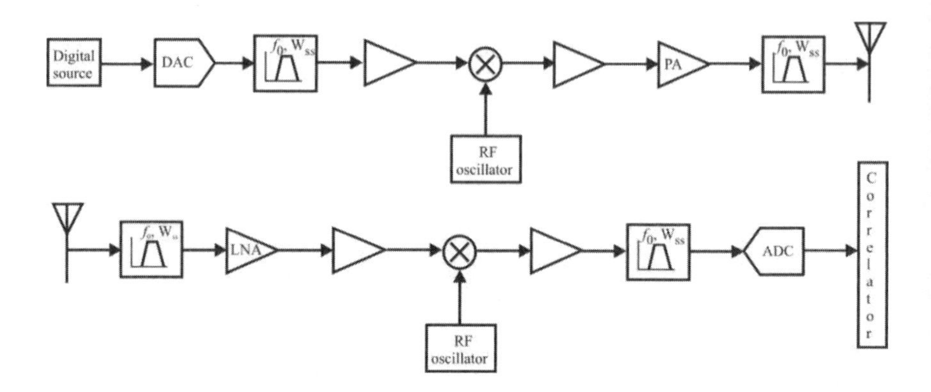

Figure 3.67 Traditional IF sampling transceiver architecture.

impressing the information stream onto the carrier, such as OFDM and BPSK, are certainly possible.

In UWB PPM communications, information is conveyed with narrow pulses (monocycles), so the RF and analog circuitry is reduced to a wideband low-noise amplifier, a correlator and data converters. No up/downconversion or mixer are needed, which results in substantial reduction of transceiver size and power consumption. A comparison of a direct IF sampling transceiver and a UWB transceiver architecture is shown is Figures 3.67 and 3.68. It indicates that a UWB transceiver can be implemented with considerably fewer components than the IF sampling receiver and can be implemented as a simple chipset with very few off-chip components.

In PPM, typically ultrashort "Gaussian" monocycle pulses with carefully controlled pulse-to-pulse intervals are transmitted. A Gaussian monocycle is the first derivative of the Gaussian pulse shown in Figure 3.69 [34]. The transmitter sends the appropriately delayed Gaussian pulse to the antenna. The function of the transmitter antenna is to formulate the first derivative of the Gaussian pulse yielding the Gaussian monopulse as the actual waveform transmitted shown in

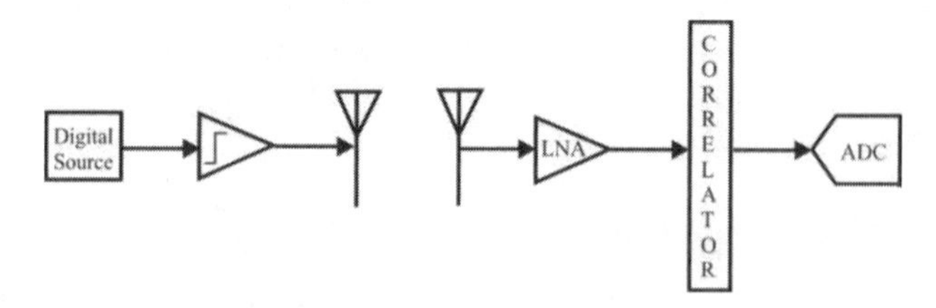

Figure 3.68 UWB transceiver architecture is much simpler than a superheterodyne receiver.

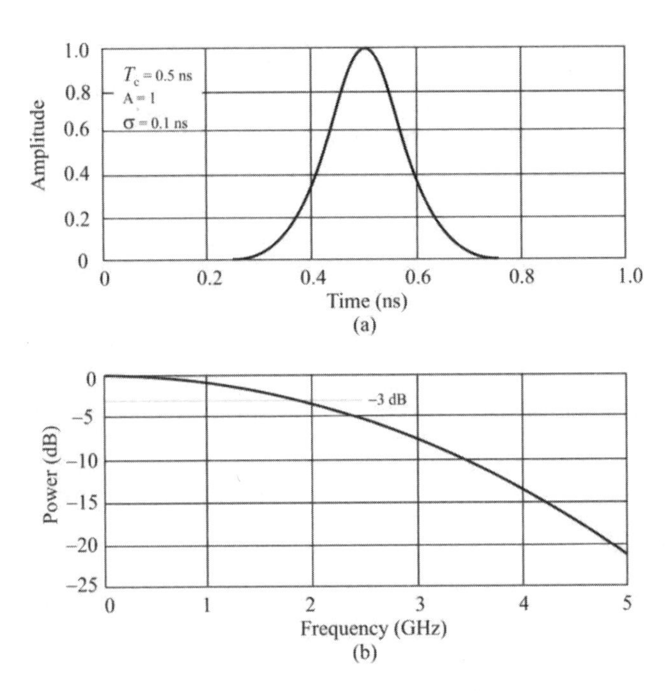

Figure 3.69 (a) Waveform and (b) PSD of Gaussian pulse. (Source: [34]. © 2002 IEEE. Reprinted with permission.)

Figure 3.70. At the receive antenna, the function of the antenna is also to find the first derivative of the received signal, and that is what is sent to the receiver.

The widths of these pulses is typically in the 200 ps to 1.5 ns range, while the time between data pulses is on the order of 100 ns to 1 μs. With these pulse rates, data rates on the order of 1 Mbps to 10 Mbps are possible. These short monocycle pulses are inherently wideband. For example, a pulse width of 200 ps corresponds to a bandwidth of 5 GHz. Typical time-domain and the corresponding frequency

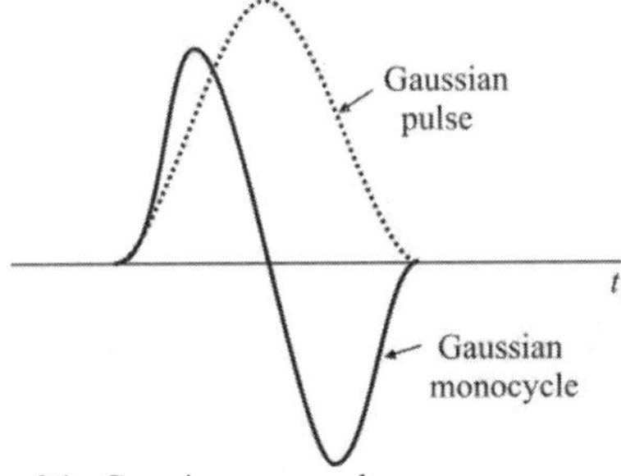

Figure 3.70 Time waveform of the Gaussian monocycle.

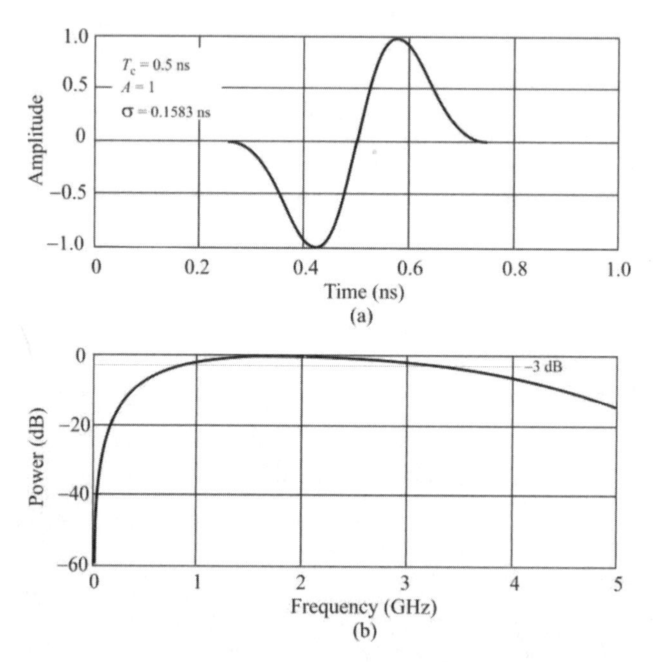

Figure 3.71 (a) Waveform and (b) PSD of Gaussian monocycle. (Source: [34]. © 2002 IEEE. Reprinted with permission.)

domain performances are illustrated in Figure 3.71.

We will focus on PPM here as the type of modulation that appears to be the most flexible and exhibits most of the desirable characteristics of UWB systems. The information is conveyed by the position of the pulse, as illustrated in Figure 3.66. In this case, if a 0 is to be received then the pulse will occur at the receiver at time t. If a 1 is to be received, the pulse occurs at $t + \delta T_{\mathrm{p}}$. As will be explained below, δ is a fraction between 0 and 1, and T_{p} is the pulse width.

The UWB signaling emitted by the kth transmitter using encoded PPM can be expressed as [35]

$$s_{\mathrm{tr}}^{(k)}(t^{(k)}) = \sum_j w_{\mathrm{tr}}(t^{(k)} - jT_{\mathrm{f}} - c_j^{(k)} T_{\mathrm{c}} - \delta d_{\lfloor j/N_{\mathrm{s}} \rfloor}^{(k)}) \tag{3.151}$$

where $w_{\mathrm{tr}}(t)$ is the transmitted monocycle waveform, T_{f} is the pulse repetition time, which is generally much larger than the monocycle width, and $\lfloor x \rfloor$ means the integer part of x. $\delta d_{\lfloor j/N_{\mathrm{s}} \rfloor}^{(k)}$ represents the data modulation. Each information bit from the binary sequence, $d^{(k)}_{\lfloor j/N_s \rfloor}$, is encoded in the pulse train by delaying N_s monopulses by an additional amount, which can be written as

$$\text{delay} = \begin{cases} 0 & \text{if } d^{(k)}_{\lfloor j/N_s \rfloor} = 0 \\ \delta & \text{if } d^{(k)}_{\lfloor j/N_s \rfloor} = 1 \end{cases} \tag{3.152}$$

In the binary case, detection of these transmitted bits is accomplished by correlating the received signal with a template signal for a single bit duration. The received signal, $r(t)$, is given by [36]

$$r(t) = \sum_{j=1}^{N_u} A_j s^{(j)}(t - \tau_j) + n(t) \tag{3.153}$$

where N_u represents the number of users in the channel, A_j is the gain (propagation loss included) of the jth user, τ_j is the random time variable representing the asynchronous relationship between user j and the desired signal, and $n(t)$ represents AWGN. For simplicity we assume that $A_j = A$ for all j. For the binary receiver, the template waveform used in the correlator for the qth bit, $v(t)$, is formed by the difference between two waveforms,

$$v_{\text{bit}}(t) = p_{\text{bit}}(t) - p_{\text{bit}}(t - \delta) \tag{3.154}$$

where $p_{\text{bit}}(t)$ is given by

$$p_{\text{bit}}(t) = \sum_{n=qN_{si}}^{(q+1)N_{si}-1} p\left(t - nT_f - c_n^{(i)} T_c + \varepsilon_n\right) \tag{3.155}$$

where ε_n is a zero mean normal random variable representing jitter error in the timing estimate.

We assume that the receiver is selecting the ith desired transmitter and that $\tau_i = 0$ for this case. As was the case for the transmitter, the receiver clock for p_{bit} is modified for timing error with the independent random variable ε_n. Again this error is modeled as a zero mean normally distributed random variable. The binary bit decision for the qth data bit made at the correlator output is given by

$$d_q^{(i)} = \begin{cases} 0 & \text{if } \int_{t\in\{\Im_q\}} r(t)v_{\text{bit}}(t)dr > 0 \\ 1 & \text{if } \int_{t\in\{\Im_q\}} r(t)v_{\text{bit}}(t)dr \le 0 \end{cases} \quad (3.156)$$

where $\{\Im_q\}$ is the set of disjoint time intervals corresponding to the qth bit.

Information bits are normally transmitted with several (many) pulses, so that a system with a PPM pulse rate of 10 Mbps (10 Mcps, T_f = 100 ns) and an information exchange rate of 10 kbps transmits N_s = 1,000 bits per source symbol.

As in CDMA DSSS, each link is assigned a distinct, pseudo-random code. In this case, however, distinct from CDMA, the code is used to implement time hopping. The code $c_j^{(k)}$ and $c_j^{(k)}T_c$ defines a pseudo-random additional time shift to the jth monocycle, where T_c is the chip duration.

At the receiver, when N_u links are active, then the received signal can be expressed as [36]

$$r(t) = \sum_{n=1}^{N_u} A_k r_{\text{rec}}^{(n)}(t-\tau_n) + n(t) \quad (3.157)$$

Here A_n models the attenuation of the nth transmitter signal path and τ_n represents time asynchronism between the clocks of the nth transmitter and the receiver. $r_{\text{rec}}^{(n)}(t-\tau_n)$ represents the nth signal received and $n(t)$ represents AWGN.

In AWGN, the optimum receiver for a single bit of a binary PPM signal is a correlator that implements the decision rule (assuming without loss of generality that signal 1 is the signal of interest) [see Chapter 4]

$$d_0^{(1)} = \begin{cases} 0, & \text{when} \sum_{j=0}^{N_s-1} \int_{\tau_1+jT_f}^{\tau_1+(j+1)T_f} r(t)v(t-\tau_1 - jT_f - c_j^{(1)}T_c)dt > 0 \\ 1, & \text{otherwise} \end{cases} \quad (3.158)$$

where $v(t) = w_{\text{rec}}(t) - w_{\text{rec}}(t-\delta)$ is the correlation template (3.154). In a multiuser environment, it is reasonable to approximate the combined effect of the other users' dehopped interfering signals as a Gaussian random process, and thus the same reception algorithm can be used [35].

The PSD of a UWB PPM signal is illustrated in Figure 3.72. We can see that the envelope of the PSD follows (3.165), but the actual PSD consists of impulse functions at the multiples of the symbol rate. These impulse functions occur at the

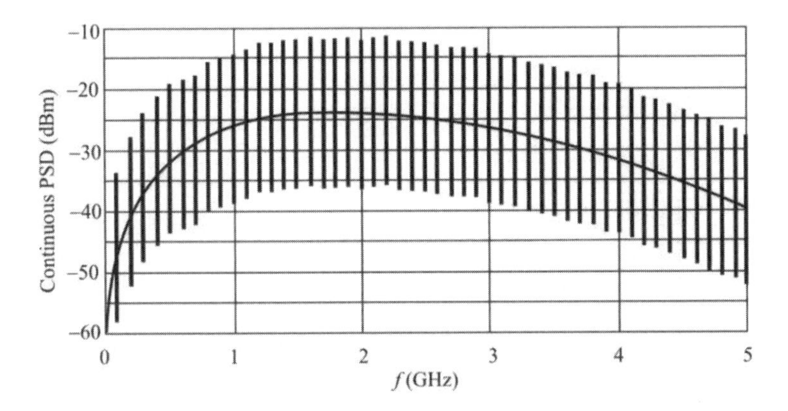

Figure 3.72 Continuous PSD of the Gaussian monocycle for a total UWB signal power of 10 dBm. The pulse duration is 0.5 ns, $M = 10$, $\varepsilon_c = 10$ ns, $R_d = 10$ Mbps.

harmonics of the data rate. The impulses can be minimized by dithering the data signal [37].

3.13.6.5 Regulatory Emission Limits

Since UWB technology is, by definition, wideband, it has the potential to interfere with many other RF signals if the power is high enough. Therefore, in 2002 in the United States, the FCC developed a power level mask that defines, depending on the frequency, how powerful a UWB single can be. That mask is shown graphically in Figure 3.73 along with several previously approved services. It is these services that dictated the structure of the mask. Because the signal must traverse at least one wall to propagate from inside a building to outside, there is a different mask depending on whether the UWB system is operating inside or outside a building. These two masks are illustrated in Figure 3.73, along with a similar mask adopted by CEPT in Europe.

3.13.6.6 Monocycle Shapes for UWB Communications

Unfortunately, the Gaussian monopulse does not satisfy the FCC limit. This is illustrated in Figure 3.74, where the spectrum of the monopulse is compared with the outdoor limit. Therefore, other possible monocycle shapes were investigated.

UWB is a baseband technology, and thus the choice of the monocycle shape will affect the performance. Several possible monocycles (duration $T_p = 0.5$ ns for specificity) for UWB are discussed in this section.

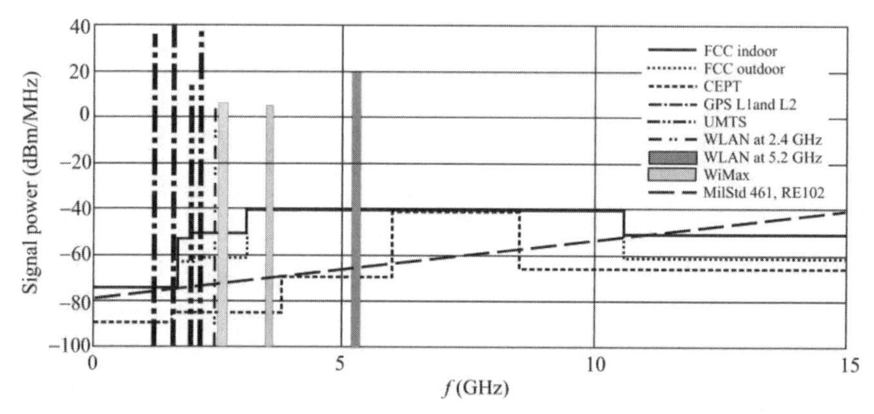

Figure 3.73 U.S. and European UWB emission masks and services.

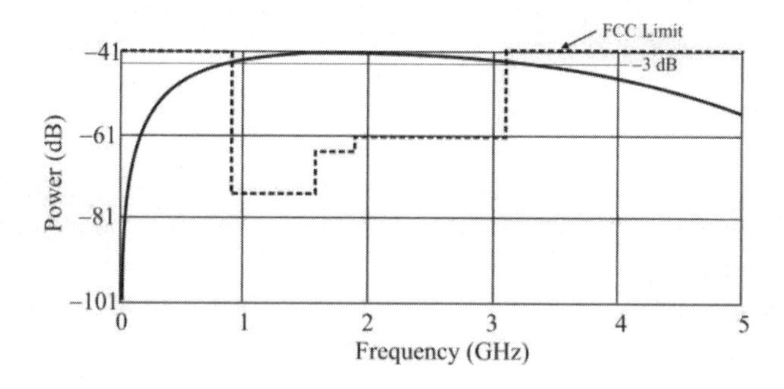

Figure 3.74 PSD of the Gaussian monocycle with limit (outdoor).

Gaussian Pulse and Monopulse

The impulse waveform used for analysis of UWB communication systems is frequently the Gaussian monocycle shown in Figure 3.71. This monocycle is the first derivative of the Gaussian pulse, where the Gaussian pulse is given by

$$w(t) = g(t) = \frac{A}{\sqrt{2\pi\sigma^2}} e^{-\frac{1}{2}\left(\frac{t-\bar{t}}{\sigma}\right)^2} \tag{3.159}$$

where A is the pulse amplitude, $\bar{t}$ is the mean, and σ^2 the variance. σ is referred to as the *pulse shape parameter* given by

$$\sigma = \frac{\sqrt{\ln 2}}{\sqrt{2}W_{-3\text{dB}}} = \frac{0.5887}{W_{-3\text{dB}}} \tag{3.160}$$

where $W_{-3\text{dB}}$ is the –3dB bandwidth of the Gaussian monocycle. It is the pulse half duration at the $1 / e$ points. The waveform and PSD of the Gaussian pulse are shown in Figure 3.70. The PSD of the Gaussian pulse is given by

$$G(\omega) = Ae^{-\frac{\omega^2}{2(1/\sigma)^2}} \tag{3.161}$$

The pulse width T_p is related to the variance as

$$T_\text{p} = 2\pi\sigma \tag{3.162}$$

The nominal center frequency and bandwidth of this monocycle depend on the pulse width. Bandwidth is approximately 116% of the center frequency, so the UWB criteria on bandwidth is clearly met. The spectrum of the time hopped UWB signal is asymmetrical. As seen, the monocycle has a single zero crossing. If additional derivatives of the Gaussian function are used, the relative bandwidth decreases and the center frequency increases for fixed time delay.

The transmit antenna for such wideband signals serves as a highpass filter and as such essentially differentiates the transmitted impulses [37, 38]. The number of differentiations depends on the type of antenna. If the Gaussian monopulse is used as the transmit impulse function, then the second derivative of the Gaussian function is what is received. If the third derivative is used as the impulse function, then the received impulse is the fourth derivative, and so on.

If the transmitter produces the Gaussian pulse and sends it to the antenna, the antenna effectively computes the first derivative of the pulse, and transmits that signal. The Gaussian monocycle is given by the first derivative of the Gaussian pulse and is

$$w(t) = g'(t) = -A\frac{t - T_c}{\sqrt{2\pi\sigma^3}} e^{-\frac{(t-T_c)^2}{2\sigma^2}} \tag{3.163}$$

The kth order derivative of the Gaussian pulse is given by [39]

$$w^{(k)}(t) = -\frac{n-1}{\sigma^2} w^{(k-2)}(t) - \frac{t}{\sigma^2} w^{(k-1)}(t) \tag{3.164}$$

The corresponding kth derivative normalized PSD is given by [39]

$$|P_k(\omega)| \triangleq \frac{|X_k(\omega)|^2}{|X_k(\omega_M)|^2} = A_{\max}(\omega\sigma)^{2k} \frac{e^{-(\omega\sigma)^2}}{k^k e^{-k}} \tag{3.165}$$

Note that there is a phase factor in $P_k(\omega)$ due to the delay of duration T_c in $w(t)$ that does not show up in $|P_k(\omega)|$. Also note that T_c does not show up at all in $|P_k(\omega)|$, so charts that illustrate the PSD later are invariant to it. The PSD is a function of T_p however, because σ is a function of it.

In theory, the Gaussian pulse and all of its derivatives are infinite in duration ($T_p \to \infty$). However, current dogma defines the pulse width, T_p, to be the duration of the pulse that contains 99.99% of the energy in the pulse. In that case, $T_p \sim 7\sigma$ in the first derivative.

For a constant energy signal at a UWB receiver, as the width of the pulse decreases, the peak power must increase because

$$E_b = P_{peak} T_p \tag{3.166}$$

where E_b is the energy per bit and P_{peak} is the peak power of the UWB signal. This peak power occurs over a very short period of time, however, so the time-averaged power is quite low. The spectrum is more or less constant over a considerable range. Typical frequencies are 1.5 GHz with greater than 25% fractional bandwidth, as previously mentioned.

For realistic values of the parameters, the communication range of UWB technologies is relatively small, especially for ground-to-ground based applications (multiple-mile ranges have been achieved). Signals above 1 GHz require antennas with considerable gain to extend beyond a few kilometers due to the R^{-4} propagation characteristics close to the earth and the inverse dependency of received power on frequency. Considerable data rates are possible, making such technology potentially applicable to ad hoc, multinode communication networks such as those for sensor field communications, for example, or video transmission over short ranges. However, air-to-ground or air-to-air communications would enjoy much better range performance where the propagation is proportional to R^{-2}.

The relationship between the pulse's center frequency and its duration is

$$T_{\text{au}} = \frac{1}{\pi f_{\text{c}}} \tag{3.167}$$

where T_{au} represents the time between the maximum and minimum amplitudes. The half power bandwidth is 116% of the center frequency and since f_{c} defines the pulse width, then the pulse width specifies both the center frequency and bandwidth. In practice, the center frequency of a pulse is approximately the reciprocal of the pulse's length and the bandwidth is approximately equal to the center frequency. Thus, for the 0.5 ns pulse shown in Figure 3.71, the center frequency and the half power bandwidth are approximately 2 GHz.

The 3 dB bandwidth of the UWB signal is given by

$$W = 2\frac{f_{\text{H}} - f_{\text{L}}}{f_{\text{H}} + f_{\text{L}}} \tag{3.168}$$

The time bandwidth product of a signal is on the order of unity, that is,

$$T_{\text{p}} W \approx 1 \tag{3.169}$$

where T_{p} is the pulse width and W is its bandwidth. Thus, when $f = 2$ GHz, a 25% fractional bandwidth is 500 MHz, and $T_{\text{p}} \approx 0.5$ ns. The spectra of a Gaussian monopulse of a few typical durations are shown in Figure 3.75.

The First Seven Derivatives. The time waveforms of the first seven derivatives of the Gaussian pulse are illustrated in Figure 3.76, while the corresponding PSDs are shown in Figure 3.77 along with the FCC mask. As is evident, the seventh derivative is the minimum order derivative that satisfies the outdoor mask while the fifth derivative satisfies the indoor mask. It is also evident that as the order

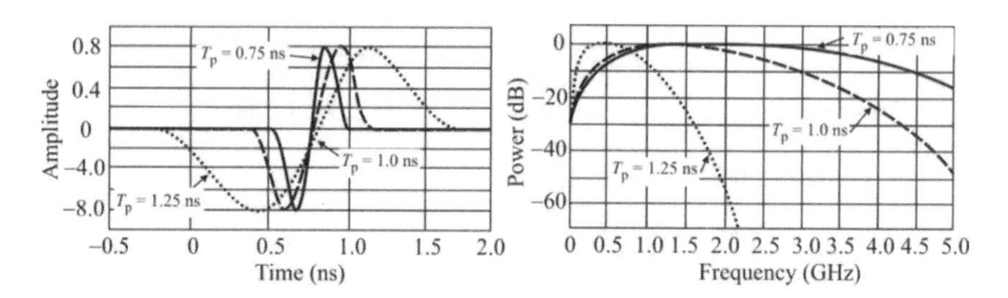

Figure 3.75 Typical Gaussian monopulses and associated spectrums. The shorter the pulse, the wider the frequency spectrum occupied.

increases, the bandwidth is reduced somewhat. In addition, as the order increases, the number of zero crossings increases by one for each increase in order. This implies that the frequency is increasing, which is evident from Figure 3.77.

The value of σ varies according to the derivative order as shown in Table 3.8. This table also lists the important frequency information of the first seven derivatives.

Other Possible Waveforms

The Gaussian pulse and Gaussian monocycle are arguably the most prolific waveforms used in UWB PPM communication systems. Some others are also useful, however, and a few are summarized in Table 3.9 [34]. Quickly plotting the FCC limits on the PSDs listed in Table 3.9 shows that there are several regions for each of the PSDs that exceed the limits.

Waveforms other than the Gaussian monocycle are also possible candidates for UWB communication systems, especially if the FCC mask is not a factor. In this section we present a few and compare their BER performance with simulations. The waveforms along with their PSDs are illustrated in Table 3.9.

Scholtz Monocycle. The Scholtz monocycle is very similar to the second derivative of the Gaussian monocycle so its performance would be expected to be similar.

Manchester Monocycle. The Manchester monocycle has amplitude A during the first half of the monocycle width and has amplitude $-A$ during the latter half.

RZ-Manchester Monocycle. The return-to-zero Manchester monocycle has amplitude A and $-A$ for only a portion of each half monocycle width.

Sine Monocycle. The sine monocycle is simply just one period of a sine wave.

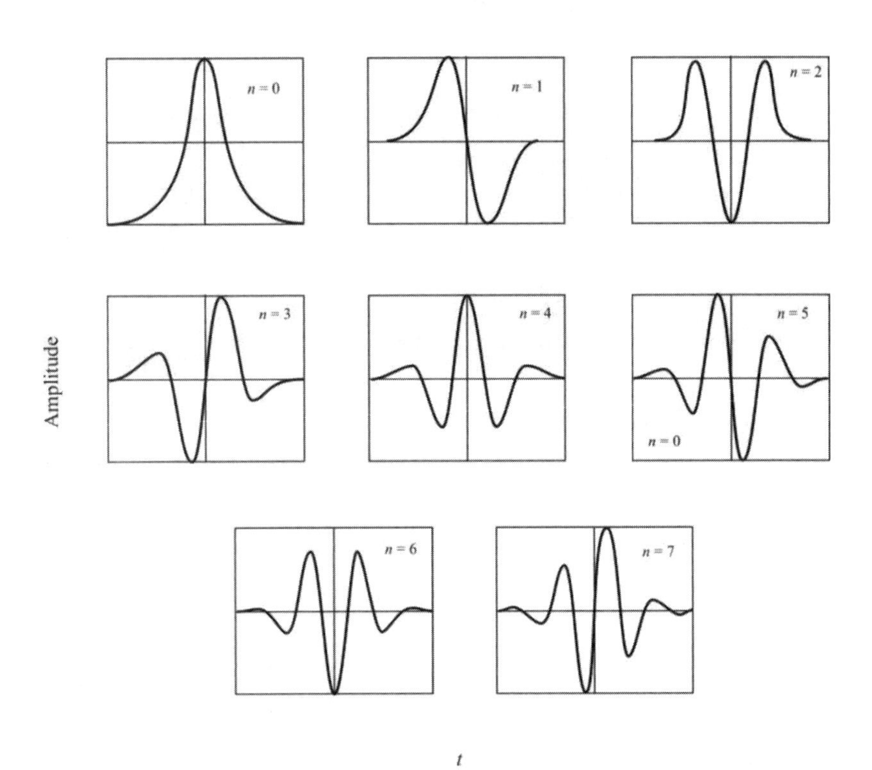

Figure 3.76 UWB Gaussian time pulse derivatives.

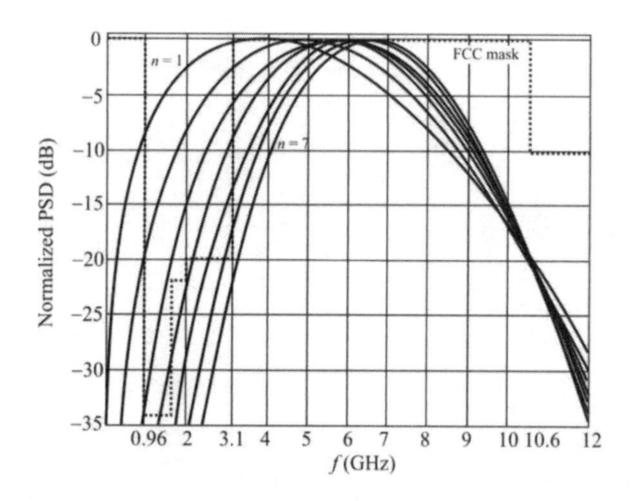

Figure 3.77 Spectrums for UWB Gaussian pulse derivatives. FCC mask is for outside. 0 dB is –41.3 dBm.

Table 3.8 Properties of the Gaussian Pulse Derivative Spectra

n	σ (ps)	f_L (GHz)	f_H (GHz)	f_M (GHz)	W_{3dB} (GHz)
1	33	2.31	7.84	4.79	5.53
2	39	3.57	8.33	5.78	4.76
3	44	4.33	8.60	6.34	4.28
4	47	4.85	8.79	6.72	3.93
5	51	5.25	8.92	7.01	3.67
6	53	5.57	9.03	7.23	3.46
7	57	5.83	9.12	7.42	3.29

Source: [40].

Table 3.9 Some Fast Monocycle Waveforms and Their PSDs

Pulse Name	Waveform	Waveform and PSD
Scholtz Monocycle	$w(t) = A\left[1-4\pi\left(\frac{t-T_c}{\sigma}\right)^2\right]e^{\left[-2\pi\left(\frac{t-T_c}{\sigma}\right)^2\right]}$	Amplitude / Time; Power / Frequency; −3 dB
Manchester Monocycle	$w(t)=\begin{cases}A, & T_c/2 \le t \le T_c \\ -A, & T_c < t \le T_c + T_c/2 \\ 0, & \text{otherwise}\end{cases}$	Amplitude / Time; Power / Frequency; −3 dB
Rectangle Monocycle	$w(t)=\begin{cases}A, & T_c/2 \le t \le T_c + T_c/2 \\ 0, & \text{otherwise}\end{cases}$	Amplitude / Time; Power / Frequency; −3 dB

Source: [34].

Table 3.9 (continued)

Pulse Name	Waveform	Waveform and PSD
RZ-Manchester Monocycle	$$w(t)=\begin{cases}A, & T_c/2\le t\le T_c/2-pT_c\\ -A, & T_c/2\le t\le T_c-pT_c\\ 0, & \text{otherwise}\end{cases}$$	Amplitude / Time; Power / Frequency, −3 dB
Sine Monocycle	$$w(t)=\begin{cases}A\sin 2\pi f_c t, & T_c/2\le t\le T_c+T_c/2\\ 0, & \text{otherwise}\end{cases}$$	Amplitude / Time; Power / Frequency, −3 dB
Wavelet Monocycle	$$w_{a,b}(t)=\frac{C}{\sqrt{a}}e^{-\frac{(t-b)^2}{2a^2}}\cos\left[2\pi f_0\left(\frac{t-b}{a}\right)\right]$$	Amplitude / Time; Power / Frequency, FCC mask

Rectangle Monocycle. The rectangle monocycle is a pulse that has uniform amplitude A during the whole pulse width.

Wavelet-Based Monocycle. Liang and Zhou [40] described a monocycle based on the Marlet mother wavelet given by

$$w_{a,b}(t) = \frac{C}{\sqrt{a}} e^{-\frac{(t-b)^2}{2a^2}} \cos\left[2\pi f_0\left(\frac{t-b}{a}\right)\right] \tag{3.170}$$

where a is the scale factor and b is the shift factor. C is a constant adjusted in order to meet the FCC mask amplitude. The corresponding normalized PSD is given by

$$|P(f)| = \exp\left[-(2\pi af - 2\pi f_0)^2\right] \tag{3.171}$$

Performance

Simulations of the different monocycle shapes for UWB help to illustrate the trade-offs among the effects of the monocycle shapes on the design of filters, the choice of receiver bandwidth, the antenna design, the bit error rate performance, and the multipath performance. The spectrum characteristics of these monocycles and their bit error rate performances in an AWGN are compared by simulation. The spectrum characteristics for these monocycles are illustrated in Table 3.9. In the bit error rate simulations, every monocycle width is chosen to be 0.5 ns, and the ratio of T_f and T_m is set to be 1,024.

Monocycle Spectrum Characteristics. Unlike the conventional narrowband communication systems, it is desirable for UWB signaling to spread the energy as widely in frequency as possible to minimize the PSD. As shown in Table 3.9, Gaussian pulse and rectangle monocycle have DC components, which may reduce the antenna radiating efficiency. The Scholtz monocycle, the Gaussian monocycle and the RZ-Manchester monocycles have wider 3 dB bandwidth than others.

BER Performances in AWGN Channel. As discussed previously, the optimum receiver in AWGN is a correlation receiver. As obvious from (3.158), the modulation parameter δ will affect the BER, since δT_m is the length of the delay between a zero and one in PPM (see Figure 3.66). The optimal choice of δ for a given received monocycle on a single link is given by [35]

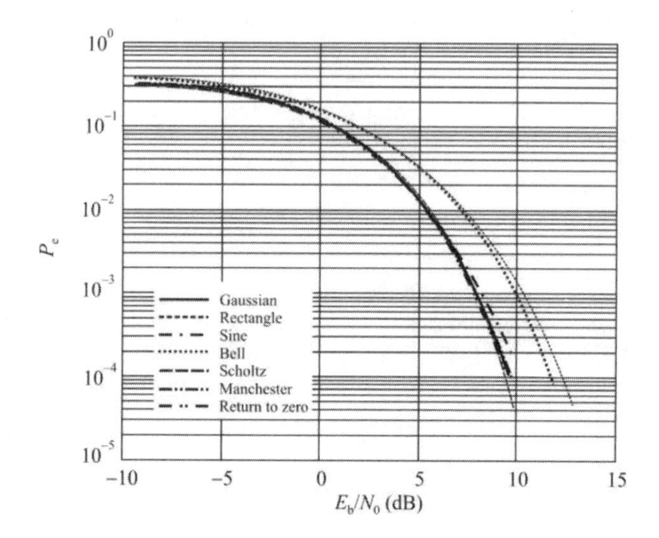

Figure 3.78 UWB BER performance in AWGN. (Source: [34]. © 2002 IEEE. Reprinted with permission.)

$$\delta_{opt} = \arg\min_{\delta} \int_{-\infty}^{\infty} w_{rec}(u) w_{rec}(u-\delta) du \tag{3.172}$$

Here $0 \le \delta \le 1$, where T_m is the monocycle duration. Note that when δ is greater than one, $w_{rec}(t)$ and $w_{rec}(t - \delta)$ are orthogonal. Therefore, δ_{opt} can be chosen as one. To compare the BER performance, δ_{opt} for the Gaussian pulse, the Gaussian monocycle, the Scholtz monocycle, the Manchester monocycle, the RZ Manchester monocycle, the sine monocycle, and the rectangle monocycle are first determined by using (3.172). In this case, $\delta_{opt}T_m$ are given by 0.5 ns (δ_{opt} = 1), 0.1917 ns (δ_{opt} = 0.3834), 0.125 ns (δ_{opt} = 0.025), 0.25 ns (δ_{opt} = 0.5), 0.25 ns (δ_{opt} = 0.5), 0.25 ns (δ_{opt} = 0.5), and 0.5 ns (δ_{opt} = 1), respectively. The simulation results are illustrated in Figure 3.78 [40]. It shows that the performances of the Scholtz, Manchester, RZ-Manchester, sine and Gaussian monocycles are somewhat better than those of the Gaussian pulse and rectangle monocycle, although the results are within 2 dB of each other.

Summary. In this section, we investigated several candidate monocycle shapes for UWB. Their single-link performances are simulated and compared. The results show that monocycle shape affects the PPM BER; however, the performance in all cases is very similar. The monocycle shape also determines the spectrum characteristics of UWB signals.

Of those monocycles we discussed, only the seventh-order derivative of the Gaussian monocycle complies with the FCC mask. All the other monocycles exceed the mask in most of the spectrum.

3.13.6.7 Jam Resistance and Processing Gain

Processing gain is a measure of the communication system's resistance to interference and UWB has a huge processing gain. One definition of processing gain is the ratio of the bandwidth of the signal to the bandwidth of the information signal. For example, IS-95 spread spectrum system with 8 kHz of information bandwidth and a 1.25 MHz channel bandwidth yields a processing gain of 156 (22 dB). An impulse system transmitting the same 8 kHz information bandwidth and a 2 GHz channel bandwidth the processing gain is 250,000 or 54 dB.

Other ways of calculating the processing gain for a UWB signal are:

- The duty cycle of the transmission (e.g., a 1% duty cycle yields a process gain of 20 dB);
- The effect of integrating over multiple pulses to recover the information (e.g., integrating energy over 100 pulses to determine one digital bit yields a process gain of 20 dB);
- The total process gain is then 40 dB.

Thus, a 2 GHz/10 Mpps link transmitting 8 kbps would have a process gain of 54 dB, because it has a 0.5 ns pulse width with a 100 ns pulse repetition interval = 0.5% duty cycle (23 dB) and 10 Mpps/8,000 bps = 1,250 pulses per bit (another 31 dB).

3.13.6.8 Multipath and Propagation

Multipath fading is less of a problem for UWB systems than for conventional communication systems. In fact, Rayleigh fading, so noticeable in cellular communications, is a continuous wave phenomenon, not an impulse communications phenomenon. In an impulse system, in order for there to be multipath effects, one of the following special conditions must exist:

- The path length traveled by the multipath pulse must be less than the pulse width times the speed of light. For a 2 GHz pulse, that equals 0.15 m or about 1/2 ft, that is, [1/2 ns] × [300,000,000 meters/second]. This is

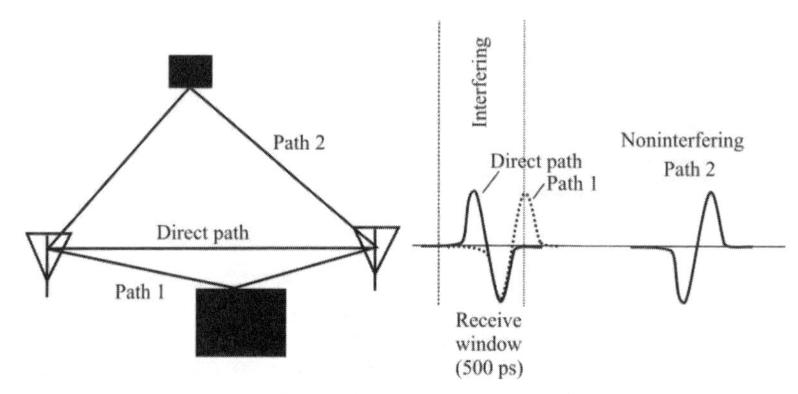

Figure 3.79 UWB multipath. A multipath problem occurs when the reflecting path is short as in path 1.

illustrated in Figure 3.79, in the case where the pulse traveling path 1 arrives one half a pulse width after the direct path pulse.

- The multipath pulse travels a distance that equals multiples of the interval of time between pulses multiplied by the speed of light. For a 1 Mpps system, that would be equal to traveling an extra 300, 600, and 900 meters. However, because each individual pulse is subject to the pseudo-random dither, these pulses are decorrelated.

Pulses traveling between these intervals do not cause self-interference. In Figure 3.79, this is illustrated by the pulse traveling path 2, while pulses traveling grazing paths, as illustrated in Figure 3.79 by path 1, create impulse radio multipath effects. Figure 3.80 shows how easy it is to resolve multipath impulse signals in the time domain. Note that the receive antenna's characteristics convert the Gaussian monocycle into a "W" waveform as seen in Figure 3.80. The first arriving pulse (between 3 ns and 6 ns) is of a lower amplitude because it traveled through more walls than some later arriving pulses (the pulse arriving at between 8 ns and 11 ns).

UWB PPM communication systems have a far superior performance in high multipath environments than narrowband systems, as well as DSSS systems. Multipath is a critical problem for most communication systems in urban environments because there are many objects that reflect RF signals. In these environments, UWB systems can operate at high data rates with lower bit error rates and with lower transmit powers.

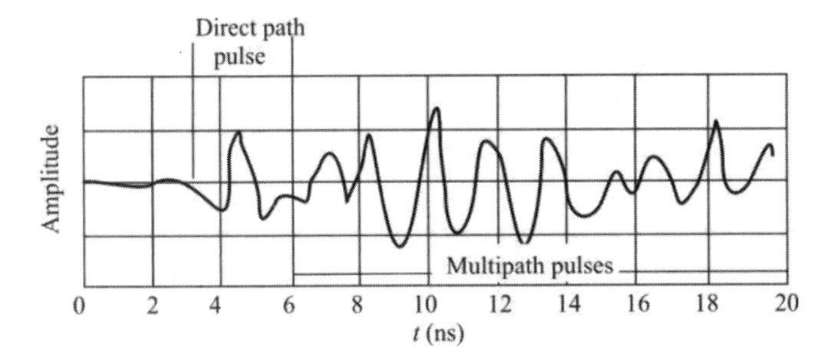

Figure 3.80 UWB multiple multipath examples.

3.14 Orthogonal Frequency Division Multiplexing

This section will review the fundamental characteristics of *orthogonal frequency division multiplexing* (OFDM) techniques, and also discuss common impairments and how, in some cases, OFDM mitigates their effect. Where applicable, the impairment effects and techniques will be compared to those in a single carrier system.

3.14.1 Single-Carrier Modulation System

A typical single-carrier modulation spectrum is shown in Figure 3.81. A single carrier system modulates information onto one carrier using frequency, phase, or amplitude adjustment of the carrier. For digital signals, the information is in the form of symbols that are modulated onto the carrier. As higher data rates are used, the duration of one bit or symbol of information becomes shorter. The system becomes more susceptible to loss of information from impulse noise, signal

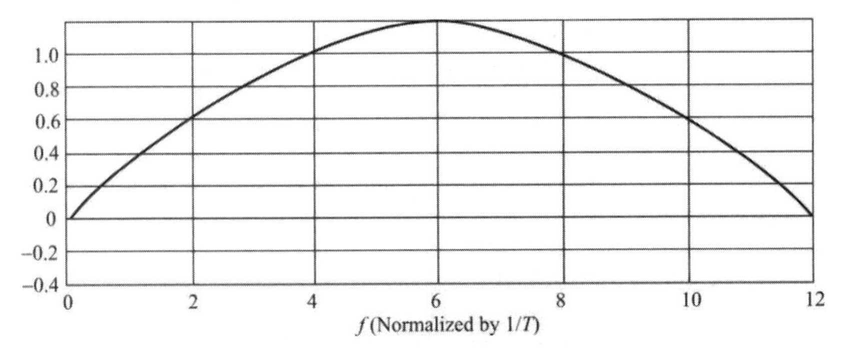

Figure 3.81 Single carrier spectrum.

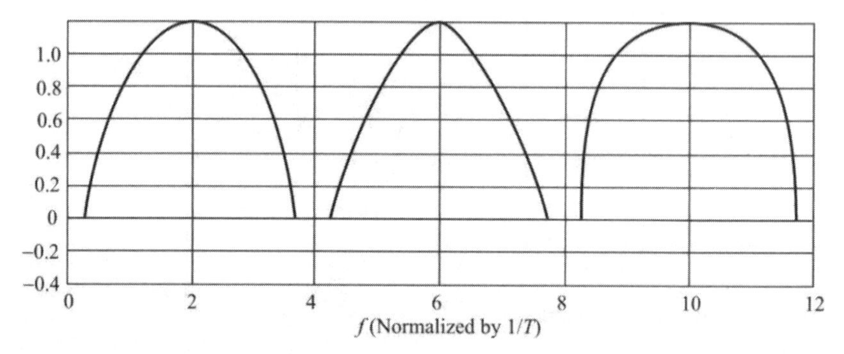

Figure 3.82 Frequency division multiplexing spectrum.

reflections, and other impairments. These impairments can impede the ability to recover the information sent. In addition, as the symbol rate increases, ceteris paribus, the bandwidth used by a single carrier system, increases, and the susceptibility to interference from other continuous signal sources becomes greater.

3.14.2 Frequency Division Multiplexing Modulation System

Frequency division multiplexing (FDM) extends the concept of single carrier modulation by using multiple subcarriers within the same single channel. This combination of subcarriers is frequently moved in frequency as a group. The total data rate to be sent in the channel is divided between the various subcarriers. The data do not have to be divided evenly, nor do they have to originate from the same information source or even have similar statistical properties. Advantages include using separate modulation/demodulation customized to a particular type of data, or sending out banks of dissimilar data that can be best sent using multiple, and possibly different, modulation schemes.

FDM systems, an example of a spectrum for which there are three subcarriers, as illustrated in Figure 3.82, usually require a guard band between modulated subcarriers to prevent the spectrum of one subcarrier from interfering with another. These guard bands lower the system's effective information rate when compared to a single carrier system with similar modulation because the spectral efficiency is less.

3.14.3 Orthogonality and OFDM

If the FDM system shown in Figure 3.82 used a set of subcarriers that were orthogonal to each other, a higher level of spectral efficiency could be achieved.

The guard bands that were necessary to allow individual demodulation of subcarriers in an FDM system would no longer be necessary. The use of orthogonal subcarriers would allow the subcarriers' spectra to overlap, thus increasing the spectral efficiency.

As pointed out in (3.11), if the dot product of two deterministic signals is equal to zero, these signals are said to be orthogonal to each other. Similarly, if two stochastic processes are uncorrelated, then they are orthogonal. OFDM is usually implemented in practice using the *discrete Fourier transform* (DFT). The sinusoids of the DFT form an orthogonal basis set, and a signal in the vector space of the DFT can be represented as a linear combination of these orthogonal sinusoids. One view of the DFT is that the transform essentially correlates its input signal with each of the sinusoidal basis functions. If the input signal has some energy at a certain frequency, there will be a peak in the correlation of the input signal and the basis sinusoid at that frequency. This transform is used at the OFDM transmitter to map an input signal onto the orthogonal basis functions of the DFT, which form a set of orthogonal subcarriers. Similarly, the transform is used again at the OFDM receiver to process the received subcarriers. The signals from the subcarriers are then combined to form an estimate of the source signal from the transmitter. Since the basis functions of the DFT are uncorrelated, the correlation performed in the DFT for a given subcarrier only sees energy for that corresponding subcarrier. The energy from other subcarriers does not contribute because it is uncorrelated. This separation of signal energy is the reason that the OFDM subcarriers' spectra can overlap without causing interference. Figure 3.83 illustrates an OFDM spectrum where there are nine subcarriers. Notice that where each of the spectra for the subcarriers peaks, all the other spectra are zero. This is a feature of the orthogonality of the subcarriers. Each of these subcarriers carries its own information stream that may or may not be related to the information streams in the other subcarriers.

3.14.4 Implementation of OFDM Systems

The discrete Fourier transform (DFT) and the *inverse discrete Fourier transform* (IDFT) are used in OFDM systems because they can be viewed as mapping data onto orthogonal subcarriers. For example, the IDFT is used to take in frequency-domain data and convert it to time-domain data. In order to perform that operation, the IDFT correlates the frequency-domain input data with its orthogonal basis functions, which are sinusoids at certain frequencies. This correlation is equivalent to mapping the input data onto the sinusoidal basis functions.

In practice, the DFT and IDFT are implemented with the fast Fourier transform (FFT) and inverse fast Fourier transform (IFFT) blocks. An OFDM system treats the source symbols (e.g., the QPSK or QAM symbols that would be present in a single carrier system) at the transmitter as though they are in the

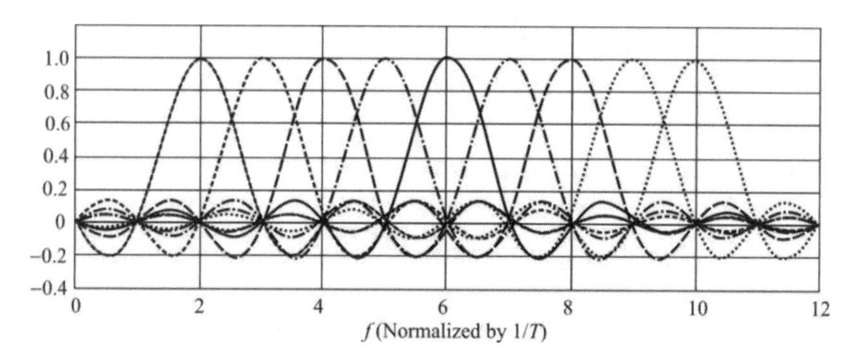

Figure 3.83 OFDM spectrum.

frequency domain. These symbols are used as the inputs to an IFFT block that converts the signal into the time domain. The IFFT takes in N symbols at a time where N is the number of subcarriers in the system. Each of these N input symbols has a symbol period of T seconds. Recall that the basis functions for an IFFT are N orthogonal sinusoids. These sinusoids each have a different frequency and the lowest frequency is DC. Each input symbol acts like a complex weight for the corresponding sinusoidal basis function. Since the input symbols are complex, the value of the symbol determines both the amplitude and phase of the sinusoid for that subcarrier. The IFFT output is then summed over all N sinusoids. Thus, the IFFT block provides a simple way to modulate data onto N orthogonal subcarriers. The block of N output samples from the IFFT make up a single OFDM symbol. The length of the OFDM symbol is NT, where T is the IFFT input symbol period mentioned above.

The time domain signal that results from the IFFT is transmitted across the channel to the receiver over at least one path. At the receiver, an FFT block is used to convert it back into the frequency domain. Ideally (lacking significant noise, interference, and ISI), the FFT output will be the original symbols that were sent to the IFFT at the transmitter. When plotted in the complex plane, the FFT output samples will form a constellation, such as 16-QAM shown in Figure 3.84. The block diagram in Figure 3.84 illustrates the conversions between the frequency domain, the time domain, and back again in an OFDM system and shows in this case how the 16 QAM signal is reconstructed.

3.14.5 Multipath Channels and the Use of Cyclic Prefix

A major problem in most wireless communication systems is the presence of at least one multipath channel. In a multipath environment, the transmitted signal

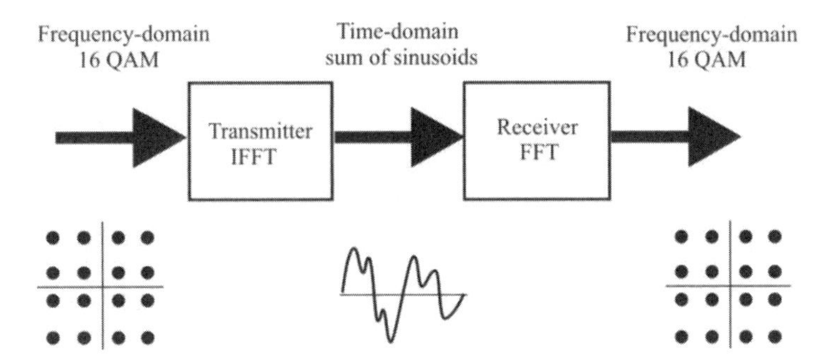

Figure 3.84 Simplified OFDM system.

reflects off of several objects as it traverses from the transmitter to the receiver(s). As a result, multiple delayed versions of the transmitted signal arrive at the receiver. The multiple versions of the signal can cause the received signal to be distorted.

A multipath channel will cause two problems for an OFDM system. The first problem is ISI, illustrated in Figure 3.85. This problem occurs when the received OFDM symbol is distorted by the previously transmitted OFDM symbol. The effect is similar to the ISI that occurs in a single-carrier system. However, in that case, the interference is usually due to several previous symbols instead of just the previous symbol; the symbol period in single carrier systems is typically much shorter than the time span of the channel (the delay time of the longest significant delayed component), whereas the typical OFDM symbol period is much longer than the time span of the channel. The second problem is unique to multicarrier systems and is called the *intrasymbol interference*. It is the result of interference among a given OFDM symbol's own subcarriers.

3.14.5.1 Intersymbol Interference

Perhaps the biggest advantage of OFDM is its ability to essentially remove ISI. Assume that the time span of the channel is L samples long. Instead of a single carrier with a data rate of R sps, an OFDM system has N subcarriers, each with a data rate of R / N sps. Because the data rate is reduced by a factor of N in each subcarrier, the OFDM symbol period is lengthened by a factor of N. By choosing an appropriate value for N, the length of the OFDM symbol becomes longer than the time span of the channel. Because of this configuration, the effect of ISI is the distortion of the first L samples of the received OFDM symbol. An example of this effect is shown in Figure 3.85. By noting that only the first few samples of the symbol are distorted, we can use a guard interval to remove the effect of ISI. For

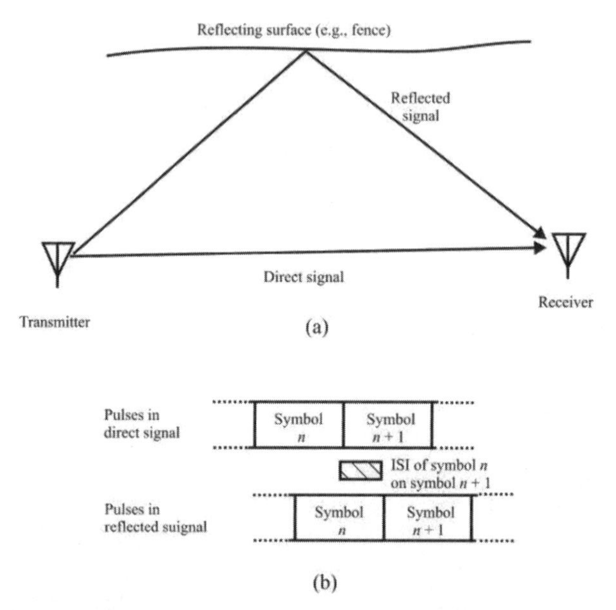

Figure 3.85 Example of intersymbol interference. (a) Scenario with multipath, and (b) symbol streams exhibiting ISI caused by the time delay of reflected signal at the receiver.

example, the guard interval could be a section of all zero samples transmitted in front of each OFDM symbol. Since it does not contain any useful information, the guard interval would be discarded at the receiver. If the length of the guard interval is properly chosen such that it is longer than the time span of the channel, the OFDM symbol itself will not be distorted. Thus, by discarding the samples from the guard interval, the effects of intersymbol interference are eliminated as well.

In OFDM an all-zero guard interval is not really used because it is impractical. Instead, the *cyclic prefix* is inserted during this time interval. Its samples are still thrown away at the receiver, but it provides some useful properties that an all-zero guard interval does not have.

3.15 Concluding Remarks

The most common forms of spread spectrum AJ communication techniques, DSSS, FHSS, and THSS, were discussed in this chapter. These were developed primarily to, if not defeat, then certainly to mitigate the effects of jamming. This was the motivation of the military for such development, but the advantages of, in particular, DSSS for providing CDMA, and FHSS for GSM, to commercial cellular communications has become obvious. Thus, the capabilities of DSSS and FHSS have found a niche in the commercial telephone market and THSS is emerging. Methods to counter the advantages of SS are the subject of the remainder of this book.

In this chapter we discussed the performance of the main types of modulation used in SS systems. This analysis resulted in expressions for the BER (P_e) as a function of the SNR of the signal. It was shown that for the postdetection SNRs expected to be achieved in these communications systems (>10 dB) the BER of BPSK is 10^{-5} or better while that for the higher modulations, 4QAM and 16QAM, the BER is about 10^{-3} or better.

References

[1] Peterson, R. L., R. E. Ziemer, and D. E. Borth, *Introduction to Spread Spectrum Communications*, Upper Saddle River, NJ: Prentice Hall, 1995, p. 7.

[2] Peterson, R. L., R. E. Ziemer, and D. E. Borth, *Introduction to Spread Spectrum Communications*, Upper Saddle River, NJ: Prentice Hall, 1995, p. 386.

[3] Torrieri, D. J., "The Information-Bit Error Rate for Block Codes," *IEEE Transactions on Communications*, Vol. COM-32, No. 4, April 1984, pp. 474–476.

[4] Nicholson, D. L., *Spread Spectrum Signal Design,* Rockville, MD: Computer Science Press, 1988, pp. 207–208.

[5] Torrieri, D. J., *Principles of Secure Communication Systems*, 2nd ed., Norwood, MA: Artech House, 1992, p. 54.

[6] Torrieri, D. J., *Principles of Secure Communication Systems*, 2nd ed., Norwood, MA: Artech House, 1992, p. 72.

[7] Trumpis, B. D., "On the Optimum Detection of Fast Frequency Hopped MFSK Signals in Worst Case Jamming," TRW Internal Memorandum, June 1981.

[8] Torrieri, D. J., *Principles of Secure Communication Systems*, 2nd ed., Norwood, MA: Artech House, 1992, p. 35.

[9] Ziemer, R. E., and W. H. Tranter, Principles of Communication Systems, Modulation, and Noise, 5th ed., New York: John Wiley & Sons, 2002, p. 398.

[10] Simon, M. K., J. K. Omura, R. A. Scholtz, and B. K. Levitt, *Spread Spectrum Communications Handbook,* New York: McGraw Hill, 1994, pp. 668–672.

[11] Ziemer, R. E., and W. H. Tranter, *Principles of Communication Systems, Modulation, and Noise*, 5th ed., New York: John Wiley & Sons, 2002, p. 399.

[12] Proakis, J. G., *Digital Communications*, 3rd ed., New York: McGraw-Hill, 1995, pp. 269–274.

[13] Simon, M. K., J. K. Omura, R. A. Scholtz, and B. K. Levitt, *Spread Spectrum Communications Handbook,* New York: McGraw-Hill, 1994, p. 671.

[14] Ziemer, R. E., and W. H. Tranter, *Principles of Communications,* 5th ed., New York: John Wiley & Sons, 2002, p. 356.

[15] Ziemer, R. E., and W. H. Tranter, *Principles of Communications Systems, Modulation, and Noise*, 5th ed., New York: John Wiley & Sons, 2002, pp. 358–359.

[16] Zeng, M., and Q. Wang, "On the Probability Distribution of Differential Phase Perturbed by Tone Interference and Gaussian Noise," *IEEE Transactions on Communications*, Vol. 47, No. 4, April 1999, pp. 508–510.

[17] Torrieri, D. J., *Principles of Secure Communication Systems*, 2nd ed., Norwood, MA: Artech House, 1992, pp. 13–23.

[18] Torrieri, D. J., *Principles of Secure Communication Systems*, 2nd ed., Norwood, MA: Artech House, 1992, p. 276.

[19] Torrieri, D. J., *Principles of Secure Communication Systems*, 2nd ed., Norwood, MA: Artech House, 1992, p. 275.

[20] Torrieri, D. J., *Principles of Secure Communication Systems*, 2nd ed., Norwood, MA: Artech House, 1992, p. 19.

[21] Ziemer R. E. and W. H. Tranter, *Principles of Communications*, 5th ed., New York: John Wiley & Sons, 2002, p. 413.

[22] Torrieri, D. J., *Principles of Secure Communication Systems*, 2nd ed., Norwood, MA: Artech House, 1992, p. 22.

[23] Simon, M. K., J. K. Omura, R. A. Scholtz, and B. K. Levitt, *Spread Spectrum Handbook,* New York: McGraw-Hill, 1994, p. 177.

[24] Noneaker, D. L., "Bandwidth-Efficient Modulation for Improved ISI Protection in High-Data-Rate SFHSS Spread-Spectrum Communications," *Proceedings IEEE MILCOM*, 2001.

[25] Kennedy, M. P., R. Rovatti, and G. Setti, (eds.), *Chaotic Electronics in Telecommunications,* Boca Raton, FL: CRC Press, 2000.

[26] Pecora, L. M., and T. L. Carroll, "Synchronization in Chaotic Systems," *Physical Review Letters,* Vol. 64, No. 8, February 1990, pp. 821–824.

[27] Leuciuc, A., "Information Transmission Using Chaotic Discrete-Time Filter," *IEEE Transactions on Circuits and Systems, Part I,* Vol. 47, No. 1, January 2000, pp. 82–88.

[28] Tam, W. M., and C. K. Tse, "Multi-User Detection Techniques for Multiple Access Chaos-Based Digital Communication Systems," *International Symposium on Nonlinear Theory and its Applications*, October 7–11, 2002, pp. 503–506.

[29] Rulkov, N. F., and L. S. Tsimring, "Synchronization Methods for Communication with Chaos over Bandlimited Channels," *International Journal of Circuit Theory and Applications,* Vol. 27, 1999, pp. 555–567.

[30] Volkovskii, A. R., S. C. Young, L. S. Tsimring, and N. F. Rulkov, "Multi-User Communication Using Chaotic Frequency Modulation," *IEEE Transactions on Circuits and Systems–I: Fundamental Theory and Applications,* Vol. 48, No. 12, December 2001.

[31] Tang, S., H. F. Chen, S. K. Hwang, and J. M. Liu, "Message Encoding and Decoding Through Chaos Modulation in Chaotic Optical Communications," *IEEE Transactions on Circuits and Systems–I: Fundamental Theory and Applications,* Vol. 49, No. 2, February 2002, pp. 163–169.

[32] Xie, Q., and G. Chen, "Hybrid Chaos Synchronization and Its Application in Information Processing," *Mathematical and Computer Modeling*, Vol. 35, 2002, pp. 143–163.

[33] Lau, F. C. M., M. Ye, C. K. Tse, and S. F. Hau, "Anti-Jam Performance of Chaotic Digital Communication Systems, "*IEEE Transactions on Circuits and Systems–I: Fundamental Theory and Applications*, Vol. 49, No. 10, October 2002, pp. 1486–1494.

[34] Chen, X., and S. Kaiei, "Monocycle Shapes for Ultra Wideband System," *Proceedings IEEE MILCOM*, 2002, pp. I-597–600.

[35] Win, M. Z., and R. A. Scholtz, "Impulse Radio: How It Works," *IEEE Communications Letters*, Vol.2, No.2, January 1998, pp. 10–12.

[36] Scholtz, R. A., "Multiple Access with Time-Hopping Impulse Modulation," *Proceedings MILCOM 1993*, October 11–14, 1993.

[37] Zhang, Q, D. L. Goeckel, J. Burkhart, B. K. Mui, N. Merrill, M. Carrier, and R. Jackson, "FSR-UWB (TR-UWB without the Delay Element): Effect of Impulse Dithering and Experimental Results," www-unix.ecs.umass.edu/~goeckel/promotion_files/fsr_uwb_prototype.pdf.

[38] Franschetti, G., and C. H. Papas, "Pulsed Antennas," *IEEE Transactions on Antennas and Propagation,* Vol. AP-22, 1974, pp. 651–661.

[39] Sheng, H., P. Orlik, A. M. Haimovich, L. J. Cimini, and J. Zhang, "On the Spectral and Power Requirements for Ultra-Wideband Transmission," *Proceedings IEEE International Conference on Communications*, 2003.

[40] Liang, A., and Z. Zhou, "The Performance of UWB System-Influence on Monocycle Shape and Synchronization Error," *ECTI Transactions on Electrical Engineering, Electronics, and Communications*, Vol. 3, No. 1, February 2005, pp. 44–49.

Chapter 4

Antijam Signal Detection

4.1 Introduction

In many cases of practical utility, the presence of signals must be established before jamming is initiated. There are several reasons for this to include avoiding possible friendly fratricide, to conserve energy in the jammer, and to ascertain that the target has not moved in frequency. Determining such presence in this context is called *detection*. We specifically address AJ signal detection in this chapter.

There are several authoritative sources of information on SS signal detection. Dillard and Dillard [1] focus specifically on the topic and present many design charts and equations that can be used to ascertain detection capability. The handbook by Simon et al. [2] provides much signal design information if you are interested in designing SS communication systems. Other excellent sources for SS system design are the books by Peterson et al. [3] and Torrieri [4], the latter of which also includes some intercept and EW information. Dixon [5] was one of the first to write a book on the topic and remains an excellent source. A book covering only CDMA DSSS systems was written by Viterbi [6]. Nicholson [7] prepared a book describing how to design signals to be LPI. Since there are several excellent sources, the information contained in this chapter is sufficient to introduce the reader to the subject, but is not a complete treatise. Also presented are the results of an extensive simulation on the detectability of such signals.

This chapter is organized as follows. We begin with an introduction to the types of receivers typically used for the detection of AJ signals. We then discuss detection of DSSS signals by various means, some of which are specific to a type of signal while some are more general (such as the radiometer discussed Chapter 2). Detection techniques for THSS/UWB signals are presented next. Detectors and their performance for FHSS systems are then presented. Finally, the results of a performance simulation are presented where the target environment consisted of several networks of SFH signals.

4.2 Signal Detection

Signal detection is for the purpose of controlling the jammer. As such, parameters associated with signals are estimated in the detection process. The frequency, time of intercept, and power (or amplitude) of the signals being detected are determined. In addition, the type of modulation of the signals is a parameter that is important in order to properly match the jamming waveform to the signals.

How signals are detected depends on the characteristics of the signal. Optimal strategies to detect analog signals are different from those for digital signals. Strategies for nonspread signals are different from those for spread signals. If spread, the strategies are different for DSSS signals and FHSS signals [1, 8–10].

Simply detecting the presence of an FHSS signal in a crowded spectrum is rarely sufficient. The theoretical calculations on the ability to detect targets form bounds on possible performance, however. It is almost always necessary to perform measurements on detected signals to sort them and attempt to associate one hop with previous ones. Therefore, except for the simplest cases when BBN or PBN jamming can be used, aspects of follower jamming are almost always implemented in practice.

4.3 Receivers

RF receivers are used for the detection of AJ communication signals. They are used to convert the frequency of the intercepted signal to some lower frequency range where the signal can be processed. In some cases, the receiver itself provides demodulation and detection functions. Sometimes these are provided by modules external to the receiver. The particular receiver employed will vary in architecture depending on the types of signals to be detected. Such receivers usually either have a wide instantaneous bandwidth in order to receive a significant portion of the signals, or they must scan the frequency spectrum very fast, essentially emulating wideband receivers. Some of the general characteristics of receivers are covered here.

The role of receivers in EW, in particular ES, is to *detect* or *intercept* targets, in the shortest possible time. However, the bandwidth in which targets operate, especially AJ targets, may span a wider bandwidth than is practically instantaneously covered by a receiver. In order to cope with this very large bandwidth, a widely favored receiver architecture employs a receiver of a more modest bandwidth with an agile center frequency. The frequency-swept superheterodyne receiver is an example of this sort of receiver.

There are two general types of receivers used for these functions. Due to the wideband characteristic of AJ signal, be they DSSS, THSS, or FHSS, the receivers must be able to tune, usually rapidly, over that bandwidth. They can also be staring receivers where a significant portion of the entire bandwidth is within the passband of the receiver.

4.3.1 Staring Receivers

Staring receivers do not change their frequency, but stay tuned to the same place. While their input is normally fairly wideband, that need not be the case. FHSS systems, for example, could hop into the bandwidth of a relatively narrow bandwidth receiver. Such detection performance, however, is neither reliable nor robust.

Digital receivers for SS signal detection, which are typically staring over limited bands, were examined extensively by Tsui [11]. A single digital receiver with substantial dynamic range, for example, can cover the low VHF range. Wide dynamic range is important so that close, usually friendly, transmitters do not saturate the receiver, thereby precluding the reception of weak, distant targets.

4.3.2 Scanning Receivers

Scanning receivers measure energy in a (relatively) narrow bandwidth while being tuned over a wider bandwidth. There are two types of these receivers: scanning superheterodyne and compressive. The former typically has a much narrower instantaneous bandwidth than the latter and therefore a finer frequency resolution. On the other hand, the latter are considerably faster than the former and therefore possess a much higher probability of detection, when sensitivity and dynamic range are not issues.

4.3.2.1 Superheterodyne Receiver

One method to measure the energy of an FHSS signal and determine the hop frequencies is with a scanning narrowband superheterodyne receiver. This receiver will not detect every hop, but it has been shown that several hops in a 5 second transmission of an FHSS transmitter can be detected [12]. A simplified block diagram of such a configuration is shown in Figure 4.1. Such a system linearly changes its frequency with time, and the output of an envelope detector is used to estimate the location of the frequency of energy incident to the receiver [13]. When the input signal is given by

$$r(t) = \sqrt{2R}\cos(2\pi f_0 t) \tag{4.1}$$

where R is the average power in the signal, the peak voltage at the output of the envelope detector is given by [14]

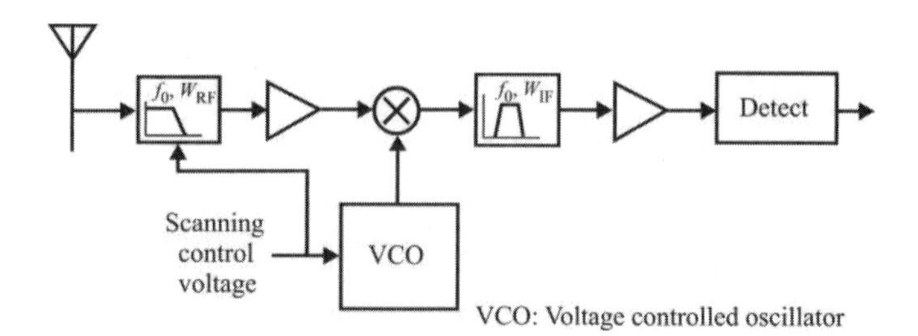

Figure 4.1 Block diagram of a scanning superheterodyne receiver.

$$\alpha = \left(1 + 0.195\frac{\mu^2}{W^4}\right)^{-1/4} \tag{4.2}$$

where μ is the scan rate in Hz/sec, and W is the 3 dB bandwidth of the associated filter. The frequency resolution in hertz is given by [14]

$$\delta = W\left(1 + 0.195\frac{\mu^2}{W^4}\right)^{1/2} \tag{4.3}$$

while the sweep speed is given by [15]

$$\mu = \left(\frac{\delta_f^2}{5.282 T_s^2 W^4} + 1\right)^{-1/4} \tag{4.4}$$

when δ_f is the sweep width and T_s is the time for one sweep.

Many modern superheterodyne receivers are digitally tuned, which means that a digital command rather than a voltage level as shown in Figure 4.1 tunes the local oscillator. This configuration is shown in Figure 4.2. A receiver such as this dwells at a frequency channel, the width of which is determined by the bandpass

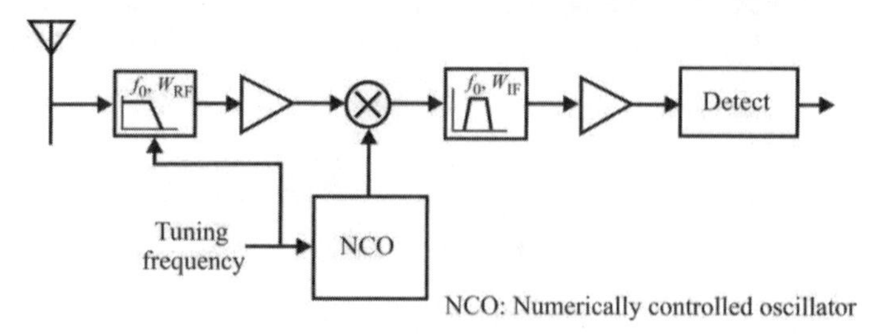

Figure 4.2 Digitally controlled superheterodyne receiver that can scan the frequency spectrum.

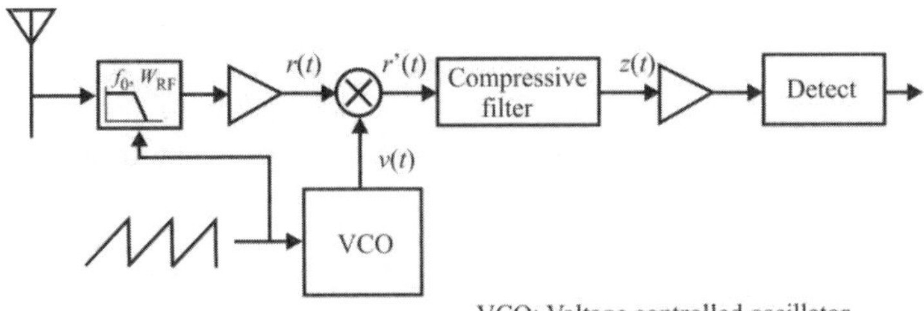

Figure 4.3 Simplified block diagram of a compressive receiver.

filter. A typical dwell might be 10 ms, or it could match the hop rate of a specific target. Digitally controlled synthesizers (NCO) such as the one shown here can tune from one frequency to the next rapidly, especially if the next frequency channel is adjacent to the current one. In fact, with two digitally controlled synthesizers, one could be tuning to the next frequency while the other one is used to collect data at the current channel. In that case the tuning time is the time it takes to switch between oscillators and the settling time of the bandpass filter—typically approximately $1 / W_F$ when W_F is the width of the filter.

A narrowband receiver system is simulated in [12], for example, where the narrowband receiver is scanned across the FHSS system's bandwidth to ascertain detectability of an FHSS system.

4.3.2.2 Compressive Receiver

A compressive receiver performs a continuous fast scan over a frequency range of interest. In the case of SS systems that region of interest would be W_{ss} or a substantial fraction of it. These receivers can scan several hundreds of megahertz in microseconds, facilitating a high probability of detection performance. One such configuration is shown in Figure 4.3. The incoming signal $r(t)$ is mixed with a VCO signal whose output is given by

$$v(t) = \cos(2\pi f_0 t - \mu t^2), \qquad 0 \le t \le T_s \tag{4.5}$$

When T_s is the time for one sweep. Thus,

$$r'(t) = r(t)\cos(2\pi f_0 t - \mu t^2), \qquad 0 \le t \le T_s \tag{4.6}$$

The time delay of the compressive filter is inversely matched to that of the sweep and the impulse response is given by

$$h(t) = W(t)\cos(2\pi f_0 t + \mu t^2), \qquad 0 \le t \le T_c \tag{4.7}$$

$W(t)$ is a mathematical weighting function that lowers the peaks of the sidelobes in the frequency response, which, however, also widens the width of the main lobe. T_c is the compression time and typically $T_c = 2T_s$.

The compressive filter output is therefore

$$\begin{aligned} z(t) &= \int_0^{T_c} r(t-\tau)h(\tau)d\tau \\ &= \int_0^{T_c} r(t-\tau)\cos[2\pi f_0(t-\tau) - \mu(t-\tau)^2]W(t)\cos[2\pi f_0(\tau) + \mu^2(\tau)]d\tau \end{aligned} \tag{4.8}$$

The output of the compressive filter has energy at a time corresponding to the location in frequency of the input signal. Thus, a time to frequency transformation has occurred. By measuring the time from the start of the sweep to when that energy appears, an estimate of the frequency can be determined along with an estimate of the energy in the signal [12, 16–21].

Compressive receivers have been analyzed for the intercept of frequency hopping targets [17, 20] and for the intercept of DSSS systems [21]. Results are shown in Figure 4.4 for the former and in Figure 4.5 for the latter. The SNR in

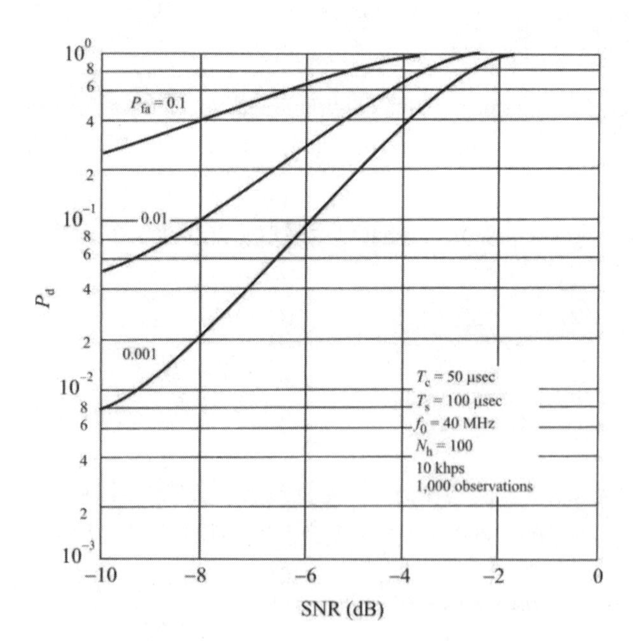

Figure 4.4 Compressive receiver performance against FHSS example. Note that P_d for the compressive receiver is much higher than the radiometer.

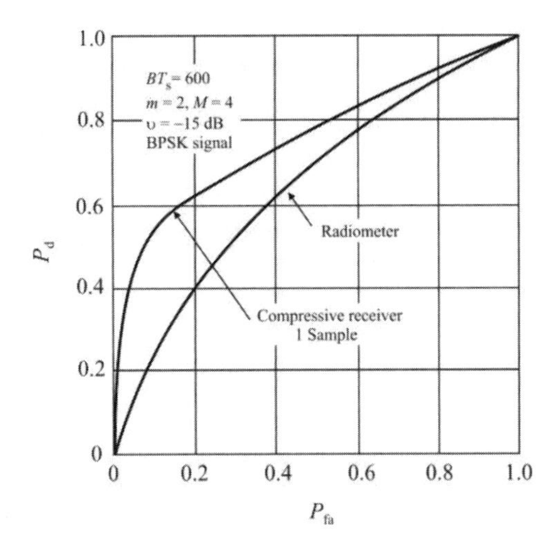

Figure 4.5 ROC for DSSS signal detection. (Source: [21]. © IEEE 1991. Reprinted with permission.)

both these charts is based on all of W_{ss} as the noise source. An SNR of –15 dB over W_{ss} = 50 MHz corresponds to an SNR of –15 + 33 = 18 dB over a channel bandwidth of 25 kHz, for example. In Figure 4.4, if T_b is the data bit time, then $T_1 = mT_b$ is the duration of the compressive filter impulse response, $T_s = MT_b$ is the scan period, and υ is the SNR. BT_s is the product of the bandwidth of the compressive filter and the scan period.

Notice that in Figure 4.4, the results are based on 1,000 observations. Since the compressive receiver scans every 100 μs in this example, the data applies to 100 ms of signal samples, implying a maximum target hop rate of 10 hps. Compressive receivers can handle faster hop rate targets, but the performance is less than that shown in Figure 4.4 because of the fewer samples available.

4.3.3 Detectors

Detectors are those parts of receivers that perform the function of determining the presence of specific signals. The two kinds of detectors considered here are matched filters with envelope detection and radiometers. In addition, the filter bank combiner that uses them is discussed.

4.3.3.1 Matched Filter

A matched filter followed by an envelope detector is frequently implemented in SS detectors [22]. Matched filters are optimum detectors. A filter is matched to a signal $r(t)$, $0 \le t \le T$ if its impulse response is given by

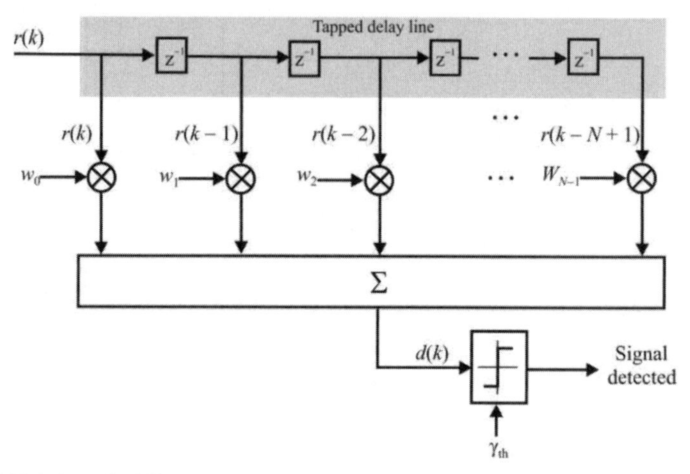

Figure 4.6 Digital matched filter.

$$h(t) = r(T - t) \tag{4.9}$$

The output of such a filter in response to $r(t)$ at its input has a peak when $t = T$, and the value of that peak is $R_{xx}(0)$ [23].

One of the advantages of matched filters in detectors is their processing gain. Denote the output SNR of the filters as υ_o and that at the input υ_i. Then

$$\upsilon_o = 2TW_F\upsilon_i \tag{4.10}$$

where W_F is the noise bandwidth of the filter. Thus, the output SNR can be larger than that at the input by the factor $2TW_F$. For measuring for the presence of DSSS signals, for example, T and W_F can be quite large, say, $T = 10$ seconds and $W_F = 10$ MHz, yielding a 83 dB processing gain. Practical matters would limit this to something less than 83 dB; nevertheless, the gain can be substantial. Note that W_F should not be larger than that of the signal for which detection is desired. If it is, then more noise and possibly other signals are admitted into the detection process than need to be.

Matched filters can be implemented in analog or digital form. *Surface acoustic wave* (SAW) devices are popular analog device choices for matched-filter construction, but with the progress in high-speed digital components, digital implementations are the method of choice when possible. Digital implementations allow easily changing the configuration as technology progresses, whereas a SAW implementation would require changing the hardware. One configuration for a digital matched filter is illustrated in Figure 4.6. The delayed input signals $r(k - j)$

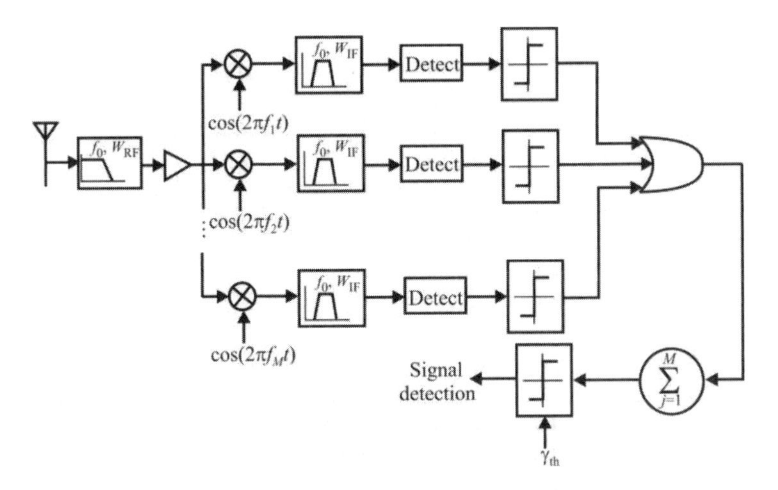

Figure 4.7 Filter bank combiner. γ_{th} is the detection threshold that is normalized to the output standard deviation. (Source: [25]. © Artech House 1989. Reprinted with permission.)

are each multiplied by a weight w_j, which is then summed and compared with a threshold. The filter output is thus

$$\vec{d}(k) = \vec{w}^T \vec{r}(k) = \sum_{i=0}^{N-1} w_i r(k-i) \tag{4.11}$$

The values of the weights are determined by the signal to which the filter is matched.

Since matched filters are designed to match specific signals about which all is known, they are of limited use for detection of generally unknown signals in general-purpose, noncooperative EW systems. Of course, in those cases where the signal specifics are totally known, the matched filter is the optimum way to detect the signal. In many cases, enough is known about the SOI that some of the matched filter concepts can be employed, however. In addition, detailed knowledge is required on how matched filters work for EA purposes, discussed elsewhere in this book. A much more promising candidate for the general-purpose EW system is the energy detector, also known as a radiometer, which we discussed in Chapter 2.

4.3.4 Filter Bank Combiner

When several parallel channels of radiometers or matched filters are implemented, then the configuration is called a *filter bank combiner* (FBC), one configuration of which is shown in Figure 4.7 [24–26]. There are N_c paths in the FBC, each tuned to the center frequency of a channel where N_c is the number of channels in W_{ss}.

Thus,

$$N_{\mathrm{c}} = \frac{W_{\mathrm{ss}}}{W_{\mathrm{c}}} \tag{4.12}$$

Each such path could consist of a radiometer followed by a threshold detector, a matched filter followed by an envelope detector and threshold detector, or any other mechanism for energy detection. If the threshold is exceeded in a path, the logical output of the threshold detector is a 1, and the OR output is 1 whenever any of the inputs is a 1. After a measurement interval defined by M dwells, a detection is declared if the output of the summing device exceeds the detection threshold γ_{th}.

When there is a path for each possible channel in W_{ss}, then the system shown in Figure 4.7 is called an FBC. When there are fewer than that, it is called a *partial band FBC* (PB FBC). Due to implementation complexity, in most cases of practical interest, the PB FBC is implemented [1].

Usually the overall performance of a detector would be specified in terms of P_{d} and P_{fa}—typically in the form of performance versus SNR. With this information, the per-channel detection and false alarm probabilities can be obtained from [28]

$$\begin{aligned} P_{\mathrm{d}_i} &\approx \frac{P_{\mathrm{d}} N_{\mathrm{c}}}{W_{\mathrm{ss}} T_{\mathrm{M}}} \\ P_{\mathrm{fa}_i} &\approx \frac{P_{\mathrm{fa}}}{W_{\mathrm{ss}} T_{\mathrm{M}}} \end{aligned} \tag{4.13}$$

when $\gamma_{\mathrm{th}} = 1$ and N_{c} and N_{h} (N_{h} are the number of hops in a message) are adequately large, where T_{M} is the message duration, P_{d} is the overall probability of detection, and P_{fa} is the overall probability of false alarm. The SNR required to achieve these performance levels is given by [26].

$$\frac{R}{N_0} = \kappa d \sqrt{\frac{W_{\mathrm{ss}}}{N_{\mathrm{c}}}} \tag{4.14}$$

$$\upsilon = \frac{R}{P_{\mathrm{N}}} = \frac{R}{N_0 W_{\mathrm{c}}} = \frac{1}{W_{\mathrm{c}}} \kappa d_{\mathrm{I}} \sqrt{\frac{W_{\mathrm{ss}}}{N_{\mathrm{c}}}} \tag{4.15}$$

where

κ = a correction factor that accounts for differences in Gaussian and chi-square statistics [26]

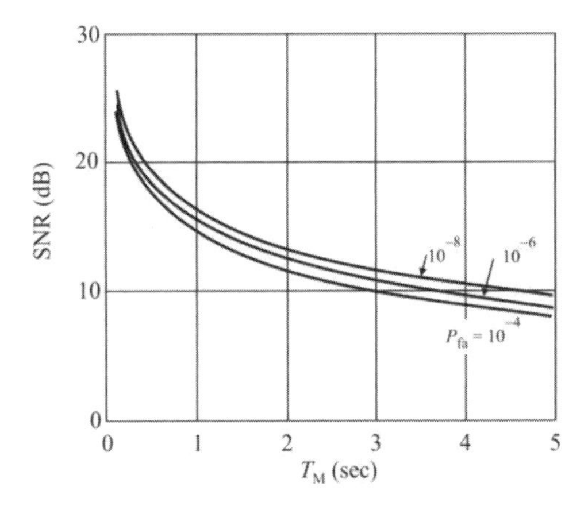

Figure 4.8 SNR required to obtain P_{fa} as indicated with $P_d = 0.9$ for radiometer implementation of an FBC. In this example, $\kappa = 0.5$, $\gamma_{th} = 1$, hop rate = 100 hps, $W_{ss} = 60$ MHz, and $W_h = 25$ kHz.

$$d = Q^{-1}(P_{fa_1}) - Q^{-1}(P_{d_1})$$

W_c = bandwidth of each channel
N_c = number of channels in W_{ss}

The last step follows because $N_c = W_{ss}/W_h$. The SNR is plotted in Figure 4.8 with parameters compatible with the low VHF band. In this case $\gamma_{th} = 1$, so that a single hit is all that is necessary to declare detection. As the length of the message increases past 3 seconds or so, the SNR required to achieve $P_d = 0.9$ and $P_{fa} = 10^{-4}$ gets smaller than 10 dB, even for a single hit requirement. For these numbers to apply, however, there can be only a single signal present in W_{ss} so, again, it describes a bounding case that is unlikely to occur in practice.

For an optimum FBC, $M = N_h$, where N_h is the number of hops in the message. Obviously, then, a PB FBC is a suboptimal detector.

FBC performance is illustrated in Figure 4.9 compared to a compressive receiver for the same parameters. The FBC is only slightly better than the compressive receiver in this case, indicating that the realizable compressive receiver is a good approximation to the FBC.

4.3.4.1 Stepping Wideband Radiometer

Radiometers were discussed at length in Chapter 2. We briefly discuss their application to receivers here.

A simplified block diagram of a receiver with an embedded radiometer is shown in Figure 4.10. Up until the last stages of the IF chain the receiver is the same as described previously. The last stage of the IF chain provides for noise limiting to constrain the amount of noise input to the radiometer. After such filtering, the signal is squared and then integrated over the interval $(0, T)$,

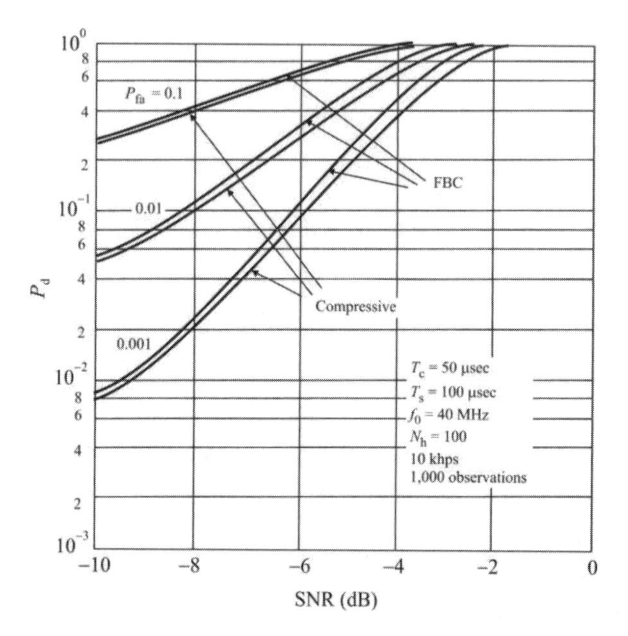

Figure 4.9 Compressive receiver and filter bank combiner (FBC) performance against FHSS example. (Source: [22]. © IEEE 1994. Reprinted with permission.)

effectively determining the energy present over the interval. After T seconds the output of the integrator is sampled and compared to the threshold γ_{th}. As above, if the integrator output is above the threshold at that time, then a signal presence is declared; otherwise, no signal present.

Krasner formulated the wideband radiometer detection performance evaluation as a nonlinear programming problem [27]. His analysis assumed that a wideband radiometer was stepped across a frequency span, dwelling at a channel for a period of time and that P_d and P_{fa} are specified, for each dwell (need not be the same for each dwell). The goal was to find the optimal dwell times for each

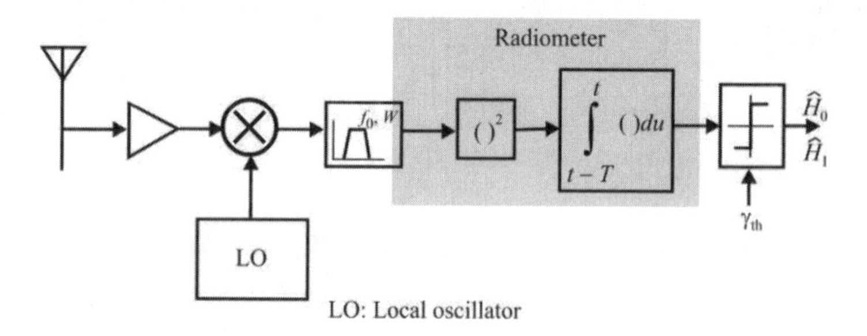

Figure 4.10 A radiometer detector embedded in a receiver.

dwell, assumed to be different, in general. We will summarize that analysis in this section.

We assume that a filter of bandwidth W_F is stepping through a band of total width W in search of a signal of bandwidth W_s. In our applications, W is fixed but larger than W_s, and the filter and signal have ideal rectangular shapes in magnitude versus frequency. We further assume that there is a total of M steps to be searched within the band W. These steps may either be disjoint or may overlap. If the signal to be detected may occur in any location within the input passband, then the adjacent steps should overlap by at least W_s to ensure that in one of the steps the signal is wholly contained within the stepping filter's passband. This will maximize the probability of detecting the signal. Overlap of adjacent steps by W_s implies a number of steps equal to $(W - W_s)/(W_F - W_s)$, assuming that this quantity is an integer. Note that there are no repeats of dwell bands. W is searched from one end to the other one time. What happens if repeat sweeps are allowed, which is a much more probable scenario?

For simplicity we assume that the SOI has energy in only one of the M steps. This is a fair approximation for the case in which $W >> W_s$ and adjacent steps are overlapped by W_s. Then the probability that the signal of interest straddles two steps is small.

The basic parameters are the dwell times of the steps, which we denote by T_i, $i = 0, \ldots, M - 1$. In general, $T_i \neq T_j$, although equality applies in some cases. We want to find those T_i such that the probability of detection is maximized, with a fixed P_{fa} (CFAR detection), and subject to the total time constraint T as

$$\sum_{i=0}^{M-1} T_i = T \tag{4.16}$$

Total time T is measured in "steps." We assume that the dwell time at each step is sufficiently long that square-law detection followed by integration at each step yields an output statistic that is Gaussian. As such a dwell step might be on the order of 1 ms, the total time of 10,000 units, for example, would then correspond to 10 seconds. As mentioned, however, the dwell time of 1 ms is the *average* time since we are seeking the optimum combination of dwell times to maximize the detection performance.

For DSSS we assume that the input SNR, SNR_i, measured in bandwidth W, is very small compared to unity. That is, $SNR_i << 0$ dB.

For a given frequency step, suppose that the input SNR has a specified functional form (the form need not be the same for each step). The output SNR of the square-law detector followed by the integrator is given by

$$\upsilon_{o,i} = K \frac{R_{s,i}^2}{N_{0,i}^2 W^2 + 2R_{s,i} N_{0,i} W} T_i W \tag{4.17}$$

$$\upsilon_{o,i} \triangleq K \frac{\gamma_i^2}{1 + 2\gamma_i} T_i W \tag{4.18}$$

$$= \Gamma_i T_i \tag{4.19}$$

where

$$\Gamma_i = K \frac{\gamma_i^2}{1 + 2\gamma_i} W \tag{4.20}$$

$$\gamma_i^2 = \frac{R_{s,i}^2}{N_{0,i}^2 W^2} \tag{4.21}$$

and where the subscript i denotes the ith frequency step, and $R_{s,i}$ and $N_{0,i}$ are the signal power and single-sided noise power density at the ith step, yielding γ_i as the input SNR in step i.[1] The constant K is on the order of unity.

The probabilities of detection and false alarm for a single step are

$$P_{fa,i} = Q(\gamma'_{th,i}) \tag{4.22}$$

$$P_{d,i} = Q\left[\gamma'_{th,i} - \upsilon_{o,i}^{1/2}\right] \tag{4.23}$$

where $Q(x)$ is the Q-function (Appendix A), and where $\gamma'_{th,i}$ is a detection threshold normalized by the detector output standard deviation with noise only at the input. $\gamma'_{th,i}$ is chosen to yield CFAR detection. Suppose we require that $P_{fa,i}$ be the same for each step and denote that value by P_{fa}. Let $P_{fa,M}$ denote the composite false alarm probability. Then

$$P_{fa,M} = 1 - (1 - P_{fa})^M \tag{4.24}$$

and

$$P_{d,i} = Q\left(\alpha - R_{s,i}^{1/2} T_i^{1/2}\right) \tag{4.25}$$

[1] The output SNR is defined as the ratio of the square of the mean shift due to the signal present to the output noise variance with the signal present.

where

$$\alpha = Q^{-1}\left[1-(1-P_{\text{fa,M}})^{1/M}\right] \tag{4.26}$$

In the above, $P_{\text{d},i}$ represents the probability of detection given that the signal is received in the ith frequency step. Note that, in general, different thresholds are used for different steps to maintain CFAR detection per step. The overall probability of detection for uniform a priori probability distribution on the T_i's is

$$P_{\text{d}} = \frac{1}{M}\sum_{i=0}^{M-1} Q\left(\alpha - \Gamma_i^{1/2} T_i^{1/2}\right) \tag{4.27}$$

Our problem then is to maximize (4.27) with respect to T_i, $i = 0, 1, \ldots, M-1$ given that (4.16) is obeyed and all the T_i's are nonnegative. This is a constrained nonlinear programming problem.

Now consider when a probability distribution is assigned to the SNR at each frequency step. That is, let us assume that $p_i(.)$ is the PDF corresponding to the probabilistic description of $\Gamma_i^{1/2}$ [which from (4.20) we see that in the low SNR regime is proportional to input SNR]. Then, to obtain the detection characteristics, we need to integrate our previous results with respect to the density function. That is, the optimal values for T_i are the solutions to the following maximization problem

$$\underset{T_i}{\text{maximize}}\, P_{\text{d}}(\bar{T}) \triangleq \frac{1}{M}\sum_{i=0}^{M-1}\int_{-\infty}^{\infty} Q(\alpha - uT_i^{1/2}) f_i(u)du \tag{4.28}$$

$$\text{subject to} \sum_{i=0}^{M-1} T_i = T; \qquad 0 \le T_i \le T, i = 0,1,\ldots,M-1 \tag{4.29}$$

Performance

Reference [27] provides an algorithm for optimally assigning the time to dwells that we won't discuss here. We will, however, show some of the results obtained by simulating the algorithm.

Figure 4.11 shows the probability of detection when the maximum total search time was 250,000 with the other parameters shown in the caption. For short T, the detection probability is quite low. The optimum dwell times, however, increase P_{d} much faster than when the dwell times were all set equal. For longer total times, say, beyond about 125,000, the detection performance is almost perfect.

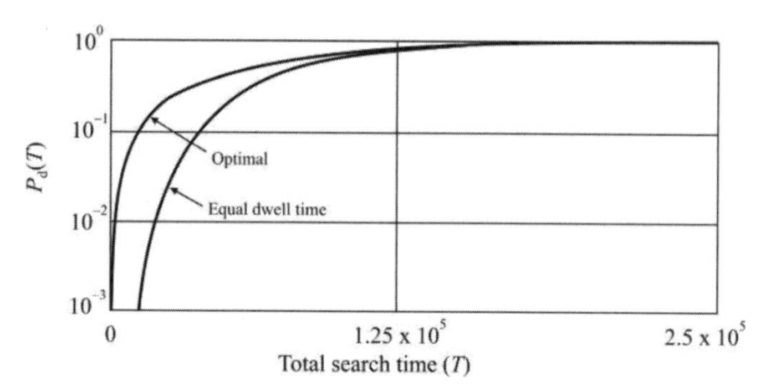

Figure 4.11 P_d for optimal dwell times and equal dwell times. Input SNR = –10 dB, a priori probability = 0.04, total search time = 250,000, $\alpha = 5$ yielding $P_{fa} = 10^{-5}$.

It is interesting to note that in this example, when T is small, the total time gets distributed to few dwells [27], but as it gets larger, the algorithm tends to distribute the time fairly evenly across the dwells. This performance can be seen as the two curves in Figure 4.11 coalesce at about 100,000 steps.

4.4 DSSS Signal Detection

We will discuss several techniques for detecting DSSS signals in this section [28–30]. To detect DSSS signals, the signals typically must be extracted from below the noise floor, or at least from one another as in CDMA. Wideband receivers are typically used and suppression of narrowband interference sources is usually required.

DS signals are normally detected with one of the energy detectors discussed in Section 4.2. Since DS signals do not change frequency, a starting receiver can look at the appropriate part of the spectrum for whatever length of time is required. For hybrid SS signals, where both FHSS and DSSS are used, this is not the case, however, and the receiver must be able to detect both modulation types.

For tactical communications, one of the issues that must be addressed by an interceptor is interference. The VHF and UHF spectrums are very crowded and narrowband interference is very likely to be present. In most countries, for example, the frequency of the lowest TV channel is around 50 MHz, in the middle of the low VHF military band. One method to mitigate such interference is with notch filters [31].

4.4.1 Signal Specific Detection

There are techniques for detection of DSSS signals that depend on the structure of the specific signals. Some of those techniques are presented in this section.

4.4.1.1 Squaring for BPSK

A BPSK modulated carrier signal can be expressed as

$$r(t) = \begin{Bmatrix} +1 \\ -1 \end{Bmatrix} \sin(2\pi f_c t + \phi_c) \tag{4.30}$$

with the +1 or –1 depending on the results of the product of the data signal and the spreading signal. If $r(t)$ is squared,

$$\begin{aligned} r^2(t) &= \begin{Bmatrix} +1 \\ -1 \end{Bmatrix}^2 \sin^2(2\pi f_c t + \phi_c) \\ &= \sin^2(2\pi f_c t + \phi_c) \\ &= \frac{1}{2} - \frac{1}{2}\cos(2\pi 2 f_c t + 2\phi_c) \end{aligned} \tag{4.31}$$

The effects of the digital modulation are removed and the square of the signal has a significant signal component at twice the carrier frequency. This characteristic can be used to detect whether a BPSK signal is present. Of course, the original carrier frequency must be known or estimated. Alternately, the squared spectrum can be examined for energy peaks. Where such peaks occur indicates a possible DSSS signal at half that frequency.

4.4.1.2 2^n PSK

A similar characteristic is observed for higher forms of PSK. However, for DSSS it is frequently the case that the signal is below the noise level, and thus $\upsilon < 1$. In that case raising the exponent reduces the SNR more because

$$\begin{aligned} \upsilon_{\text{dB}}^n &= 10\log\left(\upsilon^n\right) \\ &= 10n\log\left(\upsilon\right) \end{aligned} \tag{4.32}$$

but since $\upsilon = R / P_{\text{N}} < 1$, $\log(\upsilon) < 0$. Thus, the larger the value of n, the smaller the SNR becomes.

4.4.1.3 Cyclostationary Characteristics

Many data modulations for DSSS have statistical characteristics that vary in a periodic sense. The mean, variance, and so forth for these processes repeat

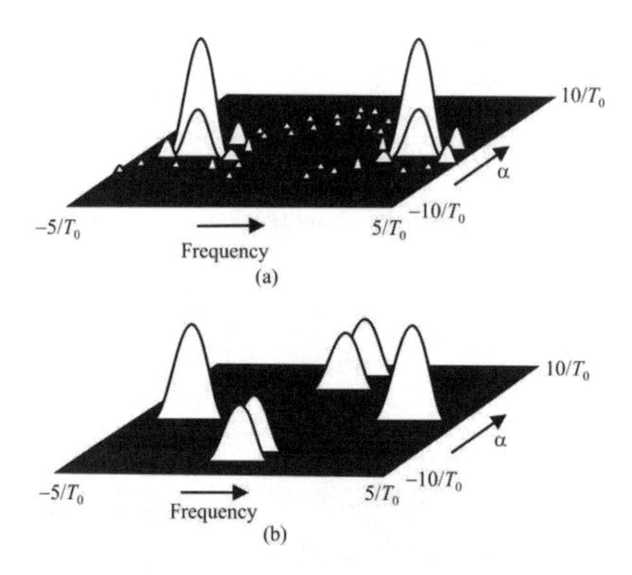

Figure 4.12 Spectrum of signals that exhibit cyclostationary statistical properties, such as most digital modulations. (a) QPSK and (b) MSK.. (Source: [32]. © 1988 IEEE. Reprinted with permission.) $T_0 = 1/f_0$ with f_0 the center frequency and α is the cyclic frequency.

themselves. This gives rise to the possibility of exploiting these characteristics to ascertain both the presence of the signal (detection) as well as determination of the type of signal (classification). An example of this is illustrated in Figure 4.12 [32]. The cyclic spectrum of QPSK is shown in Figure 4.12(a), while that for MSK is shown in Figure 4.12(b). Measuring these spectra in the parameter space shown would allow for successfully distinguishing between these two signals.

This, then, is a type of detection of DSSS signals that relies on characteristics of the signals themselves.

4.4.2 Nonsignal Specific Detection

Specific characteristics of the modulations were exploited in the last section to facilitate DSSS signal detection. It is possible to filter the signal at the transmitter to minimize these features, however, and thereby making detection using these techniques less reliable.

In addition to this shortcoming, such techniques are only useful when searching for specific signals—the so-called *directed search*. When *general search* for DSSS signals is required, when the types of signals in the spectrum are not necessarily known a priori, these approaches are inappropriate. General search is accomplished without reliance on signal structure. Two such approaches are presented in this section: the radiometer and a technique based on time series processing.

4.4.2.1 Radiometer Detection

The radiometer discussed in Section 2.7 is perhaps the most common way to detect the presence of DSSS signals. The measurement of the energy in a passband, which may or may not contain a signal, is performed. The resulting energy level is compared with a threshold, and, if larger, signal detection is declared. Values below the threshold are interpreted as noise, and values above the threshold are interpreted as signal presence. The threshold level is therefore critical for proper performance of a radiometer for DSSS signal detection. However, as discussed in Section 2.7, long-term variation in the background noise dictates that the threshold be adjusted if that background changes. Since the characteristics of the signal are not specifically exploited, the radiometer can be used against a wide variety of signals to detect energy.

The radiometer system is shown in Figure 4.10. The input bandpass filter selects the center frequency, f, and bandwidth, W, of interest. In general, these parameters are varied in a search for the correct W, and f, of the target signal. In practice, this signal is an IF signal where the bandwidth and frequency have already been accounted for in the filtering and mixing that precedes the radiometer. The squaring device follows the input filter to give a measure of received energy. Finally, the integrator integrates over the observation interval, T, for the energy measurement.

The output of the integrator is compared to a threshold and if the output exceeds the threshold, detection is declared. A false alarm occurs when a threshold crossing occurs with no signal present. With a deterministic signal and AWGN, the output statistics are characterized by a chi-square distribution. However, for large W_sT, the output statistics are approximately Gaussian. The Nieman Pearson test is parameterized by the probability of false alarm and the probability of correct detection.

The target signal is assumed to be a DSSS signal of the form

$$r(t) = A\sum_i d_i c(t - iT_c)\cos(\omega t) \tag{4.33}$$

where A is the amplitude, $\{d_i\}$ is a sequence of independent ± 1 r.v.s representing the data bits, $c(t)$ is the chip sequence, T_c is the pulse repetition period, and ω is the center frequency. The data rate is given by R_d in bits per second. The chip rate is given by $R_c = 1/T_c$ in chips per second. The ratio of bit energy to chip energy is the processing gain

$$G_p = \frac{R_c}{R_d} \tag{4.34}$$

This is the amount by which the power of jamming/interfering signals are reduced during the despreading process at the target receiver.

Performance Model

The performance model for the radiometer can be simplified by assuming that the radiometer output is Gaussian, which is the assumption often made. The approximation of the radiometer is based on the mean value

$$\mu = \begin{cases} \mu_{s+n} & \text{signal present} \\ \mu_n & \text{noise only} \end{cases} \tag{4.35}$$

and the standard deviation

$$\sigma = \begin{cases} \sigma_{s+n} & \text{signal present} \\ \sigma_n & \text{noise only} \end{cases} \tag{4.36}$$

For low input SNR, the means and variances in (4.35) and (4.36) are close to the same for either hypothesis; signal present or noise only. The signal is detectable if, after sufficient integration time, the difference between μ_{s+n}, and μ_n is greater than the standard deviation $\sigma_{s+n} = \sigma_n$. Thus, the radiometer performance can be given by the output SNR

$$\upsilon_o = \frac{(\mu_{s+n} - \mu_n)}{\sigma_{s+n}^2} \tag{4.37}$$

The approximation can be tailored to the DSSS signal by including the signal spectral terms.

4.4.2.2 Delay and Multiply Receiver

A code clock estimator is depicted in Figure 4.13. The technique used for the extraction of the chip rate is called "delay and mix," or "delay and multiply," with the resulting architecture referred to as the *delay and multiply receiver* (DMR). The idea is to delay the spreading sequence by a fraction of a chip duration (½ of a chip is illustrated in Figure 4.13) and multiply it by the undelayed sequence. This delay and multiply operation produces a spectral line at the bit rate that can be detected by a radiometer and/or tracked by a PLL. Once again, this detection method is effective on the binary DSSS systems because of the binary nature of their spreading sequences. However, since chaotic spreading sequences are not

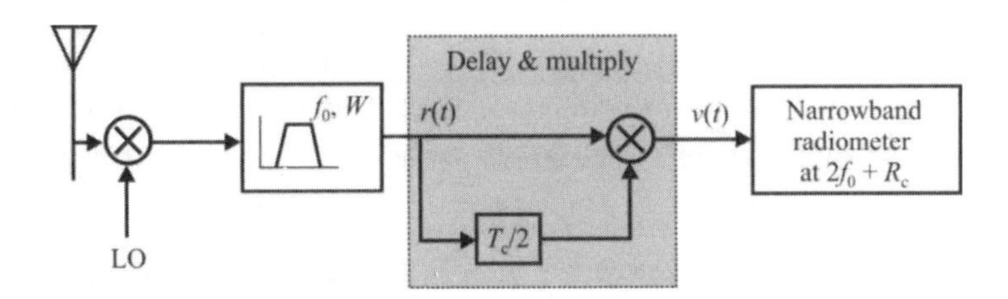

Figure 4.13 Delay and multiply receiver.

binary, it can analytically be shown that the chaotic DSSS system is resistant to this method of detection as well.

The delay and multiply detector, like most other suboptimal DSSS detectors, does not operate well when the SNR is negative [33]. When the noise power is greater than the signal power, producing a negative SNR, the operation of multiplication actually reduces the resulting SNR even further, as explained in Section 4.4.1.2.

The output bandpass filter of the DMR selects the rate tone. If the actual rate tone is not known exactly, a search in the center frequency, $f_r = 1/T_c$, must be performed, which makes the DMR more complex than the radiometer. The number of rate-tone frequencies that must be searched depends on the uncertainty of f_r and the bandwidth of the output filter which is assumed to be $1/T_i$, where T_i is the integration, or observation, time.

The output of the multiplier can be divided into four components: rate tone, self noise, signal-times-noise, and noise-times-noise. For the case of very weak signals, as DSSS signals generally are, all but the first and last terms may be ignored. This leads to an expression for the SNR at the output of the DMR to be [34]

$$\upsilon_o = \frac{(E_c / N_0)^2 P_S}{P_{nn}} \tag{4.38}$$

where E_c is the energy of one chip in the spread-spectrum signal, N_0 is the one-sided noise spectral density at the input, P_s is the normalized power of the rate tone, and P_{nn} is the normalized background noise power at the output filter of the DMR.

4.4.2.3 Comparison of the DMR and Radiometer Intercept Receiver in Nonstationary Environments

The radiometer is a standard for detection of unknown signals in noise [35], and is optimal when the noise is AWGN. In ideal cases, it can detect arbitrarily weak signals by integrating received energy for a sufficiently long time. Detectors that take advantage of the cyclostationarity of digitally modulated signals can have superior performance in a nonstationary noise environment [36]. This phenomenon is discussed here with simple models used to show the relationships between spread spectrum processing gain (PG), detection performance, and noise fluctuation.

The source of noise fluctuation may be temperature drift in the receiver (warmer receiver front-ends will be noisier), changes in external noise due to antenna pointing or otherwise changing in the physical environment, or fluctuation of interfering signals. This fluctuation in received energy, not caused by the signal of interest, can cause false detections or make it difficult to set the detection threshold.

The quadratic detector considered here is the DMR discussed in Section 4.4.2.2. This circuit, used as a detector, can be used to overcome the problem with nonstationary noise. This is done with added complexity at the intercept receiver. The DMR may be countered with proper signal design [37]. This is also considered in the comparison of the radiometer and DMR.

Stationary Noise

We now compare the performances of the radiometer and DMR for the case of stationary noise which is the ideal case. The performances are investigated and the effects of detector parameters are considered. Simple models are used to find the operating threshold of each detector.

Radiometer

The radiometer performance is very similar to that of the DMR. For very low input υ_j, the output SNR is given by (4.38) with appropriate modification of P_s, and P_{nn}. Thus, the radiometer SNR, is also proportional to $(E_c / N_0)^2$. The performance depends on the prefilter bandwidth, W, as does that of the DMR. One of the advantages of the radiometer is that there is no delay to set, so its performance is simpler to quantify.

The sensitivity of performance to bandwidth is important because it determines how the detection threshold affects the searches in center frequency and chip rate. In Figure 4.14 we plot the product $\upsilon_o \times W_n$ versus WT_c, assuming a fixed target signal chip rate. Here $W_n = T_c/T_i$ is the normalized bandwidth of the integrator. Notice that WT_c can range from 0.25 to 2.0 with only a 3 dB loss in normalized SNR. The optimum WT_c is approximately 0.5 where the bandpass radiometer bandwidth is matched to the chip rate. The optimum $\upsilon_o \times W_n$ is –23 dB for $E_c/N_0 = -10$ dB. For other signal levels, assuming they are less than 0 dB, we can approximate υ_o by

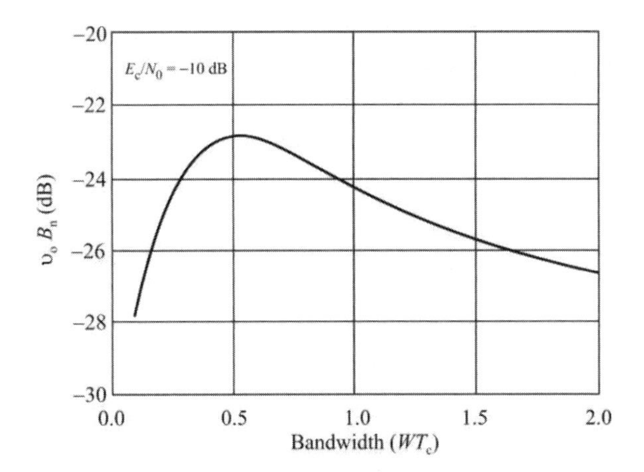

Figure 4.14 Normalized SNR versus normalized bandwidth. (Source: [38]. © IEEE 1989. Reprinted with permission.)

$$\upsilon_o W_n = -23 + 2\left(\frac{E_c}{N_0} + 10\right) \text{ dB} \tag{4.39}$$

As predicted in [37] with υ_o = 10 dB and $P_{fa} = 10^{-3}$, the probability of detection is $P_d = 0.8$. Using this detection criteria the input CNR threshold can be found. First E_c/N_0 is solved for in (4.39) letting υ_o = 10 dB. Because of the bandwidth factor of 2 between R_c and the null-to-null BPSK bandwidth, the CNR is 3 dB less than E_c/N_0. Finally, letting $W = 1/T_j$, we have $W_n = 1/(R_c T_j)$. The resulting threshold for radiometer detection is

$$CNR_t = 3.5 - 5\log_{10}(R_c T_i) \text{ dB} \tag{4.40}$$

Thus, given the chip rate and the integration time, (4.40) is the lowest CNR for which the detection criteria can be met.

DMR

The DMR multiplies the signal by a delayed version of itself, which produces the rate line and provides the chip rate of the target signal. When the exact signal bandwidth and R_c are not known, a variable bandwidth and center frequency can be incorporated in the DMR. However, the delay also affects performance and it must be chosen to give a good performance over a range of uncertainty in R_c.

The DMR performance is given by the equations in [39]. These are used to plot υ_o versus WT_c in Figure 4.15 where graphs are provided for $T_d W$ = 0.3, 0.5, 0.7, and 0.9. The delay is set relative to the prefilter bandwidth W. We note that if

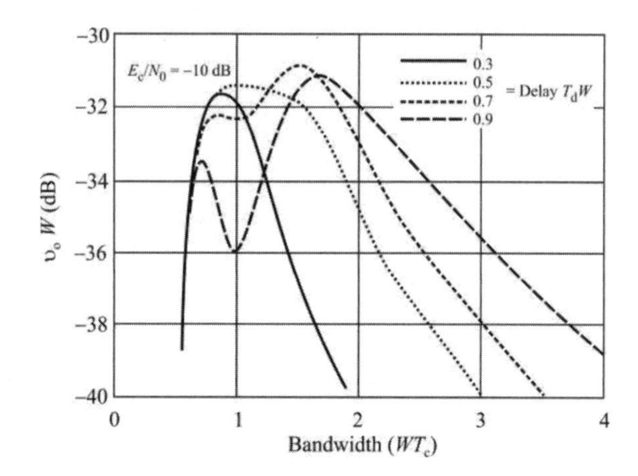

Figure 4.15 DMR performance versus normalized bandwidth. (Source: [38]. © 1989 IEEE. Reprinted with permission.)

T_d is chosen too small, the DMR has a much higher sensitivity to bandwidth than the radiometer. The curve for $T_dW = 0.5$ shows the least bandwidth sensitivity without a ripple. The curve for $T_dW = 0.9$ shows a better performance at large bandwidths, but has a ripple of variation in υ_o. In this curve, we see the effects of the bandwidth and delay. The bandwidth causes the first peak in υ_o and the delay causes the second.

Once the delay is set, we see that the DMR bandwidth can vary from WT_c equal to 0.6 to 2.25. Thus, the DMR is more sensitive to bandwidth mismatch than is the radiometer. The optimum bandwidth gives $\upsilon_o W = -31$ dB. This is 8 dB worse than the radiometer, but because of the square relationship between input and output SNR, this means a 4 dB difference in CNR threshold.

The detection performance can again be extrapolated giving the approximate υ_o as

$$\upsilon_o W = -31 + 2\left(\frac{E_c}{N_0} + 10\right) \text{ dB} \tag{4.41}$$

Using the detection criteria of $\upsilon_o = 10$ dB as before, we find the input CNR threshold to be given by

$$CNR_t = 7.5 - 5\log_{10}(R_c T_i) \text{ dB} \tag{4.42}$$

With the CNR thresholds from (4.40) and (4.42), we have the relationships between spreading rate, integration time, and CNR. We see from (4.40) and (4.42) that the CNR_t decrease as the square root of the chip rate. Also, the detectability threshold can be lowered by increasing the integration time. The threshold decreases as the square root of the integration time.

The target signal can lower the CNR by increasing G_p. This is done by keeping the data rate, R_d, constant and increasing the chip rate, R_c, according to (4.34). Assuming that the target system needs CNR = 9 dB after despreading, the CNR in the spread bandwidth is given by

$$\text{CNR}_{\text{RF}} = 9 - 10\log_{10}\left(\frac{R_c}{R_d}\right) \text{dB} \tag{4.43}$$

So, for a fixed data rate, the target signal can reduce CNR proportionally to the increase in R_c. Thus, the target signal has this advantage over both the radiometer and DMR.

The detectability threshold for a radiometer and achievable CNR to spread bandwidth are shown in Figure 4.16 where two example data rates are shown: 300 bps and 16 kbps. Detection threshold curves are shown for the radiometer and the DMR with 1 second integration time. Notice that any pair of target CNR_{com} line and CNR_t line have an intersection point. To the left of this point are the spread bandwidths for which the target signal can be detected. To the right of the intersection point are the spreading bandwidths for which the target signal cannot be detected with sufficient quality at the given integration time.

We see that for a 300 bps data rate and the maximum spreading bandwidth, the radiometer cannot detect the target signal with a 1 second integration time. However, for a 16 kbps data rate, the target signal can be detected using its maximum spreading rate with the DMR.

Nonstationary Noise Environment

In the previous section there were no interference fluctuations. If interference fluctuations are present during detection, the performance of the detectors deteriorates. This section considers the effects of interference fluctuations on detection performance.

First, consider the effects of interference fluctuations on radiometer performance. The radiometer detects the presence of a signal by the change in received background energy. Without signal interference this background energy is due to the various noise sources. This level changes very slowly and is considered to be constant. When a signal other than noise is received, the received energy level is changed and the increase is detected. With interference fluctuations present, the background energy will change with these fluctuations.

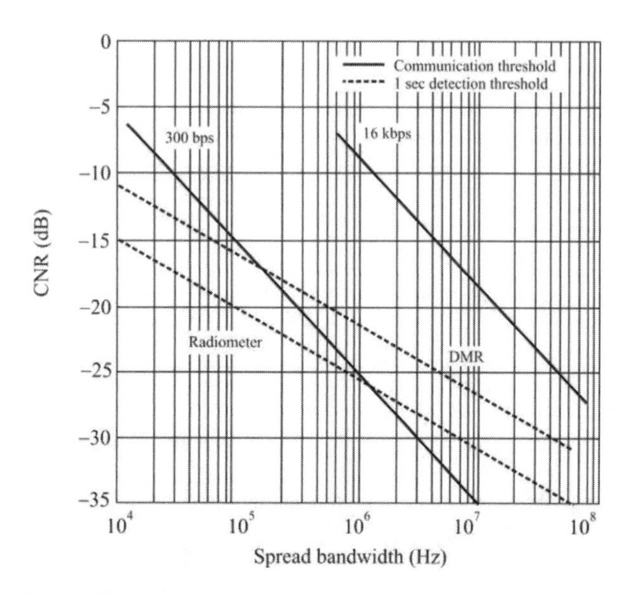

Figure 4.16 Stationary noise thresholds. (Source: [38]. © IEEE 1989. Reprinted with permission.)

The background energy received by a radiometer may vary unpredictably in real signal environments. This variation can be modeled as an additive r.v. on the mean output of the radiometer. Thus,

$$\mu = \mu_0(1+\varepsilon) \tag{4.44}$$

where ε is a zero mean r.v. with standard deviation σ_ε. With this model for the interference fluctuations we can derive a new approximation for the radiometer's output SNR. The interference variation adds to the output variance, μ_o, in (4.37). Combining this interference fluctuation model with (4.37) gives

$$\upsilon_o = \frac{(E_c / N_0)^2}{\sigma_n^2 + \sigma_\varepsilon^2} \tag{4.45}$$

This is an approximation for the case of low input SNR and no bandwidth mismatch. This approximation leads to a different detection threshold, CNR_t, for the radiometer with interference fluctuations

$$CNR_t = 3.5 + 5\log_{10}\left[\frac{1}{R_c T_i} + \frac{\sigma_\varepsilon^2}{2}\right] \tag{4.46}$$

Of course, the detection threshold with interference fluctuations is the same as before when $\sigma_e = 0$ (no interference fluctuations). When $\sigma_e \neq 0$, on the other hand, there is a lower bound on the detection threshold. That is, after a point we cannot increase the detection probability by increasing the integration time. Thus, the interference fluctuations can limit the performance of the radiometer.

Next we consider the effect of interference fluctuations on DMR performance. The interference fluctuations do not affect the strength of the rate line amplitude with the DMR. The interference fluctuations may increase or vary the background noise level, but increasing the integration time in the DMR overcomes this problem. Thus, the detection threshold is not changed significantly by interference fluctuations.

There is another consideration, however, with the use of the DMR when interference fluctuations are present. By design, the DMR produces tones, or rate lines, from received signals and the interference fluctuations may include some digitally modulated carriers. In that case, the DMR output will have rate lines at the symbol rate of the interference signals as well. Also, if unmodulated carriers are present in the interference, there will be tones at the beat frequencies (i.e., difference frequencies). These added rate lines and beat tones can easily be confused with the actual rate line of the signal of interest.

In the worst case, the interference could add a spectral line at the DMR output that would cover up the target signal's rate line. However, with a 1 second integration time, the interfering tone would have to be within 1 or 2 Hz of the target signal chip rate. Since this is unlikely, the interference fluctuations are assumed to have no effect on the DMR detection threshold. In addition, if the interference rate line is equal to that of the SOI, then the chip rate will be properly determined even though it may not belong to the SOI.

If filtering is used by the communicator in order to defeat the DMR, the performance of the DMR may be severely degraded. For example, filtering may reduce the rate tone SNR by 20 dB. In this case, the detection threshold will be raised by 10 dB.

Finally, we consider the detection thresholds of the radiometer and the DMR with interference fluctuations. These are depicted in Figure 4.17 where CNR_t is plotted versus spread bandwidth for the radiometer and the DMR. Both use a 1 second integration time. The radiometer threshold is plotted for three values of interference fluctuation: σ_e = 0.01, 0.1, and 1%. Notice that the DMR performs better than the radiometer for any amount of interference fluctuation if the input bandwidth is large enough. Even with only 1% variation in the background power level, the DMR performs better than the radiometer for the spread bandwidths of 200 kHz or greater. However, for the case of signal pulse filtering, labeled DMR/LPI, the radiometer has better performance for spread bandwidths up to 10 MHz.

We compared the performances of the radiometer and DMR intercept receivers in this section. The DMR can give superior performance in an

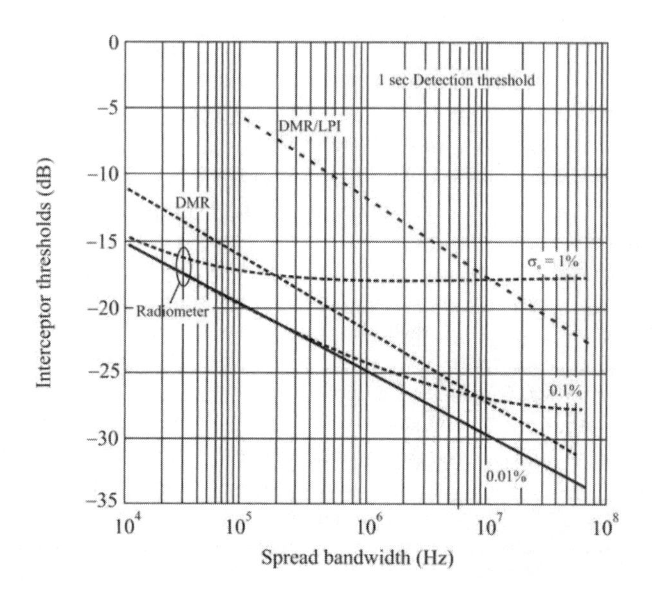

Figure 4.17 Interceptor thresholds with nonstationary noise. DMR/LPI represents signal pulse filtering be the target communicator. (Source: [38]. © IEEE 1989. Reprinted with permission.)

environment with nonstationary interference. This is achieved at the cost of higher complexity at the detector (adding a delay module). Some of this performance gain can be nullified if the target uses filtering to defeat the DMR. Which detector is most appropriate depends on the target signal and the cost complexity trade space.

4.4.2.4 Autoregressive-Based DSSS Detection

As above, the estimation problem associated with detection of DSSS signals is to determine whether a noise signal $r(t) = n(t)$ or a signal plus noise signal, $r(t) = s(t) + n(t)$ is present [40]. Denote the noise process by $\{n_i\}_1^n$ and the signal process by $\{s_i\}_1^n$.

Suppose that $\{y_i\}_{-\infty}^{\infty}$ represents a stationary random process. For every sample, $\mathcal{E}\{y_i\} = 0$ and $\text{var}\{y_i\} = \sigma_0^2$. This process is an *autoregressive* (AR) process of order p if the samples can be represented as

$$r_n = \sum_{j=1}^{p} a_j r_{n-j} + e_n \tag{4.47}$$

where the coefficients a_j are the autoregressive coefficients, e_n is a sample from a wide sense stationary Gaussian process with zero mean, and covariance σ_p^2 called the *innovation* of the random process. That is, sample y_n can be determined from the previous p samples plus a sample from the Gaussian innovations process. Gaussian noise alone is represented by an AR(0) model, where the p AR coefficients are all 0. The innovation process can be considered an error term in representing y_n by a linear combination of the previous p values

$$e_n = r_n - \sum_{j=1}^{p} a_j r_{n-j} \tag{4.48}$$

Any discrete time sequence can be perfectly represented by an AR(p) process for some p, which may be infinite. For time sequences that require a large value of p, often a smaller number of values are included and the subsequent modeling error is accepted. Therefore, sampled DSSS signals can be represented by an AR process. The coefficients a_i and the samples e_i are unknown, and the autoregressive method of DSSS signal detection relies on determining whether the AR(p) model is present or the AR(0) model is present. Therefore, the null hypothesis is

$$H_0 : r_i = n_i, \qquad i = 1, 2, \cdots, n$$

where n_i are samples from a zero mean Gaussian noise process, and the alternate hypothesis is

$$H_1 : r_i = s_i + n_i, \quad i = 1, 2, \cdots, n$$

where r_i is a sample of an order p AR process.

The likelihood ratio in this case is given by

$$z = \frac{\hat{\sigma}_0^2}{\hat{\sigma}_p^2} \tag{4.49}$$

where $\hat{\sigma}_0^2$ is determined by the previous n sample data and $\hat{\sigma}_p^2$ is an estimate of the variance of the pth order innovations. The numerator is estimated by

$$\hat{\sigma}_0^2 = \frac{1}{n} \vec{r}^{\mathrm{H}} \vec{r} \tag{4.50}$$

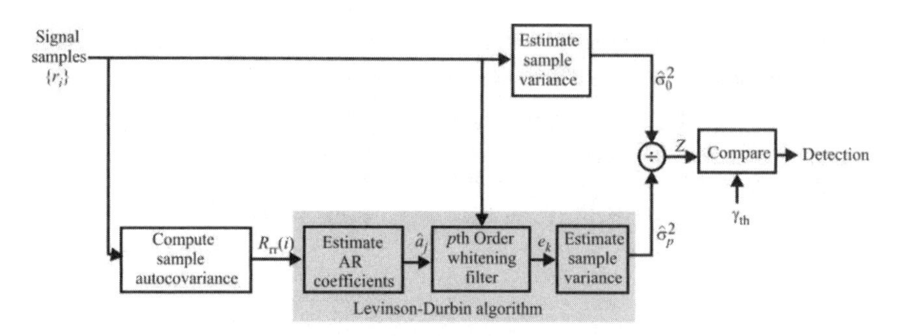

Figure 4.18 AR algorithm for DSSS signal detection. (Source: [41].)

when the n values of the sample process are represented as a vector $\vec{r} = (r_1, r_2, \cdots, r_n)^{\mathrm{T}}$. The denominator is calculated by first computing lags 0 through p of the sample autocorrelation function of r

$$R_{\mathrm{rr}}(i) = \frac{1}{n-p} \sum_{k=0}^{n-p} r_{\mathrm{k}}^{*} r_{\mathrm{i+k}} \tag{4.51}$$

and applying an appropriate algorithm to find $\hat{\sigma}_p^2$ from these lags. The technique is summarized in the flow diagram depicted in Figure 4.18 [41].

The algorithm for computing $\hat{\sigma}_p^2$ shown in Figure 4.18 is the Levinson-Durbin algorithm that is an efficient algorithm for solving the Yule-Walker equations associated with this AR model. The Yule-Walker equations are given by [42]

$$R_{\mathrm{rr}}(k) = -\sum_{j=1}^{n} a_{\mathrm{j}} R_{\mathrm{rr}}(k-j), \quad k \geq 1 \tag{4.52}$$

$$R_{\mathrm{rr}}(0) = \sigma^2 - \sum_{j=1}^{n} a_j R_{\mathrm{rr}}(-j) \tag{4.53}$$

Therefore, if the n + 1 values of $R_{\mathrm{rr}}(i)$ are known or can be estimated then, the n + 1 model parameters a_j and σ^2 can be found from these n + 1 equations. However, the Levinson-Durbin algorithm is more efficient than solving these n + 1 equations directly.

This algorithm recursively calculates a_n and σ_n^2 from a_{n-1}, $a_{n-2}, \ldots, a_1$. The initial values are [42]

$$a_1(1) = -\frac{R_{rr}(1)}{R_{rr}(0)} \tag{4.54}$$

$$\sigma^2(1) = [1 - a_1^2(1)]R_{rr}(0) \tag{4.55}$$

The nth values are

$$a_n(n) = -\frac{1}{\sigma^2(n-1)}\left[R_{rr}(n) + \sum_{q=1}^{n-1} a_q(n-1)R_{rr}(n-q)\right] \tag{4.56}$$

$$\sigma^2(n) = \left[1 - a_n^2\right]\sigma^2(n-1) \tag{4.57}$$

$$a_j(n) = a_j(n-1) + a_n(n)a_{n-j}(n-1), \quad j = 1,2,3,\cdots,n-1 \tag{4.58}$$

The value of use in (4.49) is

$$\hat{\sigma}_p^2 = \sigma^2(p) \tag{4.59}$$

when p is the order of the AR model. This technique is modulation independent, and does not require any particular structure to be present in the signal.

One advantage of this approach over the radiometer is its relative insensitivity to long-term variations in the background noise (stationarity). This insensitivity is illustrated in Figure 4.19, which illustrates P_{fa} as the noise variance estimate error increases away from nominal (at zero variance). The probability of detection as the error in this estimate changes is illustrated in Figure 4.20. For the radiometer, whereas the probability of detection increases as the error increases, the probability of false alarm does as well. This feature is not present in the AR-based detection process.

4.4.2.5 Fluctuations of the Autocorrelation

Burel, Quinquis, and Azou [43] presented a method of not only detecting the presence of a DSSS signal, but, determining the symbol period also based on the eigendecomposition of the autocorrelation matrix of the received signal. We present a summary of their approach in this section.

It is assumed that the DSSS signal uses short codes so that a full period of the spreading sequence occurs in each symbol. It is based on the behavior of DSSS statistical properties versus those of random thermal noise.

The DSSS signal is given by

$$s(t) = \sum_{k=-\infty}^{\infty} d_k h(t - kT_s) \tag{4.60}$$

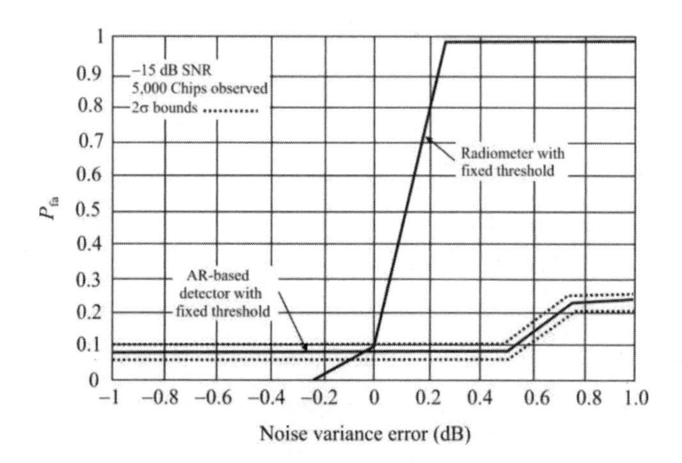

Figure 4.19 Probability of false alarm variation of the AR-based DSSS detector versus that of a radiometer, as the long-term background noise level changes. (Source: [41].)

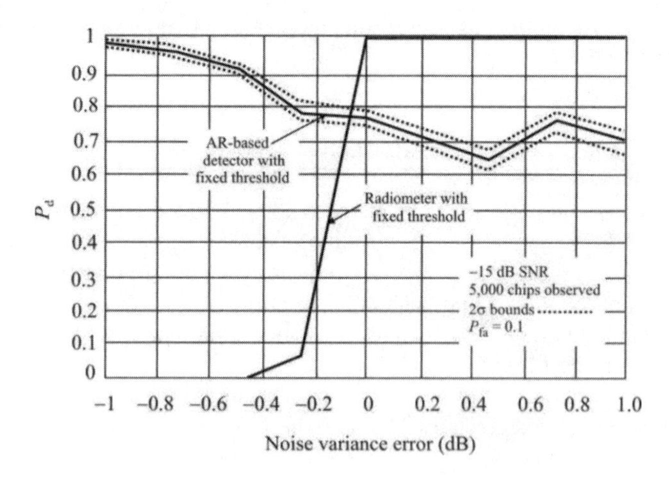

Figure 4.20 Probability of detection variation of the AR-based DSSS detector versus that of a radiometer, as the long-term background noise level changes. (Source: [41].)

where $h(t)$ is the convolution of the pseudo-random sequence with the transmitter filter (and with the channel echoes if there are any). We make the following assumptions:

- The symbols are centered and uncorrelated.
- The noise at the input of the receive filter, $n_c(t)$, is white, Gaussian, centered, and uncorrelated with the signal. Its PSD is $N_0/2$.
- The signal is hidden in the noise. [The signal-to-noise ratio (in decibels) at the output of the receiver filter is negative.]

The signal received at the EW receiver is given by

$$r(t) = s(t) + n(t) \tag{4.61}$$

where $n(t)$ represents the noise.

Signal Detection

First, a technique is presented for detection of the presence of a DSSS signal that is below the noise based on computation of the first- and second-order statistics of the autocorrelation estimate. The received signal is divided into M nonoverlapping windows of duration T, which contains a few symbols or more. Within each of these windows the autocorrelation estimate is found as

$$\hat{R}_{rr}^{(m)}(\tau) = \frac{1}{T}\int_0^T r_m(t) r_m^*(t-\tau)\,dt \tag{4.62}$$

where m is an index on the windows. The second order moment estimate is calculated based on (4.62) as

$$\rho(\tau) = \frac{1}{M}\sum_{m=1}^{M}\left|\hat{R}_{rr}^{(m)}(\tau)\right|^2 \tag{4.63}$$

which is an indication of the variability of the autocorrelation estimate. Under the assumption of a flat frequency channel response, with no signal present, then the expected value and standard deviation of ρ are given by

$$\mathcal{E}\{\rho\} = \mu_\rho^{(n)} = \frac{1}{TW}\sigma_n^4 \tag{4.64}$$

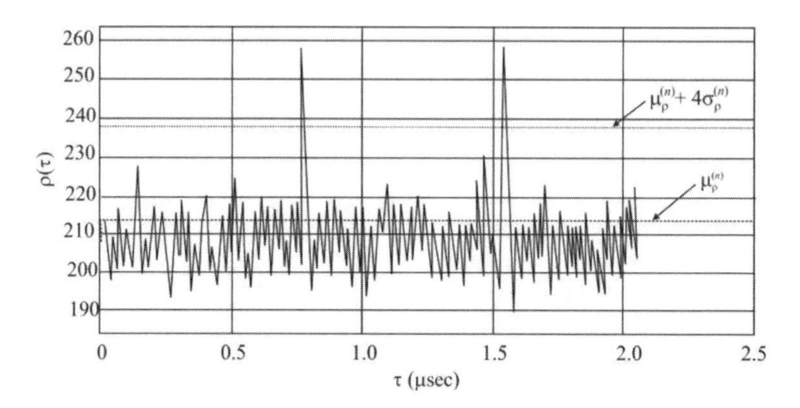

Figure 4.21 Detector output when a DSSS signal is present 8 dB below the noise floor. The two horizontal lines show $\mu_\rho^{(n)}$ and $\mu_\rho^{(n)} + 4\sigma_\rho^{(n)}$. ($\mu_\rho$ = fluctuations mean, σ_ρ = fluctuations standard deviation.) The peaks occur at multiples of the symbol period. (Source: [43]. © IEEE 2002. Reprinted with permission.)

(σ_n is the noise standard deviation) and

$$\sigma_\rho^{(n)} = \frac{\mu_\rho^{(n)}}{\sqrt{M}} \tag{4.65}$$

respectively. With a signal present below the noise floor, its contribution to the fluctuation is small except in regions around τ, which are multiples of the symbol period. In those regions that are multiples of the symbol period, the expected value becomes

$$m_\rho^{(\mathrm{S})} = \frac{T_s}{T}\sigma_s^4 \tag{4.66}$$

(σ_s is the signal standard deviation). Then

$$\frac{m_\rho^{(\mathrm{S})}}{\sigma_\rho^{(n)}} = \sqrt{\frac{M}{2}T_s W}\frac{\sigma_s^4}{\sigma_n^4} \tag{4.67}$$

A representative example of $\rho(\tau)$ is plotted versus τ in Figure 4.21. If the peaks are present, then there is a signal hidden in the noise. From this computation, an estimate T_s of the symbol period is available as the distance between the peaks in the curve. In this example, $\mu_\rho^{(n)} \approx 213$ and $\mu_\rho^{(n)} + 4\sigma_\rho^{(n)} \approx 238$, while the estimated symbol time, $T_s \approx 0.8\,\mu\text{s}$ implying that the symbol rate is 1.25 Mbps.

Determining the Threshold

To determine the detection threshold, we assume that noise only is present and compute the theoretical average value and standard deviation of $\rho(\tau)$ given by (4.63). The theoretical average value of the fluctuations is given by

$$\mu_\rho^{(n)} = \mathcal{E}\{\rho(\tau)\} = \mathcal{E}\left\{\left|\hat{r}_{\text{nn}}(\tau)\right|^2\right\} \tag{4.68}$$

This is the average power of the estimated autocorrelation signal. Thus

$$\mu_\rho^{(n)} = \int_{-\infty}^{\infty} p_{\hat{R}}(u)du \tag{4.69}$$

where $p_{\hat{R}}(u)$ is the PSD of $\hat{R}$. If the duration of the analysis window, T, is not too small,

$$p_{\hat{R}}(u) = \frac{1}{T}\left|p_{\text{n}}(u)\right|^2 \tag{4.70}$$

and

$$\mu_\rho^{(n)} = \frac{1}{T}\int_{-\infty}^{\infty} \left|p_{\text{n}}(u)\right|^2 du \tag{4.71}$$

The PSD of the noise at the output of the receiver filter is given by

$$p_\mu(\nu) = \left|G(\nu)\right|^2 \frac{N_0}{2} \tag{4.72}$$

where $G(\nu)$ is the Fourier transform of the receiver filter impulse response, $g(t)$.The standard deviation of the fluctuations is given by

$$\sigma_\rho^{(n)}(\tau) = \sqrt{\text{var}\{\rho(\tau)\}} \tag{4.73}$$

Since the windows are independent, using (4.63), we obtain

$$\sigma_{\text{n}}^{(n)} = \sqrt{\frac{1}{N}\text{var}\left\{\left|\hat{r}_{\text{nn}}(\tau)\right|^2\right\}} \tag{4.74}$$

where

$$\mathrm{var}\left\{\left|\hat{r}_{\mathrm{nn}}(\tau)\right|^2\right\} = \mathcal{E}\left\{\left|\hat{r}_{\mathrm{nn}}(\tau)\right|^4\right\} - \left[\mu_\rho^{(n)}\right]^2 \tag{4.75}$$

The statistical behavior of $\hat{r}_{\mathrm{nn}}(\tau)$ is essentially Gaussian, because it is the average of a large number of independent random variables. Furthermore, except for low values of τ, its average value is zero. Hence,

$$\mathcal{E}\left\{\left|\hat{r}_{\mathrm{nn}}(\tau)\right|^4\right\} = 3\left[\mu_\rho^{(n)}\right]^2 \tag{4.76}$$

Note that this result does not depend on τ. Then we obtain

$$\sigma_\rho^{(n)} = \sqrt{\frac{2}{N}}\mu_\rho^{(n)} \tag{4.77}$$

Therefore, based on the preceding analysis, determining the threshold requires the following steps:

1.Compute the PSD, $p_{\mathrm{n}}(\nu)$, of the signal at the output of the receiver filter assuming only noise is present.
2.Compute the theoretical average value $\mu_\rho^{(n)}$, of the fluctuations (4.71).
3.Compute the theoretical standard deviation $\sigma_\rho^{(n)}$ of the fluctuations (4.77).
4.Compute the theoretical approximate upper bound of the fluctuations: $\mu_\rho^{(n)} + 4\sigma_\rho^{(n)}$.

Detection

Figure 4.21 shows the curve $\rho(\tau)$ computed by the data fluctuations and the threshold. If no signal is present, the curve remains under the threshold with a high probability. If a signal is hidden in the noise, the curve goes above the threshold for multiples of the symbol period.

High values of $\rho(\tau)$ are obtained for every τ multiple of the symbol period T_s. This can be seen from the following. From (4.60) and (4.62) we can show that:

$$\hat{r}_{\mathrm{ss}}(T_s) = \frac{1}{T}\sum_{k=-\infty}^{\infty} d_k d_{k-1}^* \int_0^T \left|q(t-kT_s)\right|^2 dt \tag{4.78}$$

where

$$q(t) = g(t) * h(t) \tag{4.79}$$

is the convolution of $h(t)$ with the receiver filter impulse response $g(t)$. With some algebra, the average value of its square modulus reduces to

$$\mu_\rho^{(S)} = \mathcal{E}\left\{\left|\hat{r}_{ss}(\tau)\right|^2\right\} = \frac{T_s}{T}\sigma_s^4 \tag{4.80}$$

This is the contribution of the noise-free signal.

We assume a rectangular filter frequency response on [$-W/2$, $+W/2$]. The contribution of the noise is found from (4.71). We can show that:

$$\mu_\rho^{(n)} = \frac{1}{TW}\sigma_n^4 \tag{4.81}$$

Hence, from (4.77), the standard deviation of the fluctuations is

$$\sigma_\rho^{(n)} = \sqrt{\frac{2}{N}\frac{1}{TW}}\sigma_n^4 \tag{4.82}$$

then

$$\frac{\mu_\rho^{(S)}}{\sigma_\rho^{(n)}} = \sqrt{\frac{N}{2}}T_s W \frac{\sigma_s^4}{\sigma_n^4} \tag{4.83}$$

This is the ratio between the mean value of the peaks created by the DSSS signal (if there is one such signal hidden in the noise), and the standard deviation of the noise. However, (4.83) is not easy to use because σ_n and σ_s depend on W.

We can show that the optimal value of the receiver filter bandwidth is approximately $1/T_c$. In that case, we have

$$\frac{\mu_\rho^{(S)}}{\sigma_\rho^{(n)}} = \sqrt{\frac{N}{2}}M \frac{\sigma_s^4}{\sigma_n^4} \tag{4.84}$$

where M is the length (in bits) of the pseudo-random sequence.

Example: Suppose

$$\frac{\sigma_s^2}{\sigma_n^2} = 0.05 \text{ (input } \upsilon = -13\text{dB)}$$

$$N = 1{,}000 \text{ (size of analysis windows)}$$

$$M = 255$$

The output SNR is 14.2 (+11.5 dB).

Performance

Figure 4.21 shows an example of the detector output. The horizontal axis represents τ (in μs) while the curve represents $\rho(\tau)$ (i.e., the estimated fluctuations of the autocorrelation estimator). The horizontal lines represent the theoretical mean fluctuation and the approximate upper bound. We can see the peaks extending above the upper bound. Furthermore, we can see that these peaks are located at multiples of a given period. This means that a DSSS signal is hidden in the noise, which is the case. The parameters of this signal are: input $\upsilon = -10$ dB, $M = 31$, and $N = 255$.

Summary

DSSS signals are, by design, difficult to detect. Indeed, they are often transmitted below the noise level. Furthermore, a DSSS signal is especially built to be similar to a noise, in order to have a low probability of detection and interception. The autocorrelation of a spread spectrum signal is close to a Dirac delta function, as well as the autocorrelation of white noise (this is due to the pseudo-random sequence). The detector described here is based on the fluctuations of autocorrelation estimators, rather than on the autocorrelation itself. Although the autocorrelation of a DSSS signal is similar to the autocorrelation of noise, we have shown that the fluctuations of the second order estimators when a signal is buried in the noise versus just noise alone are different. The proposed method is able to estimate the symbol period of the DSSS signal. It should be noted, however, that it was assumed that the spreading code is short so that at least one complete cycle through the code is contained within every symbol.

4.5 Frequency-Hopping Spread Spectrum Signal Detection

4.5.1 Introduction

The FBC and PB FBC discussed in Section 4.3 are the most common methods to detect the presence of an FHSS signal. As mentioned, however, simply detecting

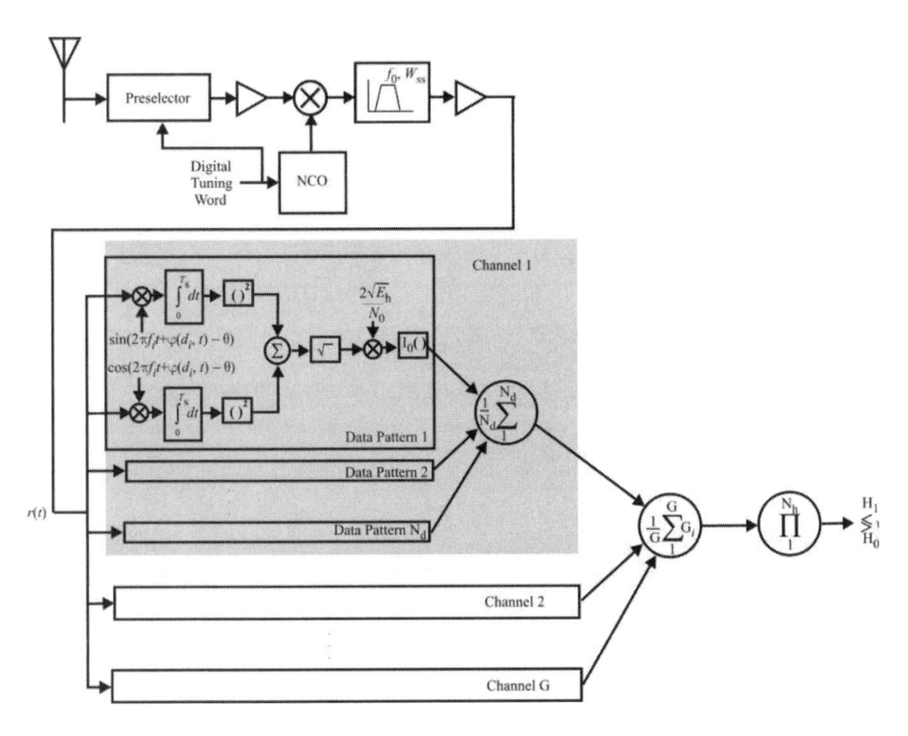

Figure 4.22 Optimal receiver architecture for noncoherent SFHSS signals. (Source: [44]. © IEEE 1994. Reprinted with permission.)

the presence is rarely sufficient in crowded spectrums. Since the purpose of detection is to indicate to a jammer whether a signal is present, it is almost always necessary to measure parameters based on the energy in the channel to perform hop association.

Measuring appropriate parameters depends on somehow determining the frequency to which the transmitter has hopped. The entire frequency band over which the targets hop can be simultaneously monitored with one of the receiver types discussed in Section 4.3. Once new energy is detected at some frequency, then measurements must be made to determine if that is the target of interest because, typically, several new energy alarms will occur if there are very many targets present, and they are frequently changing frequency. Thus, signal sorting becomes important.

Optimum receiver structures have been determined for the interception of FHSS signals. The specific architectures for these receivers depend on the type of signal and detection goal. The optimum receiver architecture for SFHSS targets (multiple bits per hop) is illustrated in Figure 4.22 while that for FFHSS signals (multiple hops per bit) is shown in Figure 4.23 [45]. After appropriate filtering and

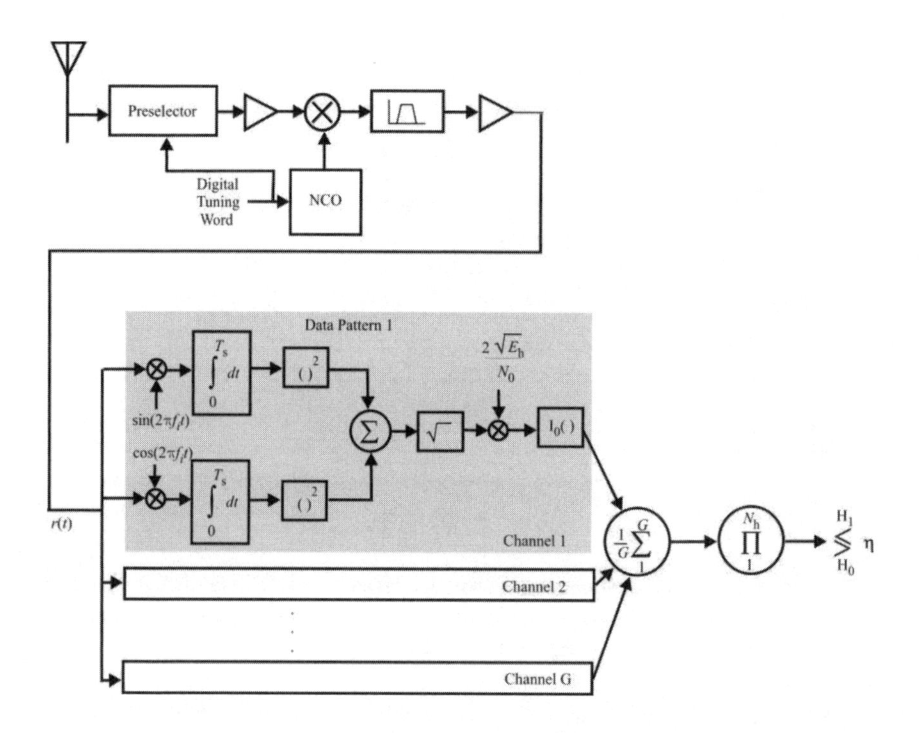

Figure 4.23 Architecture for optimum FFHSS receiver for noncoherent detection. (Source: [45]. © IEEE 1992. Reprinted with permission.)

frequency downconversion, an FBC is used to ascertain whether the signal is present or not.

For SFHSS signals, optimum detection requires searching over all the possible data patterns, since many data bits are contained within each dwell. On the other hand, for FFHSS, there are many dwells per data bit and it is only necessary to search over the possible dwell frequencies. Neither of these architectures is practically realizable in general. They do, however, form the limiting cases of what performance is possible. These architectures assume that the timing epoch information is available at the receiver and the receiver integration times are aligned to it.

4.5.2 Searching for FHSS Targets

Described in this section is a technique for searching for FHSS targets. In addition, a method for estimating the time to first intercept for each target is presented. We assume that the target emitters hop in a (pseudo) random fashion from one frequency in their hop set to the next. The search strategy employed by the receiver, on the other hand, is to step from one frequency channel to the next in a linear fashion—the receiver does not hop around looking for energy.

4.5.2.1 Linear/Periodic Search Strategy

It is necessary to repeatedly retune the center frequency of the receiver in order to maintain surveillance over the entire search bandwidth. The sequence and timing of changes to the center frequency constitute a search strategy. This is a sensor scheduling problem. Typically, the search bandwidth is divided equally by the bandwidth of the receiver and the strategy involves tuning the receiver into each of the smaller bands in some sequence. In each band, we say the receiver *dwells* for a certain period before retuning to a different band.

An obvious and widely used strategy is the *periodic strategy*. Here, the receiver simply steps through each of the bands sequentially with equal dwell periods (although as discussed in Section 4.3.4.1, allocating time in equal dwells may not be the best strategy). Once the receiver steps to the last band in the sequence, it begins again from the first. We call the *sweep period* the time to complete a sequence.

On the other hand, FHSS emitters typically employ a random (more accurately, a pseudo-random) hopping strategy. The set of frequencies used by an emitter constitutes its hop set and may consist of all the frequency channels in a band or some subset of them.

The goal of the receiver is to try to detect radiation from the emitter in the shortest possible time. For detection to occur, the receiver must be dwelling in the "right" band (i.e., the one on which the emitter operates, at the right time).

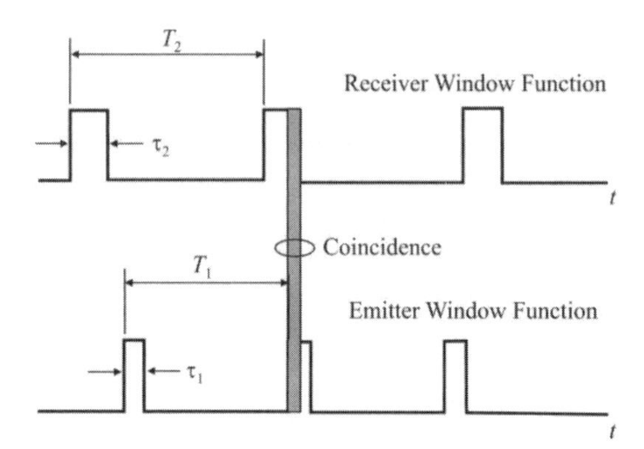

Figure 4.24 Coincidence of two window functions.

The fundamental trade-off being examined in this technique is the scan rate of the EW receiver versus receiver bandwidth. For a constant signal level and bandwidth, the greater the extent of the receiver bandwidth beyond the bandwidth of the target signal, the lower the receive SNR since more noise is permitted into the receiver while more signal energy is not. Therefore, the wider the bandwidth the fewer the number of targets that can be detected. However more channels are covered in a given period of time, thus searching faster. On the other hand, the narrower the bandwidth, the slower the scan rate and therefore the fewer the number of channels covered during a given time period, taken as 1 ms and 10 ms for now. In this case, the SNRs are higher, however.

4.5.2.2 Window Functions, Pulse Trains, and the Interception Process

We model the process of detection or interception of an emitter by a receiver in terms of *window functions* or *pulse trains* [46]. A window function is binary, "on" or "off," and is a function of time. Interception is said to occur when two independent window functions are simultaneously "on." One window function represents the times at which the receiver is dwelling on a channel in which the emitter operates. The other window function represents the times at which the emitter is in the channel of the receiver. The situation is depicted in Figure 4.24. Interception/detection is said to occur when these pulses coincide. The receivers synchronously and repetitively linearly scan a bandwidth W from a low-frequency channel centered on f_L, to a maximum frequency channel centered on f_M. If energy is detected in the currently selected bandwidth, W, a detection is declared.

One additional assumption is worth mentioning. We assume that in the event of a coincidence of the two window functions, there is enough of an overlap that the energy received from the target at the receiver is sufficient to produce an adequate SNR to produce usable detection results.

Markov-Chain Model for FHSS Hopping Characterization

The hopping characteristics of the emitters are modeled as a *discrete-time Markov chain* (DTMC), which is a statistical construct for modeling certain kinds of stochastic processes. One of the processes well modeled by a Markov chain is the hop sequence of a frequency hopping emitter as long as the hop pattern is assumed to be truly random. We will make that assumption here. Since the Markov chain is the mathematical construct that we will use, we will first review the characteristics of such chains that are pertinent to our discussions. For those readers who would like to delve further into Markov chains, [47] is recommended reading.

Markov Chains

A Markov chain is a *stochastic process with state space* $\mathcal{S}$, that is, it is a sequence $\{X_t\}$ of random variables defined on some probability space and taking values in $\mathcal{S}$. The variable t represents time. For a DTMC, the time variable t takes on integer values, whereas in a *continuous-time Markov chain* (CTMC), it takes real values. We assume that the set $\mathcal{S}$ is finite (i.e., we consider only *finite* Markov chains). Without loss of generality, its elements are the numbers 1, ..., $|S|$.[2] We briefly review the properties of discrete- and continuous-time Markov chains that are relevant here.

Our approach in this section is to examine a linear receiver search strategy targeted against FHSS targets whose hop pattern is based on a DTMC. We seek to find the smallest maximum time (minimax) to first intercept of a target.

Discrete-Time Markov Chains

For a DTMC, the process obeys the *discrete-time Markov property* that

$$\Pr\{X_{n+1} = x_{n+1} | X_n = x_n, X_{n-1} = x_{n-1}, \ldots\} = \Pr\{X_{n+1} = x_{n+1} | X_n = x_n\} \quad (4.85)$$

for all $x_i \in \mathcal{S}$ and all integer n. In other words, this means, given X_n, that the next state X_{n+1} is conditionally independent of $X_{n-1}, X_{n-2}, \ldots$. The next state does not depend on how we got to the current state. A DTMC is called *time invariant* if the right hand side (RHS) of (4.85) is independent of n. A (time-invariant) DTMC can be conveniently defined in terms of its transition matrix **P** with elements

$$p_{ij} = \Pr\{X_{n+1} = j | X_n = i\} \quad (4.86)$$

---

[2] $|\mathcal{S}|$ denotes the cardinality of set $\mathcal{S}$.

Note that the column vector $\vec{1}$, all of whose elements are 1, is a right eigenvector of $\mathbf{P}$ corresponding to the eigenvalue 1.

Under certain conditions which need not be described here, a Markov chain may have a *stationary distribution* $\vec{\pi}$ (i.e., a distribution $\Pr\{X_n = i\} = \pi_i$) which is independent of n. It is important to note that a stationary distribution, when one exists, is a left eigenvector of $\mathbf{P}$ corresponding to the eigenvalue 1.

The *first-passage time* between states i and j when $i \neq j$ is the amount of time, given that, at time t, the Markov chain is in state i, it takes until the chain enters state j for the first time. The mean first-passage time between i and j, when it exists, can be found by forming a new matrix $\hat{\mathbf{P}}$ from $\mathbf{P}$ by deleting the jth row and column. Computation of

$$\vec{m}_j = (\mathbf{I} - \hat{\mathbf{P}})^{-1}\vec{1} \tag{4.87}$$

then yields a vector $\vec{m}_j$ of the mean first-passage times to state j from state i in an ascending order of index, excluding the mean first-passage time from state j. (The matrix $\mathbf{I}$ is the appropriately sized identity matrix.)

The ES receiver is assumed to progress linearly from some initial frequency channel in succession up to some maximum frequency channel, and then is reset and starts over at the initial channel again.

Time to First Intercept

Our goal is to determine the maximum time to the first intercept of a frequency hopping target and to minimize it. The spectrum characteristics of a target then is modeled as a Markov chain. The intercept receiver is modeled as above—a sweeping receiver stepping from one frequency channel to the next in a linear fashion, dwelling on each channel sufficiently long to reliably measure the energy present on that channel.

Suppose that an FHSS emitter with a random hop pattern begins operating in channel i. Without a loss of generality, suppose that it begins emitting its first (detectable) pulse at time 0. Let $T_i(\tau)$ be the time at which our linear-based search strategy first intercepts a pulse; the intercept time τ is the total scan time of the receiver (from f_L to f_M). The *intercept time* is the first time after (or at) time 0 that the receiver visits band i when a pulse is emitted. The intercept time is an r.v. so we can define a mean or expected intercept time $\mathcal{E}\{T_i(\tau)\}$. Moreover, the expected intercept time can be maximized over the pulse width of the emitter pulse train. We call this the *maximum expected intercept time* or *MEIT*, $f_i(\tau)$. The intercept time is always maximized when the pulse width is 0. Let $\vec{f}(\tau)$ denote the vector of these MEITs in each channel.

The maximum expected intercept time is determined with the following property (suitably modified to reverse the characteristics of the target and receiver).

Denote by R_i the event that the ES receiver is tuned to channel i and by T_j the event that the transmitter is emitting in channel j. Then *coincidence* occurs when the events R_i *and* T_i occur. Denote this event by R_iT_i. Note that we are *not* assuming here that the receiver and transmitter are synchronized (transmitter epoch timing is known at the receiver). As mentioned, we *are* assuming that when R_iT_i occurs there is enough time overlap that there is enough signal received by the receiver to produce sufficient energy for detection purposes.

The simultaneous operation of the receiver and transmitter form a DTMC. The transition matrix for this DTMC can be written as follows. From this description of the DTMC we can say that if the current state of the DTMC is R_iT_j, then the next state can be any state from the set

$$S_{i+1|i} = \{R_{i+1}T_k \,|k \in \{1,2,\ldots,N_{ch}\}, k \neq j\} \tag{4.88}$$

where N_{ch} is the number of channels that the transmitter has available to use. For the problem described here, the receiver has the same number of channels to scan. For example, for the low VHF range from 30–90 MHz, there are N_{ch} = 2,400 channels available. (For simplicity, we are assuming that the hop set for the transmitter consists of all these channels. Extension to when the hop set is a subset is straightforward.)

The targets of interest here for the most part form PTT networks, and in such networks the emitters are typically off considerably more than they are on. This must be taken into account in the modeling. We will assume that the average duty cycle of the targets is the same and we will denote it by χ. Palpably, a target must be on in order to be detected.

First, let's examine what happens when N_{ch} = 2. The transition matrix in this case is

P =

		0 $R_1T_1T_{on}$	1 $R_1T_2T_{on}$	2 $R_2T_1T_{on}$	3 $R_2T_2T_{on}$	4 $R_1T_1T_{off}$	5 $R_1T_2T_{off}$	6 $R_2T_1T_{off}$	7 $R_2T_2T_{off}$
0	$R_1T_1T_{on}$	0	0	0	χ	0	0	0	$(1-\chi)$
1	$R_1T_2T_{on}$	0	0	χ	0	0	0	$(1-\chi)$	0
2	$R_2T_1T_{on}$	0	χ	0	0	0	$(1-\chi)$	0	0
3	$R_2T_2T_{on}$	χ	0	0	0	$(1-\chi)$	0	0	0
4	$R_1T_1T_{off}$	0	0	0	χ	0	0	0	$(1-\chi)$
5	$R_1T_2T_{off}$	0	0	χ	0	0	0	$(1-\chi)$	0
6	$R_2T_1T_{off}$	0	χ	0	0	0	$(1-\chi)$	0	0
7	$R_2T_2T_{off}$	χ	0	0	0	$(1-\chi)$	0	0	0

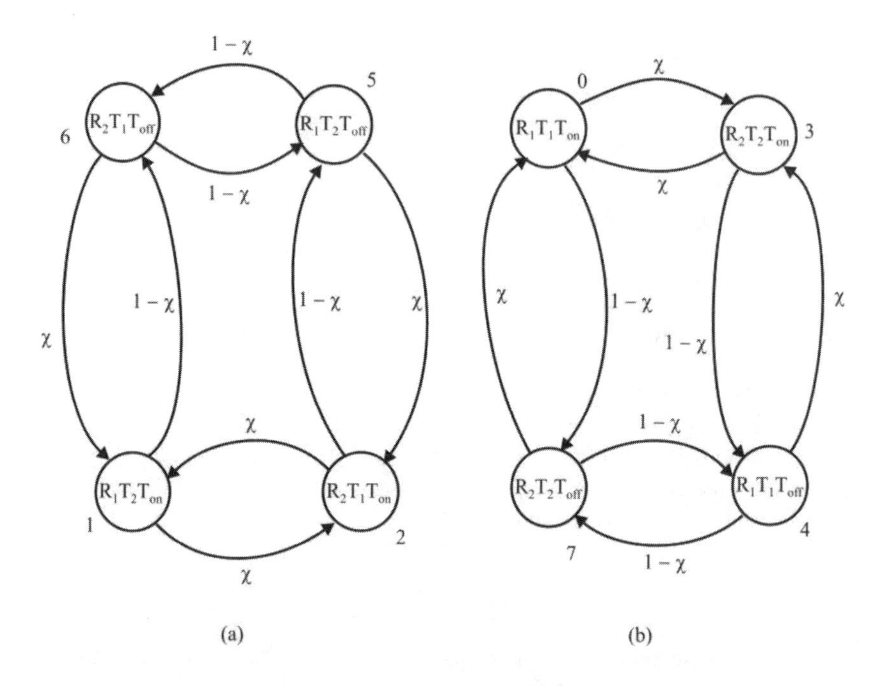

Figure 4.25 Markov transition diagram: (a) corresponds to being synchronized so that no detections ever occur while (b) corresponds to a detection every time that the target is on.

The transition diagram corresponding to this transition matrix is shown in Figure 4.25. As we see from either **P** or Figure 4.25, the Markov chain consists of two subchains that do not communicate with each other. Depending on the initial state, the chain remains in one or the other subchain. Figure 4.25(a) corresponds to when the receiver and target are "synchronized." There is never a detection no matter how long the Markov chain operates. Figure 4.25(b) corresponds to continuous detection. There is detection on every step, as long as the target is emitting.

Now suppose $N_{\text{ch}} = 3$. Taking the duty cycle into consideration, the transition matrix is comprised of four parts as

$$\mathbf{P} = \begin{bmatrix} \mathbf{P}_{\text{on,on}} & \mathbf{P}_{\text{on,off}} \\ \mathbf{P}_{\text{off,on}} & \mathbf{P}_{\text{off,off}} \end{bmatrix} \tag{4.89}$$

where the elements in $\mathbf{P}_{\text{on,on}} = \left[p_{ij} \right]_{\text{on,on}}$ correspond to states where the transmitter is on at state i and it is also on at state j. The elements in $\mathbf{P}_{\text{on,off}} = \left[p_{ij} \right]_{\text{on,off}}$ correspond to states where the transmitter is on at state i and off at state j. The other two submatrices are defined similarly. Define

$$\alpha = \frac{1}{N_{\text{ch}} - 1}\chi \tag{4.90}$$

and

$$\beta = \frac{1}{N_{\text{ch}} - 1}(1 - \chi) \tag{4.91}$$

Then, based on (4.88), these matrices are given by

$$\mathbf{P}_{\text{on,on}} = \begin{array}{c|ccccccccc} & R_1T_1 & R_1T_2 & R_1T_3 & R_2T_1 & R_2T_2 & R_2T_3 & R_3T_1 & R_3T_2 & R_3T_3 \\ R_1T_1 & 0 & 0 & 0 & 0 & \alpha & \alpha & 0 & 0 & 0 \\ R_1T_2 & 0 & 0 & 0 & \alpha & 0 & \alpha & 0 & 0 & 0 \\ R_1T_3 & 0 & 0 & 0 & \alpha & \alpha & 0 & 0 & 0 & 0 \\ R_2T_1 & 0 & 0 & 0 & 0 & 0 & 0 & 0 & \alpha & \alpha \\ R_2T_2 & 0 & 0 & 0 & 0 & 0 & 0 & \alpha & 0 & \alpha \\ R_2T_3 & 0 & 0 & 0 & 0 & 0 & 0 & \alpha & \alpha & 0 \\ R_3T_1 & 0 & \alpha & \alpha & 0 & 0 & 0 & 0 & 0 & 0 \\ R_3T_2 & \alpha & 0 & \alpha & 0 & 0 & 0 & 0 & 0 & 0 \\ R_3T_3 & \alpha & \alpha & 0 & 0 & 0 & 0 & 0 & 0 & 0 \end{array}$$

and

$$\mathbf{P}_{\text{on,off}} = \begin{array}{c|ccccccccc} & R_1T_1 & R_1T_2 & R_1T_3 & R_2T_1 & R_2T_2 & R_2T_3 & R_3T_1 & R_3T_2 & R_3T_3 \\ R_1T_1 & 0 & 0 & 0 & 0 & \beta & \beta & 0 & 0 & 0 \\ R_1T_2 & 0 & 0 & 0 & \beta & 0 & \beta & 0 & 0 & 0 \\ R_1T_3 & 0 & 0 & 0 & \beta & \beta & 0 & 0 & 0 & 0 \\ R_2T_1 & 0 & 0 & 0 & 0 & 0 & 0 & 0 & \beta & \beta \\ R_2T_2 & 0 & 0 & 0 & 0 & 0 & 0 & \beta & 0 & \beta \\ R_2T_3 & 0 & 0 & 0 & 0 & 0 & 0 & \beta & \beta & 0 \\ R_3T_1 & 0 & \beta & \beta & 0 & 0 & 0 & 0 & 0 & 0 \\ R_3T_2 & \beta & 0 & \beta & 0 & 0 & 0 & 0 & 0 & 0 \\ R_3T_3 & \beta & \beta & 0 & 0 & 0 & 0 & 0 & 0 & 0 \end{array}$$

$$\mathbf{P}_{\text{off,on}} = \mathbf{P}_{\text{on,on}} \qquad \mathbf{P}_{\text{off,off}} = \mathbf{P}_{\text{on,off}} \tag{4.92}$$

Note that these notions are consistent with how FHSS transmitters actually operate. When not transmitting, their clocks are still running and internally the transmitters are still changing frequencies according to the *m*-sequence coding. So to change states to one that corresponds to a transmitter being off makes practical sense.

This DTMC is periodic by definition (receiver repeats as does the emitter). The transition diagram is not shown here as it is too large. Therefore, this DTMC is an ergodic, cyclic DTMC [48] and several important results that apply to regular DTMCs [49] apply to this problem, in particular, the mean time to first passage and their variances [50]. This is the form of the transition matrix irrespective of the size of the problem.

Since we are assuming the transmitter frequency pattern is determined by an *m*-sequence[3], there are no repeated channels until they have all been visited by the transmitter.

As indicated in [49], the matrix $\mathbf{M} = [m_{ij}]$ whose entries contain the mean time to first passage for a finite, ergodic, possibly cyclic DTMC is calculated with

$$\mathbf{M} = (\mathbf{I} - \mathbf{Z} + \mathbf{E}\mathbf{Z}_{\text{dg}})\mathbf{D} \tag{4.93}$$

where

$\mathbf{I}$ is the appropriately sized identity matrix;
$\mathbf{Z} = [\mathbf{I} - (\mathbf{P} - \mathbf{A})]^{-1}$ is the *fundamental matrix* for $\mathbf{P}$ [48];
$\mathbf{A}$ is the *limiting matrix* for P $\left(\mathbf{A} = \lim_{n\to\infty} \mathbf{P}^n\right)$;
$\mathbf{E}$ is an appropriately sized matrix of all ones;
$\mathbf{Z}_{\text{dg}}$ is a diagonal matrix containing the diagonal elements of Z on the diagonal;
$\mathbf{D}$ is the diagonal matrix with diagonal elements $d_{ii} = 1/\pi_i$.

m_{ij} represents the average time until state j is reached assuming the DTMC begins operation in state i.

Also from [49], the matrix of the variances of the time to first passage can be determined using

$$\mathbf{M}_2 = \mathbf{M}(2\mathbf{Z}_{\text{dg}}\mathbf{D} - \mathbf{I}) + 2[\mathbf{Z}\mathbf{M} - \mathbf{E}(\mathbf{Z}\mathbf{M})_{\text{dg}}] \tag{4.94}$$

Example: FHSS Searching

Suppose that we search the lower VHF range for short duration targets (FHSS). As an example, suppose we search 30–90 MHz with a scanning receiver with an instantaneous bandwidth of W = 20 MHz, followed by a 800 bin FFT, yielding 25 kHz search channels. Thus N_{ch} = 3. We assume that the average duty cycle

[3] We will discuss *m*-sequences in Chapter 6.

for each transmitter is 5 / 372. The matrices in this case are as in (4.92), with $\chi = 5/372,\ \alpha = \chi/2,$ and $\beta = (1-\chi)/2.$

Detection occurs when the receiver channel and target channel coincide, and when the target is on, which is state 0 $(R_1\ T_1)_{on}$, 4 $(R_2T_2)_{on}$, and 8 $(R_3T_3)_{on}$. The vectors of mean time to first passage (in steps) corresponding to (4.93) and these states are given by

$$\vec{m}_0 = \begin{bmatrix} 452 \\ 452 \\ 452 \\ 450 \\ 452 \\ 452 \\ 452 \\ 450 \\ 450 \end{bmatrix} \qquad \vec{m}_4 = \begin{bmatrix} 1.7\times10^4 \\ 1.7\times10^4 \\ 1.7\times10^4 \\ 1.7\times10^4 \\ 1.7\times10^4 \\ 1.7\times10^4 \\ 1.7\times10^4 \\ 1.7\times10^4 \\ 1.7\times10^4 \end{bmatrix} \qquad \vec{m}_8 = \begin{bmatrix} 452 \\ 452 \\ 450 \\ 450 \\ 450 \\ 452 \\ 452 \\ 452 \\ 452 \end{bmatrix}$$

(differences in these numbers are due to round-off errors). We can see that the largest number of steps to first passage occurs for any initial state and state 4, that is $(R_2T_2)_{on}$, with 1.7×10^4 steps. If the targets are FHSS emitters with a nominal hop rate of 100 hps (10 ms dwell), then the longest mean time until first detection is 170 seconds.

The variance vectors for the elements of concern here, (R_1T_1), (R_2T_2), and (R_3T_3), are given by

$$\vec{\sigma}_0^2 = \begin{bmatrix} 4.1\times10^5 \\ 4.1\times10^5 \\ 4.1\times10^5 \\ 4.1\times10^5 \\ 4.1\times10^5 \\ 4.1\times10^5 \\ 4.1\times10^5 \\ 4.0\times10^5 \\ 4.0\times10^5 \end{bmatrix} \quad \vec{\sigma}_4^2 = \begin{bmatrix} 5.7\times10^8 \\ 5.7\times10^8 \\ 5.7\times10^8 \\ 5.7\times10^8 \\ 5.7\times10^8 \\ 5.7\times10^8 \\ 5.7\times10^8 \\ 5.7\times10^8 \\ 5.7\times10^8 \end{bmatrix} \quad \vec{\sigma}_8^2 = \begin{bmatrix} 4.1\times10^5 \\ 4.1\times10^5 \\ 4.1\times10^5 \\ 4.1\times10^5 \\ 4.1\times10^5 \\ 4.1\times10^5 \\ 4.1\times10^5 \\ 4.1\times10^5 \\ 4.1\times10^5 \end{bmatrix}$$

The standard deviation of the time to first passage (or first detection) corresponding to the largest mean time to first passage (R_2T_2) is $\sqrt{5.7\times10^8} = 23{,}400$ steps, or 234 seconds; therefore, 99.7% of the targets are detected in 170 + 3×234 = 932 seconds.

The mean times for subsequent detections are also available from these vectors. Considering state 0 (R_1T_1), the time between visits to the state is given by m_{00}. In this case it is T_0 = 452 steps or 4.5 seconds. The mean time to revisit state 4 (R_2T_2) is given by m_{44} or 17,000 steps corresponding to 170 seconds. Finally, the mean time to revisit state 8 (R_3T_3) is given by m_{88} or 452 steps, corresponding to 4.5 seconds. So the model indicates that, on average, every target will be detected at least every 170 seconds.

$\chi = 5/372$ corresponds to the average time a network member other than the *net control station* (NCS) transmits in each frame in the simulation presented later in this chapter. As described there, the NCS actually transmits once for each of the other network members, of which there are four on each network. The duty cycle of the NCS is therefore 20/372. The maximum mean time to first detection for the NCSs is therefore 1,122 steps, or 4.2 seconds, with a variance of 2.7×10^4 steps^2, or a standard deviation of 164 seconds.

4.6 FHSS Signal Sorting

In a follower jammer, it is necessary to determine the frequency to which the target has hopped since it is assumed herein that a priori information about this is unavailable to the jamming system. Thus, measurements of the spectrum are necessary and there are parameters determined that are used for such discrimination. These parameters are random variables and so the performance of the jamming system can only be described statistically.

Typical parameters for tracking frequency-hopping targets are azimuth, hop phase, and amplitude. *Azimuth* is the angle of arrival of a signal at the jammer and will be a constant, subject, of course, to statistical variability and systematic errors, as long as the jammer and target are not moving. Even if they are moving, however, azimuth may still be usable since the azimuth will not vary much for subsequent hops for most situations. *Hop phase* is the expected time that the next hop will start based on the timing of the current hop. As long as the hop rate is a constant, then this can be predicted. One of the ECCM techniques for hopping systems, however, is to vary the hop timing, so this parameter is not always useful.

Amplitude is the absolute amplitude of the signal at the jammer. Variations in propagation phenomena as well as antenna gain can make it difficult to use amplitude as a sort parameter, however. This is especially true for ground-based platforms. Airborne jammers experience less variation in propagation so they may be better able to take advantage of the amplitude as a sort parameter.

Tracking frequency-hopping targets based on these parameters requires establishing gates which determine the limits within which the parameters must be within in order to declare a hop as belonging to a particular target or not. The gates have a specified minimum and maximum. The gate for azimuth, for example, might be $\pm 10^{\circ}$, depending on the accuracy of the instrument used to measure the azimuth. The gates are based on the initial detection of a potential target and the measurements made thereon. Normally the gates would be continuously updated as the collection process proceeds.

4.6.1 Sort Parameters

Details of the aforementioned frequency-hopping sort parameters are discussed in this section.

4.6.1.1 Hop Phase

Hop phase is defined as the time that a hop begins (or ends) relative to the time that a previous hop began (or ended). The resolution on measuring the hop phase is limited by, inter alia, the revisit time of a scanning receiver. For example, for the low VHF frequency range, 30 to 90 MHz is the spectrum of concern, and if the jammer receiver is using a 4 MHz instantaneous bandwidth, then N_c = 60 MHz/4 MHz = 15, and is dwelling for 200 μs on each frequency, a revisit to any particular channel is given by 15 × 200 μs = 3 ms. This is the limit on the ability to measure hop phase. In this example then, for 100-hps targets, the resolution using only hop phase is given by 3 ms/10 ms or 0.33. Therefore, on average, only about three targets can be sorted based on hop phase alone.

When a new net member starts communicating, it most likely has a new azimuth associated with it, whereas when tracking the same target, azimuth is an important track parameter, and this situation must be accommodated. Hop phase is critical in this situation.

4.6.1.2 Azimuth

The azimuth angle of arrival is the direction from which the target signal originates. Since the target of the jammer is the receiver to which the target transmitter is sending information, not the transmitter, the azimuth can still be used as a sort parameter. It cannot, in all cases, be used to point a directional jamming antenna, however. Such would be the case for small-scale operations.

The smaller the azimuth gates are, the fewer measured azimuths occur within the gates, which results in fewer hops being detected. On the other hand, the wider the gates are, the higher the probability that more than one target will appear within the gates, leading to confusion over which hop belongs to the correct target. The better the instantaneous DF accuracy, the smaller the azimuth gate can be. Unfortunately, the limits on DF accuracy are often not imposed by the EW system itself, but are caused by external factors such as multipath reflections, cochannel interference, and high noise levels.

If υ represents the received SNR, T_d is the sample time, and B_n is the noise bandwidth (25 kHz in the low VHF range), then the Cramer-Rao bound limits the ability to measure azimuths to

$$\sigma \geq \frac{1}{\sqrt{\upsilon T_d B_n}} \tag{4.95}$$

where σ is the standard deviation of the measurement. Thus, the larger any of these parameters is the better the azimuth can be measured. The Cramer-Rao bound is a theoretical bound assuming that the variations in the parameters are normally distributed. Being a lower bound on performance, the accuracy will be worse than that indicated by this bound.

4.6.1.3 Amplitude

Amplitude information is, in general, not very reliable when the entire low VHF spectrum is used for hopping. The received power levels of signals in free space are given by the equation (more on this in Chapter 5)

$$P_R = \frac{P_T G_{TR} G_{RT} \lambda^2}{(4\pi D_{TR})^2} \tag{4.96}$$

where P_T is the transmitted power, G_{TR} is the transmit antenna gain in the direction of the receiver, G_{RT} is the receive antenna gain in the direction of the transmitter, D_{TR} is distance between the transmitter and the receiver, and λ is the wavelength. Theoretically, then, the power levels will decrease by 6 dB for every octave increase in frequency. For the low VHF, the frequencies are varied between 30 and 90 MHz, which represent 1.5 octaves. Therefore, signal levels should not vary by more than about 9 dB across the band in free space. Close to the ground propagation can best be described by setting the exponent in the denominator in this expression to four rather than two, which yields a 12 dB loss per octave distance. Again, over the low VHF range, the total variability of the amplitude would be about 18 dB for stationary targets. In fading channels, this variability will be larger.

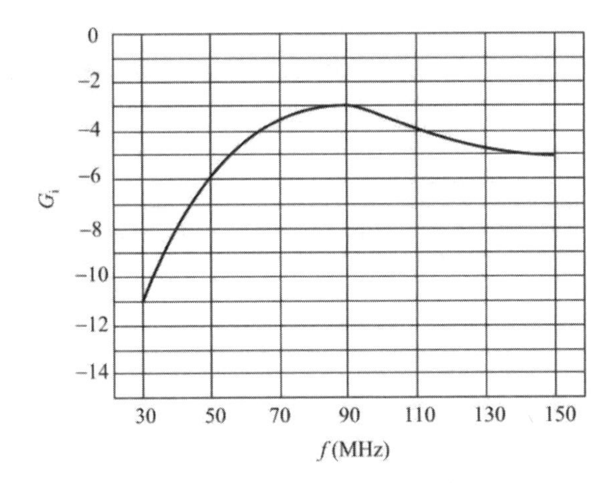

Figure 4.26 Antenna gain response in the horizontal plane used in the intercept system simulation.

The receive antenna gain influences the amount of received power. The antenna gain shown in Figure 4.26 is typical for a ground-based antenna in this frequency range, tuned to a half wavelength at 60 MHz. Considerable variation can be seen over the lower VHF range that complicates the utilization of signal amplitude for sorting.

There are some special cases, however, when amplitude sorting can be used. In standoff situations, that is, the collection systems are some distance from the targets, amplitude can be used to eliminate friendly communications since they will typically be much stronger. Another case where amplitude can be used is when the collection systems are much closer to the targets than friendly communicators, which could be the case for UASs, for example. Again, however, amplitude could be used primarily to sort friendly from target communications since in this case, the target signals will be much stronger.

4.6.1.4 Summary

As an example of the above discussion, suppose there are two parameters: azimuth and hop phase with probabilities P_a and P_h, respectively. Further suppose, for discussion purposes, that these parameters have a uniform probability distribution, that is, any value within appropriate ranges is equally likely. In addition, suppose that gating is used. That is, a measured parameter is compared with hard limits within the tracker. If the sort azimuth gates are set at 10° and the time resolution is 500 μs, then the probability of a target being within the azimuth gates is $P_a = 20^\circ / 360^\circ = 0.055$, assuming omnidirectional antenna coverage in the horizontal plane. The probability of measuring two hops with the same hop phase is $P_h = 0.5$ ms/10 ms = 0.05 assuming 100 hops per second targets. The probability of confusing hops is then $P = P_a P_h = 0.055 \times 0.05 = 0.00275$. The theoretical upper bound on size of the target environment that can be dealt with without ambiguity is given by the reciprocal of P. In this example the bound is 364.

4.6.2 Tracking Metrics

There are two ways discussed here for measuring the closeness of a set of measured parameters to another set. The first is based on an approach that assumes that the error sources are Gaussian. The second is based on a weighted Euclidean distance measurement.

If there are M sort parameters and they are independent random variables, then the probability of getting targets confused can be expressed as

$$P = \prod_{k=1}^{M} P_k \tag{4.97}$$

where P_k is the probability associated with the occurrence of sort parameter k.

Upon initial detection, the parameters are measured and assigned to trackers. Upon subsequent detections and assignment of hops, the parameters are compared to statistical values related to these parameters and, if sufficiently close, the hop is associated with the transmitting net being tracked.

Suppose that the statistics associated with measuring the sort parameters are Gaussian. The mean and standard deviation values of differences in measured and expected parameters are calculated for the target. The joint probability density function of these parameters is given by

$$p(\vec{x}) = \prod_{j=1}^{M} \frac{1}{\sqrt{2\pi}\sigma_j} e^{\frac{(x_j - m_j)^2}{2\sigma_j^2}} \tag{4.98}$$

where the vector $\vec{x}$ is given by

$$\vec{x} = (x_1, \ x_2, \ \cdots, \ x_M) \tag{4.99}$$

m_j is the mean of x_j and σ_j is the standard deviation. Of the signals detected at any instant, the one that maximizes this probability density should be assigned to the target being tracked, subject to being a certain minimum value.

Taking the natural logarithm of this equation yields

$$\ln p(\vec{x}) = -\frac{1}{2}\sum_{j=1}^{M} \ln 2\pi - \sum_{j=1}^{M} \ln \sigma_j + \sum_{j=1}^{M} \frac{(x_j - m_j)^2}{2\sigma_j^2} \tag{4.100}$$

This equation yields a statistic for measuring a closeness of fit given by

$$\beta = -\sum_{j=1}^{M} \ln \sigma_j + \sum_{j=1}^{M} \frac{(x_j - m_j)^2}{2\sigma_j^2} \tag{4.101}$$

since the first term in (4.100) is a constant that can be absorbed in β. The smaller the value of β, the better the fit to the statistics, that is, the larger $p(\vec{x})$ will be. The minimum threshold value of $p(\vec{x})$ is reflected in (4.101) by a threshold value of β, given by β_{th}. If $\beta \le \beta_{\text{th}}$ for one or more detected signals, then that signal with the smallest β is associated with the target. If $\beta > \beta_{\text{th}}$ for all detected targets, then no hop is associated with the target and that hop is not jammed.

The second metric used for tracking a target from one hop to the next, as well as tracking nets when a new net member starts communicating, is a weighted Euclidean distance measurement of the measured parameters versus the tracker values. This can be expressed as

$$D = \sqrt{w_1 x_1^2 + w_2 x_2^2 + \cdots + w_M x_M^2} \tag{4.102}$$

where the w_k are the weights.

The weights are selected so that selective importance is assigned to all the terms in (4.102). No hop is detected when D exceeds a predefined threshold. The distance measurement is calculated on a hop-to-hop basis. Once no hop is detected for the wait time (defined by hop dwells), then that target is declared "down", and the operator is free to be assigned to a new target. The smaller this value is, the more the amount of time waiting for the next transmission on a net is reduced. This decreases the probability of being able to track communications on a particular net, but may increase the likelihood of being able to detect more targets, regardless of the net.

The Euclidean distance method for tracking targets from one hop to the next is certainly heuristic as opposed to the Gaussian-based distance that is optimum, as long as the statistics involved are normal.

The net effect of including the probability of detecting the correct hop is to decrease the probability of a bit error in the targeted communication system. The probability of being confused as to the identity of the correct hop, denoted here as P_{confused}, is found as follows. Denote the total number of possible targets or transmitters by N_{T}. The probability of exactly n events happening out of a possible N_{T} is given by the binomial distribution

$$P(n) = \binom{N_{\text{T}}}{n} p^n (1-p)^{N_{\text{T}}-n} \tag{4.103}$$

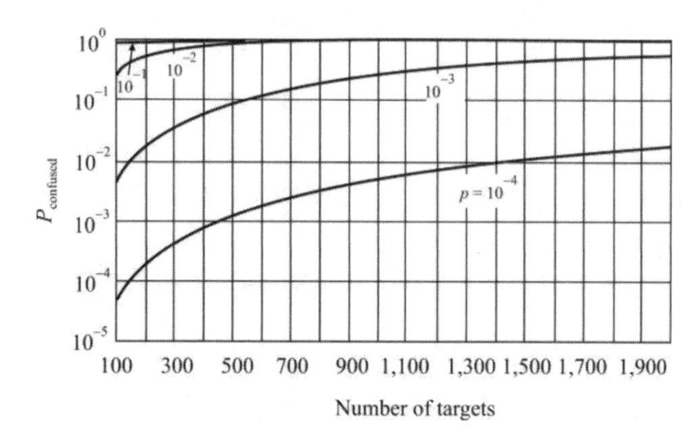

Figure 4.27 Probability of getting hops confused versus the number of potential targets with the probability of a single event as a parameter.

where p is the probability of a single event happening and $\binom{a}{b}$ represents the bth binomial coefficient given by

$$\binom{a}{b} = \frac{a!}{(a-b)!b!} \tag{4.104}$$

The tracking will get confused if there are more than one possible targets that pass the gate testing described above. Denoting by p the probability of a single target passing the gates, then

$$\begin{aligned} P_{\text{confused}} &= P(n > 1) = 1 - P(n \leq 1) \\ &= 1 - \binom{N_\text{T}}{0} p^0 (1-p)^{N_\text{T}} - \binom{N_\text{T}}{1} p(1-p)^{N_\text{T}-1} \end{aligned} \tag{4.105}$$

In this case, with M sort parameters, P is given by (4.97). Thus

$$P_{\text{confused}} = 1 - (1-p)^{N_\text{T}} - N_\text{T} p (1-p)^{N_\text{T}-1} \tag{4.106}$$

This expression is plotted in Figure 4.27 for several values of N_T and p.

As an example suppose there are two parameter gates, azimuth and hop phase. Using the numbers above, suppose the azimuth gate is set at $\pm$ 10° and the hop phase gate is set at $\pm$ 0.25 ms. Then

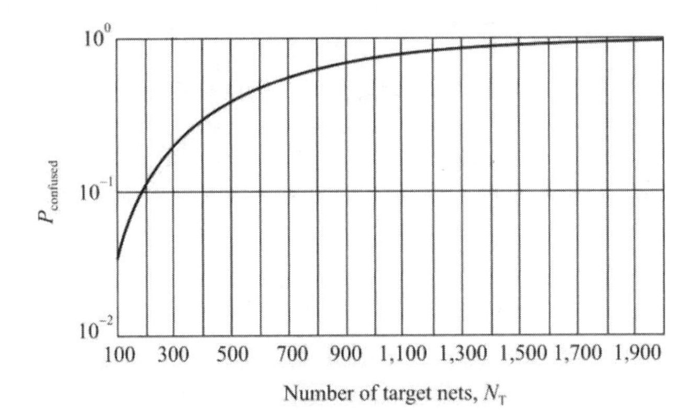

Figure 4.28 Probability of being confused for the example when $p = 0.0024$.

$$p = \frac{20^\circ}{360^\circ}\frac{0.5\text{ms}}{10\text{ms}} = 0.0028 \tag{4.107}$$

for a 100-hps target system. This function is plotted in Figure 4.28. It only takes an environment of about 200 targets to cause confusion 1/10 of the time. Above about 1,000 targets, confusion is almost always assured.

The probability of bit error calculated earlier can now be modified to include the effects of tracking the frequency hopping targets. Frequency-hopping communication system jammers track a target until some time expires from when the last hop from the target transmitter was detected. Call the number of hops over this time N_H. Let event A = *this hop passes the tracking gate and it is the only one* and event B = *at least one hop passed the tracking gates on the last N_H hops*. If it is assumed that the trackers and their gates have already been assigned to a target, then the probability that this hop will pass the tracking gates and the jammer will not be confused (although it could still jam the wrong target by these criteria because a wrong target could be the one that passed the tracking criteria) is given by $P\{A|B\}$.

The conditional probability is necessary because, if the target has been detected and assigned to a tracker, then there is a very high probability that the target will again pass the tracking gates. The binomial distribution described above would yield the probability of passing the tracking gates if the events were independent. Now

$$P\{A|B\} = 1 - \Pr\{\text{no hops passed the tracking gates on the last } N_H \text{ hop intervals}\}$$

This expression can be written in terms of the binomial distribution as

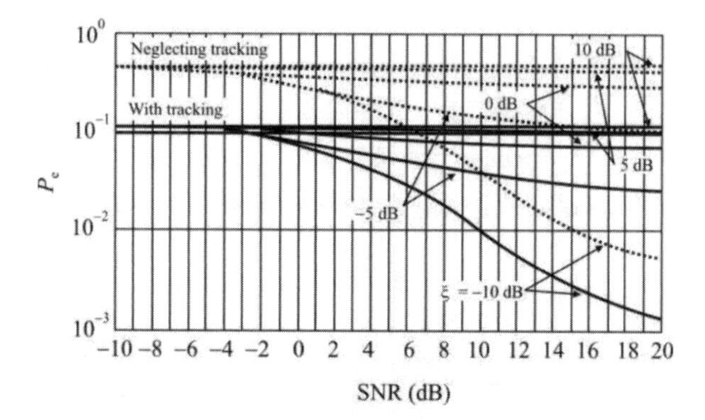

Figure 4.29 Probability of bit error including the effects of tracker errors for $N_H = 10$.

$$
\begin{aligned}
P\{A|B\} &= 1 - \overbrace{\binom{N_H}{0} p^0 (1-p)^{N_H - 0}}^{N_H \text{ times}} = \overbrace{(1-p)^{N_H}}^{N_H \text{ times}} \\
&= (1-p)^{N_H^2}
\end{aligned}
\tag{4.108}
$$

This function is plotted in Figure 4.29 for $N_H = 10$ hops and in Figure 4.30 for $N_H = 100$ hops. The smaller the number of hops before which the target is declared Gone, the easier it is for the jammer to make a tracking error. Conversely, the longer the interval over which no hops make it through the tracking gates, the more likely a hop will occur within the tracking space.

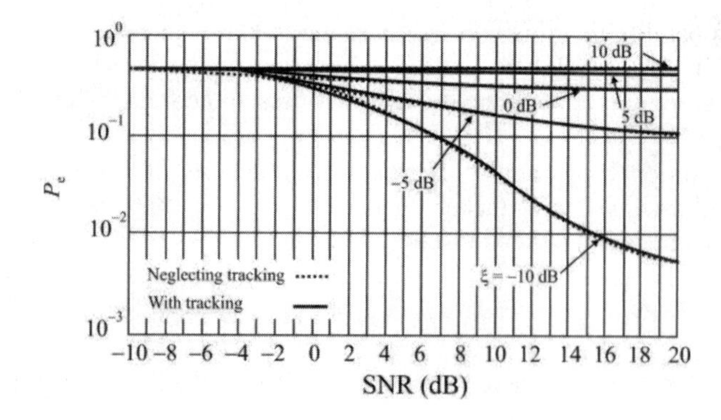

Figure 4.30 Probability of bit error including tracking errors for $N_H = 100$.

4.7 Performance Simulation

This section presents the results of a computer simulation of an ES system collecting FHSS AJ targets in support of the EA function, based on the discussions in Sections 4.1through 4.6.

Realistic simulation of the operation of an ES system interacting with a target environment is a multidimensional problem; the number of parameters is very large. Nevertheless, the analysis was performed to ascertain the collection (detection, identification, and location) performance of such a system measured in terms of probabilities of collecting (detecting and identifying) and locating targets in these environments. Only a few of the various parameters were varied, made necessary by practical computation limitations.

4.7.1 Simulation Description

The computer models of the target environment and the collection systems were essentially autonomous. The only common factors were the total simulation time and the simulation time increment per pass. One hour of simulated time was used for all cases considered. This value was selected primarily to keep the computer time to within reasonable limits. The time increment per pass (epoch) was 0.01 seconds for all cases. Ten values of seed random variables were used for each case and the results were averaged.

The Longley-Rice propagation model [51] (see Chapter 5) was used to compute all signal levels in the simulation, to include the target signal levels at the target receiver as well as the jammer signal level at the receiver. Model parameters used were typical of a region consisting of rolling hills and not too good but not too bad ground conditions. All transmissions were assumed to be vertically polarized.

4.7.1.1 Target Environment

The target environment consisted of a variable number of communication networks, ranging from 12 to 96, where each net had five members. The first 12 target nets are illustrated in Appendix B based on the nodes indicated in Figure 4.31. The remainder of the target nets was randomly placed within the target area shown by first placing the *net control station* (NCS) for each. The remainder of a net was then deployed randomly within a 14 km radius of the respective NCS.

The targets were all assumed to be hopping at a rate of 100 hps over the low VHF frequency range of 30–90 MHz. The effects of transmitter switching times were not considered. Each transmitter had an ERP of 50 watts and the transmit antenna heights were all 5 m.

Over the 1 hour of simulated time, it was assumed that the targets were stationary. The transmissions on the nets consisted of the NCS transmitting to a

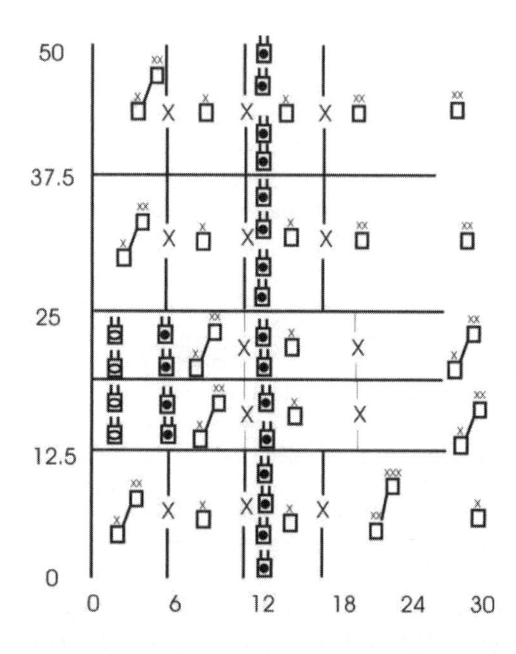

Figure 4.31 Target scenario for the simulation.

net member for 5 seconds and then that net member responded with a 5 second transmission. The NCS then transmitted to the next net member and that net member responded with a 5 second transmission. Between each of these transmissions, the radios were off for 2 seconds. This continued until the NCS had called all stations on its net and they had all responded. The network was then inactive for an average time of 300 seconds. All of these time variables were randomly varied with approximately normal distributions.

4.7.1.2 ES Systems

Two ES system configurations were examined: UAS mounted receiving systems and ground standoff platforms. For the UAS configuration, the collection receivers were mounted in two UASs and were deployed at an altitude of 1,000 m, 8 km across the *forward line of own troops* (FLOT). The systems were netted together for obtaining target fixes. Implicit in this scenario, then, was a data link to ground processors from the UASs. The UASs loitered in the same position for the 1 hour simulation time. Each ground station contained two operators for a total of four, two per UAS.

In the ground standoff configuration, the receiving equipment was contained within the same shelter as the operators. The systems were deployed 5 km behind the FLOT and the receive antenna heights were 10 m. Three such systems were included in the simulation, with two collection operators in each, for six operators. Other than this, the scenario was the same as the UAS configuration.

The collection system operation was simulated as a scanning wideband receiver with *direction finding* (DF) capability. The instantaneous bandwidth of the receivers was a simulation variable and was set at 4 MHz, 12 MHz, and 24 MHz. A filter bank channelized to 25 kHz followed this receiver. The process of target collection started with scanning the frequency spectrum looking for energy. If energy was detected at a channel, an attempt to compute the location of the emitter generating the energy was made. If a location was computed, then that target was assigned to an operator for copy. After this assignment, copy proceeded by that operator based on hop phase, instantaneous azimuth, and power level measurements as well as minimum signal level requirements.

Each operator was assigned a geographic area defined as a rectangle. The rectangles were obtained by dividing the target area from top to bottom into equal areas. This area was then used to assign targets to operators: four areas for the UAS configuration and six areas for the ground standoff. These rectangular areas were derived by placing horizontal slices across the target area shown in Figure 4.31, equally spaced top to bottom.

As mentioned, targets were assigned to collection operators based on their geographic locations. If the conglomerate ensemble of collection systems could compute a target fix [a minimum of two *lines of bearing* (LOBs) were necessary], and the collection operator that was assigned to the area where the fix is in was not busy prosecuting another target, that operator was assigned to copy the new target. That operator continued to copy the target net until no hops matched the tracking criteria for a prescribed number of hops called herein the *wait time*. If a hop reappeared that did satisfy the tracking criteria, then this time clock was reset regardless of whether the hop was associated with the correct target. Once no target satisfied the tracking criteria for this number of hops, that operator was proclaimed not busy anymore and was available to be assigned to the next target that appeared within his/her assigned area.

The scan rate of the receivers was varied between 20 GHz per second to 120 GHz per second. Furthermore, the scanning receivers were time-synchronized between collection systems so that the same frequencies were scanned at the same time by all receivers. This scan rate is determined by the instantaneous bandwidth of the wideband receiver and the sample (receiver dwell) time. To put these numbers in perspective, suppose that the receiver had a 2.5 MHz instantaneous bandwidth (IBW) and dwelled at a tuned frequency for 100 μs. The search rate in this case was 25 GHz per second. The channel resolution in this case was 1/100 μs = 10 kHz, but with a signal processing smearing factor (defined later) of 2.5, the net resolution was approximately 25 kHz, the assumed width of a VHF channel.

In such a system configuration, when the scanning receiver detected energy at a channel and subsequent measurements allowed that channel to be declared associated with the last hop, a separate, narrowband but fast tuning receiver would be tuned quickly to that frequency. There would be one such hand-off receiver for each operator.

The receiver was tuned to a frequency and the energy within that channel (25 kHz wide) was measured. A target location was automatically obtained on each cell that contained sufficient energy, as long as adequate data link capacity was present.

Intersystem Data Link

The rate of the data link that interconnects the systems limited the number of LOBs that could be transferred among systems. Since all the systems were time synchronized and covered the same instantaneous frequency band at the same time, a target detected at one system at a frequency would also be detected at the same time at all collection systems, subject to signal level constraints. That is, the signals must have been at an adequate level as well as have an adequate SNR to be detected. If too many signals were detected at a time for the data link to transfer the LOBs to the other stations, the remaining signals were considered not detected for the purpose of computing locations. The data link data rate used for all the results herein was 200 kbps at a frequency of 1.7 GHz. The SNR characteristics of the data link at this frequency were included.

In the event of copying a target, as opposed to searching for targets, the hop phase and instantaneous LOB at the copying site were used along with the signal amplitude so the data link is not a constraint in this case.

Signal Detection

Typically, for single hop detection, a 15–20 dB SNR is necessary to detect BFSK signals when $P_{fa} \approx 10^{-3}$ and $P_d \approx 0.99$. P_{fa} refers to the probability of declaring a hop present when none was and P_d refers to the probability of declaring a hop present correctly. These results assume that an FFT channelized receiver architecture is implemented and is based on a statistical estimation likelihood theory leading to optimum results (as long as the noise present is Gaussian.) For these results to apply, however, the targets must be fast frequency hopping. For the SFHSS cases herein, although not necessarily optimum, the detection results are applicable. A P_{fa} of 10^{-3} produces about one error every 10 seconds per channel. A P_d of 0.99 ensures that 99% of the time a hop is detected in the channel when one is there. Typically, data rates around 16 kbps are used, leading to approximately 160 data bits or BFSK tones in a 10 ms hop. They are not optimum in the sense that not all of the information in the signal is exploited—in particular, the data sequence.

This leads to the second assumption about the thresholds: that the FFT sample windows coincide precisely with the period of the sine wave input signal. When misalignment of the window occurs, spectral splattering results, pushing signal energy into adjacent FFT bins [52]. For this analysis, it was assumed that the alignment of FFT points and target tones occurred and they were correctly combined so that FFT points were an accurate representation of the energy in the signal.

The third assumption is that the hop epoch information was completely known. That is, it was assumed that the target hop time was known to the receiver. If this information is not known, then the statistical derivation leading to the exact optimal results is an approximation and some method of establishing (measuring) the timing information is required. Establishing the optimum thresholds then becomes a matter of simulation because some of the statistics involved are correlated [52].

The fourth simplifying assumption, made for simulation feasibility, is that the hopping targets all changed channels at the same time. This occurred once per pass through the simulation at a rate of 100 hps.

The signal must have been detected within the first 30% of the hop. If it was, the hop was *detected*. One event that could have led to failure to collect 70% of a hop was due to the scanning receiver. If the instantaneous frequency of an operator's target was not covered by the receiver until 30% or more of the hop was transmitted, then the hop was not detected. This could have happened at all scan rates less than 20 GHz/sec. Because achieving a scan rate of 20 GHz/sec is not a major challenge with modern technology, all the scan rates considered herein were at or above this rate. Therefore, this is not a limitation. Another reason why an individual hop may not have been detected is that the SNR was too low. The noise sources simulated consisted of RF background noise (galactic, atmospheric, and man-made) and system noise, as represented by the system noise figure. A high rural noise environment was assumed [51].

Another case was when the signal level was too low. The receiver dynamic range was a simulation variable and close interferers combined with this dynamic range set the minimum signal levels that could be detected. Lastly, the copying operator's measured LOB to the target may have been consistently too far away from what the tracker expected. How these parameters were factored into the detection performance is described later.

A target transmitter is considered *identified* if the transmission was detected within the first 2 seconds. Seventy percent or more of the hops associated with a transmission must have been detected in order to proclaim the collection a success. When this occurred, the transmission was said to have been *copied*.

Noise Figure

The system noise figure was varied between 0 and 10 dB. Noise is added to a signal by the electronics of the system, primarily by the front end. The noise figure is a measure of this noise. For a well-designed system in this frequency range,

10 dB is achievable but with difficulty. With cooled receiver front ends, noise figures lower than this are possible. It can be argued, however, that the external noise level is higher than this, however.

Minimum Detectable Signal

In dense target environments, almost every channel is occupied by a signal—either friendly or hostile (and often both). With stand-off platforms, friendly transmitters are closer and therefore higher levels of their signals are received while target signals are further away and weaker. In the case of UAS platforms, the targets are closer than friendly interferers so the dynamic range is not as much of a concern. We must design for the case where weak target signals are in channels adjacent to powerful signals and not masked by the sidelobe response of the receiver chain.

The dynamic range and signal level of the closest interfering transmitter set the *minimum detectable signal* (MDS) level given by

$$\text{MDS}_{\text{dBm}} = P_{\text{I,dBm}} - \text{Dynamic Range}_{\text{dB}} \tag{4.109}$$

where P_{I} is the signal level of the interfering signal at the collection site. The sidelobes generated by any signal processing system that samples the signal can mask signals of interest, which frequently are substantially weaker than interfering signals. The level of the highest of these sidelobes is given by the first sidelobe dynamic range, as given above. Therefore, for a signal to be detected, its power level must have been above the MDS. Three values of receiver dynamic range were used in the analysis: 50 dB, 72 dB, and 96 dB. These levels are achieved by weighting the sidelobe response at the expense of widening the width of the main lobe. This is called *smearing*. In all cases in this simulation, the closest interferer was located 0.5 km away from the collection system and its ERP was 30 watts. The MDS values corresponding to the above conditions are –110 dBm for a 96 dB dynamic range, –86 dBm for a 72 dB dynamic range, and –64 dBm for a 50 dB dynamic range.

Rejection of Friendly Transmitters

It was assumed that friendly transmissions could be automatically detected, sorted, and rejected as signals of interest. This is a somewhat optimistic assumption against modern frequency hopping targets. With the older version of some radios, there was a 150 Hz squelch tone that could be used for this purpose. This tone was transmitted to reduce the noise in the receiver as a signal is received intermittently. Many new radios also have this feature so its utility for sorting is limited. However, other parameters, such as unique hop dwell times, are sometimes used for this purpose.

Table 4.1 Target Network Capacity

IBW	Azimuth Gate	Number of Target Nets
4 MHz	1°	1,377
	10°	138
12 MHz	1°	4,545
	10°	455
24 MHz	1°	9,090
	10°	909

4.7.2 Statistics Collected

The statistical parameters consisted of the following: (1) probability of copying a target at least once (P_c), (2) probability of identifying a target at least once (P_{it}), (3) probability of copying an NCS at least once (P_{cn}), (4) probability of identification of an NCS at least once (P_{in}), (5) probability of locating an NCS at least once (P_{ln}), and (6) probability of locating a target at least once (P_{lt}). A target in these definitions refers to any transmitter, including the NCS. Copying and identification are as defined earlier.

All of the results herein are based approximately on the equivalent of 10 dB amplitude gates resulting in an approximate resolution of 20 / 50 = 0.4. Table 4.1 shows the upper bound on the size of the target environment based on the parameters used in this analysis with the assumptions mentioned. To ascertain the effects of the wait time before declaring the signal gone, three values were evaluated: 20 hops equating to 0.2 second; 200 hops equating to 2 seconds; and 500 hops equating to 5 seconds.

4.7.3 Results

The simulation results are presented in two sections, with each section referring to the particular platform configuration. First, the results for the UAS configuration are presented, followed by those for the ground stand-off configuration.

4.7.3.1 UAS

All of the results for the UAS configuration used the weighted Euclidean metric for sorting and tracking.

Shown in Figure 4.32 is P_c plotted versus the number of target nets when IBW = 4 MHz and the dynamic range is 72 dB. P_c decreases approximately linearly as the number of nets increased and waiting only 20 hops before moving on the next target is noticeably better than waiting 200 hops, while 200 hops is better than 500 hops. The trade-off on the number of hops to wait is whether the goal is to copy a net or, as this statistic measures, to copy as many individual

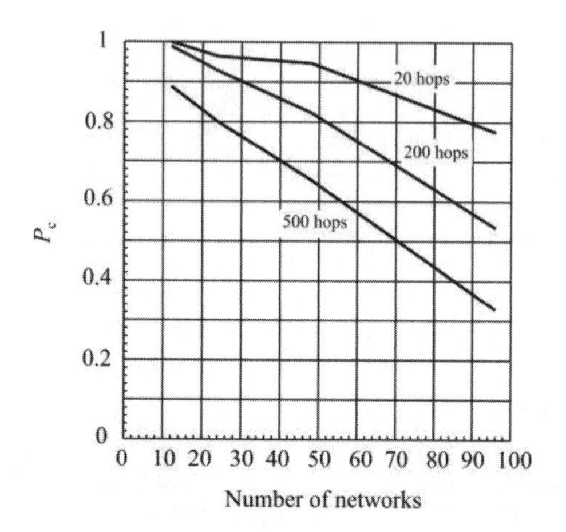

Figure 4.32 Probability of copying a target at least once, 4 MHz, 72 dB, UAS, weighted Euclidean metric.

transmissions as possible, although it is not obvious a priori which approach yields better performance. P_c is based on the latter, whereas the former likely produces more information. If the goal is to copy as many transmissions as possible, then setting the number of hops to wait as short as possible is reasonable. Figure 4.33 shows P_{it} as a function of the number of nets. Again, the linear decrease is present.

Waiting only 20 hops before moving on to the next target produced P_c above 0.8 for up to 90 nets, while P_{it} remained above 0.7 for these same conditions. These are relatively high probabilities and indicate that signal levels are probably not the limiting parameters. Timing of intercept operations caused some targets to

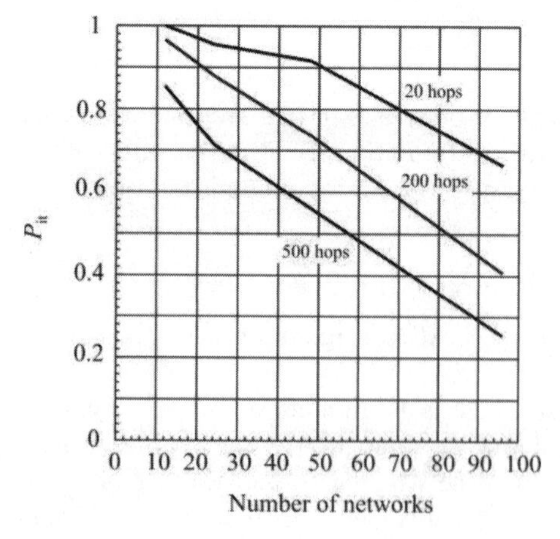

Figure 4.33 Probability of identification of a target at least once; 4 MHz, 72 dB, UAS, and weighted Euclidean metric.

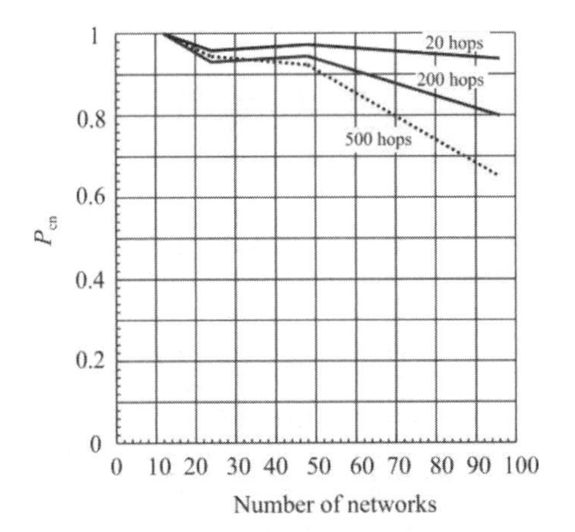

Figure 4.34 Probability of copying an NCS at least once, 4 MHz, 72 dB, UAS, and weighted Euclidean metric.

be missed, although identifying 70% of the targets indicates that not many were.

Shown in Figure 4.34 is P_{cn}, the probability of copying an NCS at least once while Figure 4.35 shows P_{in}, the companion statistic that illustrates the ability to identify an NCS. These probabilities remained reasonably high as the number of nets increased with no substantial decrease until the number of nets exceeded 44. In fact, for a wait time of 20 hops, P_{cn} and P_{in} remained relatively flat regardless of the number of nets.

Even when the wait time was allowed to extend to 200 and 500 hops, these statistics remained relatively high. The NCS transmitted the most compared to other members on the nets, which mostly accounts for this. The wait time is the

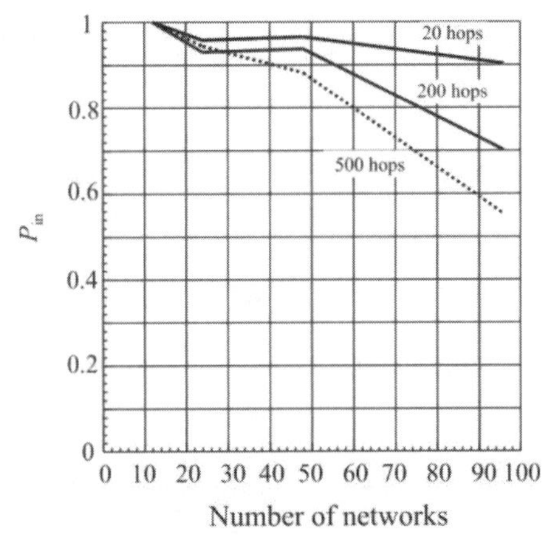

Figure 4.35 Probability of identification of an NCS at least once, IBW = 4 MHz, 72 dB dynamic range, UAS, and weighted Euclidean metric.

delay before declaring a target as gone. Since the NCS transmitted after each of the other nodes, the longer the wait time, the higher the probability of capturing the NCS. The tradeoff is waiting for the next node to transmit versus covering all the necessary frequencies.

P_{lt} and P_{ln} equaled unity for the above cases, indicating that a system with the performance described has no problem locating targets. Recall that the target location process was independent of what the system operators were doing; therefore, there is no variation with the wait time. The only limits on target location performance, as represented by these probabilities, were the data link capacity and target detectability (a target must be detected before it can be located). This is not to say, however, that the DF performance is not reflected in the other results. Target locations, to include the effects of DF accuracy, are critical elements of the copy performance due to the method of tasking operators with detected targets and the subsequent target tracking.

Varying the dynamic range, IBW, azimuth gates, noise figure, and detection threshold had little effect for the UAS configuration. For the cases considered here, achieving a 15 dB SNR at UAS collection systems was never a problem since the UASs flew very close to the targets. In fact, the SNRs were always 20 dB or better, considerably exceeding the required 15 dB. The target signal levels were sufficient so that the MDS levels were always met or exceeded. This is illustrated in Figure 4.36 where P_c is shown for when the dynamic range was reduced to 50 dB. There are no significant differences from when the dynamic range was 72 dB.

Changing the azimuth gates had no effect because of the statistics captured. Concentrating on collection of a single target instead of a target net negates some of the requirement for more accurate azimuth measurements.

Comparing Figure 4.37 for 12 MHz and Figure 4.38 for 24 MHz with Figure 4.36 shows that widening the IBW and thereby increasing the scan rate had no effect on the results. This indicates that widening the IBW beyond 4 MHz had no impact on the ability to capture transmissions in adequate time for target identification. It further indicates that the limit on hop phase resolution imposed by the IBW (about 3 ms) was adequate. The ability to copy targets was, however, subject to the assumptions given earlier (such as 70% of a hop). If these assumptions are violated, then a wider bandwidth may very well be necessary.

In summary, for cross-FLOT UAS collection platforms a dynamic range of 50 dB was adequate as was a 4 MHz IBW scanning receiver. Adequate SNRs and signal levels were available to provide reliable detection, collection and identification performance.

4.7.3.2 Ground Stand-Off

Both the Gaussian and the weighted Euclidean tracking metrics were examined for the ground stand-off configuration. Ground stand-off platforms suffer primarily

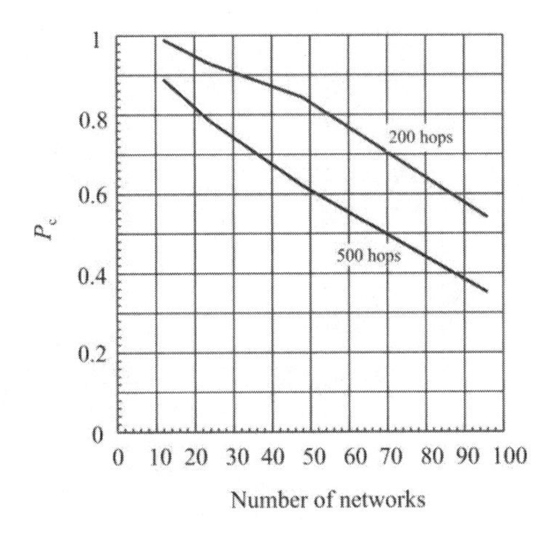

Figure 4.36 Probability of copy of a target at least once; 4 MHz, 50 dB, UAS, and weighted Euclidean metric.

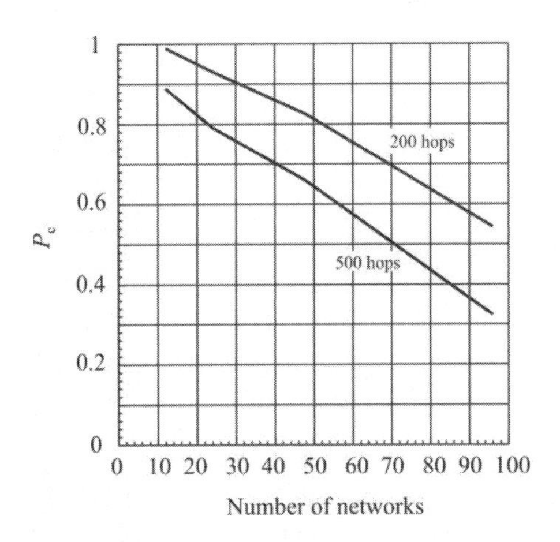

Figure 4.37 Probability of copy of a target at least once, UAS, 72 dB dynamic range, 15 dB threshold, 10 dB NF, 12 MHz IBW, and weighted Euclidean metric.

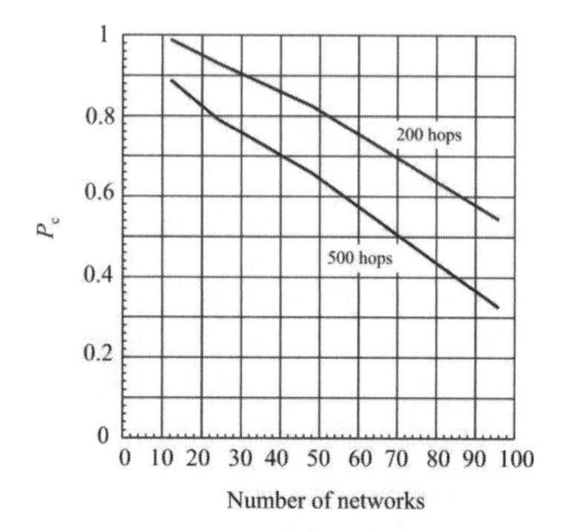

Figure 4.38 Probability of copy of a target at least once, UAS, 15 dB threshold, 10 dB NF, 24 MHz IBW, 72 dB dynamic range, and weighted Euclidean metric.

from the signal propagation characteristics discussed in the next chapter. The signal loss is much greater when the collection site is close to the ground. There are two fundamental limitations: noise induced onto the signal by external and internal means and the desensitizing effects of close (friendly) interfering transmitters. The effects of these limitations are shown in this section.

Euclidean Metric

The probability of copy of a target at least once, when the system noise figure is 10 dB, the SNR threshold was 15 dB, representing a type of radiometer detector, and the dynamic range was 72 dB, is shown in Figure 4.39. The 50 dB dynamic range performance (not shown) produced very poor results because of the signal propagation effects discussed previously. The signal levels of close transmitters produced too much interference and distant targets could not be detected. The 72 dB dynamic range did not improve this much. Detection and collection probabilities were less than about 0.2, irrespective of the number of nets. Since the target networks were placed uniformly throughout the target area, as more were added, a smaller percentage of them fell within the detection range of the ground, standoff collection systems. The probability of identification of a target is shown in Figure 4.40, which is essentially identical to Figure 4.39, indicating that for when the number of target nets was greater than 25, $P_{it} < 0.2$. Therefore, if a transmission was copied, it was identified, at least once, although the fraction of targets for which this happened was relatively low.

The probability of copy of an NCS is shown in Figure 4.41, with similar results. Few of the control nodes were copied. Likewise, the probability of

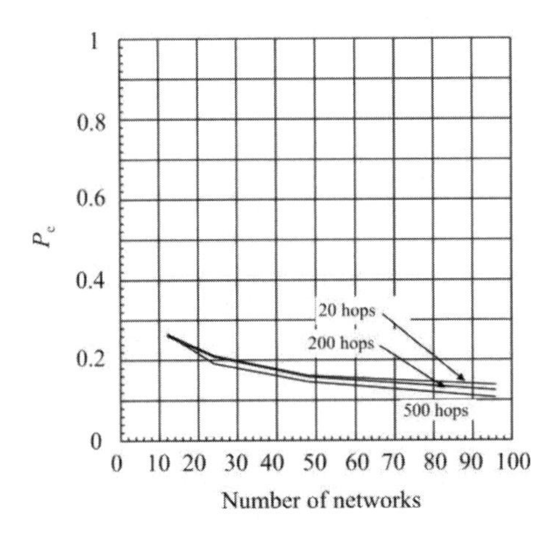

Figure 4.39 Probability of copying a target at least once, ground stand-off, 4 MHz IBW, 10 dB NF, 15 dB threshold, 72 dB dynamic range, and weighted Euclidean metric.

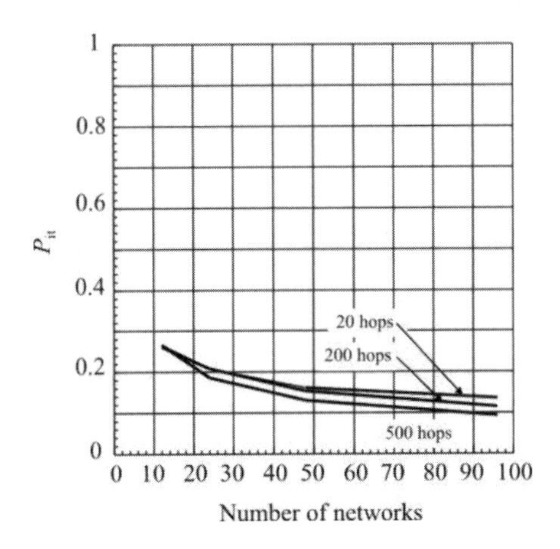

Figure 4.40 Probability of identifying a target at least once, ground stand-off, 4 MHz IBW, 10 dB NF, 15 dB threshold, 72 dB dynamic range, and weighted Euclidean metric.

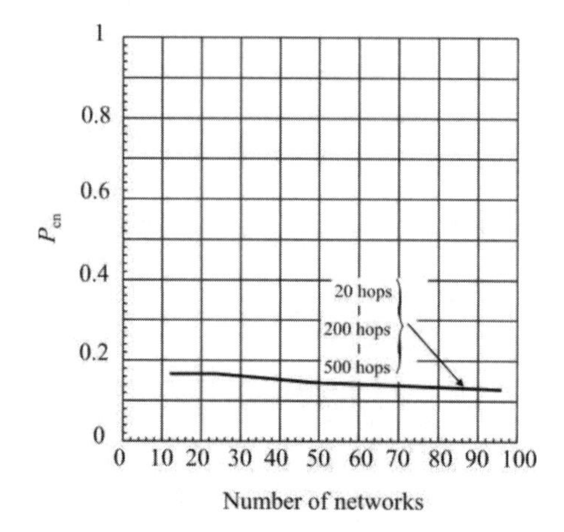

Figure 4.41 Probability of copy of an NCS at least once, ground stand-off, 4 MHz IBW, 10 dB NF, 15 dB threshold, 72 dB dynamic range, and weighted Euclidean metric.

identification of an NCS is shown in Figure 4.42 with the same results. If an NCS was copied, it was identified. The low results are due to the relatively few control nodes within range of the ground stand-off collection systems.

The probability of locating a target is shown in Figure 4.43, again indicating the relatively low fraction of targets located. Somewhat more overall targets were located than NCS nodes, however, the latter of which is shown in Figure 4.44.

Less than 0.2 of the number of control nodes were located for any number of networks.

Typical SNR densities for 12 and 96 target nets are shown in Figure 4.45 and

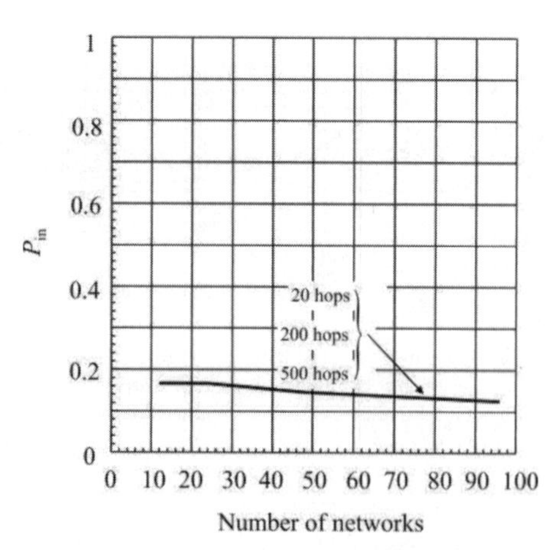

Figure 4.42 Probability of identification of an NCS at least once: ground stand-off, 4 MHz IBW, 10 dB NF, 15 dB threshold, 72 dB dynamic range, and weighted Euclidean metric.

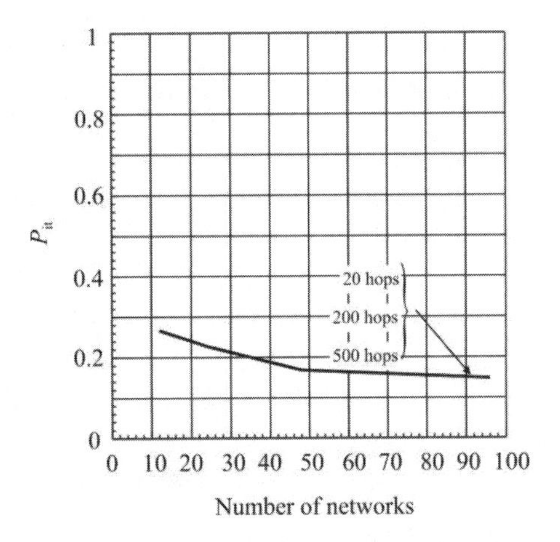

Figure 4.43 Probability of location of a target at least once, ground stand-off, 4 MHz IBW, 10 dB NF, 15 dB threshold, 72 dB dynamic range, and weighted Euclidean metric.

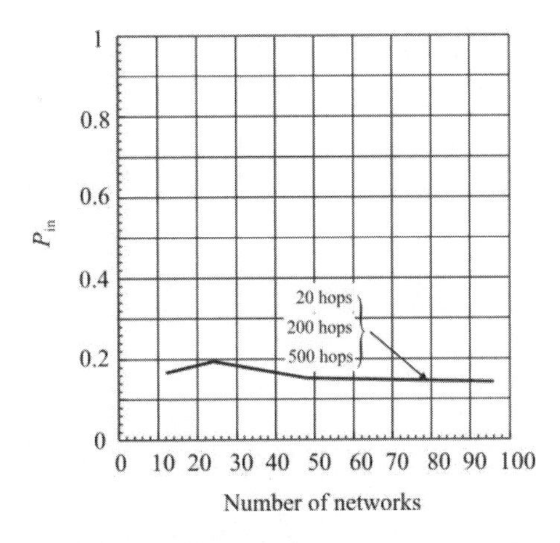

Figure 4.44 Probability of location of an NCS at least once, ground stand-off, 4 MHz IBW, 10 dB NF, 15 dB threshold, 72 dB dynamic range, and weighted Euclidean metric.

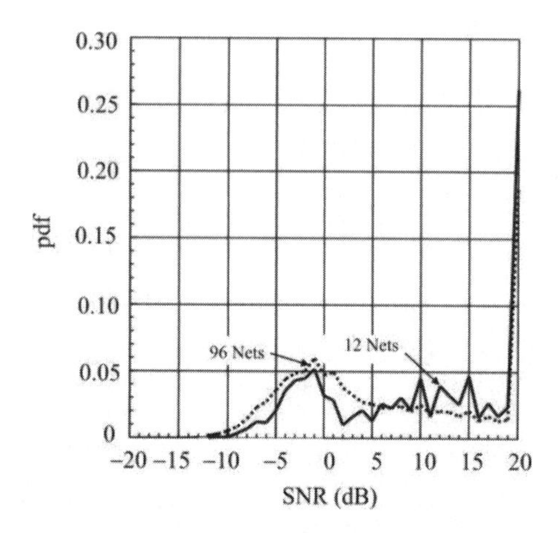

Figure 4.45 Typical SNR probability density for ground stand-off, 10 dB NF.

the corresponding distribution functions are shown in Figure 4.46. Recall that the SNRs calculated here incorporate the system noise figure. That is, it is the SNR expected at the input to the detector. For 12 nets, about 65% of the received SNRs were below the necessary 15 dB, while for 96 nets, about 75% were smaller. Reducing the noise figure by 10 dB (to 5 dB) increased the SNR by the same amount in these figures, which would effectively move the abscissa the same amount to the right, yielding about 35% for 12 nets and 55% for 96 nets with SNRs below the detection threshold. Therefore, one would expect a substantial improvement, other factors being the same, if the noise figure could be reduced. It will be shown below that this is precisely what happened. It is also clear from these figures that for the standoff case with limited dynamic range and sensitivity,

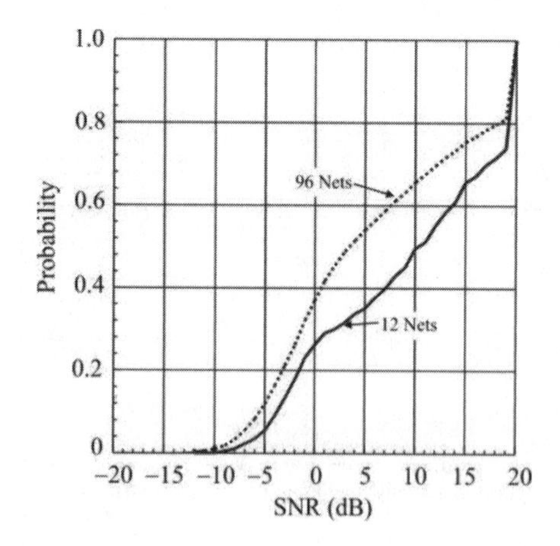

Figure 4.46 Typical SNR probability distribution for ground stand-off, 10 dB NF.

the wait time did not matter. This is because the collection systems only saw a limited set of targets; those they could hear, they collected.

Although the fraction of targets collected and located was low in all cases for ground stand-off collection systems, as more target nets were included, in general the ES results decreased. This was caused by two factors. The first were the timing impacts even though the number of target networks within range was limited. The operators were busier handling the additional targets and some were undetected. The second was, as more targets were added, they were added uniformly in the target array. Since the ground assets could only see a limited amount of this target area, the majority of the new target nets added was not within their field of view. These factors were true both for collection as well as target location.

Gaussian Metric

All of the above results are based on the heuristic Euclidean distance metric discussed earlier. The effects of employing the Gaussian based metric did not change performance from the Euclidean case. Either distance metric may be used. Target location performance obviously is not affected. These results do indicate, however, that these two tracking metrics can be used to track frequency hopping targets.

The dynamic range is not the only limiting factor, as increasing it by 24 dB did not change the results. It can therefore be concluded that the noise, as represented by the noise figure and detection threshold, was also a prominent cause of lack of collection from ground stand-off platforms. Reducing the noise figure to 0 dB yields about two to four times improvement for all statistics over the 10 dB noise figure case.

Some limits are imposed by the dynamic range. Increasing the dynamic range to 96 dB from 72 dB at a 0 dB noise figure produced about a double increase in performance. Some performances, however, were worse using the larger dynamic range, in particular, for the longer wait times and larger target nets. This is because the collection systems were hearing substantially more targets.

For the UAS case there was no variation in results when the azimuth gate parameter was varied. Although this parameter was not varied for the ground stand-off case, we would expect a greater effect since the LOB accuracy for each LOB obtained is inversely proportional to $\sqrt{\upsilon}$, and the SNRs were substantially less in many cases.

For the ground stand-off case, then, in summary, the system noise figure and the dynamic range were critical factors affecting collection performance. With a 72 dB dynamic range, a 10 dB noise figure, and a 15 dB SNR detection threshold, collection performance was fairly poor, typically below 20% to 30% of all targets. Increasing the dynamic range to 96 dB while decreasing the system noise figure to 0 dB substantially improved performance.

4.8 Concluding Remarks

The results presented in this chapter point out that a UAS is an excellent platform for collection of low VHF frequency-hopping communication signals. The system parameters necessary to be effective are quite modest. A ground stand-off platform on the other hand, has a very limited capability with technology that is available today. The performance can be substantially improved with some developments in the areas of noise reduction and detection algorithms.

The statistics used for this evaluation were more appropriate for target identification and tracking than for information collection (e.g., situation assessment). For the latter function, more than two collection systems would be needed and different statistical criteria would be more appropriate. Nevertheless, the results presented do indicate that significant signal collection performance can be achieved.

References

[1] Dillard, R. A., and G. M Dillard, *Detectability of Spread Spectrum Signals,* Norwood, MA: Artech House, 1989.
[2] Simon, M. K., J. K. Omura, R. A. Scholtz, and B. K. Levitt, *Spread Spectrum Communications Handbook,* New York: McGraw-Hill, 1994.
[3] Peterson, R. L., R. E. Ziemer, and D. E. Borth, *Introduction to Spread Spectrum Communications,* Upper Saddle River, NJ: Prentice Hall, 1995.
[4] Torrieri, D. J., *Principles of Secure Communication Systems,* 2nd ed., Norwood, MA: Artech House, 1992.
[5] Dixon, R. C., *Spread Spectrum Systems,* 2nd ed., New York: Wiley, 1984.
[6] Viterbi, A. J., *CDMA Principles of Spread Spectrum Communication,* Reading, MA: Addison-Wesley, 1995.
[7] Nicholson, D. L., *Spread Spectrum Signal Design,* Rockville, MD: Computer Science Press, 1984.
[8] Torrieri, D. J., *Principles of Secure Communication Systems* 2nd ed., Norwood, MA: Artech House, 1992, Ch. 4.
[9] Simon, M. K., J. K. Omura, R. A. Scholtz, and B. K. Levitt, *Spread Spectrum Communications Handbook,* New York: McGraw-Hill, Part 5, Ch. 4, 1994.
[10] Peterson, R. L., R. E. Ziemer, and D. E. Borth, *Introduction to Spread Spectrum Communications,* Upper Saddle River, NJ: Prentice Hall, 1995, Ch. 10.
[11] Tsui, J., *Digital Techniques for Wideband Receivers,* Norwood, MA: Artech House, 1995.
[12] Poisel, R. A., *Introduction to Communication Electronic Warfare Systems*, 2nd ed., Norwood, MA: Artech House, 2008, Ch. 10.
[13] Torrieri, D. J., *Principles of Secure Communication Systems*, 2nd ed., Norwood, MA: Artech House, 1992, pp. 323–324.
[14] Torrieri, D. J., *Principles of Secure Communication Systems*, 2nd ed., Norwood, MA: Artech House, 1992, p. 324.
[15] Erst, S. J., *Receiving Systems Design,* Artech House: Dedham, MA, 1984, p. 183.
[16] Torrieri, D. J., *Principles of Secure Communication Systems*, 2nd Ed., Norwood, MA: Artech House, 1992, pp. 328–335.

[17] Snelling, W. E., and E. Geraniotis, "Analysis of Compressive Receivers for the Optimal Interception of Frequency-Hopped Waveforms," *IEEE Transactions on Communications,* Vol. 42, No. 1, January 1994, pp. 127–134.

[18] Jack, M. A., P. M. Grant, and J. H. Collins, "Theory, Design, and Applications of Surface Wave Fourier-Transform Processors," *Proceedings of the IEEE,* Vol. 68, No. 4, April 1980, pp. 450–464.

[19] Zeal, A. J., I. S. Simi, and A. Petrolia, "Mismatched Compressive Receiver with Rejected Sidelobes," *ISIT* 1998.

[20] Harms, B. K., and D. R. Hummels, "Calculation of Detection Probability for Frequency Compressive Receivers," *IEEE Transactions on Aerospace and Electronic Systems,* Vol. AES-21, No. 1, January 1985, pp. 106–116.

[21] Li, K. H., and L. B. Milstein, "On the Use of a Compressive Receiver for Signal Detection," *IEEE Transactions on Communications,* Vol. 39, No. 4, April 1991, pp. 557–566.

[22] Turin, G. L., "An Introduction to Digital Matched Filters," *Proceedings of the IEEE,* Vol. 64, No. 7, July 1976, pp. 1092–1112.

[23] Torrieri, D. J., *Principles of Secure Communication Systems*, 2nd ed., *Second Edition,* Norwood, MA: Artech House, 1992, p. 151.

[24] Dillard, R. A., and G. M. Dillard, *Detectability of Spread Spectrum Signals,* Norwood, MA: Artech House, 1989, pp. 24–32.

[25] Simon, M. K., J. K. Omura, R. A. Scholtz, and B. K. Levitt, *Spread Spectrum Communications Handbook,* New York: McGraw-Hill, Vol. III, 1994, p. 1043.

[26] Edell, J. D., *Wideband, Noncoherent, Frequency-Hopped Waveforms and Their Hybrids in Low-Probability-of-Intercept Communications*, Naval Research Laboratory, Report 8025, November 8, 1976.

[27] Krasner, N. F., "Efficient Search Methods Using Energy Detectors—Maximum Probability of Detection," *IEEE Journal on Selected Areas in Communications*, Vol. SAC-4, No. 2, March 1986, pp. 273–279.

[28] Tsatsanis, M. K., and G. B. Giannakis, "Blind Estimation of Direct Sequence Spread Spectrum Signals in Multipath," *IEEE Transactions on Signal Processing*, Vol. 45, No. 5, May 1997, pp. 1241–1252.

[29] Burel, G., "Detection of Spread Spectrum Transmissions Using Fluctuations of Correlation Estimators," *Proceedings IEEE ISPACS*, Honolulu, HI, November 5–8, 2000.

[30] Krasner, N. F., "Maximum Likelihood Parameter Estimation for LPI Signals*," Proceedings IEEE MILCOM*, 1982, pp. 2.3-1–2.3-4.

[31] Milstein, L. B., "Interference Refection Techniques in Spread Spectrum Communications," *Proceedings of the IEEE,* Vol. 76, No. 6, June 1988, pp. 657–671.

[32] Gardner, W. A., "Signal Interception: A Unifying Theoretical Framework for Feature Detection," *IEEE Transactions on Communications,* Vol. 36, No. 8, August 1988, pp. 897–906.

[33] Holmes, J. K., *Coherent Spread Spectrum Systems*, New York: Wiley, 1982, pp. 583–584.

[34] Reed, D. E., "The Performance of Rate-Line Generation Circuits for Determining the Symbol Rate of Weak and Bandlimited Digitally Modulated Signals," Ph.D. Dissertation, University of Colorado at Colorado Springs, 1981.

[35] Urkowitz, H. "Energy Detection of Unknown Deterministic Signals," *IEEE Proceedings*, Vol. 55, April, 1968, pp. 523–531.

[36] Liebetreu, J. M., and D. E. Reed, *Detection of Spread-Spectrum Signals; Communications Study*, Technical Report CD-1011-1005A, January 1984.

[37] Reed, D. E., and M.A. Wickert, "Minimization of Symbol-Rate Spectral Lines by Delay-and-Multiply Receivers," *IEEE Transactions on Communication*, Vol. 36, January 1988, pp. 118–120.

[38] Reed, D. E., "Comparison of Symbol-Rate Detector and Radiometer Intercept Receiver Performances in a Nonstationary Environment," *Proceedings MILCOM,* 1989, pp. 19.5.1–19.5.5.

[39] Reed, D. E., "The Performance of Rate-Line Generation Circuits for Determining the Symbol Rate of Weak and Bandlimited Digitally Modulated Signals," Ph.D. Dissertation, University of Colorado at Colorado Springs, 1988.

[40] Gardner, *Introduction to Random Processes with Applications to Signals and Systems,* New York: Macmillan, 1986, p. 279.

[41] Sousa, M. J., and J. W. Betz, *Robust Detection of Wideband Signals*, MITRE Unpublished Research Report, 1991.

[42] Gardner, W. A., *Introduction to Random Processes with Applications to Signals and Systems,* New York: Macmillan, 1986, p. 280.

[43] Burel, G., A. Quinquis, and S. Azou, "Interception and Furtivity of Digital Transmissions," *Proceedings IEEE Communications 2002*, Bucharest, Romania, December 5–7, 2002.

[44] Levitt, B. K., U. Cheng, A. Polydoros, and M. K. Simon, "Optimum Detection of Slow Frequency Hopped Signals," *IEEE Transactions on Communications,* Vol. 42, No. 2/3/4, February/ March/April 1994, pp. 1990–2000.

[45] Levitt, B. K., and U. Cheng, "Optimum Detection of Frequency Hopped Signals," *Proceedings IEEE MILCOM, 1992*.

[46] Schleher, D. C., *Electronic Warfare in the Information Age*, Norwood, MA: Artech House, 1999, pp. 388–389.

[47] Kemeny, J. G., and J. L. Snell, *Finite Markov Chains*, New York: Springer-Verlag, 1976.

[48] Kemeny, J. G., and J. L. Snell, *Finite Markov Chains*, New York: Springer-Verlag, 1976, Ch. V.

[49] Kemeny, J. G., and J. L. Snell, *Finite Markov Chains*, New York: Springer-Verlag, 1976, Ch. IV.

[50] Kemeny, J. G., and J. L. Snell, *Finite Markov Chains*, New York: Springer-Verlag, 1976, pp. 101–102.

[51] Hufford, G., *A Guide to the Use of the ITS Irregular Terrain Model in the Area Prediction Mode*, U.S. Department of Commerce, National Telecommunications and Information Administration, NTIA Report 82-100, April 1982.

[52] Tsui, J., *Digital Techniques for Wideband Receivers,* Norwood, MA: Artech House, 1995, pp. 84–85.

Chapter 5

Radio Signal Propagation

5.1 Introduction

Signals propagate from the transmit antenna to the receive antenna in all RF communication systems in the form of *electromagnetic* (EM) waves. How well they do this largely determines the signal strength at the receiver. The propagation effects are the same for the signal from the intended transmitter as well as for the jammer-to-receiver link, although the specifics may vary somewhat because the propagation paths are normally different. Any obstacles that are between the transmitter and the receiver would probably not be in the path from the jammer to the receiver. The fundamentals, however, are the same.

The other factor that limits how well a signal is received is the noise in the channel as perceived by the receiver. The *signal-to-noise ratio* (SNR) at the receiver is the overarching determinant of how well a signal is received. Noise arises externally to the receiver and within the receiver. Both limit performance, and normally they are additive. Noise effects were discussed in Chapter 2.

Modern communication systems can be found in most frequency bands. The ones considered here, however, will be limited to the HF (0.5–30 MHz), VHF (30–300 MHz), and UHF (300–3,000 MHz) frequency bands. For HF, both sky-wave and ground-wave propagation systems can be used for communications.

We cover radio signal propagation in general in this chapter. This material describes the basic propagation mechanisms found in most places (under water is an exception to this that is not covered). Because of its increased importance, we devote Chapter 16 specifically to signal propagation in urban terrain, and cover the frequency range associated with the most prolific modern signals in this setting.

All of the important communication systems as of this writing, forming the targets for our EW systems, are implemented digitally so those are the only signals of interest. There are some exceptions to this, primarily in the urban environment, such as CB radios. We will not delve into the details of these radios, however.

It is the BER that specifies how well a digital communication system works and is a function of the SNR at the receiver RF input. An EW system tries to decrease the SNR at the target receiver by raising the noise level. The usual jammer performance in such instances is specified by the *jam-to-signal ratio* (JSR) at the receiver input. We will discuss many facets of EW systems and how, and how well, the SNR can be decreased.

We begin this chapter with a discussion of the fundamental characteristics of VHF and above signal propagation and, in particular, the effects of the Earth since we are most concerned with signals propagating close to the Earth. Following that, we enter into a discussion of HF signal propagation and present two ways to model nonionospheric signal propagation: a simple model and a complex model suitable for computer implementation. Lastly, HF sky-wave propagation is discussed.

5.2 Propagation of Signals in the VHF Range and Above

5.2.1 Introduction

It can be argued that the low VHF frequency range is the most heavily used for communications by land-mobile military forces. That is certainly where the majority of the tactical CNRs have been built. It offers reasonable propagation ranges coupled with acceptable bandwidths, along with reasonable noise levels. Lower than this, in the HF range, the signals will normally propagate further for a given amount of *effective radiated power* (ERP), but the bandwidth is smaller. Higher than this, larger bandwidths are possible but the signals do not propagate as far, typically. In this section, the discussion about signal propagation will begin with the VHF range and higher. HF signal propagation is presented after that [1–3].

5.2.2 Free-Space Propagation

From basic physics, as a radio wave travels away from an isotropic antenna,[1] its energy expands spherically. The power per unit area on the surface of the sphere thus decreases. The power field associated with the propagating wave is called its *power density* and is given in units of W m^{-2}. The *electric field* associated with a propagating signal is denoted by E and is given in units of V m^{-1}. If there are no obstructions to distort the wave, the power density decreases as the square of the

[1] An *isotropic antenna* is an imaginary antenna where energy leaves equally in all directions.

distance it travels. This is the case when the radio wave is far from the surface of the Earth, such as in outer space.

The equation that dictates the amount of power received by a receive antenna with gain[2] G_{RT} in the direction of the transmitter is known as the Friis equation and is given by

$$P_R = G_{TR} P_T G_{RT} \frac{\lambda^2}{(4\pi D_{TR})^2} \tag{5.1}$$

In this equation, G_{TR} is the transmitter antenna gain in the direction of the receiver, P_T is the amount of power that is entering the transmit antenna, and λ is the wavelength of the signal, given by $\lambda = c / f$ when f is the frequency (in hertz). The constant c is the speed of propagation, normally accepted as the speed of light for RF propagation in free space. D_{TR} is the distance between the transmitter and receiver, in units consistent with λ. The product $G_{TR}P_T$ is often referred to as the ERP.

While communications close to the earth, described next, do not normally follow (5.1), at altitudes such as those typified by air-to-air communications, this expression does apply. Furthermore when the jamming target is close to the ground and the jammer is airborne, and vice versa, the jamming link may follow (5.1). Thus, the links between the transmitter and jammer and jammer and receiver may have entirely different characteristics.

5.2.3 Propagation Close to the Earth

Expression (5.1) dictates that the power received decreases as the square of the distance traveled. When close to the Earth or other large object, this dependence on the square of the distance is no longer valid. It actually normally falls faster than the square of the distance, and the rate depends on the propagation medium and the nature of any obstructions. A common approximation that is useful in many cases is a model that assumes the decrease in power density is proportional to D_{TR}^{-n}, where the value of n varies with the particular situation. Outdoors, in reasonably flat Earth surface conditions, $n = 4$ is frequently used. Inside buildings and factories it can be more than or less than 4. In fact, n has been measured to vary over the range of 1.6 to over 5 [4].

---

[2] The *gain* of an antenna in a given direction is the degree to which the antenna prefers to emit (or receive) energy in that direction. It is usually specified relative to an isotropic antenna where it is denoted G_i. Thus, an antenna that emits twice the amount of energy in some direction has a gain of 2 in that direction. Since $G_{dB} = 10 \log_{10} G$, a factor of two is about 3 dB.

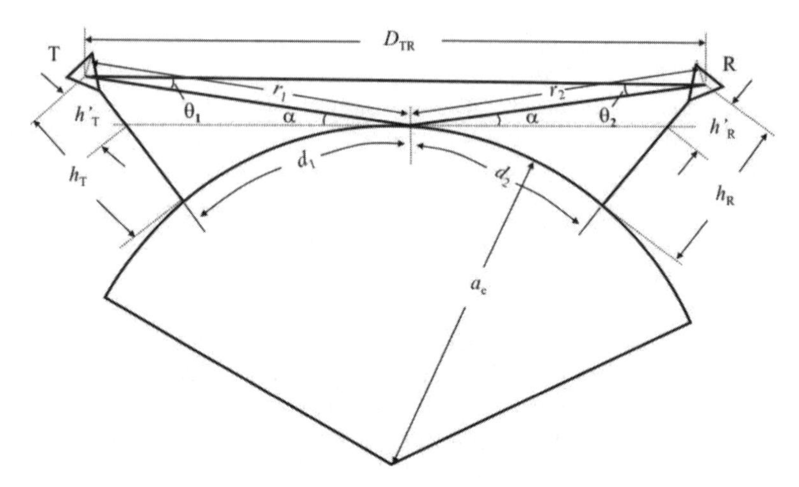

Figure 5.1 Direct wave and reflection off the Earth. (Source: [5]. © IRE 1979. Reprinted with permission.)

The discussions in this section apply to both the communication path as well as the transmitter-to-jammer path and the jammer to receiver path, as long as these links are all close to the Earth.

When the transmit and receive antennas are high enough and/or the propagation path short enough, then the propagation mechanism consists of a *direct wave* component as well as a component that is reflected off the Earth at some point between the antennas. If the Earth is smooth at the point of reflection, then the reflected component will be *specular* with significant deterministic characteristics. If the Earth is very rough at the point of reflection, then the reflected component will be largely *diffuse* (also referred to as *scattered*) and spatially uniform with reflection angle. Figure 5.1 shows the parameters involved [5].

5.2.4 Smooth Earth

In general, the electric field at the receiver is given by

$$E_p = \frac{E_0}{D_{\text{TR}}}\left[1+\rho_k e^{j\delta} + (1-\rho_k)Ae^{j\delta}\right] \tag{5.2}$$

where $p \in \{\text{h, v}\}$ for horizontal and vertical polarizations, respectively, and δ is the phase difference at the receiver between the direct path signal and the reflected signal[3]. E_0 is the free-space field strength when there are no ground effects. Parameter ρ_k is referred to as the *reflection coefficient* and, in general, is a complex number. The first term in brackets represents the direct wave; the second term is the *reflected component*, while the third represents the *surface wave*. A represents the attenuation of the surface wave, which, as mentioned above, depends on the Earth's surface conditions.

At the point of reflection, the wave leaves with a different phase angle that depends on the polarization of the signal at this point, the angle of incidence, and the ground characteristics. For antennas close to the ground (relative to a wavelength), ρ_k approaches –1 and δ approaches 0, causing the first two terms to cancel, leaving only the third, surface wave, term. For antennas that are elevated further relative to a wavelength, the surface wave component is negligible and only the direct wave and ground reflected wave need be considered. Thus, for VHF frequencies and above, where the antennas are typically elevated, the third term in (5.2) is small compared to the other two and is assumed to be 0. For the HF range, and in particular when $f < 10$ MHz, the antennas under consideration here are relatively close to the ground and the first two terms cancel, leaving only the third term. Furthermore, there can only be direct and reflected components when the transmitter and receiver are within radio line of sight of each other, whereas the surface wave can propagate much further than that, depending on surface conditions.

When the signal at the receiver is composed of the sum of the direct wave and the reflected wave, the reflected path will impose a certain amount of additional phase delay. In addition to this phase difference at the receiver, an additional phase term is added due to the complex nature of the reflection coefficient at the point of reflection.

The surface wave travels along the surface of the Earth from the transmitter to the receiver. The magnitude of this component of the signal is strongly dependent on the characteristics of the Earth between the transmitter and receiver. When good ground conditions for propagation (wet, for example, or sea water) are present, the surface wave component can extend to 100 km or more.

First, consider the VHF case when there is no significant surface wave. Close to the earth, surface waves propagate poorly in the VHF and above ranges. Vertically polarized EM waves propagate better than those with horizontal polarization. For reliable propagation via the surface wave, the frequency must be limited to a range of about 1.5 MHz to 5 MHz.

---

[3] The *polarization* of a propagating wave is the orientation of the E-field vector relative to something. Close to the Earth, that something is usually taken to be the unit vector tangent to the Earth's surface. Not close to the Earth (in space, for example), the polarization is specified relative to something—usually an antenna.

The reflection coefficients ρ_v and ρ_h are comprised of three parts denoted $r_{v,h}$, $g(v)$, and D as

$$\rho_{v,h} = r_{v,h} \times g(\chi_P, \chi_{P'}) \times D \tag{5.3}$$

These components are described next.

The first part consists of the reflection coefficient as if the reflection were off an infinite plane. For a horizontally polarized wave with an angle of incidence given by ϕ [6],

$$r_h = \frac{\sin\phi - \sqrt{\left(\varepsilon_r - j\dfrac{\sigma}{2\pi f \varepsilon_0}\right) - \cos^2\phi}}{\sin\phi + \sqrt{\left(\varepsilon_r - j\dfrac{\sigma}{2\pi f \varepsilon_0}\right) - \cos^2\phi}} \tag{5.4}$$

where ε_r is the *relative dielectric constant* (also called the permittivity) of the Earth at that point, ε_0 is the dielectric constant of a vacuum, σ is the conductivity of the Earth at that point, and f is the frequency. For vertical polarization,

$$r_v = \frac{\left(\varepsilon_r - j\dfrac{\sigma}{2\pi f \varepsilon_0}\right) - \sqrt{\left(\varepsilon_r - j\dfrac{\sigma}{2\pi f \varepsilon_0}\right) - \cos^2\phi}}{\left(\varepsilon_r - j\dfrac{\sigma}{2\pi f \varepsilon_0}\right) + \sqrt{\left(\varepsilon_r - j\dfrac{\sigma}{2\pi f \varepsilon_0}\right) - \cos^2\phi}} \tag{5.5}$$

The permittivity and conductivity actually vary considerably with frequency and soil type as shown in Figure 5.2 [7]. The variability with frequency over the typical land mobile communication system range (low VHF) is not so great. In the cellular bands around 900 MHz and 1,800 MHz, however, the ground conditions significantly affect the permittivity and conductivity.

The vertical polarization reflection coefficient, r_v, is plotted in Figure 5.3 for when the frequency is mid-range through the low VHF, 60 MHz. The curves do not change much for anywhere in the range of 30 MHz to 90 MHz, however. Figure 5.3(a) is the magnitude, while Figure 5.3(b) is the phase. The three curves correspond to poor soil, average soil, and wet (good) soil with parameters given in Table 5.1.

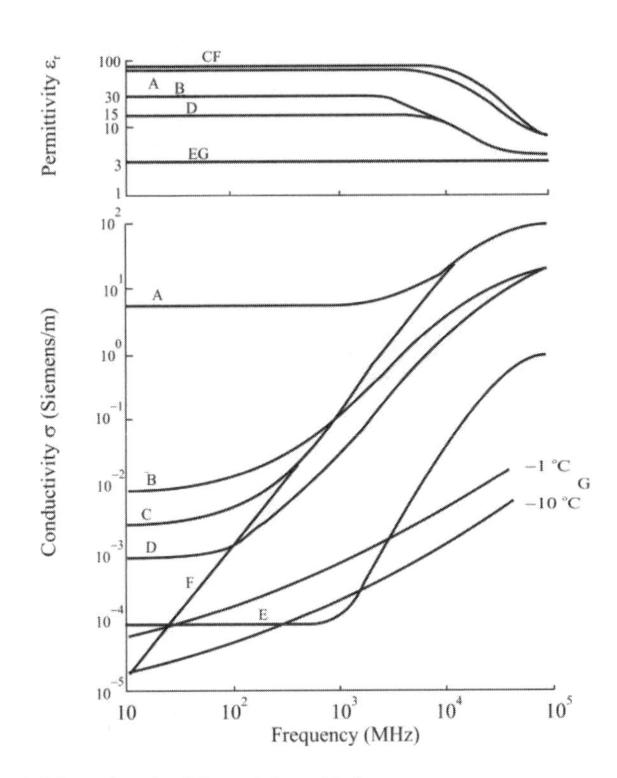

Figure 5.2 Permittivity and conductivity variation with frequency.

A: Sea water, 20°C	C: Fresh water, 20°C	E: Very dry ground	G: Ice
B: Wet ground	D: Medium dry ground	F: Pure water, 20°C	

(Source: [7]. © IRE 1979. Reprinted with permission.)

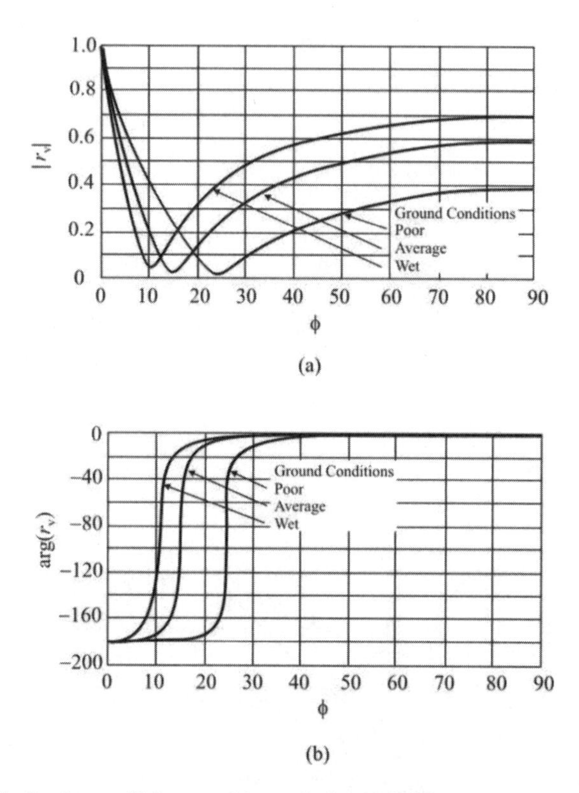

Figure 5.3 Vertical reflection coefficient r.v.: (a) magnitude and (b) phase.

Table 5.1 Ground Conditions for Three Soil Conditions in Figures 5.3 and 5.4

Ground Conditions	**Conductivity, σ (Siemens)**	**Dielectric Constant, ε**
Poor	10^{-3}	5
Average	5×10^{-3}	15
Wet	2×10^{-2}	30

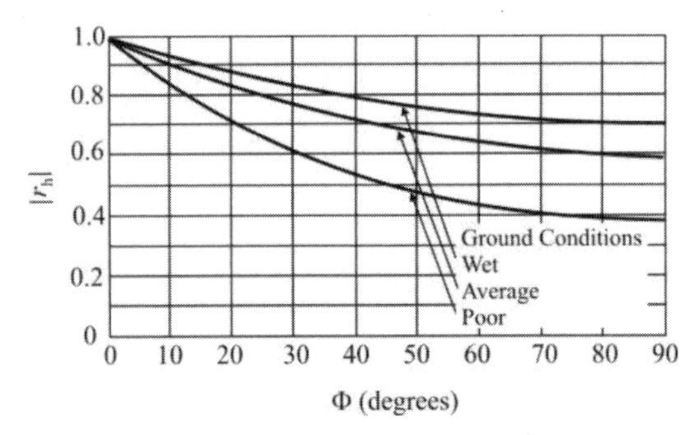

Figure 5.4 Horizontal polarization refection coefficient amplitude. The phase of the reflection coefficient is 180° for all of the conditions indicated.

The angle where the minimum in the amplitude occurs is called the *pseudo-Brewster angle*. This angle decreases with increasing soil conductivity being the least for salt water.

The phase of the horizontal reflection coefficient, r_h, is 180° for all of the conditions considered here, while the amplitude is shown in Figure 5.4.

In summary, horizontally polarized waves always undergo an 180° phase reversal at the point of reflection. Vertically polarized waves undergo this phase reversal at reflection angles below the pseudo-Brewster angle but not above it, where the phase shift is less.

The second component in the reflection coefficient in (5.3) is a factor $g(\chi_P, \chi_{P'})$, which accounts for the finite dimensions of the reflecting plane. This function is given by

$$g(\chi_P, \chi_{P'}) = g(\chi_P) + g(\chi_{P'}) - 1 \tag{5.6}$$

where

$$\begin{aligned} \chi_P &= -\frac{d_P}{\sin(\alpha)}\sqrt{\frac{d_1 + d_2}{2d_1 d_2 \lambda}} \\ \chi_{P'} &= -\frac{d_{P'}}{\sin(\alpha)}\sqrt{\frac{d_1 + d_2}{2d_1 d_2 \lambda}} \end{aligned} \tag{5.7}$$

and

$$G(\chi_P, \chi_{P'}) = -20\log_{10}[g(\chi_P, \chi_{P'})] \tag{5.8}$$

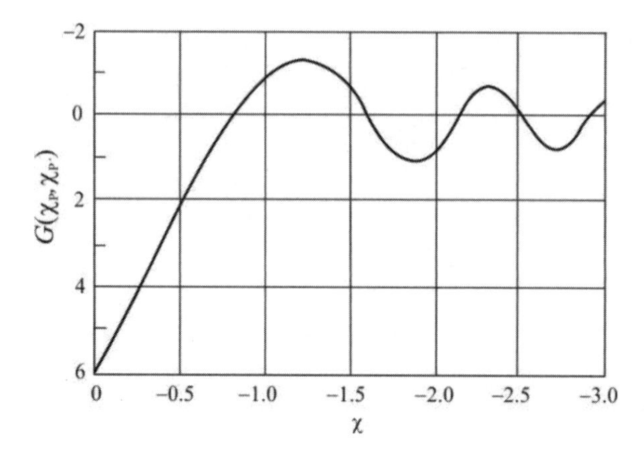

Figure 5.5 Fresnel-Kirchoff function for a knife-edge diffraction. (Source: [8]. © IRE 1979. Reprinted with permission.)

is the value of the Fresnel-Kirchoff function shown in Figure 5.5.

Distances d_P and $d_{P'}$ in these expressions are derived from analysis of the Fresnel zones. See [9] for a complete derivation. Figure 5.6 depicts the case of a reflection off the ground somewhere between the transmitter and receiver. The rays that travel between T and R that have equal time delays define the ellipsoid. When the path length is an integral multiple of $\lambda/2$, the ellipsoid is a cross section of the Fresnel zone corresponding to that multiple. Distance d_P is the length of PC and $d_{P'}$ is the length of P'C. In general $d_P \neq d_{P'}$, but they are equal if $d_1 = d_2$ in Figure 5.1.

The number of Fresnel zones that must be included in this analysis depends on the size of the reflecting area. This, in turn, depends on the height of the

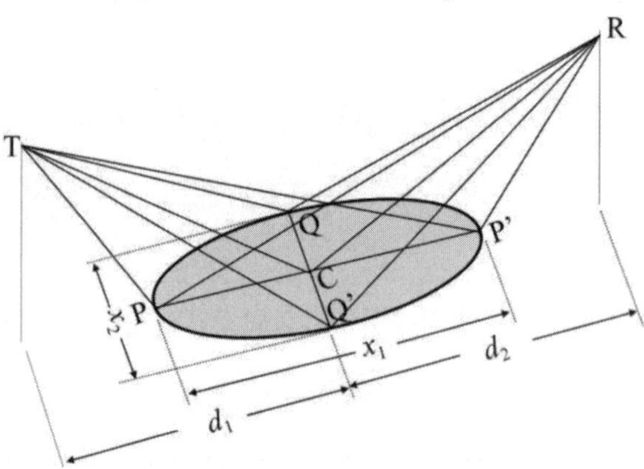

Figure 5.6 Reflection ellipsoid between transmitter (T) and receiver (R). (Source: [9]. © IRE 1979. Reprinted with permission.)

antennas and the terrain between the antennas. The dimensions shown in Figure 5.6 when the path length difference between the direct wave and the reflected wave is $n\lambda/2$ are

$$x_2 = 2R_n \tag{5.9}$$

where R_n is the radius of the nth Fresnel zone

$$R_n = \sqrt{\frac{n\lambda d_1 d_2}{d_1 + d_2}} \tag{5.10}$$

In addition, x_1 is given by

$$x_1 = \frac{x_2}{\sin(\alpha)} \tag{5.11}$$

The third component in (5.3) is a *divergence factor*, denoted by D. This parameter accounts for curvature of the Earth at the point of reflection. For smooth Earth D is given by [10].

$$D = \sqrt{1 + \frac{2d_1 d_2}{a_e(h_T' + h_R')}} \tag{5.12}$$

The parameters in this equation are as shown in Figure 5.1.

When the distances involved are short, the Earth can be modeled as a flat plane. In those cases, a propagating wave can be modeled as a direct wave and one interfering wave reflected off the Earth's plane as seen in Figure 5.7. The Earth at the point of reflection can be modeled as smooth, where the wave is perfectly reflected, or rough. In the latter case, the wave after reflection is better modeled with random parameters. The method to discern between the two is presented here.

The phase difference between the direct wave and reflected wave at the receiver can be expressed as [11]

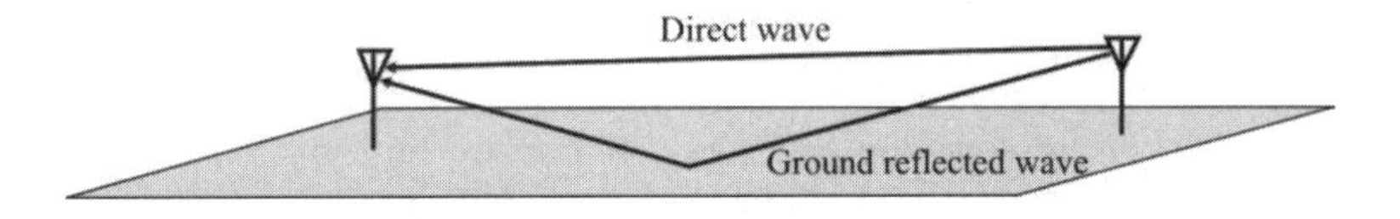

Figure 5.7 The *two-ray* model consists of a direct wave with a single reflected wave.

$$\delta = \frac{2\pi}{\lambda}\left(\sqrt{D_{\text{TR}}^2 + (h_{\text{T}} + h_{\text{R}})^2} - \sqrt{D_{\text{TR}}^2 + (h_{\text{T}} - h_{\text{R}})^2}\right) \tag{5.13}$$

where h_{T} and h_{R} are the heights (above ground) of the transmit and receive antennas, respectively, λ is the wavelength, and D_{TR} is the distance between the transmitter and receiver. Since

$$\sqrt{1+x} \approx 1 + \frac{x}{2} \tag{5.14}$$

when the distance D_{TR} is large then

$$\delta \approx \frac{4\pi}{\lambda}\frac{h_{\text{T}} h_{\text{R}}}{D_{\text{TR}}} \tag{5.15}$$

In addition, the received power is

$$P_{\text{R}} = \frac{\lambda^2}{(4\pi D_{\text{TR}})^2}\left[2\sin\frac{2\pi}{\lambda}\frac{h_{\text{T}} h_{\text{R}}}{D_{\text{TR}}}\right]^2 G_{\text{TR}} P_{\text{T}} G_{\text{RT}} \tag{5.16}$$

where G_{TR} is the gain of the transmit antenna in the direction of the receiver, P_{T} is the transmitter power, and G_{RT} is the gain of the receive antenna in the direction of the transmitter. The distance

$$D_{\text{TO}} = \frac{4}{\lambda} h_{\text{T}} h_{\text{R}} \tag{5.17}$$

is called the *turnover distance,* in kilometers. Beyond that point

$$P_{\text{R}} \approx \frac{(h_{\text{T}} h_{\text{R}})^2}{D_{\text{TR}}^4} G_{\text{TR}} P_{\text{T}} G_{\text{RT}} \tag{5.18}$$

The received power at large distances (beyond D_{TO}) therefore falls as the fourth power of that distance. In addition, it increases as the square of the heights of the product of the antennas. Lastly, it is independent of frequency. In the sequel, this propagation model will be called the *two-ray model.*

This function is plotted in Figure 5.8 out to 100 km for some typical parameters. In this case, D_{TO} is as shown in Table 5.2 so the graphs for ranges less

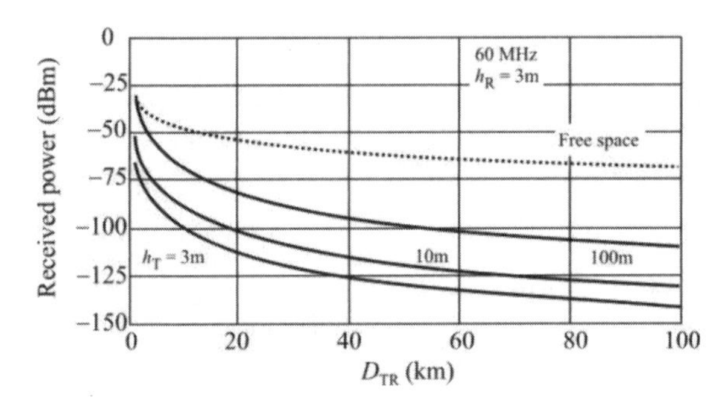

Figure 5.8 Power received based on the ground reflection model of signal propagation. In this case D_{TO} is given in Table 5.2. $P_T = 10$ W, $G_{TR} = G_{RT} = 1$.

than this are not accurate. Also plotted is the received power due to free space for comparison.

The free-space model (5.1) prescribes that the signal level falls as the square of the frequency. Thus, if the frequency is doubled, the signal level at a given distance is 6 dB less. This model indicates that beyond D_{TO}, the signal level is independent of frequency. Thus, there is a fundamental difference between free-space propagation and propagation when there is a ground reflected wave involved.

Parameter A in (5.1) is the Earth attenuation factor. It is given approximately by [6]

$$A \approx \frac{-1}{1 + j\dfrac{2\pi D_{TR}}{\lambda}(\sin\alpha + z)^2} \tag{5.19}$$

where

$$z = \frac{\sqrt{\varepsilon_0 - \cos^2\alpha}}{\varepsilon_0}, \qquad \text{vertical polarization}$$

$$z = \sqrt{\varepsilon_0 - \cos^2\alpha}, \qquad \text{horizontal polarization}$$

Table 5.2 Values of Turnover Distance for the Antenna Heights in Figure 5.11 (h_R = 3m)

h_T (m)	D_{TO} (km)
3	5.4
10	8
100	80

α = angle between reflected ray and the ground
= 0 for antennas at ground level;
ε = dielectric constant of the ground relative to unity in free space;
σ = conductivity of the ground in mhos m^{-1}.

5.2.5 Effective Antenna Height

Terrain conditions near the transmit and receive antennas cause the effective height of these antennas to be different from their physical values in the VHF range. The *effective antenna heights* are given by [12].

$$h'_{\rm T} = \sqrt{h_{\rm T}^2 + h_0^2} \tag{5.20}$$

and

$$h'_{\rm R} = \sqrt{h_{\rm R}^2 + h_0^2} \tag{5.21}$$

where

$$h_0 = \left(\frac{\lambda}{2\pi}\right)\left[(\varepsilon_{\rm r} + 1)^2 + (60\lambda\sigma)^2\right]^{1/4} \tag{5.22}$$

for vertical polarization and

$$h_0 = \left(\frac{\lambda}{2\pi}\right)\left[(\varepsilon_{\rm r} - 1)^2 + (60\lambda\sigma)^2\right]^{-1/4} \tag{5.23}$$

for horizontal polarization. In these expressions ε_r and σ are the permittivity and conductivity of the soil in the area given in Figure 5.2 and h_T and h_R are the physical heights of the phase centers of the antennas. The ground characteristics do not change the antenna heights much above the VHF range. Table 5.3 illustrates h_0 for a few conditions. The effective antenna height for horizontal polarizations for all practical purposes is the same as the physical height for land mobile communications. For vertical polarizations in the low VHF range, the effective height can be considerably higher than the physical height. The effective height is increased more over water than over ground and is increased more over wet, or better conducting, ground than over poorly conducting, dry ground.

Table 5.3 Values of h_0 (in m) for a Few Conditions of Interest for Land Mobile Communications

Soil Type	**Frequency**		
Vertical Polarization	30 MHz	100 MHz	1,000 MHz
Sea Water (20°C)	87	14	0.53
Wet Ground	8.9	5.7	0.27
Fresh Water	14	4.3	0.43
Medium Dry Ground	6.4	1.9	0.19
Very Dry Ground	3.2	0.96	0.095
Pure Water	14	4.30	0.430
Ice	3.2	0.96	0.095
Horizontal Polarization			
Sea Water (20°C)	0.03	0.02	0.004
Wet Ground	0.30	0.09	0.009
Fresh Water	0.18	0.05	0.005
Medium Dry Ground	0.42	0.13	0.013
Very Dry Ground	1.13	0.34	0.034
Pure water	0.18	0.05	0.005
Ice	1.12	0.34	0.034

The jamming antenna also has an effective height that would typically be different from the transmitter and receiver. This would be especially true if the jammer were airborne.

5.2.6 Surface Roughness

The two-ray model described above assumes that the Earth is a smooth plane. In fact, this is rarely the case and surface roughness can impact on the reflection. If σ represents the standard deviation of the irregularities of the Earth's surface relative to the mean height in a region, Parsons [13] derives the following as the *Rayleigh criterion*. This criterion, borrowed from the field of optics, was put forth by Sir Walter Rayleigh as the point when two diffracted light beams could be resolved. Considering Figure 5.9, the two beams on the left are not resolvable, the ones in the middle are just resolvable, and those on the right are over-resolved. In this figure, θ is the phase difference of the two beams. Rayleigh's criterion says that the resolution limit is where the null of one beam is placed at the maximum of the other. This occurs at $\theta = \pi/5$.

The same signal reflected from two surfaces a distance Δh apart, as illustrated in Figure 5.10, defines the roughness of a surface. The lower signal travels a distance of $2x_1$ more than the upper signal. By simple trigonometry,

$$2x_1 = 2\Delta h \sin\alpha \tag{5.24}$$

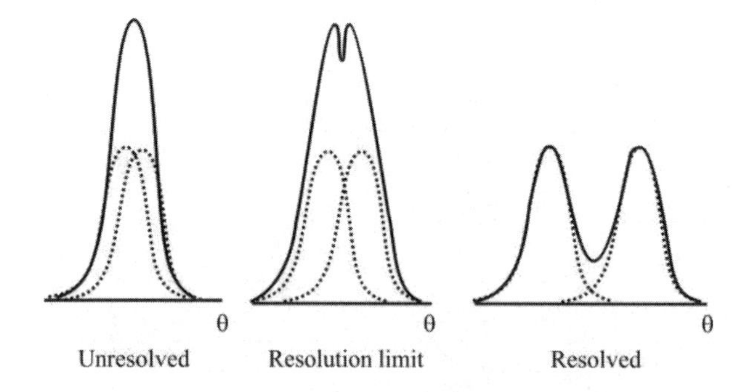

Figure 5.9 Establishing the Rayleigh criterion for resolving two diffracted light beams.

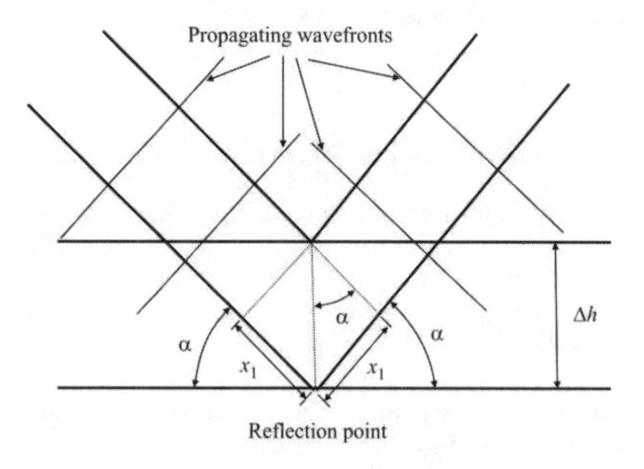

Figure 5.10 Details of the reflection of a signal from two planes.

so

$$\sin\alpha = \frac{x_1}{\Delta h} \tag{5.25}$$

The difference in phase of these two signals caused by traveling this additional distance is given by

$$\begin{aligned}\Delta\phi &= \frac{2\pi(2x_1)}{\lambda} \\ &= \frac{2\pi}{\lambda}2\Delta h\sin\alpha\end{aligned} \tag{5.26}$$

Therefore, the Rayleigh limit is given by

$$\frac{4\pi\Delta h}{\lambda}\sin\alpha = \frac{\pi}{2} \tag{5.27}$$

or

$$\Delta h = \frac{\lambda}{8\sin\alpha} \tag{5.28}$$

Some values for Δh are given in Table 5.4.

Hall reports that the RMS specular component of the reflection coefficient at a point on the Earth is the product of that for smooth Earth times the terrain roughness factor $f(\sigma_i)$ [14]. Thus,

$$\rho_{\text{specular}} = \rho_{\text{smooth}} f(\sigma_i) \tag{5.29}$$

with the terrain roughness factor given by

$$f(\sigma_i) = e^{-\frac{1}{2}\left(\frac{4\pi\sigma_i\sin(\alpha)}{\lambda}\right)^2} \tag{5.30}$$

where σ_i is the standard deviation of the irregularities. From these expressions, it can be seen that for very irregular terrain when σ_i is large, the roughness factor gets small and the reflection is mostly due to the diffuse component. Conversely,

Table 5.4 Rayleigh Criterion, Δh (m), for Some Conditions

$f =$ $\lambda =$	**30 MHz** **10 m**	**100 MHz** **3 m**	**1,000 MHz** **0.3m**
α^{o}			
0.1	716	215	21
0.5	143	43	4.3
1.0	72	21	5.1

when the surface is smooth, $\sigma_i \approx 0$, the $f(\sigma_i) \approx 1$, and the reflection is mostly specular. This terrain roughness factor is plotted in Figure 5.11.

5.2.7 Diffraction Loss

When a propagating signal encounters an obstacle with a sharp edge, signal *diffraction* occurs. Diffraction is a bending of the wave and accounts for the ability to receive signals behind mountains, for example. The scenario is shown schematically in Figure 5.12. In that case, define [15]

$$v = h_{\text{m}} \sqrt{\frac{2}{\lambda}\left(\frac{1}{D_{\text{T}}} + \frac{1}{D_{\text{R}}}\right)} \tag{5.31}$$

then the additional loss due to the diffraction, over and above the free-space path loss, is given by [11, 16]

$$L_{\text{Diff,dB}} = \begin{cases} 0, & v < 0 \\ 6 + 9v - 1.27v^2, & 0 < v < 2.4 \\ 13 + 20\log_{10} v, & v > 2.4 \end{cases} \tag{5.32}$$

This function is plotted in Figure 5.13. There are also approximations to account for multiple such sharp peaks [17, 18].

If the edge is not peaked, but rounded, then the losses are normally greater than those predicted by this expression. The more rounded the peaks, the higher the diffraction loss. If the sharp peak in the knife-edge diffraction is replaced by a cylinder of radius r corresponding to approximately the roundness of the peak, then an excess loss is experienced given by [19]

$$L_{\text{round,dB}} = 11.7\left(\frac{\pi r}{\lambda}\right)^{1/2} \alpha \tag{5.33}$$

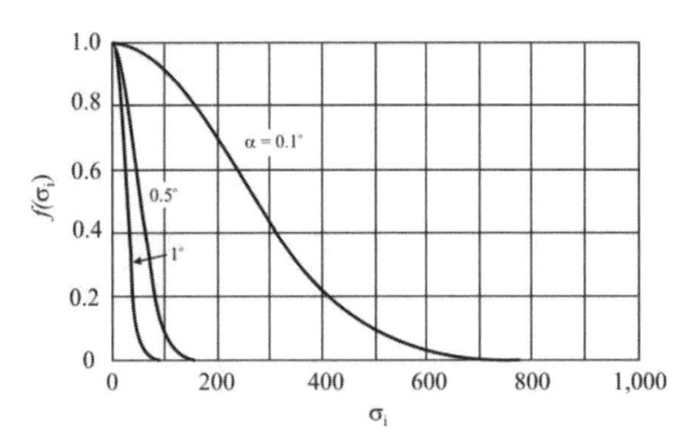

Figure 5.11 Terrain roughness factor when $f = 60$ MHz, so $\lambda = 5$ m.

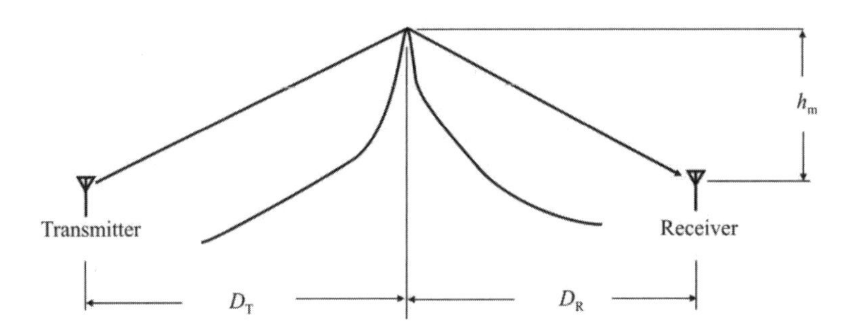

Figure 5.12 Diffraction at a sharp edge.

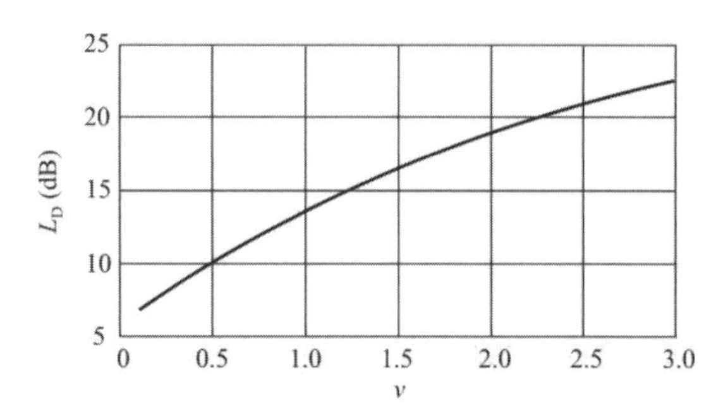

Figure 5.13 Diffraction loss as a function of *v*.

where α denotes the angle between the ray paths as they cross over the peak. See Figure 5.14. The amount of this excess loss is shown in Figure 5.15 for several values of r / λ. Parson reports that if the peak is irregular, covered with trees for example, then the loss is only about 65% of the value computed with this equation.

5.2.8 Terrain Irregularity

The micro-effects of terrain irregularity on reflected wave propagation were discussed in Section 5.2.6. Larger-scale terrain irregularity affects all the propagation models discussed here. Let Δh denote the terrain irregularity parameter that is the interdecile range of the heights of the land above some level. Then one experiment produces the standard deviation of the field strength (E in $\text{dB}_{\mu\text{V/m}}$) as ($\Delta h / \lambda < 5{,}000$) [20].

$$\sigma_L = 6 + 0.69\sqrt{\frac{\Delta h}{\lambda}} - 0.0063\left(\frac{\Delta h}{\lambda}\right) \tag{5.34}$$

This function is plotted in Figure 5.16. Up to about $\Delta h / \lambda = 3{,}000$ or so, an increase of terrain variability Δh relative to a wavelength causes an increase in the variability of the received signal strength.

5.2.9 Attenuation Due to Woodlands

Trees and other vegetation in dense woodlands will attenuate the signal level. The effect is seen in Figure 5.17 [21]. Denser and/or moister vegetation can increase these levels even more. Note that vegetation attenuates vertically polarized signals more than horizontally polarized ones.

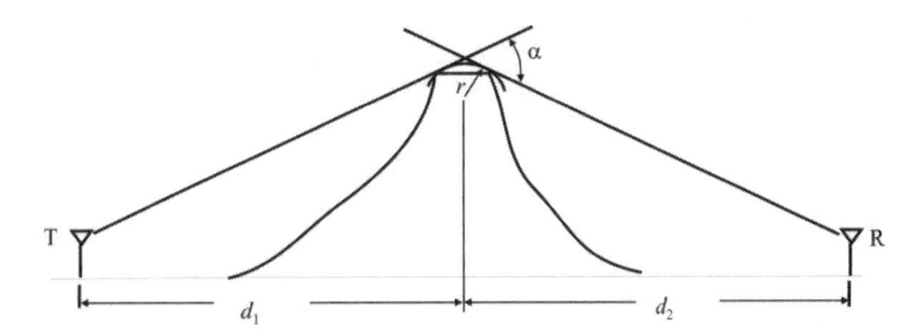

Figure 5.14 Diffraction over an obstacle other than a knife edge.

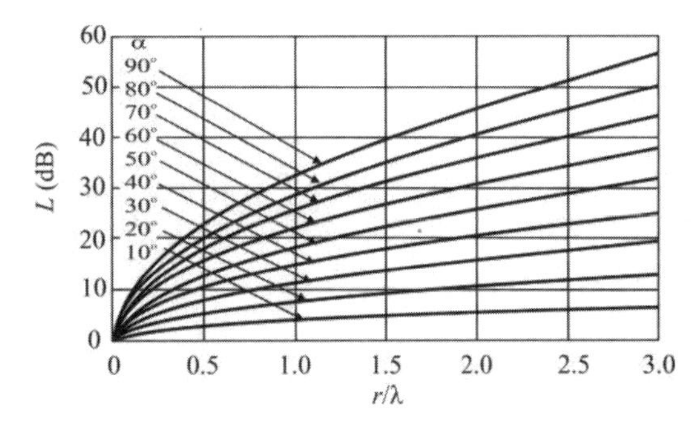

Figure 5.15 Excess losses over a knife edge when the obstruction is not sharp.

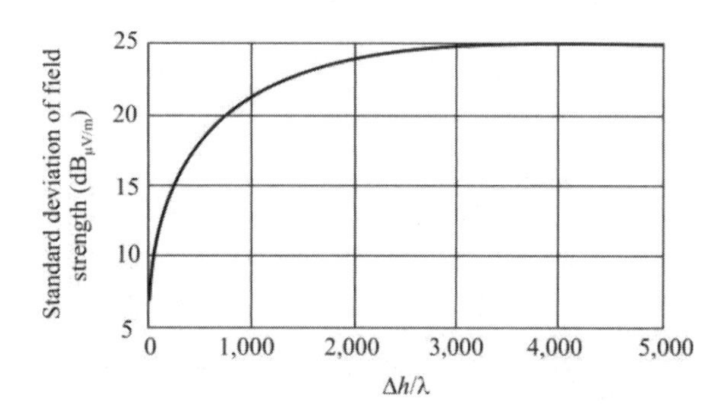

Figure 5.16 Standard deviation of the field strength.

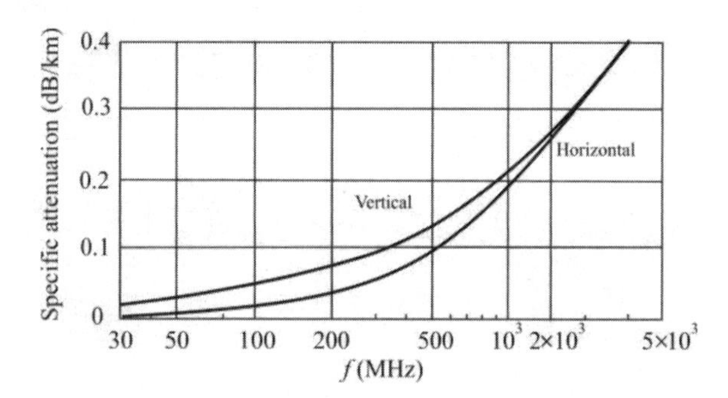

Figure 5.17 Specific attenuation caused by vegetation in woodlands. (Source: [21]. © IRE 1979. Reprinted with permission.)

This curve applies when the transmitter and receiver are both in and close to the woods. If they are well outside the woods with the woods between them, then the woods forms a diffracting obstacle and the discussion in Section 16.4.2 applies.

5.2.10 Multipath Reception

In the VHF and above frequency range, signals normally arrive at a receiver from multiple directions, whether that receiver is the intended one or the jammer receiver. One of these could be the direct path from the transmitter or not. This phenomenon is caused by reflections of signals off multiple objects within the range of the communication system. In some situations, such as cellular phone systems in cities, multipath reflections may be the only components present—a direct path between the mobile handset and a base station does not exist. Moreover, in mobile communication situations, which are normally the case for land-mobile tactical communications, the arrival direction of these signals varies as the vehicles move. Such movement causes changing multipath effects. The faster the movement of the receiver and/or transmitter, the more rapid the fluctuations in the received signal.

Multipath effects can include signal enhancement or signal degradation, depending on the phases and amplitudes of the reflected signals as they add at the receiver. Furthermore, for digital modulations *intersymbol interference* (ISI) occurs. This is when one or more symbols preceding the current symbol being received interfere with the current symbol due to one or more of these reflections because the distance traveled by the reflected signal is larger. Symbol overlap of 10% or greater is usually enough to cause significant degradation in the communication system. The *delay spread* is a statistical measure of the time delays of the various multipath components.

Two statistical approaches are usually used to model multipath reception: Ricean or Rayleigh. If there is no strong direct wave component, then the signal is diffuse and the Rayleigh distribution more accurately depicts the statistics of the received signals. If there is a strong direct wave component, then the signal is largely specular and the Ricean distribution is more accurate. A third approach that is more accurate than either of these and, in fact, includes the Ricean and Rayleigh channels as special cases is the Nakagami distribution.

5.2.10.1 Ricean Fading

Ricean fading occurs when there is a strong specular component present along with diffuse components. The joint *probability density function* (PDF) for amplitude a and phase θ in this case is given by [11]

$$p_{A,\theta}(a,\theta)=\frac{a}{2\pi\sigma^2}e^{-\frac{a^2-2c_\sigma a\cos\theta+c_\sigma^2}{2\sigma^2}} \tag{5.35}$$

where σ^2 is the local mean scattered power and $c_\sigma^2/2$ is the power of the dominant wave. By integrating over θ, assuming that θ is uniformly distributed over $[-\pi, \pi)$, the PDF of the amplitude only can be found to be

$$\begin{aligned}p_A(a)&=\int_{-\pi}^{\pi}p_{A,\theta}(a,\theta)d\theta\\&=\frac{a}{\sigma^2}e^{-\frac{a^2+c_\sigma^2}{2\sigma^2}}I_0\left(\frac{c_\sigma a}{\sigma^2}\right)\end{aligned} \tag{5.36}$$

where $I_0(x)$ is the modified Bessel function of the first kind and zeroth order (see Appendix A).

The *Ricean K-factor* is defined to be the ratio of the dominant signal component to the random multipath components power

$$K=\frac{c_\sigma^2}{2\sigma^2} \tag{5.37}$$

Using this factor, the PDF of the amplitude is given by [11]

$$p_A(a)=(1+K)e^{-K}\frac{a}{\bar{a}}e^{-\frac{1+k}{2\bar{A}}a^2}I_0\left(\sqrt{\frac{2K(1+K)}{\bar{a}}}a\right) \tag{5.38}$$

where $\bar{a}$ is the local-mean power. This function is plotted in Figure 5.18. As K increases, the mean of the density gets larger and the maximum amplitude decreases.

5.2.10.2 Rayleigh Fading

Rayleigh fading occurs when there is no dominant signal, but reception is via one or more reflected and/or scattered components. As the receiver moves, the amplitude and phases of these components change, which causes varying reception. The amplitude of the received signal can be statistically described by the Rayleigh density function given by

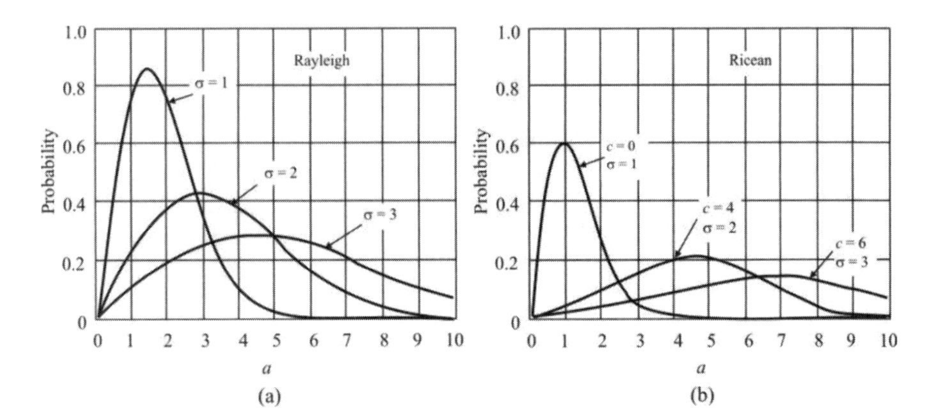

Figure 5.18 (a) Rayleigh and (b) Ricean amplitude densities. In the case of the Ricean density as c increases relative to σ, the mean of the density gets larger.

$$p_A(a) = \frac{a}{\sigma^2} e^{-\frac{a^2}{2\sigma^2}} \tag{5.39}$$

where σ^2 is the mean power and $a^2 / 2$ is the short-term signal power [22]. This function is also plotted in Figure 5.18.

5.2.10.3 Nakagami Fading

Nakagami fading can be a more accurate model than either the Ricean or Rayleigh models. The Nakagami PDF was developed to model the fading on HF channels and is based on fitting the density to measured data. It more closely models the amplitude fading that occurs when there is no direct component better than the Rayleigh density function. The PDF is given by [23]

$$p_m(x) = \frac{2m^m x^{2m-1}}{\Gamma(m)\Omega_{\mathrm{p}}^m} e^{-\frac{mx^2}{\Omega_{\mathrm{p}}}}, \quad m \geq \frac{1}{2} \tag{5.40}$$

where $\Omega_{\mathrm{p}} = \mathcal{E}\{m\}$ and $\Gamma(x)$ is the gamma function [24]. This density function is plotted in Figure 5.19 for several values of m and $\Omega_{\mathrm{p}} = 1$.

5.2.11 Doppler Shift

When a signal is sent between a transmitter and receiver and the receiver is moving with a velocity v there is a change in the frequency of the signal that

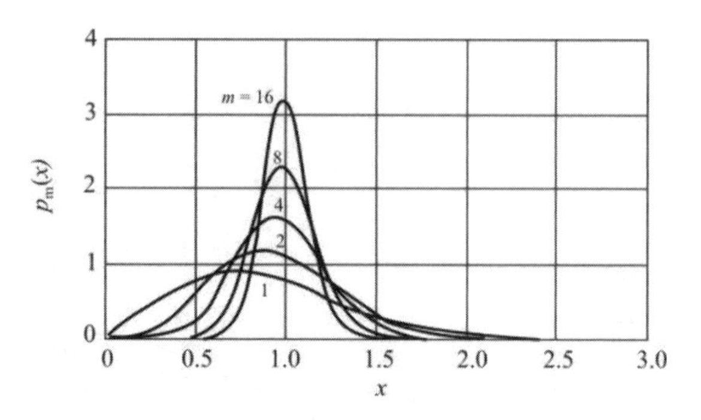

Figure 5.19 Nakagami probability density function when $\Omega_p = 1$.

occurs. This phenomenon is called the *Doppler shift*, named after the Austrian physicist Christian Doppler, who quantified it in 1845. Specifically, the frequency is changed by an amount

$$f_D = \frac{v}{\lambda}\cos(\alpha) \tag{5.41}$$

where $\lambda = c / f_c$ is the wavelength and α is the angle of the incoming signal at the receiver relative to the direction of the receiver. In mobile communication situations typically many of these signals arrive at the same time at differing angles. That means the relationship between the amplitudes and phase angles is constantly changing as well. The maximum shift occurs when the receiver is moving directly away from or toward the transmitter where $\alpha = \pm 1$. That maximum frequency shift is given by

$$f_m = \frac{v}{c} f_c \tag{5.42}$$

Each of these signals arrives with a phase angle uniform over $(0, 2\pi)$ which leads to a Doppler spectrum function [25]

$$S(f) = \frac{1}{4\pi f_m} \frac{1}{\sqrt{1 - \frac{(f - f_c)^2}{f_m^2}}} \tag{5.43}$$

shown in Figure 5.20 for various velocities. In this chart the power in the reflected waves was assumed to be 1. The region between $-f_c - f_m$ and $+f_c + f_m$ is called the

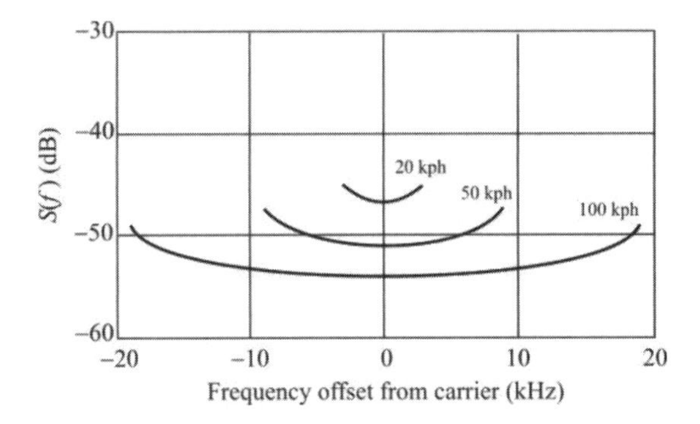

Figure 5.20 Doppler spectrums for three relatively slow velocities, 20 kph, 50 kph, and 100 kph ,when $f_c = 60$ MHz.

Doppler spread and $S(f)$ gives the density at any given frequency in this range. From the figure, it is clear that as the velocity increases, the density gets smaller and broader.

5.2.11.1 Autocorrelation with Distance

As a receiver moves, the phase and amplitudes of the multipath rays hitting the antenna change. The amount of change varies depending on the distance traveled. The covariance with distance $L(D)$ is proportional to [26]

$$L(D) = J_0\left(2\pi f_m \frac{D}{\lambda}\right) \tag{5.44}$$

where J_0 is the zeroth-order Bessel function of the first kind. This function is plotted in Figure 5.21. For short distances, the received signals are highly correlated, but that correlation falls rapidly with distance.

5.2.11.2 ISM at 5.8 GHz

There are three *instrumentation, scientific, and medical* (ISM) frequency bands in North America, Europe, and Japan, the characteristics of which are given in Table 2.3. Transmissions in these bands are unlicensed (but still restricted) by the federal governments of their respective countries so anyone can use them, subject to certain emission constraints (such as power level). The bands were established to allow for experimentation into radio technologies. They also provide useful ranges for expanding commercial cellular and PCS spectra.

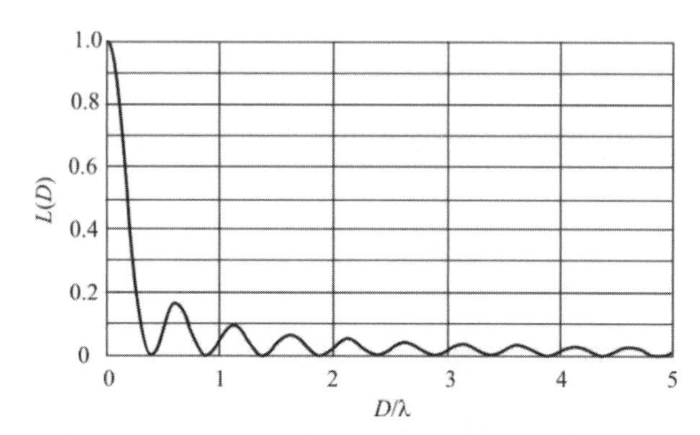

Figure 5.21 Autocorrelation of signals with distance traveled.

DSSS system can operate in the presence of other narrowband signals without much degradation to either. Therefore, the frequency ranges are of interest to DSSS system designers. The band at 5.8 GHz has developed considerable interest for in-building communication for both personal phone systems and *wireless local area networks* (WLANs).

Outdoor propagation at 5.4 GHz follows the same laws of physics as other frequency ranges in outdoor conditions. All of the models discussed apply as appropriate. Indoor propagation is different, although the same laws are followed. More on indoor propagation is presented later.

5.2.12 Oxygen Absorption at 60 GHz

The frequency band around 60 GHz has been proposed for use in outdoor microcellular systems, as well as indoor PCS systems. The primary advantage of this band is the absorption of RF signals by oxygen in the atmosphere that discourages interference of and by neighboring systems. A system on one floor of an office building, for example, would not interfere with that on another. The absorption of RF energy by oxygen and water can be seen in Figure 5.22 [27]. In addition to the absorption characteristics in the higher-frequency ranges, there is substantial bandwidth available at these frequencies. This facilitates considerably more simultaneous users.

The absorption of water is an important factor in outdoor microcellular applications because of the effects of rain. When raining, the excessive absorption around 22 GHz and above 60 GHz makes these frequencies undesirable for outdoor applications. In indoor settings, picocells, for example, 22 GHz may be a good frequency to use, although 60 GHz provides substantially more attenuation.

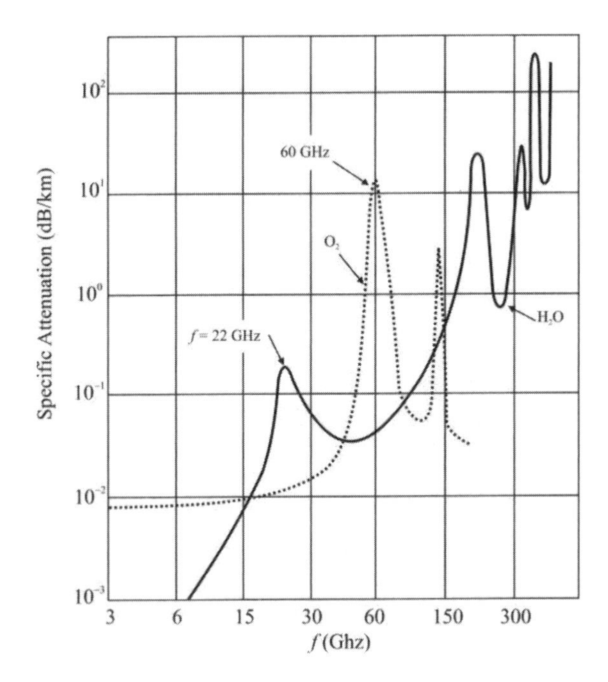

Figure 5.22 Specific attenuation at the higher frequencies. (Source: [27]. © John Wiley & Sons 1992. Reprinted with permission.

Indoor delay spreads are on the order of 15 to 50 ns. To avoid intolerable distortion due to ISI, a maximum of 10% overlap in received symbols is required, limiting the minimum symbol time to 150–500 ns. This corresponds to a symbol rate of 2 Mbps–6.6 Mbps.

5.2.13 Propagation Models

Four models for calculating the losses encountered by VHF signals and above are presented in this section.

5.2.13.1 Egli Propagation Model

In 1957, John Egli developed a propagation model that was based on the above theoretical arguments, but introduced another factor that brings the frequency back into the mix. Since propagation tends to be frequency-dependent, he added a multiplicative factor to account for this [28]. The reported limits on this model are $1 < D_{TR} < 50$ km and 30 MHz $< f <$ 1 GHz. The median (exceeded 50% of the time) received power is given by [29]

$$P_{\mathrm{R}} = \left(\frac{40}{f}\right)^2 \frac{(h_{\mathrm{T}} h_{\mathrm{R}})^2}{D_{\mathrm{TR}}^4} P_{\mathrm{T}} G_{\mathrm{TR}} G_{\mathrm{RT}} \tag{5.45}$$

where the frequency f is in megahertz, h_{T} is the height of the transmit antenna, h_{R} is the height of the receive antenna, P_{T} is the transmit power, G_{TR} is the gain of the transmit antenna in the direction of the receiver, G_{RT} is the gain of the receive antenna in the direction of the transmitter, and D_{TR} is the distance between the transmitter and receiver in units consistent with h_{T} and h_{R}. Expression (5.45) applies with the proper substitution of subscripts to the jammer links as well. This function is consistent with the two-ray model with $n = 4$ given previously. A plot of this function out to 100 km at 60 MHz is shown in Figure 5.23 for three transmit antenna heights (could be either antenna) compared with free-space propagation.

It has also been reported that the standard deviation of the received signal power varies with frequency, as we would expect. This variation is described by the empirical equation [29]

$$\sigma = 5\log_{10} f_{\mathrm{MHz}} + 2 \text{ dB} \tag{5.46}$$

with f in megahertz. This function is plotted in Figure 5.24. The standard deviation varies from about 9 to 12 dB, increasing with frequency.

5.2.13.2 R^n Model

The R^n propagation model is given by

$$P_{\mathrm{R}} = \frac{P_{\mathrm{T}} G_{\mathrm{TR}} G_{\mathrm{RT}}}{L(D)} \tag{5.47}$$

where

$$L(D) = L(D_0) + 10n\log_{10}\left(\frac{D}{D_0}\right) \tag{5.48}$$

$L(D)$ is the loss with distance D. D_0 is a reference distance at which the propagation loss is known. $L(D_0)$ could be known by measurement, or it can be approximated with

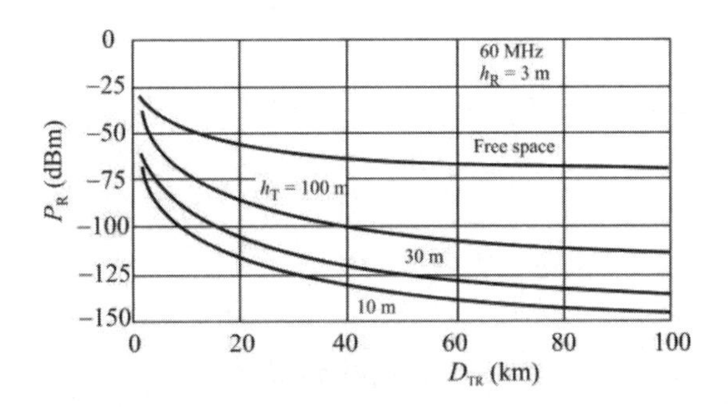

Figure 5.23 Egli's propagation model results at f = 60 MHz and h_R = 3 m. For this example G_{TR} = G_{RT} = 1 and P_T = 10 watts.

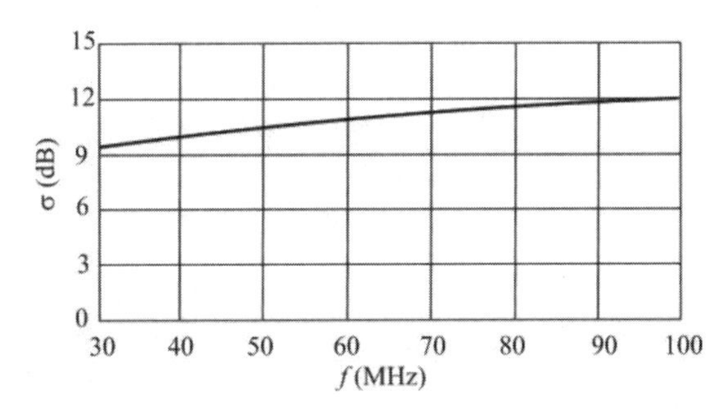

Figure 5.24 Standard deviation of the received power in Egli's model as a function of frequency.

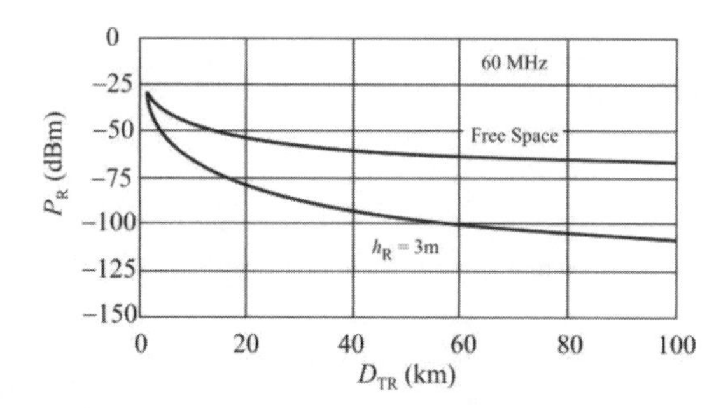

Figure 5.25 Received power as predicted by the R^n model compared with free space when $n = 4$, $f = 60$ MHz, $h_R = 3$ m, $h_T = 3$ m, and $P_T = 10$ W.

$$L(D_0) = 20\log_{10}\left(\frac{4\pi D_0}{\lambda}\right) \tag{5.49}$$

where λ is the wavelength of the signal. As mentioned previously, n is a constant that typifies the type of propagation conditions being considered. The R^n model is applicable for frequencies in the VHF range and above. This function is plotted in Figure 5.25 for $n = 4$ at 60 MHz. Again, (5.49) applies to the jammer links as well.

Shown in Figure 5.26 is the SNR versus distance between the transmitter and a receiver, either the intended receiver or the receiver of the jammer, for three transmitter power levels. These SNR values are based on the R^n propagation model. The frequency for the curves in Figure 5.26 was chosen to be midway through the low VHF range at 60 MHz. At that frequency, the noise is approximated according to the curves shown in Figure 2.10, where a suburban environment was assumed. In that case, the external noise is about $30\,\text{dB}_{k_BTW}$, or 30 dB above a value of k_BT_0W. The receiver noise figure also adds to the noise level at the receiver as explained in Section 2.5.1. For the example considered here, it was assumed to be 15 dB, substantially less than the external noise, making this term negligible. It was further assumed that $G_T = G_R = 1$ (0 dB).

Whereas the received signal level decreases, the noise level does not. It is frequently assumed constant, independent of spatial parameters. This, of course, is not always true especially when considering MMN, such as that from vehicle ignitions and welding machines. Thus, in this example, with a transmitter power of 1 W, the SNR value of 8.9 dB or greater, corresponding to a BER of 10^{-2}, is achieved at ranges less than about 20 km with no fade margin. If the transmitter power is 10 W then 8.9 dB is achieved at about 35 km and for 100 W at about 55 km.

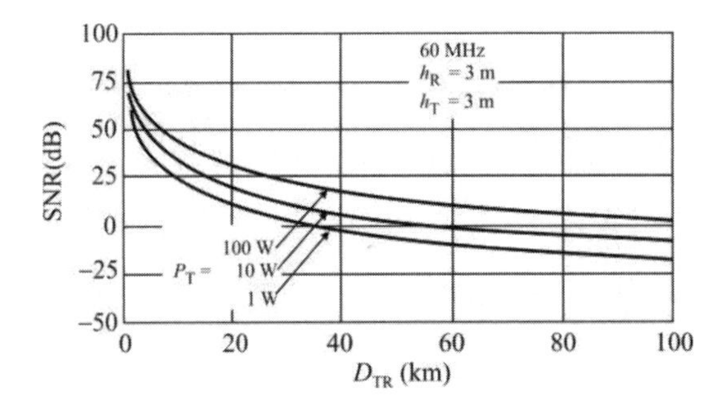

Figure 5.26 SNR decreases as the distance between the transmitter and receiver increases. In this R^n model example, $f = 60$ MHz.

5.2.13.3 Nicholson Propagation Model

In 1988 David L. Nicholson presented a propagation model [30]. In that model, the received power is given by

$$P_R = \frac{1}{L(D, f)} P_T G_{TR} G_{RT} \tag{5.50}$$

The denominator is the propagation loss and is given by

$$L(D, f)_{dB} = C(f) + L_{fs}(f) + 40\log\left(\frac{R}{R_0}\right) \tag{5.51}$$

which, with close inspection, can be seen to be the same as the R^n model with $n = 4$ and different constants. $L_{fs}(f)$ is the free space propagation loss given by

$$L_{fs}(f) = 37 + \log(f_{MHz}) \tag{5.52}$$

and $C(f)$ is a constant given by Table 5.5. R_{min} and R_{max} refer to the minimum and maximum ranges over which the numbers are accurate.

Applicability to the jammer links applies for the Nicholson model; however, attention to the ranges in Table 5.5 the altitude of the jammer is important as well.

5.2.13.4 Longley-Rice Propagation Model

This is a computer model developed in the 1960s for calculating propagation losses close to the Earth's surface [31]. It is an empirical model with parameters

Table 5.5 Parameters for the Nicholson Propagation Model

Frequency (MHz)	One Antenna on Ground, the Other Height in Feet	$C(f)$ (dB)	R_{min} (km)	R_{max} (km)
30	0	28	0.2	33
30	100	15	0.7	25
60	0	34	0.2	33
60	100	15	0.8	17
150	0	42	0.2	17
150	100	15	0.2	10
300	10	35	0.2	17

based on field measurements. These measurements were taken at frequencies between 20 MHz–20 GHz.

In the point-to-point mode, when the communication is between a single transmitter and single receiver (as opposed to the mobile, broadcast mode), the propagation path is determined to be free-space, line-of-sight, diffraction, or troposcatter, as illustrated in Figure 5.27, and the appropriate model parameters are adjusted accordingly. The model calculates the additional loss imposed over and above the free-space loss as a function of path length. This additional loss is denoted A and is composed of two components: the reference attenuation, denoted as A_{ref}, which is attenuation due to frequency and distance, and V_{med}, which is an adjustment to A_{ref} that accounts for time, location, and situation variability. In the point-to-point mode, the location variability is not a factor. Therefore,

$$A = A_{ref} + V_{med} \tag{5.53}$$

The time variability takes into account nine radio climates that vary throughout the year. V_{med} is the mean additional attenuation and adjusts A_{ref} for the median. Therefore, the resulting total loss is the loss that can be expected 50% of the time.

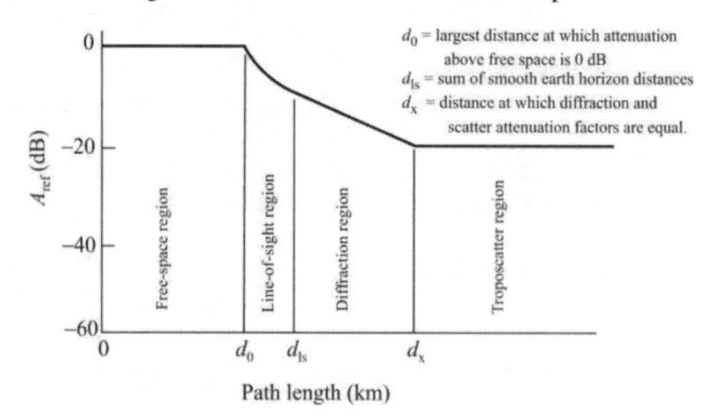

Figure 5.27 Regions in the Longley-Rice propagation model.

The total loss is then calculated as

$$L_T = L_{fs} + A_{tot} \tag{5.54}$$

where

L_{fs} = Calculated free-space loss

Free-space loss is calculated as

$$L_{fs} = 10\log_{10}\left(\frac{4\pi f D}{c}\right)^2 \tag{5.55}$$

where c is the speed of light, f is the frequency, and D is the distance, all in compatible units.

Extensive documentation for the Longley-Rice model is available at [32] along with a software listing in FORTRAN that implements it. It also contains the algorithm in pseudo-code so that adaption to other programming languages is straightforward.

The Longley-Rice model applies to the jammer links as well. When used for placement of a jammer, the area mode is more applicable since the target locations are not normally known in advance.

5.3 HF Signal Propagation

5.3.1 Introduction

HF signals propagate via two fundamental mechanisms: a ground-wave phenomenon and a sky-wave mode. While the overall characteristics are presented here, [33] is an excellent source for more detailed information on the HF signal propagation modes.

5.3.2 Noise

Just as in the VHF range and higher, there are two fundamental sources of noise: external and internal. Below about 10 MHz, however, the external noise is usually much greater than the internal noise. In this frequency range, atmospheric noise is by far the dominating source. This is noise caused by lightning in thunderstorms and sandstorms. It can travel great distances by the same mechanisms that signals propagate in the ionosphere.

Figure 2.9 illustrates the levels of atmospheric noise in the frequency range below 100 MHz. Even though the product $k_B TW$ represents a power level, the value of F_a in these charts can be used to predict the noise electrical field strength. The RMS field strength is related to F_a, which is the noise level relative to $k_B T_0 W$, by

$$E_n = F_a + 20\log_{10} f_{MHz} - 65.5 \tag{5.56}$$

where E_n is for a 1 kHz bandwidth in decibels above 1 μV m^{-1}. If bandwidths other than 1 kHz are under consideration then $20\log_{10} W_{kHz}$ must be added. For example, if the bandwidth is 10 kHz, then 20 dB more noise will be present. If the bandwidth is 500 Hz then $20\log_{10}(0.5) = -6$ dB must be added. For different conditions such as time of year, or time of day, these levels would be different; those shown are representative. The noise power in decibels above $k_B TW$ is given by

$$P_N(\text{dBm}) = F_a + 20\log f_{MHz} - 174 \tag{5.57}$$

There is variability to these parameters since they represent statistical processes. A reasonable value to use in many cases for this variability is 10 dB [34]. Thus, to ensure system performance, 10 dB must be added to the ordinate. This factor represents a reasonable estimate for a fade margin in this frequency range.

MMN is also shown in Figure 2.9. This is noise from arc welders, neon signs, vehicle ignitions and the like, that are products of humans. These noise sources can be significant in the HF frequency range, especially since an advancing brigade is not exactly quiet. We discussed MMN in depth in Chapter 2.

5.3.3 Ground Wave

Reliable ground-wave propagation is limited to the frequency range of 1.5 to 10 MHz. The quality of such communications is highly dependent upon the Earth characteristics over which the signal propagates. Used for communication by ships at sea where the sea provides favorable conditions for propagation, ranges up to 1,000 km can be accommodated with the ground wave at relatively low transmitted power.

Two models for HF ground EM wave propagation are presented in this section. The first is considerably simplified, which makes quick calculations easy. The second, due to Bremmer, is considerably more complicated, as well as accurate, and is suitable for computer implementation. As explained later, the Bremmer model is not suitable for sky-wave propagation.

5.3.3.1 Simplified Model

From (5.5)

$$E = \frac{E_0}{D_{TR}}\left[1+\rho e^{j\delta}+(1-\rho)Ae^{j\delta}\right] \tag{5.58}$$

where ρ is the reflection coefficient at the point of reflection and δ is the phase difference at the receiver between the direct and reflected wave. Close to the Earth when the heights of the antennas are λ/4 or less, which is normal for such antennas in the HF range, $\alpha \approx 0$, $\rho \approx -1$, and $\delta \approx 0$, so the first two terms cancel and

$$E_R \approx \frac{2E_0}{D_{TR}} A \tag{5.59}$$

The term $2E_0$ is given by

$$\begin{aligned} 2E_0 &= 300\sqrt{P_T G_{TR}} \ \ \text{mV m}^{-1} \\ &= 3\times10^5 \sqrt{P_T G_{TR}} \ \ \mu\text{V m}^{-1} \end{aligned} \tag{5.60}$$

when P_T is in kilowatts. Therefore

$$E_R = \frac{3\times10^5}{D_{TR}} A\sqrt{P_T G_{TR}} \ \ \mu\text{V m}^{-1} \tag{5.61}$$

Braun defines four conditions for which calculation of the propagation loss factor A is different, as follows [35]:

1. A_v, where $D_{TR} < D_{CR}$, for vertical polarization and the distance is short enough so the curvature of the Earth can be ignored;
2. A_h, where $D_{TR} < D_{CR}$, same as 1 for horizontal polarization;
3. A_v, where $D_{TR} > D_{CR}$, for vertical polarization and the distance is such that the curvature of the Earth must be taken into consideration;
4. A_h, where $D_{TR} > D_{CR}$, same as 3. for horizontal polarization.

The critical distance D_{CR} is given by

$$D_{CR} = \frac{80}{\sqrt[3]{f}}, \text{ km} \tag{5.62}$$

when f is expressed in megahertz. This function is plotted in Figure 5.28. It is assumed here that the ground is homogeneous between the transmitter and receiver. *Homogeneous* in this case means that the soil conditions are the same all along the path between the transmitter and receiver. In addition to that, it assumes that the conditions are the same for considerable depth into the soil. At HF frequencies the signals penetrate the Earth as a function of the soil conditions approximately as shown in Table 5.6 [36]. Since HF signals can propagate for considerable distances, this assumption is only valid for short paths. For longer paths, the propagation must be evaluated for each segment. This is illustrated in Figure 5.29.

For all practical purposes, horizontally polarized surface waves do not propagate well in the HF range so this polarization will not be considered further.

Define

$$p = 1.745 \times 10^{-4} \frac{f \cos(\beta)}{\sigma} \frac{D_{TR} \times 10^3}{\lambda} \tag{5.63}$$

$$\beta = \tan^{-1}\left(\frac{(\varepsilon + 1)f}{1.8 \times 10^4 \sigma}\right) \tag{5.64}$$

where f is in megahertz, D_{TR} is in kilometers, λ is in meters, ε is the dielectric constant of the Earth, and σ the conductivity in S m^{-1}. For $\beta \leq 90^\circ$ then

$$A_v = \frac{2 + 0.3p}{2 + p + 0.6p^3} - \sqrt{\frac{p}{2}} e^{-1.44 p \log(\varepsilon)} \sin(\beta) \tag{5.65}$$

The signal strength using this function is plotted in Figure 5.30 for medium soil conditions at 20°C. For this curve $P_T = 50$ W, $G_{TR} = 0$ dB and $h_T = h_R = \lambda / 4$.

Curves for the other soil conditions are similar to Figure 5.30. Salty seawater facilitates ground-wave propagation better than fresh water. Fresh water and moist ground characteristics are similar.

Receiver sensitivities of 15 $\text{dB}_{\mu V/m}$ are typical in the HF range. Therefore, all conditions considered here allow C2 of land forces at tactical ranges (< 30 km).

These calculations and curves apply, of course, to the transmitter-to-receiver path, the transmitter to jammer path and the jammer-to-receiver path. The signal level, however, is only half the story. Reception quality also depends on the amount of noise present. Define a measure of signal-to-noise ratio as

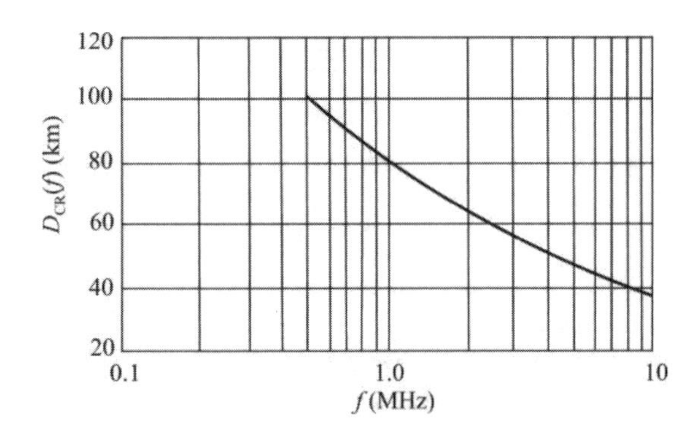

Figure 5.28 Critical distance defined by (5.62) beyond which the calculation of the surface-wave attenuation factor must take the curvature of the Earth into consideration.

Table 5.6 Penetration Depths of EM Waves into the Earth

	Penetration Depth (m)			
Frequency (MHz)	0.5	1	5	10
Sea Water (20°C) Average Salt Content	0.35	0.25	0.1	0.075
Fresh Water (20°C)	18	15	12	8.5
Moist Soil	7	5.5	3.8	1.5
Medium Soil	28	23	18	16
Very Dry Soil	90	90	90	90

Source: [36].

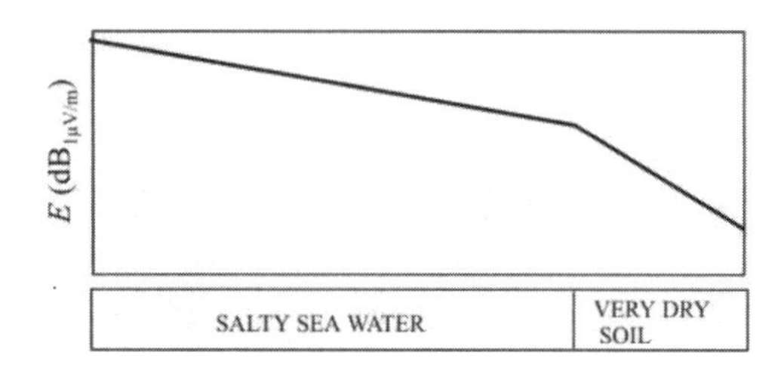

Figure 5.29 Signal strength in two nonhomogeneous regions.

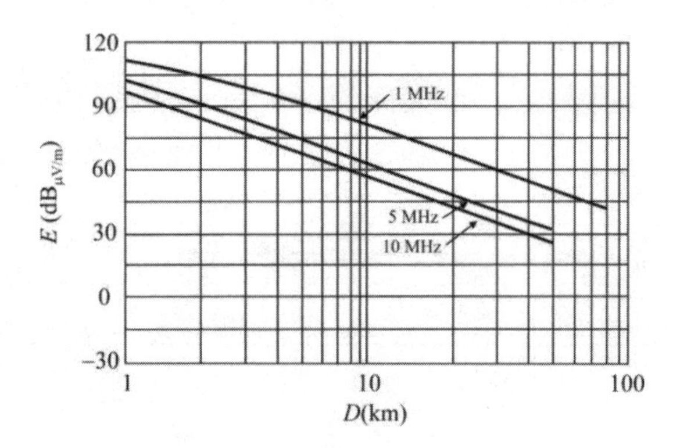

Figure 5.30 Signal strength for medium soil ($\varepsilon = 15$, $\sigma = 10^{-3}$) in decibels relative to 1 μV/m.

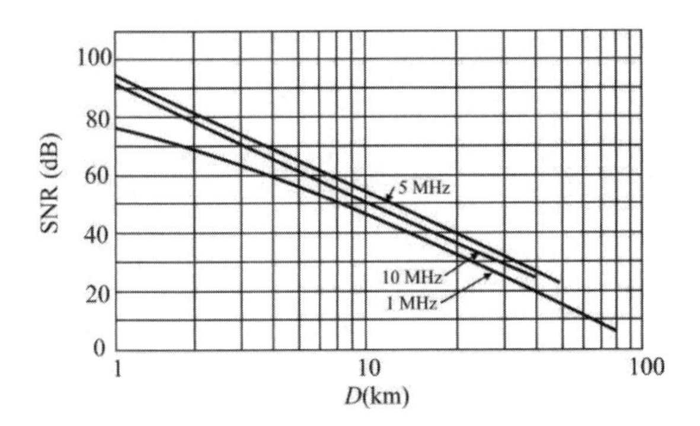

Figure 5.31 SNR for medium soil conditions.

SNR = E_R/E_n as determined by (5.56) and (5.61). Then Figure 5.31 shows the SNR for medium soil conditions when P_T = 50 W, G_{TR} = 0 dB, $h_R = h_T = \lambda/2$, and the noise level at 1 MHz is 90 dB in Figure 2.9. The bandwidth of the signal for this chart was 3 kHz, typical for voice communications in the HF region. Then at 1 MHz, F_a = 50 dB, at 5 MHz F_a = 30 dB, and at 10 MHz, F_a = 30 dB. Note from Figure 2.9 that at these frequencies and grades of service, the MMN at a quiet location is on the same order as the atmospheric noise. Thus, the amount of noise is approximately twice that of each source alone, or about 3 dB higher.

An SNR of 10 dB would be considered enough for relatively clear reception as long as it was at this level constantly. Charts similar to Figure 5.31 show that with a 50 dB noise level at 1 MHz, at link distances typical of land force C2, the signal produced by the ground-wave is large enough out to at least 40 km to facilitate communication.

When the diffraction angle $\alpha < \lambda / 2\pi R_e$ radians, where R_e is the radio radius of the Earth (4/3 its actual radius assuming it is a sphere), then there is no appreciable direct wave nor can there be a reflected wave. Diffraction is the predominant cause of propagation in this region and the loss settles to a constant $0.62/\lambda^{1/3}$ dB km^{-1}. The critical angle is plotted in Figure 5.32. The loss in this region is shown in Figure 5.33 for a few frequencies.

5.3.3.2 Bremmer Model

Bremmer documented an HF ground-wave propagation model [37] that was subsequently modified somewhat by Fulks for the Defense Nuclear Agency (DNA) [38]. It is widely believed that in the event of a nuclear explosion in the

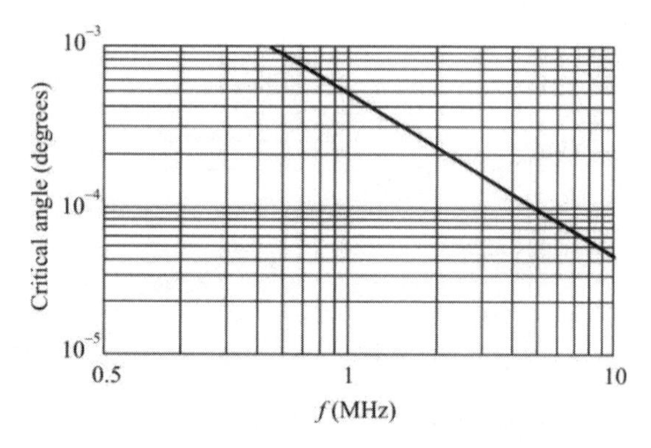

Figure 5.32 The diffraction angle below which propagation is due to only diffraction.

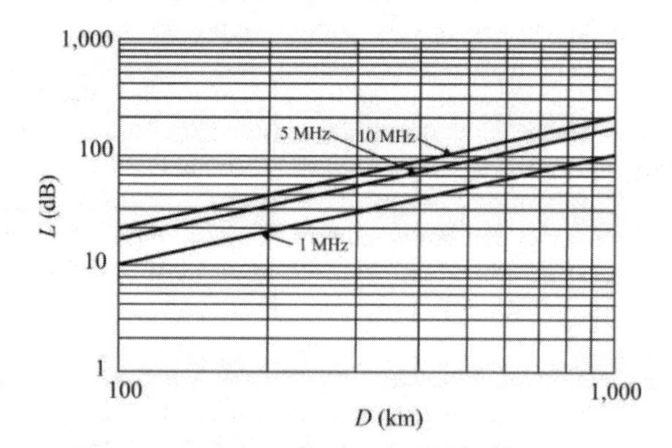

Figure 5.33 Attenuation in the region beyond that defined by the diffraction angle.

atmosphere, the D-layer (explained later) will be so disturbed and loaded with extra charges that sky wave HF propagation will not be possible as the D-layer would not let any wavefronts through (in either direction). HF ground-waves would not be affected, however. Therefore, the DNA was interested in how well HF ground-waves propagate.

In any case, the model estimates the electric field emitted from a short vertical dipole antenna driven by a 1 kW transmitter through lossless cables at distances over an assumed smooth earth. As modified by Fulks (and corrected here), it is given by the expression

$$E = \left[346.4 \sqrt{(h_1 + a)^2 + (h_2 + a)^2 - 2(h_1 + a)(h_2 + a)\cos\left(\frac{d_{\text{km}}}{a}\right)} \right] \times \left| \sum_{i=0}^{\infty} f_i(h_1) f_i(h_2) \frac{\exp(j\tau_i \chi)}{2\tau i - \frac{1}{\delta^2}} \right| \quad (5.66)$$

where

d_{km} = path length between the transmitter and receiver along the surface of the Earth, km;
$\chi = 0.053693\, d_{\text{km}} / \lambda^{1/3}$;
λ = wavelength, m;
$f_i(h_1)$ = height gain factor given by

$$f_i(h_k) = \left(\frac{\chi_1^2 - 2\tau_i}{-2\tau_i} \right)^{1/2} \frac{H_{1/3}^{(1)}\left[\frac{1}{3}\left(\chi_1^2 - 2\tau_i\right)^{3/2} \right]}{H_{1/3}^{(1)}\left[\frac{1}{3}\left(-2\tau_i\right)^{3/2} \right]}, \quad k = 1,2 \quad (5.67)$$

where $H_n^{(1)}[x]$ is the Hankel function of the first kind and order n, sometimes called Bessel functions of the third kind. The Hankel function of the first kind is given by

$$H_n^{(1)}(x) = J_n(x) + jY_n(x) \quad (5.68)$$

where $J_n(x)$ and $Y_n(x)$ are Bessel functions of the first and second kinds, respectively. For large x and $x >> n$, this function can be approximated by

$$H_n^{(1)}(x) \approx \sqrt{\frac{2}{\pi x}} \exp\left[j\left(x - \frac{\pi}{4} - \frac{n\pi}{2} \right) \right] \tag{5.69}$$

These functions were so approximated by Bremmer, yielding

(a) for $h_k > 50\lambda^{2/3}$, $k = 1, 2$:

$$f_i(h_k) = \frac{A_i}{\delta_e \sqrt[4]{\chi_k^2 - 2\tau_i}} \left\{ \begin{array}{l} \exp\left[-j\frac{\pi}{4} + j\frac{1}{3}\left(\chi_1^2 - 2\tau_i\right)^{3/2} \right] \\ \quad \times \left[1 - j\frac{0.2038}{\left(\chi_k^2 - 2\tau_i\right)^{3/2}} - \frac{0.3342}{\left(\chi_k^2 - 2\tau_i\right)^{3}} \right] \\ - \exp\left[j\frac{\pi}{4} - j\frac{1}{3} \right]\left(\chi_k^2 - 2\tau_i\right)^{3/2} \\ \quad \times \left[1 + j\frac{0.2083}{\left(\chi_k^2 - 2\tau_i\right)^{3/2}} \right] \end{array} \right\} \tag{5.70}$$

where

$$\begin{aligned} A_0 &= 0.3582 e^{j120^o} \\ A_1 &= 0.3129 e^{-j60^o} \\ A_2 &= 0.2903 e^{j120^o} \\ A_3 &= 0.2760 e^{-j60^o} \\ A_i &= 0.3440 \frac{(-1)^{i+1}}{\left(i + \frac{3}{4}\right)^{1/6}} \exp\left(-j\frac{\pi}{3} \right), \quad i > 3 \end{aligned} \tag{5.71}$$

(b) for $h_k < 50\lambda^{2/3}$, $k = 1, 2$

$$f_i(h_k) = 1 + 6.283\left(\frac{1}{x^{1/3}\delta_e} - \frac{1}{x} \right)\frac{h_k}{\lambda} - 39.48\frac{1 - x^{2/3}\delta_e\tau_i}{x^{4/3}\delta_e}\left(\frac{h_k}{\lambda} \right)^2 \tag{5.72}$$

where

$$x = \frac{4\times 10^7}{\lambda} \tag{5.73}$$

In both (5.70) and (5.72) the following definitions apply

$$\delta_e = K_e^{j(135^o - \varphi_e)} \tag{5.74}$$

$$K_e = 0.002924\lambda^{1/3} \frac{\sqrt{\varepsilon^2 + 3.6\times 10^{25}\sigma_e^2\lambda^2}}{\sqrt[4]{(\varepsilon-1)^2 + 3.6\times 1025\sigma_e^2\lambda^2}} \tag{5.75}$$

$$\varphi_e = \tan^{-1}\left(\frac{\varepsilon}{6\times 10^{12}\sigma_e\lambda}\right) - \frac{1}{2}\tan^{-1}\left(\frac{\varepsilon-1}{6\times 10^{12}\sigma_e\lambda}\right) \tag{5.76}$$

$$\tau_i = \mathrm{Re}(\tau_i) + j\,\mathrm{Im}(\tau_i) \tag{5.77}$$

When K_e is small

$$\begin{aligned}\mathrm{Re}(\tau_0) = 0.928 + K_e\cos(45^\circ + \varphi_e) + 1.237K_e^3\cos(75^\circ + 3\varphi_e)\\ -0.5K_e^4\cos(4\varphi_e) - 0.2755K_e^5\cos(75^\circ - 5\varphi_e)\end{aligned} \tag{5.78}$$

$$\begin{aligned}\mathrm{Re}(\tau_1) = 1.622 + K_e\cos(45^\circ + \varphi_e) + 2.163K_e^3\cos(75^\circ + 3\varphi_e)\\ -0.5K_e^4\cos(4\varphi_e) - 8.422K_e^5\cos(75^\circ - 5\varphi_e)\end{aligned} \tag{5.79}$$

$$\begin{aligned}\mathrm{Re}(\tau_2) = 2.191 + K_e\cos(45^\circ + \varphi_e) + 2.921K_e^3\cos(75^\circ + 3\varphi_e)\\ -0.5K_e^4\cos(4\varphi_e) - 15.36K_e^5\cos(75^\circ - 5\varphi_e)\end{aligned} \tag{5.80}$$

$$\begin{aligned}\mathrm{Re}(\tau_3) = 2.694 + K_e\cos(45^\circ + \varphi_e) + 3.592K_e^3\cos(75^\circ + 3\varphi_e)\\ -0.5K_e^4\cos(4\varphi_e) - 23.227K_e^5\cos(75^\circ - 5\varphi_e)\end{aligned} \tag{5.81}$$

$$\begin{aligned}\mathrm{Re}(\tau_k) = 1.116\left(k + \frac{3}{4}\right)^{2/3} + K_e\cos(45^\circ + \varphi_e)\\ +1.488\left(k + \frac{3}{4}\right)^{2/3} K_e^3\cos(75^\circ + 3\varphi_e)\\ -0.5K_e^4\cos(4\varphi_e)\\ -3.987\left(k + \frac{3}{4}\right)^{4/3} K_e^5\cos(75^\circ - 5\varphi_e), \qquad k > 3\end{aligned} \tag{5.82}$$

$$\begin{aligned}\mathrm{Im}(\tau_0) = 1.607 - K_e\sin(45^\circ + \varphi_e) - 1.237K_e^3\cos(75^\circ + 3\varphi_e)\\ +0.5K_e^4\cos(4\varphi_e) - 0.2755K_e^5\sin(75^\circ - 5\varphi_e)\end{aligned} \tag{5.83}$$

$$\begin{aligned}\mathrm{Im}(\tau_1) = 2.810 - K_\mathrm{e}\sin(45^\circ + \varphi_\mathrm{e}) - 2.163K_\mathrm{e}^3\cos(75^\circ + 3\varphi_\mathrm{e}) \\ +0.5K_\mathrm{e}^4\cos(4\varphi_\mathrm{e}) - 8.422K_\mathrm{e}^5\sin(75^\circ - 5\varphi_\mathrm{e})\end{aligned} \tag{5.84}$$

$$\begin{aligned}\mathrm{Im}(\tau_2) = 3.795 - K_\mathrm{e}\sin(45^\circ + \varphi_\mathrm{e}) - 2.921K_\mathrm{e}^3\cos(75^\circ + 3\varphi_\mathrm{e}) \\ +0.5K_\mathrm{e}^4\cos(4\varphi_\mathrm{e}) - 15.36K_\mathrm{e}^5\sin(75^\circ - 5\varphi_\mathrm{e})\end{aligned} \tag{5.85}$$

$$\begin{aligned}\mathrm{Im}(\tau_3) = 4.663 - K_\mathrm{e}\sin(45^\circ + \varphi_\mathrm{e}) - 3.592K_\mathrm{e}^3\cos(75^\circ + 3\varphi_\mathrm{e}) \\ +0.5K_\mathrm{e}^4\cos(4\varphi_\mathrm{e}) - 23.227K_\mathrm{e}^5\sin(75^\circ - 5\varphi_\mathrm{e})\end{aligned} \tag{5.86}$$

$$\begin{aligned}\mathrm{Im}(\tau_k) = 1.932\left(k + \frac{3}{4}\right)^{2/3} - K_\mathrm{e}\sin(45^\circ + \varphi_\mathrm{e}) \\ -1.488\left(k + \frac{3}{4}\right)^{2/3} K_\mathrm{e}^3\cos(75^\circ + 3\varphi_\mathrm{e}) \\ +0.5K_\mathrm{e}^4\cos(4\varphi_\mathrm{e}) \\ -3.987\left(k + \frac{3}{4}\right)^{4/3} K_\mathrm{e}^5\cos(75^\circ - 5\varphi_\mathrm{e}), \qquad k > 3\end{aligned} \tag{5.87}$$

For large K_e

$$\begin{aligned}\mathrm{Re}(\tau_0) = 0.4043 + 0.6183\frac{\cos(15^\circ - \varphi_\mathrm{e})}{K_\mathrm{e}} - 0.2364\frac{\sin(2\varphi_\mathrm{e})}{K_\mathrm{e}^2} \\ -0.0533\frac{\cos(15^\circ + 3\varphi_\mathrm{e})}{K_\mathrm{e}^3} + 0.00226\frac{\cos(60^\circ - 4\varphi_\mathrm{e})}{K_\mathrm{e}^4}\end{aligned} \tag{5.88}$$

$$\begin{aligned}\mathrm{Re}(\tau_1) = 1.288 + 0.194\frac{\cos(15^\circ - \varphi_\mathrm{e})}{K_\mathrm{e}} - 0.0073\frac{\sin(2\varphi_\mathrm{e})}{K_\mathrm{e}^2} \\ +0.0120\frac{\cos(15^\circ - 3\varphi_\mathrm{e})}{K_\mathrm{e}^3} - 0.00160\frac{\cos(60^\circ - 4\varphi_\mathrm{e})}{K_\mathrm{e}^4}\end{aligned} \tag{5.89}$$

$$\mathrm{Re}(\tau_k) = 1.116\left(k + \frac{1}{4}\right)^{2/3} + \frac{0.2241}{\left(k + \frac{1}{4}\right)^{2/3}}\frac{\cos(15^\circ - \varphi_\mathrm{e})}{K_\mathrm{e}}, \qquad k > 1 \tag{5.90}$$

$$\begin{aligned}\mathrm{Im}(\tau_0) = 0.7003 - 0.6183\frac{\sin(15^\circ - \varphi_\mathrm{e})}{K_\mathrm{e}} + 0.2364\frac{\cos(2\varphi_\mathrm{e})}{K_\mathrm{e}^2} \\ -0.0533\frac{\sin(15^\circ + 3\varphi_\mathrm{e})}{K_\mathrm{e}^3} - 0.00226\frac{\sin(60^\circ - 4\varphi_\mathrm{e})}{K_\mathrm{e}^4}\end{aligned} \tag{5.91}$$

$$\text{Im}(\tau_1) = 2.232 - 0.1940\frac{\sin(15^\circ - \varphi_e)}{K_e} + 0.0073\frac{\cos(2\varphi_e)}{K_e^2}$$
$$+0.0120\frac{\sin(15^\circ + 3\varphi_e)}{K_e^3} + 0.00160\frac{\sin(60^\circ - 4\varphi_e)}{K_e^4} \quad (5.92)$$

$$\text{Im}(\tau_k) = 1.932\left(k + \frac{1}{4}\right)^{2/3} - \frac{0.2241}{\left(k + \frac{1}{4}\right)^{2/3}}\frac{\sin(15^\circ - \varphi_e)}{K_e}, \qquad k > 1 \quad (5.93)$$

For these expressions to apply

$$d_{\text{km}} > 5\lambda^{1/3} \quad (5.94)$$

These equations will also work for horizontal polarized ground-wave HF signals provided δ_h is substituted for δ_e where

$$\delta_h = K_h e^{j(45^\circ + \varphi_h)} \quad (5.95)$$

$$K_h = 0.002924\frac{\lambda^{1/3}}{\sqrt[4]{(\varepsilon - 1)^2 + 3.6\times10^{25}\sigma_e^2\lambda^2}} \quad (5.96)$$

$$\varphi_h = \frac{1}{2}\tan^{-1}\left(\frac{\varepsilon - 1}{6\times10^{12}\sigma_e\lambda}\right) \quad (5.97)$$

The surface conductivity in these expressions is in e.m.u units. To convert σ to S m^{-1}, use

$$\sigma_e(\text{e.m.u.}) = 10^{-11}\sigma_\Omega(\text{S m}^{-1}) \quad (5.98)$$

One particular example of this model is illustrated in Figure 5.34. In this case, $\sigma = 4\times10^{-11}$ e.m.u. (4 S m^{-1}) and $\varepsilon = 80$.

This model accurately predicts field strengths beyond the HF range, through the VHF range.

5.3.4 Sky Wave

The ionosphere is the region of the Earth's atmosphere that extends from approximately 50 km to 600 km altitude. It is located just above the stratosphere that extends from about 10 km to 50 km altitude. The stratosphere, in turn, is

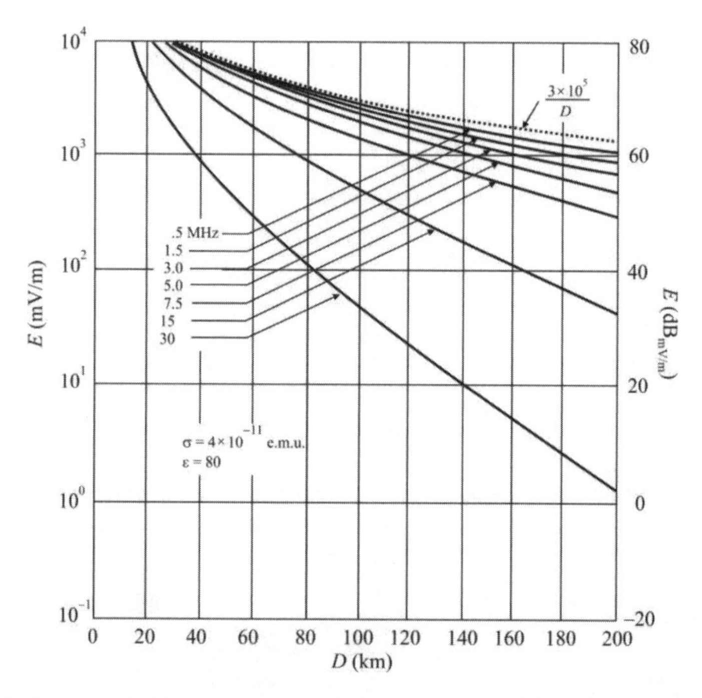

Figure 5.34 Example of field strengths computed with Bremmer's model. These ground constants are for sea water. In addition, the transmitter power is 1 kW, vertical polarization, and neither antenna is elevated. (Source: [38].)

situated just above the troposphere that extends from the Earth's surface to an altitude of about 10 km.

Except for Earth to space links, all VHF and above propagation occurs within the troposphere and so propagation characteristics are determined mostly by this region. HF and lower communications can transpire within the troposphere as well, as in the case of the ground-wave discussed above. There is another mode of propagation, however, that is supported by the ionosphere. This region will reflect (refract, actually) HF and lower-frequency signals back to the Earth. The Earth, in turn, will sometimes reflect the signals back up toward the ionosphere. Very long-distance communication is possible this way with multiple such refractions/reflections.

Propagation of HF signals via the sky-wave is complicated to treat theoretically. The interaction of the Earth's magnetic field with electrons in the ionosphere and the ionization of the ionosphere by the sun lead to many complex properties. Propagation via ionospheric refraction is made possible by ultraviolet rays from the sun energizing atoms in the ionosphere that cause free electrons, thus creating positive ions. The electron density in the ionosphere tends to be formed into regions at specific heights, which vary according to the sunspot activity that frees the electrons. The interaction of the traveling waves with these ions and free electrons facilitates propagation. The ion density in the ionosphere can reach as high as 10^{12} ions (or electrons) per cubic meter.

During the daytime, there are four identifiable layers in the ionosphere: D, E, F1, and F5. The E, F1, and F2 layers will refract EM waves, while the D layer does not. The D layer, however, can significantly attenuate signals passing through it. The D layer forms at altitudes of about 60 to 90 km with a *half thickness*[4] of typically 10 km. As mentioned above, in the event of a nuclear explosion, it is believed that the D layer will become so ionized that the traversal of HF EM waves through it will be impossible or at least severely attenuated.

The maximum ionization of the E layer normally occurs at about 120 km with a half thickness of 20 to 25 km. The F1 layer forms about 170–220 km with a half thickness of about 50 km. The F2 layer forms from 225 to 400 km altitude with a half thickness of 100 to 200 km. In addition, there is occasionally an additional layer called sporadic E, denoted E_s, which forms at an altitude of 120 km with a very small half-thickness of 300 m to 1 km. The concentration of electrons in the E_s layer is much higher than the other regions and therefore higher frequencies are refracted in them. At night, the D and E layers completely disappear and the two F layers combine into a single layer. Figure 5.35 depicts the characteristics of the ionosphere.

As the sun traverses the sky, the layers are formed and changed. The concentration of the D layer is highest at midday. Because of the changing layer electron densities and half thickness, the frequencies that are usable by refraction change as the day goes on. Furthermore, when the day to night and night to day transitions (called the *day/night terminators*) occur, there is considerable turmoil in the ionosphere. Thus, it is normally necessary to monitor the usable frequencies and adjust the one in use. In addition, the usable frequencies change with the season and latitude.

Figure 5.36 shows some of the possible paths HF sky waves can take as they propagate. A single wave can be refracted by any of the refractive layers and reflected by the Earth several times.

The ionosphere will refract a maximum frequency at any given moment. There is also a minimum. If the frequency is too high, it will pass through without being refracted. If it is too low, the D-layer absorption will prevent it from reaching the refracting layers. The maximum frequency is called the *maximum usable frequency* (MUF), while the minimum frequency is called the *lowest usable frequency* (LUF). These frequencies are different for the E layer, F1 layer, and F2

[4] The half thickness of a layer is the thickness where the density falls to one-half of its maximum value.

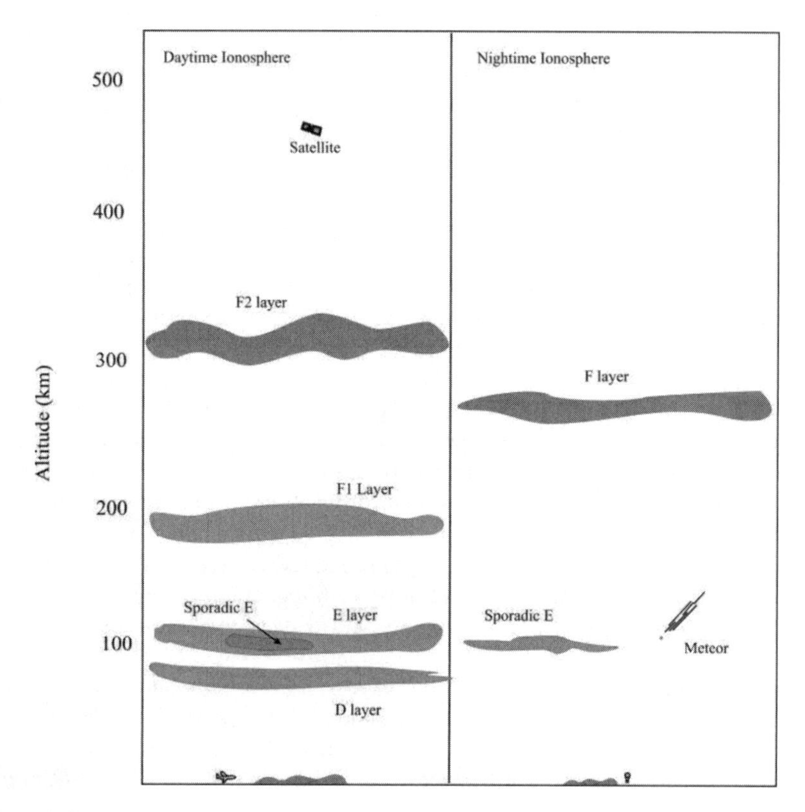

Figure 5.35 Depiction of the ionosphere.

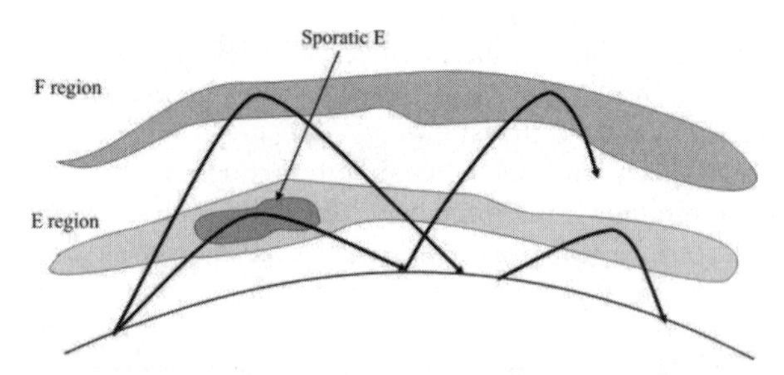

Figure 5.36 A few of the many possible paths that HF sky-wave signals can take. The F region could be either the F1 or F2 or just the nighttime F.

layer because they are largely determined by the actual electron density near the refraction region.

The *critical frequency* (also called the *ionospheric plasma frequency*) is the highest frequency that the F layer will refract. Ionospheric sounders normally are used to determine this frequency. A *vertical sounder* radiates signals directly upward toward the ionosphere. The frequency is swept and the return time and amplitude are measured. The frequency where the signal no longer returns is the critical frequency. More accurate ionospheric information is obtained with an *oblique sounder*, which attempts to measure the characteristics near the region where refraction occurs. It does this by aiming its swept signal toward that region. This is possible in a cooperative communication situation, but is not likely to be possible for uncooperative EW or EA.

The amount of bending due to refraction in the ionosphere is due primarily to two factors: the wavelength (and therefore frequency) of the signal and the ion density in the ionospheric layer. The higher the density, the more bending occurs, and the longer the wavelength (lower the frequency), the more the bending. An additional factor called the *critical angle* is also a determinant of whether a signal is refracted enough by the ionosphere to return to Earth. At a given frequency and ionospheric layer with a given density, the critical angle is that angle at which if the signal impinges on the layer at a higher angle (more toward zenith), the signal will pass through and not be refracted by that layer. With (im)proper antenna orientation it is possible for the sky wave to interfere with the ground-wave. Normally there is, by design, a skip zone, which is the region between where the ground-wave ends and the sky wave returns to Earth the first time. In this region, there is no appreciable signal strength. By orienting the antenna so that the sky-wave signal takes off close to zenith, however, the sky wave will return virtually straight back down, forming a cone around the transmitter. This is called the *near vertical incidence sky-wave* (NVIS) and is useful for relatively short-range communications in mountainous terrain, for example. The ground-wave can interfere with the NVIS signal, however, and vice versa.

As noted, it is possible that an EM wave can be refracted by the ionosphere and reflected by the Earth many times. The wave propagation mode is indicated by the notations 1F1, 1F2, 2F2, and so forth. The first number is the number of hops the wave makes. The second letter/number combination is the ionospheric layer of refraction. Because of the altitudes of the refracting layers, very long communication paths are possible using HF sky-wave links. One-hop F layer refractions can propagate 3,000 km or more, while multihop signals can propagate around the world. One-hop E-layer signals can propagate up to around 2,000 km.

The signal arriving at the receiver, then, consists of potentially several signals refracted off the ionosphere as well as a ground-wave component. Let $G_j(A_j,\nu_j,\sigma_j,t)$ denote the gain of path j where A_j denotes the attenuation factor of the path, ν_j is the Doppler shift, and σ_j is the spread of the frequency shift. Then the signal electrical field strength due to path j is given by

$$E_{\mathrm{R}_j} = G_j(A_j, \nu_j, \sigma_j, t) E(t - \tau_j) \tag{5.99}$$

where τ_j is the delay in the path. The total signal then arriving at the receiver is given by

$$E_{\mathrm{R}}(t) = A_{\mathrm{g}} E(t - \tau_{\mathrm{g}}) + \sum_{j=1}^{K} G_j(A_j, \nu_j, \sigma_j, t) E_j(t - \tau_j) \tag{5.100}$$

where K represents the number of such paths. The first term in (5.100) is due to the ground-wave.

A closed-form expression for determining the signal level for sky-wave signals in the HF range is currently unavailable. Calculation of the performance of HF sky-wave links is based on measured parameters of the ionosphere such as the length. Braun [39] describes a procedure for calculating the expected performance of sky-wave communication links. A similar procedure is described by Davies [40].

Phenomena known as *traveling ionospheric disturbances* (TIDs) frequently occur. These are uneven shapings of one or more of the layers and can cause unexpected reflection angles from the layer. Signals can be focused or defocused by the TIDs. This focusing is illustrated in Figure 5.37, which can cause signals to arrive stronger or weaker. A TID can last for 10 minutes or longer and travel through a layer. Their horizontal speed is on the order to 5–10 km per minute.

Primarily because closed-form equations for the prediction of HF sky-wave signal propagation characteristics are not available, a modeling program for worldwide HF propagation has been developed over several years by the U.S. Department of Commerce, National Telecommunications and Information Administration (NTIA), Institute for Telecommunications Sciences (ITS) and its predecessors. The program is available free on the Internet [41]. It is called the Ionospheric Communication Enhanced Profile Analysis and Circuit (ICEPAC) prediction program. This model is very flexible and provides data such as the MUF, expected SNRs at the receiver using two points or in an area mode, and field strength at the receiver using two points or in an area mode. Many graphical presentations are available from the model—example outputs are shown in Figures 5.38 and 5.39 (the actual displays are, of course, in color).

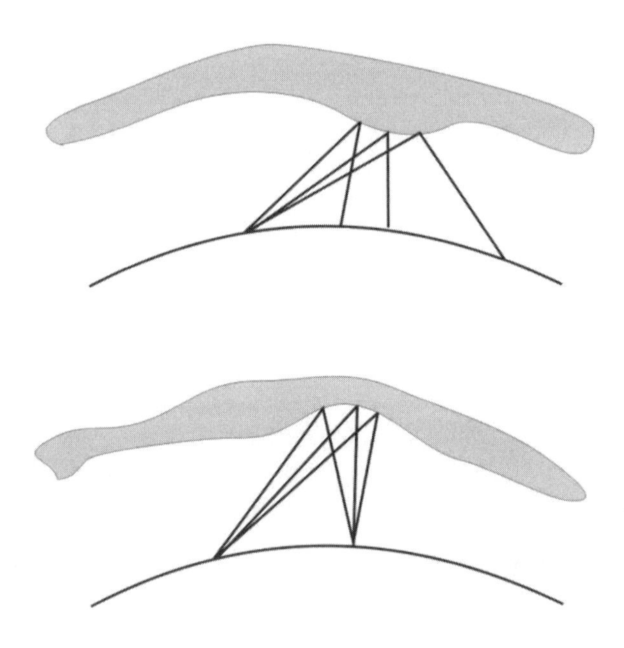

Figure 5.37 TIDs. TIDs can cause dispersion or concentration due to their focusing effects.

Jammers for the sky-wave modes of HF propagation are subject to the same limitations as the communicator. The implications of the MUFs and LUFs are the same, but it is important to keep in mind that the propagation paths between the transmitter and receiver are generally going to be dramatically different from the jammer links.

5.4 Concluding Remarks

The fundamentals of RF signal propagation important to understanding spread spectrum systems were presented in this chapter. The vast majority of spread spectrum systems for land mobile forces use the VHF and higher-frequency spectrum, but some use the HF range. Therefore, salient characteristics of both were presented.

The bandwidth available in the VHF range and above is the primary reason for the interest in that frequency spectrum. In the range below about 30 MHz, by international agreement, the bandwidths are narrow—typically 3 kHz or so. In the VHF range, they are normally 25 kHz, and in the UHF range they are wider than that. With newer communication technology, these bandwidths are becoming

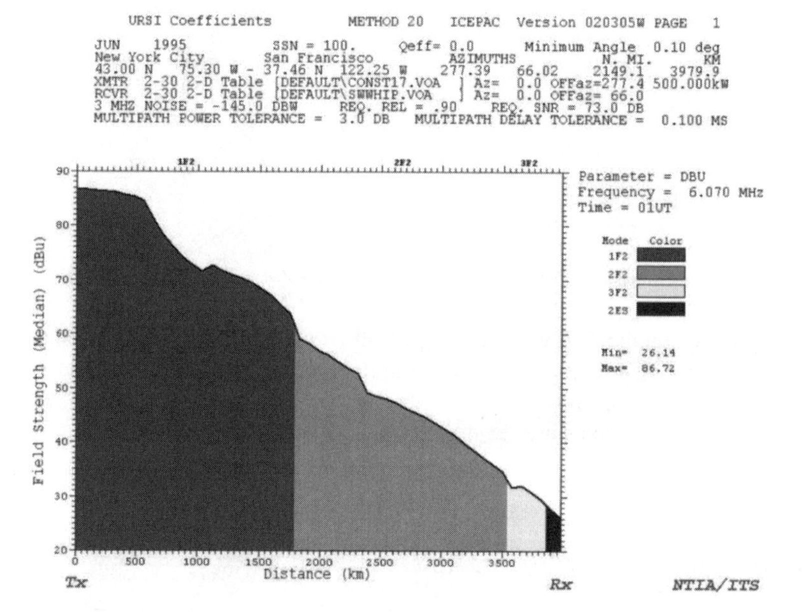

Figure 5.38 ICEPAC output graphic showing median field strength versus distance at 6.070 MHz on the date specified between New York City and San Francisco.

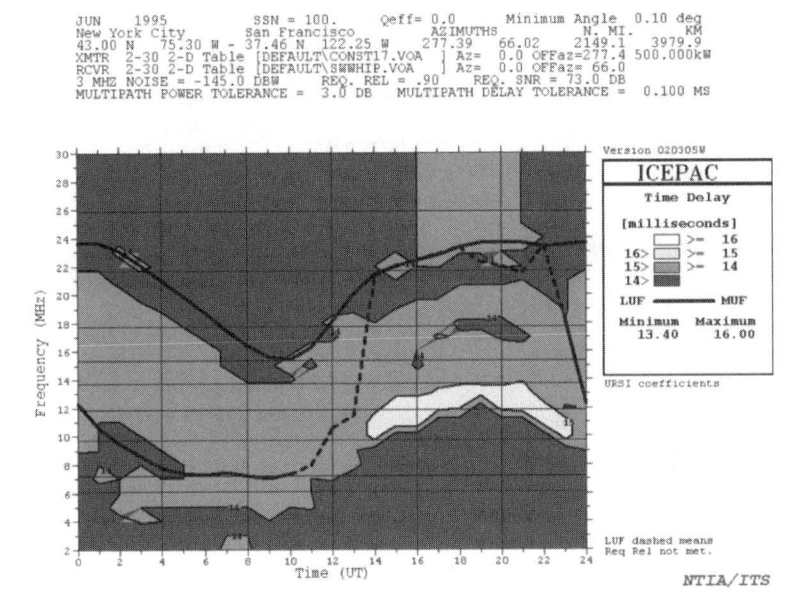

Figure 5.39 ICEPAC graphical output showing time delay.

narrower, however, and the same or more information can be sent over such channels.

The attenuation of RF signals with distance is probably the most important propagation effect. In clear space, when there are no obstacles to impact the propagating sphere of RF energy, the attenuation increases as $1 / R^5$. In most cases of C2 of military ground forces, it is more appropriate to use $1 / R^4$, however. If propagation into and out of buildings is important, a factor higher than 4 is necessary.

Direct wave propagation in the HF range follows these same physical laws. Ionospheric, long-distance HF communication, however, is governed by factors that are more complicated and it is normally necessary to measure the attenuation.

References

[1] Safer, H., "Propagation Measurement-Based Probability of Error Predictions for the Tactical VHF-Range," *Proceedings IEEE MILCOM*, 1999, http://www.greenhouse.com /society/TacCom/milcom_99_papers.html.

[2] Foran, R. A., T. B. Welch, and M. J. Walker, "Very Near Ground Radio Frequency Propagation Measurements and Analysis for Military Applications," *Proceedings IEEE MILCOM*, 1999, http://www.greenhouse.com/society/TacCom/milcom_99_papers.html.

[3] Hande, P., L. Tong, and A. Swami, "Channel Estimation for Frequency Hopping Systems," *Proceedings IEEE MILCOM*, 1999, http://www.greenhouse.com/society /TacCom/ milcom_99_ papers.html.

[4] Poisel, R. A., *Introduction to Communication Electronic Warfare Systems*, 2nd ed, Norwood, MA: Artech House, 2008, pp. 65–66.

[5] Hall, M. P. M., *Effects of the Troposphere on Radio Communications,* London: Institution of Electrical Engineers, 1979, p. 84.

[6] Parson, D., *The Mobile Radio Propagation Channel,* New York: John Wiley & Sons, 1992, p. 20.

[7] Hall, M. P. M., *Effects of the Troposphere on Radio Communications,* London: Institution of Electrical Engineers, 1979, p. 89.

[8] Hall, M. P. M., *Effects of the Troposphere on Radio Communications,* London: Institution of Electrical Engineers, 1979, p. 80.

[9] Hall, M. P. M., *Effects of the Troposphere on Radio Communications,* London: Institution of Electrical Engineers, 1979, p. 91.

[10] Hall, M. P. M., *Effects of the Troposphere on Radio Communications,* London: Institution of Electrical Engineers, 1979, p. 92.

[11] http://wireless.per.nl/multimed/cdrom97/pel.htm, 10/2/2001.

[12] Hall, M. P. M., *Effects of the Troposphere on Radio Communications,* London: Institution of Electrical Engineers, 1979, p. 93.

[13] Parson, D., *The Mobile Radio Propagation Channel,* New York: John Wiley & Sons, 1992, p. 26.

[14] Hall, M. P. M., *Effects of the Troposphere on Radio Communications,* London: Institution of Electrical Engineers, 1979, p. 86.

[15] Parson, D., *The Mobile Radio Propagation Channel,* New York: John Wiley & Sons, 1992, p. 39.

[16] Hall, M. P. M., *Effects of the Troposphere on Radio Communications,* London: Institution of Electrical Engineers, 1979, p.155.
[17] Parson, D., *The Mobile Radio Propagation Channel,* New York: John Wiley & Sons, 1992, pp. 48–53.
[18] Hall, M. P. M., *Effects of the Troposphere on Radio Communications,* London: Institution of Electrical Engineers, 1979, Ch. 7.
[19] Parson, D., *The Mobile Radio Propagation Channel,* New York: John Wiley & Sons, 1992, p. 46.
[20] Hall, M.P.M., *Effects of the Troposphere on Radio Communications,* London: Institution of Electrical Engineers, 1979, p. 161.
[21] Hall, M. P. M., *Effects of the Troposphere on Radio Communications,* London: Institution of Electrical Engineers, 1979, p. 165.
[22] Parson, D., *The Mobile Radio Propagation Channel,* New York: John Wiley & Sons, 1992, p. 120.
[23] Stuber, G. L., *Principles of Mobile Communication,* Boston, MA: Kluwer Academic Publishers, 1996, p. 48.
[24] Abramowitz, M., and I. A. Stegun, (eds.), *Handbook of Mathematical Functions*, New York: Dover, 1965, Ch. 6.
[25] Parson, D., *The Mobile Radio Propagation Channel,* New York: John Wiley & Sons, 1992, p. 117.
[26] Stuber, G. L., *Principles of Mobile Communications*, Boston: Kluwer Academic Publishers, 1996, p. 40.
[27] Parson, D., *The Mobile Radio Propagation Channel,* New York: John Wiley & Sons, 1992, p. 8.
[28] Egli, J. J., "Radio Propagation Above 40 Mc Over Irregular Terrain," *Proceedings of the IRE,* Vol. 45, No. 10, 1957, pp. 1383-1391.
[29] Parson, D., *The Mobile Radio Propagation Channel,* New York: John Wiley & Sons, 1992, p. 68.
[30] Nicholson, D. L., *Spread Spectrum Signal Design LPE and AJ Systems,* Rockville, MD: Computer Science Press, 1988, pp. 20–25.
[31] Longley, A. G., and P. L. Rice, *Prediction of Tropospheric Radio Transmission Loss Over Irregular Terrain, A Computer Method*, ESSA Technical Report ERL 79–ITS 67, NTIS Access No. 676–874, 1968.
[32] Hufford, G., "The ITS Irregular Terrain Model, Version 1.5.2, The Algorithm," http://flattop.its.bldrdoc.gov/itm.html.
[33] Braun, G, *Planning and Engineering of Shortwave Links,* New York: John Wiley & Sons, 1986.
[34] *Reference Data for Radio Engineers*, New York: Howard W. Sams, 1975, p. 28-18.
[35] Braun, G., *Planning and Engineering of Shortwave Links,* New York: John Wiley & Sons, 1986, p. 260.
[36] Braun, G., *Planning and Engineering of Shortwave Links,* New York: John Wiley & Sons, 1986, p. 259.
[37] Bremmer, H., *Terrestrial Radio Wave/Theory of Propagation,* New York: Elsevier Publishing Company, Inc., 1949.
[38] Fulks, G. J., *HF Ground-wave Propagation Over Smooth and Irregular Terrain*, Final Report for Contract DNA 001-80-C-0022, April 1981.
[39] Braun, G., *Planning and Engineering of Shortwave Links,* New York: John Wiley & Sons, 1986, pp. 140–143.
[40] Davies, K., *Ionospheric Radio,* London, U.K.: Peter Peregrinus, Ltd., 1989, pp. 454–459.
[41] ICEPAC, http://elbert.its.bldrdoc.gov/hf.html.

Chapter 6

Feedback Shift Registers and Recursive Sequences

6.1 Introduction

SS systems use pseudo-random sequences as the underlying codes to spread the data streams. These sequences are normally, but not always, generated with shift registers. This chapter introduces shift registers and how they are used to generate such sequences. The specific types of shift registers considered here *linear feedback shift registers* (LFSRs), which have properties that are useful for generating *linear recursive sequences* (LRSs), and *nonlinear feedback shift registers* (NLFSRs), which generate *nonlinear recursive sequences* (NLRSs). LRS is another appellation for a pseudo-random sequence. Before discussing these topics, however, a discussion of Galois fields is presented for those not familiar with the topic. FSR operation and RS properties are derived from the Galois field theory.

This chapter is included to provide an introduction to this basic technology and how such registers are used in AJ systems. Linear sequences are relatively easy to exploit, however.

The structure of this chapter is as follows. We start with a review of Galois fields, as they form the mathematical underpinnings for shift registers. We then discuss the basic characteristics of shift registers. Maximal sequences are introduced after that followed by a discussion on correlation functions. The properties of *m*-sequences as they are used in modern communication systems is then covered. An introduction to product codes, such as Gold codes and Kasami sequences, is then provided. Some linear feedback shift register fundamental design issues are then presented followed by a discussion of some applications of LFSRs in modern communication systems, particularly LPI systems. Lastly, we briefly discuss nonlinear feedback shift registers, although most of the theoretical properties of these are still unknown.

6.2 Galois Fields

A *field* is a mathematical construct comprised of a closed set of elements $X = \{X_i\}$, possibly infinite, and two operations denoted $*$ and $\bullet$. The operations $*$ and $\bullet$ are called *multiplication* and *addition*, by convention. *Closed* here means that if X_i and X_j are in the field then so are $X_i * X_j$ and $X_i \bullet X_j$. In a field there exists an *additive identity element* denoted by 0 such that for every element X_i in X there exists another element in X, called the additive inverse and denoted by $-X_i$, such that $X_i \bullet (-X_i) = 0$. In a field there exists a *multiplicative identity element* denoted by 1 such that for every element X_i in X, except 0, there exists another element in X, denoted X_i^{-1}, such that $X_i * X_i^{-1} = 1$.

In addition to these characteristics, for X to be a field, $*$ and $\bullet$ must be *commutative*:

$$\begin{aligned} X_i \bullet X_j &= X_j \bullet X_i \\ X_i * X_j &= X_j * X_i \end{aligned} \tag{6.1}$$

Also, $*$ is *distributive* over $\bullet$:

$$X_i * (X_j \bullet X_k) = X_i * X_j \bullet X_i * X_k \tag{6.2}$$

The last requirement is that $*$ and $\bullet$ are *associative*:

$$\begin{aligned} X_i \bullet (X_j \bullet X_k) &= (X_i \bullet X_j) \bullet X_k \\ X_i * (X_j * X_k) &= (X_i * X_j) * X_k \end{aligned} \tag{6.3}$$

where the parentheses mean that the operation contained within them is performed first.

A familiar example of a field is the set of all real numbers, both positive and negative, including 0; $*$ is the arithmetic multiplication, and $\bullet$ is arithmetic addition. The multiplicative identity element is the number 1 while the additive identity element is the number 0. The set of integers is not a field since for any element other than 1, there is no integer multiplicative inverse.

LRS theory is based on Galois fields. A *Galois field* is one that is limited in size (not infinite) and the number of elements in the field is a prime number, p, or an integer power of a prime number. Galois fields are denoted by GF(p). It can be shown that for any prime number p and integer m, GF(p^m) is also a field. GF(p^m) is called an *extension field* of p.

Table 6.1 Modulo-2 Addition and Multiplication

		Modulo-2 Addition (⊕)
x_1	x_2	
0	0	0
0	1	1
1	0	1
1	1	0
		Modulo-2 Multiplication (⊗)
0	0	0
0	1	0
1	0	0
1	1	1

For the analysis of LFSRs for AJ communications, particular Galois fields are of interest. These have elements from 2^m for some integer m and they are denoted by GF(2^m). The binary field is perhaps the simplest to understand. It consists of two elements, 0 and 1, and is denoted as GF(2). The operations in this field are the *exclusive OR*, denoted by ⊕, which is analogous to the addition function in the field of real numbers but without carry, and multiplication, denoted by ⊗, which is analogous to the multiplication function in the field of real numbers. The definition of these operations is shown in Table 6.1. Herein, when there is no confusion that would result, modulo-2 multiplication, in conformance with normal practice, will be denoted simply by placing variables next to each other; thus, $a \otimes b = ab$. Note that multiplication modulo-2 is the same function as normal integer multiplication. It is also equivalent to the binary logical AND function.

GF(2^m) consists of all m-tuples of binary digits, that is, from {0, 1}. Thus, when $m = 4$, for example, GF(2^4) consists of the elements given in Table 6.2. The arithmetic in GF(2^m) is called *modulo-2 arithmetic*.

Modulo-2 arithmetic is used in GF(2^m). This arithmetic is similar to binary arithmetic except there is no carry function. That is,

$$\begin{array}{r} 01 \\ \oplus\, 11 \\ \hline 10 \end{array}$$

and the 1 that would be carried from the first column on the right in binary arithmetic is not carried in modulo-2 arithmetic. Also, in modulo-2 arithmetic, each element in GF(2^m) is its own additive inverse because $0 \oplus 0 = 0$, $1 \oplus 1 = 0$, while $1 \oplus 0 = 1$ and $0 \oplus 1 = 1$. Also, subtraction of b from a, defined as $a + (-b)$, is accomplished by the addition operator; thus, *addition and subtraction are equivalent*.

Table 6.2 Elements of GF(2^4)

0000	1000
0001	1001
0010	1010
0011	1011
0100	1100
0101	1101
0110	1110
0111	1111

6.2.1 Polynomials

Polynomials of the form

$$f(D) = f_0 + f_1 D + f_2 D^2 + \cdots + f_m D^m \tag{6.4}$$

where the coefficients f_i are selected from GF(2) and D is an *indeterminate*, are useful for analyzing LFSRs. The nature of D is not significant for our purposes. The mathematical operations that are defined for when the coefficients are real numbers apply in this case as well, except the mathematics is carried out modulo-2. Thus, addition, subtraction, multiplication, and division are defined in the familiar way.

The Euclidean division algorithm [1] states that any polynomial $f(D)$ of degree m over GF(2) can be divided by another polynomial, $g(D)$ over GF(2), of degree at least 1 as

$$f(D) = g(D)h(D) + r(D) \tag{6.5}$$

where $h(D)$ is the *quotient* and $r(D)$ is the *remainder*, and $r(D)$ has a maximum degree of $m - 1$. As an example, suppose $f(D) = D^4 + D^2 + 1$ and $g(D) = D^3 + 1$. Then

$$\begin{array}{rl} & D \\ D^3 + 1 \big) & \overline{D^4 + D^2 \qquad + 1} \\ & \underline{D^4 \qquad\quad + D \quad} \\ & \qquad D^2 + D + 1 \end{array}$$

so $h(D) = D$ and $r(D) = D^2 \oplus D \oplus 1$.

If $r(D)$, = 0 then $g(D)$ is said to *divide* $f(D)$. If $g(D)$ divides $f(D)$, then $g(D)$ is said to be a *factor* of $f(D)$. Most of the characteristics of interest for LRS generation and analysis are associated with the remainder, $r(D)$.

These polynomials have roots just as polynomials over the real field do. A *root* of $f(D)$ is any value of D for which $f(D) = 0$. Thus, $D = 1$ is a root of $D^3 \oplus 1$ because letting $D = 1$, $1^3 \oplus 1 = 1 \otimes 1 \otimes 1 \oplus 1 = 0$. The roots of a polynomial need not necessarily be in GF(2), just as the roots of a polynomial over the real field may be complex and therefore not part of the real field. In addition, if $f(D)$ is factorable, then it is said to be *reducable*; otherwise, it is said to be *irreducible*. The factorization of $f(D)$ is unique as well. An irreducible polynomial is also known as a *primitive polynomial*. From above, $(D - 1)$ is a factor of $D^3 \oplus 1$. This can be verified by

$D - 1 = D \oplus 1$ (Subtraction and addition are equivalent since $1 - 1 = 0$ and $1 \oplus 1 = 0$)

$$
\begin{array}{r|l}
 & D^2 \oplus D \quad \oplus 1 \\
\hline
D \oplus 1 & D^3 \qquad\quad \oplus 1 \\
 & \underline{D^3 \oplus D^2} \\
 & \quad D^2 \quad \oplus 1 \\
 & \quad \underline{D^2 \oplus D} \\
 & \qquad D \oplus 1 \\
 & \qquad \underline{D \oplus 1} \\
 & \qquad\quad 0
\end{array}
$$

Thus, $r(D) = 0$ and $D^3 \oplus 1 = (D \oplus 1)(D^2 \oplus D \oplus 1)$.

The extension field of GF(2) of degree m, GF(2^m), has 2^m elements in it. There are 2^m distinct polynomials over GF(2) of degree $m - 1$. Therefore, these polynomials can represent each of the elements of GF(2^m) on a one-to-one basis. Addition of the polyonmials occurs in the normal way and each polynomial is its own additive inverse. The product of two polynomials of degree m (all polynomials are of degree m as long as m is equal to or greater than the largest degree of all the polynomials under consideration, with the appropriate coefficients set to 0, as necessary, in some of the polynomials) is defined as the remainder after the two polynomials are multiplied normally, and the product divided by the irreducible polynomial, $h(D)$, used to define multiplication. Thus, the multiplication is defined modulo-$h(D)$. Therefore, the elements of GF(2^m) can be described by their corresponding polynomial.

Polynomials defined over GF(2) form what is called an *Abelian* (commutative) *group* with polynomial addition as the operator. A group is a mathematical structure with fewer restrictions than a field; that is, a field is an

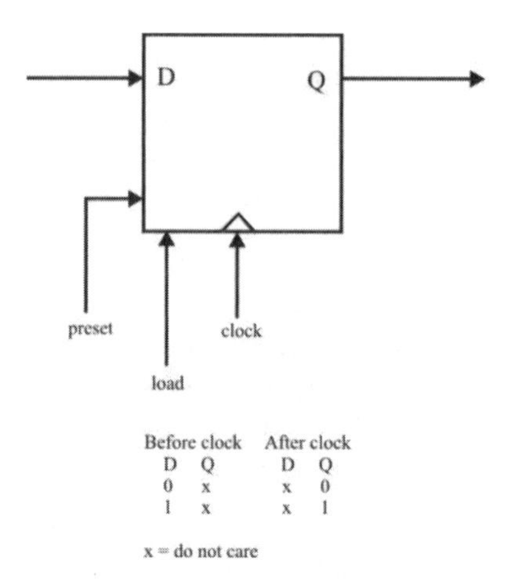

Figure 6.1 Delay element.

Abelian group, but the inverse is not necessarily true. Polynomial multiplication is communicatve and associative, and obeys the distributive law, but polynomials defined over a field do not form a field. What is missing is a multiplicative inverse of $f(D)$, $f^{-1}(D)$, so that $f(D)f^{-1}(D) = 1$.

6.3 Shift Registers

Shift register sequences are the mechanism to control the frequency in an FHSS system and to determine the bit sequence in DSSS systems. Short sequences are more readily exploited than long sequences, mainly because the latter take longer to repeat. Short sequences used by FHSS targets facilitate discovering the hopping frequencies in use and their pattern, and therefore predictive jamming can be used, which is the most efficient and effective kind.

A shift register consists of delay elements and digital logic. The delay elements are depicted as in Figure 6.1. Delay elements are also called *one-bit registers* and *flip-flops*. Whatever digital bit is present on the input (D) is stored when the clock signal occurs. The stored bit is available on the output (Q). This bit is stored indefinitely until either the power is removed (assuming volatile implementations) or another clock signal occurs. Most delay elements can also be preloaded with binary digits as well. Whatever value is on the preset line when the load signal occurs is stored in the element.

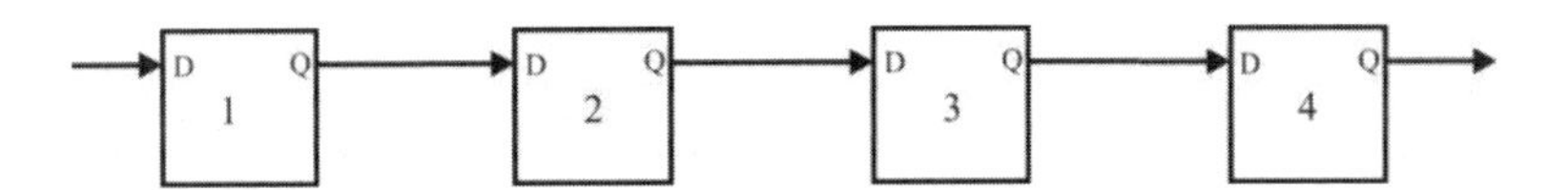

Figure 6.2 Four-stage shift register.

In the remainder of this book, for clarity we will not show the clock, preset, or load inputs to the delay elements. It is important to remember, however, that they are involved in the proper operation of shift registers.

A shift register is configured with one or more delay elements connected one after the other. A four stage shift register is depicted in Figure 6.2. The output of one element forms the input to the next. The serial output of the register is usually taken as the output of the last stage. It is also possible to use all or some of the stages as simultaneous, parallel outputs.

Suppose this shift register is preloaded with 1001 and the sequence 0100, in succession, was applied to the input to the first stage. The bits would shift through the register on each clock pulse following the pattern shown in Figure 6.3. At the end of $m = 4$ clock pulses, the shift register contains exactly what comprises the input sequence, with the first bit in the last stage, the second bit in the third stage, the third bit in the second stage, and the last bit in the first stage.

Note that the register can also be preloaded by inputting the initial sequence through the first stage serially and clocking the register four times. Parallel initialization such as indicated here is faster, if that is an issue, which it normally is not for spread spectrum applications.

Now suppose that the last stage of the shift register is fed back to the input stage as shown in Figure 6.4. If the register is preloaded with 1001, the register will continue to generate this sequence after every four clock pulses.

Linear recursive sequences are formed with shift registers by combining the outputs of one or more stages and feeding them back to the input stage. By including these elements, sequences longer than m can be generated with an m-

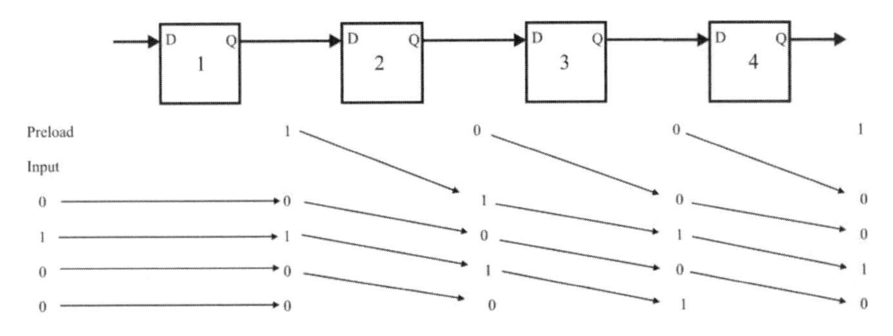

Figure 6.3 Four-stage shift register sequence.

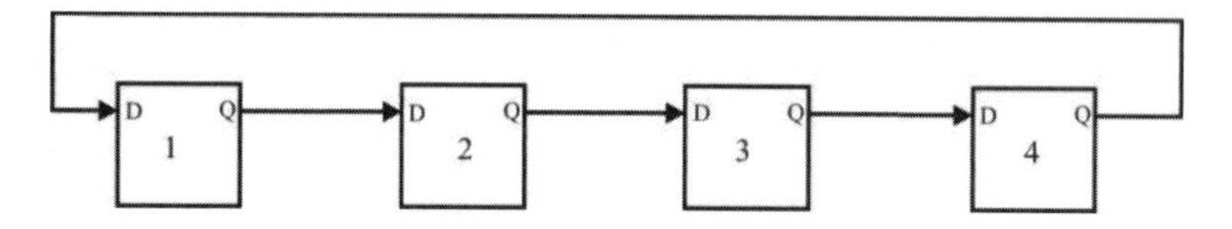

Figure 6.4 Linear shift register with the output stage fed back to the input stage.

stage register. The logic element used to perform this combining is the XOR. Its truth table is given in Figure 6.5(a). The XOR is sometimes referred to as a *half adder* because its output is the binary sum of the inputs with no carry function implemented (see Figure 6.6). Note that the truth table in Figure 6.5 is the same as the operation performed in GF(2^m) by the addition operator.

Suppose that stages 3 and 4 are combined and fed back to the input of the first stage as illustrated in Figure 6.7. Then with the initial load of 1001 the sequence shown in Table 6.3 is generated. The two rows in bold are the same, indicating the sequence is starting over. A length 15 sequence is thus generated with a four-stage register. This is the maximum length, as it is equal to $N = 2^m - 1$ with $m = 4$ in this case. Such a sequence is called a *maximal sequence*, or *m-sequence*.

By using different feedback taps in an LFSR, different sequences are generated even with the same initial load; however, some of these sequences may not be maximal. A different initial load of an LFSR that generates an m-sequence will generate the same m-sequence but with a different phase—that is, the m-sequence

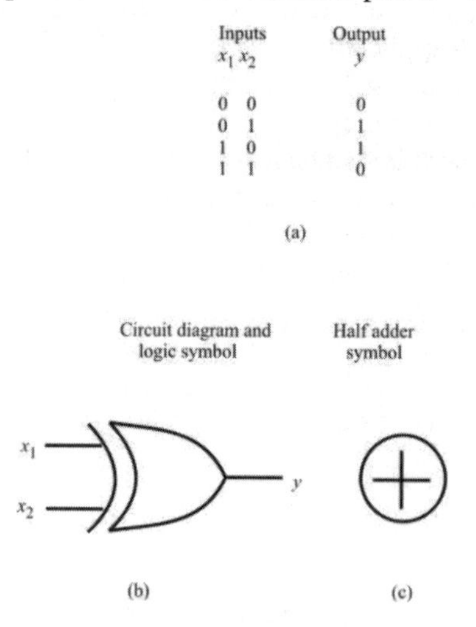

Inputs x_1 x_2	Output y
0 0	0
0 1	1
1 0	1
1 1	0

Figure 6.5 The XOR element: (a) truth table, (b) circuit diagram, and (c) half-adder symbol.

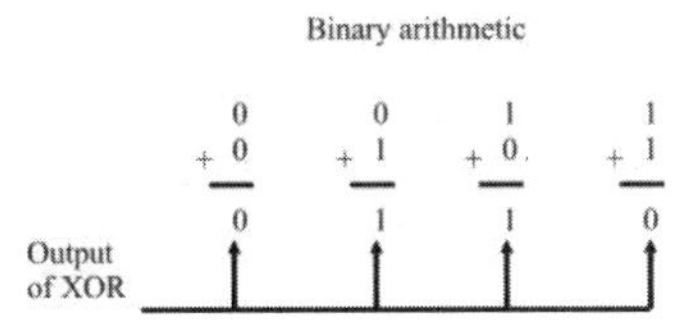

Figure 6.6 Half adder.

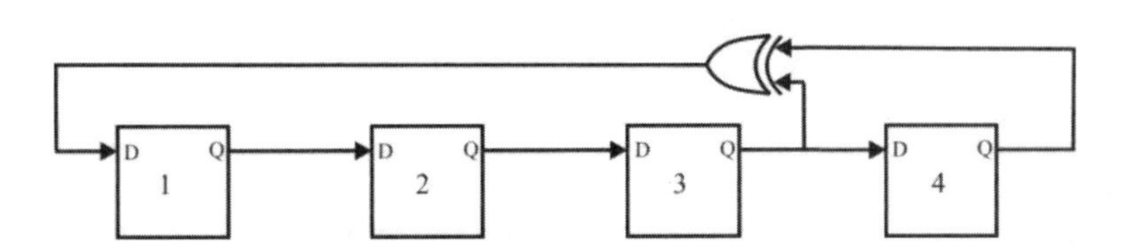

Figure 6.7 Linear feedback shift register.

Table 6.3 Linear Sequence of Shift Register Shown in Figure 6.7

		Stage				
Clock Pulse		**1**	**2**	**3**	**4**	**XOR Output**
	Preload	**1**	**0**	**0**	**1**	**1**
1		1	1	0	0	0
2		0	1	1	0	1
3		1	0	1	1	0
4		0	1	0	1	1
5		1	0	1	0	1
6		1	1	0	1	1
7		1	1	1	0	1
8		1	1	1	1	0
9		0	1	1	1	0
10		0	0	1	1	0
11		0	0	0	1	1
12		1	0	0	0	0
13		0	1	0	0	0
14		0	0	1	0	1
15		**1**	**0**	**0**	**1**	**1**

starts at a different point but the bit pattern is the same.

Careful inspection of Table 6.3 reveals that every 4-bit binary number except 0 is displayed in one of the rows and each occurs only once. This is characteristic of maximal sequences. If the shift register ever contains all 0's, it will remain in that state forever. Therefore, all 0's in the shift register is not part of the sequence generated nor can it be an initial load.

The LRS generator shown in Figure 6.7 has the XOR in the feedback path. Such a configuration is sometimes referred to as an *S-type* LFSR. Another name for this architecture is the *Fibonacci configuration* [2]. It is also possible to put them in the feedforward path between the register stages as shown in Figure 6.8. The sequence generated by this register is given in Table 6.4. Notice that in this case the sequence generated is not maximal. The pattern repeats itself after only three clock pulses, illustrated by the bold rows in Table 6.4.

Feedforward shift registers can also be configured to generate maximal sequences, however, as seen in Figure 6.9 and Table 6.5. Here, the m-sequence is generated with period $N = 2^m - 1 = 15$. Feedforward LRS generators are also sometimes referred to as an *M-type* LFSR. Another name for this type of shift register is the *Galois configuration*. The Galois configuration is usually preferred

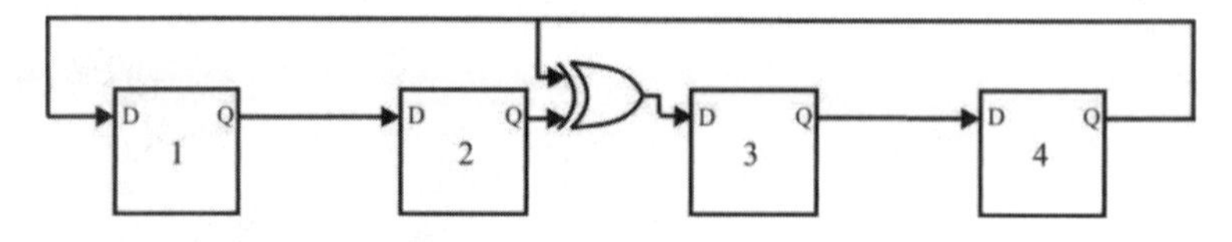

Figure 6.8 Feedforward shift register.

Table 6.4 A Non-Maximal Sequence

		Stage				
Clock Pulse		**1**	**2**	**3**	**4**	**XOR Output**
	Preload	1	0	0	1	1
1		1	1	1	0	1
2		0	1	1	1	0
3		1	0	0	1	1
4		1	1	1	0	1
5		0	1	1	1	0
6		1	0	0	1	1
7		1	1	1	0	1
8		0	1	1	1	0
9		1	0	0	1	1
10		1	1	1	0	1
11		**0**	**1**	**1**	**1**	**0**
12		**1**	**0**	**0**	**1**	**1**
13		**1**	**1**	**1**	**0**	**1**
14		0	1	1	1	0
15		1	0	0	1	1

when implemented in hardware because there are fewer logic elements in the feedback path and therefore the LFSR is faster. This is less of a factor when the LFSR is implemented in software.

The general forms for the two types of LFSRs are shown in Figures 6.10 and 6.11. The multiplication factors g_i determine whether stage i in the LFSR is used in the feedback or feedforward paths.

The following analysis of the mathematics of the LFSR are taken from [3] and follows that development closely. From Figure 6.10, the sequence x_i is given by the expression

$$x_i = g_1 x_{i-1} \oplus g_2 x_{i-2} \oplus \cdots \oplus g_{m-1} x_{i-(m-1)} \oplus g_m x_{i-m} = \sum_{k=1}^{m} g_k x_{i-k} \tag{6.6}$$

where the variables are selected from {0, 1} and the mathematics is carried out modulo-2, according to Table 6.1. The *generating function* of this sequence is given by

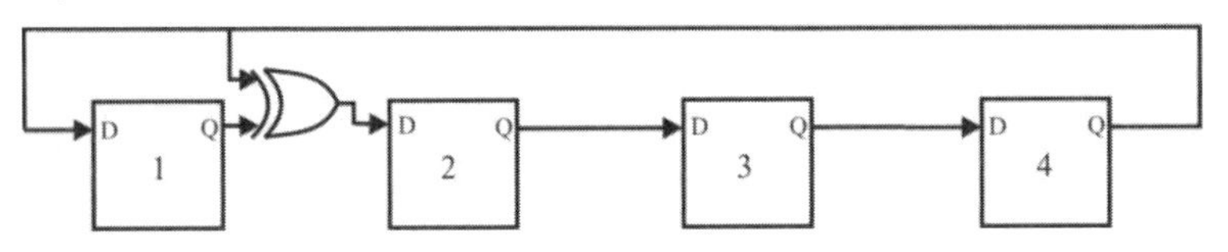

Figure 6.9 An M-type LFSR generator that generates a maximal sequence.

Table 6.5 *M*-Sequence Generated by the LFSR in Figure 6.9

		Stage				
Clock Pulse		**1**	**2**	**3**	**4**	**XOR Output**
	Preload	**1**	**0**	**0**	**1**	**0**
1		1	0	0	0	1
2		0	1	0	0	0
3		0	0	1	0	0
4		0	0	0	1	1
5		1	1	0	0	1
6		0	1	1	0	0
7		0	0	1	1	1
8		1	1	0	1	0
9		1	0	1	0	1
10		0	1	0	1	1
11		1	1	1	0	1
12		0	1	1	1	1
13		1	1	1	1	0
14		1	0	1	1	0
15		**1**	**0**	**0**	**1**	**0**

$$g(D) = x_0 \oplus x_1 D \oplus x_2 D^2 \oplus \cdots = \sum_{i=0}^{\infty} x_i D^i \qquad (6.7)$$

where the D denotes delay as in Figure 6.10.

This sequence is *periodic* with period N if $x_{i+N} = x_i$. If $g(D)$ is periodic, then

$$\begin{aligned}
i = 0: \quad & x_0 \oplus x_1 D \oplus x_2 D^2 \oplus \cdots \oplus x_{N-1} D^{N-1} \\
i = 1: \quad & x_0 D^N \oplus x_1 D^{N+1} \oplus x_2 D^{N+2} \oplus \cdots \oplus x_{N-1} D^{N+(N-1)} \\
& = D^N (x_0 \oplus x_1 D \oplus x_2 D^2 \oplus \cdots \oplus x_{N-1} D^{N-1}) \\
i = 2: \quad & x_0 D^{2N} \oplus x_1 D^{2N+1} \oplus x_2 D^{2N+2} \oplus \cdots \oplus x_{N-1} D^{2N+(N-1)} \\
& = D^{2N} (x_0 \oplus x_1 D \oplus x_2 D^2 \oplus \cdots \oplus x_{N-1} D^{N-1}) \\
\cdots
\end{aligned} \qquad (6.8)$$

So clearly $g(D)$ can be written as

$$\begin{aligned}
g(D) &= \sum_{i=0}^{\infty} D^{iN} (x_0 \oplus x_1 D \oplus x_2 D^2 \oplus \cdots \oplus x_{N-1} D^{N-1}) \\
&= (x_0 \oplus x_1 D \oplus x_2 D^2 \oplus \cdots \oplus x_{N-1} D^{N-1}) \sum_{i=0}^{\infty} D^{iN}
\end{aligned} \qquad (6.9)$$

But

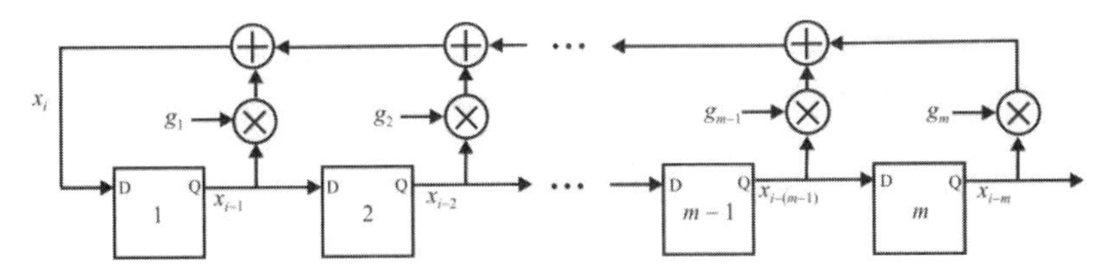

Figure 6.10 General form of Fibonacci LFSR.

$$\sum_{i=0}^{\infty} y^i = \frac{1}{1-y}, \qquad -1 < y < 1 \tag{6.10}$$

so

$$g(D) = \frac{(x_0 \oplus x_1 D \oplus x_2 D^2 \oplus \cdots \oplus x_{N-1} D^{N-1})}{1 \oplus (-D^N)} \quad \text{for } -1 < D^N < 1 \tag{6.11}$$

Since

$$g(D) = \sum_{i=0}^{\infty} x_i D^i \tag{6.12}$$

and

$$x_i = \sum_{k=1}^{m} g_k x_{i-k} \tag{6.13}$$

then

$$g(D) = \sum_{i=0}^{\infty} \left(\sum_{k=1}^{m} g_k x_{i-k} \right) D^i \tag{6.14}$$

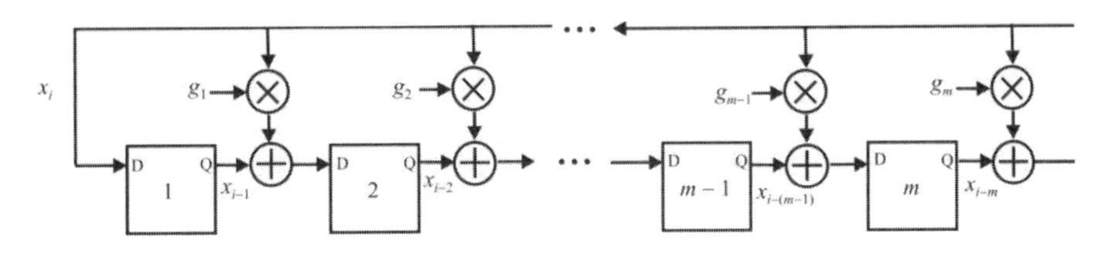

Figure 6.11 General form of Galois LFSR.

But the order of the summations can be interchanged. Also, a D^k is factored out of the second sum yielding

$$g(D) = \sum_{k=1}^{m} g_k D^k \sum_{i=0}^{\infty} x_{i-k} D^{i-k} \tag{6.15}$$

but

$$\begin{aligned} \sum_{i=0}^{\infty} x_{i-k} D^{i-k} &= x_{-k} D^{-k} \oplus x_{1-k} D^{1-k} \oplus x_{2-k} D^{2-k} \\ &\quad \oplus \cdots \oplus x_{-2} D^{-2} \oplus x_{-1} D^{-1} \oplus x_0 D^0 \oplus x_1 D^1 \oplus \cdots \\ &= x_{-k} D^{-k} \oplus x_{1-k} D^{1-k} \oplus x_{2-k} D^{2-k} \\ &\quad \oplus \cdots \oplus x_{-2} D^{-2} \oplus x_{-1} D^{-1} \oplus g(D) \end{aligned} \tag{6.16}$$

so

$$\begin{aligned} g(D) &= \sum_{k=1}^{m} g_k D^k [x_{-k} D^{-k} \oplus x_{1-k} D^{1-k} \oplus x_{2-k} D^{2-k} \\ &\quad \oplus \cdots \oplus x_{-2} D^{-2} \oplus x_{-1} D^{-1} \oplus g(D)] \\ &= \sum_{k=1}^{m} g_k D^k [x_{-k} D^{-k} \oplus x_{1-k} D^{1-k} \oplus x_{2-k} D^{2-k} \\ &\quad \oplus \cdots \oplus x_{-2} D^{-2} \oplus x_{-1} D^{-1}] \oplus g(D) \sum_{k=1}^{m} g_k D^k \end{aligned} \tag{6.17}$$

In order to remove the term with $g(D)$ in it on the right side of this expression, it is sufficient to simply add the term to both sides. Thus,

$$\begin{aligned} g(D)\left(1 \oplus \sum_{k=1}^{m} g_k D^k\right) &= \sum_{k=1}^{m} g_k D^k [x_{-k} D^{-k} \oplus x_{1-k} D^{1-k} \oplus x_{2-k} D^{2-k} \\ &\quad \oplus \cdots \oplus x_{-2} D^{-2} \oplus x_{-1} D^{-1}] \end{aligned} \tag{6.18}$$

so

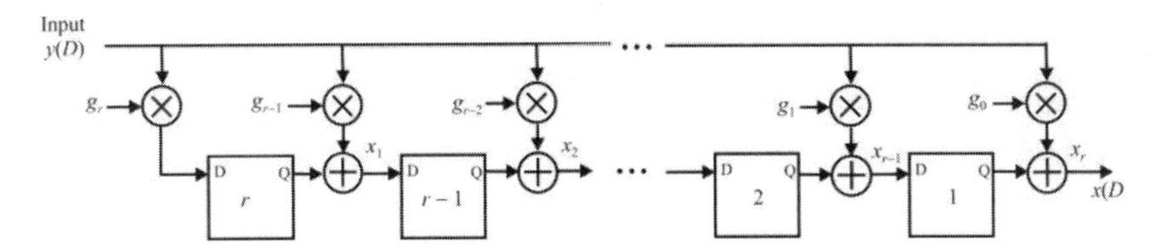

Figure 6.12 Shift register that multiplies two polynomials, given by $y(D)$ and $g(D)$.

$$g(D) = \frac{\sum_{k=1}^{m} g_k D^k \begin{pmatrix} x_{-k} D^{-k} \oplus x_{1-k} D^{1-k} \\ \oplus x_{2-k} D^{2-k} \oplus \cdots \oplus x_{-2} D^{-2} \oplus x_{-1} D^{-1} \end{pmatrix}}{1 \oplus \sum_{k=1}^{m} g_k D^k}$$

$$= \frac{g_0(D)}{f(D)} \tag{6.19}$$

The denominator in this ratio is called the *characteristic polynomial* for the LFSR and depends only on the interconnections as represented by the g_i. The numerator depends on the initialization load for the LFSR and determines the phase of the sequence.

If the sequence corresponding to $g(x)$ is an m-sequence, that is, it has a period N, then $f(x)$ must divide $1 \oplus D^N$. In addition, for $g(x)$ to generate an m-sequence, it is necessary (but not sufficient) that $f(x)$ be primative, which, as discussed above, is true if and only if $f(x) \neq f_1(x)f_2(x)$ for any $f_1(x)$ and $f_2(x)$.

The XOR logic element is the one used for the combination of the stages in an LFSR, however, other logic elements are also used. In an LFSR they are not used for feedback or feedforward purposes, but to form the the output sequence. We will see later that they are used for these purposes in nonlinear shift registers. AND, OR, NAND, and NOR combinational logic elements can be used to advantage to form sequences with particular properties.

Shift registers can be used to multiply and divide polynomials whose coefficients are from GF(2) as described in Section 6.1.1. The diagram shown in Figure 6.12 multiplies the two polynomials $y(D)$ and $g(D)$ [4, 5], $x(D) = y(D)g(x)$. This can be seen by:

$$\begin{aligned} x(D) &= x_r(D) \\ &= x_{r-1}(D)D \oplus g_0 y(D) \\ &= x_{r-2}(D)D^2 \oplus g_1 y(D)D \oplus g_0 y(D) \end{aligned}$$

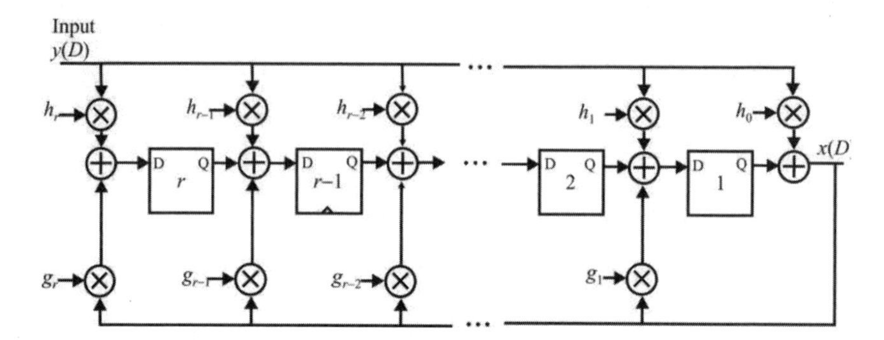

Figure 6.13 An LFSR that multiplies the input by $h(D)$ and divides it by $g(D)$, simultaneously.

$$\begin{aligned}&= g_r y(D)D^r \oplus g_{r-1} y(D)D^{r-1} \oplus \cdots \oplus g_0 y(D)\\ &= g(D)y(D)\end{aligned} \tag{6.20}$$

The multiplication here is not the polynomial modulo-multiplication discussed above but normal polynomial multiplication using modulo-2 operations.

The LFSR shown in Figure 6.13 performs two functions simultaneously: it multiplies the input $y(D)$ by $h(D)$ and divides it by $g(D)$ [6]; thus,

$$x(D) = \frac{h(D)}{g(D)} y(D) \tag{6.21}$$

6.4 Maximal Sequences

Two m-sequences are *cyclically equivalent* if they are time shifted versions of one another. If two m-sequences are not cyclically equivalent, then they are called *cyclically distinct*. Table 6.6 gives the number of cyclically distinct m-sequences for some values of N [6]. The number of primitive polynomials of degree m is given by [6]

$$N_p(m) = \frac{2^{m-1}}{m} \prod_{i=1}^{K} \frac{\pi_i - 1}{\pi_i} \tag{6.22}$$

where π_i, $i = 1,\ldots,K$ is the *prime decomposition* of $2^m - 1$ given by

Table 6.6 The Number of Cyclically Distinct *m*-Sequences

L	$N_p(L)$	L	$N_p(L)$
2	1	16	2,048
3	2	17	7,710
4	2	18	7,776
5	6	19	27,594
6	6	20	24,000
7	18	21	84,672
8	16	22	120,032
9	48	23	356,962
10	60	24	276,480
11	176	25	1,296,000
12	144	26	1,719,900
13	630	27	4,202,496
14	756	28	4,741,632
15	1,800	29	18,407,808

Source: [6]

$$2^m - 1 = \prod_{i=1}^{K} \pi_i^{n_i} \tag{6.23}$$

where π_i is prime number i in the decomposition and n_i is the degree of that prime number. Prime number π_i is unique by the unique factorization theorem of arithmetic.

A sequence generated with an LFSR of length N is referred to as an RN code. Thus, when $N = 7$, an R7 code is generated.

6.5 Correlation Functions

The *correlation* of two sequences is a measure of how similar the two are at bit offsets from one another. A sequence that is not offset at all from a copy of itself will have maximum correlation since the bit values are identical. The correlation is given by the sum of the products of the two sequences taking each bit position one at a time. Since $1 \times 1 = -1 \times -1 = 1$, when they are identical, the correlation value will equal $N = 2^m - 1$. When the two sequences are offset from one another, approximately half of the values in the correlation sum will be +1 and half will be −1. The sums are therefore near zero.

The *autocorrelation* function of a sequence $x(k)$ consisting of elements from GF(2) is given by

$$R_{xx}(k)=\sum_{i=0}^{N-1} z(i)z(i+k), \quad k=0,1,\ldots,N-1 \tag{6.24}$$

where $z(k)=1-2x(k)$ (this changes the sequence of 0s and 1s to +1s and −1s, respectively). An autocorrelation function is calculated graphically in Figure 6.16.

The *cross correlation* function between two sequences $x(k)$ and $y(k)$ consisting of elements from GF(2) is given by

$$R_{xy}(k)=\sum_{i=0}^{N-1} z_1(k)z_2(i+k), \; k=0,1,\ldots,N-1 \tag{6.25}$$

where $z_1(k)=1-2x(k)$ and $z_2(k)=1-2y(k)$.

The cross-correlation of two 7-bit sequences that are not alike is shown in Figure 6.15. Note that even with these short sequences, the correlation was higher when the sequences were alike in Figure 6.14 at a 0-bit lag.

In spread spectrum communication systems, it is important to have sequences where the autocorrelation function is large at zero lag so that synchronization can be accomplished. At lags other than zero it is desirable that the autocorrelation be small so the false synchronization lock can be avoided. In addition, the cross correlation between two sequences being used by two different communication systems should be low even at zero lag so that a false correlation indication of the two signals is avoided.

The properties mentioned above apply when the entire period of a sequence is available for processing. This is not always the case, especially when N is large. In those cases, large values of the correlation function of a sequence may very well occur at other than zero lag and this must be taken into consideration.

In addition, these properties assume that the signals are time-synchronized, that is, the bit timing is the same for the signals being considered. Synchronizing these signals is more difficult (we discuss a technique for chip-timing synchronizing in otherwise asynchronous systems in Chapter 7). Sometimes GPS signals are used to accomplish this time synchronization. GPS timeticks are accurate to within 100 ns or so and very accurate bit clocks can be maintained in the mobile phones.

An effective jamming technique that will be discussed at length later is to attack the synchronization process in both DSSS and FHSS systems. If the targets can be prevented from synchronizing they cannot communicate at all.

There is a lower bound on the cross-correlation function between two m-sequences with period N, which was discovered by Welch. It is given by [3, 7]

$$R_{xy}(\tau)\geq N\sqrt{\frac{m-1}{mN-1}}\approx\sqrt{N} \tag{6.26}$$

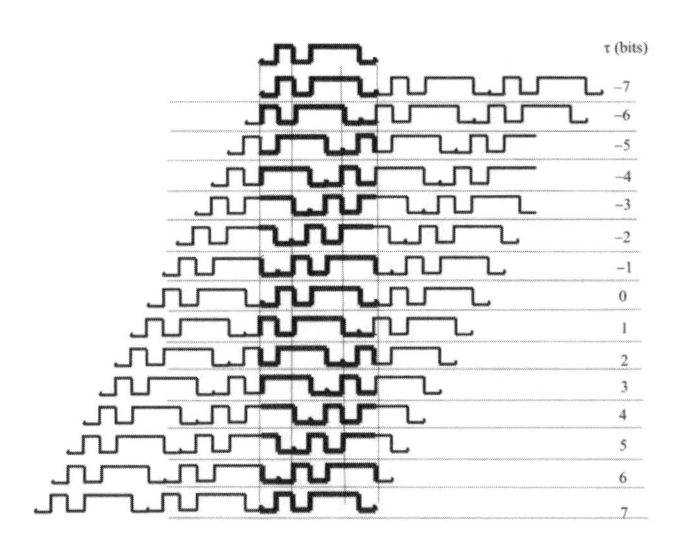

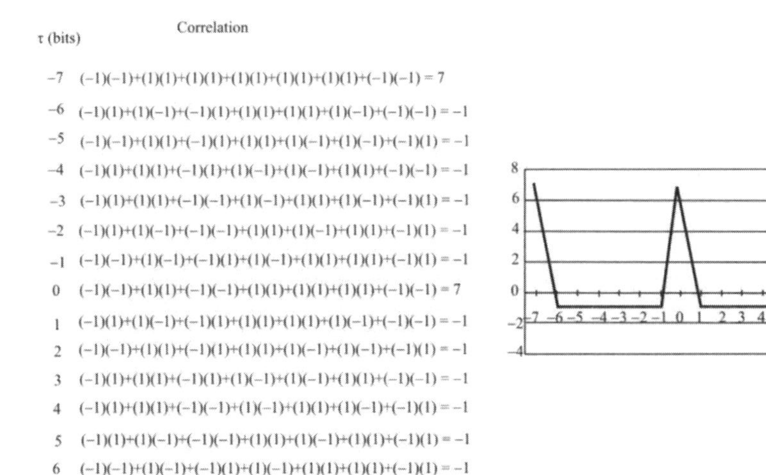

Figure 6.14 Correlation of a 7-bit *m*-sequence.

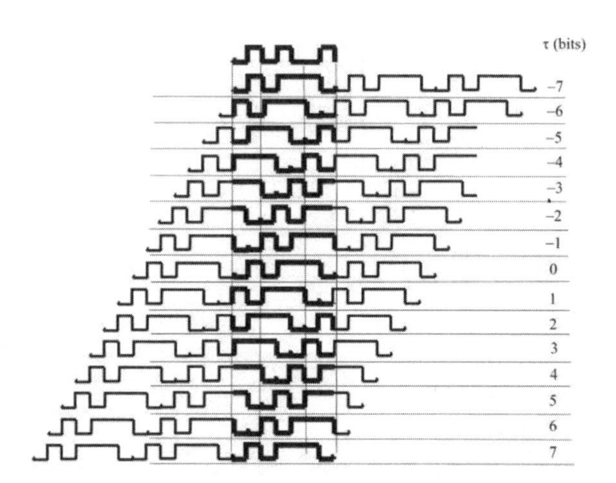

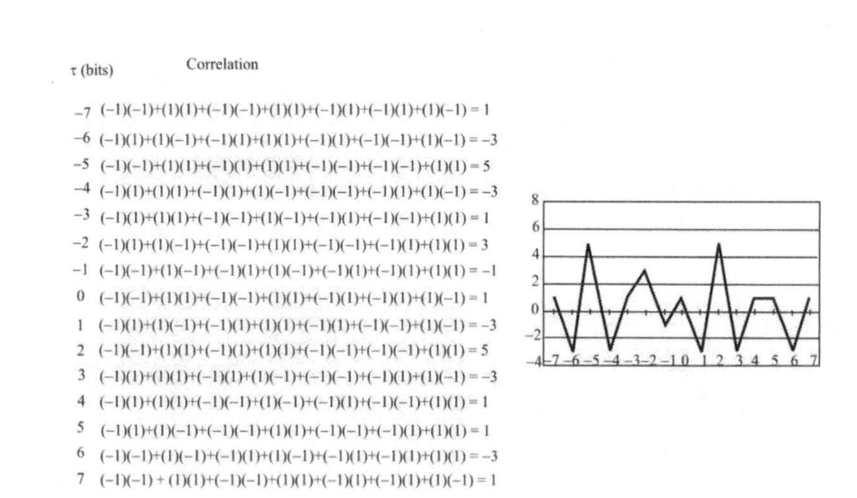

Figure 6.15 Correlation function of two dissimilar sequences when $N = 7$.

6.6 Properties of *m*-Sequences

This section presents some properties of *m*-sequences that make them particularly useful in SS systems.

Property 1: There is only one value of correlation between two phases of an *m*-sequence that is other than $1/N$. Thus, for sequence x:

$$R_{xx}(t-\tau) = \begin{cases} \dfrac{1}{N} & \tau \neq 0 \\ N & \tau = 0 \end{cases} \tag{6.27}$$

The correlation function repeats itself every $2^m - 1$ bits. This is particularly useful for synchronizing the locally generated sequence to that of the incoming signal.

The bit offset of the local sequence is frequently determined by sliding the two sequences past each other. When there is no bit offset between the two, then $\tau = 0$ and the correlation peak occurs. This is illustrated in Figure 6.16. If

$$q(\tau) \equiv \begin{cases} 1 - \dfrac{|\tau|}{T_c}, & |\tau| \leq T_c \\ 0, & \text{otherwise} \end{cases} \tag{6.28}$$

then when $N >> 1$ this autocorrelation function can be expressed as

$$R_{xx}(\tau) = N \sum_{i=-\infty}^{\infty} q(\tau - iNT_c) \tag{6.29}$$

Property 2: As illustrated in Table 6.3, an *m*-sequence generates all possible m bit binary numbers except for the all zero tuple. Furthermore there is one run of 1s of length m, one run of 0s of length $m - 1$, two runs of 1s of length $m - 2$, two runs of 0 of length $m - 2$ and, in general, n runs of 1s of length $m - n$, and n runs of 0 of length $m - n$, where $n = 1$ to m.

Property 3: Combining two phase offset versions of the sequence yields another phase offset version of the same sequence. Combining in this case refers to the XOR function or modulo-2 addition.

Property 4: An *m*-sequence contains one more one than zero. The number of 1s will equal 2^{m-1}, while the number of 0s will equal $2^{m-1} - 1$.

Property 5: In any two instantiations of an m-sequence (same sequence but shifted in offset), the number of 1s is about equal to the 0s in both sequences.

Property 6: An LFSR of size m produces a sequence with period N.

6.7 Product Codes

Product codes are codes that are generated by combining two (or more) m-sequences with an XOR as shown in Figure 6.16. The constituent m-sequences are called *factors* since they, in a sense, are multiplied together to generate the product code and they are factors just as 2 and 3 are factors of 6. Product codes generated this way are not necessarily maximal codes. One family of important product codes are the Gold codes.

6.7.1 Gold Codes

Gold colds are product codes made from two maximal sequences of the same length. Certain m-sequences, when offset and linearly combined, generate another m-sequence. Such m-sequences are called *preferred pairs*. Preferred pair Gold codes have three-valued cross correlations as shown in Table 6.7 [8], and these three values occur at the frequency indicated in the fourth column.

Fold codes have some very desirable correlation properties. Gold codes can be generated by initially loading one of the LFSRs in Figure 6.16 with any value other than 0, and seeding the other shift register with any of the numbers between 0 and $2^m - 1$. The resultant sequences are the Gold code sequences. Another code is generated by seeding the first LFSR with all 0s, and the second LFSR will generate the initial m-sequence. Thus, there are $2^m + 1$ Gold codes available

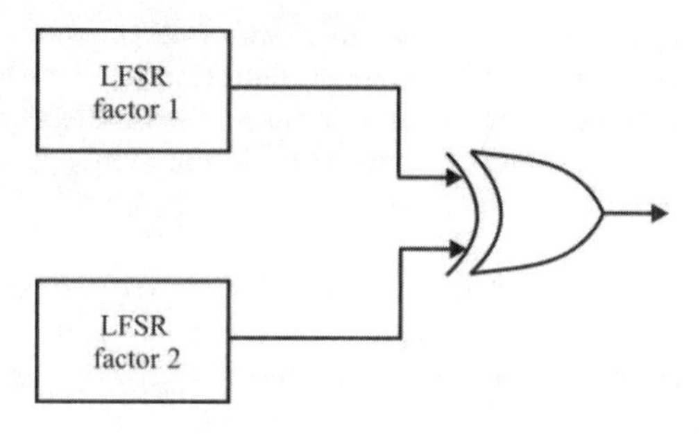

Figure 6.16 Product codes are generated by combining two m-sequences with an XOR.

Table 6.7 Cross Correlation Function of Gold Codes

LFSR Length, m	Code Length	Cross Correlation	Frequency of Occurrence
m odd	$N=2^m-1$	-1	~50%
		$-(2^{(m+1)/2}+1)$	~25%
		$+(2^{(m+1)/2}-1)$	~25%
m even and not divisible by 4	$N=2^m-1$	-1	~75%
		$-(2^{(m+2)/2}+1)$	~12.5%
		$+(2^{(m+2)/2}-1)$	~12.5%

Source: [8].

associated with a particular LFSR structure. Gold codes of length 1,023 (m = 10) are used in GPS.

6.7.2 Kasami Codes

Kasami codes also exhibit correlation functions with very desirable properties. In fact, their cross-correlations can reach the Welch lower bound in (6.26). If s denotes an m-sequence of degree m, then s^* is derived from s by taking every rth bit of s. Sequence s^* is sometimes denoted by $s[r]$. When $r = 2^{m/2} + 1$, $s[r]$ is periodic with a period of $2^{m/2} - 1$. If $s[r]$ is repeated r times, another sequence s_1 is obtained. When s is added to the $2^{m/2} - 2$ cyclically shifted versions of s_1, and including s and s_1, the Kasami sequences result. The correlation function values for Kasami sequences are the same as for Gold codes given in Table 6.7.

6.8 LFSR Design

Galois field theory is useful for determining the taps for LRS generation. A *tap* is used in the LFSR if its coefficient α_i = 1 and is a 0 otherwise. These α_i are the coefficients of the generator polynomial for that LFSR:

$$g(x) = \alpha_m x^m \oplus \alpha_{m-1} x^{m-1} \oplus \alpha_{m-2} x^{m-2} \oplus \cdots \oplus \alpha_2 x^2 \oplus \alpha_1 x \oplus \alpha_0 \qquad (6.30)$$

Recall that $\alpha_m \equiv 1$ and $\alpha_0 \equiv 1$ for LFSRs. For example, the generating polynomial for the LFSR in Figure 6.17 is

$$g_{4.17}(x) = x^3 \oplus x \oplus 1 \qquad (6.31)$$

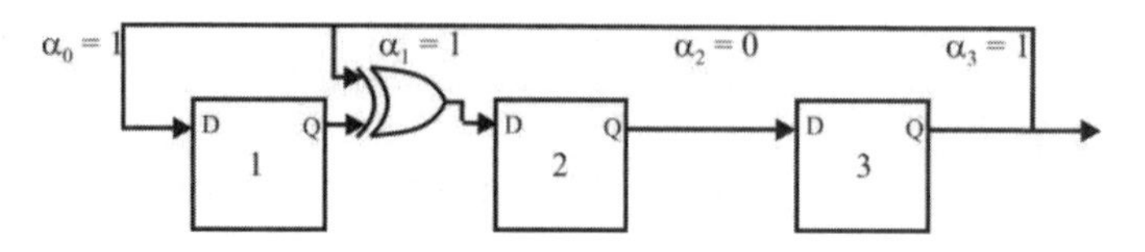

Figure 6.17 LFSR corresponding to $g_{6.14} = x^3 + x + 1$.

In addition to being prime, a primitive polynomial must be a factor of $x^N \oplus 1$, where $N = 2^m - 1$. It is such a factor if it divides $x^N \oplus 1$ with no remainder. The generator polynomial $g_{6.17}(x)$ is sometimes denoted as $[4, 1]_{g_{4.17}}$. The LFSR corresponding to primitive polynomials generate m-sequences.

Generator polynomial $g_{6.17}$ cannot be factored, and therefore it is prime, and it divides $x^7 \oplus 1$ evenly because

$$
\begin{array}{r|l}
 & x^4 \oplus x^2 \oplus x \oplus 1 \\
\hline
x^3 \oplus x \oplus 1 & x^7 \oplus 1 \\
 & x^7 \oplus x^5 \oplus x^4 \\
\hline
 & x^5 \oplus x^4 \oplus 1 \\
 & x^5 \oplus x^3 \oplus x^2 \\
\hline
 & x^4 \oplus x^3 \oplus x^2 \oplus 1 \\
 & x^4 \oplus x^2 \oplus x \oplus 1 \\
\hline
 & x^3 \oplus x \oplus 1 \\
 & x^3 \oplus x \oplus 1 \\
\hline
 & 0
\end{array}
$$

Therefore, the corresponding primitive polynomial will generate a maximal sequence. The sequence for $g_{6.17}$ is given in Table 6.8.

6.8.1 Synthesis of *m*-Sequence LFSRs

It is possible to determine all of the LFSRs of a given size m that will generate m-sequences. From the above, it is known that the corresponding polynomials must divide $x^m + 1$. It is simply necessary to factor $x^m + 1$ to determine all its factors. Those factors with the same order as the specified LFSR size will correspond to m-sequence-generating LFSRs. For example, suppose $m = 3$. Then $N = 2^3 - 1 = 7$ and the factors of $x^7 \oplus 1$ must be found. These are

$$x^7 \oplus 1 = (x \oplus 1)(x^3 \oplus x \oplus 1)(x^3 \oplus x^2 \oplus 1) \tag{6.32}$$

Table 6.8 Sequence for $g_{6.17}$

Clock Pulse		Stage 1	2	3	XOR Output
	Preload	1	0	0	1
1		**0**	**1**	**0**	**0**
2		0	0	1	1
3		1	1	0	1
4		0	1	1	1
5		1	1	1	0
6		1	0	1	0
7		1	0	0	1
8		**0**	**1**	**0**	**0**

and the two factors $x^3 \oplus x \oplus 1$ and $x^3 \oplus x^2 \oplus 1$ satisfy the criteria and therefore corresond to LFSRs that generate *m*-sequences. One of these generators is shown in Figure 6.17. The other is shown in Figure 6.18. Table 6.9 shows the sequence for the LFSR in Figure 6.18.

6.8.2 Other Ways to Generate *m*-Sequences

There are other ways to generate LRSs than with LFSRs. The two that are mentioned here are memory-based and counter-based. Both of these methods of generating *m*-sequences can be used to generate any sequence, since they are totally general.

6.8.2.1 Memory-Based Generator

With the desired sequence stored in a memory, using a simple address generator to access the applicable memory location would produce the desired sequence. Such an architecture is shown in Figure 6.19. With the development of large memories on a single integrated circuit, such a configuration is very attractive from a cost and size point of view. The address generator could be nothing more than the appropriately sized counter.

The memory could be *read-only memory* (ROM) or read-write memory, also called *random access memory* (RAM). The former can be packed more densely but is limited in flexibility. Once the ROM is programmed, the sequence cannot be changed until the memory is replaced. The latter type of memory provides the flexibility to change the sequence generated but requires more area on an integrated circuit chip.

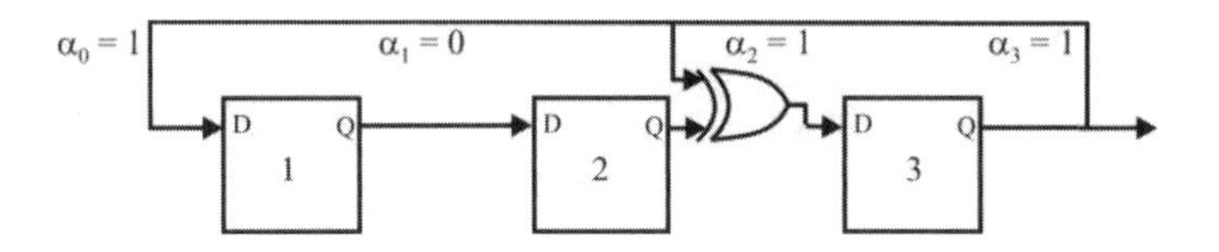

Figure 6.18 LFSR corresponding to $x^3 + x^2 + 1$.

Table 6.9 Maximal Sequence Corresponding to $x^3 \oplus x^2 \oplus 1$

		Stage			
Clock Pulse		**1**	**2**	**3**	**XOR Output**
	Preload	1	0	0	0
1		**0**	**1**	**0**	**1**
2		0	0	1	1
3		1	0	1	1
4		1	1	1	0
5		1	1	0	1
6		0	1	1	0
7		1	0	0	0
8		**0**	**1**	**0**	**1**

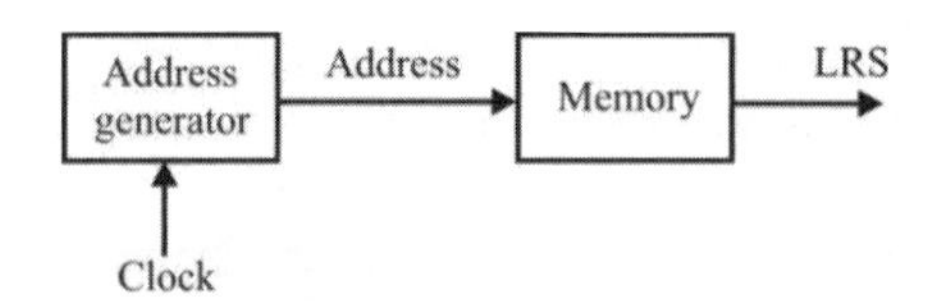

Figure 6.19 Memory-based LRS generator.

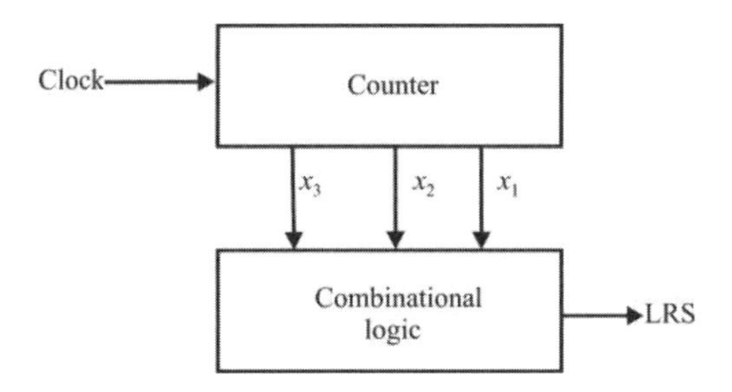

Figure 6.20 Counter-based LRS generator.

6.8.2.2 Counter-Based Generator

A simple binary counter, coupled with Boolean logic elements (AND, OR, NOT) can generate an arbitrary sequence. This configuration is shown in Figure 6.20 for a simple three-stage counter that implements an LRS. The logic for even this simple counter sequence generator is substantial and becomes quite complicated for large m.

6.9 Applications

LFSRs are used in DSSS, FHSS, and THSS AJ communication systems. While they, in general, are multiple-use finite state machines [9], their typical use in these systems is for generation of the sequence used to code the data sequence. Two typical examples of how this can be done are described in this section.

6.9.1 FHSS

The parallel output configuration could be used, for example, to select the frequencies in FHSS systems. The frequencies would appear to be pseudorandomly selected. All numbers are in such a sequence eventually, however. There are about 2,400 channels in the lower VHF frequency band of 30–90 MHz. To be able to address all of these, 12 bits are necessary because 2^{11} = 2,048 and 2^{12} = 4,096. Frequency-hopping systems typically do not use every channel all the time but rather a selected subset called the hop set. The arrangement shown in Figure 6.21 could be used with such systems, where the channel mapping component is where the hop set is defined. In this architecture, many of the random m-sequence symbols would be mapped to the same RF channel.

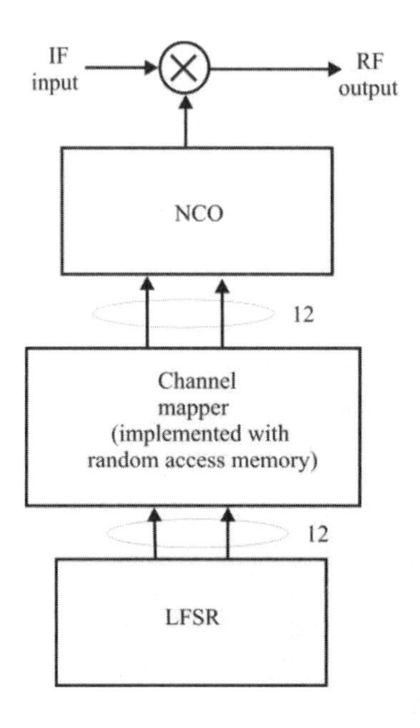

Figure 6.21 Generation of the frequency hopping hop set with an LFSR.

6.9.2 DSSS

A DSSS data modulator can be configured with an LFSR using the serial output of the last stage in the register as shown in Figure 6.22. The XOR effectively multiplies the data sequence by the much higher bit rate LFSR sequence, thereby generating the DSSS chipped signal. The output of this stage is then sent to the RF modulator to move the signal to the appropriate frequency for transmission.

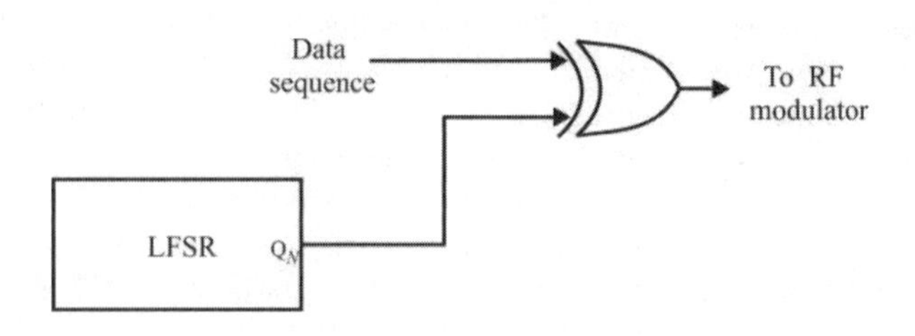

Figure 6.22 DSSS modulator.

6.10 Nonlinear Feedback Shift Registers

While LFSRs with maximal sequences can be, and have been, used in generating spreading codes, they are susceptible to relative ease in determining the taps, thus making them vulnerable to rather casual exploitation. To address this problem, *nonlinear feedback shift registers*[1] (NLFSRs) were developed whose sequences are considerably more difficult to determine. We will discuss them in this section.

It is important to note that avoiding exploitation is not always the reason for spreading a signal. A good example of this is PCS and cellular phone systems that employ CDMA modulation. In that case, CDMA is used to facilitate spectrum sharing rather than avoiding exploitation. Secrecy is accomplished in other ways in these systems. The codes used to spread the signals for channel access are not only linear, but they are published in the open literature.

It should be noted that a general procedure for implementing NLFSRs is, at the time of this writing, an ongoing research problem. Only solutions for special cases have been devised [10–16].

As was established earlier in this chapter, sequences of length $2^n - 1$ can always be obtained from an n-stage shift register by means of a feedback logic consisting entirely of modulo-2 additions. The number of modulo-2 unique configurations yielding the maximum length of $2^n - 1$ is given by $\phi(2^n - 1)/n$, where ϕ is Euler's function [17]. To a first approximation, the number of linear configurations yielding maximum length is $2^n / n$. Sometimes a linear configuration for maximum length involves only two taps.

The removal of the restriction that the feedback configuration be linear increases the number of maximum length shift register codes of degree n from less than $2^n/n$ to $2^{2^{n-1}} / 2^n$. This enormous increase in the number of maximal length codes makes determining the particular code in use considerably more difficult to ascertain, thereby making exploitation much more difficult. In spread spectrum applications this is a very desirable quality, especially in military settings.

While there are several general methods available to implement NLFSRs we will briefly discuss three of the most common methods:

- Nonlinear combination of the outputs of two or more LFSRs, which we will assign the appellation nonlinear combination generators;

- Nonlinear combination of several bits from a single LFSR, which here we will call a nonlinear filter;

---

[1] Linear logic is when the logic is comprised of only modulo-2 adders, also known as exclusive-OR networks. Nonlinear logic adds the other logic components (AND, OR, NAND, NOR) to the mix. Any Boolean function is a candidate to use for the feedback function.

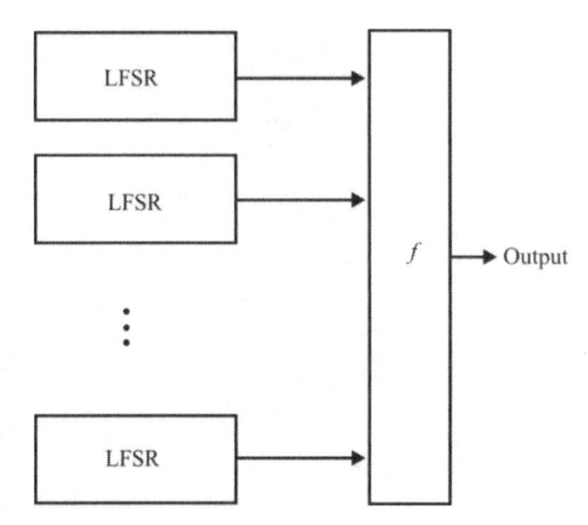

Figure 6.23 Combining generator.

- Variable clock.

6.10.1 Nonlinear Combination Generators

A nonlinear combination generator uses several maximum-length LFSRs. The sequence is generated as a nonlinear Boolean function f of the outputs of these LFSRs. The function f is called the *combining function*. If n maximum-length LFSRs with lengths l_1, l_2, ..., l_n (in general $l_i \neq l_j$) are used together with the Boolean function f, the linear complexity of the sequence is

$$f(l_1, l_2, \ldots, l_n) = a_0 + a_1 l_1 + \ldots a_n l_n + a_{12\ldots k} l_1 l_2 \cdots l_k + \ldots + a_{12\ldots n} l_1 l_2 \cdots l_n, \quad k < n \tag{6.33}$$

where a_0, a_1, ... are the coefficients $(a_i \in \{0,1\})$ in the algebraic normal form of f. Thus, it is desirable to use a combining function with a high nonlinear order. Several feedback shift registers operate in parallel and their output is combined using a suitable function f (see Figure 6.23).

6.10.2 Nonlinear Filter Generators

Another way to implement NLFSRs is to use a nonlinear filtering function f with one output and n inputs, which uses a single maximum-length LFSR (see Figure

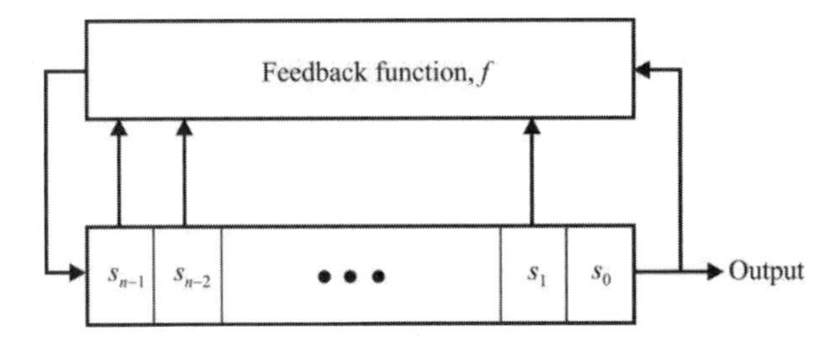

Figure 6.24 Nonlinear filter. The notation is opposite what is normally used with LFSRs (n – 1)st stage as the output, (zeroth stage as the input) by tradition.

6.24), where n is the number of stages in the LFSR. This configuration is also known as the *Fibonacci NLFSR* [18]. When less than all of the n (say, k) stages are used as inputs for the feedback function it is called an (n, k) NLFSR [19].

A nonlinear filter generator uses a single maximum-length LFSR, and the sequence is generated as a nonlinear function f of the state of the LFSR. The function f is called the filtering (or feedback) function. If the LFSR has length n and f has a nonlinear order m, the linear complexity is at most

$$L_m = \sum_{i=1.}^{m} \binom{n}{i} \tag{6.34}$$

Note the two fundamental differences between the nonlinear combination generator and the nonlinear filter generator:

- In the former the SRs are LFSRs and the nonlinear output function is determined by the output stage of these registers.
- In the latter the feedback function of a single SR is where the nonlinear function is placed.

6.10.3 Variable Clock

There is another way to remove the linearity property from LFSRs. That is to vary the clock rate. This can be accomplished by having one LFSR control the clock rate of one or more other LFSR. This is illustrated in Figure 6.25 [17]. In this case, $LFSR_0$ controls whether $LFSR_1$ or $LFSR_2$ is clocked. The outputs of $LFSR_1$ and $LFSR_2$ are XORed and the resulting output is thus randomized. When the receiver contains this same module, is synchronized, and is loaded with the same initial seed, the receiver will recover the sequence and will despread the signal.

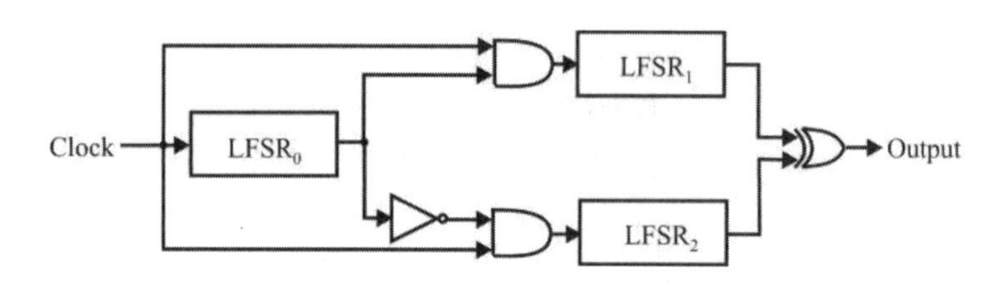

Figure 6.25 Variable clock. (Source: [16]. © 2008 World Academy of Sciences. Reprinted with permission.)

6.11 Concluding Remarks

This chapter presented a very brief introduction to FSRs and their properties. Also presented were some of the useful characteristics of the FSR that such devices generate. These characteristics are exploited in spread spectrum communication systems, primarily to generate the codes used to spread the data.

It should be noted that linear shift registers are fairly easy to exploit, given a sequence of the code. Nonlinear shift registers are much more difficult and therefore are more prolific. Nonlinear registers were only briefly discussed in this chapter, as it is intended to be an introduction and/or review of shift registers. For those interested in pursuing the nonlinear shift register theory further, [10] is an excellent reference.

The characteristics of FSRs are based on Galois field mathematical theory, and in particular GF(2^m), the extension field of the prime number 2, consisting of m-tuples of binary digits.

References

[1] Gallager, R. G., *Information Theory and Reliable Communication,* New York: John Wiley & Sons, 1968, p. 216.

[2] Simon, M. K., J. K. Omura, R. A. Scholtz, and B. K. Levitt, *Spread Spectrum Communications Handbook,* New York: McGraw-Hill, 1994, p. 276.

[3] Dinan, E. H., and B. Jabbari, "Spreading Codes for Direct Sequence CDMA and Wideband CDMA Cellular Networks," *IEEE Communications Magazine,* September 1998, pp. 48–56.

[4] Simon, M. K., J. K. Omura, R. A. Scholtz, and B. K. Levitt, *Spread Spectrum Communications Handbook,* New York: McGraw-Hill, 1994, p. 100.

[5] Peterson, W. W., and E. J. Weldon, *Error Correcting Codes,* Cambridge, MA: MIT Press, 1972.

[6] Simon, M. K., J. K. Omura, R. A. Scholtz, and B. K. Levitt, *Spread Spectrum Communications Handbook,* New York: McGraw-Hill, 1994, p. 286.

[7] Welch, L. R., "Lower Bounds on the Maximum Cross-Correlation of Signals," *IEEE Transactions on Information Theory,* Vol. IT-20, May 1974, pp. 397–399.

[8] Holmes, J. K., *Coherent Spread Spectrum Systems,* New York: John W. Wiley & Sons, 1982, p. 553.
[9] Golomb, S. W., *Shift Register Sequences,* Laguna Hills, CA: Aegean Park Press, 1982, Ch. 2.
[10] Golomb, S., *Shift Register Sequences*, Aegean Press, 1982.
[11] Mykkeltveit, J., M.-K. Siu, and P. Tong, "On the Cycle Structure of Some Nonlinear Shift Register Sequences," *Information and Control*, Vol. 43, No. 2, 1979, pp. 202–215.
[12] Robshaw, M. J. B., "On Binary Sequences with Certain Properties," Ph.D. Dissertation, University of London, 1992.
[13] Linardatos, D., and N. Kalouptsidis, "Synthesis of Minimal Cost Nonlinear Feedback Shift Registers," *Signal Processing*, Vol. 82, No. 2, 2002, pp. 157–176.
[14] Ahmad, A., M. J. Al-Mushrafi, and S. Al-Busaidi, "Design and Study of A Strong Crypto-System Model for E-Commerce," *ICCC 2002*, *Proceedings of the 15th International Conference on Computer Communications*, Washington D.C., International Council for Computer Communications, 2002, pp. 619–630.
[15] Janicka-Lipska, J. S. I., "Boolean Feedback Functions for Full-Length Nonlinear Shift Registers," *Telecommunications and Information Technology*, Vol. 5, 2004, pp. 28–29.
[16] Win, M. S. M., "A New Approach to Feedback Shift Registers," *World Academy of Science, Engineering, and Technology*, Vol. 48, 2008, pp. 185–189.
[17] Golomb, S. W., *Shift Register Sequences,* Revised Edition, Laguna Hills, CA: Aegean Park Press, 1982, p. 119.
[18] Dubrova, E., M. Teslenko, and H. Tenhunnen, "On Analysis and Synthesis of (n, k)-Non-Linear Feedback Shift Registers," *EDDA* 2008.

Chapter 7

Synchronization and Tracking in Spread Spectrum Systems

7.1 Introduction

All forms of SS communication systems considered here have the requirement to keep the transmitter and receiver synchronized. Synchronization of frequency, chip time, and code are necessary. These variables are generally initially unknown. Furthermore, Doppler shifts in frequency due to motion of the transmitter, receiver, or both frequently occur. Other frequency offsets are due to drifting of the transmitter and receiver oscillators. The initial acquisition of the correct chip time and phase offset is referred to as *coarse acquisition*. The subsequent tracking, once coarse acquisition has been achieved, is sometimes referred to as *fine acquisition*. Both terms will be used herein.

Coarse code acquisition consists of searching the time/frequency space illustrated in Figure 7.1. Each combination of phase offsets and frequency is called a *cell*. In general, both the time and frequency dimensions must be searched for coarse acquisition. In many practical cases, however, one of these can be readily estimated, and the searching is only needed over the other dimension. In DSSS, for example, coarse acquisition attempts to adjust the phase offset of the locally generated pseudo-random sequence to within a large fraction of one chip time, T_c. In DSSS the carrier frequency is generally known in advance. Tracking in DSSS attempts to keep the phase offset within a small percentage of T_c so that correlation can be maintained. For FHSS, coarse acquisition consists of finding where in the hop sequence the system is at, although maintaining the correct timing is also important.

In general, the methods to acquire synchronization and facilitate tracking are different, depending on whether the communication system is employing DSSS or FHSS. In this chapter, DSSS acquisition and tracking will be discussed first, followed by that for FHSS.

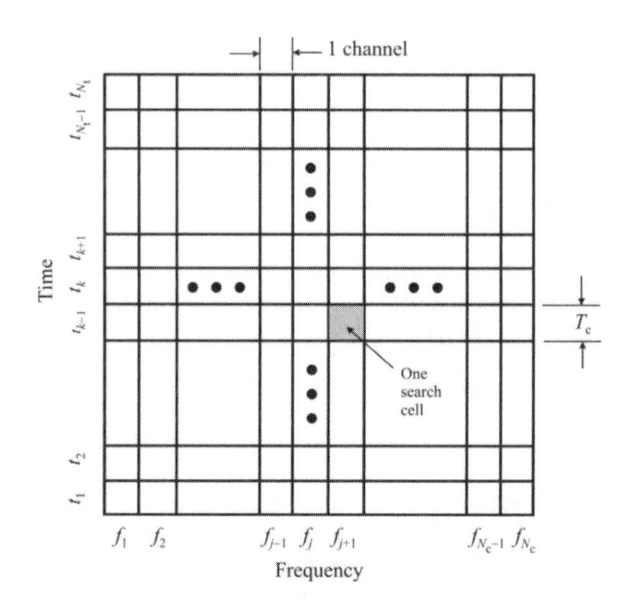

Figure 7.1 Time/frequency search space associated with coarse sync acquisition.

7.2 DSSS Synchronization

DSSS systems use coding to generate the spreading sequence. If the transmitted code and the locally generated code at the receiver are not synchronized, then the data bits will not be decoded properly. The sequences must be well within a bit duration to function [1, 2].

We consider in this section two types of synchronizations and one type of tracking, since the latter has few variations. For synchronization, we first must accomplish blind chip timing synchronization. This lines up the chip transition times among all the incoming CDMA signals so asynchronous DSSS systems become synchronous. Second is synchronization of the codes of the received signal and the locally generated signal

Consider the simple case illustrated in Figure 7.2. We have represented the code in the received signal as a single 1 and all the rest as zeros. The known local code is shown at the bottom and it exhibits both timing offset and code offset. It also has a single 1 and all the rest are zero. Note that in the cooperating (communicating) case, the code is known. In the noncooperating (EW) case, the code may not be known, so a search for the code may also be required. Searching for these offsets and potentially codes is frequently done at least partially in parallel.

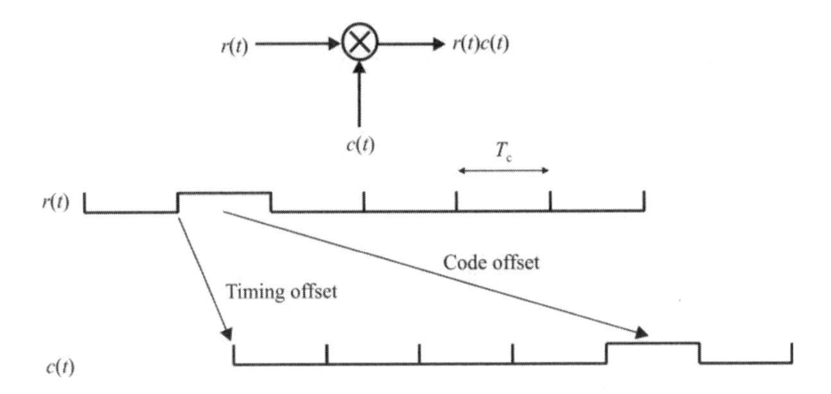

Figure 7.2 There are two types of offsets associated with DSSS synchronization—both are shown here. The first involves a timing offset, while the second is due to a code offset.

7.2.1 DSSS Code Synchronization

Many coarse acquisition methods are based on the block diagram shown in Figure 7.3. The received signal is multiplied by the locally generated pseudo-random sequence at some phase offset, which is estimated. If the offset corresponds to the same offset in the received signal, then the signal at the output of the mixer will be the narrowband data signal generated at the transmitter. Significant energy will emerge from the bandpass filter and be subsequently detected by the energy detector. The energy will exceed the properly selected threshold and coarse acquisition will have been achieved. The receiver will then engage the fine acquisition/tracking mode.

If the locally generated phase offset does not coincide with the offset of the received signal, the output of the mixer will still be a wideband signal and little energy will pass through the bandpass filter. The energy detected will then be low and not exceed the threshold. In that case, the phase will be adjusted forward or backward by some mechanism. This is usually accomplished by skipping one chip period of the clock driving the generator. This process will proceed until the correct phase offset is found. It is generally not necessary to search the entire phase offset space because it is typically known or can be estimated what the maximum offset is due to other considerations, for example, the length of time the local oscillator could have drifted or the deployment geometry.

There are several forms of energy detectors that can be used in Figure 7.2. Matched filters and radiometers are two of them. The matched filter will be analyzed in detail here [3]. The radiometer was examined in Chapter 2.

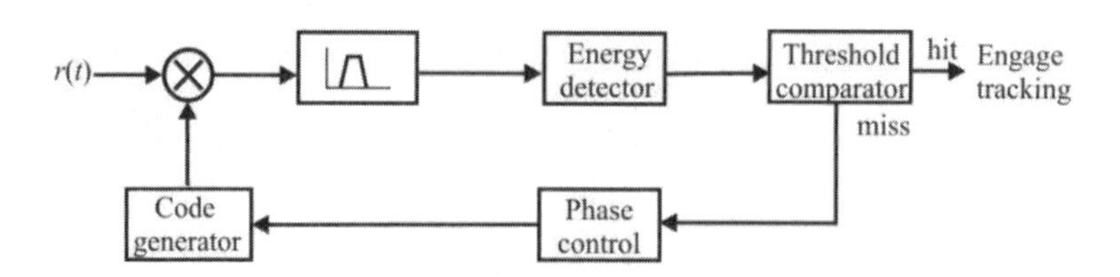

Figure 7.3 Coarse acquisition flow diagram.

7.2.1.1 Matched Filter

A matched filter combines the implementation of the bandpass filter with the energy detector in Figure 7.3. Matched filters can be implemented either in analog or digital form. An extensive and thorough treatment of digital matched filter performance is provided by Turin [4]. For convenience it is assumed that the DSSS signal is BPSK. Extension to higher forms of PSK is straightforward. Suppose the signal at the transmitter is given by

$$s_{\mathrm{T}}(t) = \sqrt{2S_{\mathrm{T}}}x(t) \tag{7.1}$$

where S_{T} is the average power in the transmitted signal and $x(t)$ is the spread signal given by

$$x(t) = \sum_{k=-\infty}^{\infty} x_k p_{\mathrm{T}}(t - kT_{\mathrm{c}}) \tag{7.2}$$

when T_{c} is the chip time. The coefficients $x_k \in \{-1, +1\}$ determine the phase of the BPSK signal. Assuming, temporarily, that the transmitted signal is the only signal impinging on the receiver in bandwidth W_{ss} and neglecting noise, then the received signal is given by

$$r(t) = \sqrt{2R}x(t - \delta) \tag{7.3}$$

where δ is the propagation and processing delay and R is the average power in the received signal, substantially less than S_{T}. Neglecting mixing losses or gains, the signal emerging from the mixer (multiplier) in Figure 7.3 is

$$r_1(t) = \sqrt{2R}x(t - \delta)x(t - \hat{\delta}) \tag{7.4}$$

where $\hat{\delta}$ is the estimate of the phase offset. Because this is a coarse acquisition, it is only necessary to determine δ to within the chip time T_c, and therefore $\hat{\delta} = mT_c$ for integer m. This signal, $r_1(t)$, is integrated for $T = kT_c$ seconds and the decision statistic for matched filtering is

$$\begin{aligned} z &= \frac{1}{2T^2}\left|\int_{t-T}^{t} \sqrt{2R}x(t-\delta)x(t-\hat{\delta})dt\right|^2 \\ &= \frac{R}{T^2}\left|\int_{t-T}^{t} x(t-\delta)x(t-\hat{\delta})dt\right|^2 \\ &= \frac{R}{T^2}R_{xx}^2(\tau) \end{aligned} \tag{7.5}$$

where $R_{xx}(\tau)$ is the autocorrelation function of $x(t)$ and $\tau = \delta - \hat{\delta}$ is the relative offset. If $x(t)$ is an m-sequence with period N, then from

$$q(\tau) = \begin{cases} 1 - \dfrac{|\tau|}{T_c}, & |\tau| \le T_c \\ 0, & \text{otherwise} \end{cases} \tag{7.6}$$

then

$$R_{xx}(\tau) = N\sum_{i=-\infty}^{\infty} q(\tau - iNT_c) \tag{7.7}$$

When the decision threshold is set at

$$\gamma_{th} = \left(\frac{N-1}{N}\right)^2 \frac{R}{4} \tag{7.8}$$

a hit is determined by $z > \gamma_{th}$. This value of γ_{th} will ensure that approximately half of the peak autocorrelation value or more is achieved because

$$z > \gamma_{th} \Rightarrow \frac{R}{T^2}R_{xx}^2(\tau) > \left(\frac{N-1}{N}\right)^2 \frac{R}{4} \Rightarrow R_{xx}(\tau) > \frac{T}{2}\left(\frac{N-1}{N}\right) \tag{7.9}$$

When the effects of noise and possibly jamming are included, false and missed locks can occur. Assume that only noise is present and the transmitted signal is the only one impinging on the receiver. In that case the received signal is given by

$$r(t) = \sqrt{2R}x(t-\delta) + n(t) \tag{7.10}$$

where it is assumed that $n(t)$ is AWGN. The decision statistic is then

$$\begin{aligned} z &= \frac{1}{2T^2}\left|\int_{t-T}^{t} r(t)x(t-\hat{\delta})dt\right|^2 \\ &= \frac{1}{2T^2}\left|\int_{t-T}^{t} \sqrt{2R}x(t-\delta)x(t-\hat{\delta})dt + \int_{t-T}^{t} n(t)x(t-\hat{\delta})dt\right|^2 \\ &= \frac{1}{2T^2}\left|\sqrt{2R}R_{xx}(\tau) + \int_{t-T}^{t} n(t)x(t-\hat{\delta})dt\right|^2 \\ &= \left|\frac{\sqrt{R}}{T}R_{xx}(\tau) + \frac{1}{\sqrt{2}T}\int_{t-T}^{t} n(t)x(t-\hat{\delta})dt\right|^2 \end{aligned} \tag{7.11}$$

When the phases do not match $(\delta \neq \hat{\delta})$, $R_{xx}(\tau) \approx 0$, the output of the integrator is totally decorrelated, and any signal present is due to noise only. The hypothesis in this case is then

$$H_0 : z = \left|\frac{1}{\sqrt{2}T}\int_{t-T}^{t} n(t)x(t-\hat{\delta})dt\right|^2 \tag{7.12}$$

When the phases match, assume they match perfectly and $R_{xx}(\tau) = T$. In this case

$$H_1 : z = \left|\sqrt{R} + \frac{1}{\sqrt{2}T}\int_{t-T}^{t} n(t)x(t-\hat{\delta})dt\right|^2 \tag{7.13}$$

A *false alarm* occurs when a match is declared when, in fact, $\delta \neq \hat{\delta}$. The probability of false alarm, P_{fa}, is given by

$$P_{\text{fa}} = \Pr(z > \gamma_{\text{th}} | H_0) = 1 - F_Z(z | H_0) \tag{7.14}$$

where $F_Z(z| H_0)$ is the cumulative probability distribution function for z under H_0. A *miss* occurs when, in fact, $\delta = \hat{\delta}$ but the decision is made that there is no match. Such situations are caused by excessive noise in the detection process. Denoting the probability of a miss by P_{m}, then

$$P_{\text{m}} = \Pr(z \le \gamma_{\text{th}} | H_1) = F_Z(z | H_1) \tag{7.15}$$

where $F_Z(z| H_1)$ is the cumulative probability distribution function of z under H_1.

It can be shown that [5]

$$P_{\text{fa}} = e^{-\frac{\gamma_{\text{th}} T}{N_0}} \tag{7.16}$$

where $N_0/2$ is the two-sided noise PSD. In addition

$$P_{\text{m}} = \int_0^{\sqrt{\frac{2\gamma_{\text{th}} T}{N_0}}} u \exp\left\{ -\frac{1}{2}\left[u^2 + 2\left(\frac{RT}{N_0} \right)^2 \right] \right\} I_0\left(\sqrt{2}\frac{RT}{N_0} u \right) du \tag{7.17}$$

In this expression, I_0 is the modified Bessel function of the first kind and zeroth order (see Appendix A). The probability of detection is given by

$$\begin{aligned} P_{\text{d}} &= 1 - P_{\text{m}} \\ &= Q\left(\sqrt{2}\frac{RT}{N_0}, \sqrt{-2\ln P_{\text{fa}}} \right) \end{aligned} \tag{7.18}$$

where $Q(a, b)$ is Marcum's Q-function (Appendix A).

A *receiver operating characteristic* (ROC) is a graphical representation of P_{d} plotted versus P_{fa}. Any stochastic signal detector has an associated ROC curve, and, in general, they are different depending on the type of detector. Of course, the matched filter is the optimum detector structure when the noise in question is Gaussian. An example of such a curve is shown in Figure 7.4 for the matched filter for a few typical valuse of the time-bandwidth product TW_{ss} when the signal-to-noise ratio, $\upsilon = -20$ dB.

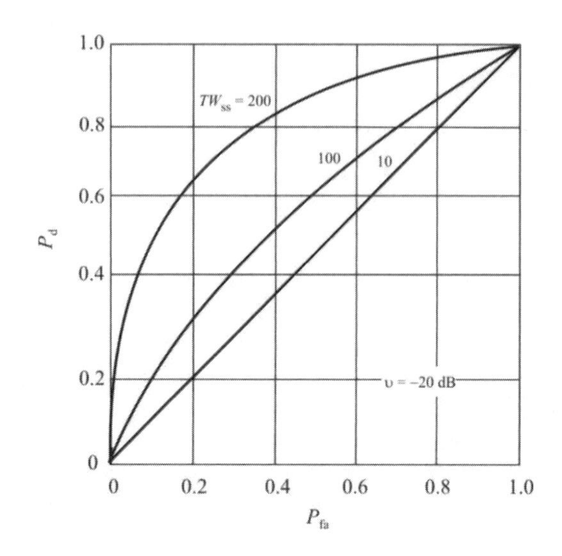

Figure 7.4 Matched filter ROC when υ = –20 dB.

7.2.1.2 Code Acquisition

Irrespective of the code used, the code space must be searched in some fashion to find the correct phase offset. There are several ways to accomplish this. The simplest is a serial approach where one phase offset at a time is attempted and the comparison with the threshold is made. If the sequence length, N, is large, however, this approach can be very slow if bounds on the search space are not available.

On the other hand, a fully parallel implementation is possible that would be the fastest way. In this architecture, N parallel matched filters would simultaneously search the code space, one offset for each matched filter. The filter with the largest output would correspond to the correct phase offset. If N is large, implementation would be prohibitive, however.

Between these two approaches, fully serial and fully parallel, there are compromises that can be made. Instead of a fully parallel implementation, for example, some smaller number of parallel matched filters could be included and they would be time shared.

One technique that is reasonably efficient from both an acquistion time point of view and hardware complexity is the multidwell approach, shown in Figure 7.5 [6]. The first energy detector implements a relatively low threshold with a short integration time. Its purpose is to quickly eliminate offsets that are not acceptable. This stage would have a relatively high false alarm rate but a corresponding high probability of detection. The parameters for the second energy detector would be

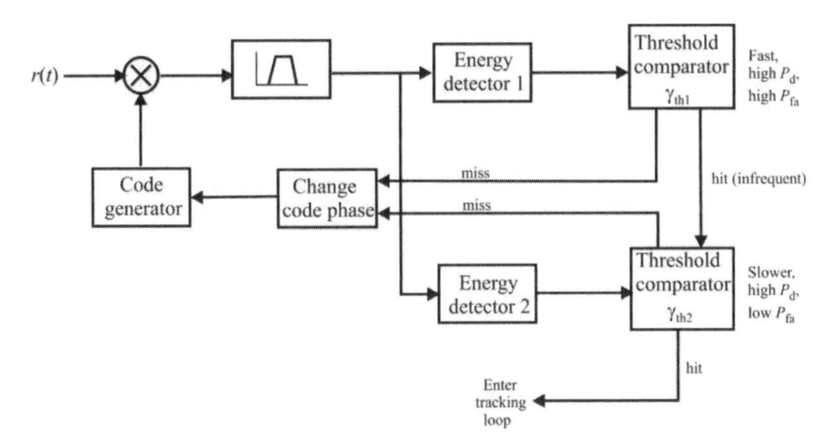

Figure 7.5 Multidwell detection flow diagram.

selected to implement a low false alarm rate and a low probability of miss (higher probability of detection), and therefore a longer acquisition time. The goal is to have the first stage hand off to the second infrequently so that the overall acquisition time is minimized.

Acquisition Time

A figure of merit for any DSSS signal acquisition approach is the average time it takes to achieve synchronization. This time is a random variable because of the underlying randomness in the pseudo-random codes. A transmission can occur at any time. Synchronization, however, is a function of the receiver. The receiver must find where the code is in the PN sequence to synchronize with it. Because of this random nature of the acquisition time, it can only be described statistically with moments and distribution functions.

The least desirable situation is *false synchronization*, also called false alarm or false lock. When this occurs, the receiver thinks it is synchronized when it is not, so it incorrectly decodes the incoming signal, generating bad or no data until synchronization is again attempted.

Serial searching is the most common form of DSSS search method. The average acquisition time for serial search is given by [7]

$$\mathcal{E}\{T_\text{s}\} = (C-1)T_\text{da}\left(\frac{2-P_\text{d}}{2P_\text{d}}\right) + \frac{T_\text{i}}{P_\text{d}} \tag{7.19}$$

Table 7.1 Barker Code Words

N	**Code Word**
2	10 or 11
3	110
4	1011 or 1001
5	11101
7	1110010
11	11100010010
13	1111100110101

where

$T_{da} = T_i + T_{fa} P_{fa}$;
T_i = evaluation time for each cell;
T_{fa} = time required to reject an incorrect cell;
$C = \Delta T / \Delta t$ (assumed to be an integer);
Δt = phase step size;
ΔT = time within which the correct phase occurs.

In the case when parallel search is executed, these variables can be adjusted to coincide with each of the parallel search paths in a straightforward manner.

The standard deviation of the acquisition time is given by

$$\sigma_{T_s}^2 = \left[\frac{C^2 - 1}{12} - \frac{(C-1)^2}{P_d} + \frac{(C-1)^2}{P_d^2} \right] T_{da}^2 \\ +(2C-1)\frac{1-P_d}{P_d^2} T_i^2 + 2(C-1)\frac{1-P_d}{P_d^2} T_i T_{fa} P_{fa} \\ -(C-1)\frac{2-P_d}{2P_d} T_{fa}^2 P_{fa}^2 + (C-1)\frac{2-P_d}{2P_d} T_{fa}^2 P_{fa} \tag{7.20}$$

Barker Codes

Barker codes are useful in spread spectrum systems for synchronization. They form a set of PN codes that is very small with short periods, the longest sequence having 13 bits. The code words are shown in Table 7.1.

The autocorrelation of the Barker 7 sequence is illustrated in Figure 7.6. It is typical of the Barker correlation functions in that the value of the correlation function is one of –1, 0, 1, or *N*, with *N* only occuring for 0 offset.

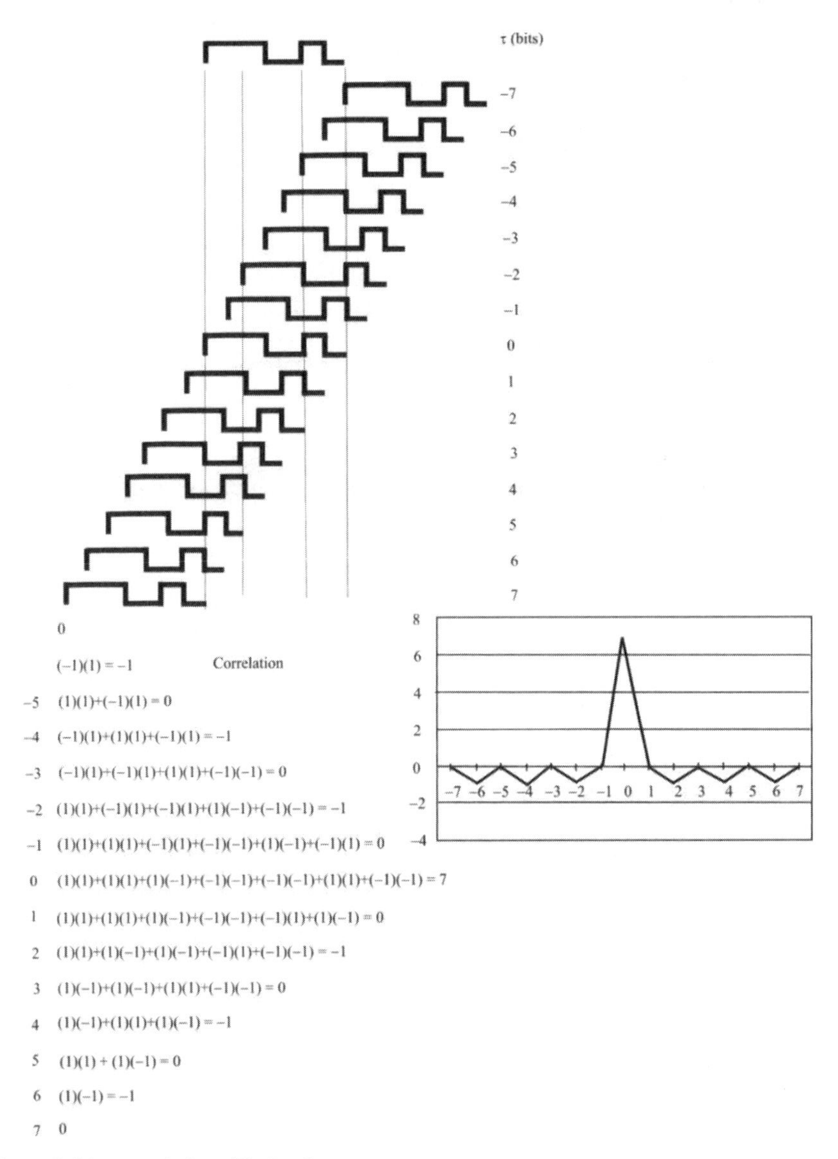

Figure 7.6 Autocorrelation of Barker 7.

These functions are normally used in a "one-shot" fashion in that they are sent through the correlator only once. They are used at system initialization where the Barker correlator looks for the sequence before any data is added to the chip sequence. In that way, accidentally sending the sequence during data transmission can be avoided because the chip sequence can be controlled.

7.2.2 Chip Timing Synchronization

Many of the algorithms discussed in the sequel assume that the CDMA communication system is operating in a synchronous mode. This is synonymous with saying that the relative timing between all the terminals are all known.

To address this issue, a method was devised by Strom et al. to essentially synchronize CDMA systems [8]. Albeit this method was for the purpose of communication system synchronization, it also applies to synchronization at an EW system attempting to intercept CDMA signals. We discuss the technique in this section.

The standard method of code acquisition (initial propagation delay estimation, typically before data transmission) is the sliding correlator and modifications thereof [9]. Just like the standard detector, these methods are single-user algorithms and work reasonably well in a multiuser environment if the received powers are similar, but fail in a near-far environment [10, 11]. Another approach to code acquisition is to transmit a known bit sequence during acquisition, but these cannot be used for tracking (propagation delay estimation during data transmission).

Strom et al. dealt with the full problem, that is, simultaneous estimation of the propagation delay, phase, and amplitude for all users in a DS-CDMA system. Most of the algorithms proposed, however, dealt with a propagation delay estimation. Obtaining accurate propagation delay estimates is a difficult problem and there exist a multitude of algorithms for (suboptimally) estimating the remaining parameters given reliable estimates of the propagation delays. Furthermore, no knowledge of the data sequences is required by the proposed algorithms. This means that the algorithms are applicable to both the acquisition and tracking problems.

7.2.2.1 Target Signal Model

The system under consideration is modeled as an asynchronous K-user DS-CDMA system operating over an additive white Gaussian noise (AWGN) channel (as mentioned in Chapter 2, in the frequency range of the UMTS systems, this is a reasonable assumption since the fractional bandwidth is low). The modulation scheme is BPSK with bit duration T and chip duration $T_c = T/N$, where N is an integer. The code waveforms are assumed to be periodic with period T. As a general rule, a subscript l implies that the subscripted quantity is due to the lth user. For instance, a period of the lth user's code waveform is denoted by $b_l(t)$, where $b_l(t) = 0$ for $t \notin [0,T]$.

The baseband signal, $s_l(t)$, is formed by pulse amplitude modulating the data stream, $d_l(m) \in \{+1,-1\}$, with a period of the code waveform, that is,

$$s_l(t) = \sum_{m=-\infty}^{\infty} d_l(m)c_l(t-mT) \tag{7.21}$$

The signal that is transmitted is formed by multiplying $s_l(t)$ with the carrier $\sqrt{2S_l}\cos(\omega_c t + \theta_l')$, where S_k is the average power and θ_k' is the random carrier phase uniformly distributed in $[0,2\pi)$. We assume, without loss of generality, that T_c has been normalized so that $T_c = 1$.

7.2.2.2 EW Receiving System

The signal received by the EW system may be written as

$$r(t) = \text{Re}\left\{\sum_{l=1}^{L} s_l(t-\tau_l)\sqrt{2R_l}\exp[j(\omega_c t + \theta_l)]\right\} + n(t) \tag{7.22}$$

where $\tau_l \in [0,T)$ is the unknown propagation delay and $\theta_l = \theta_l' - \omega_c \tau_l$. The noise waveform, $n(t)$, is a white Gaussian noise waveform with two-sided power spectral density $N_0 / 2$.

Following appropriate filtering, amplification, and probably frequency down-conversion, the receiver consists of a standard IQ-mixing stage followed by an integrate-and-dump section as shown in Figure 7.7. The integration time, T_i, is defined as $T_\text{i} = T_\text{c} / Q_\text{c}$ where Q_c is an integer and referred to as the *oversampling factor*. Nyquist theory dictates that Q_c must be at least 2 to avoid aliasing the signals.

Ignoring double-frequency terms, the equivalent complex received sequence, $r(k) = r^\text{I}(k) + jr^\text{Q}(k)$, can be expressed as

$$r(k) = n(k) + \sum_{l=1}^{L} R_l \exp(j\theta_l)\frac{1}{T_i}\int_{(k-1)T_\text{i}}^{kT_\text{i}} s_l(t-\tau_l)dt \tag{7.23}$$

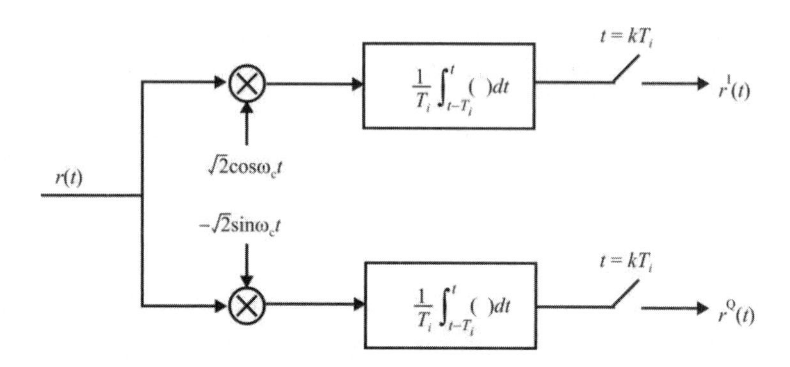

Figure 7.7 Receiver detector. (Source: [8]. © 1996 IEEE. Reprinted with permission.)

where $n(l)$ is a zero-mean white complex Gaussian sequence with variance $\sigma^2 = \mathcal{E}\{|n(l)|^2\} = N_0 / T_\mathrm{i} = N_0 Q_\mathrm{c} N / E_{\mathrm{b},1}$ and where $E_{\mathrm{b},1} = Q_\mathrm{c} N T_\mathrm{i}$ is the energy per bit for the first user. We assume without loss of generality that $R_1 = 1$.

Let the received vector during the mth bit interval, $\vec{r}(m) \in \mathbb{C}^{QN}$, and the noise vector, $\vec{n}(m) \in \mathbb{C}^{QN}$, be defined as

$$\vec{r}(m) = \left[r(mQ_\mathrm{c}N + Q_\mathrm{c}N) \quad \cdots \quad r(mQ_\mathrm{c}N + 1)\right]^\mathrm{T} \tag{7.24}$$

$$\vec{n}(m) = \left[n(mQ_\mathrm{c}N + Q_\mathrm{c}N) \quad \cdots \quad n(mQ_\mathrm{c}N + 1)\right]^\mathrm{T} \tag{7.25}$$

It is easy to show that the noise vector is a zero-mean complex Gaussian random vector with second moments[1]

$$\mathcal{E}\{\vec{n}(p)\vec{n}^*(q)\} = \sigma^2 \mathbf{I}_{Q_\mathrm{c}N} \vec{\delta}_K(p-q), \qquad \mathcal{E}\{\vec{n}(p)\vec{n}^\mathrm{T}(q)\} = \mathbf{0} \tag{7.26}$$

After some straightforward calculations, we can formulate the contribution from the lth user to $\vec{r}(m)$

$$\vec{r}_l(m) = \left[\vec{a}_{2l-1} \quad \vec{a}_{2l}\right] \begin{bmatrix} \beta_l & 0 \\ 0 & \beta_l \end{bmatrix} \begin{bmatrix} z_{2l-1}(m) \\ z_{2l}(m) \end{bmatrix} \tag{7.27}$$

where

---

[1] $\delta_K(p)$ denotes the Kronecker delta function, $\delta_K(p) = 1$ for $p = 0$ and $\delta_K = 0$ otherwise. $\mathbf{I}_r$ is the $r \times r$ identity matrix.

$$z_{2l-1}(m) = 1/2[d_l(m) + d_l(m-1)]$$
$$z_{2l}(m) = 1/2[d_l(m) - d_l(m-1)]$$
$$\beta_l = \sqrt{S_l}\exp(j\theta_l)$$

As seen from this equation, the lth user contributes $d_l(m)\beta_l\vec{a}_{2l-1}$, if $d_l(m) = d_l(m-1)$, or $d_l(m)\beta_l\vec{a}_{2l}$, if $d_l(m) = -d_l(m-1)$. The vectors $\{\vec{a}_{2l-1}, \vec{a}_{2l}\}$ are defined by the lth user's propagation delay and code waveform

$$\vec{a}_{2l-1} = \left[\frac{\delta_l}{T_i}\mathbf{D}(p_l+1,1) + \left(1-\frac{\delta_l}{T_i}\right)\mathbf{D}(p_l,1)\right]\vec{\chi}_l \tag{7.28}$$

$$\vec{a}_{2l} = \left[\frac{\delta_l}{T_i}\mathbf{D}(p_l+1,-1) + \left(1-\frac{\delta_l}{T_i}\right)\mathbf{D}(p_l,-1)\right]\vec{\chi}_l \tag{7.29}$$

where $\tau_l = p_l T_i + \delta_l$, such that p_l is an integer and $\delta_l \in [0, T_i)$, and $\vec{\chi}_l \in \mathbb{R}^{QN}$ is defined as

$$\vec{\chi}_l = \left[\chi_l(Q_c N) \quad \chi_l(Q_c N-1) \quad \cdots \quad \chi_l(1)\right]^{\mathrm{T}} \tag{7.30}$$

$$\chi_l(k) = \frac{1}{T_i}\int_{(k-1)T_i}^{kT_i} c_l(t)dt \tag{7.31}$$

The permutation matrix $\mathbf{D}(r,\alpha) \in \mathbb{R}^{Q_c N \times Q_c N}$ is defined in block form as

$$\mathbf{D}(r,\alpha) = \begin{bmatrix} \mathbf{0} & \mathbf{I}_{Q_c N-r} \\ \alpha\mathbf{I}_r & \mathbf{0} \end{bmatrix} \tag{7.32}$$

We adopt the conventions that $\mathbf{D}(0,\alpha) = \mathbf{I}_{Q_c N}$ and $\mathbf{D}(Q_c N,\alpha) = \alpha\mathbf{I}_{Q_c N}$.

The expression for $\vec{r}(m)$ can be written in a more compact form

$$\vec{r}(m) = \vec{n}(m) + \sum_{l=1}^{L}\vec{r}_l(m) = \mathbf{A}(\vec{\tau})\mathbf{B}(\vec{\gamma},\vec{\theta})\vec{z}(m) + \vec{n}(m) \tag{7.33}$$

where[2]

$$\mathbf{A}(\vec{\tau}) = \begin{bmatrix}\vec{a}_1 & \vec{a}_2 & \cdots & \vec{a}_{2L}\end{bmatrix} \in \mathbb{R}^{Q_c N \times 2l};$$
$$\mathbf{B}(\vec{\gamma},\vec{\theta}) = \text{diag}(\beta_1,\beta_1,\beta_2,\beta_2,\cdots,\beta_L,\beta_L) \in \mathbb{C}^{2L\times 2L};$$
$$\vec{z}(m) = \begin{bmatrix}z_1(m) & z_2(m) & \cdots & z_{2L}(m)\end{bmatrix}^{\mathrm{T}} \in \{-1,0,+1\}^{2L}.$$

We observe that $\mathbf{A}(\vec{\tau})$ and $\mathbf{B}(\vec{\gamma},\vec{\theta})$ are functions of the vectors $\vec{\tau},\vec{\gamma},\vec{\theta} \in \mathbb{R}^L$

$$\vec{\tau} = \begin{bmatrix}\tau_1 & \tau_2 & \cdots & \tau_L\end{bmatrix}^{\mathrm{T}} \tag{7.34}$$
$$\vec{\theta} = \begin{bmatrix}\theta_1 & \theta_2 & \cdots & \theta_L\end{bmatrix}^{\mathrm{T}} \tag{7.35}$$
$$\vec{\gamma} = \begin{bmatrix}\gamma_1 & \gamma_2 & \cdots & \gamma_L\end{bmatrix}^{\mathrm{T}} \tag{7.36}$$

where $\gamma_l = \sqrt{R_l}$. The explicit dependence of **A** and **B** on the parameter vectors are dropped for notational convenience. We will assume that **A** and **B** have full rank. **A** will have full rank if and only if $\{\vec{a}_1,\vec{a}_2,\cdots,\vec{a}_{2L}\}$ are linearly independent for all possible values of $\vec{\tau}$. This is obviously desirable for a DS-CDMA system since users otherwise can cancel each other's transmissions. **B** will be full rank if $S_l > 0$ for all l, which obviously is the case. The final complex discrete-time signal flow diagram is depicted in Figure 7.8, where the process for a single user is detailed. The flow shown is repeated for each possible received signal.

We will consider the parameter vectors $\vec{\tau}$, $\vec{\theta}$, and $\vec{\gamma}$ to be unknown and deterministic. Furthermore, it is assumed that the data streams consist of equally likely, independent bits and that the noise is independent of the data streams. The correlation matrix $\mathbf{R}_{rr} = \mathcal{E}\{\vec{r}(m)\vec{r}^{\mathrm{H}}(m)\}$ for $\vec{r}(m)$ is then

$$\begin{aligned}\mathbf{R}_{rr} &= \mathbf{A}\mathbf{B}\mathcal{E}\{\vec{z}(m)\vec{z}^{\mathrm{H}}(m)\}\mathbf{B}^{\mathrm{H}}\mathbf{A}^{\mathrm{H}} + \sigma^2\mathbf{I}_{Q_c N}\\ &= \mathbf{A}\boldsymbol{\Sigma}\mathbf{A}^{\mathrm{H}} + \sigma^2\mathbf{I}_{Q_c N}\end{aligned} \tag{7.37}$$

where

[2] diag($\alpha_1, \alpha_2, \ldots, \alpha_L$) denotes the $L \times L$ diagonal matrix whose lth diagonal element is α_k, i.e., (diag($\alpha_1, \alpha_2, \ldots, \alpha_L$))($l, l$) = α_l. On the other hand, diag(**A**) is a diagonal matrix of same dimension and diagonal as **A**, that is, (diag(**A**))(l, l) = **A**(l, l).

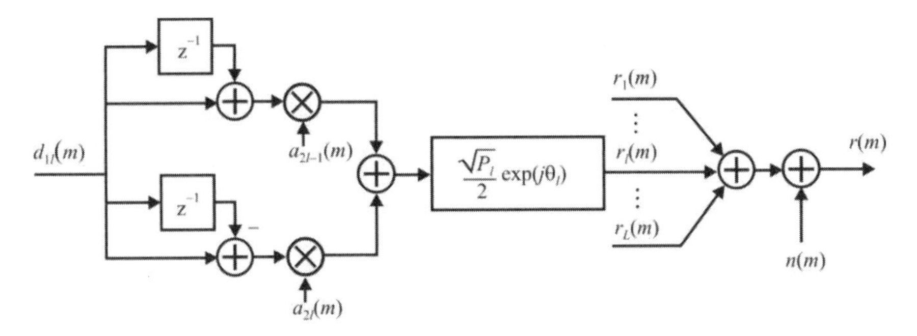

Figure 7.8 Complex discrete-time vector model of the communication system. Signal flow diagram. (Source: [8]. © IEEE 1996. Reprinted with permission.)

$$\begin{aligned}\mathbf{\Sigma} &= \mathbf{B}\mathcal{E}\{\vec{z}(m)\vec{z}^{\mathrm{H}}(m)\}\mathbf{B}^{\mathrm{H}} \\ &= \frac{1}{2}\mathrm{diag}(S_1, S_1, \cdots, S_L, S_L) \in \mathbb{R}^{2L\times 2L}\end{aligned} \tag{7.38}$$

We note that $\mathbf{R}_{\mathrm{rr}}$ is symmetric and is positive definite if $\sigma^2 > 0$.

Since $\mathbf{A\Sigma A}^*$ is real symmetric and has rank $2K$, there is an eigenvalue decomposition of $\mathbf{A\Sigma A}^*$ such that

$$\mathbf{A\Sigma A}^{\mathrm{H}} = \begin{bmatrix}\mathbf{E}_{\mathrm{s}} & \mathbf{E}_{\mathrm{n}}\end{bmatrix}\begin{bmatrix}\tilde{\mathbf{\Lambda}} & \mathbf{0} \\ \mathbf{0} & \mathbf{0}\end{bmatrix}\begin{bmatrix}\mathbf{E}_{\mathrm{s}}^{\mathrm{H}} \\ \mathbf{E}_{\mathrm{n}}^{\mathrm{H}}\end{bmatrix} \tag{7.39}$$

where $\mathbf{E}_{\mathrm{s}} \in \mathbb{R}^{QN\times 2L}$ and $\mathbf{E}_{\mathrm{n}} \in \mathbb{R}^{Q_cN\times(Q_cN-2L)}$ are such that $\begin{bmatrix}\mathbf{E}_{\mathrm{s}} & \mathbf{E}_{\mathrm{n}}\end{bmatrix} \in \mathbb{R}^{Q_cN\times Q_cN}$ is orthogonal, and where $\tilde{\mathbf{\Lambda}} = \mathrm{diag}(\lambda_1, \lambda_2, \cdots, \lambda_{2L}) \in \mathbb{R}^{2L\times 2L}$ is a diagonal matrix of the $2L$ nonzero positive eigenvalues of $\mathbf{A\Sigma A}^{\mathrm{H}}$. We define the *signal subspace* to be the subspace spanned by the columns of $\mathbf{A}$, denoted by range($\mathbf{A}$). The *noise subspace* is defined as the orthogonal complement to the signal subspace. It follows that range($\mathbf{A}$) = range($\mathbf{A\Sigma A}^{\mathrm{H}}$) = range($\mathbf{E}_{\mathrm{s}}$) and that the noise subspace consequently is range($\mathbf{E}_{\mathrm{n}}$).

From (7.37) we see that an eigenvector of $\mathbf{A\Sigma A}^{\mathrm{H}}$ (with eigenvalue λ) is also an eigenvector of $\mathbf{R}_{\mathrm{rr}}$ (with eigenvalue $\lambda + \sigma^2$). Thus, we can write an eigenvalue decomposition of $\mathbf{R}_{\mathrm{rr}} = \mathbf{A\Sigma A}^{\mathrm{H}} + \sigma^2\mathbf{I}_{Q_cN}$ as

$$\mathbf{R}_{rr} = \begin{bmatrix} \mathbf{E}_s & \mathbf{E}_n \end{bmatrix} \begin{bmatrix} \mathbf{\Lambda}_s & \mathbf{0} \\ \mathbf{0} & \mathbf{\Lambda}_n \end{bmatrix} \begin{bmatrix} \mathbf{E}_s^H \\ \mathbf{E}_n^H \end{bmatrix} \tag{7.40}$$

where $\mathbf{\Lambda}_s = \tilde{\mathbf{\Lambda}} + \sigma^2 \mathbf{I}_{2L} = \text{diag}(\lambda_1 + \sigma^2, \cdots, \lambda_{2L} + \sigma^2)$ and $\mathbf{\Lambda}_n = \sigma^2 \mathbf{I}_{Q_c N - 2L}$.

Since the noise is complex Gaussian, the (conditional) log-likelihood function is[3]

$$\ln L(r) = \text{constant} - MQ_c N \ln \sigma^2 - \frac{1}{\sigma^2} \sum_{m=1}^{M} \left\| \vec{r}(m) - \mathbf{AB}\vec{z}(m) \right\|^2 \tag{7.41}$$

7.2.2.3 Estimation Algorithms

Strom et al. [8] proposed four algorithms for estimating the parameter vector: an ML estimator, an approximate ML estimator, an estimator based on MUSIC, and an estimator based on a sliding correlator approach. The first two are detailed here. Let $\hat{\vec{\varphi}}$ denote an unbiased estimator of the vector of deterministic unknown parameters $\vec{\varphi} = \begin{bmatrix} \sigma^2 & \vec{\gamma}^T & \theta^T & \tau^T \end{bmatrix}^T$ so that $\mathcal{E}\{\hat{\vec{\varphi}}\} = 0$.

Maximum Likelihood

The first algorithm is a maximum likelihood estimator where the log-likelihood function (7.41) is maximized over $\vec{\varphi}$. As seen from (7.41), the optimization problem is separable with respect to σ^2. That is, we can fix σ^2 and maximize the log-likelihood function with respect to $\vec{\tau}$, $\vec{\theta}$, and $\vec{\gamma}$.

Maximizing the log-likelihood function is equivalent to minimizing the negative log-likelihood function. Thus, the maximum likelihood (ML) estimates $\hat{\vec{\gamma}}^{\text{ML}}, \hat{\vec{\theta}}^{\text{ML}}, \hat{\vec{\tau}}^{\text{ML}}$, and $\hat{\vec{z}}^{\text{ML}} = \begin{bmatrix} \vec{z}^T(1) & \cdots & \vec{z}^T(M) \end{bmatrix}^T$ can be written as

$$\begin{bmatrix} \hat{\vec{\gamma}}^{\text{ML}} & \hat{\vec{\theta}}^{\text{ML}} & \hat{\vec{\tau}}^{\text{ML}} & \hat{\vec{z}}^{\text{ML}} \end{bmatrix}^T = \arg\min_{\vec{\gamma},\vec{\theta},\vec{\tau},\vec{z}} \sum_{m=1}^{M} \left\| \vec{r}(m) - \mathbf{AB}\vec{z}(m) \right\|^2 \tag{7.42}$$

[3] We denote the 2-norm of a vector by $\|\vec{v}\| = \sqrt{\vec{v}^* \vec{v}}$.

Note that the maximum-likelihood algorithm estimates $\vec{\tau}$, $\vec{\theta}$, $\vec{\gamma}$, and $\vec{z}$ simultaneously. It is important to realize that the log-likelihood function is conditioned on $\vec{z}$, which, in turn, is defined by the transmitted bits from all users. Therefore, in order to use the maximum-likelihood algorithm as stated in (7.42) we need to maximize (7.42) for *all possible* bit sequences. This estimator exploits the full structure of the problem and will achieve the CRB.

The maximization over $\vec{z}$ is unfortunate since this is a mixed type optimization problem with both continuous and discrete parameters. The problem is solvable in principle, but the complexity becomes overwhelming as the number of users or the observation interval grows.

Approximate Maximum Likelihood

As mentioned above, the problem with the ML estimator is that we need to test all possible transmitted bit sequences in order to perform a true maximum-likelihood estimation of the unknown parameters. However, we can make an approximation that in essence allows us to "forget" that we have a mixed optimization problem. If we define

$$\vec{s}(m) = \mathbf{B}\vec{z}(m) \in \mathbb{C}^{2K} \tag{7.43}$$

then the received vector may be written as

$$\vec{r}(m) = \mathbf{A}\vec{s}(m) + \vec{n}(m) \tag{7.44}$$

If we assume that $\mathbf{s}(m)$ are unknown, continuous and deterministic for $m = 1, \ldots,$ M, then the ML estimate of $\mathbf{s}(m)$ is $\hat{\mathbf{A}}^{\dagger}\vec{r}(m)$. Here $\hat{\mathbf{A}}^{\dagger}$ denotes the left pseudo inverse of $\hat{\mathbf{A}} = \mathbf{A}(\hat{\vec{\tau}})$, which, since $\hat{\mathbf{A}}$ has full rank, can be written as

$$\hat{\mathbf{A}}^{\dagger} = (\hat{\mathbf{A}}^{*}\hat{\mathbf{A}})^{-1}\hat{\mathbf{A}}^{*} \tag{7.45}$$

We can now formulate an approximate ML (AML) estimate of $\vec{\tau}$

$$\begin{aligned}\hat{\vec{\tau}}^{AML} &= \arg\min_{\vec{\tau}} \sum_{m=1}^{M} \left\| \vec{r}(m) - \mathbf{A}\mathbf{B}\vec{z}(m) \right\|^2_{\mathbf{B}\vec{z}(m)=\hat{\mathbf{A}}^{\dagger}\vec{r}(m)} \\ &= \arg\min_{\boldsymbol{\tau}} \operatorname{trace}\left(\mathbf{P}^{\perp}_{\hat{\mathbf{A}}}\hat{\mathbf{R}}_M\right)\end{aligned} \tag{7.46}$$

where $\mathbf{P}_{\hat{\mathbf{A}}}^{\perp} = \mathbf{I}_{QN} - \hat{\mathbf{A}}\hat{\mathbf{A}}^{\dagger}$ is the orthogonal projection matrix onto range $(\hat{\mathbf{A}})$ and $\hat{\mathbf{R}}_M$ is the sample correlation matrix defined as

$$\hat{\mathbf{R}}_M = \frac{1}{M}\sum_{m=1}^{M}\vec{r}(m)\vec{r}^{\mathrm{H}}(m) \tag{7.47}$$

Since we are not using the full structure of $\mathbf{s}(m)$ we should expect (7.46) to yield poorer estimates than (7.42). The AML algorithm finds estimates of all the delays simultaneously.

Performance

The simulated system was a 10-user system with $N = 31$ chips per bit and $T_c = 1$ Gold code sequences [8] generated by the polynomials $g_1(x) = x^5 + x^2 + 1$ and $g_2(x) = x^5 + x^4 + x^3 + x^2 + 1$. Oversampling was not used (i.e., $Q_c = 1$).

Each Monte Carlo run represents a particular realization of the noise and data sequences. Throughout all the simulations, the delays, received amplitudes and phases were fixed. The measure of performance is the sample standard deviation, std($\hat{\tau}_1$). A total of 500 Monte Carlo runs were done for each simulation. The absolute performance of the timing estimators (in terms of standard deviation) is, of course, dependent on the particular τ for which the simulations were done. However, the relative performance of the estimators proved to be roughly independent of the delays.

The near-far ratio is defined as R_2 / R_1, where all interfering users had the same received power, $R_l = R_2$ for $l = 3, 4, \ldots, 7$. The Cramer-Rao bound serves as a lower bound on the standard deviation. For our simulated case, the actual CRB was very close to the asymptotic expression for an observation interval $M > 20$. Furthermore, the estimators turned out to be approximately unbiased, and hence the sample standard deviation is an appropriate measure of performance.

Since we are searching for the global minimum, we may sometimes find an estimate that is more than one chip duration away from the true delay. Such an estimate is called an *outlier* and was excluded from the data before the standard deviation was calculated. This was done in order to make the comparison with the CRB meaningful. An outlier error is, of course, disastrous in the acquisition phase; however, in the tracking phase, we will probably limit the search of the cost function to be in the vicinity (say, $\pm T_c / 2$) of the previous estimate. This is reasonable since the propagation delays typically are slowly varying.

The relative number of outliers are listed in Table 7.2. For the MUSIC algorithm, the number of outliers decrease as the observation interval or SNR

Table 7.2 Percentage of Outliers for MUSIC and Correlator Algorithms, No Outliers Were Detected for the Correlator When $P_2 / P_1 = 0$ dB and $\text{SNR}_1 = 5, 10, 15$ dB

	MUSIC						**Correlator**		
R_2/R_1	**0**			**20**			**20**		
υ_1 **(dB)**	5	10	15	5	10	15	5	10	15
M **= 100**	39.0	0.8	0	99.8	84.2	0	54.4	53.8	50.6
200	10.2	0	0	99.6	25.4	0	30.2	29.8	34.4
300	3.8	0	0	98.8	3.6	0	18.6	18.2	16.4
400	2.4	0	0	94.8	0.4	0	12.8	12.8	13.4

increases; the outlier frequency is only weakly dependent on the near-far ratio for medium to high SNR and sufficiently large M. The correlator had no detected outliers when $R_2 / R_1 = 0$ dB and $\upsilon_1 = 2E_{b,1} / N_0 = 5, 10, 15$ dB. However, for $R_2 / R_1 = 20$ dB, the number of outliers was high and decreased only slowly with M. Hence, the correlator has fewer outliers than MUSIC when we have perfect power control or low SNR. In case of the AML algorithm, the optimization was carried out by a numerical search started at the true delay. Therefore, we seldom encountered an outlier since we were likely to find a local minimum close to the

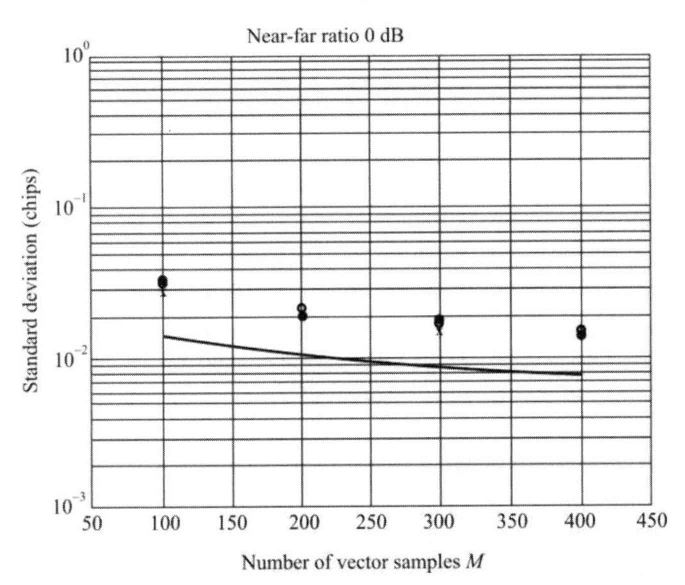

Figure 7.9 Standard deviation of $\hat{\tau}_1$ as a function of the number of observed vectors for AML (x), MUSIC (o), and correlator (•) for different near-far ratios. The solid line is the Cramer-Rao bound. $\upsilon_1 = 15$ dB. In this case the near-far ratio is 0 dB. (Source: [8]. © 1996 IEEE. Reprinted with permission.)

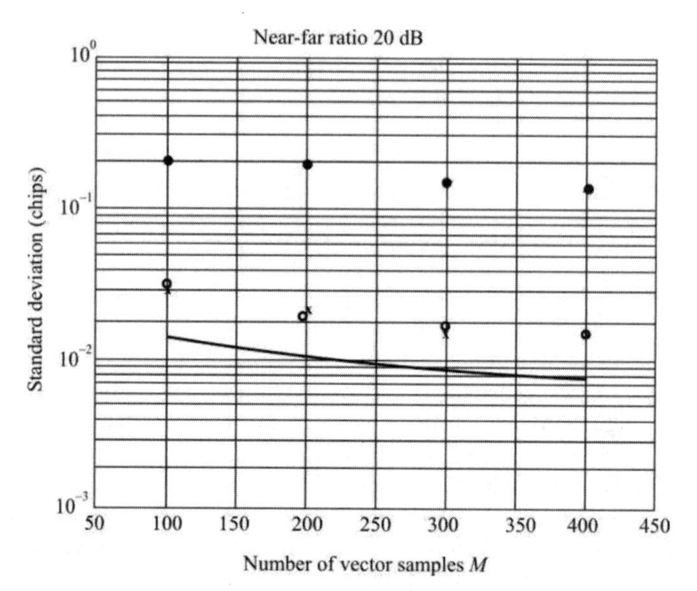

Figure 7.10 Same as Figure 7.9 except that the near-far ratio is 20 dB. Note the severe degradation of the performance of the correlator when the near-far ratio is increased. (Source: [8]. © 1996 IEEE. Reprinted with permission.)

starting point in this case. We did not attempt to further quantify the outlier frequency for the AML since, as discussed below, the AML has little practical value for acquisition.

In Figures 7.9 and 7.10 [8], the standard deviation is plotted as a function of the number of observed vectors for the AML, MUSIC, and correlator algorithms for near-far ratios 0 and 20 dB. Figure 7.11 shows the standard deviation for the estimators as a function of the near-far ratio. For both figures, the signal-to-noise ratio was υ_1 = 15 dB. Note that the MUSIC and AML algorithms performed similarly and in a near-far resistant manner. However, the correlator performance degraded rapidly with the near-far ratio. The CRB is seen to be invariant to the near-far ratio; furthermore, the CRB was not attained by any of the algorithms.

7.2.2.4 Summary and Conclusions

The standard algorithm for estimating propagation delays in an asynchronous DS-CDMA system works satisfactory when the multiuser interference is low. However, in a near-far situation, the performance is severely degraded. The structure of the asymptotic CRB indicates that it is possible to find good near-far resistant propagation delay estimators. The two proposed algorithms, AML and

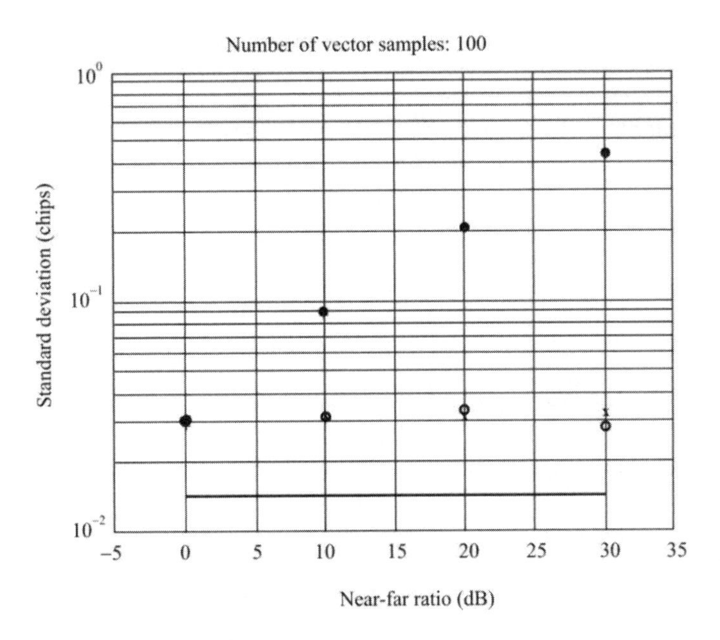

Figure 7.11 Standard deviation of $\hat{\tau}_1$ as a function of the near-far ratio for AML (x), MUSIC (o), and correlator (•) for different near-far ratios. The solid line is the Cramer-Rao bound. The number of observed vector samples is $M = 100$. Note that the AML and MUSIC estimators are unaffected by the near-far ratio while the correlator shows an increasing standard deviation as the near-far ratio is increased. (Source:[8] © 1996 IEEE. Reprinted with permission.)

MUSIC, are shown to be robust against multiuser interference. Furthermore, they place no constraints on the data sequence and can therefore be used for tracking as well. Recursive formulations are also possible, which lower the computational complexity. The MUSIC estimates are found one by one, whereas the AML algorithm computes all estimates simultaneously. Obviously, if only one or a few propagation delays are of interest, the MUSIC algorithm is preferable since, for instance, it only requires knowledge of the code sequences of the users whose delays are to be estimated. Moreover, the AML cost function is highly nonlinear with many local minima and is therefore sensitive to correct initialization. Thus, the AML is of limited practical value for the acquisition problem.

7.2.3 Non-Gaussian Noise

As we pointed out in Chapter 2, over significant portions of some of the frequency bands of interest to us, up to about 1 GHz, in fact, the predominant external noise sources are not Gaussian as we have been assuming up to now. In the HF range, below about 30 MHz, atmospheric noise dominates the external noise. Above that, from about 30 to 300 MHz, galactic noise sources (the stars) produce the largest average external noise. From about 300 MHz to about 1 GHz, man-made noise is the main culprit. It isn't until about 1 GHz that the assumption of AWGN is valid.

PTT CNRs are the predominant mode of communication in the low VHF range (30–90 MHz). The external noise here is predominantly galactic and man-made.

We present in the section an analysis of the effects of non-Gaussian noise [12] on code acquisition. Atmospheric noise is very impulsive, being caused by lightning. In the lower-frequency ranges (HF) signals can travel long distances due to the reflections off the ionosphere as discussed in Chapter 5. Being very powerful generators, lightning thus can be heard a long way from its point of origin. Typical MMN is also very impulsive in nature, being caused by mechanical and electrical devices that turn on and off frequently.

We will use the observation model described in Chapter 4, but repeated here for convenience (with different noise characteristics, however). In a DSSS system, the received signal can be expressed as

$$r(t) = \sqrt{2R}d(t-\tau T_c)c(t-\tau T_c)\cos(\omega_c t+\phi)+n(t) \tag{7.48}$$

where $R = E/T_s$ is the average power in the signal; $d(t)$ is the data sequence; T_c is the chip duration; $c(t)=\sum_{k=-\infty}^{\infty} c_k p_{T_c}(t-kT_c)$ with $c_k \in \{-1,+1\}$, the kth chip of a PN code sequence with period L and $p_{T_c}(t)$ as the PN code waveform defined as a unit rectangular pulse over $[0, T_c]$; τ is the time delay normalized to T_c; ω_c is the carrier angular frequency; ϕ is the phase distributed uniformly over $(0, 2\pi]$; and $n(t)$ is the ambient non-Gaussian channel noise. The level of $n(t)$ can be estimated with Figures 2.9 and 2.10.

We assume that there is a preamble for acquisition so that no data modulation is present during acquisition. This may seem overly restrictive; however, the UMTS PCS standards all have a code acquisition scheme that can be modeled this way.

A typical structure of a PN code acquisition module with a noncoherent I-Q correlator is shown in Figure 7.12. We consider the serial search scheme with a single dwell [13]. The kth sampled I-Q components are r_k^{I} and r_k^{Q} and can be obtained as, for $k = 1, 2, \ldots, K$

$$R_k^{\mathrm{I}} = \int_{t_k-T_c}^{t_k} r(u)c(u-\hat{\tau}T_c)\sqrt{2}\cos(\omega_c u)du \tag{7.49}$$

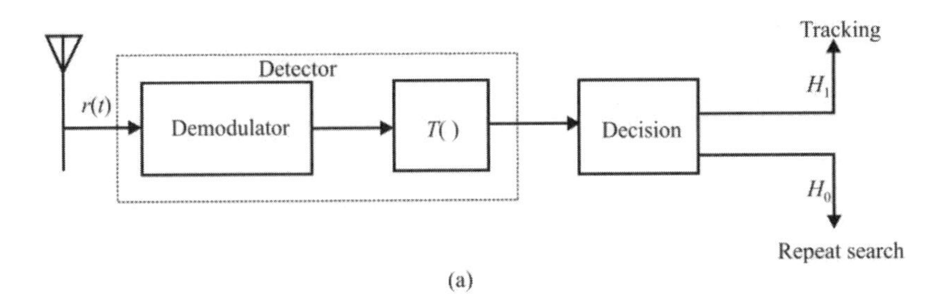

(a)

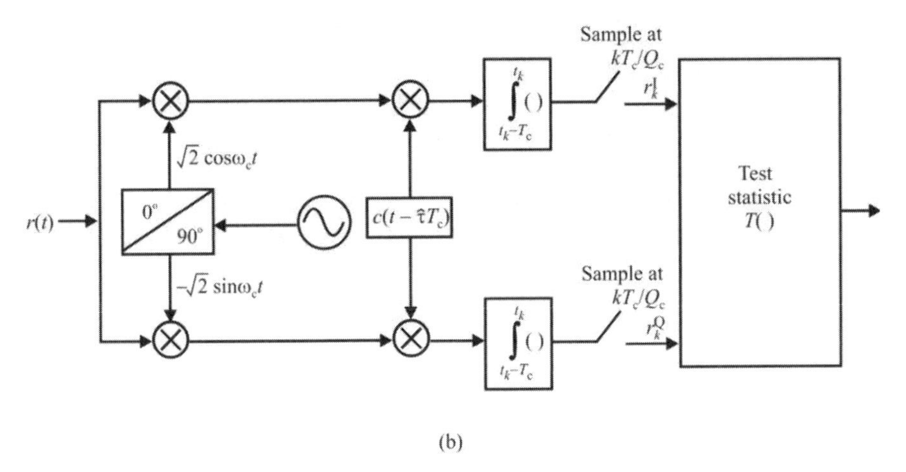

(b)

Figure 7.12 Structure of: (a) PN code acquisition system and (b) detail on the detector in (a).

$$R_k^Q = -\int_{t_k - T_c}^{t_k} r(u)c(u-\hat{\tau}T_c)\sqrt{2}\sin(\omega_c u)du \tag{7.50}$$

respectively, where K is the correlation length, $\hat{\tau}$ is the time delay (normalized to the chip duration) of the locally generated PN code, and $t_k = t_0 + kT_c$. Here, t_0 is the initial time. A test statistic is evaluated with the vectors $\vec{r}^{\,I} = \left(r_1^I, r_2^I, \ldots, r_K^I\right)^T$ and $\vec{r}^{\,Q} = \left(r_1^Q, r_2^Q, \ldots, r_K^Q\right)^T$, and then compared with a threshold. The test statistic is, for example, the DSSS statistic

$$T_{SS}(\vec{r}^{\,I}, \vec{r}^{\,Q}) = \left(\sum_{k=1}^{N} r_k^I\right)^2 + \left(\sum_{k=1}^{N} r_k^Q\right)^2 \tag{7.51}$$

We can regard the PN code-acquisition problem as a hypothesis testing problem: given $\vec{r}^{\mathrm{I}}$ and $\vec{r}^{\mathrm{Q}}$, a decision is to be made between the null hypothesis H_0 and alternative hypothesis H_1, where

$$H_0 : |\tau - \hat{\tau}| \geq 1 \tag{7.52}$$

and

$$H_1 : |\tau - \hat{\tau}| < 1 \tag{7.53}$$

Under H_1, each sampled correlation value between the locally generated and received PN codes is $\sqrt{E}(1-\delta)$, where δ is the residual shift (normalized to T_c) between the two PN codes, with the value ranging in the interval (–1, +1). For simplicity, we assume that the system is chip synchronous (that is, $\delta = 0$). Thus, each sampled correlation value is $\sqrt{E}$. On the other hand, each sampled correlation value is +1 or –1 with equal probability, and the mean value of the sampled correlation is 0 under H_0. From these results and (7.49) and (7.51), we can alternatively express H_0 and H_1 as

$$H_0 : \left(r_k^{\mathrm{I}} = n_k^{\mathrm{I}}, r_k^{\mathrm{Q}} = n_k^{\mathrm{Q}}\right), \qquad k = 1,2,\ldots,K \tag{7.54}$$

$$H_1 : \left(r_k^{\mathrm{I}} = \chi\cos\phi + n_k^{\mathrm{I}}, r_k^{\mathrm{Q}} = \chi\sin\phi + n_k^{\mathrm{Q}}\right), \qquad k = 1,2,\ldots,K \tag{7.55}$$

or simply as

$$H_0 : \chi = 0 \tag{7.56}$$

$$H_1 : \chi > 0 \tag{7.57}$$

In (7.54)–(7.57), $\chi = \sqrt{E}$ is the signal strength parameter, and $\vec{n}^{\mathrm{I}} = \left(n_1^{\mathrm{I}}, n_2^{\mathrm{I}}, \ldots, n_K^{\mathrm{I}}\right)^{\mathrm{T}}$ and $\vec{n}^{\mathrm{Q}} = \left(n_1^{\mathrm{Q}}, n_2^{\mathrm{Q}}, \ldots, n_K^{\mathrm{Q}}\right)^{\mathrm{T}}$ are the I-Q noise sample vectors, respectively. As argued in Chapter 2, we characterize $\vec{n}^{\mathrm{I}}$ and $\vec{n}^{\mathrm{Q}}$ with SαS distributions [14–17]. We let

$$n_k^{\mathrm{I}} \sim x_1 = x_k^{\mathrm{I}} - \chi\cos\phi \tag{7.58}$$

and

$$n_k^{\rm Q} \sim x_2 = x_k^{\rm I} - \chi \sin\phi \tag{7.59}$$

so that, from (2.183), the joint PDF, $p(\chi;i)$, of $r_i^{\rm I}$ and $r_i^{\rm Q}$ as a function of χ for α = 1 is

$$p(\chi;k) = \frac{\gamma}{2\pi\left[\left(x_k^{\rm I} - \chi\cos\phi\right)^2 + \left(x_k^{\rm Q} - \chi\sin\phi\right)^2 + \gamma^2\right]^{3/2}} \tag{7.60}$$

The joint PDF of the $2N$ sampled in-phase and quadrature observations $\{r_k^{\rm I}, r_k^{\rm Q}\}, k = 1,\ldots,K$, is then

$$p_{R^{\rm I},R^{\rm Q}}(x^{\rm I}, x^{\rm Q}) = \mathcal{E}_\phi\left\{\prod_{k=1}^{K} p(\chi;k)\right\} \tag{7.61}$$

under the assumption that the samples $\{r_k^{\rm I}, r_k^{\rm Q}\}$ of the bivariate noise process form a sequence of independent random vectors for given ϕ, where $\mathcal{E}_\phi\{\cdot\}$ denotes the expectation over ϕ.

A performance comparison of the non-Gaussian detection probability versus using the AWGN assumption is shown in Figure 7.13 for some values of the SNR/chip. This chart was based on a Monte Carlo simulation using as a test statistic [14]

$$T(r^{\rm I}, r^{\rm Q}) = \frac{1}{p_{R^{\rm I},R^{\rm Q}}(r^{\rm I}, r^{\rm Q})\Big|_{\chi=0}} \left.\frac{d^v p_{R^{\rm I},R^{\rm Q}}(r^{\rm I}, r^{\rm Q})}{d\chi^v}\right|_{\chi=0} \tag{7.62}$$

where v is the order of the first nonzero derivative of $p_{R^{\rm I},R^{\rm Q}}(r^{\rm I}, r^{\rm Q})$ at $\theta = 0$.

In Figure 7.13 [12], the SNR per chip is given by

$$\upsilon = \frac{1}{2C_g}\left(\frac{\sqrt{E}}{S_0}\right)^2 \tag{7.63}$$

where

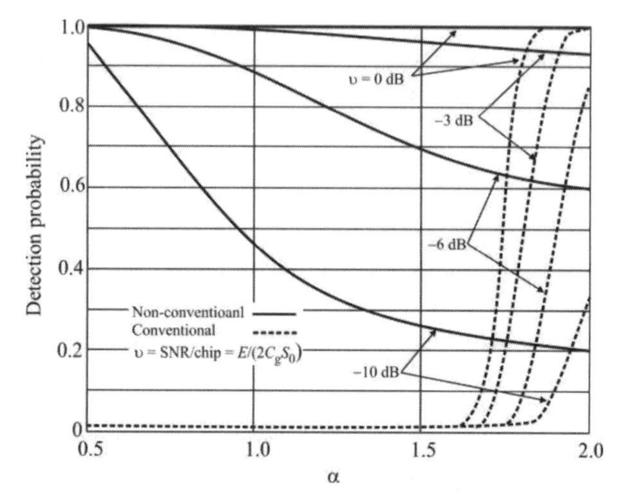

Figure 7.13 Detection probability versus alpha. $\gamma = 1$, $L = 1{,}023$, primitive polynomial $= 1 + z^3 + z^{10}$, $K = 64$, $P_{\text{fa}} = 10^{-2}$. *Conventional* here means that the noise is assumed to be AWGN at the detector while *nonconventional* assumes the noise is modeled as SαS processes. (Source: [12]. © 2004 IEEE. Reprinted with permission.)

$$S_0 = \frac{(C_g \gamma)^{1/\alpha}}{C_g} \tag{7.64}$$

and where $C_g \sim 1.78$ is the exponent of Euler's constant

$$C_g = \lim_{m \to \infty} \left[\sum_{k=1}^{m} \frac{1}{k} - \ln m \right] \tag{7.65}$$

Normalizing the SNR by $1 / C_g$ ensures that the definition of the SNR is the same in both the non-Gaussian noise and the Gaussian noise cases. In this case $\gamma = 1$.

7.2.4 Code Synchronization for Band-Limited DSSS Systems

Benedetto and Giunta documented the results of a development that determines the phase offsets of asynchronous CDMA signals when the communication channels and therefore the digital signals are band-limited [15]. Most analyses of DSSS systems and technologies assume that the PN code sequence consists of rectangular pulses, where in reality they never are. Truly rectangular pulses at the receiver implies that there is infinite bandwidth over which the signal travels. This is never the case, and band-limiting tends to round the corners of the pulses, sometimes severely. We will briefly describe the approach in this section.

Most asynchronous CDMA systems need to search for the timing (phase) offset of the PN code sequence in order to align the signals. The codes used in CDMA systems are only truly orthogonal when they are aligned in phase. Such systems employ a type of detection system that typically looks like the one in Figure 7.12. The incoming signal is first amplified and filtered (not shown) and then applied to a detector. The detector typically is a demodulator, in our case normally a correlator, and the results of correlating the incoming signal with a locally generated version of the code are applied to a threshold comparing device. If the correlation results are above a threshold (normally set by the amount of false alarms permitted), decision logic is employed as described above. The resulting decision is declared that the correct phase offset is present (H_1) if the correlation is above a threshold or declared that the phase offset incorrect (H_0) if the comparison is small. If the signal is declared present, then the receiver enters the tracking mode; if not, the search is repeated with the next candidate phase offset.

7.2.4.1 Receiver Model

The block scheme of the conventional acquisition receiver based on a matched filter is depicted in Figure 7.14 [19] where the I and Q samples at time k, R_k^{I} and R_k^{Q}, are obtained from the matched filter in Figure 7.12. At the output, we have a chip waveform, with raised-cosine spectrum and roll-off ζ, and defined as

$$h_{T_c}(t) = \frac{\sin\left(\frac{\pi}{T_c}t\right)\cos\left(\gamma\frac{\pi}{T_c}t\right)}{\frac{\pi}{T_c}t\left[1-4\zeta^2\left(\frac{t}{T_c}\right)^2\right]} \tag{7.66}$$

Now consider N samples of the complex envelope of the received signal after the matched filter $\{r(kT_c/2 - \tau T_c), k = 1,...N\}$, with the normalized timing offset τ, being $\tau = 0$ in the chip-synchronous case while randomly distributed over $-0.5 \leq \tau \leq 0.5$ in the chip-asynchronous case, being independent of the sampling times (either integer or half-integer multiples of the chip period T_c) [9].

The power detector first estimates the cross-correlation $\rho_w(\tau)$ between $r(t)$ and the code candidate $c(t)$, shifted by $(kT_c/2 + \delta T_c)$, where δ is the normalized local PN code delay and is assumed to be an integer [15]:

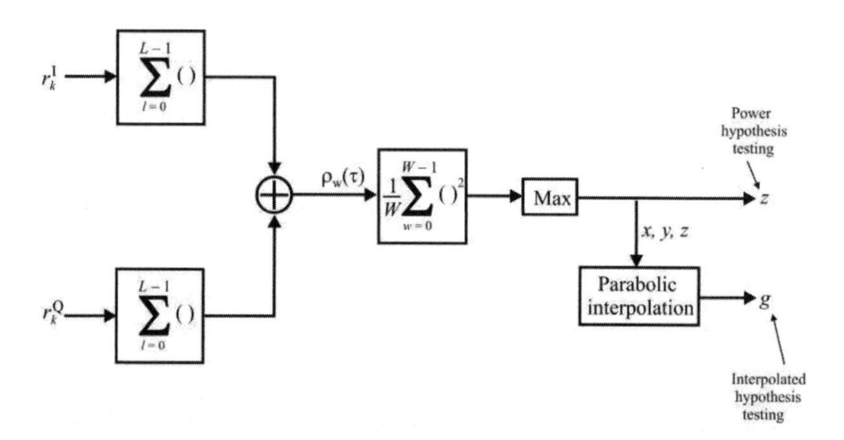

Figure 7.14 Conventional acquisition receiver based on matched filter. (After: [15]. © 2009 IEEE. Reprinted with permission.)

$$\begin{aligned}\rho_w(\delta,\tau) &= \frac{1}{N}\left[\sum_{k=1}^{N} r^{\mathrm{I}}\left(k\frac{T_{\mathrm{c}}}{2}-\tau T_{\mathrm{c}}\right)c\left(k\frac{T_{\mathrm{c}}}{2}+\delta T_{\mathrm{c}}\right)+\sum_{k=1}^{N} jr^{\mathrm{Q}}\left(k\frac{T_{\mathrm{c}}}{2}-\tau T_{\mathrm{c}}\right)c\left(k\frac{T_{\mathrm{c}}}{2}+\delta T_{\mathrm{c}}\right)\right]\\ &= \frac{1}{N}\left[\sum_{k=1}^{N} r^{\mathrm{I}}\left(k\frac{T_{\mathrm{c}}}{2}-\tau T_{\mathrm{c}}\right)+jr^{\mathrm{Q}}\left(k\frac{T_{\mathrm{c}}}{2}-\tau T_{\mathrm{c}}\right)\right]c\left(k\frac{T_{\mathrm{c}}}{2}+\delta T_{\mathrm{c}}\right)\end{aligned}$$

so that

$$\rho_w(\delta,\tau) = \frac{1}{N}\sum_{k=1}^{N} r\left(k\frac{T_{\mathrm{c}}}{2}-\tau T_{\mathrm{c}}\right)c\left(k\frac{T_{\mathrm{c}}}{2}+\delta T_{\mathrm{c}}\right) \tag{7.67}$$

Then, the power detector accumulates W blocks of the squared magnitude of the estimated cross-correlation $\rho_{\mathrm{w}}(\tau)$ to get

$$z_k(\tau) = \frac{1}{W}\sum_{w=1}^{W}\left|\rho_w(\tau)\right|^2 \tag{7.68}$$

We then compares the (kth) currently examined decision variable, $z_k(\tau)$, to a preselected threshold γ_{th}

$$z_k(\tau) \underset{H_0}{\overset{H_1}{\gtrless}} \gamma_{th} \tag{7.69}$$

As $W \to \infty$, the testing variable $z_k(\tau)$ is asymptotically Gaussian because of the central limit theorem, assuming the statistical independence among the W data blocks. The threshold γ_{th} can be determined by repeated evaluation of the Gaussian integral for a fixed probability of false alarm (CFAR detection). The CFAR test is accomplished in two successive parts. First, the mean $\mathcal{E}\{z_k(\tau)|H_0\}$ and the variance $\mathrm{var}\{z_k(\tau)|H_0\}$ are evaluated under the null hypothesis to compute the threshold γ_{th}. This threshold is determined to limit the false-alarm probability

$$P_{fa} = \Pr\{z_k(\tau) \geq \gamma_{th} |H_0\} \tag{7.70}$$

at a given reduced value (the *size* of the test). Second, the probability of detection

$$P_d = \Pr\{z_k(\tau) \geq \gamma_{th} |H_1\} \tag{7.71}$$

(the *power* of the test) is evaluated for the threshold previously determined. These steps are repeated as necessary to arrive at the final, acceptable values.

7.2.4.2 Interpolation Technique

Referring to Figure 7.14, we introduce the following quantities

$$x = z_k\left(\tau - \frac{1}{2}\right), \quad y = z_k\left(\tau + \frac{1}{2}\right), \quad z = z_k(\tau) \tag{7.72}$$

where z is the same as expressed by (7.68), while y and x are the detector's outputs evaluated at $T_c / 2$ before and after z, respectively. This detector searches for the apex of the parabolic function $g(\tau) = a\tau^2 + b\tau + c$ fitted over the three detectors' outputs x, y, and z expressed by (7.72). The parameters of the parabola fitting the measured correlation are

$$a = 2\frac{x - 2z + y}{T_c^2}, \qquad b = \frac{y - x}{T_c}, \qquad c = z \tag{7.73}$$

The coordinates of the apex of the parabola are

$$\text{abscissa: } G_k = -\frac{b}{2a} \qquad \text{ordinate: } g(G_k) = -\frac{b^2}{4a} + c \tag{7.74}$$

Using (7.73), (7.74) can be expressed in terms of x, y, and z as

$$G_k = f(x, y, z) = -\frac{T_c}{4}\frac{x - y}{x - 2z + y} \tag{7.75}$$

$$g(G_k) = g(x, y, z) = z - \frac{1}{8}\frac{(x - y)^2}{x - 2z + y} \tag{7.76}$$

We use a parabolic interpolation to approximate the ambiguity function because we are dealing with band-limited signals. Unlike rectangular chips, the auto-correlation functions of band-limited chip waveforms (such as the ones with square-root raised-cosine spectrum) are well matched by parabolic functions in the neighborhood of its maximum, according to their Taylor's expansion truncated up to the second order [15].

7.2.4.3 Performance

In this section, we evaluate the performance of the new acquisition method based on the parabolic interpolation in this section. We need to point out that, in real cases, the algorithm considers as possible candidates the only testing variables for which z is greater than x and y simultaneously.

The best-case results are illustrated in Figure 7.15. The best case is defined in Figure 7.16(a); it is where the timing offset $\tau = 0$. The worst case performance, defined in Figure 7.16(b), is illustrated in Figure 7.17. The worst case is when the timing offset is $\tau = T_c/4$. We can see that the power test is affected somewhat more by the timing offset. However, it is also clear that the interpolation test procedure performs about as well as the power test.

7.3 DSSS Tracking

Tracking DSSS signals is the process of keeping the receiver's locally generated code sequence in phase with the incoming code sequence. This is necessary for the receiver to correctly decorrelate the signal to extract the data. It also spreads any interfering signals, although proper synchronization is not necessary to accomplish

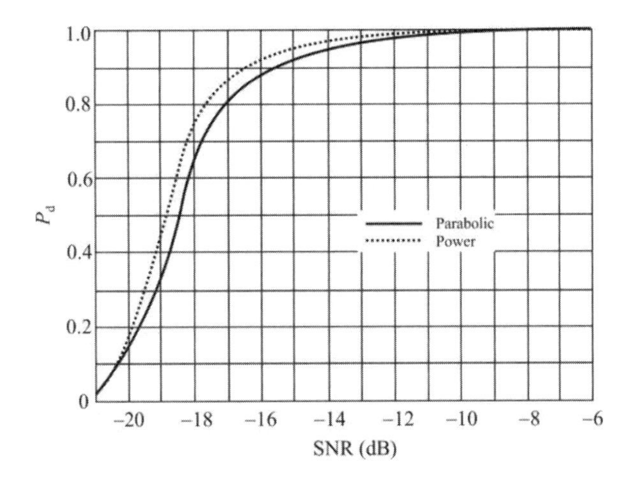

Figure 7.15 Parabolic interpolation best case. (Source: [15]. © 2009 IEEE. Reprinted with permission.).

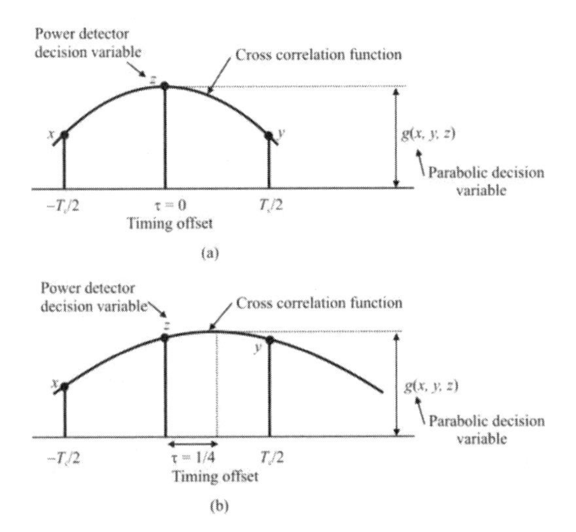

Figure 7.16 Parabolic interpolation timing offsets. (a) Best case and (b) worst case.

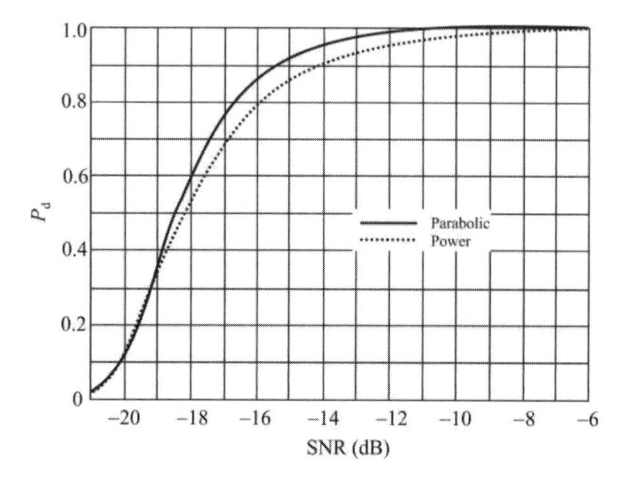

Figure 7.17 Parabolic interpolation worst case. (Source: [15]. © 2009 IEEE. Reprinted with permission.)

that. While there are numerous techniques for implementing DSSS tracking, most are based on the *delay lock loop* (DLL).

The DLL relies on the theoretical triangular shape of the correlation function of *m*-sequences. An example of such architecture is shown in Figure 7.18. The locally generated code sequence is sent to a three-stage shift register. The last stage, which represents a delay of three chips in the sequence, is the late code. The code sequence from the middle stage is used to decorrelate the incoming signal, and the sequence from the first stage is the early code. The incoming code sequence is modulo-2 added to the early local PN sequence in the upper path of Figure 7.18(a). If they are synchronized then the correlation function shown at the top of Figure 7.18(b) is generated by the early correlator. The incoming code sequence is modulo-2 added to the late local code sequence in the late correlator in Figure 7.18(a). The correlation function shown in the middle of Figure 7.18(b) is implemented through this path.

The difference in the two correlation functions, when in sync, is shown at the bottom of Figure 7.18(b), and there is a balance point that always occurs at a fixed point in time after a chip starts. If this point occurs at some other time, depending on which side of the in-sync point it occurs, it indicates that the clock should be delayed or advanced.

Another example of a correlator that would work in this architecture is shown in Figure 7.19. This is a straightforward parallel implementation of a correlator with shift registers and exclusive ORs.

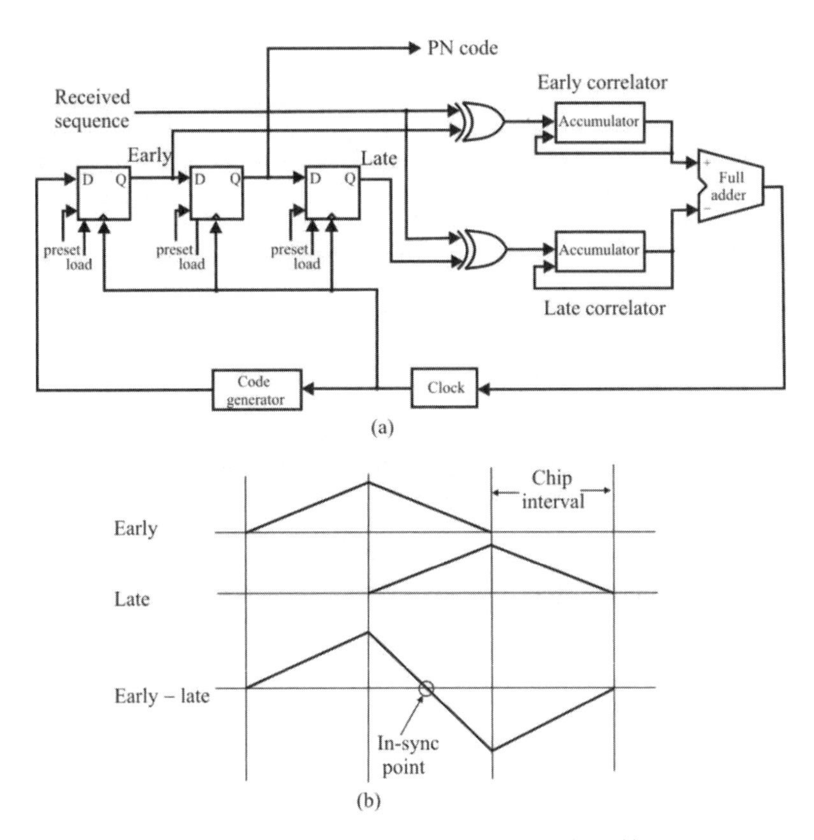

Figure 7.18 (a, b) Simplified architecture of a DLL used for DSSS code tracking.

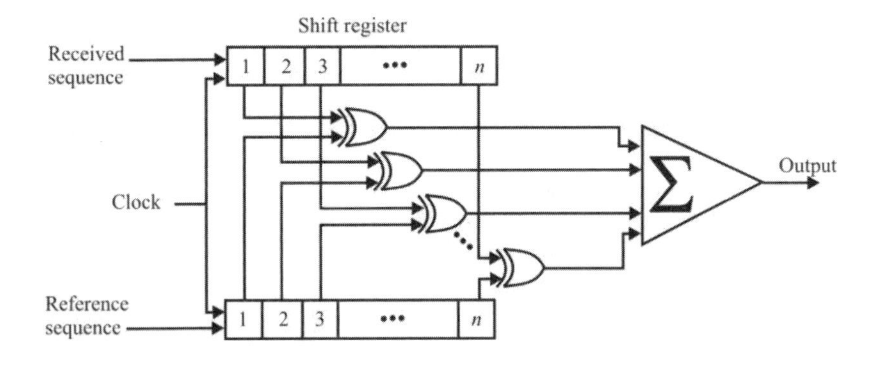

Figure 7.19 Parallel correlation of two n-bit sequences.

7.4 FHSS Synchronization

This section presents discussions of synchronization of FHSS AJ communication systems. The basic principles are similar to those for DSSS, however, the details are different. We examine three techniques for coarse acquisition of FHSS signals. This discussion follows that in [16] closely.

If an FHSS network fails to synchronize, it cannot communicate. A jammer can try to attack the synchronization process, but it must know when synchronization is occurring to be successful. As will be shown, however, such attacks can be very productive.

The environment in which these techniques are compared consists of push-to-talk networks for which there may be extended periods of no communications. In this time, the clocks in the transmitter and receiver may have drifted apart, thus causing the PN codes to be out of sync. Therefore, before each transmission the transmitter and receiver are synchronized.

Although, in general, a two-dimensional search is required over both frequency and code, in this analysis the frequency search will be ignored. In many cases of practical interest, such as ground-deployed PTT communication networks, this is a safe assumption. It may not be a safe assumption for high-velocity aircraft communications.

With this technique, a *leader* is transmitted prior to sending any data from the transmitter to the receiver. This leader is comprised of several passes through the hop set with a known data sequence attached. In general, if the synchronization search fails on the first pass through the search space, additional passes can be attempted, although normally rapid synchronization is required to maximize the data throughput. The number of hops in the synchronization sequence here is denoted H. The received signal during the kth hopping interval of the synchronization process can be expressed as

$$r_k(t) = \sqrt{2R}\cos(2\pi f_k t + \theta_k) + n_k(t) \tag{7.77}$$

where R is the average power in the signal, f_k is the frequency, θ_k is the carrier phase, and $n_k(t)$ is a Gaussian noise process with variance σ^2 (also the power of the noise when it has zero mean, which is assumed here). The probability of detection of this signal is given by

$$\begin{aligned} P_\mathrm{d} &= P\{z > \gamma_\mathrm{th}\} \\ &= Q_\mathrm{H}(\sqrt{\upsilon}, \sqrt{\gamma'}) \end{aligned} \tag{7.78}$$

where z is the test statistic,

$$\upsilon = \sum_{k=1}^{H} \frac{2R_k}{\sigma^2} \tag{7.79}$$

is a measure of the SNR, and

$$\gamma' = \frac{\gamma_{\text{th}}}{\sigma^2} \tag{7.80}$$

is a threshold normalized to the noise variance that establishes the false alarm rate. The probability of miss is given by the complementary function

$$P_{\text{NoSync}} = 1 - Q_H(\sqrt{\upsilon}, \sqrt{\gamma'}) \tag{7.81}$$

The analysis in [16] includes communications in a Ricean fading channel with potential interfering emitters, including tone jammers, impinging on the receiver. The effects of tone jammers will be discussed in Chapters 8 and 10–14. In this environment during the hop interval k, the received signal can be expressed as

$$r_k(t) = \sqrt{2a_k R}\cos(2\pi f_k t - \theta_i) + \sum_j c_j \sqrt{2R}\cos(2\pi f_k t - \theta_j) + n_k(t) \tag{7.82}$$

where f_k is the frequency and $n_k(t)$ is a Gaussian noise process with a two-sided spectral density $N_0/2$ and variance σ^2, where $\sigma^2 = N_0 B_{\text{IF}}$, with B_{IF} as the approximate noise bandwidth. For comparative analysis of the three approaches analyzed, the sum of the powers in the specular (direct) component and scatter (reflections) component to an average power over all the hops are assigned as R. The factor a_i accounts for the effects of the random amplitude fluctuations in the specular component in the Ricean channel model. The c_j factors account for the random fluctuations in the reflected components and there are several of them, indexed by j. Thus,

$$\alpha + b = 1 \tag{7.83}$$

where

$$\alpha = \frac{1}{H}\sum_k a_k \tag{7.84}$$

and

$$b = \sum_j c_j^2 \tag{7.85}$$

Thus, α is averaged over H hops and b is summed over several multipath channels.

The number of channels available to the FHSS system is denoted by F and it is assumed that h of these contain interfering tones, all of which have power equal to that of the intended signal, independent of each other, and with uniform random phases. While these are gross assumptions for cochannel interference, they do facilitate an approximate analysis of the effects of interfering signals. Furthermore, they are very good assumptions when the interference emanates from a multitone jammer discussed in Chapter 8.

Since the probability that *no* interfering signal is present in a frequency channel during a single hop is $(1 - 1/F)^h$, then the probability that *at least one* of the synchronization channels is occupied with an interfering signal is given by

$$p_h = 1 - \left(1 - \frac{1}{F}\right)^h \tag{7.86}$$

and the probability that k channels have an interfering tone is given by

$$p(k, H, p_\text{h}) = \binom{H}{k} p_\text{h}^k (1 - p_\text{h})^{H-k} \tag{7.87}$$

when the synchronization interval is over H hops. The probability of declaring synchronization on a single pass when, in fact, the synchronization sequence is not present (false alarm) is given by

$$P_\text{fa} = \sum_{k=0}^{H} p(k, H, p_\text{h}) Q_H\left(\sqrt{R_k}, \sqrt{\gamma'}\right) \tag{7.88}$$

where

$$R_k = 2k\iota\upsilon$$

ι is the interference or jammer tone to signal power ratio and $\upsilon = R / P_\text{N}$ is the SNR. P_N is the total noise power given by

$$P_{\mathrm{N}} = \sigma^2 + bR \tag{7.89}$$

Assuming that $W_F = 1/T_h$, and $Q_H(a, b)$ is the generalized Q-function. When the leader contains m sequences through the synchronization dwells, the overall probability of synchronization increases as reflected in a reduction in the probability of missed synchronization.

The probability of not achieving synchronization when the sync sequence is present on a single attempt is given by

$$P_{\mathrm{NoSync}_k} = \sum_{k=0}^{H} p(k, H, p_{\mathrm{h}})\left[1 - Q_H\left(\sqrt{R_{\mathrm{d}}}, \sqrt{\gamma''}\right)\right] \tag{7.90}$$

where

$$R_{\mathrm{d}} = 2\upsilon \frac{k\iota + H\alpha}{1 + b\upsilon}$$

$$\gamma'' = \frac{\gamma'}{1 + b\upsilon}$$

7.4.1 Matched Filter

The first technique for FHSS code synchronization employs a matched filter shown in Figure 7.20. H matched filters are implemented in parallel. The detector is normally implemented at each of H successive frequencies, but, in fact, they need not be sequential if the delays are correctly selected. The output of these detectors are delayed appropriately from 1 to H hop intervals, and the outputs of the delays are added. If this sum is above threshold, then sync lock is declared. In this case the probability of false lock is given by (7.88) and the probability of missing synchronization is given by (7.90) with $H = P$ and $\gamma' = \gamma_{\mathrm{th}_1}$.

This method of synchronization is faster than the serial search technique discussed next, but the probability of false lock is higher. The matched filter method, since several parallel channels are working at the same time, requires substantially more components than the serial search scheme. The performance of the matched filter code acquisition system with $h = 5$ interfering signals are shown in Figure 7.21 for some representative values of the parameters. For this there were no jammers considered, but the interfering signals from other transmitters were at the same level as the signal (ISR = 0 dB). P_{fa} varies with the SNR, but the threshold was varied to keep it near 10^{-4}.

The average time to achieve sync is given by

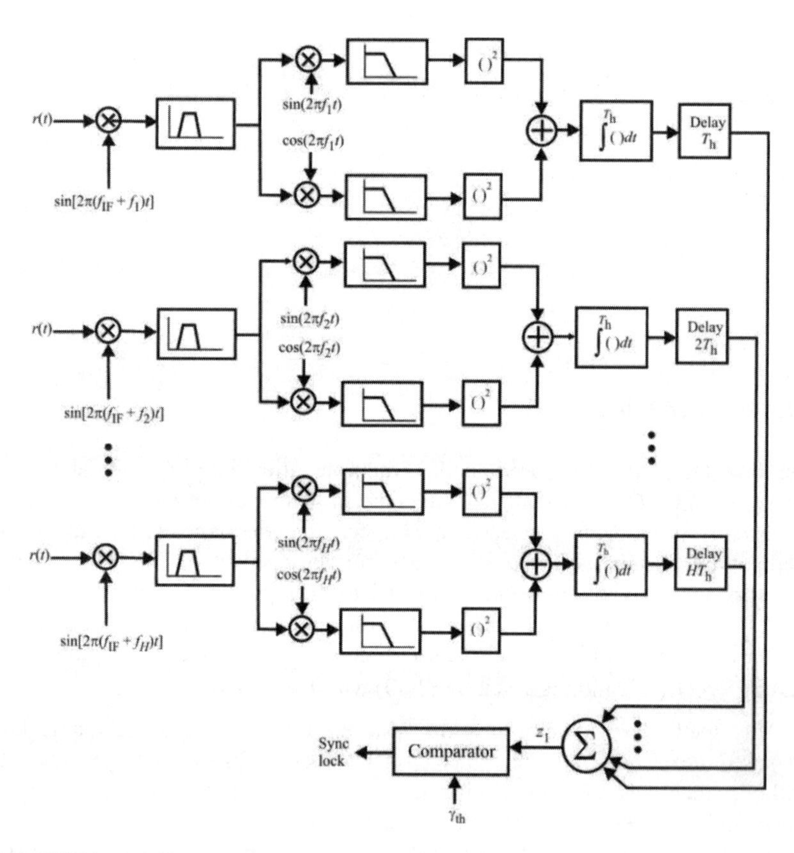

Figure 7.20 Matched filter correlator.

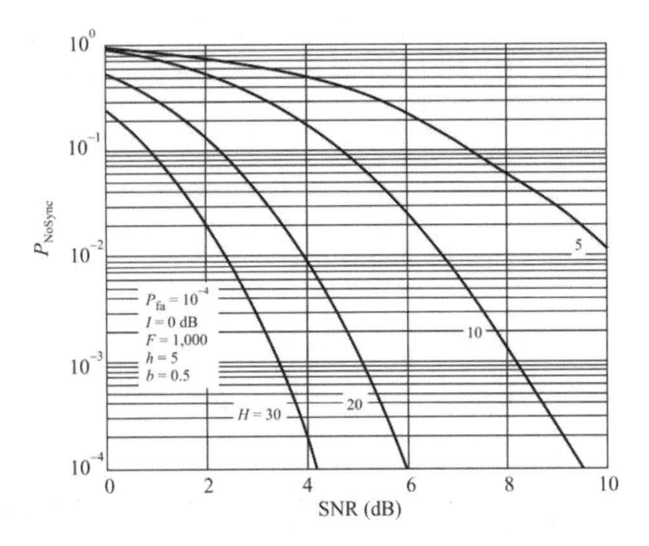

Figure 7.21 Probability of missed code synchronization for a matched filter (passive) code acquisition with $h = 5$ interfering signals all at the same power as the signal. The thresholds were $\gamma'_5 = 35$ (15 dB), $\gamma'_{10} = 50$ (17 dB), $\gamma'_{20} = 80$ (19 dB), and $\gamma'_{30} = 105$ (20 dB).

$$T_s = (N_c + H)T_h \tag{7.91}$$

7.4.2 Serial Search

The second technique employs an active correlator for serial search. Such a correlator is shown in Figure 7.22. The integration time is set at an integer number of hop intervals, $T_s = HT_h$. If the receiver is in sync with the incoming signal, then all H hops periods will contribute to the test statistic z. When z rises above the threshold level, sync lock is declared. If some of the receiver hop frequencies do not coincide with the incoming hop sequence, then during those periods with no match, the output of the squaring envelope detector will be due to noise only. If a sufficient number of hops do not match, then z will not be above the threshold and the control will hold back the clock for one period, thus slipping the code sequence by one chip. This process is repeated until the correct code phase is found.

Generally, declaring synchronization on a single successful attempt will yield poor results and too many false locks. Typically, m sequential successful synchronization indications are needed. For stepped serial synchronization the probability of false hit on a single attempt is given by (7.88) with $H = A$ and

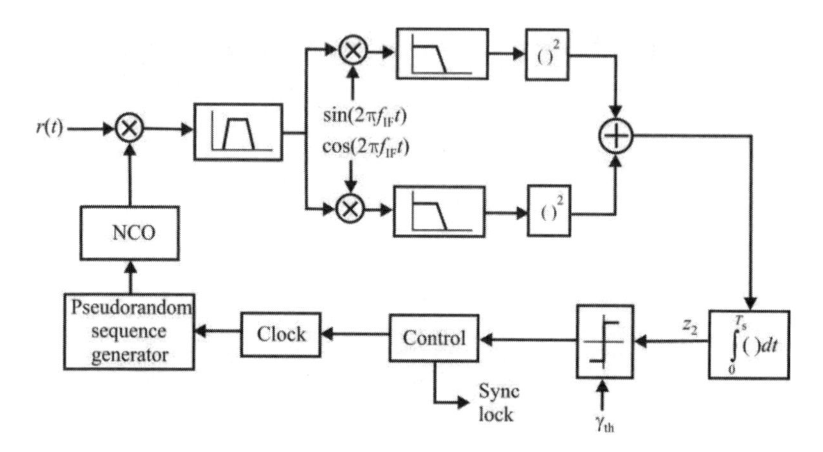

Figure 7.22 Block diagram of an active correlator implementing a serial search scheme for FHSS code synchronization.

$\gamma' = \gamma_{\text{th}_2}$, and the probability of not achieving synchronization on a single attempt is given by (7.90). The overall probability of false alarm is then given by

$$P_{\text{fa}} = P_{\text{F}}^m \tag{7.92}$$

where P_{F} is given by (7.88). The overall probability of missing code acquisition is given by

$$P_{\text{NoSync}} = 1 - P_{\text{H}}^m \tag{7.93}$$

where $P_{\text{H}} = 1 - P_{\text{NoSync}_k}$ with P_{NoSync_k} from (7.90).

Note that the receiver knows the hop sequence (code), but does not know where in the sequence the incoming signal is located within that sequence. A typical hop set would be comprised of, say, 256 frequencies, so a search over 256 hop intervals should suffice to find where the sequence is in its pattern.

The serial search scheme is slow but it has a high probability of finding the code sequence. That is, the probability of false lock is low. In the noise-free case, it is guaranteed to find the correct code location.

The probability of missed sync for the serial search technique assuming $m = 2$ sequential hits are required is illustrated in Figure 7.23. As for matched filtering, the thresholds were adjusted to keep $P_{\text{fa}} \approx 10^{-4}$.

The average time to achieve sync lock is given by

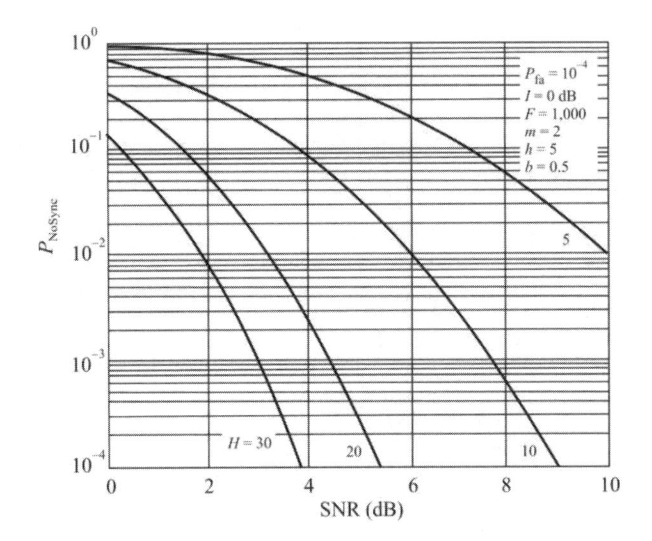

Figure 7.23 Probability of missed sync lock for the serial search approach to coarse code acquisition. When there were $h = 5$ interfering signals, the thresholds were $\gamma'_5 = 23$ (14 dB), $\gamma'_{10} = 35$ (15 dB), $\gamma'_{20} = 60$ (18 dB),and $\gamma'_{30} = 85$ (19 dB). m is the number of sequential successful synchronization indications.

$$T_s = (mN_c + 2)HT_h \tag{7.94}$$

where

n = number of cells searched per chip;
N_c = maximum delay difference between the received and locally generated code sequence.

7.4.3 Two-Step Synchronization

Because of the sporadic characteristics of PTT communication, rapid and accurate techniques are required to facilitate resynchronization upon every transmission event. That is the motivation for the third form of FHSS code synchronization approach discussed here.

This two-level code synchronization approach was first analyzed by Rappaport and Schilling [6]. This is a combination of the serial search and matched filter. A block diagram of the approach is shown in Figure 7.24. The leader transmits a short segment of H hop frequencies. The first stage is comprised of a passive correlator (matched filter), the output of which, if above threshold

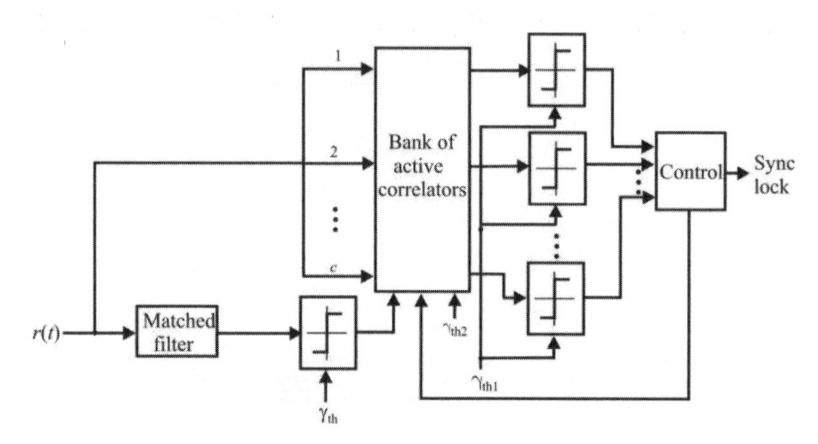

Figure 7.24 Two-step FHSS code synchronization architecture. (Source: [16]. © IEEE 1983. Reprinted with permission.)

γ_{th_1}, triggers the second stage, which is a bank of active correlator detectors. The matched filter is designed to search P of the H frequencies in this sync prefix. When the match filter exceeds its threshold, it is an indication that a possible sync prefix has been detected. In that case one of the c active correlators is started (if one is available) to look for the remainder A ($P + A = H$) of the hop set. At the end of the H hops, if the active correlator has detected enough of the hops, its output will exceed the second threshold indicating that sync lock has been achieved. If it does not exceed the second threshold then that correlator is made available to the pool for subsequent assignment by the matched filter. If there is no available active correlator when the first step indicates the presence of the sync prefix, then that notification is ignored.

There are P passive matched filters in the first stage and A active correlators in the second. The relatively low reliability but fast performance of the matched filter is used to presort through the possible code offsets before the time-consuming, but high reliability, active correlator technique is employed. The rationale for such an approach is that the passive scheme is fast but hardware-intense, whereas the active scheme is slower but hardware light. The effect of interfering signals and jamming tones is essentially to increase the number of false detections in the first step of the two-step process. This increases the load on the correlators for the second step, increasing the blocking probability. It also, however, increases the probability of correctly detecting the beginning of a correct synchronization sequence. The false detection and false dismissal probabilities are given by (7.88) and (7.90), respectively, with $H = P$ and $\gamma = \gamma_{\text{th}_1} / P_{\text{N}}$ for the first step and $H = A$ and $\gamma = \gamma_{\text{th}_2} / P_{\text{N}}$ for the second step.

The probability of a missed synchronization on a single pass through the synchronization process is given by

$$P_{\text{NoSync}_k} = P_{\text{NoSync}_1} + (1 - P_{\text{NoSync}_1})\{B(c,a) + [1 - B(c,a)]P_{\text{NoSync}_2}\} \quad (7.95)$$

where P_{NoSync_1} is the probability of the passive correlator incorrectly dismissing the in-sync condition when it is present, P_{NoSync_2} is the same probability for the active correlator, and $B(c, a)$ is the Erlang B formula given by

$$B(c,a) = \frac{\dfrac{a^c}{c!}}{\displaystyle\sum_{k=0}^{c} \frac{a^k}{k!}}, \; c = 1,2,\cdots \quad (7.96)$$

with $a = P_{\text{fa}_1} A$. The Erlang blocking formula (7.96) reflects the likelihood of a call getting blocked in a phone system, assuming the call attempts arrive according to a random process characterized by the Poisson distribution, and is used here as an indication of whether an active correlator is available. In this formula, c represents the number of processing assets available and a represents the rate at which these assets are tasked to perform a function.

Equation (7.95) is illustrated in Figure 7.25 for the two-step code acquisition process for several values of c. Note that the curves turn upward as the SNR increases past some point for $c < 7$. This is because the Erlang blocking probabilities increase as the SNR increases. It is ameliorated as the number of second-stage correlators increases, and for $c \geq 7$ for this example and for the ranges displayed the characteristic disappears. Although the false alarm probability varies with the SNR, the threshold values were adjusted for these curves to make P_{fa} approximately 10^{-5} at $\upsilon = 0$ dB.

The false alarm probability of the two-step technique is given by

$$P_{\text{fa}} = P_{\text{f}_1} P_{\text{f}_2} \left[1 - B\left(c, P \times P_{\text{f}_1}\right)\right] \quad (7.97)$$

Whereas for the serial search scheme alone, several passes through the sync frequencies could and should be attempted before declaring code lock, which is not possible in the two step technique. Multiple attempts can and should be made

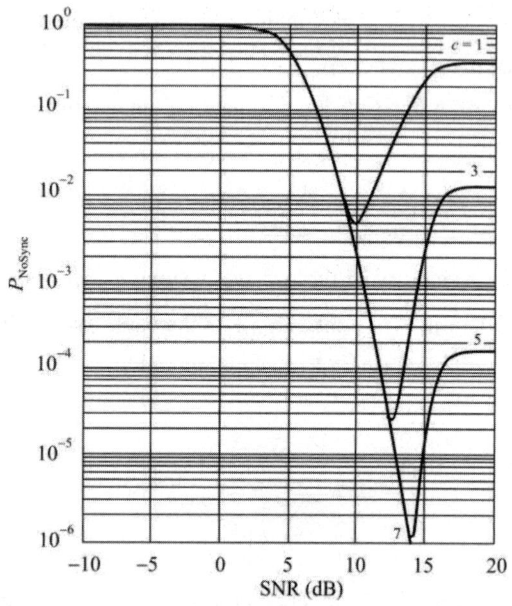

Figure 7.25 Probability of missing synchronization for the two-step code acquisition technique. In this example $b = 0.5$, $P = 10$, $A = 100$, $F = 1{,}000$, $\gamma_1 = 80$ (19 dB), $\gamma_2 = 30$ (15 dB), $I = 0$ dB, and $h = 7$.

using both steps, however. The probability of no sync acquisition, P_{NoSync}, for m passes through the hop set in the leader is given by

$$P_{\text{NoSync}} = 1 - P_{\text{H}_k}^m \tag{7.98}$$

where P_{H_k} is the probability of a hit on each individual pass, given by

$$P_{\text{H}_k} = 1 - P_{\text{NoSync}_k} \tag{7.99}$$

where P_{NoSync_k} is given by (7.95).

The time required to search the code offset space for two-level synchronization is given by

$$T_{\text{s}} = (N_{\text{c}} + H)T_{\text{h}} \tag{7.100}$$

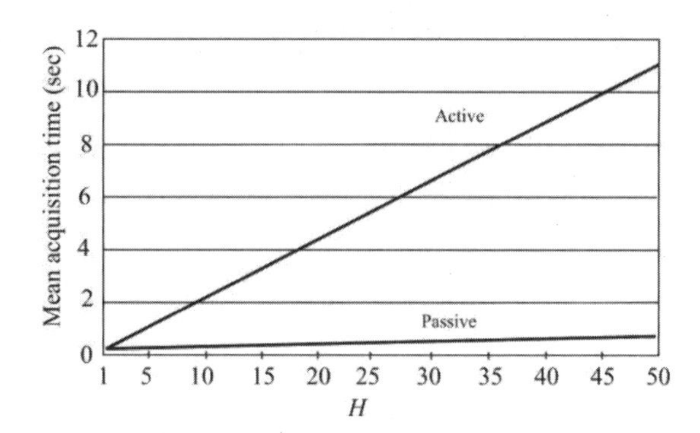

Figure 7.26 Comparison of the mean synchronization acquisition times for active and passive correlators.

7.4.4 Comparison of Mean Acquisition Times

The mean acquisition times for the serial search (active correlator) and the matched filter (passive correlator) are shown in Figure 7.26. Clearly, the passive approach is much faster than the active approach by a considerable amount for reasonable sizes of the correlators. The active approach may require many seconds. This chart assumes that $m = 1, N_c = 20$ for a 100-hps FHSS system.

Similar data for the two-level approach is shown in Figure 7.27. Relying on the presorting possible with the matched filter significantly shortens the code acquisition time. These times are short enough to facilitate PTT communications.

7.5 FHSS Tracking

Just as for DSSS, after coarse acquisition establishes code synchronization to within a single dwell time or code symbol, handoff to the tracking mode occurs. Coarse acquisition establishes this synchronization to typically half to one-quarter of a chip interval. The tracking function is to maintain the code synchronization to within a small fraction of a single chip.

There are two goals of the tracking circuit. The first is to maintain code synchronization. In addition to this, if the coarse synchronization has established a track incorrectly, the second goal of the tracking function is to recognize this and return to coarse synchronization as quickly as possible.

Perhaps the most common architecture for FHSS tracking is based on the early-late gate. That will be described here.

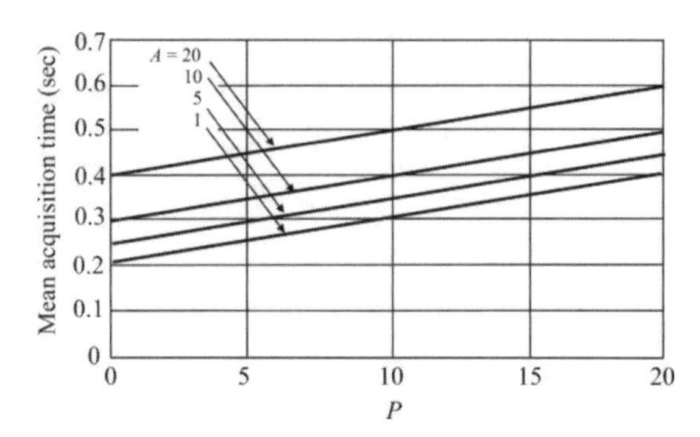

Figure 7.27 Mean synchronization acquisition time for two-level code acquisition.

7.5.1 Early-Late Gate Tracking

One implementation of an early-late tracking gate is shown in Figure 7.28. When the receiver is perfectly synchronized to the incoming chip signal, the early and late gates, which are of duration $T_c/2$ each, occupy the first half and last half of the chip duration, respectively, as shown in Figure 7.28(b). When synchronized, the output of the early gate integrator increases to a fixed level during the first half, and that level is maintained during the second half. During the second half of the chip duration, the late integrator also rises to that same level. At the end of the chip, the two integrator outputs are compared (by subtraction in this case) and the output will be 0 V. This output is sent to a voltage-controlled clock, which slows the clock rate with a positive input voltage and increases the rate for a negative input.

If the receiver is early (the gates occur too early compared to the chip), then at the end of the chip duration, there is a negative value out of the comparator as shown in Figure 7.28(c), indicating to the voltage controlled clock to shift later in time. If the receiver is too late, then at the end of the chip there is a remaining positive value out of the comparator as shown in Figure 7.28(d), indicating to the clock to adjust the time so the early and late gates occur earlier.

This description assumed that the chip was a logical 1. Additional logic is required to handle the case when the chip is a logical 0, but the modification is straightforward. This architecture, then, can be used to move the timing of the locally generated pseudo-random sequence back and forth within the duration of a chip.

There is always some residual phase jitter in the early-late tracking loop due to the uncertainty in clocks matching in the transmitter and receiver. The variance of this jitter is given by [6]

$$\sigma_{\mathrm{T}}^2 = \frac{T_{\mathrm{c}}^2}{8N_{\mathrm{h}}\dfrac{E_{\mathrm{c}}}{N_0}} \tag{7.101}$$

where N_{h} is the number of hops in the loop integration period.

Putnam, Rappaport, and Schilling reported on the analysis of an early-late gate tracking loop similar to that shown in Figure 7.28 [17]. This loop is shown in Figure 7.29 where the coarse loop and fine (tracking) loop are combined into a single module.

In hop interval k, the received signal is given by

$$r_k(t) = \sqrt{2R}\cos(2\pi f_k(t - kT_{\mathrm{c}}) + \theta_k) + n_k(t - kT_{\mathrm{c}}) \tag{7.102}$$

where R is the average power in the signal and T_{c} is the chip (dwell) time. The noise is given by $n_k(t - kT_{\mathrm{c}})$, which is assumed to be Gaussian with one-sided power density N_0 and variance σ^2. If the loop integration is over M chips and with only thermal noise present, the probability of detection is given by

$$P_{\mathrm{d}} = Q(\sqrt{A}, \sqrt{\gamma'}) \tag{7.103}$$

where

$$A = 2M\frac{R}{\sigma^2} \tag{7.104}$$

and

$$\gamma' = \gamma / \sigma^2 \tag{7.105}$$

when γ is the detection threshold. That is, if z is the test statistic, then

$$P_{\mathrm{d}} = \Pr\{z > \gamma\} = \int_{\gamma}^{\infty} p_U(u)du \tag{7.106}$$

The tracking loop integration time is MT_{c}, and γ_{h} is a measure of the threshold to the noise density given by

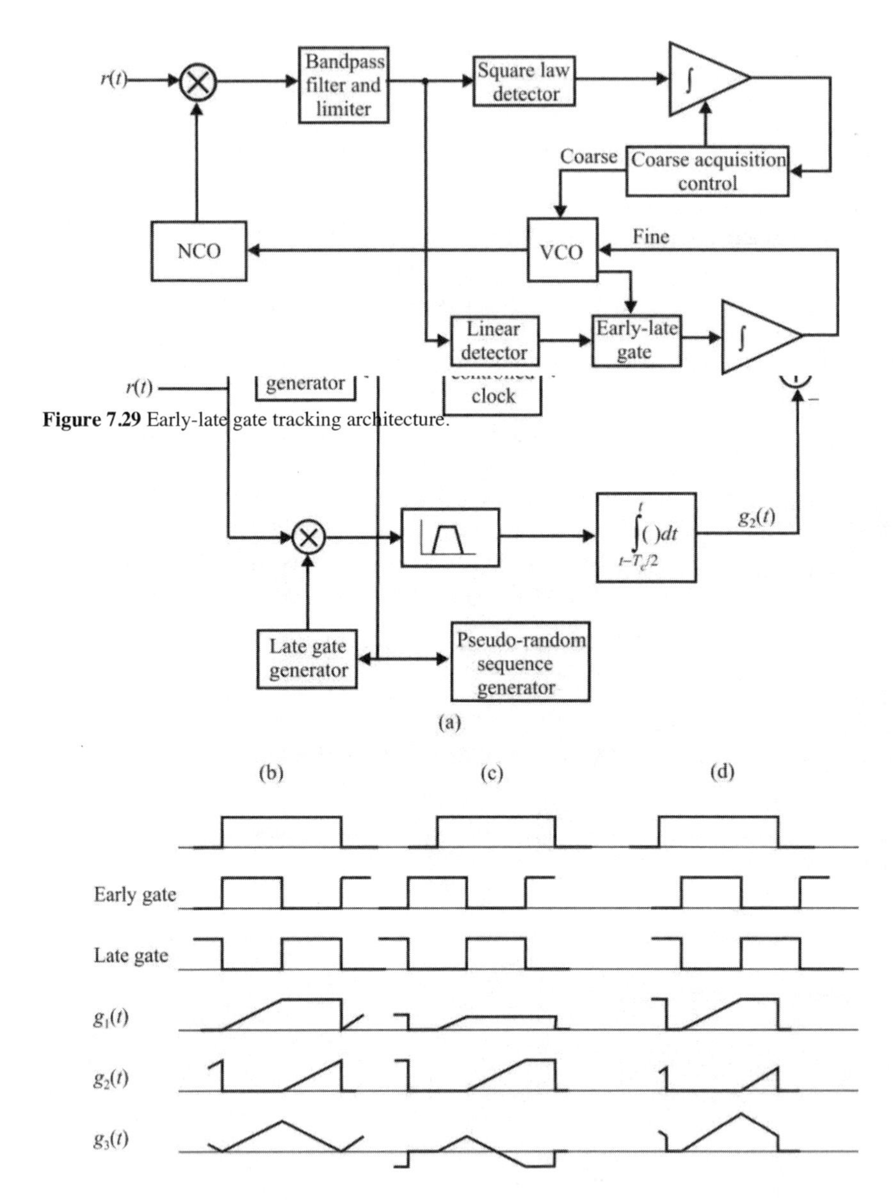

Figure 7.29 Early-late gate tracking architecture.

Figure 7.28 Early-late tracking loop for FHSS signals: (a) block diagram of tracking loop; (b) in synchronization; (c) receiver is early; and (d) receiver is late.

$$\gamma_{\mathrm{h}} = \frac{\gamma_{\mathrm{th}}}{N_0} \tag{7.107}$$

when γ_{th} is the tracking loop threshold.

The probability of miss in the tracking loop is given by

$$P_{\mathrm{m}} = 1 - P_{\mathrm{d}} \tag{7.108}$$

and the probability of false alarm is given by

$$P_{\mathrm{fa}} = \lim_{a \to 0} Q(a, \gamma_{\mathrm{h}}) \tag{7.109}$$

As explained in [6], the tracking loop phase jitter will reduce the correlation peak because the correlation is not perfect. Assuming that the phase jitter is Gaussian, 99.7% of the time it will be within 3σ of the mean so the correlation peak is reduced by a factor d given by

$$d = \left(1 - \frac{3\sigma_{\mathrm{T}}}{T_{\mathrm{c}}}\right)^2 \tag{7.110}$$

which can be included in the power as a multiplicative factor as

$$A = \frac{2MdR}{W_{\mathrm{IF}} N_0} \tag{7.111}$$

since $\sigma^2 = N_0 W_{\mathrm{IF}}$.

There are a wide variety of control processes that can be implemented. Recall that one of the goals of tracking is to return to coarse synchronization as quickly as possible when that function has incorrectly established what it thought was synchronization. The architecture of the control system for the tracking loop establishes the performance of the loop.

One possibility is to return to coarse acquisition when a single miss in the tracking loop is detected. This is perhaps the fastest way to correct incorrect coarse synchronization, but will also lead to a higher level of correct tracking being rejected. It is also possible to have m out of n type of control, where $m < n$ detections of sync-loss out of n passes are required before returning to coarse synchronization. A refinement of this second approach could be to return to $m = 0$

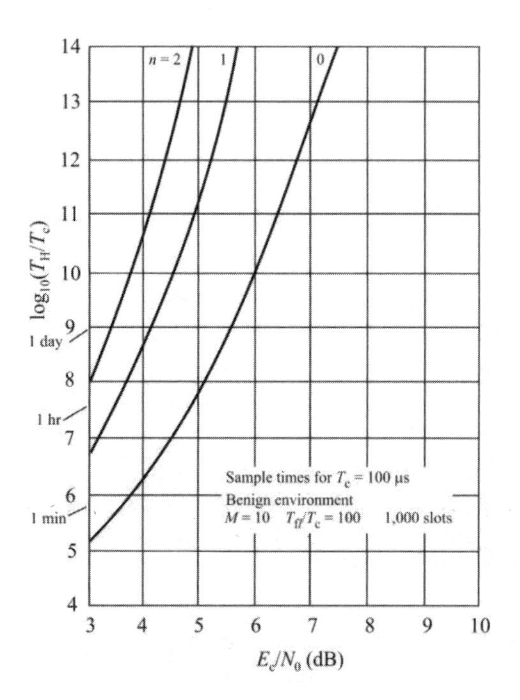

Figure 7.30 Normalized mean hold time for T_c = 100 μs. (Source: [17]. © IEEE 1983. Reprinted with permission.)

upon any detection of correct tracking. For this last approach, the mean time to loss of synchronization, denoted T_L, is calculated to be [6]

$$T_L = MT_c \frac{1-q^{n+1}}{(1-q)q^{n+1}} \tag{7.112}$$

When $q = P_m$, then T_L is the mean time to loss of synchronization for this approach, also known as the *mean hold time* (T_H), when the code epoch has been correctly determined. When $q = 1 - P_{fa}$, the mean hold time for when the code epoch is not established is given, also known as *mean time for false lock* (T_{fl}).

The same Ricean fading model, as described previously by (7.82), was used to describe the signal propagation environment. The normalized mean hold time is shown in Figure 7.30 (normalized by the chip time) performance versus the chip energy SNR when there are neither interferers nor tone jammers. The parameter n is the number of times the tracking detector must detect loss of sync before returning to the coarse acquisition process. For example, for an SNR of 5 dB and

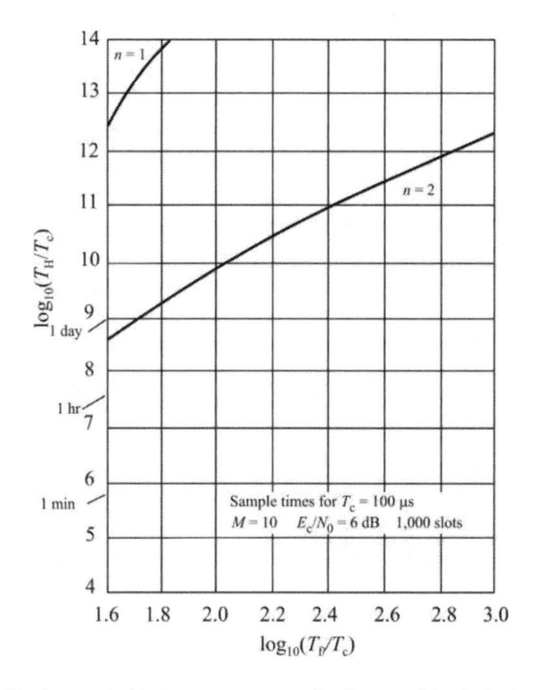

Figure 7.31 Normalized mean hold time versus normalized mean false lock time. (Source: [17]. © 1983. Reprinted with permission.)

with one detection of loss of sync adequate to reinitialize coarse acquisition, the mean hold time ratio is

$$\log\left(\frac{T_{\mathrm{H}}}{T_{\mathrm{c}}}\right) \approx 11 \tag{7.113}$$

so

$$T_{\mathrm{H}} \approx 10^{7}\ \mathrm{sec} = 115\ \mathrm{hours} \tag{7.114}$$

Thus, synchronization will be maintained for an average of 115 hours in this case.

Figure 7.31 illustrates the normalized mean hold time versus the normalized mean false lock time. For these example parameters, the recovery time from a false lock condition is essentially instantaneous. For example, when $\log_{10} T_{\mathrm{FL}} / T_{\mathrm{c}} = 2$, $T_{\mathrm{FL}} = 0.01$ second and for this value of the abscissa and $n = 2$, $\log_{10} T_{\mathrm{H}} / T_{\mathrm{c}} \approx 10^{6}$ second.

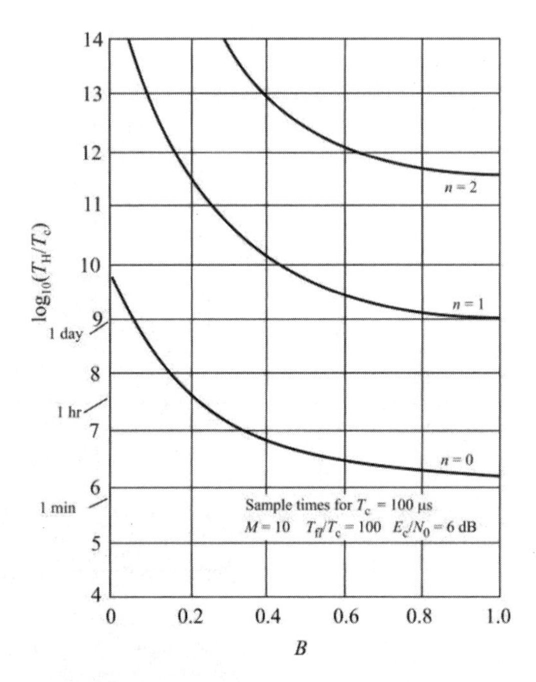

Figure 7.32 Normalized mean hold time versus the strength of the scatterers. (Source: [17]. © IEEE 1983. Reprinted with permission.)

The normalized mean false hold time is illustrated in Figure 7.32 versus the relative strength of the scattered signals at the receiver, as expressed by B. As more power shifts into the reflected components from the direct component (B increases), the hold time tends to approach asymptotic values for each n. Likewise, for small amounts of power in the scattering components, the hold time decreases relatively rapidly as B increases.

The normalized mean hold time is shown in Figure 7.33 versus the number of other users of the spectrum as heard by the receiver. As opposed to coarse acquisition, where interfering signals have a profound impact on synchronization, the hold time is not particularly sensitive to the number of potential interfering users for the parameters indicated. The effects of interfering signals on coarse acquisition are discussed earlier.

7.6 Concluding Remarks

The fundamental properties of synchronization and tracking in DSSS and FHSS systems are presented in this chapter. PN code synchronization and tracking are

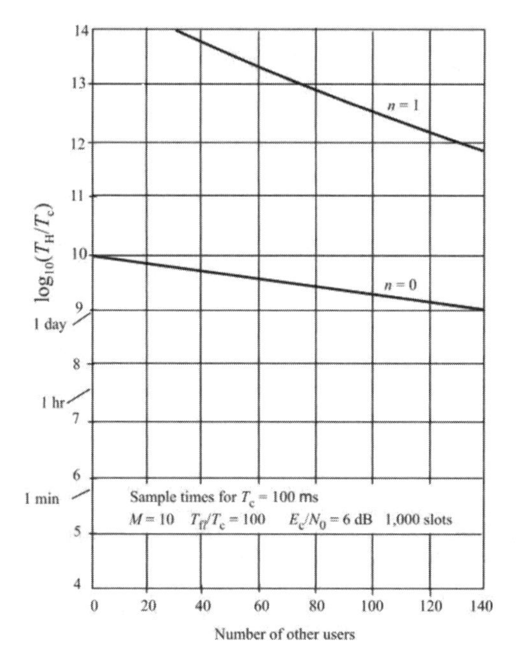

Figure 7.33 Normalized mean hold time versus the number of other users of the spectrum. (Source: [17]. © IEEE 1982. Reprinted with permission.)

requirements of all AJ systems. Synchronization is the process of initially determining where in the PN sequence the received signal is located. Once this is established, handoff to the tracking function is accomplished. The tracking function attempts to keep the receiver PN code generator within a small fraction of a chip time.

This discussion is included here to establish the baseline for effective jamming techniques against these functions to be discussed next in Chapter 8.

References

[1] Noneaker, D. L., "The Performance of Serial Matched Filter Acquisition in Direct Sequence Packet Radio Communications," *Proceedings IEEE MILCOM*, 2001.

[2] Curry, S. J., "State Transition Assisted Code Synchronization," *Proceedings IEEE MILCOM* 1999, http://www.agreenhouse.com/society.TacCom/milcom_99_ papers.shtml.

[3] Yoon, S, I. Song, S. R. Park, S. Y. Kim, and J. Lee, "A Code Acquisition Scheme for DS/CDMA Systems," *Proceedings IEEE MILCOM*, 1997, http://www.agreenhouse.com /society.TacCom/milcom_99_papers.shtml.

[4] Turin, G. L., "An Introduction to Digital Matched Filters," *Proceedings of the IEEE*, Vol. 64, No. 7, July 1976, pp. 1092–1112.

[5] Wong, T. F., "Spread Spectrum & CDMA," http://www.dsp.ufl.edu /~twong/notes1.html.

[6] Rappaport, S. S., and D. L. Schilling, "A Two-Level Code acquisition Scheme for Spread Spectrum Radio," *IEEE Transactions on Communications*, Vol. COM-28, No. 9, September 1980, pp. 1734–1742.

[7] Peterson, R. L., R. E. Ziemer and D. E. Borth, *Introduction to Spread Spectrum Communications*, Upper Saddle River, NJ: Prentice Hall, 1995, pp. 225–230.

[8] Strom, E. G., S. Parkvall, S. L. Miller, and B. E. Ottersten, "Propagation Delay Estimation in Asynchronous Direct Sequence Code-Division Multiple Access Systems," *IEEE Transactions on Communications*, Vol. 44, No. 1, January 1996, pp. 84–93.

[9] Pickholtz, R. L., D. L. Schilling, and L. B. Milstein, "Theory of Spread Spectrum Communications—A Tutorial," *IEEE Transactions on Communications*, Vol. 30, No. 5, May 1982, pp. 855–884.

[10] Moon, T. K., R. T. Short, and C. K. Rushforth, "Average Acquisition Time for SSMA Channels," *Proceedings IEEE MILCOM*, 1991, pp. 1042–1046.

[11] Rappaport, S. S., and D. M. Grieco, "Spread-Spectrum Signal Acquisition: Methods and Technology," *IEEE Communications Magazine*, Vol. 22, No. 6, June 1984, pp. 621–635.

[12] Yoon, S., I. Song, and S. Y. Kim, "Code Acquisition for DSSS Communications in Non-Gaussian Impulsive Channels," *IEEE Transactions on Communications*, Vol. 52, No. 2, February 2004.

[13] Rick, R. R., and L. B. Milstein, "Parallel Acquisition in Mobile DS-CDMA Systems," *IEEE Transactions on Communications*, Vol. 45, November 1997, pp. 1466–1476.

[14] Song, I., J. Bae, and S. Y. Kim, *Advanced Theory of Signal Detection*, Berlin, Germany: Springer-Verlag, 2002.

[15] Benedetto, F., and G. Giunta, "A Self-Synchronizing Method for Asynchronous Code Acquisition in Band-Limited Spread Spectrum Communications," *IEEE Transactions on Communications*, Vol. 57, No. 8, August 2009, pp. 2410–2417.

[16] Putman, C. A., S. S. Rappaport, and D. L. Schilling, "A Comparison of Schemes for Coarse Acquisition of Frequency-Hopped Spread-Spectrum Signals," *IEEE Transactions on Communications*, Vol. COM-30, No. 2, February 1983, pp. 183–187.

[17] Putman, C. A., S. S. Rappaport, and D. L. Schilling, "Tracking of Frequency-Hopped Spread Spectrum Signals in Adverse Environments," *IEEE Transactions on Communications*, Vol. COM-31, No. 8, August 1983, pp. 955–964.

Chapter 8

Jamming Techniques

8.1 Introduction

In this chapter we discuss types of jamming strategies that a jammer could use against AJ targets. The coverage is not all-inclusive, but most of the more common approaches are presented. A jammer can use several possible strategies—each has its own advantages and disadvantages and may be the "best" strategy against a particular set of targets.

First, noise jamming is presented. Within this category are wideband noise, partial-band noise, and narrowband noise. The first two could apply to a nonagile jammer where the jamming signals occupy a portion of or the entire spectrum in use by the AJ system, but the signals stay in one place in the spectrum. The narrowband noise case is associated with follower jamming where attempts to follow a frequency-hopping AJ target are made. If the signal can be correctly found quickly enough then jamming energy can be applied and communication denial can be attempted at the new frequency.

Next tone jamming is discussed. Tone jammers can be applied to both DSSS and FHSS AJ signals. A single tone can be used against DSSS or FHSS when the latter is attempted with the follower strategy. Multiple tones can be applied to both types of AJ signals as well. Multiple-tone jamming is also referred to as multitone, or MT, here.

Swept jamming is presented next, which takes one of the basic techniques and scans the waveform across all or a significant portion of the spectrum of interest. This sweeping can be accomplished with analog "scanning" exciters or digital "stepping" exciters. In the former case the jamming waveform is only at a given channel for an "instant." In the latter, the jamming signal can dwell at a channel for arbitrary (and mixed) times.

Pulse jamming is the appellation applied to intermittently transmitting high-power energy. It is the DSSS analog to partial-band noise jamming against FHSS AJ signals and the results are similar.

The last jamming technique presented is smart jamming, where particular parts of a signal are targeted. Much must be known about the particular target signal to apply this technique. The ability to synchronize the jamming waveform with the target signal is also required with smart jamming. This problem is exacerbated by, typically, the flight time of the jamming signal, which is difficult to predict. The jamming signal must be synchronized with the target signal at the *target receiver*.

The chapter concludes with a discussion of *jam-to-signal ratio* (JSR) calculations with the models presented in Section 2.1. The JSR is the parameter that indicates whether jamming is effective. The required JSR depends on the particular type of signal being considered. When referring to JSRs herein, it is implicitly meant to refer to these signal levels at the target receiver. Occasionally, to reinforce this point, the received jammer power will be referred to as J to distinguish it from the jammer ERP at the transmitter, denoted as J_T here and previously. S and R are only used to refer to the average signal power at the receiver depending on whether the signal is specified as $s(t)$ or $r(t)$, while P_T refers to the target net transmitter ERP.

8.2 Jamming Strategies

It has been shown that jamming approximately 30% or more of a voice transmission degrades the intelligibility significantly enough to deny an effective transfer of information [1]. Therefore, denying 30% of transmissions represents a reasonable goal to strive for and serves as a useful threshold to achieve for such communications. Jamming of substantially less than 30% is effective against coded or uncoded digital data communications, as will be shown. Jamming effectiveness criteria for digital communications will be substantially different from that for analog, voice communications. All important SS communication systems utilize digital signaling techniques.

There are several possible strategies that a jammer can employ against a communication system to include AJ systems. Some techniques are more effective than others, and a successful strategy depends on the particular type of AJ employed.

Two fundamental waveforms are typically used against AJ communication systems. A carrier signal centered on the transmitting frequency is unmodulated, modulated with one or more tone signals, or modulated with a noise signal. The bandwidth of the noise can be varied. When the carrier is not modulated, the jamming waveform is a single tone. When modulated with more than one tone, then multiple tones are emitted by the jammer. Usually, the placement of these tones is based on some knowledge of the target or targets to be jammed. Noise is used to raise the background noise in the spectrum in which the AJ system is operating.

The taxonomy of jamming strategies consists of how the available jammer power is distributed with frequency, the type of modulation transmitted (if any), the time-sharing of the jammer among several targets, and power sharing, where the instantaneous power can be distributed among several targets.

8.2.1 Partial Dwell Jamming of FHSS Systems

When jamming FHSS systems using a follower jamming technique, only a portion of each dwell is jammed. There is a finite amount of time to ascertain if newly detected energy belongs to the correct signal to jam. It is not normally possible to jam the entire hop dwell at the receiver, but only a fraction of it, denoted here by γ.

Likewise, when jamming using a partial-band technique, with either noise or multiple tones, only a portion of the dwells is jammed. The FHSS system can hop into a band or tone that contains a jamming signal or not, depending on the placement of the jamming signal and the hop sequence.

The overall probability of jamming a dwell is given by the probability of a symbol error when the jammer is not present and P_e is determined by the thermal background noise, and probability of a symbol error when the jammer is present, where the BER is determined by the composite effects of the jammer energy combined with the thermal noise. For simplicity, in this section it is assumed that the FHSS system is using BFSK modulation. Extensions to higher orders of FSK modulation are straightforward. The probability of a bit error occurring when the jammer is not present is due to the background noise only and is given by (3.93) with $P_n = 0$ as

$$P_{e_1} = \frac{1}{2}\exp\left(-\frac{1}{2}\frac{R}{P_N}\right) \tag{8.1}$$

where R is the average power in the signal, P_N is the thermal noise power, and it has been assumed that $R_c = W_F$.

Let P_{e_2} denote the probability of a symbol error with the jammer present. Then without diversity, that is, for SFHSS,

$$P_e = (1-\gamma)P_{e_1} + \gamma P_{e_2} \tag{8.2}$$

The particular expression for P_{e_2} depends on the individual jamming technique invoked and is discussed in Chapter 11.

8.2.2 Noise Jamming

For *noise jamming*, the jamming carrier signal is modulated with a random noise waveform. The intent is to disrupt the AJ communication waveform by inserting noise into the receiver. The bandwidth of the signal can be as wide as the entire spectrum width used by the AJ system or much narrower, occupying only a single channel. The effects are different depending on the details of the implementation.

The noise is generally assumed to be Gaussian, and therefore this is the type of noise sources designed into jamming systems. Theoretical Gaussian noise has an infinite frequency extent. Colored Gaussian noise is Gaussian noise that has been subject to filtering and is the appropriate type to use in situations where the filtering effects are important.

8.2.2.1 Broadband Noise Jamming

Broadband noise (BBN) jamming places noise energy across the entire width of the frequency spectrum used by the target communication systems. It is also called *full band* jamming and is sometimes called *barrage jamming*. This latter appellation, however, also refers to cases where less than the full band is jammed. This type of jamming is useful against all forms of AJ communications. It is generally useful for coverage of an area for screening purposes as well. In this EP role, the jammer is placed between an adversary's ES system(s) and friendly communications. Directional antennas are, of course, required to be pointed in the direction of the ES systems if friendly fratricide is to be minimized. If used correctly, this prevents the adversary from intercepting friendly communications at least for a time.

Since BBN jamming generates signals that are similar to broadband noise, the level of jamming power is sometimes referred to as J_0 and is measured in watts/hertz just as background noise is specified. The primary limitation of BBN jamming is that it results in low J_0 as limited jammer power is spread very wide. It is not as effective as partial-band jamming can be against BFSK. The spectrum of BBN jamming is illustrated in Figure 8.1(b).

This type of jamming essentially raises the background (thermal) noise level at the receiver, creating a higher noise environment for the AJ system. Noise is the nemesis for any communication system and if the noise level can be increased, it makes it more difficult for the communication system to operate. At the very least it decreases the range over which the communication system is effective. In some cases, this effect is adequate to accomplish the goals of ECM. In other cases, communication can be totally denied.

BBN jamming is a direct assault on the channel capacity of a communication system. This capacity, assuming the noise is Gaussian, was first investigated by Shannon in 1948 [2]. It expresses the maximum data rate that the channel can

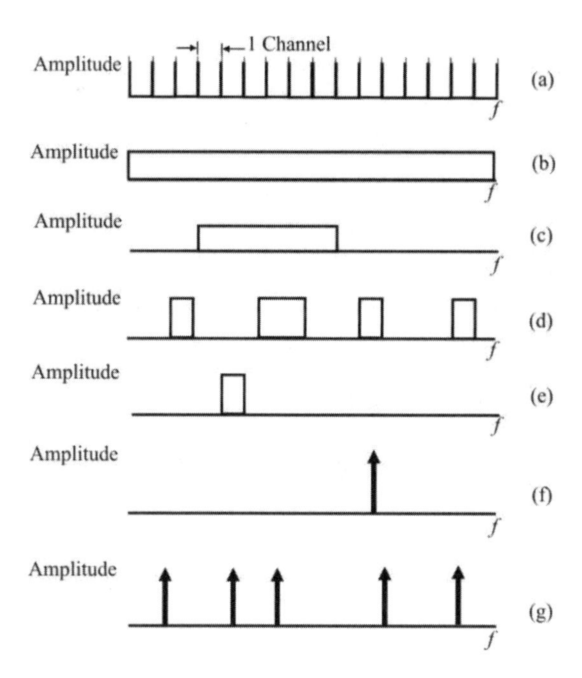

Figure 8.1 Possible strategies a jammer may use based on the channelized spectrum shown in (a) and (b) full-band jamming (BBN), (c) contiguous partial-band jamming (PBN), (d) noncontiguous partial-band jamming (PBN), (e) narrowband noise jamming (NBN), (f) single-tone (ST) jamming, and (g) multitone (MT) jamming.

carry with an arbitrarily small error rate. If an attempt is made to transmit a digital signal through the channel with a higher bit rate than that given by the capacity, then errors are assured in the received signal. The capacity of a channel corrupted by such noise is given by

$$C = W_{ss} \log_2 \left(1 + \frac{R}{P_T}\right) \tag{8.3}$$

where W_{ss} is the bandwidth of the signal, R is the average power of the signal, and P_T is the total average noise present given by $P_T = W_{ss}(N_0 + J_0)$. Clearly, as the noise level is raised by intentionally inserting Gaussian noise into the channel, the SNR decreases, thus decreasing the channel capacity. BBN jamming of FHSS systems utilizing BFSK produces the P_e given in (3.89).

BBN Jamming Effects on Synchronization

PTT FHSS systems typically must maintain tracking on every transmission attempt. The tracking consists of aligning the time epoch of the transmitter at the receiver and maintaining the PN code sequence. Generally maintaining coarse synchronization can be accomplished by keeping the transmitter and receiver clocks/oscillators running even when not transmitting. The fine tracking, however, must be adjusted on each attempt. Tracking causes the receiver to hop to the same frequencies as the transmitter and ensures that these hops occur at the same time. If the receiver is precluded from tracking the transmitter, then communication is denied. The effects of BBN jamming when FHSS systems are trying to synchronize and track are addressed in this section.

BBN jamming raises the background noise levels and, as such, can be used to attack the synchronization and tracking processes. The arguments in Section 7.4.1 apply directly here with the higher noise level. Thus,

$$\begin{aligned}\xi' &= \sum_{k=1}^{H} \frac{2R_k}{\sigma^2 + J} \\ &= \sum_{k=1}^{H} \frac{1}{\dfrac{\sigma^2}{2R_k} + \dfrac{J}{2R_k}} \\ &= \sum_{k=1}^{H} \frac{1}{\upsilon_k + \xi_k}\end{aligned} \tag{8.4}$$

where υ_k is the SNR during hop interval k and ξ_k is the JSR during the same interval. Furthermore,

$$T' = \frac{\gamma_{\text{th}}}{\sigma^2 + J} \tag{8.5}$$

The probability of false alarm is given by (7.84) and the probability of missed sync is given by (7.86) with $\upsilon = \upsilon'$ and $T = T'$.

Matched Filter Coarse Acquisition. When the matched filter approach to achieving coarse acquisition is used, then (7.84) and (7.86) are used directly. The results for an example when υ = 10 dB are shown in Figure 8.2.

As noted in Chapter 4, well-designed FHSS receivers employ limiters prior to their detectors to eliminate excessive jamming signal energy. As seen in Figure

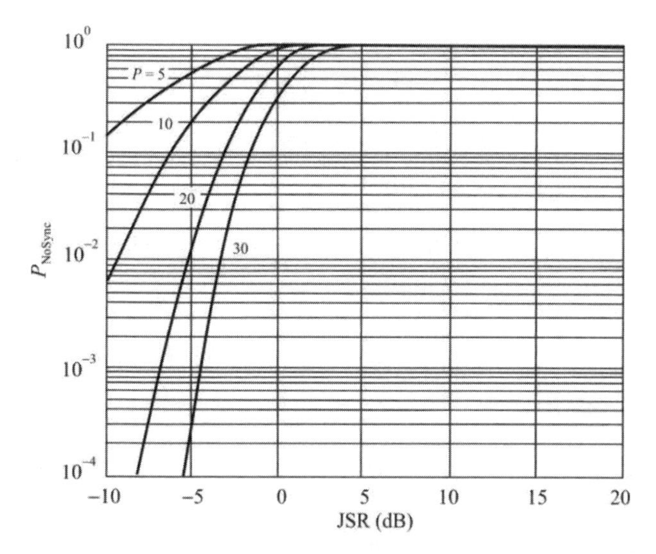

Figure 8.2 Probability of not achieving coarse code acquisition with BBN jamming of FHSS parallel synchronization (matched filter) when υ = 10 dB. The threshold ratios were adjusted to keep $P_{\text{fa}} \approx 10^{-4}$ yielding: P = 5, $\gamma_1 = 35(15\text{ dB})$, P = 10, $\gamma_1 = 22(13\text{ dB})$, P = 20, $\gamma_1 = 80(19\text{ dB})$, and P = 30, γ_1 = 105 (20 dB). (P is the number of passive correlators.)

8.2, however, for matched filter jamming, the JSR has little additional effect above $\xi \approx 0$ dB.

Stepped Serial Coarse Code Acquisition. Just as for matched filter jamming, the analysis in Section 7.4.1 applies here as well since the noise level is raised. For serial search code acquisition, one technique is to require that some number of successive hits is required before declaring that coarse synchronization has been achieved. Another is to require some number of hits, and with a miss, the system drops back one stage. This is illustrated with the finite state machine in Figure 8.3. When the strategy is that m consecutive hits must occur to declare synchronization, then

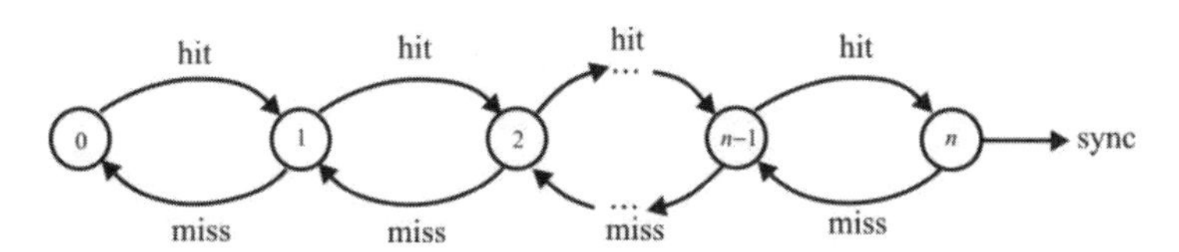

Figure 8.3 Finite state machine representation of strategy of going back one stage at a time when a missed sync indication occurs.

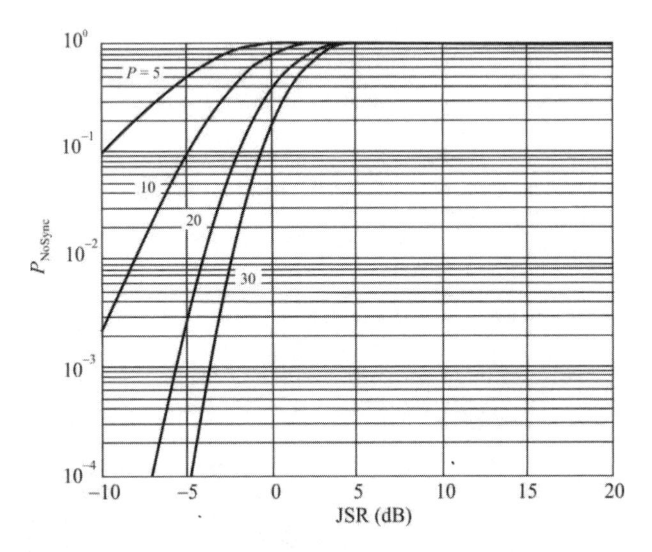

Figure 8.4 Probability of missed coarse code acquisition in serial search in BBN jamming when SNR = 10 dB. The threshold ratio was adjusted to keep $P_{fa} \approx 10^{-4}$ yielding: $A = 5, \gamma_1 \approx 23$ (13 dB), $A = 10, \gamma_1 \approx 35$ (15 dB), $A = 20, \gamma_1 \approx 60$ (18 dB), and $A = 30, \gamma_1 \approx 85$ (19 dB).

$$P_{\text{NoSync}} = 1 - P_{\text{H}_k}^m \tag{8.6}$$

where P_{H_k} is given by $1 - P_{\text{NoSync}_k}$ with P_{NoSync_k} given by (7.86). Likewise,

$$P_{\text{fa}} = P_{\text{fa}_k}^m \tag{8.7}$$

where P_{fa_k} is the false alarm probability per pass, given by (7.84). These results are shown in Figure 8.4 for $n = 3$. For the parameters used in this illustration, when $P_{\text{NoSync}} > 0.3$, the results are shown in Table 8.1. These results indicate that attacks on the synchronization process are potentially very lucrative.

Table 8.1 JSR Values for $P_{\text{NoSync}} > 0.3$

P	$\xi_{P_{\text{NoSync}>0.3}}$ (dB)
5	−7
10	−3
20	−0.5
30	1

Two-Step Coarse Acquisition. When the acquisition function is split into two steps as described in Section 7.4.1.3, the first step is passive correlation with a matched filter and the second step is active correlation with a stepped serial correlator. Again, the analysis in Section 7.4.1 applies with suitably modified noise levels.

The total SNR received over the synchronization interval is given by (7.75) where $K = P$ for the first, passive step and $K = A$ for the second, active step. That due to step 1 is given by

$$\upsilon_1 = \frac{P}{P+A}\upsilon \tag{8.8}$$

while that due to step 2 is given by

$$\upsilon_2 = \frac{A}{A+P}\upsilon \tag{8.9}$$

These same comments apply to the JSR. Therefore,

$$\upsilon_1' = \frac{1}{\dfrac{1}{\dfrac{P}{A+P}\upsilon} + \dfrac{P}{A+P}\xi} \tag{8.10}$$

and

$$\upsilon_2' = \frac{1}{\dfrac{1}{\dfrac{A}{A+P}\upsilon} + \dfrac{A}{A+P}\xi} \tag{8.11}$$

where P dwell frequencies are examined during the parallel (passive) phase and $A = H - P$ frequencies are examined during the serial (active) phase. The probability of false alarm is given by (7.84) and the probability of missing synchronization is given by (7.86) with these modifications.

The BBN jamming performance is illustrated in Figure 8.5 where the probability of not acquiring synchronization is depicted for a few values of SNR. In this example, jamming is very effective at precluding synchronization at relatively high SNRs. The threshold of $P_{\text{NoSync}} > 0.3$ is reached at $\xi \approx -4$ dB at $\upsilon = 10$ dB, $\xi \approx -1$ dB for $\upsilon = 20$ dB, and $\xi \approx -2$ dB for $\upsilon = 30$ dB.

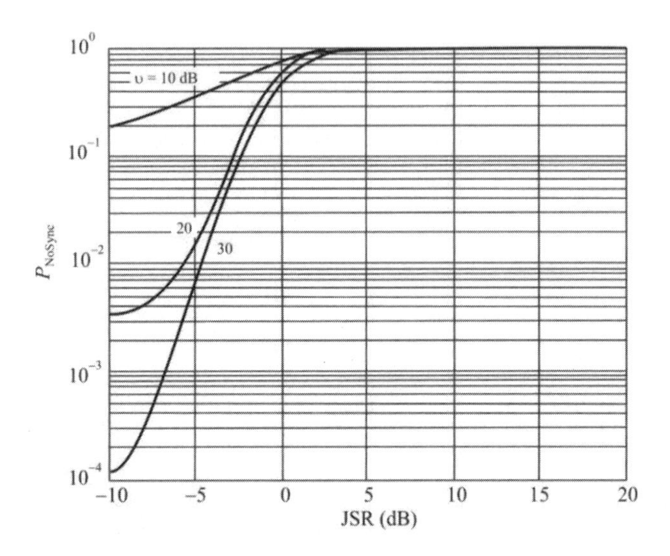

Figure 8.5 BBN jamming performance example for FHSS two-step coarse code acquisition, c = 3, A = 20, P = 10, I = 0 dB, b = 0.5, m = 2, υ = 5. The thresholds were varied to maintain $P_{fa} \approx 10^{-4}$ so for υ = 10 dB, γ_1 = 60 (18 dB), γ_2 = 20 (13 dB); for υ = 20 dB, γ_1 = 70 (18 dB), γ_2 = 30 (15 dB); and for υ = 30 dB, γ_1 = 80 (19 dB), γ_2 = 40 (16 dB).

8.2.2.2 Partial-Band Noise Jamming

Partial-band noise (PBN) jamming places noise-jamming energy across multiple, but not all, channels in the spectrum used by the targets. These channels may or may not be contiguous. The spectrum for PBN jamming is illustrated in Figure 8.1(c, d); Figure 8.1(c) illustrates contiguous channels while Figure 8.1(d) illustrates noncontiguous channels.

Let ν denote the fraction of the target net band instantaneously covered by the jammer. It will be shown later that optimal PBN jamming is more successful than BBN jamming. Optimal in this sense means that there is a "best" value of ν. The performance of PBN jamming is given by (8.2) where P_{e_2} is given by (3.93) with $P_n = J$

$$P_{e_2} = \frac{1}{2}\exp\left(-\frac{1}{2}\frac{R}{P_N + J}\right) \tag{8.12}$$

Therefore,

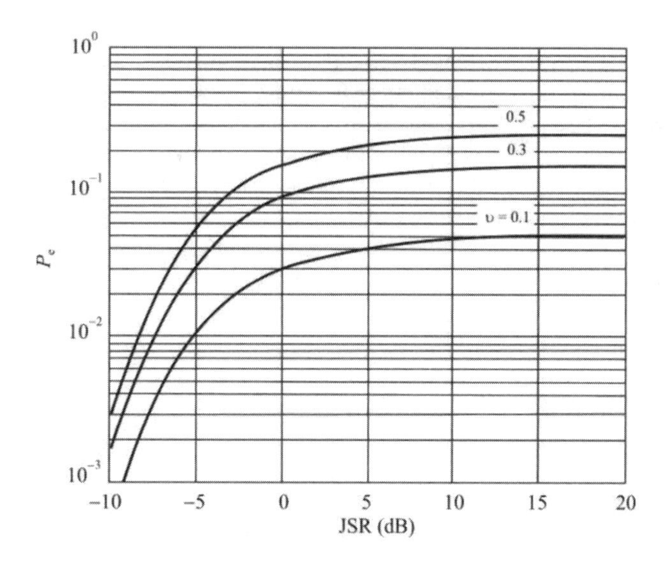

Figure 8.6 PBN jamming performance for BFSK.

$$P_e = (1-\nu)\exp\left(-\frac{1}{2}\upsilon\right) + \frac{\nu}{2}\exp\left(-\frac{1}{2}\frac{1}{\frac{1}{\upsilon}+\xi}\right) \tag{8.13}$$

This function is plotted in Figure 8.6 for some values of ν. For $\nu \geq 0.3$, $P_e > 10^{-1}$ for $\xi \geq 0$ dB indicating good jamming performance. Below about $\nu = 0.3$, $P_e < 10^{-1}$ for $\xi \geq 0$ dB so about one-third of the channels must be jammed.

8.2.2.3 Narrowband Noise Jamming

Narrowband noise (NBN) jamming places all of the jamming energy into a single channel. The bandwidth of this energy injection could be the whole width of the channel or it could be only the data signal width or the complementary signal width. In the former, when the target is a digital signal, BFSK, for example, both the mark and the space tone frequencies receive jamming energy. Narrowband noise jamming is illustrated in Figure 8.1(e). P_e for jamming both data paths is given by the same expression as for BBN as (3.96) with $P_n = J$ assuming follower jamming is employed (the only situation where NBN jamming of FHSS signals makes sense). This also ignores tracking issues discussed later.

When there is a noise signal of average power J injected into the data channel only, then P_e can be calculated by considering the data channel as receiving a signal that is Rayleigh distributed while the complementary channel is receiving a signal that is Ricean distributed. In a Rayleigh distributed channel the received signal consists of several randomly distributed specular components and no direct component while the received signal in a Ricean channel has a significant direct component in addition to the specular components. In that case, the PDF of the signals in the two filters is given by [3]

$$p_R(r_1) = \frac{r_1}{P_N + J} \exp\left(-\frac{1}{2} \frac{r_1^2 + R}{P_N + J} \right) \times I_0 \left(\frac{\sqrt{R} r_1}{P_N + J} \right) \tag{8.14}$$

$$p_R(r_2) = \frac{r_2}{P_N} \exp\left(-\frac{1}{2} \frac{r_2^2}{P_N} \right) \tag{8.15}$$

and the overall BER is calculated by

$$P_{e_2} = \Pr\{r_2 > r_1\} = \int_0^\infty p(r_1) \left[\int_{r_1}^\infty p(r_2) dr_2 \right] dr_1 \tag{8.16}$$

When the above are substituted into this expression and the integration carried out, P_{e_2} results, given by

$$P_{e_2} = \frac{P_t}{2P_N + J} \exp\left(-\frac{R}{2P_N + J} \right) \tag{8.17}$$

where P_N is the thermal noise power and J is the average jammer noise power.

8.2.3 Tone Jamming

In *tone jamming*, one or more jammer tones are strategically placed in the spectrum. Where they are placed and their number affect the jamming performance. Two types of tone jamming are illustrated in Figure 8.1. Single-tone jamming places a single tone where it is needed and is illustrated in Figure 8.1(g). Multiple-tone jamming distributes the jammer power among several tones and is illustrated in Figure 8.1(g).

The phase of the jammer tone relative to the target signal can be an important parameter [4, 5]. Suppose that the data symbol is a mark. Further, assume the jammer signal level is sufficiently greater than the noise that it can be neglected. When n = 1, in a jammed channel the single jammer tone present is either at the

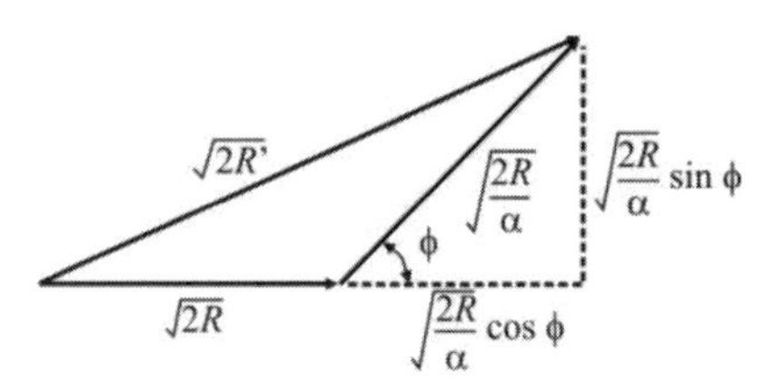

Figure 8.7 Phase relationships between the jammer and the symbol tone. (Source: [5] © IEEE 1985. Reprinted with permission.)

mark or the space frequency. If it is at the mark frequency, then the phase can present a problem if the jammer tone is sufficiently out of phase as explained below. If it is at the space frequency, then if the JSR is large enough, the symbol is jammed independent of the phase relationship.

When $n > 2$ and the tones are present in a contiguous band of channels, then jamming performance, the same as that for PBN jamming is achieved. For PBN jamming, it makes no theoretical difference whether the jammer power that makes it through the detector filters is a tone with or without noise modulation. Thus, we have partial-band multitone jamming. Realistically, a tactical jammer is limited to about 100 watts ERP for a UAS jammer to about 5 kW ERP for a large ground-based jammer. In these cases, n would be limited to on the order of 10 for the jammer to be effective, thus, for the low VHF (30 to 90 MHz), $\gamma \leq 0.004$ or so.

For $n = 2$ there will be a symbol error if the channel is jammed at all and the jammer tone power in the complementary channel is greater than the power in the symbol channel, which consists of the symbol tone and one of the jammer tones. The power in the symbol channel depends on the phase relationship between the symbol signal and the jammer tone.

To determine the effects of the phase relationship, consider the phasor diagram shown in Figure 8.7 [5]. Assume that the phase of the jammer tone relative to that of the symbol tone is uniformly distributed over $(0, 2\pi)$; thus, $p(\phi) = 1/2\pi$ as shown in Figure 8.8. For the jammer to be effective in a jammed channel then the jammer wants to optimize a parameter α so that

$$J > \frac{R}{\alpha} \tag{8.18}$$

The total signal at the symbol frequency that also has the jammer tone is thus given by

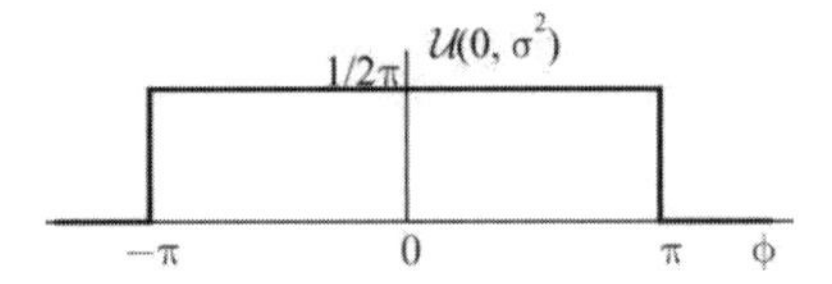

Figure 8.8 Uniform probability density function of ϕ.

$$\sqrt{2R}\sin\left(2\pi f_i t\right)+\sqrt{\frac{2R}{\alpha}}\sin(2\pi f_i t+\phi)=\sqrt{2R'}\sin(2\pi f_i t+\phi') \tag{8.19}$$

From the phasor diagram, the total power at that tone frequency is then given by

$$\begin{aligned} R' &= R\left[\left(1+\frac{1}{\sqrt{\alpha}}\cos\phi\right)^2+\left(\frac{1}{\sqrt{\alpha}}\sin\phi\right)^2\right] \\ &= R\left(1+\frac{2}{\sqrt{\alpha}}\cos\phi+\frac{1}{\alpha}\right) \end{aligned} \tag{8.20}$$

The power in the jammed complementary channel is just $J = S/\alpha$ so there will be a symbol error if $R' < J$, or

$$\begin{aligned} P_s &= \Pr\{R' < J\} \\ &= \Pr\left\{R\left(1+\frac{2}{\sqrt{\alpha}}\cos\phi+\frac{1}{\alpha}\right)<\frac{R}{\alpha}\right\} \\ &= \Pr\left\{\cos\phi < -\frac{\sqrt{\alpha}}{2}\right\} \\ &= \Pr\left\{\phi < \cos^{-1}\left(-\frac{\sqrt{\alpha}}{2}\right)\right\} \\ &= \int_0^{\cos^{-1}\left(\sqrt{\alpha}/2\right)} p(\phi)d\phi \\ &= \int_0^{\cos^{-1}\left(\sqrt{\alpha}/2\right)} \frac{1}{2\pi}d\phi \end{aligned}$$

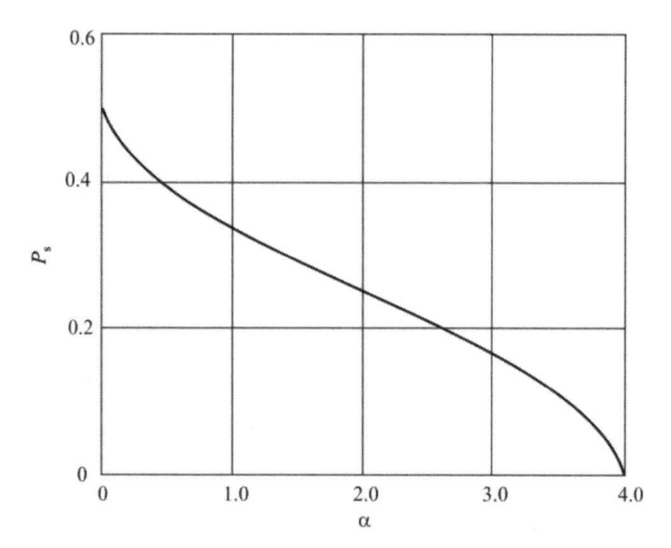

Figure 8.9 Probability of a symbol error caused by $n = 2$ and one jammer tone hits the correct symbol.

$$= \frac{1}{\pi}\cos^{-1}\left(\frac{\sqrt{\alpha}}{2}\right) \tag{8.21}$$

since $\cos^{-1}(x)$ is an even function of x.

Note that α is upper bounded by four, as any larger value would produce an argument for the $\cos^{-1}$ function that is greater than 1. In addition, it is lower-bounded by zero, since at 0 it implies that the jammer is infinitely more powerful than the signal. Since $J = R/\alpha$, this expression indicates that J can be as much as 6 dB less than R and still accomplish jamming, but the phase relationship must be correct. P_s is plotted in Figure 8.9.

The probability of the channel being jammed at all is given by γ. Because these are independent events,

$$P_s' = \gamma P_s \tag{8.22}$$

Therefore, when the jammer tones enter both detector channels with equal power given by J, then the probability of a symbol error is given by

$$P_e = P_s' P_{sb}$$

$$= \frac{\gamma}{\pi}\cos^{-1}\left(\frac{\sqrt{\alpha}}{2}\right)P_{\text{sb}} \tag{8.23}$$

where [6]

$$P_{\text{sb}} = \frac{1}{2\pi}\int_0^{2\pi}\left\{\begin{array}{l} Q\left[\sqrt{\frac{J}{P_{\text{N}}}}, \frac{D(x)}{\sqrt{2P_{\text{N}}}}\right] \\ \qquad -\frac{1}{2}\exp\left[-\frac{2J+D^2(x)}{2P_{\text{N}}}\right]I_0\left[\frac{\sqrt{2J}D(x)}{\sqrt{2P_{\text{N}}}}\right]\end{array}\right\}dx \tag{8.24}$$

with

$$D^2(x) = 2R + 2J + 4\sqrt{RJ}\cos(x)$$

This function is plotted in Figure 8.10 for some values of SNR versus α when the number of channels jammed is 10 out of 2,400. Over the majority of the range of α, the SNR makes little difference.

Peterson [7] argues that when the background noise is negligible, and only one jamming tone is allowed per channel, then close to the optimum value of n is given by

$$n^* = \left\lfloor \frac{J}{R} \right\rfloor \tag{8.25}$$

where $\lfloor x \rfloor$ stands for the integer part of x, and in this case, J stands for the total average jammer power arriving at the receiver. The rationale for this is that if n is larger than this, the jammer power per channel is less than the signal power in that channel and (neglecting noise effects) the signal in the channel will be detected properly. Likewise, if n is smaller than this value, the jammer power per channel is more than is needed to accomplish the jamming, and jamming power is wasted and more channels could be jammed with the same total jammer power.

8.2.3.1 Single Tone

A jamming signal transmitted at a single frequency was shown in Figure 8.1(f). Thus, the jamming signal is a CW tone placed at a single frequency. Single-tone jamming is also called *spot jamming*. It will be shown in later chapters that CW

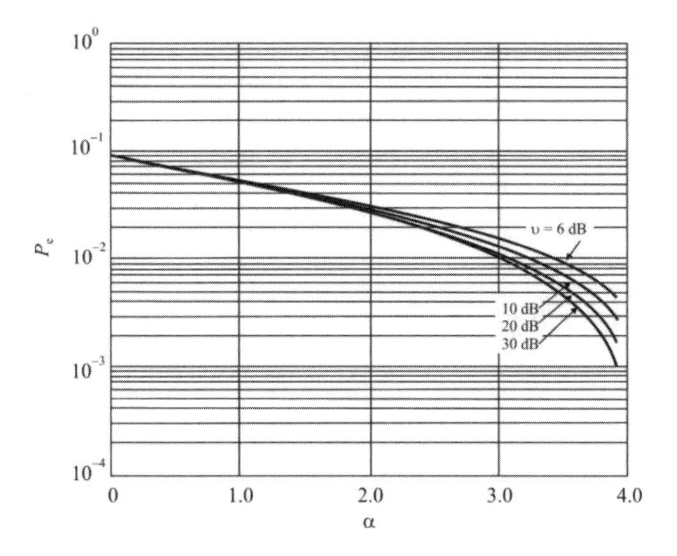

Figure 8.10 BER versus α for several SNRs for $n = 2$ multitone jamming when one of the jamming tones hits the symbol tone.

jamming of FHSS systems is unsuccessful. It has been successfully applied to very narrow targets such as OOK telegraphy that is not changing frequency, however.

Single-tone jamming may be useful, for example, against DSSS AJ systems by overcoming the processing gain of such systems at the receivers and causing deleterious effects at the despread level. When the total jammer power is fixed, more power can be placed in a single tone than in each of multiple tones, thereby increasing the probability of overcoming the processing gain.

When there is a single tone and it is placed in the data channel, while the complementary channel has only thermal noise, then the jammer can enhance the ability of the receiver to correctly decode the data bit, depending on the phase relationship between the interfering tone and the data tone and the relative magnitudes of the tones. Based on (3.35), the probability of bit error in this case, is, however,

$$P_e = \frac{1}{2} Q\left(\sqrt{\frac{J}{P_N}}, \sqrt{\frac{R}{P_N}} \right) \tag{8.26}$$

which is

$$P_e = \frac{1}{2} Q\left(\sqrt{\xi \upsilon}, \sqrt{\upsilon}\right) \tag{8.27}$$

When the single tone is correctly placed in the complementary channel, P_e is given by (3.102) [8]

$$P_e = \frac{1}{2} \frac{P_N + J}{2P_N + J} \exp\left(-\frac{R}{2P_N + J}\right) \tag{8.28}$$

8.2.3.2 Multiple Tones

The jammer could emit $L > 1$ tones, randomly placed, or placed at specific frequencies. If a particular target AJ communication system is vulnerable to specific tones and the jammer knows that, then it may be more prudent to use tones at those specific frequencies rather than placing them randomly.

When the tones are in consecutive channels, this is called *comb jamming*. Herein, no matter which tone jamming strategy is being discussed, it is tacitly assumed that the tone is placed precisely at a frequency in the spectrum so that the jammer tone passes through the receiver filters without distortion or attenuation. Multitone jamming is illustrated in Figure 8.1(g).

When the interference is jammer tones only into both the data channel and the complementary channel with no noise, then P_e is given by (3.99)

$$P_e = \frac{1}{2\pi} \int_0^{2\pi} \left\{ \begin{array}{l} Q\left[\sqrt{\dfrac{J}{P_N}}, \dfrac{D(x)}{\sqrt{2P_N}}\right] \\ -\dfrac{1}{2} \exp\left[-\dfrac{2J + D^2(x)}{4P_N}\right] I_0\left[\dfrac{\sqrt{2J} D(x)}{2P_N}\right] \end{array} \right\} dx \tag{8.29}$$

where

$$D^2(x) = 2R + 2J + 4\sqrt{RJ} \cos(x)$$

This expression must be evaluated numerically since there is no known closed-form solution.

Confined to Within the Same Channel, $L \leq M$

The L tones can be confined to be in the same channel or they can be distributed through the spectrum. It is assumed that each of the jamming tones is at the same power level as the others since the jammer does not know which of the M data tones is present and has no basis upon which to apply higher power to one of the tones over another. The goal when constrained to within the channel is to place tones at two or more locations where data tones may be located.

If an FFHSS AJ system hops to a channel where such a jammer is located, one of M tones will be transmitted representing a single symbol. By jamming with two or more tones, the jammer is assured of hitting at least one incorrect tone and maybe two. It is also possible for the jammer to reinforce the correct symbol tone with one of its own tones. Therefore, whether the jammer is effective or not will depend on: (1) if the correct symbol tone is not hit then the JSR at the tone frequencies, and (2) if the correct symbol tone is hit with one of the jammer tones, then depending on the phase relationship between the jammer tone and data tone and the JSR, the receiver may correctly decide which symbol was sent and the jammer will be ineffective or the jammer could be successful.

If an SFHSS AJ system hops to such a channel then some of the symbol tones could be received in error depending on the JSR. At those tone frequencies, representing symbols that were not sent the jammer signal must be larger than the signal frequency to be effective. At those symbol times when the jammer hits the symbol tone that was sent, then the receiver may correctly decide the transmitted symbol since the jammer reinforces the symbol energy, but this depends on the phase relationship.

If there are L tones all placed in the same channel, and left there then this jamming technique would be ineffective against FHSS AJ communication networks because the target would have to hop to that channel to be jammed. Therefore, a better strategy is for the jammer to be swept or hopped randomly to attempt to hit as many channels as possible.

This technique could be used against DSSS AJ systems, but jammer performance is critically important for effective jamming. The best jammer performance is obtained by placing the multiple tones at integer multiples of the data rate. The tone jammer performance against DSSS is illustrated in Figure 8.11 and 8.12. In Figure 8.11, a single tone is placed at $\Delta f = 1/2T_c$ away from the carrier with the curves based on exact calculations. The curves in Figure 8.12 are based on an example of five tones placed at $\Delta f = \pm 1/T_c, \pm 1/2T_c$, and 0. These curves are based on an approximate bound. Also shown in Figure 8.12 is the communication system performance in the presence of noise only (no jammer).

These charts demonstrate that tone jamming, whether single or multiple tones are used, is relatively ineffective for even moderate values of the SNR when $P_e \geq 10^{-1}$ is the performance criteria. Typical reliable communication system performance would implement SNRs higher than those shown in Figure 8.11 and Figure 8.12.

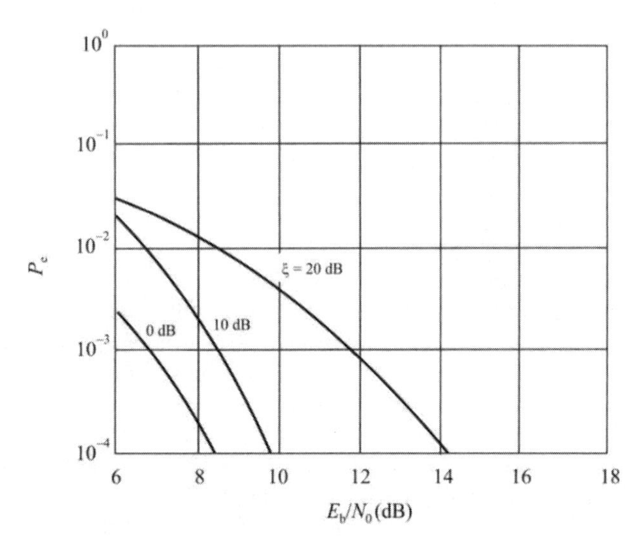

Figure 8.11 Tone jammer performance against DSSS systems. In this case the tone is placed at $\Delta f = 1/2T_c$ away from the carrier frequency of the DSSS system. (Source: [9]. © IEEE 1982. Reprinted with permission.)

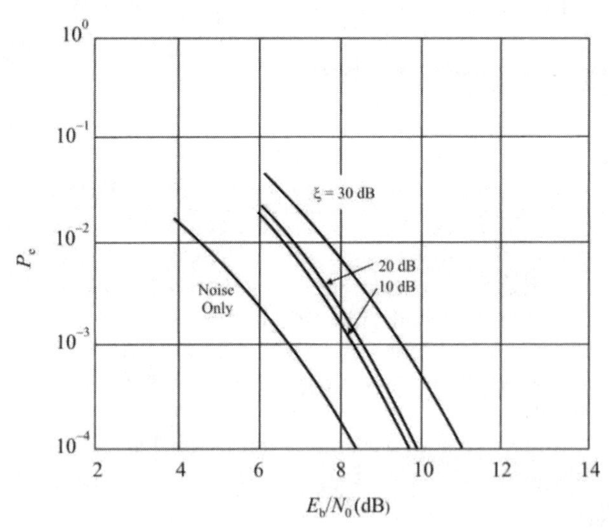

Figure 8.12 Multitone jammer performance against DSSS systems. Here five tones are placed at $\Delta f = \pm\ 1/T_c$, $\pm 1/2T_c$, and 0 away from the carrier frequency of the DSSS system. (Source: [9]. © IEEE 1982. Reprinted with permission.)

Not Confined to Within the Same Channel

The jammer tones can be placed throughout the spectrum of operation of the communication system. In that case, a frequency-hopping system may hop to one of the tones and send data at that tone or it may not. A DSSS system may also encounter one of the tones. A hybrid system may hop to one of the tones, but the effect will be different from the FHSS case because of the direct sequence spreading.

Multiple Tone Effects on Synchronization

Some synchronization processes for FHSS were presented in Section 7.4.1. The FHSS analysis included the effects of thermal noise and interfering signals on the performance of three schemes for synchronization. It was assumed that interferers arrived at the target receiver at the same signal level as the intended signal. Those interferers could just as easily be considered multiple tone jammers. Presented here are the performances of multiple tone jammers against these synchronization techniques.

To consider the tone jammers, it is only necessary for

$$J = P_{\mathrm{I}} \tag{8.30}$$

$$\xi = I \tag{8.31}$$

in that analysis. The number of jamming tones is given by h.

It is assumed here that the jammer has no knowledge of the hop sequence. Thus, the only alternative is to place the jammer tones randomly throughout W_{ss}. If the jammer knew the hop sequence, then it would be perfectly effective, limited only by the amounts of power involved. That is not the problem analyzed here.

It is assumed that the FHSS receiver implements limiters prior to the FHSS detectors. In that case, the optimum jamming power per tone is slightly greater than the maximum signal level because less than this and the receiver is not jammed, while with jamming power greater than this, the limiters remove the excess jamming energy and is therefore wasted. Hence, when N_{J} is the number of jammer tones present, since

$$J = N_{\mathrm{J}} P_{\mathrm{i}} = N_{\mathrm{J}} R \tag{8.32}$$

then

$$N_{\mathrm{J}} = J / R = \xi \tag{8.33}$$

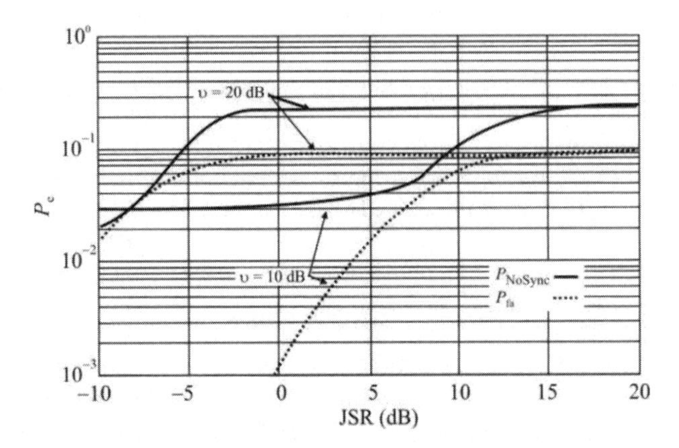

Figure 8.13 Multiple-tone jamming performance against two-step coarse synchronization. For this example $\gamma_1 = 60$ (18 dB), $\gamma_2 = 20$ (13 dB), $b = 0.5$, $c = 3$, $P = 10$, $A = 10$, and $h = 20$.

Multiple-tone jamming performance against two-step synchronization is shown in Figures 8.13 and 8.14. The jamming effectiveness as the JSR is increased is shown in Figure 8.8. T_1 would normally be somewhat higher than T_2 because in two-step synchronization the first stage has lower reliability but is faster, while the second stage has higher reliability but can be somewhat slower. As there are more correlators available for the second step in the synchronization technique, weaker jammer signals have less effect on the received signal. Therefore, the jammer performance improves as the JSR is increased when there are more second-step correlators available.

The effects of increasing the number of tones are shown in Figure 8.14. Higher SNR means that in step 1 there are more hits that are false. With few correlators for the second step, many of the valid hits are discarded so the BER is higher than for the lower SNR. This is evident in Figure 8.14 by comparing like SNR curves for $c = 3$ and $c = 7$.

In this analysis $W_{IF} = 1/T_h$ so it is applicable to FFHSS systems. For SFHSS systems, this bandwidth has insufficient width to communicate much data; for example, at 100 hps, $W_{IF} = 1/10$ ms = 100 Hz. Instead $W_{IF} \approx 1/T_b$, yielding typical VHF bandwidths of 25 kHz for a 20 kbps data rate.

As mentioned in Section 7.5.1, with early-late gate *tracking* of FHSS systems, successful communications is not substantially affected by interfering tones. That is not the case here for coarse synchronization. Therefore, it is important for successful ECM to deny the initial synchronization of a communication network.

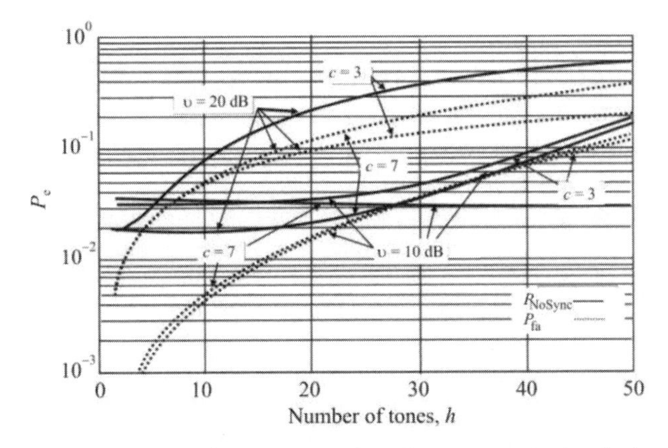

Figure 8.14 Multiple-tone jamming performance example against two-step synchronization acquisition as the number of tones, h, is varied. In this case $\gamma_1 = 60$ (18 dB), $\gamma_2 = 20$ (13 dB), $b = 0.5$, $P = 10$, $\xi = 0$ dB, $m = 1$, and $A = 10$.

8.2.4 Swept Jamming

A concept similar to broadband or partial-band noise jamming is *swept jamming*. This is when a relatively narrowband signal, which could be as narrow as a tone but more often is a PBN signal, is swept or scanned in time across the frequency band of interest. At any instant in time, the jammer is centered on a specific frequency and the only portion of the spectrum being jammed is in a narrow region around this frequency. However, since the signal is swept, a broad range of frequencies can be jammed in a short period. When implemented digitally, for example, the jammer may spend 100 μs at any one frequency before moving on to the next band to be jammed. Normally these bands would be consecutive, but do not have to be—the bands could actually be selected randomly with digital synthesizers generating the jamming waveform. In this way this jammer could cover the whole 30–90 MHz band in about 240 ms.

The net effect of such a jamming strategy is similar to a barrage jammer except that the full power of the jammer is employed at each dwell bandwidth. It is also possible to sectorize the jamming strategy and avoid certain bands that might be in use by friendly forces. This is true only when the timing is tailored to the target receivers so that the jamming signal is present at the receiver for an adequate dwell time. The characteristics of the target receivers must be taken into consideration for swept jamming to be effective.

It is also important to note that the characteristics of the target receiver are important for evaluating the effectiveness of swept jamming. The filtering process in the receiver has considerable impact on the dwell duration required. All the analyses herein are predicated on the jammer dwell being long enough so the jammer effects are not hindered by such concerns.

The main purpose for sweeping the partial-band noise waveform is to ensure that the jammer enters the frequency spectrum where an FHSS target net is located. Normally FHSS AJ networks do not use every channel from, say, 30 MHz to 90 MHz, but only a portion, called the hop set. These hop sets need not be that large to be effective. It could be that a partial-band jammer permanently situated in a portion of the spectrum does not cover any of the hop set frequencies, or perhaps just a few of them. This renders the partial-band jammer ineffective. By sweeping the jamming waveform over the whole range, then the jammer is ensured to jam at the entire set of hop frequencies.

Timing is one of the more important parameters for a swept jammer. The sweeping must be fast enough to ensure that the whole band is covered in a sufficiently short period or hops will occur for which no jamming signal is present. On the other hand, sweeping cannot be so fast that when a hop is jammed, an inadequate fraction of the hop is jammed. A BER of 10^{-1} means that it is necessary to jam 1 bit out of 10, or for an AJ system that is sending data at 20 kbps, 2,000 bits must be jammed to produce this BER. If this system is an SFHSS network at 100 hps, each hop will contain around 200 bits (this discussion ignores the finite time it takes to go from hop to hop). Therefore, at least 10 hops per second must be jammed. Since these hops can be anywhere in the spectrum from the point of view of the jammer, at least 10 sweeps per second are required. To cover the 60 MHz in the low VHF range 10 times per second requires a sweep rate of 600 MHz per second. Each channel is 25 kHz, so this equates to 24,000 channels per second, or a maximum of 42 μs on each channel if the jamming waveform is one channel wide (25 kHz). If it is 10 channels wide (250 kHz), then 4.20 μs per channel is allowed and if it is 100 channels wide (2.5 MHz) then 4.2 ms can be spent on each channel.

The trade-off with these numbers is that as the instantaneous bandwidth is increased, the power per channel decreases if the total jammer power remains fixed. Obviously, when the instantaneous bandwidth gets large enough to be equal to the total target system bandwidth, the results are the same as BBN jamming. In that case, there is no point in sweeping at all.

8.2.5 Pulse Jamming

Pulse jamming is similar in concept to partial-band noise jamming. The fraction, ν, in this case corresponds to the portion of time that the jammer is on relative to off, whereas for PBN jamming the fraction corresponded to the portion of the spectrum covered at a time. The statistics work out to be the same, however. Short pulses have broad spectral content and therefore are similar to broadband noise when they are on.

Pulse jamming can have lower average power than some of the other jamming techniques discussed herein, and be just as or more effective. The duty cycle determines the relationship between the average power and peak power. The jamming effects depend on the peak power and how often that signal returns to the receiver.

8.2.6 Follower Jamming

A *follower jammer* attempts to locate the frequency to which the frequency-hopping transmitter went, identify the signal as the one of interest (the target), and jam at the new frequency. This jamming waveform could be in the form of tones or it could modulate the tones with, say, noise using FM modulation. Follower jamming is also referred to as *responsive jamming, repeater jamming*, and *repeat-back jamming*.

Some basic limitations to employing follower jammers against FHSS targets were determined by Torrieri [10].[1] These limits are related to the placement of the communicators relative to the jammer that imposes timing constraints.

The task of the ES system is to acquire where an FHSS signal has moved in frequency so the EA system can place the jamming signal at that frequency. This is typically done by measuring the spectrum for new energy loss or gain and making appropriate measurements on those channels. A loss of energy implies that the target has moved to a new frequency. A gain of energy in a channel implies that a new signal is present, which may or may not be the target. ES to find the new frequency can be very difficult to do accurately and rapidly. Even when the channel structure of the spectrum is limited, when measuring just energy, most targets may look similar. It requires an examination of the details of each new signal to determine if it is the target.

The timing of a follower jammer must satisfy certain constraints due to the finite propagation and processing time of the signals. In general, to meet the timing requirements, the jammer must lie anywhere on or within the ellipse shown in Figure 8.15 based on [10].

$$\frac{D_{\mathrm{TJ}} + D_{\mathrm{JR}}}{c} + T_{\mathrm{J}} \leq \frac{D_{\mathrm{TR}}}{c} + \gamma T_{\mathrm{d}} \tag{8.34}$$

[1] The limitations are not truly fundamental in a scientific sense because several approximations were used in the derivations.

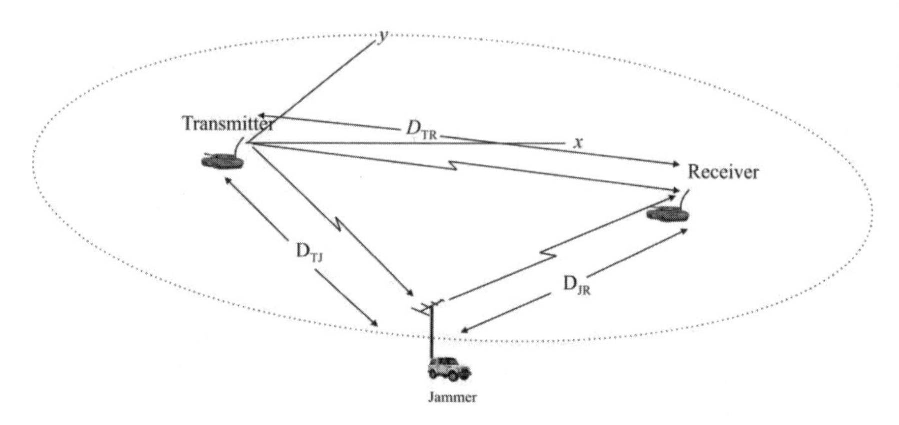

Figure 8.15 Ground-based jamming scenario.

where the distances are as shown in the figure, T_J is the processing time at the jammer, T_d is the dwell time of the target signal, ν is the fraction of the dwell that must be jammed to be effective, and c is the speed of propagation. The transmitter and receiver are the foci of this ellipse. This equation can be manipulated to

$$\frac{4(x-D_{TR})^2}{(D_{TR}+c\nu T_d)^2}+\frac{4y^2}{(D_{TR}+c\nu T_d)^2-D_{TR}^2}=1 \tag{8.35}$$

The timing of such is shown in Figure 8.16. T_δ represents the difference in propagation time of the transmitter signal at the receiver and the jammer. This time may actually be negative if the jammer is closer to the transmitter than the receiver. The jammer takes a certain amount of time to process the signal and this is represented by T_{proc}. The jammer signal takes a certain amount of time to propagate to the receiver once the jammer signal is started. This is represented by T_{prop}. For the majority of the cases of interest herein, the propagation times are negligible compared with the other times involved.

The dwell fraction within which follower jamming will be effective depends on the path length difference between the path between the transmitter and receiver versus the path from the transmitter to the jammer and the jammer to the receiver. It is also affected by the amount of processing time required at the jammer to make a jamming decision. This is illustrated in Figure 8.17 as an example. In this case, it is assumed that the jammer processing time is 10% of the dwell time. The data shown in Figure 8.17 corresponds to the fraction of a dwell that is available for the jammer to jam. A dwell of 10^{-5}, for example, if L_F = 4,

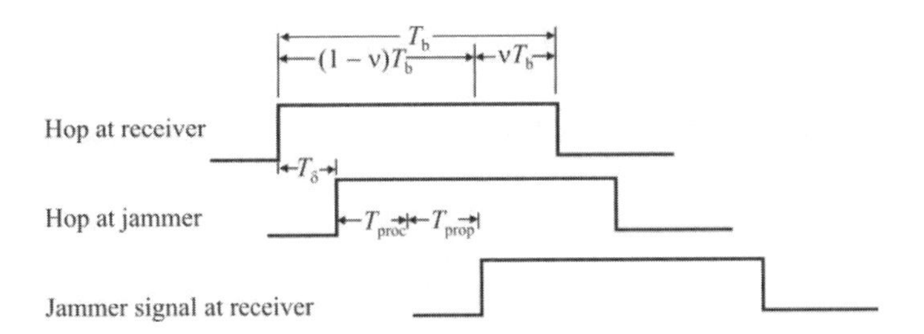

Figure 8.16 FHSS follower jamming timing.

would correspond to a bit rate of 25 kbps, about that required to transmit uncoded voice traffic.

When one adjacent channel is jammed, the full power of the jammer is placed into that channel while when two or more channels are jammed, the jammer power is divided among those channels, leaving the power per channel less than what it is otherwise. For BFSK, on the other hand, if only one adjacent channel is jammed, the jammer does not know if it is the complementary channel so the probability of jamming the hop is half of what it is otherwise.

It may be required to jam both adjacent channels. To do this without jamming the data channel adds complexity to the jammer, so therefore the jammer may choose to jam all three channels, thus ensuring that the complementary channel is jammed. This does, of course, decrease the power per channel by a factor of 3 (4.8 dB). Figure 8.18 illustrates jamming all three consecutive channels. If it is decided to jam just one adjacent channel, P_e is decreased by half in that case because either adjacent channel could be the correct one and it will be jammed half the time.

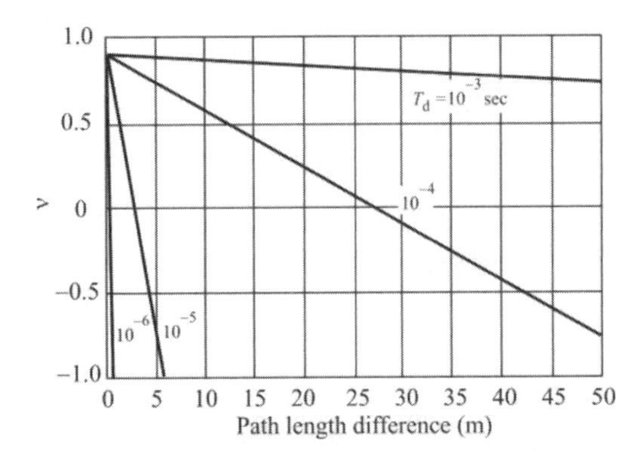

Figure 8.17 Dwell fraction as a function of the path length difference.

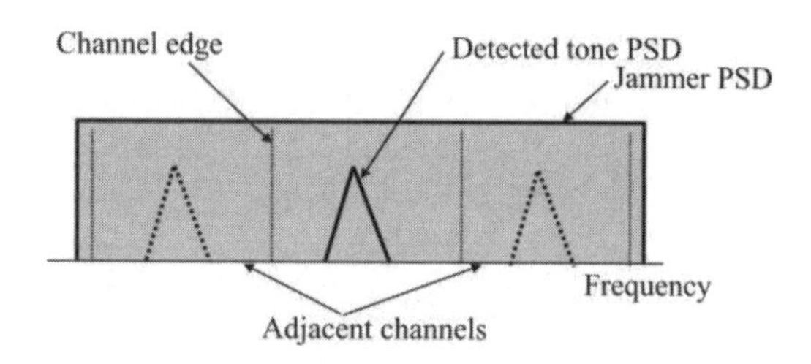

Figure 8.18 Follower jammer jamming three contiguous channels.

The data shown in Figure 8.19 compares the jamming effectiveness when one complementary channel is jammed versus when all three are jammed when $\upsilon = 20$ dB. For the parameters used in this example, jamming one adjacent channel is better even though some (on average, half) of the symbols are not jammed due to ignorance of the correct complementary channel.

The requirement to jam the current symbol is restrictive relative to the placement of the jammer. For example, for SFHSS and digitized voice communications, the data rate would be on the order of 20 kbps so the symbol time is about 500 μs. The amount of the symbol necessary to jam depends on the JSR relative to the SNR since the decision as to whether a mark or space was sent is based on the energy detectors at the output of the bandpass filters. Suppose that a mark was sent (see Figure 8.20). A space will be decided if the jammer catches the last $\gamma \times 500$ μs of the symbol in this case. The allowable path-length differences are shown in Figure 8.21 for three times corresponding to $\gamma = 0.4, 0.45$, and 0.5.

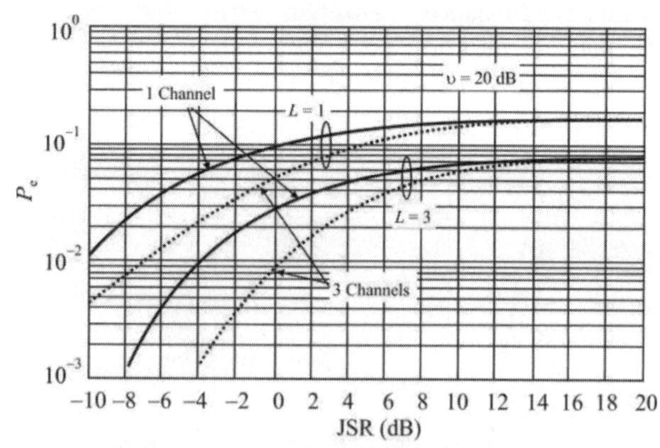

Figure 8.19 Comparison of jamming three consecutive channels versus only one adjacent channel, $\nu = 0.3$ and $\upsilon = 20$ dB. L represents the diversity of the modulation.

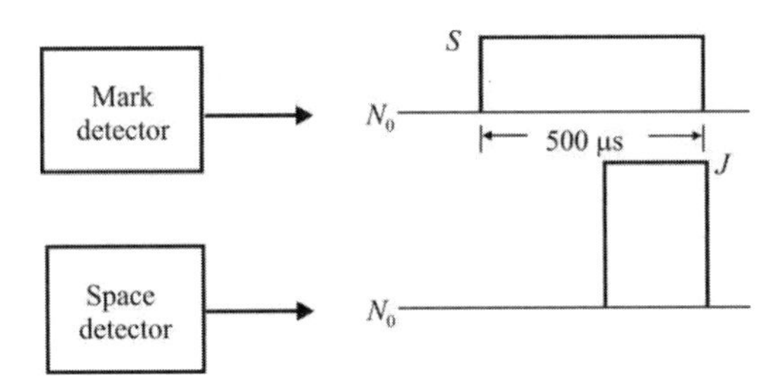

Figure 8.20 Example of follower jamming timing for BFSK.

8.2.6.1 Follower Jamming with Narrowband Noise

Receiver detection performance is based on the SNR in the channel after the detector filters have removed out-of-channel noise. Therefore, placing a noise waveform in the channel will affect the ability of the receiver to properly detect the tone.

In this section, the analytical results are first presented. They are compared in the last section.

Narrowband Noise Jamming of the Data Channel

When jamming the data channel, P_e is given by (3.93) with $P_n = J$ [11]

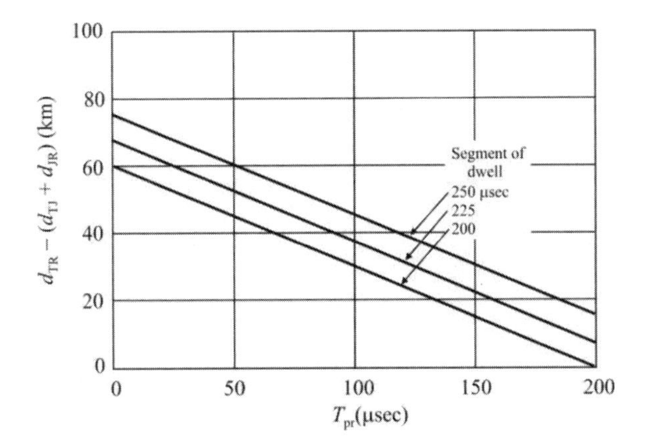

Figure 8.21 Allowable differences in the path lengths $d_{TR} - (d_{TJ} + d_{JR})$.

$$P_{e_2} = \frac{P_N}{2P_N + J} \exp\left(-\frac{R}{2P_N + J}\right) \tag{8.36}$$

From this, the optimum jammer power can be easily calculated to be

$$J = \begin{cases} R - 2P_N, & \upsilon > 3 \text{ dB} \\ 0, & \upsilon < 3 \text{ dB} \end{cases} \quad \text{dB} \tag{8.37}$$

Thus, for $\upsilon >> 3$ dB, the optimum jammer power is approximately equal to the signal power ($R - 2P_N \approx R$, $\xi \approx 0$ dB), while at low SNRs it is better not to jam at all. A BER of 10^{-1} or higher is achieved by the noise alone in that case.

When the optimum jammer power is used, then the BER is given by

$$P_{e_2} = \begin{cases} \dfrac{P_N}{eR}, & \upsilon > 3 \text{ dB} \\ \dfrac{1}{2}\exp\left(-\dfrac{1}{2}\dfrac{R}{P_N}\right), & \upsilon < 3 \text{ dB} \end{cases} \tag{8.38}$$

When $L = 1$ and $\gamma = 1$, this BER at the optimum jamming power is shown in Figure 8.22, which exhibits an inverse linear characteristic beyond $\upsilon = 3$ dB.

From (8.1), (8.2), and (8.36), the overall probability of a symbol error on a dwell for narrowband noise jamming of the data channel only is

$$P_e = (1-\gamma)\frac{1}{2}\exp\left(-\frac{1}{2}\frac{R}{P_N}\right) + \gamma\frac{P_N}{2P_N + J}\exp\left(-\frac{R}{2P_N + J}\right) \tag{8.39}$$

This function is plotted in Figure 8.23 for representative values.

Narrowband Noise Jamming of the Complementary Channel

Jamming the complementary channel produces a higher BER than jamming the data channel [12]. This is because P_{e_2} when jamming the complementary channel is given by (8.2) with

$$P_{e_2} = \frac{P_N + J}{2P_N + J} \exp\left(-\frac{R}{2P_N + J}\right) \tag{8.40}$$

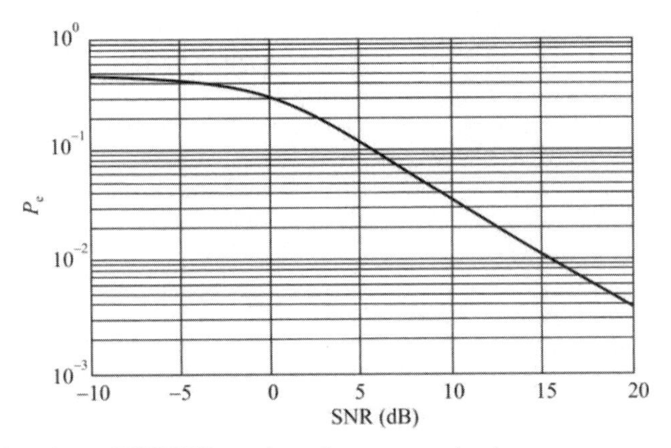

Figure 8.22 Noncoherent BFSK BER at optimum jammer power level.

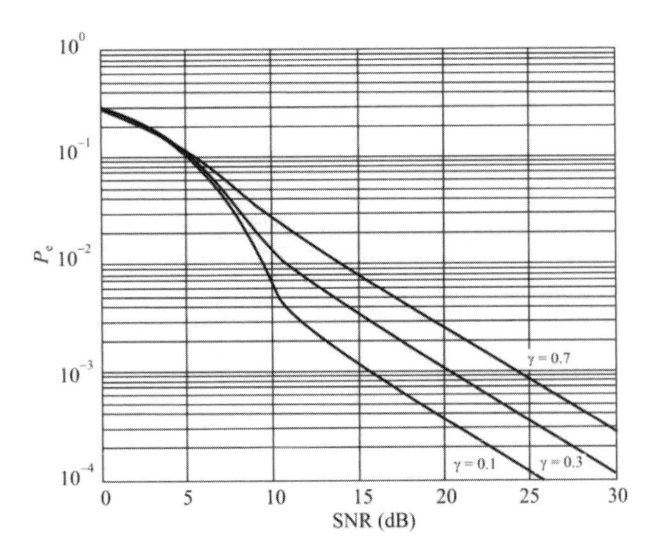

Figure 8.23 Probability of error for NBN jamming.

whereas jamming the data channel is given by (8.36), which is clearly smaller due to the lack of the J term in the numerator of the multiplying coefficient.

When picking just one of the adjacent channels shown in Figure 8.18, the probability of successfully jamming the correct channel is half of what it would be otherwise. Therefore, when including the effects of jamming, only a portion of the dwell, and using (8.40) in (8.2) yields

$$P_{\mathrm{e}}=\frac{1}{2}\left[(1-\gamma)\frac{1}{2}\exp\left(-\frac{1}{2}\frac{R}{P_{\mathrm{N}}}\right)+\gamma\frac{P_{\mathrm{N}}+J}{2P_{\mathrm{N}}+J}\exp\left(-\frac{R}{2P_{\mathrm{N}}+J}\right)\right] \tag{8.41}$$

This will be denoted as P_{comp}.

Jamming Two Contiguous Channels with Narrowband Noise

When both the data channel and one of the adjacent channels are jammed with noise signals that are wide enough to fill the bandwidth of the detector filters, then the same results as broadband jamming ensues, as given by (3.95). The jamming power per channel is reduced by half, however so $P_{\mathrm{n}}=J/2$. P_{e_2} in that case is given by [13, 14]

$$P_{\mathrm{e}_2}=\frac{1}{2}\exp\left(-\frac{1}{2}\frac{R}{P_{\mathrm{N}}+J/2}\right) \tag{8.42}$$

It is not known which adjacent channel is the correct one, so half the time, on average, the correct one will be jammed. In addition, the effects of not jamming the entire dwell must be taken into consideration as well as multiple dwells per data bit. In that case the per dwell P_{e} is given by (8.2) with (8.42)

$$P_{\mathrm{e}}=\frac{1}{2}\left[(1-\gamma)\frac{1}{2}\exp\left(-\frac{1}{2}\frac{R}{P_{\mathrm{N}}}\right)+\gamma\frac{1}{2}\exp\left(-\frac{1}{2}\frac{R}{P_{\mathrm{N}}+J/2}\right)\right] \tag{8.43}$$

which is denoted as P_2.

Jamming Both Adjacent Channels Only

When it is unknown which adjacent channel is the complementary channel when jamming two contiguous channels, the correct one is hit on average, half the time reducing the jamming performance accordingly. It is possible to place narrowband noise in both the adjacent channels without jamming the data channel. In that way, it is assured that the complementary channel will be jammed. The drawback is that, all the time, half the jammer power is wasted on the noncomplementary channel. In that case P_{e_2} is given by (3.91) with $P_n = J/2$

$$P_{e_2} = \frac{1}{2}\frac{P_N + J/2}{2P_N + J/2}\exp\left(-\frac{R}{2P_N + J/2}\right) \tag{8.44}$$

and P_e is

$$P_e = \frac{1}{2}\left[(1-\gamma)\frac{1}{2}\exp\left(-\frac{1}{2}\frac{R}{P_N}\right) + \gamma\frac{1}{2}\frac{P_N + J/2}{2P_N + J/2}\exp\left(-\frac{R}{2P_N + J/2}\right)\right] \tag{8.45}$$

which is denoted by P_{adj}.

Jamming Three Channels with Narrowband Noise

It is also possible to apply narrowband noise jamming to all three channels at the same time, thereby achieving the benefits of ensuring that the complementary channel gets jammed. The jammer power per jammed channel is reduced to one-third, but there is no loss associated with not knowing which adjacent channel is the proper one. Therefore, P_{e_2} is given by (3.91) with $P_n = J/3$

$$P_{e_2} = \frac{1}{2}\exp\left(-\frac{1}{2}\frac{R}{P_N + J/3}\right) \tag{8.46}$$

P_e in this case is

$$P_e = \frac{1}{2}\left[(1-\gamma)\frac{1}{2}\exp\left(-\frac{1}{2}\frac{R}{P_N}\right) + \gamma\frac{1}{2}\exp\left(-\frac{1}{2}\frac{R}{P_N + J/3}\right)\right] \tag{8.47}$$

which is denoted by P_3.

8.2.6.2 Follower Tone Jamming

Jamming the frequency where the tone is detected with a tone-jamming signal can enhance the intended receiver's ability to properly detect the signal as it does for

NBN jamming. Signal detection in noncoherent FSK receivers is based on measuring the energy from the channel filters. Placing additional energy at the correct FSK tone increases this energy, thus facilitating easier detection.

Again, we are faced with the quandary whether to jam one, two, or three channels.

Single Tone in Data Channel

Torrieri derived the following expression for the BER when there is a single jamming or interfering tone in the data channel [12]:

$$P_e = \frac{1}{2}\exp\left(-\frac{R+J}{2P_N}\right)I_0\left(\frac{\sqrt{RJ}}{P_N}\right) \tag{8.48}$$

Single Tone in Complementary Channel

When only the complementary channel is jammed with a tone, then P_{e_2} is given by [15] with a multiplier

$$P_{e_2} = \frac{1}{2}\left[Q\left(\sqrt{\frac{J}{P_N}},\sqrt{\frac{R}{P_N}}\right)-\frac{1}{2}\exp\left(\frac{R+J}{2P_N}\right)I_0\left(\frac{\sqrt{RJ}}{P_N}\right)\right] \tag{8.49}$$

The leading factor 1/2 is due to not knowing which adjacent channel is the correct one to jam, thereby decreasing P_e. The overall BER is given by

$$\begin{aligned}P_e &= (1-\gamma)\frac{1}{2}\exp\left(-\frac{1}{2}\frac{R}{P_N}\right)\\ &+\gamma\frac{1}{2}\left[Q\left(\sqrt{\frac{J}{P_N}},\sqrt{\frac{R}{P_N}}\right)-\frac{1}{2}\exp\left(\frac{R+J}{2P_N}\right)I_0\left(\frac{\sqrt{RJ}}{P_N}\right)\right]\end{aligned} \tag{8.50}$$

Wherein most FHSS systems are built with the data channel and the complementary channel in two contiguous channels, an effective ECCM technique is to use independent synthesizers in the FHSS system to place the mark and space center frequencies independently of each other [12]. The jammer, in this case, cannot simply jam the adjacent channel with any success.

Tones in Both the Data Channel and Complementary Channel

With tones in both the data channel and both of the adjacent channels, it is known that the jammer energy injected into the noncomplementary channel is wasted but it is also known (in the majority of cases) that both the data channel and the complementary channel are being jammed. With independent synthesizers for the mark and the space, this ECM approach is ineffective as discussed in the last section.

When both the data channel and the complementary channel are jammed with tones, then P_{e_2} is given by (3.99) but with the power half of what it is otherwise

$$P_{e_2} = \frac{1}{2\pi}\int_0^{2\pi}\left\{\begin{aligned} &Q\left[\sqrt{\frac{J/2}{P_N}}, \frac{D(\theta)}{\sqrt{2P_N}}\right] \\ &-\frac{1}{2}\exp\left[-\frac{2(J/2)+D^2(\theta)}{4P_N}\right] I_0\left[\frac{\sqrt{2(J/2)}D(\theta)}{2P_N}\right]\end{aligned}\right\} d\theta \quad (8.51)$$

and where

$$D^2(\theta) = 2R + 2(J/2) + 4\sqrt{R(J/2)}\cos(\theta) \quad (8.52)$$

and the jamming power per channel is half of what it was when a signal channel was jammed as explicitly noted in these expressions.

As long as it is known that the complementary channel is adjacent to the data channel, then P_e is reduced by a factor of 2, assuming that the mark and the space are independent. As long as the extra channel covered by the jammer is always below (or above) the detected tone frequency, then the complementary channel will be jammed half the time. Thus, $P_2 = P_e$ in (8.2) where P_{e_2} is given by (8.51).

Follower Jamming with Tones in the Two Adjacent Channels

In this case, the jammer performance is specified by the performance when the correct adjacent channel is jammed, but the jammer power per channel is half otherwise. Specifically, $P_{adj} = P_e$ in (8.2) with

$$P_{e_2} = \frac{1}{2}Q\left[\sqrt{\frac{J/2}{P_N}}, \sqrt{\frac{R}{P_N}}\right] \quad (8.53)$$

where it is explicitly noted that the power is reduced by half. The BER is given by

$$P_{\mathrm{e}} = (1-\gamma)\frac{1}{2}\exp\left(-\frac{1}{2}\frac{R}{P_{\mathrm{N}}}\right) + \gamma\frac{1}{2}Q\left[\sqrt{\frac{J/2}{P_{\mathrm{N}}}}, \sqrt{\frac{R}{P_{\mathrm{N}}}}\right] \tag{8.54}$$

Follower Jamming with Tones in All Three Channels

The most general case is when a tone is placed in both adjacent channels as well as the data channel. The correct complementary channel is assured to be hit, whereas the power per channel is reduced to one-third otherwise. Of course, the jammer power in the incorrect adjacent channel is wasted. Thus, P_{e_2} is given by (3.99)

$$P_{\mathrm{e}_2} = \frac{1}{2\pi}\int_0^{2\pi} \left\{ \begin{array}{l} Q\left[\sqrt{\dfrac{J/3}{P_{\mathrm{N}}}}, \dfrac{D(x)}{\sqrt{2P_{\mathrm{N}}}}\right] \\ -\dfrac{1}{2}\exp\left[-\dfrac{2(J/3)+D^2(x)}{4P_{\mathrm{N}}}\right] I_0\left[\dfrac{\sqrt{2(J/3)}D(x)}{2P_{\mathrm{N}}}\right] \end{array} \right\} dx \tag{8.55}$$

with

$$D^2(x) = 2R + 2(J/3) + 4\sqrt{R(J/3)}\cos(x) \tag{8.56}$$

in this case.

8.2.7 Smart Jamming

This category of jamming techniques attempts to disrupt portions of digital signals only, selecting only those portions necessary to deny communications, if possible. Some types of communication systems must be synchronized to operate properly. This synchronization is different from that described in Chapter 7 for synchronization of phase shift key signals. IS-95, for example, uses a separate coded Walsh channel for synchronization purposes. That channel alone could be attacked to degrade the synchronization process.

Likewise, some AJ signals must track the timing and phase of the transmitted signal. Sometimes separate channels are used for this purpose that are not spread. Some FHSS schemes use just a few known frequencies for acquisition. Some scheme is necessary so new members can join the network. An attack on the acquisition of the receiver of the incoming digital signal timing and framing information is thus possible. DSSS systems must establish where the transmitted signal is in the coding sequence so a search over the code space is required for acquisition. The major limitation is that the time acquisition occurs must be known if it's time related or the separate channel location if that is the approach used. Cell systems and PCS systems used out-of-band signaling to initiate calls. That channel can be jammed to preclude call set-up, to include in GSM. Some two-way pager systems use these techniques as well. *Deception jamming* is also possible. Here, false messages are sent to the receiver, such as movement orders.

Taking the notion of smart jamming to the limit, we obtain what could be called *brilliant jamming*. In this category an attempt is made to change specific bit patterns in a digital message so that an incorrect but valid message is received. The biggest limitations of this are that ultra-timing precision is required as well as significant a priori knowledge of the target signal structure.

8.3 Asset Sharing

Improving the efficiency of a jammer by increasing the number of signals it can simultaneously jam is desirable. This is possible with several techniques. Sharing of the jamming assets among several targets increases the utility of the jammer as a weapon. It is possible to share the power among targets. It is also possible to share the time that the jammer is on, switching from one target to the next in succession. This section describes some of these.

8.3.1 Look-Through

For nonspread signals look-through is often employed as a strategy to determine whether the target has moved to a new frequency or otherwise has stopped transmitting. Look-through is implemented to conserve jamming resources and/or to increase the effectiveness of the jammer by jamming more targets than otherwise possible.

Look-through is accomplished by turning the jammer transmitter off for a short period and measuring the spectrum. It generally does not work against FHSS AJ systems because the frequencies change so often when it is assumed that the jammer does not know the hop set frequencies. The same can be said for FH/DSSS hybrid AJ systems. Look-through for DSSS systems may work if the signal has sufficient strength to detect it.

8.3.2 Power Sharing

It is possible to jam more than one target at the same time. One manifestation of this is the aforementioned MT jamming. However, the signals need not be narrowband tones to implement such power sharing. Such techniques are discussed more fully in [16].

8.3.3 Time Sharing

Another technique for simultaneous coverage of more than one target is sharing the full power of the jammer on each target over time. Jamming a digital signal can be accomplished by raising the BER up to a certain level. Such signals need not be jammed 100% of the time. One bit in 10 in error is a BER of 1/10, which will cause unreliable communication in computer-to-computer links, although such a BER would probably be tolerable for voice communications. Neglecting the time it takes to switch between frequencies, one jammer could simultaneously service 10 such target nets.

As mentioned in the introduction to this chapter, it has been shown that for analog communication only about 30% of the transmission need be jammed to preclude effective voice understanding [16]. Therefore, if a scheme could be devised to move back and forth between targets, up to about three could be jammed essentially simultaneously with the same jammer. This is called *time sharing*.

8.4 Jamming Power to Signal Power Ratios

The jamming power to signal power ratio at the receiver for the most part determines the degree to which jamming will be successful. For digital signals, the jammer's goal is to raise this ratio to a level such that the BER is above a certain threshold. For analog voice signals, the goal is to reduce the articulation performance so that the signals are difficult to understand. For analog voice signals the articulation index depends on the timing of the disruptions as well as the JSR. Analog voice communication signals are not of significant interest here since all important SS systems employ digital techniques. For voice communications, the analog voice signals are converted to digital form with A/D converters for DSSS, FHSS, and THSS systems. For computer-to-computer communications such conversion is, of course, not necessary.

Actual jamming levels are only approximately predicted by propagation models discussed in Chapter 5. On the other hand, in most practical situations, actually measuring the JSR cannot reasonably be accomplished. Thus, it is wise to design jamming margins into the system. If a JSR of 6 dB is required, and the model indicates this can be accomplished with a certain jamming power level at a certain distance, then designing for a 10 dB JSR, for example, is prudent to account for the unknowns in the transition from the models to actual practice.

8.4.1 R^n Model JSR

The propagation loss with distance in this model as expressed by (5.48) is given in decibels. Therefore, the JSR can be expressed as

$$\begin{aligned}
\xi = \frac{J}{R} &= \frac{\dfrac{J_\mathrm{T} G_\mathrm{JR} G_\mathrm{RJ}}{10^{L(D_\mathrm{JR})/10}}}{\dfrac{S_\mathrm{T} G_\mathrm{TR} G_\mathrm{RT}}{10^{L(D_\mathrm{TR})/10}}} \\
&= \frac{J_\mathrm{T} G_\mathrm{JR} G_\mathrm{RJ}}{10^{\left(L(D_0)+10n\log_{10}(D_\mathrm{JR}/D_0)\right)/10}} \frac{10^{\left(L(D_0)+10n\log_{10}(D_\mathrm{TR}/D_0)\right)/10}}{S_\mathrm{T} G_\mathrm{TR} G_\mathrm{RT}} \\
&= \frac{J_\mathrm{T} G_\mathrm{JR} G_\mathrm{RJ}}{S_\mathrm{T} G_\mathrm{TR} G_\mathrm{RT}} 10^{n[\log_{10}(D_\mathrm{TR}/D_0)-\log_{10}(D_\mathrm{JR}/D_0)]} \\
&= \frac{J_\mathrm{T} G_\mathrm{JR} G_\mathrm{RJ}}{S_\mathrm{T} G_\mathrm{TR} G_\mathrm{RT}} 10^{n\log_{10}(D_\mathrm{TR}/D_\mathrm{JR})} \qquad (8.57)
\end{aligned}$$

where it has been assumed that the ground characteristics between the transmitter and the receiver are the same as those between the jammer and receiver [thus, $L(D_0)$ is the same for both paths].

This function is plotted in Figure 8.24 as a function of the ratio of the distances indicated and the power ratio of the jammer to the transmitter where it is assumed that $n = 4$ and $G_\mathrm{JR} = G_\mathrm{RJ} = G_\mathrm{TR} = G_\mathrm{RT}$.

8.4.2 Two-Ray Propagation JSR

When the antennas are elevated high enough off the ground, there could be a significant ground reflection component at the receiver. As an additional consideration, there could be a reflection from the transmitter as well as from the jammer. Here it is assumed that there are reflections due to both sources. The JSR in this case is given by

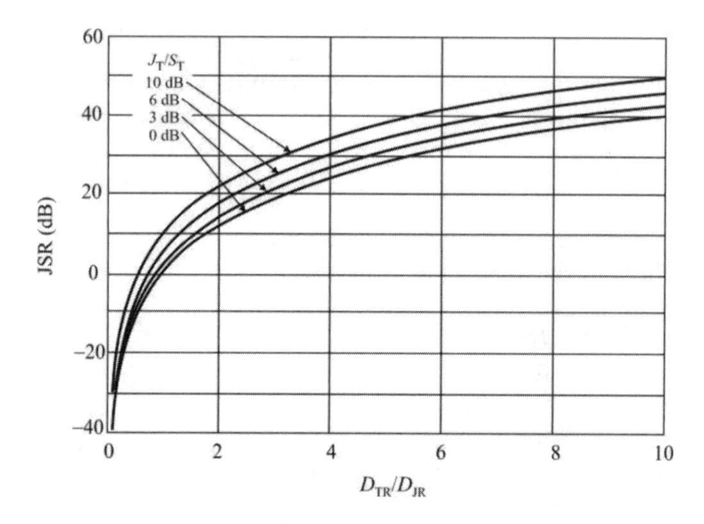

Figure 8.24 JSR versus the distance ratio D_{TR}/D_{JR} with J_T/P_T as a parameter for the R^n model.

$$\xi = \frac{J_T G_{JR} G_{RJ} \dfrac{1}{D_{JR}^4}(h_J h_R)^2}{S_T G_{TR} G_{RT} \dfrac{1}{D_{TR}^4}(h_T h_R)^2}$$

$$= \frac{J_T G_{JR} G_{RJ}}{S_T G_{TR} G_{RT}} \left(\frac{h_J}{h_T}\right)^2 \left(\frac{D_{TR}}{D_{JR}}\right)^4 \tag{8.58}$$

This equation is plotted in Figure 8.25 for two antenna height ratios. When $h_J/h_T = 1$, the results are the same as the R^n model above. When this ratio is five, the jamming performance is somewhat better as shown.

Inspection of this expression reveals that the JSR varies linearly with the ratio of the gain of the jammer antenna (toward the receiver) to that of the transmitter (also toward the receiver). It also varies linearly with the ratio of the power of the jammer to that of the transmitter. The receiver antenna gain, for most tactical AJ systems, would be the same in the directions of the transmitter and receiver, so it would have no effect. The JSR also varies as the fourth power of the ratio of the distance between the jammer and receiver to that between the transmitter and receiver. It also varies with the square of the ratio of the antenna heights. The receiver antenna height has no effect either, since it is the same for the transmitter path as for the jammer path.

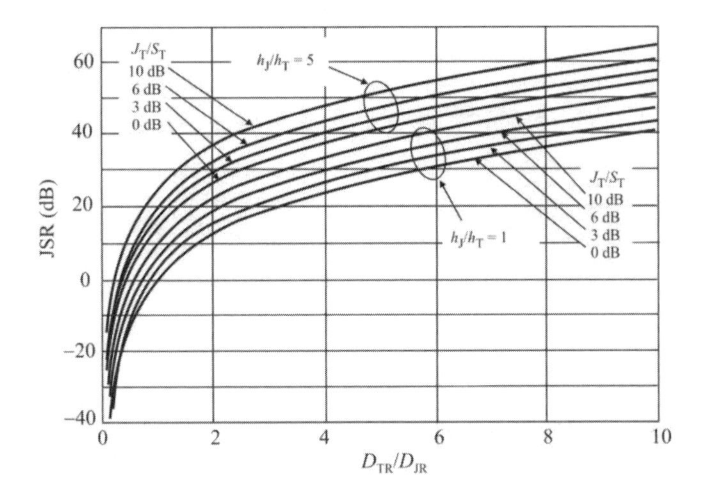

Figure 8.25 Reflection JSR levels for two antenna height ratios.

Thus, the JSR increases by 6 dB for every doubling of the jammer to transmitter antenna height ratio and –12 dB for every doubling of the jammer-receiver range compared to the transmitter-receiver range. This illustrates that jammer antenna height is important, as is the range between the jammer and the receiver. A jammer mounted on a UAS addresses both of these issues positively allowing considerable antenna heights to be achieved along with placing the jammer close to the target receiver (assuming that the approximate location of the target receiver is known).

The variation with height ratio is shown in Figure 8.26. A jammer mounted on a UAS that can get reasonably close to the target can perform very effectively denying communications. If the target receiver height is 3 m, then at $D_{TR} / D_{JR} > 0.5$, a UAS need only fly at an altitude of about 300 m to produce a JSR of 20 dB or more.

8.4.3 Nicholson JSR

Again, the JSR calculated here is

$$\xi = \frac{J}{R} = \frac{J_T G_{JR} G_{RJ}}{S_T G_{TR} G_{RT}} \frac{10^{C(f)+L_{fs}(f)+40\log\left(\frac{D_{TR}}{D_0}\right)/10}}{10^{C(f)+L_{fs}(f)+40\log\left(\frac{D_{JR}}{D_0}\right)/10}}$$

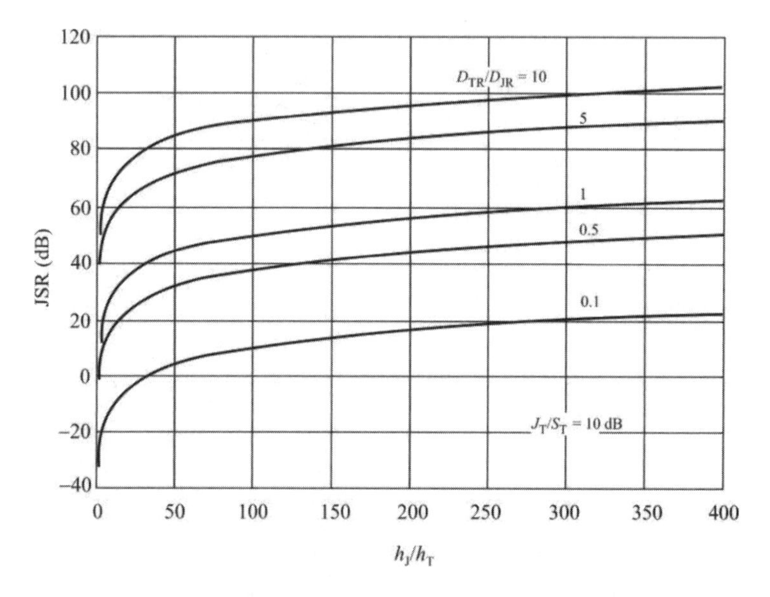

Figure 8.26 Variation of the JSR ratio with the height ratio. In this case J_T / S_T = 10 dB.

$$= \frac{J_T G_{JR} G_{RJ}}{S_T G_{TR} G_{RT}} 10^{4\log\left(\frac{D_{TR}}{D_{JR}}\right)} \tag{8.59}$$

where, as above, it is assumed that the ground characteristics between the transmitter and receiver are the same as those between the transmitter and receiver. This is the same result as for the R^n model above with $n = 4$.

8.4.4 Egli Model JSR

The JSR computed with this model yields the same expression as (8.58) since the $(40/f)^2$ terms cancel.

8.5 Jammer Platform Configurations

Jammers for land warfare can be configured in many forms whether they are AJ jammers or not. A *standoff jammer* is one that is deployed within the geography of friendly forces. A *stand-in jammer* is deployed among the adversarial forces. Due to the nature of broadband or partial band-jamming waveforms, such are not normally used in standoff configurations because of the possibility for excessive fratricide for friendly forces. The exception to this is screening jamming when the goal is to jam an adversary's ES systems to preclude interception of friendly communications.

Jammers can be vehicle mounted or dismounted. If they are vehicle mounted, they can also be implemented to operate on the move. If they are dismounted, they would remain stationary.

Airborne jammers are employed to extend the range of effectiveness as indicated above. A jammer mounted in a UAS can be placed closer to the target receivers in a stand-in configuration, thereby substantially reducing the fratricide problem.

The simplest waveform for EA against digital communications is BBN. It also is the least complicated and therefore the least expensive system to implement. Since a UAS can be a stand-in platform, BBN would typically be the waveform of choice for this application.

Various delivery mechanisms exist to deploy dismounted jammers. They can be hand-emplaced by individuals. They can be fired from artillery. UAVs or other aircraft, both manned and unmanned, can deploy them. When a UAS is used in this role, the overflight possibilities of the UAS can be favorably employed to placed the expendable jammers very close the receivers and could be deployed as arrays. Such arrays could be configured to cover a whole region with BBN jamming energy. Jammers can be deployed via missiles that travel at significant velocity. These configurations can be readily retargeted in flight and cover regions that are widely separated.

8.6 Concluding Remarks

This chapter presented the most common jamming strategies against AJ targets. The parameters involved are the bandwidth of the jamming waveform relative to that of the AJ signals, and, if tone jamming, the number and placement of the tones. The effectiveness of these jamming techniques at jamming AJ signals is presented in Chapters 10–14.

The concept of swept jamming was also presented. This is when one of the basic jamming waveforms is swept across the AJ frequency band. Its advantage is that at each individual frequency the entire jammer power is applied to the channel. Its disadvantage is that it is not present continually at any one frequency. Characteristics of the target receivers are important considerations for swept jamming. The dwell must be long enough at each frequency or frequency band to realize the effects of the jammer.

Follower jamming was also discussed, which attempts to ascertain where an FHSS AJ signal appears once it has left the last frequency. This is accomplished with a supporting ES system. There are significant timing issues associated with follower jammers that are dependent on distances and processing times.

Smart jamming was discussed as a potential jamming strategy. If significant information is known about the AJ target, then potentially advantage can be taken of it for jamming purposes. Smart jamming targets only portions of digital messages and precision timing is required to accomplish it.

The last topic presented was estimation of the JSR using the different models presented in Chapter 2. It was shown that the four models presented reduced to two when the JSR was the parameter of importance. The difference between the two models is an antenna height ratio. In addition, the models are independent of frequency because of the assumption that the propagation conditions between the transmitter and receiver are the same as between the jammer and receiver and the frequency effects cancel.

The modeling results indicated that if the jammer can get close and high in the air, the jamming could be particularly effective. This is a UAS configuration of a jammer.

Note that there are countermeasures to jamming that an AJ system can employ. Rejection of narrowband jammer signals is possible with one or more rejection filters in the receiver, which can be particularly effective against the MT jamming of DSSS systems. It is difficult to reject BBN and PBN jamming waveforms, however, as this would require excising considerable portions of the spectrum and communication would be hindered. It is also difficult to excise the effects of follower jammer waveforms on FHSS systems, as these are predicated on detecting the energy from the communication system prior to jamming, implying that the frequency is being used.

References

[1] Poisel, R. A., *Introduction to Communication Electronic Warfare Systems*, 2nd ed., Norwood, MA: Artech House, 2008, p. 35.

[2] Shannon, C. E., "A Mathematical Theory of Communication," originally published in *Bell System Technical Journal*, Vol. 27, pp. 379–423 and 623–656, July and October 1948. Also available as *The Mathematical Theory of Communication,* Urbana, IL: University of Illinois Press, 196, with some corrections and with Appendix 4 rewritten. The entire paper can be downloaded from http://cm.bell-labs.com/cm/ms/what/shannonday/paper.html.

[3] Riddle, L. P., "Performance of a Hybrid Spread Spectrum System against Follower Jamming," *Proceedings IEEE MILCOM*, 1990, pp. 18.8.1–18.8.5.

[4] Houston, S. W., "Modulation Techniques for Communication, Part 1: Tone and Noise Jamming Performance of Spread Spectrum M-ary and 2, 4-ary DPSK Waveforms," *NAECON '75 Conference Record*, 1975, pp. 51–58.

[5] Levitt, B. K., "FH/MFSK Performance in Multitone Jamming," *IEEE Transactions on Selected Areas in Communications,* Vol. SAC-3, No. 5, September 1985, pp. 627–643.

[6] Torrieri, D. J., *Principles of Secure Communication Systems,* 2nd ed., Norwood, MA: Artech House, 1992, p. 100.

[7] Peterson, R. L., R. E. Ziemer and D. E. Borth, *Introduction to Spread Spectrum Communications,* Upper Saddle River, NJ: Prentice Hall, 1985, pp. 368–373.

[8] Torrieri, D. J., *Principles of Secure Communication Systems*, 2nd ed., Norwood, MA: Artech House, 1992, p. 20.

[9] Milstein, L. B., S. Davidovice, and D. L. Schilling, "The Effect of Multiple-Tone Interfering Signals on a Direct Sequence Spread Spectrum Communication System," *IEEE Transactions on Communications,* Vol. Com-30, No. 3, March 1982.

[10] Torrieri, D. J., "Fundamental Limitations on Repeater Jamming of Frequency-Hopping Communications," *IEEE Journal on Selected Areas in Communications,* Vol. 7, No. 4, May 1989, pp. 569–575.

[11] Torrieri, D. J., *Principles of Secure Communication Systems*, 2nd ed., Norwood, MA: Artech House, 1992, p. 275.

[12] Torrieri, D. J., *Principles of Secure Communication Systems*, 2nd ed., Norwood, MA: Artech House, 1992, p. 276.

[13] Torrieri, D. J., *Principles of Secure Communication Systems*, 2nd ed., Norwood, MA: Artech House, 1992, pp. 275–278.

[14] Torrieri, D. J., *Principles of Secure Communication Systems*, 2nd ed., Norwood, MA: Artech House, 1992, p. 18.

[15] Torrieri, D. J., *Principles of Secure Communication Systems*, 2nd ed., Norwood, MA: Artech House, 1992, p. 277.

[16] Poisel, R. A., *Introduction to Communication Electronic Warfare Systems,* Norwood, MA: Artech House, 2002, Ch. 13.

Chapter 9

Blind CDMA Code Discovery

9.1 Introduction

CDMA technology specifically is for the purpose of allowing multiple communication systems to share the same spectrum simultaneously. A narrowband digital signal is spread out in the frequency spectrum by multiplication by a chip sequence that has a much higher clock rate, as illustrated in Figure 9.1. Signals spread by this method have very little energy at any narrowband portion of the RF spectrum and can be below the noise level within that band. This can make such signals very hard to detect.

In this chapter we present techniques for blindly determining the spreading codes in CDMA signals. The chapter is structured as follows. First, we briefly review what it means to blindly estimate a spreading code sequence. We then present a technique for estimating the code when it is known that there is a single DSSS signal present. We then present three algorithms that estimate the spreading codes when multiple CDMA signals could be present. The first uses subspace decomposition with a MUSIC approach. The second also uses subspace decomposition but implements an iterative procedure for estimating the code sequence. The last algorithm can be used if it is known that the codes are based on *m*-sequences using an extended Massey algorithm.

9.2 CDMA Signals

Conventional techniques for geolocation of targets emitting RF energy, such as interferometry or differential Doppler, cannot separate the energy from more than one signal simultaneously present on a frequency. These techniques assume that there is a single target present.

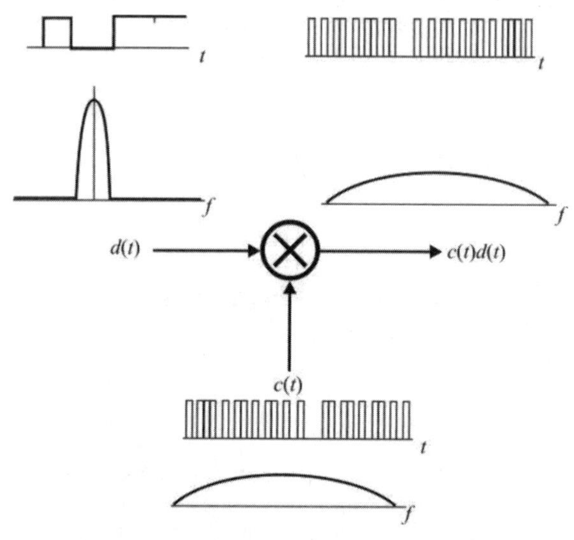

Figure 9.1 Generating CDMA signals. The data sequence is given by *d*(*t*) and *c*(*t*) is the chip sequence.

CDMA, by its very nature, imposes several broadband signals on the same portion of the frequency spectrum. These signals are separated in the receivers in the CDMA system by using codes. The receivers know what code is used by the MTs, and can process the signals received with those codes, properly decoding the signal of interest while simultaneously suppressing all other signals in the band.

A very common technique for spreading the digital signals in CDMA systems is illustrated in Figure 9.2. A short code, say, with a 15-bit PN spreading code, is applied to the symbols, separating them into I and Q channels, and the result applied to the RF phase modulators. These two channels (I and Q) are then combined in the RF modulator and transmitted.

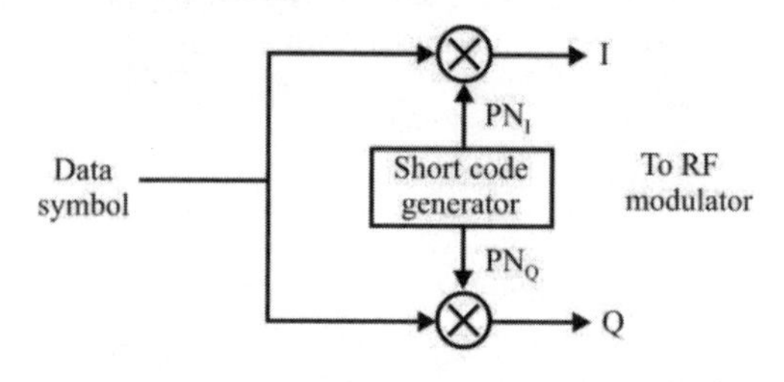

Figure 9.2 Logic for encoding data with a CDMA code.

Generally, it is not known a priori by an EW intercept system what specific CDMA codes are in use in a region. By applying blind discovery techniques, the codes can be determined, as can the power levels of the individual signals. For systems employing short codes, however, it is frequently sufficient to try all possible sequences to determine which are in use at the current time. This requires knowledge of timing offsets as will be described.

The chip sequence $c(t)$ determines the characteristics of how the channel can be shared among several users. The codes for $c(t)$ are orthogonal for all users of the channel. Therefore, their cross-correlation is low, while exhibiting a large autocorrelation value at zero lag offset.

Once the code is determined, several functions can be performed. If the code so discovered is the totality of the spreading code, then despreading is possible and the spreading code can be applied to the received signal and the despread, narrowband signal ensues. The *time difference of arrival* (TDOA), *frequency difference of arrival* (FDOA), or *direction of arrival* (DOA) methods can be applied to geolocate the transmitter. Whether or not the code is the totality of the spreading process, users can be tracked over the battlespace once they are identified by their unique code sequence.

9.3 Single Code Discovery

In this section we discuss a technique for blindly determining the spreading code of a DSSS signal when there is only a single signal present. It is based on results presented by Burel and Bouder [1]. Generally, the properties of the transmitter are unknown, including the spreading code. Also, we do not assume that we know the family of spreading codes possible. We do assume that the symbol period T_s is known; if it is not, then it can be estimated with the technique outlined in Section 4.4.2.5.

First we divide the received signal into windows, each the size of the symbol period. Each of these windows contains the end of one symbol of duration $T_s - t_0$ and the beginning of the next symbol of duration t_0, which is assumed to be unknown. Eigenanalysis of the autocorrelation of the samples in the windows produces two large eigenvalues, assumed without loss of generality to be the first two. Denote by $h(t)$ the convolution of the pseudo-random sequence with the impulse responses of all the filters in the transmission path:

$$h(t) = \sum_{l=0}^{L-1} c_l p(t - lT_c) \tag{9.1}$$

where T_c is the chip period, c_l, $l \in \{0,1,\ldots,L-1\}$, is the pseudo-random sequence, and $p(t)$ is the convolution of the transmitter filter, the channel filter characteristic, and the receiver filter [1]. Denote by $\vec{r}_m$ the contents of a window; that is, $\vec{r}_m$ is the measured received signal in that window. Then

$$\vec{r}_m = a_m \vec{h}_0 + a_{m+1} \vec{h}_{-1} + \vec{n}_m \tag{9.2}$$

where

$\vec{n}_m$ = noise;

$\vec{h}_0$ = vector containing the end of the spreading waveform $h(t)$ of duration $T_s - t_0$, followed by zeros of duration t_0;

$\vec{h}_{-1}$ = vector containing zeros of duration $T_s - t_0$ followed by the beginning of the spreading waveform $h(t)$.

Note that for the cases of concern here, the sample period, $T_e = T_c$ is known but the phase offset, t_0, in general, is not known. Here we set the sample time equal to T_c.

Therefore, the autocorrelation matrix is

$$\mathbf{R} = \mathcal{E}\{\vec{r}\vec{r}^{\mathrm{T}}\} = \mathcal{E}\left\{\|a_m\|^2\right\}\vec{h}_0\vec{h}_0^{\mathrm{H}} + \mathcal{E}\left\{\|a_{m+1}\|^2\right\}\vec{h}_{-1}\vec{h}_{-1}^{\mathrm{H}} + \sigma_n^2\mathbf{I} \tag{9.3}$$

where H indicates Hermitian (conjugate transpose). From (9.3) we can see that **R** has two eigenvalues larger than the rest. The associated eigenvectors for these eigenvalues are, within multiplicative factors, equal to $\vec{h}_0$ and $\vec{h}_{-1}$.

Equation (9.3) is computed each sample time and the averages of the first and second eigenvectors are used as estimates of the code sequence. The results of the example contained in [1] are illustrated in Figures 9.3 and 9.4. Figure 9.3 shows the average of 211 samples of the first two eigenvectors, while Figure 9.4 shows the estimated signature sequence along with the actual sequence. The agreement is very good. This example is based on a signal where $\upsilon = -9$ dB, and a Gold code sequence with length = 31.

Denote the variance of the symbols by σ_a^2, and let

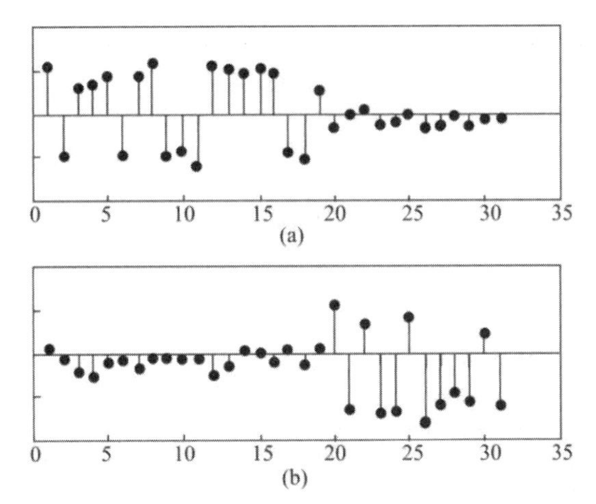

Figure 9.3 Normalized eigenvectors: (a) first eigenvector, and (b) second eigenvector. $\upsilon = -9$ dB. (Source: [1]. © 2000 IEEE. Reprinted with permission.)

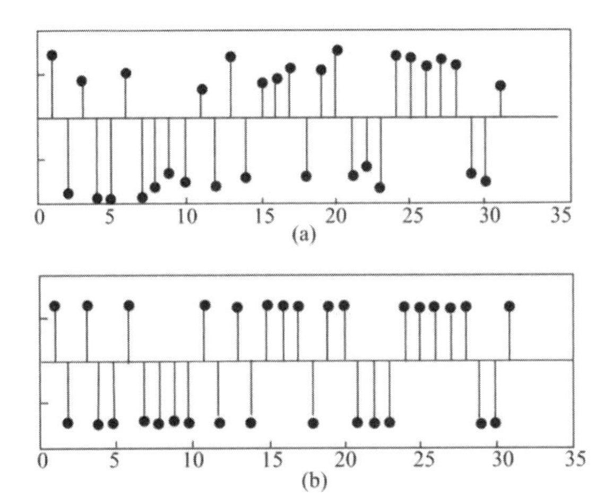

Figure 9.4 Estimated and true signature sequence: (a) estimated sequence, and (b) true sequence. $\upsilon = -9$ dB. (Source: [1]. © 2000 IEEE. Reprinted with permission.)

$$\varepsilon_{\rm h}^2 = \int_{-\infty}^{\infty} |h(u)|^2 \, du \approx T_{\rm e} \left\| \vec{h} \right\|^2 \tag{9.4}$$

then the DSSS signal variance is given by

$$\sigma_{\rm s}^2 = \frac{\sigma_{\rm a}^2 \varepsilon_{\rm h}}{T_{\rm s}} \tag{9.5}$$

and the power SNR is

$$\upsilon = \frac{\sigma_{\rm s}^2}{\sigma_{\rm n}^2} \tag{9.6}$$

When $t_0 \le T_{\rm s} / 2$, the eigenvalues of $\mathbf{R}$ are given by

$$\lambda_1 = \left(1 + \upsilon \frac{T_{\rm s} - t_0}{T_{\rm e}}\right) \sigma_{\rm n}^2 \tag{9.7}$$

$$\lambda_2 = \left(1 + \upsilon \frac{t_0}{T_{\rm e}}\right) \sigma_{\rm n}^2 \tag{9.8}$$

$$\lambda_i = \sigma_{\rm n}^2, \qquad i \ge 3 \tag{9.9}$$

An estimate of the SNR is given by (9.7) and (9.8) as

$$\hat{\upsilon} \approx \left(\frac{\lambda_1 + \lambda_2}{\sigma_{\rm n}^2} \right) \frac{T_{\rm e}}{T_{\rm s}} \tag{9.10}$$

Likewise, an estimate of t_0 is

$$\hat{t}_0 \approx \frac{T_{\rm e}}{\hat{\upsilon}} \left(\frac{\lambda_2}{\sigma_{\rm n}^2} - 1 \right) \tag{9.11}$$

When $t_0 > T_{\rm s} / 2$, the order of the first two eigenvalues is reversed.

A limitation of this method is that it assumes that short codes are used. Also, the derivation assumes that there is a single DSSS signal present in the passband of the intercept receiver. Therefore, it has EW applications for finding the codes of DSSS signals but limited utility for intercept of CDMA signals where multiple signals are expected to be the norm.

9.4 Blind Estimation of Multiple Codes with Subspace Decomposition and MUSIC

Haghighat and Soleymani [2] developed a method for blind code discovery based on the eigendecomposition of the autocorrelation matrix of the received signal and the MUSIC algorithm. We summarize their technique in this section. In this approach all possible potential spreading codes must be examined by projecting onto the noise and signal subspaces. If these potential codes are unknown, then all possible 2^{N-1} sequences must be examined.

9.4.1 Signal Model

We assume that the direct sequence CDMA communication system is synchronous. As previously mentioned, the synchronization of a PCS downlink can be accomplished at the BS. While the PCS uplink is asynchronous at RF, there are ways to synchronize the signals in the intercept system (and BS). No a priori information is assumed to be known about the target CDMA system, including the possible spreading codes.

When there are L users of a channel, the received signal prior to sampling in that channel during the time interval $(0, T)$ can be represented as

$$r(t) = \sum_{l=1}^{L} A_l d_l c_l(t) + n(t) \tag{9.12}$$

where A_l is the amplitude of the lth user, assumed unknown but constant over the time of concern, $d_l \in \{-1,+1\}$ represents the data sequence, with both values equiprobable, and $c_l(t)$ is the pseudo-random spreading sequence of the lth user. Furthermore, it is assumed that the spreading codes are short.[1] The noise in the channel is given by $n(t)$, which is assumed to be AWGN with variance σ^2.

After sampling, (9.12) can be represented by

---

[1] A short spreading sequence is one where the extent of the code is 1 bit. This can be extended to a few bits and the results herein still apply.

$$\vec{r} = \sum_{l=1}^{L} A_l d_l \vec{c}_l + \vec{n} \tag{9.13}$$

where $\vec{c}_l = (1/\sqrt{N})\begin{bmatrix} c_{l1} & c_{l2} & \cdots & c_{lN} \end{bmatrix}^{\mathrm{T}}$ is the code sequence for the lth user. N is the length of the code, given by

$$N = \frac{T_{\mathrm{s}}}{T_{\mathrm{c}}} \tag{9.14}$$

where T_{s} is the symbol time and T_{c} is the chip time. In other words, $T_{\mathrm{s}} = NT_{\mathrm{c}}$. Vector $\vec{n}$ is a zero mean white Gaussian noise vector with covariance matrix $\sigma^2\mathbf{I}$. Expression (9.13) can be written as

$$\vec{r} = \mathbf{CA}\vec{d} + \vec{n} \tag{9.15}$$

where $\mathbf{C} = \begin{bmatrix} \vec{c}_1 & \vec{c}_2 & \cdots & \vec{c}_L \end{bmatrix}$ and $\mathbf{A} = \mathrm{diag}\begin{bmatrix} A_1 & A_2 & \cdots & A_L \end{bmatrix}$.

9.4.2 Subspace Decomposition

As above, the autocorrelation matrix of $\vec{r}$ is given by

$$\mathbf{R} = \mathcal{E}\{\vec{r}\vec{r}^{\mathrm{T}}\} \tag{9.16}$$

where $\mathbf{R} \in \mathbb{R}^{N \times N}$ is real and symmetric. Substituting (9.15),

$$\mathbf{R} = \mathbf{CA}\vec{d}\vec{d}^{\mathrm{T}}\mathbf{A}^{\mathrm{T}}\mathbf{C}^{\mathrm{T}} + \sigma^2\mathbf{I} \tag{9.17}$$

$$= \mathbf{CAA}^{\mathrm{T}}\mathbf{C}^{\mathrm{T}} + \sigma^2\mathbf{I} \tag{9.18}$$

since the noise is uncorrelated with the signals and $\vec{d}\vec{d}^{\mathrm{T}} = \mathbf{I}$.

The eigenvalues and eigenvectors of $\mathbf{R}$ can be found by calculating the *eigenvalue decomposition* (EVD) of (9.18). Since $\mathbf{R}$ is real and symmetric, however, the best way to find these elements is with the *singular value decomposition* (SVD), since in that case the EVD and SVD are equivalent.[2] This yields

$$\mathbf{R} = \mathbf{U}\Sigma\mathbf{U}^{\mathrm{T}} = \begin{bmatrix}\mathbf{U}_{\mathrm{s}} & \mathbf{U}_{\mathrm{n}}\end{bmatrix}\begin{bmatrix}\Sigma_{\mathrm{s}} & \mathbf{0} \\ \mathbf{0} & \Sigma_{\mathrm{n}}\end{bmatrix}\begin{bmatrix}\mathbf{U}_{\mathrm{s}}^{\mathrm{T}} \\ \mathbf{U}_{\mathrm{n}}^{\mathrm{T}}\end{bmatrix} \tag{9.19}$$

where $\mathbf{U}$ and $\Sigma = \mathrm{diag}\begin{bmatrix}\lambda_1 & \lambda_2 & \cdots & \lambda_N\end{bmatrix}$ are the eigenvector and eigenvalue matrices, respectively.

Singular vectors spanning the signal subspace are given by $\mathbf{U}_{\mathrm{s}}$ and those spanning the noise subspace are given by $\mathbf{U}_n$. The dimension of the signal subspace is given by ρ_0 and, in this case, is the number of active users in the channel ($\rho_0 = L$). The eigenvalues in $\mathbf{\Sigma}$ can be separated into two disjoint sets: those that are larger than some threshold and those that are smaller than that threshold. The smallest eigenvalues have multiplicity $N - L$ [3].

Denoting the eigenvectors by $\vec{e}_j$, these subspaces can be represented as follows:

$\mathbf{E}_{\mathrm{s}}$ denotes the signal subspace $= \begin{bmatrix}\vec{e}_1 & \vec{e}_2 & \cdots & \vec{e}_L\end{bmatrix}$

$\mathbf{\Sigma}_{\mathrm{s}} = \mathrm{diag}\begin{bmatrix}\lambda_1 & \lambda_2 & \cdots & \lambda_L\end{bmatrix}$

$\mathbf{U}_{\mathrm{s}} = \begin{bmatrix}\vec{u}_1 & \vec{u}_2 & \cdots & \vec{u}_L\end{bmatrix}$

$\mathbf{E}_{\mathrm{n}}$ denotes the noise subspace $= \begin{bmatrix}\vec{e}_{L+1} & \vec{e}_{L+2} & \cdots & \vec{e}_N\end{bmatrix}$

$\mathbf{\Sigma}_{\mathrm{n}} = \mathrm{diag}\begin{bmatrix}\lambda_{L+1} & \lambda_{L+2} & \cdots & \lambda_N\end{bmatrix}$

$\mathbf{U}_{\mathrm{n}} = \begin{bmatrix}\vec{u}_{L+1} & \vec{u}_{L+2} & \cdots & \vec{u}_N\end{bmatrix}$

The signal and noise subspaces are orthogonal. A code sequence $\vec{c}_i$ can be projected onto these subspaces as

$$f_i = \left\|\vec{c}_i^{\mathrm{T}}\mathbf{E}_{\mathrm{n}}\right\|^2 = \left(\vec{c}_i^{\mathrm{T}}\mathbf{E}_{\mathrm{n}}\right)\left(\vec{c}_i^{\mathrm{T}}\mathbf{E}_{\mathrm{n}}\right)^{\mathrm{T}} \tag{9.20}$$

$$g_i = \left\|\vec{c}_i^{\mathrm{T}}\mathbf{E}_{\mathrm{s}}\right\|^2 = \left(\vec{c}_i^{\mathrm{T}}\mathbf{E}_{\mathrm{s}}\right)\left(\vec{c}_i^{T}\mathbf{E}_{\mathrm{s}}\right)^{\mathrm{T}} \tag{9.21}$$

[2] *Equivalent* here means that the eigenvectors are the same as the singular vectors. The eigenvalues are given by $\mathbf{\Sigma}$.

If $\vec{c}_i$ corresponds to an active user in the channel, then the code sequence lies in the signal subspace and $f_i = 0$. On the other hand if $f_i > 0$, then that user is not active at the moment. Likewise, if the ith user is active, then $g_i = 1$, and if the ith user is not active, then $g_i < 1$. This is the MUSIC algorithm.

A unique signature sequence (code) is assigned for each mobile terminal. It is generated by using a combination of a code for the manufacturer and the mobile serial number.

Thus, expressions (9.20) and/or (9.21) can be used as tests for the presence or absence of a particular code sequence. If the code sequence is known in advance, then the tests given by these functions will determine its presence. If the possible code sequences are not a priori known, then (9.20) must be tested for all possible code sequences. In all cases of practical interest, however, the number of possible code sequences are limited.

The spreading codes consist of N chips. Therefore, the examination of all possible 2^{N-1} different combinations is required to ascertain which code sequences are active at the moment. This, however, is not adequate to determine precisely which users are active since linear combinations of the code sequences can cause similar behavior for tests (9.20) and (9.21). That is, such linear combinations can produce $f_i \approx 0$ and $g_i \approx 1$. Thus, false positives can be produced, depending on which particular code sequences are active. Some method is therefore required to determine how many users are active.

Such a method is provided by decorrelating the received signal. Such a decorrelating function can be determined as [2]

$$\vec{\delta}_i = \mu_i \mathbf{U}_s (\boldsymbol{\Sigma}_s - \sigma^2 \mathbf{I})^{-1} \mathbf{U}_s^T \vec{c}_i \tag{9.22}$$

for each possible sequence $\vec{c}_i$, $1 \le i \le K'$, where K' is the maximum possible code sequences. In (9.22), μ_i is a normalizing constant given by [4]

$$\mu_i = \frac{1}{\vec{c}_i^T \mathbf{U}_s (\boldsymbol{\Sigma}_s - \sigma^2 \mathbf{I})^{-1} \mathbf{U}_s^T \vec{c}_i} \tag{9.23}$$

If $\vec{c}_i$ is not a legitimate code sequence, but a combination of two or more active sequences, $\vec{\delta}_i$ will be a linear combination of the decorrelating function corresponding to those sequences

$$\vec{c}_i = \sum_{k=1}^{K} \alpha_k \vec{c}_k \tag{9.24}$$

and

$$\vec{\delta}_i = \mu_i \mathbf{U}_s (\mathbf{\Sigma}_s - \sigma^2 \mathbf{I})^{-1} \mathbf{U}_s^{\mathrm{T}} \sum_{k=1}^{K} \alpha_k \vec{c}_k \tag{9.25}$$

On applying (9.22) to the received signal

$$z_i = \vec{\delta}_i^{\mathrm{T}} \vec{r} = \vec{\delta}_i^{\mathrm{T}} \mathbf{C}\mathbf{A}\vec{d} + \vec{\delta}_i^{\mathrm{T}} \vec{n} \tag{9.26}$$

$$= \vec{\delta}_i^{\mathrm{T}} \mathbf{C}\mathbf{A}\vec{d} + w_i \tag{9.27}$$

where w_i is white Gaussian noise with variance $\sigma_{w_i}^2 = \left(\vec{\delta}_i^{\mathrm{T}} \vec{\delta}_i\right)\sigma^2$. This represents noise enhancements on the received signal so that

$$z_i = \begin{cases} A_i d_i + w_i, & \vec{c}_i \text{ is an authentic code} \\ \sum_{k=1}^{K} \alpha_k A_k d_k + w_i, & \vec{c}_i \text{ is a combination of codes} \end{cases} \tag{9.28}$$

When $\vec{c}_i$ is an authentic code, the decorrelator output clusters around $\pm A_i$. In that case the noise component is due to background and system noise only, and interference from other users is not present. When $\vec{c}_i$ is a combination of codes, significant interference from the other codes is present and such clustering does not occur and the decorrelator output is dispersed. This is illustrated in Figure 9.5. In Figure 9.5(a) there is an authentic code while in Figure 9.5(b) there is a combination of codes that causes the PDF to spread out.

A cost function can be defined based on this clustering/disbursing characteristic as

$$J(\vec{\delta}_i) = \left| \frac{\mathcal{E}\{z_i^2\}}{\mathcal{E}^2\{|z_i|\}} - 1 \right| \tag{9.29}$$

Once the authentic codes are determined, the power of the user's signals can be determined. The total power can be determined from (9.18) as

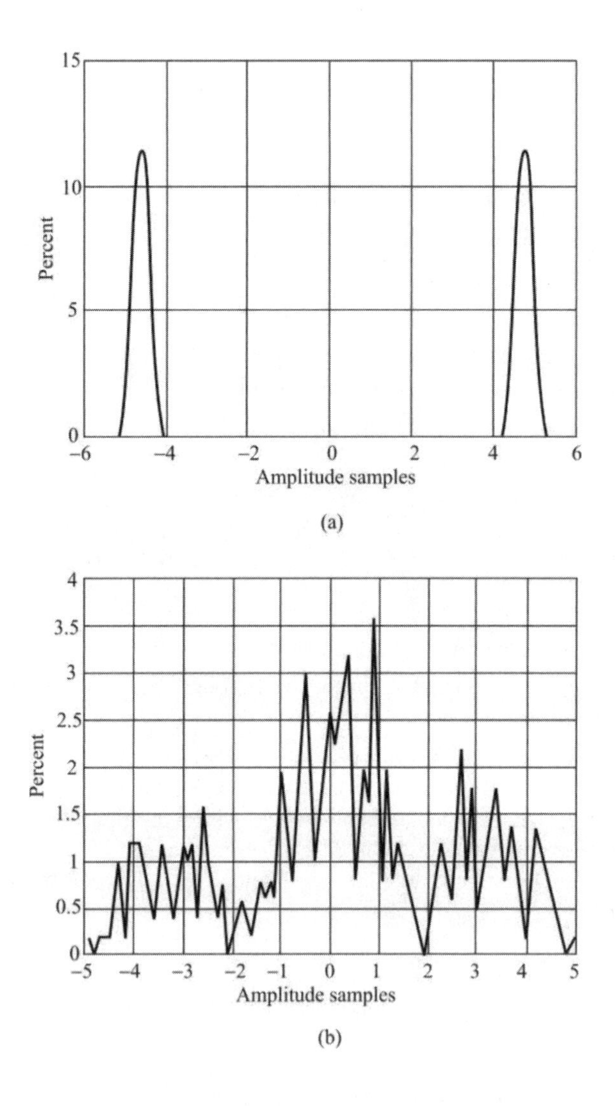

Figure 9.5(a, b) Blind multicode PDFs. (Source: [2]. © IEEE 2003. Reprinted with permission.)

$$\hat{\mathbf{A}}\hat{\mathbf{A}}^T = (\mathbf{C}^T\mathbf{C})^{-1}\mathbf{C}^{\mathrm{T}}(\mathbf{R} - \sigma^2\mathbf{I})\mathbf{C}(\mathbf{C}^{\mathrm{T}}\mathbf{C})^{-1} \tag{9.30}$$

or

$$\hat{\mathbf{A}}\hat{\mathbf{A}}^{\mathrm{T}} = N^2\mathbf{R}^{-1}\mathbf{C}^{\mathrm{T}}(\mathbf{R} - \sigma^2\mathbf{I})\mathbf{C}\mathbf{R}^{-1} \tag{9.31}$$

The noise variance σ^2 is available from the original SVD as the smallest singular value. An individual user's power can be estimated as

$$\hat{A}_i^2 = \mathcal{E}(z_i^2) - \sigma_{w_i}^2 = \mathcal{E}(z_i^2) - (\vec{\delta}_i^{\mathrm{T}}\vec{\delta}_i)\sigma^2 \tag{9.32}$$

9.4.3 Performance

The results presented above were simulated. A processing gain of N = 16 was used. When there were 10 users of the channel, the initial SVD indicated that there were 64 users present. The inverse of the cost function is illustrated in Figure 9.6. In Figure 9.6(a), all the users had equal power. In Figure 9.6(b) one user had an SNR of 6 dB, while all the others had an SNR of 26 dB. In both cases, the procedure described above found the 10 authentic code sequences.

9.4.4 Summary

While the technique presented in this section can be used to determine multiple codes as demonstrated in Figure 9.6, K' can be, and usually will be, much larger than the number of possible code sequences K. This results in a considerable computation burden, so parallel implementations are required in practice. In many cases this number can be pared down considerably.

9.5 Blind Estimation of Multiple Codes with Iterative Subspace Decomposition

9.5.1 Iterative Subspace Method

Haghighat and Soleymani [4] presented a method of blind code discovery based on subspace decomposition of the autocorrelation matrix coupled with an iterative procedure for signature sequence estimation. As above, the eigenvalue decomposition of $\mathbf{R} = \mathcal{E}\{\vec{r}\vec{r}^{\mathrm{T}}\}$, given by (9.18) and (9.19), is the starting point for

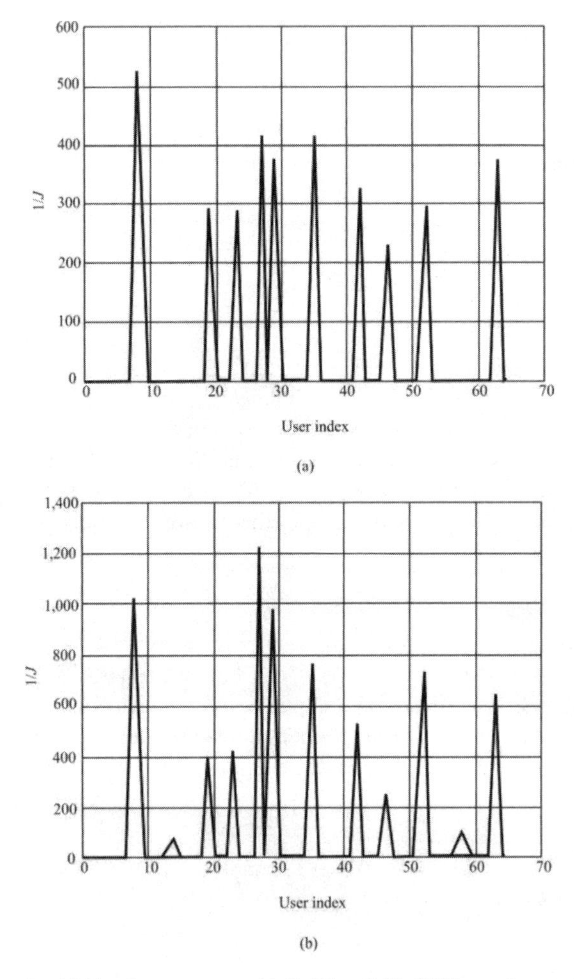

Figure 9.6 $1 / J$ results. (a) Equal power users with $E_b / N_0 = 6$ dB. (b) Unequal power users, one user with $E_b / N_0 = 6$ dB and all others with $E_b / N_0 = 26$ dB.

the method. Using (9.19) in (9.18), we can write (9.18) as

$$\mathbf{CA}(\mathbf{CA})^{\mathrm{T}} = \mathbf{U}_{\mathrm{s}}(\mathbf{\Sigma}_{\mathrm{s}} - \sigma^2\mathbf{I}_{\mathrm{L}})\mathbf{U}_{\mathrm{s}}^{\mathrm{T}} \tag{9.33}$$

where $\mathbf{I}_{\mathrm{L}}$ is the $L \times L$ identity matrix. Denote $\mathbf{Z} = \mathbf{CA}$ so that

$$\mathbf{Z} = \begin{bmatrix} \vec{z}_1 & \vec{z}_2 & \cdots & \vec{z}_{\mathrm{L}} \end{bmatrix} \tag{9.34}$$

with

$$\vec{z}_k = \sum_{l=1}^{L} w_{kl}\lambda_l \vec{u}_l \tag{9.35}$$

or, in matrix form,

$$\begin{bmatrix} \vec{z}_1 & \cdots & \vec{z}_L \end{bmatrix}_{N\times L} = \begin{bmatrix} \vec{u}_1 & \cdots & \vec{u}_L \end{bmatrix}_{N\times L} \begin{bmatrix} \lambda_1 & & 0 \\ & \ddots & \\ 0 & & \lambda_L \end{bmatrix}_{L\times L} \begin{bmatrix} \vec{w}_1 & \cdots & \vec{w}_L \end{bmatrix}_{L\times L} \tag{9.36}$$

where

$$\vec{w}_l = \begin{bmatrix} \vec{w}_{l1} & \cdots & \vec{w}_{lL} \end{bmatrix}^{\mathrm{T}} \tag{9.37}$$

contains the signature sequence information. Now

$$\mathbf{R} = \mathbf{CAA}^{\mathrm{T}}\mathbf{C}^{\mathrm{T}} + \sigma^2\mathbf{I} = \mathbf{ZZ}^{\mathrm{T}} + \sigma^2\mathbf{I} \tag{9.38}$$

or

$$\mathbf{Z} = \mathbf{U}_{\mathrm{s}}\mathbf{\Sigma}_{\mathrm{s}}\mathbf{W} \tag{9.39}$$

where **W** contains the signature sequence. The noise term in (9.38) is temporarily ignored so that we get the recursive algorithm:

$$\mathbf{R} = \mathbf{ZZ}^{\mathrm{T}} \tag{9.40}$$

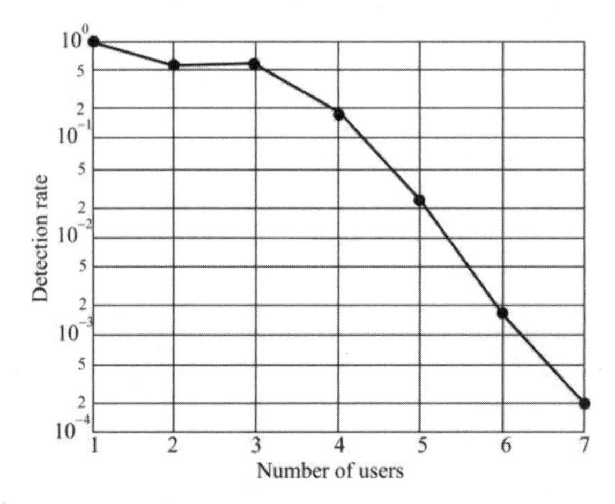

Figure 9.7 Successful identification rate versus the number of active users ($\mu = 0.05$). Detection rate is the fraction of the codes found.

The known parameters are **C**, $\mathbf{U}_s$, and $\mathbf{\Sigma}_s$, and the unknowns are **W** and **Z**. An initial guess for $\mathbf{W}_m$ is made and at the *m*th iteration the error expression

$$\Delta\mathbf{W}_m = \mathbf{\Sigma}_s^{-1}\mathbf{U}_s^{\mathrm{T}}(\mathbf{R}-\hat{\mathbf{R}}_m)\mathbf{U}_s\mathbf{\Sigma}_s^{-1}(\mathbf{\Omega}_m^{\mathrm{T}})^{-1} \tag{9.41}$$

with updating function

$$\mathbf{W}_{m+1} = \mathbf{W}_m + \mu\,\mathrm{sgn}(\Delta\mathbf{W}_m) \tag{9.42}$$

After convergence (assuming convergence), the *k*th signature sequence is given by

$$\vec{s}_k = \mathrm{sgn}(\vec{z}_k) \tag{9.43}$$

Performance of this algorithm for a particular example is shown in Figure 9.7 Note that the detection rate decreases rather rapidly with multiple users. The discovery rate falls to 50% with only two CDMA signals present and becomes virtually useless with more than three simultaneous users.

This algorithm does work when there is more than one user present, but is limited to synchronous CDMA systems. We discussed methods of forming synchronous CDMA networks from asynchronous ones in Chapter 7.

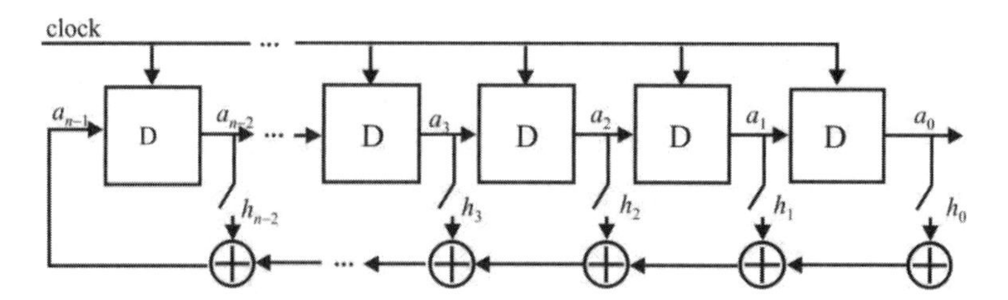

Figure 9.8 Linear feedback shift register.

9.6 Blind Discovery of DSSS *m*-Sequence Chip Codes

If the targets of concern are known to employ linear *m*-sequences, including Gold codes constructed from two *m*-sequences, then there is an alternative method to estimate the chip sequence [6] based on an extended Massey algorithm [7]. It is based on analysis of the LFSR as described in Chapter 7, one of which is shown in Figure 9.8 for convenience. The next value to enter the leftmost stage (*n*th) is given by

$$a_{n+1} = \sum_{i=0}^{n} h_i a_i \tag{9.44}$$

The process of determining what *m*-sequence is being intercepted is equivalent to determining the values of the coefficients h_i. As in the reference, the process will be illustrated with an example.

Suppose $n = 4$. Then

$$\begin{bmatrix} a_1 \\ a_2 \\ a_3 \\ a_4 \end{bmatrix} = \begin{bmatrix} 0 & 1 & 0 & 0 \\ 0 & 0 & 1 & 0 \\ 0 & 0 & 0 & 1 \\ h_0 & h_1 & h_2 & h_3 \end{bmatrix} \begin{bmatrix} a_0 \\ a_1 \\ a_2 \\ a_3 \end{bmatrix} \tag{9.45}$$

$2n$ continuous chips of the intercepted chip stream are used to find the n values of the chips and n values of the tap coefficients. (n is assumed to be known, which is likely the case in practice.) The next four chip values are

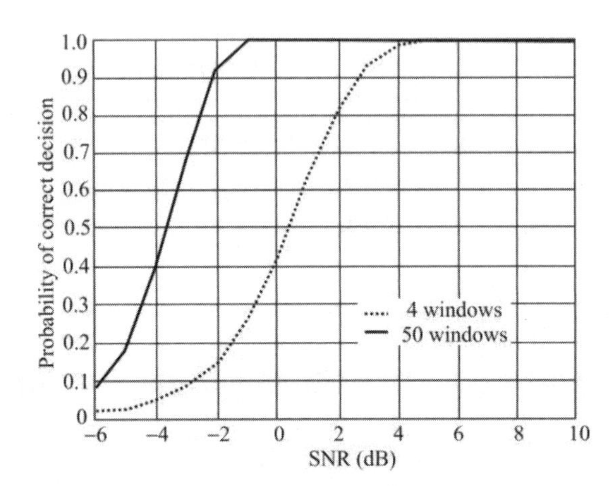

Figure 9.9 Majority voting results for a few cases: $n = 8$ taps, 4 and 50 contiguous windows, averaged over 1,000 Monte Carlo runs.

$$\begin{bmatrix} a_4 \\ a_5 \\ a_6 \\ a_7 \end{bmatrix} = \begin{bmatrix} a_0 & a_1 & a_2 & a_3 \\ a_1 & a_2 & a_3 & a_4 \\ a_2 & a_3 & a_4 & a_5 \\ a_3 & a_4 & a_5 & a_6 \end{bmatrix} \begin{bmatrix} h_0 \\ h_1 \\ h_2 \\ h_3 \end{bmatrix} \tag{9.46}$$

or

$$\vec{\tilde{a}} = \mathbf{A}\vec{h} \tag{9.47}$$

The vector $\vec{h}$ is found as

$$\vec{h} = \mathbf{A}^{-1}\vec{\tilde{a}} \tag{9.48}$$

Some method must be found that estimates the tap weights using (9.48). One method is to examine all tap set vectors and count identical ones. The vector that occurs the majority of the time is assumed to be the actual vector. The results for 4 windows and 50 windows are illustrated in Figure 9.9 for $n = 8$ taps, averaged over 1,000 simulation runs.

Gold codes are built from two or more linear feedback shift registers, with the case of two of order p_1 and p_2 illustrated in Figure 9.10. Two m-sequences modulo-2 added in this way generate the linear product code of length $p_1 + p_2$.

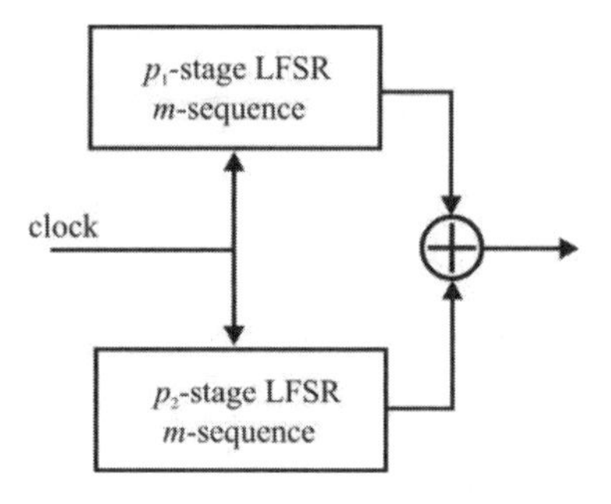

Figure 9.10 Generation of Gold codes.

This sequence can be discovered in a similar way as described above. Knowing the LFSR structure, it is possible to determine which transmitter is broadcasting.

Limited results are available based on general CDMA signaling. These results are delineated in Table 9.1 for up to three signals. When the *m*-sequence lengths are different, in some cases no solutions are available, but for the important special case of equal sequence lengths, the technique does yield viable solutions.

9.7 Concluding Remarks

We examined several techniques for estimation of the spreading codes of CDMA signals in this chapter. Since, in general, for EW applications, knowledge of the specific code in use is not generally known, methods to recover it from signals received are important.

Table 9.1 Estimated Codes

Code Condition	Estimated Code(s)
$p_1 = p_2$	$n = 2p + 1$ Inverted Gold
$p_1 = p_2$ Antiphase	$n = 2p$ Gold
$p_1 \neq p_2$	$n = p_1 + p_2 + 1$ Inverted linear product
$p_1 \neq p_2$ Antiphase	$n = p_1 + p_2$ Linear product
$p_1 = p_2 = p_3$ All phase	n is order $2p$ No sensible solution
$p_1 = p_2 \neq p_3$	n is order $2p$ Inverted linear product
$p_1 \neq p_2 \neq p_3$	No solution available

In some cases, even though the specific code is not known at the intercept receiver, it may be possible that the set of possible codes is known. This information can be exploited in those cases to optimize the search for the specific code.

References

[1] Burel, G., and C. Bouder, "Blind Estimation of the Pseudo-Random Sequence of a Direct Sequence Spread Spectrum Signal," *Proceedings IEEE MILCOM 2000*, 2000, pp. 967–970.

[2] Haghighat, A., and M. R. Soleymani, "A Blind MUSIC-Based Algorithm for User Identification in Multiuser DS-CDMA," *Proceedings IEEE GLOBECOM 2003*, 2003, pp. 2447–2452.

[3] Halford, K. W., and M. Brandt-Pearce, "New-User Identification in a CDMA System," *IEEE Transactions on Communications,* Vol. 46, No. 1, January 1998.

[4] Haghighat, A., and M. R. Soleymani, "A Subspace Scheme for Blind User Identification in Multi-User DS-CDMA," *Proceedings IEEE Wireless Communications and Networking Conference*, 2003, Vol. 1, pp. 688–692.

[5] Wang, X., and H. V. Poor, "Blind Multiuser Detection: A Subspace Approach," *IEEE Transactions on Information Theory,* Vol. 44, No. 2, March 1998.

[6] Hill, P. C. J., and M. E. Ridley, "Blind Estimation of Direct-Sequence Spread Spectrum M-Sequence Chip Codes," *IEEE Transactions 6th International Symposium on Spread-Spectrum Technology & Applications,* NJIT, Newark, NJ, September 6–8, 2000, pp. 305–309.

[7] Massey, J. L., "Shift Register Synthesis and BCH Decoding," *IEEE Transactions on Information Technology,* Vol. IT-15, No. 1, January 1969.

Chapter 10

Electronic Warfare and Direct Sequence Spread Spectrum Systems

10.1 Introduction

Performance of jamming techniques against DSSS targets is presented in this chapter. DSSS systems spread the baseband data signal over a broad bandwidth to achieve AJ protection. Frequently these signals are below the noise floor at the point of reception, so even detecting their presence can be difficult. However, the existence of the near-far problem in multiple access DSSS systems supports the notion that DSSS signals can be jammed. The near-far problem is created by one signal (DSSS or otherwise) being significantly stronger than a DSSS signal within the same bandwidth so the receiver is swamped by the first signal. This is similar to jamming.

Because they are the most prolific forms, all of the results in this chapter are presented for signals with PSK modulations, and, in particular, attention is focused on BPSK and QPSK. QPSK can be considered two independent BPSK signals at half the data rate. In addition, offset QPSK and MSK are two forms of QPSK that have desirable properties in the form of reduced sidelobes, which lower the interference in adjacent frequency channels. Although these forms of modulation imply coherent detection, noncoherent demodulation is also possible [1].

The resultant spectra of UWB signals are similar to those of DSSS; however, we believe that this topic is of sufficient importance that there is a separate chapter for it.

This chapter is structured as follows. We first define the DSSS communication system that comprises the target we are talking about. Next, the characteristics of BBN jamming of DSSS systems is addressed, limiting the discussion to BPSK and QPSK modulations as these are the most prolific modulations for DSSS systems at the time of this writing. Then, PBN jamming of DSSS is presented, followed by similar coverage for pulsed jamming. Tone jamming, both single tone and

Table 10.1 Notation

Spreading Modulation	Data Modulation	Acronym
Biphase	Biphase	BPSK/BPSK
Biphase	Quadriphase	BPSK/QPSK
Quadriphase	Biphase	QPSK/BPSK
Quadriphase	Quadriphase	QPSK/QPSK

multitone, is discussed last, covering the characteristics of both long and short spreading codes, as the jamming performance depends on code length.

10.2 DSSS Communication Systems

10.2.1 Introduction

In the DSSS transmitter, the data signal modulates a carrier, then the data-modulated signal modulates a wideband spreading code onto a wideband carrier. In the cases we consider, the phase of the carrier is the manifestation of the modulation by the spreading code. Furthermore, we only consider binary and quadriphase modulations.

At a synchronized receiver, the received signal is despread by a despreading code and demodulated by a carrier. The modulation in all the cases we are concerned with is digital phase modulation. Note that different modulation may be used for the carrier and the spreading signal.

To measure the BER, it is necessary to know the test statistics of the data, jammer, and noise signals at the output of the correlator in the receiver. It is assumed that the noise is Gaussian with variance N_0W. Throughout this section, coherent synchronization is assumed.

We will consider two types of spreading modulation, biphase and (offset) quadriphase, and two types of data modulation, biphase and quadriphase in this chapter. A word on notation is in order at this point. We will use the acronyms shown in Table 10.1 since we use the terms for the modulation types quite frequently.

10.2.2 DSSS Transmitter

Figure 10.1 [2] shows a mathematical model for a BPSK/QPSK DSSS transmitter. The in-phase signal $s^{\mathrm{I}}(t)$ is given by

$$s^{\mathrm{I}}(t) = \sqrt{S}d(t-T_0)w^{\mathrm{I}}(t)c^{\mathrm{I}}(t-T_0)\cos(\omega_0 t+\phi_{\mathrm{s}}) \tag{10.1}$$

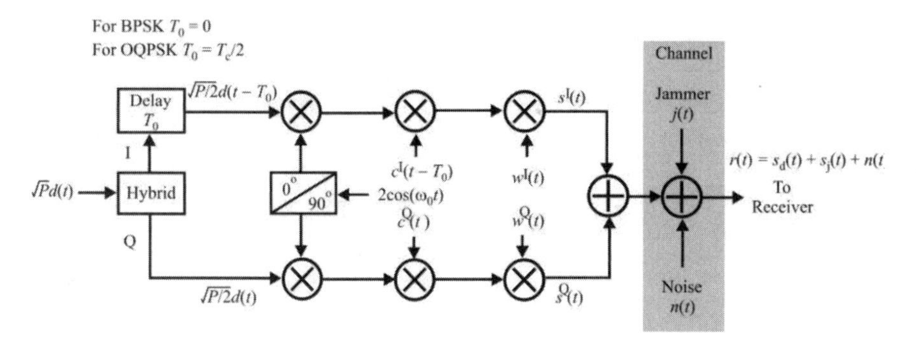

Figure 10.1 BPSK and QPSK DSSS transmitter and channel model. (Source: [2]. © 1989 IEEE. Reprinted with permission.)

and the quadriphase signal $s^Q(t)$ is given by

$$s^Q(t) = \sqrt{S}d(t)w^Q(t)c^Q(t)\sin(\omega_0 t + \phi_s) \tag{10.2}$$

where S is the average power of the data signal and $d(t)$ is the normalized data signal with values ± 1 with equal probability. $c^I(t)$ and c^Q(t) are spreading waveforms, also with values ± 1 . The jammer frequency ω_j may or may not be equal to the signal frequency ω_0. $w^I(t)$ and $w^Q(t)$ are weighting functions for the in-phase and quadrature phase paths, respectively, which are used in the model to control whether the modulation is BPSK or QPSK. T_0 represents the time offset of the code waveform for the in-phase component, relative to that for the quadrature component. For BPSK/BPSK, $T_0 = 0, w^I(t) = \sqrt{2}$, and $w^Q(t) = 0$. For QPSK/BPSK, $T_0 = T_c / 2$, where T_c is the chip duration, $w^I(t) = w^Q(t) = 1$ [3, 4]. The combined signal due to the data components, denoted by $s_s(t)$, is obtained at the output by adding the two quadrature components together:

$$s_s(t) = s^I(t) + s^Q(t) \tag{10.3}$$

10.2.3 Receiver

A mathematical model block diagram of a BPSK/QPSK DSSS receiver is shown in Figure 10.2. The received signal is split into the I and Q paths, and weighting is applied to determine the modulation type. The first mixer stages then convert the signals in the two paths to an IF frequency. The signals are then multiplied by a synchronized replica of the code sequence, bandpass filtered, and mixed at the

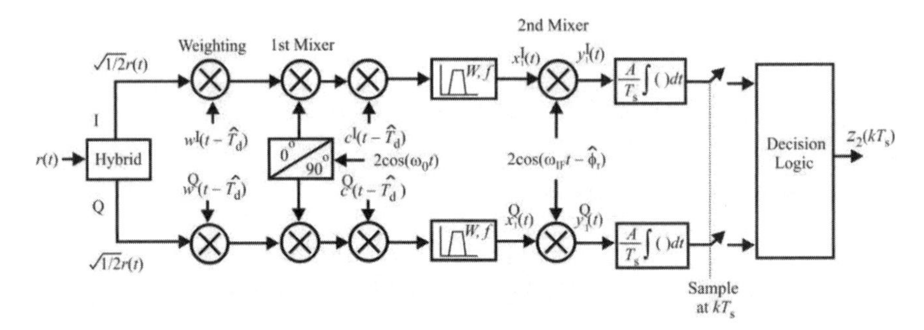

Figure 10.2 BPSK/BPSK BPSK/QPSK DSSS receiver model. (Source: [2]. © 1989 IEEE. Reprinted with permission.)

second mixing stage, the result of which is fed to an integrator (the combination of bandpass filtering and integrator is a correlator). Decision logic is at the output of the integrators to decide which symbol was sent.

10.2.3.1 DSSS Demodulator Structures

Figure 10.3 [5] illustrates the most commonly considered demodulator structures for BPSK and QPSK despreading.

We will first consider the simplest case of biphase spreading and BPSK data encoding to establish the basic parameters used in the analysis. The decision statistic is given by

$$z(\theta_{\mathrm{J}}) = \pm\sqrt{E_{\mathrm{s}}} + g + \sqrt{\frac{J}{T}} \int_{T_0}^{T_0+T_{\mathrm{s}}} c(t)\cos[(\omega_{\mathrm{J}} - \omega_{\mathrm{c}})t + \theta_{\mathrm{J}}] \tag{10.4}$$

A simplified coherent biphase spread/BPSK data demodulator implementing (10.4) is shown in Figure 10.4. The integration in (10.4) is over a bit interval and can be written as a sum of integrations over chips as

$$z = \pm\sqrt{E_{\mathrm{s}}} + g + \sqrt{\frac{J}{T}} \sum_{i=0}^{N-1} c_{i+l} \int_{T_0 + i\frac{T_{\mathrm{s}}}{N}}^{T_0 + i\frac{T_{\mathrm{s}}}{N} + \frac{T_{\mathrm{s}}}{N}} \cos(\Delta\omega t + \theta_{\mathrm{J}})dt \tag{10.5}$$

where T_0 is the bit start time, $c_i = \pm 1$ corresponds to the ith chip of the spreading code, $\pm\sqrt{E_{\mathrm{b}}}$ is the effect on z due to the signal and is positive or negative

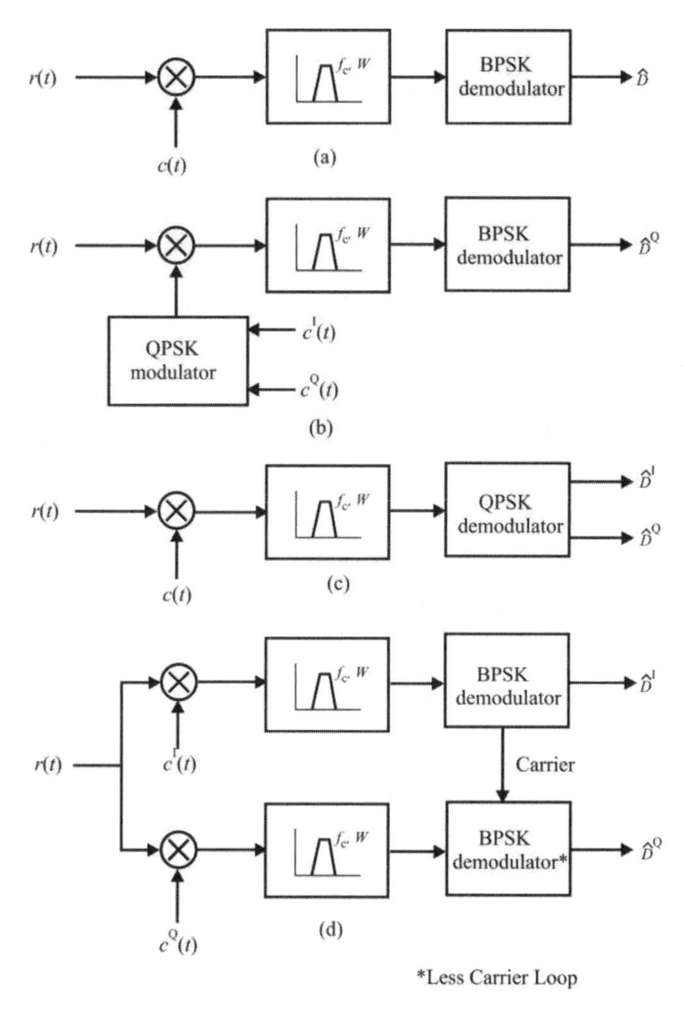

Figure 10.3 DSSS demodulators. (a) BPSK/BPSK data, (b) QPSK/BPSK data, (c) BPSK/QPSK data, and (d) QPSK/QPSK data. The bandpass filters are narrowband with bandwidth W. (Source: [5]. © 1983 IEEE. Reprinted with permission.)

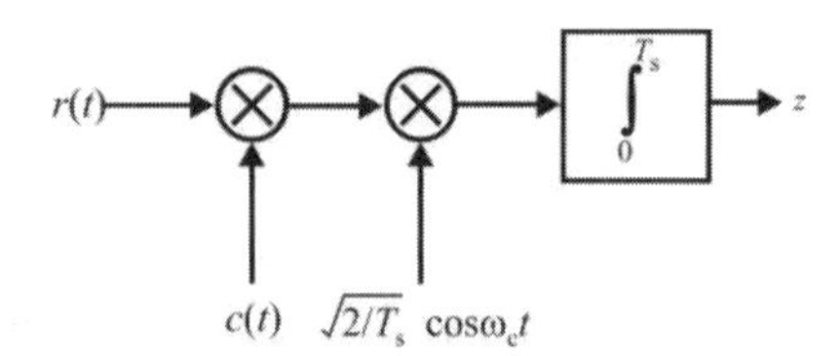

Figure 10.4 BPSK demodulator blocks in Figure 10.3.

depending on the identity of the data bit, g is a zero mean Gaussian r.v. with variance $N_0/2$ and is the response of the correlation receiver to thermal channel noise $n(t)$, and the third term is the response to the jamming signal.

The third term in (10.5) depends on a specific code sequence c_{i+l} and the jammer frequency offset Δω. In general, it is not possible to find a closed form expression for the moments of the third term in (10.5). However, there are two cases in which the third term can be simplified. The first case is when the period of the spreading code is one data bit long, that is, a short code. In this case, the third term is a deterministic quantity and not a random variable. Therefore, the jammer contribution is to the mean and not to the variance. This case has been investigated in detail for various data modulation schemes when Δω is zero [1]. A closed form solution for this case will be given when the jammer frequency offset is an integer multiple of the bit rate.

The second case is when the period of the spreading sequence is long and the number of chips per bit is large. This case is particularly interesting, as many practical systems satisfy this condition. In this case, closed form solutions can be obtained for the bit error rate using Gaussian approximations.

10.2.4 Spreading Codes

Jamming performance depends on the length of the spreading code used in DSSS systems. A short code repeats itself every one or a few data bits while a long code repeats itself over many more data bits. Hence, for the former, $L \approx N$ while for the latter $L >> N$, where $N = 10^{G_p/10}$. Since the short code sequence is repeated in every data bit, and a symbol is either a 1 or 0 for BPSK or a 00, 01, 10, or 11 for QPSK, it is relatively easy to recover the code sequence in a short code. Hence, short codes are much more vulnerable to ES, and therefore EA, than long codes. In non-adversarial situations, however, when recovery of the codes is not a concern, short codes can be used to advantage. The spreading codes for 3G+ PCS systems, for example, are published in the open commercial literature and knowledge of the one in use is not an issue at all in commercial applications.

For EW we encounter both short and long codes, and, as we will see, jamming performance is not the same for different length codes. We will consider the following two cases of interest for urban EW applications:

- Long period sequences, that is, L is finite and $L >> N$;
- Short period sequences where $L \leq N$.

10.2.4.1 Short Codes [6]

As mentioned above, the length L of a short code sequence is such that one or more cycles of the code is contained in each symbol; that is, $L \leq N$.

It can be shown that for MT jamming, the type of data modulation (BPSK or QPSK) yields the same average performance for quadriphase spreading so we will discuss only the three cases:

- Biphase spreading with BPSK and QPSK data modulation;
- Biphase spreading with QPSK data modulation;
- Quadriphase spreading with BPSK and QPSK data modulation.

As a point of interest for EW, using short codes with DSSS makes such signaling quite vulnerable. The military development of DSSS focused on long codes since their spread properties are considerably better than short codes. For example, techniques are available to automatically determine the spreading code when it is repeated within each bit several times, thereby considerably easing the difficulties involved with EW. If an EW system can determine the code, the DSSS signals can be despread within that system, just as it is despread at the target receiver. In addition, with the code known, a jammer can encode the jamming signal with that code thereby removing all of the AJ advantages of DSSS.

However, personal communication systems employing CDMA are not designed to avoid intercept or provide for jamming resistance. CDMA is used in these systems to increase the spectrum usage, thereby increasing the revenue flow for the service provider. Using CDMA with these systems provides other advantages as well, such as soft degradation as more users come online; however these advantages do not include LPI or AJ.

10.3 Spectral Characteristics of DSSS

This segment follows [7] closely. It is assumed that the received signal is synchronized in time with the local epoch and the receiver is synchronized with the spreading code. For an arbitrary jamming signal $j(t)$, its autocorrelation is given by

$$R_{\text{jj}}(\tau) = \mathcal{E}\{j(t+\tau)\,j(t)\} \tag{10.6}$$

where $\mathcal{E}\{\bullet\}$ denotes the expectation function, with a corresponding PSD given by

$$S_{\text{J}}(f) = \int_{-\infty}^{\infty} R_{\text{jj}}(\tau)\exp(-j2\pi f\,\tau)d\tau \tag{10.7}$$

Likewise, the PN sequence denoted by $c(t)$ has the autocorrelation function

$$R_{\text{cc}}(\tau) = \mathcal{E}\{c(t+\tau)c(t)\} \tag{10.8}$$

and since $c(t)$ is assumed to be a maximal sequence,

$$R_{\text{cc}}(\tau) = \begin{cases} 1 - \dfrac{|\tau|}{T_{\text{c}}}, & |\tau| \le nT_{\text{c}},\ n \text{ integer} \\ 0, & \text{otherwise} \end{cases} \tag{10.9}$$

The received signal prior to decorrelation is given by

$$r(t) = As(t - T_{\text{d}}) + n(t) + j(t) \tag{10.10}$$

where $n(t)$ is due to thermal noise, A represents the attenuation of the channel on the transmitted signal $s(t)$, and T_{d} is the delay of the signal caused by finite propagation time. Both A and T_{d} can be random variables or deterministic functions of time. Assume for the moment that the jammer signal is BBN and is much larger than the thermal noise so the latter can be ignored. In that case the decorrelator multiplies $r(t)$ by a copy of the code sequence $c(t)$ and a jammer component given by

$$n_{\text{J}}(t) = c(t)\,j(t) \tag{10.11}$$

is generated. The autocorrelation of this function is given by

$$R_{\text{n}_\text{J}\text{n}_\text{J}}(\tau) = \mathcal{E}\{n_{\text{J}}(t+\tau)n_{\text{J}}(t)\}$$

$$
\begin{aligned}
&= \mathcal{E}\{c(t+\tau)j(t+\tau)c(t)j(t)\} \\
&= \mathcal{E}\{c(t+\tau)c(t)\}\mathcal{E}\{j(t+\tau)j(t)\} \\
&= R_{\text{cc}}(\tau)R_{\text{jj}}(\tau)
\end{aligned} \tag{10.12}
$$

From linear system theory, multiplication of two time domain signals results in convolution of their PSDs in the frequency domain, so the PSD of $n_{\text{J}}(t)$ is

$$
S_{\text{n}_\text{J}}(f) = S_{\text{c}}(f) * S_{\text{J}}(f) \tag{10.13}
$$

where $*$ denotes convolution given by

$$
S_{\text{c}}(f) * S_{\text{n}_\text{J}}(f) = \int_{-\infty}^{\infty} S_{\text{c}}(u) S_{\text{n}_\text{J}}(u - f) du \tag{10.14}
$$

Then

$$
\begin{aligned}
S_{\text{n}_\text{J}}(0) &= \int_{-\infty}^{\infty} S_{\text{c}}(u) S_{\text{n}_\text{J}}(u) du \\
&\le S_{\text{c}}(0) \int_{-\infty}^{\infty} S_{\text{n}_\text{J}}(u) du \\
&= S_{\text{c}}(0) J
\end{aligned} \tag{10.15}
$$

since

$$
\int_{-\infty}^{\infty} S_{\text{n}_\text{J}}(u) du = J \tag{10.16}
$$

and

$$
S_{\text{c}}(f) \le S_{\text{c}}(0) \tag{10.17}
$$

The power spectrum of a PN sequence is [8]

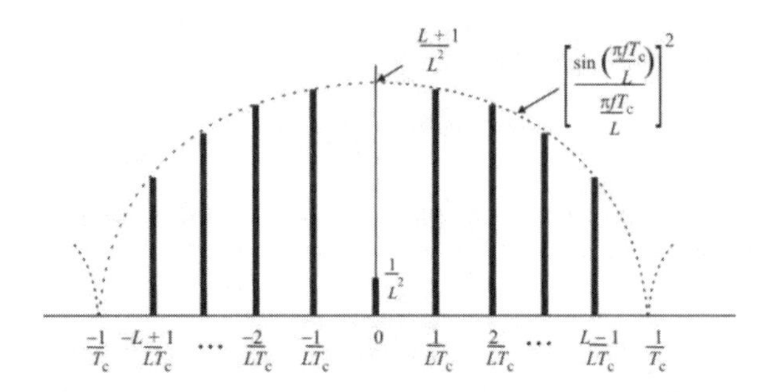

Figure 10.5 Spectrum of a maximal length PN sequence of period T_c and length L.

$$S_c(f) = \frac{1}{L^2}\delta(f) + \frac{L-1}{L^2}\sum_{\substack{i=-\infty \\ i\neq 0}}^{\infty} \text{sinc}^2\left(\frac{i}{L}\right)\delta\left(f - \frac{i}{LT_c}\right) \tag{10.18}$$

This function is plotted in Figure 10.5 where the PN sequence has maximal length with period T_c and length L chips. The resulting jammer power emerging from the decorrelation process is therefore spread over a wide bandwidth and is decorrelated with the signal. This jammer signal has the same characteristics as AWGN on the DSSS system.

The PSDs of the PN sequence and the jammer can be approximated by boxcar functions shown in Figure 10.6 where $A_J = J/W_J$ and $C \approx 1/L$. The graphical convolution of these PSDs is also shown in Figure 10.6. The resulting signal is broadband with bandwidth $W_{n_J} \approx W_{ss}$ when $W_J << W_{n_J}$ and amplitude given by $A_J C$. Therefore the despreading process produces a noise-like spectrum at the output of the decorrelator.

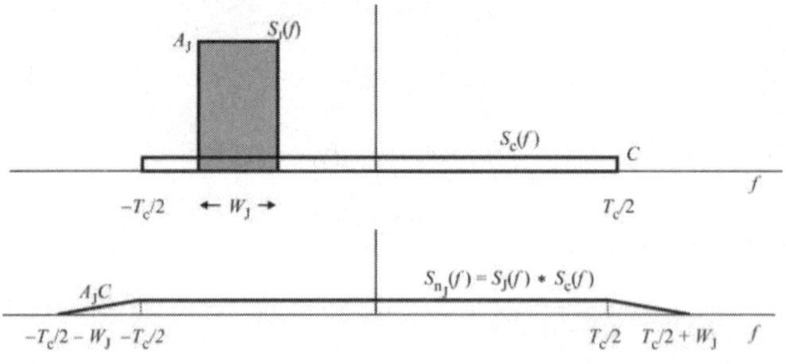

Figure 10.6 PSD of the signal resulting from the multiplication of $j(t)$ by $c(t)$. It is given by the convolution of $S_J(f)$ and $S_c(f)$.

The fundamental property of PN sequences that DSSS systems take advantage of is the effect of this spreading of unsynchronized energy in the decorrelation process, while simultaneously compressing the synchronized energy down to its original bandwidth. These effects are independent of whether the unsynchronized energy is noise, is caused by unintentional interference, or caused by intentional interference such as a jammer. It is also independent of the type of interfering signal. The signal could be BBN, PBN, NBN, or multiple tones [9, 10].

For BBN, PBN, and tone jamming, all of which are constantly present, the effect of the jamming is to raise the postdetection noise floor. In those cases, repeat coding is ineffective at reducing the BER because the coding reduces the per symbol energy as E_s / m while the postdetection jamming signal is present at all frequencies in W_{ss}. There is no benefit achieved by possibly placing a coded symbol either at a frequency or a time that does not experience the jamming signal. Thus, the effect is simply to reduce the SNR of the data bits, rendering jamming more effective and coding ineffective at defeating it. Increasing m simply increases the BER.

Repeat coding against pulsed jamming, on the other hand, can be effective. In that case there is a time when the jammer is off and there is a possibility that a coded symbol is transmitted when the jammer is off.

Although signal fading due to frequency-selective channel characteristics can affect the BER of DSSS signal, such effects are not analyzed here. The interest herein is to ascertain the effects of jamming waveforms. If desired, the effects of channel fading can be included [11, 12].

10.3.1 Signal Formats

The signal types being considered here generally have the following form

$$s(t) = \sqrt{2S}[d^{\mathrm{I}}(t)c^{\mathrm{I}}(t)w^{\mathrm{I}}(t)\cos\omega_c t + d^{\mathrm{Q}}(t)c^{\mathrm{Q}}(t)w^{\mathrm{Q}}(t)\sin\omega_c t] \qquad (10.19)$$

where $S = E_s / T_s$ is the average power in the signal, $d^{\mathrm{I}}(t)$ and $d^{\mathrm{Q}}(t)$ are baseband bipolar, (±1) data waveforms representing the in-phase and quadrature-phase data streams, respectively, for QPSK data modulation. The case when $d^{\mathrm{I}}(t) = d^{\mathrm{Q}}(t)$ represents BPSK data modulation. Similarly, $c^{\mathrm{I}}(t)$ and $c^{\mathrm{Q}}(t)$ are two ±1 sequences occurring at the chip rate (much higher than the data rate), with chip time denoted by T_c, and represent spreading codes. Biphase spreading results when $c^{\mathrm{I}}(t) = c^{\mathrm{Q}}(t)$. Quadriphase chip spreading is obtained by setting $|c^{\mathrm{I}}| = |c^{\mathrm{Q}}|$; otherwise, unbalanced quadriphase spreading is obtained. L is the cycle period of the $c^{\mathrm{I}}(t)$ and $c^{\mathrm{Q}}(t)$ chip sequences (assumed to be the same). The duration of a data symbol is T_s

and the energy per data symbol is E_s, while ω_c, is the carrier frequency. There are N chips of $c^I(t)$ and $c^Q(t)$ in a symbol duration T_s.

The signal received at a receiver is a combination of the signal of interest (SOI) to which the thermal channel noise and the jammer signal are added and is given by

$$r(t) = s(t - T_d) + n(t) + j(t) \tag{10.20}$$

The jammer and noise components are modeled as

$$j(t) = \sqrt{2J}\cos(\omega_J t + \theta_J) \qquad \text{(jammer)} \tag{10.21}$$

and

$$n(t) = n^I(t)\cos\omega_c t + n^Q \sin\omega_c t \qquad \text{(noise)} \tag{10.22}$$

where J is the average jammer power, ω_J is the jammer frequency, and θ_J is the jammer phase; $n^I(t)$ and $n^Q(t)$ correspond to in-phase and quadrature components of the noise.

We will consider four types of spreading/modulation schemes and examine EW performance against each of them. These are: (1) biphase spreading/BPSK data modulation, (2) quadriphase spreading/BPSK data modulation, (3) biphase spreading/QPSK data modulation, and (4) quadriphase spreading/QPSK data modulation. These phase spreading/modulation techniques all require coherent demodulation. While it is possible to implement DSSS systems that are non-coherent, we assume herein that all the DSSS implementations are coherent. That assumption is satisfied by all the DSSS implementations with which we are concerned.

Typically postdetection SNRs on the order of 10 dB or higher are required. When $N \sim 400$, the processing gain is about 26 dB, indicating that the RF SNR must be about 10 dBm – 26 dB = –16 dB or more. When $N \sim 10$, the processing gain is about 10 dB, indicating an RF SNR of 10 dBm – 10 dB = 0 dB is required. Between these values of N, values of required RF SNR will thus vary between –16 dB to 0 dB.

10.4 BBN Jamming of DSSS Systems

BBN jamming of DSSS signals is when the jammer signal is noise with a bandwidth approximately the same as the DSSS signal. BBN jamming is

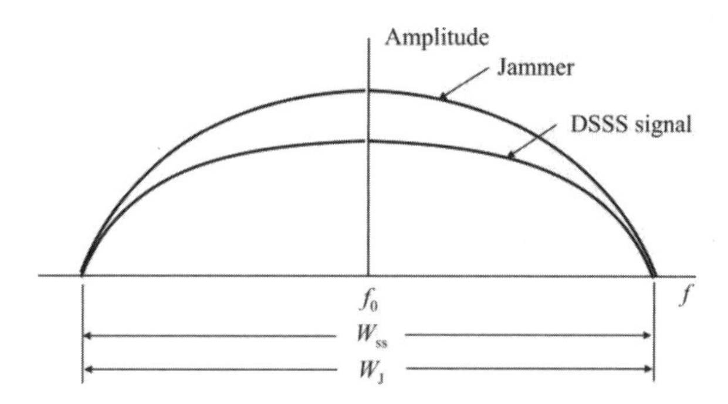

Figure 10.7 BBN jamming of DSSS.

illustrated in Figure 10.7. It is tacitly assumed that the center frequency of the jamming signal is the same as that of the DSSS target. Only hard decision decoding with no JSI is considered in this section.

With multiple users of a DSSS system within range of a single receiver, the effects of interference are very similar to AWGN. Techniques have been investigated to mitigate this interference and therefore mitigate the effects of BBN jamming [13–17].

From an intuitive point of view, at the target DSSS receiver we know that the detector output SNR, SNR_o, is increased by G_p. The output signal level is increased by G_p while the output noise is not. The decorrelation process does not affect the noise, and whatever noise exists at the input will exist at the output of the detector. Therefore to use BBN as the jamming technique, the processing gain must be overcome in order to put enough noise at the detector output to affect the detection process.

The signal-to-noise plus jamming ratio after the despreading and demodulation processes for DSSS receivers, denoted by υ_0, when the jamming waveform is BBN is given by

$$\upsilon_o = \frac{R}{N_0 W_{ss} / 2 + J} G_p \tag{10.23}$$

where R is the signal power received, N_0 is the one-sided noise spectral density, W_{ss} is the bandwidth of the signal, J is the jamming power at the receiver, and G_p is the number of chips per data bit. In this expression it is assumed that the receiver noise bandwidth is matched to that of the signal. If T_b denotes the baseband data bit duration and T_c denotes the spreading waveform chip duration, then G_p, the processing gain, is given by

$$G_{p} = \frac{T_{b}}{T_{c}} \quad (10.24)$$

Since

$$T_{s} = \frac{1}{R_{s}} \quad (10.25)$$

and

$$R_{c} = \frac{1}{T_{c}} \quad (10.26)$$

then

$$G_{p} = \frac{R_{c}}{R_{s}} \quad (10.27)$$

Also

$$G_{p,dB} = 10\log(G_{p}) \quad (10.28)$$

is the processing gain expressed in decibels. As illustrated in Figure 10.7, $W_{ss} = 2/T_c$ from null to null while the 3 dB bandwidth is $W_{ss} \approx 0.88/T_c$. It is frequently assumed that $W_{ss} = 1/T_c$ as we will in the sequel.

The ratio in (10.23) can be defined to be the input SNR

$$\upsilon_{i} = \frac{R}{N_{0}W_{ss}/2 + J} \quad (10.29)$$

so that

$$\upsilon_{o} = G_{p}\upsilon_{i} \quad (10.30)$$

Thus, the received SNR is effectively increased by the processing gain of the DSSS system. In order to disrupt the signal at the receiver, the jamming signal must somehow overcome this processing gain. In addition to the processing gain to be overcome by the jammer, the effectiveness of a jammer against DSSS depends on whether the DSSS signal is error protection coded or not.

The performance of DSSS systems that are phase modulated with BBN jamming is the same as the performance with AWGN only. Thus, [18, 19]

$$P_{\mathrm{e}} = Q\left(\sqrt{\frac{2E_{\mathrm{s}}}{N_{\mathrm{T}}}}\right) \tag{10.31}$$

where

$$N_{\mathrm{T}} = N_0 + \kappa J_0 \tag{10.32}$$

when N_0 is the one-sided thermal noise spectral density and J_0 is the jammer noise spectral density given by $J_0 = J/W_{\mathrm{ss}}$, and constant κ is determined by the method of imposing the PSK modulation onto the carrier [19]

$$\kappa = \begin{cases} 0.903, & \text{BPSK DSSS} \\ 0.995, & \text{MFSK DSSS} \end{cases} \tag{10.33}$$

Most of the time $\kappa = 1$ is assumed, and that convention will be adopted here.

The receiver architecture under consideration is essentially that shown in Figure 3.18. During the *k*th time interval, the received signal takes the form

$$r_k(t) = d_k^{\mathrm{I}}(t)c^{\mathrm{I}}(t)\cos(2\pi f_0 t) + d_k^{\mathrm{Q}}(t)c^{\mathrm{Q}}(t)\sin(2\pi f_0 t), \qquad (k-1)T_{\mathrm{c}} \le t < kT_{\mathrm{c}} \tag{10.34}$$

where the data signals are given by

$$\begin{aligned} d_k^{\mathrm{I}}(t) &= d_k^{\mathrm{I}} p(t) \\ d_k^{\mathrm{Q}}(t) &= d_k^{\mathrm{Q}} p(t) \end{aligned} \tag{10.35}$$

where the data values d_k^{I} and d_k^{Q} take on values ± 1, and where $p(t)$ determines the pulse shape. The spreading signals are given by $c^{\mathrm{I}}(t)$ and $c^{\mathrm{Q}}(t)$ for the in-phase and quadrature phase channels, respectively. T_{c} denotes the duration of a single chip during time interval k.

Schilling et al. examined the problem of noise and tone jamming of *M*-ary DSSS systems [20]. In particular they examined the effects of BBN and tone jamming on BPSK and QPSK modulated DSSS signals. They defined a parameter β as

$$\beta = \int_0^{T_b} c(t)dt \tag{10.36}$$

where T_b is the data bit time. For short codes where the length of the spreading sequence is such that the sequence repeats itself within the bit duration (short codes, $N \approx L$) then, since $c(t)$ is a maximal sequence, there is one more +1 than –1 and

$$\int_0^{T_b} c(t)dt = \sum_{i=1}^{T_s/2} (+1)\frac{T_s}{L} + \sum_{i=1}^{T_s/L-1} (-1)\frac{T_s}{L} = \frac{T_s}{L} \tag{10.37}$$

Therefore $\beta = T_s/L = T_c$. For long codes this relationship is not necessarily true, and β will depend on the particular properties of $c(t)$. There could be many more +1's than –1's for example, depending on which part of the sequence is included in the data bit. On average, however, $\beta = T_c$ even for long codes. Therefore they assumed $\beta = T_c$ and we will do that here as well.

10.4.1 BPSK and QPSK

It was established in Chapter 3 that the bit error performance for BPSK and QPSK in AWGN are the same. The power per symbol for QPSK is half that for BPSK while the symbol duration is twice as long, which ensures that the energy per bit for both approaches is the same. For rectangular pulse shapes and BBN jamming the bit error performance is given by

$$P_{e,\text{BPSK}} = Q\left\{\sqrt{\frac{2E_s}{N_T}}\right\} \tag{10.38}$$

where

$$N_T = N_0 + J_0 \tag{10.39}$$

This is the familiar result for BPSK modulation in the presence of thermal noise where the thermal noise level has been increased by the BBN level.

To put (10.38) into a form containing power ratios,

$$P_e = Q\left\{\sqrt{\frac{2E_s}{N_T}}\right\}$$
$$= Q\left\{\sqrt{\frac{2E_s}{N_0 + J_0}}\right\}$$
$$= Q\left\{\sqrt{\frac{2E_s}{N_0 + J / W_{ss}}}\right\}$$
$$= Q\left\{\sqrt{\frac{2RT_s W_{ss}}{W_{ss} N_0 + J}}\right\} \quad (10.40)$$

But $W_{ss} = 1/T_c$ and $P_N = W_{ss}N_0$ is the noise power so

$$P_e = Q\left\{\sqrt{\frac{2RT_s \dfrac{1}{T_c}}{P_N + J}}\right\} \quad (10.41)$$

but the processing gain $G_p = T_s/T_c$ so

$$P_e = Q\left\{\sqrt{\frac{2G_p}{\dfrac{1}{\upsilon} + \xi}}\right\} \quad (10.42)$$

where ξ denotes the JSR. This function is plotted in Figure 10.8 for a few values of SNR when $G_p = 100$ ($G_{p,\text{dB}} = 20$ dB).

The SNR here includes the noise in W_{ss}. The equivalent narrowband SNR can be determined if the despread bandwidth of the data symbol is known. It is given by $R_s = 1/T_s = W_F$. Suppose $W_F = 25$ kHz and $W_{ss} = 50$ MHz. If the spread SNR is –20 dB, then the despread SNR is

$$-20 + 10\log_{10}\left(\frac{50\times 10^6}{25\times 10^3}\right) \approx 13\,\text{dB} \quad (10.43)$$

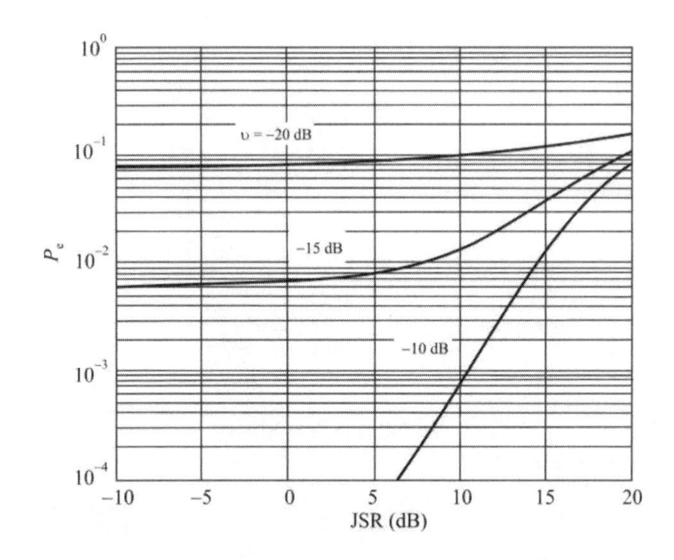

Figure 10.8 BBN jamming performance of uncoded BPSK and QPSK DSSS systems, $N = 100$.

In the case illustrated in Figure 10.8, the BBN jammer remains relatively ineffective until the processing gain ($N = 100$, 20 dB) is overcome. At that point however, $P_e \approx 10^{-1}$ or higher irrespective of the spread SNR of those considered here.

The jamming performance for BBN as the G_p is varied from $G_p = 10$ to $G_p = 1{,}000$ is shown in Figure 10.9 for $\upsilon = -20$ dB. Note that $G_p = 100$ is approximately that for some forms of 3G+ PCS. Increasing the JSR by 10 dB improves the jamming performance, as measured by the processing gain overcome, by about 3 dB at $P_e = 10^{-1}$. It is worth noting again, to be fair, that 3G+ PCS systems were not designed for antijam performance.

10.4.2 BBN Jamming of Chaotic Systems

The effects of BBN jamming on the chaotic communication systems introduced in Section 3.13.5 are similar to those discussed for BPSK and QPSK. The jammer produces the equivalent of additional background noise with which the communication system must contend. Adding BBN essentially increases the background noise throughout the whole spectrum. Chaotic signals are noiselike in character, just as DSSS signals. The effects of the various jamming techniques, including BBN, would be expected to be similar.

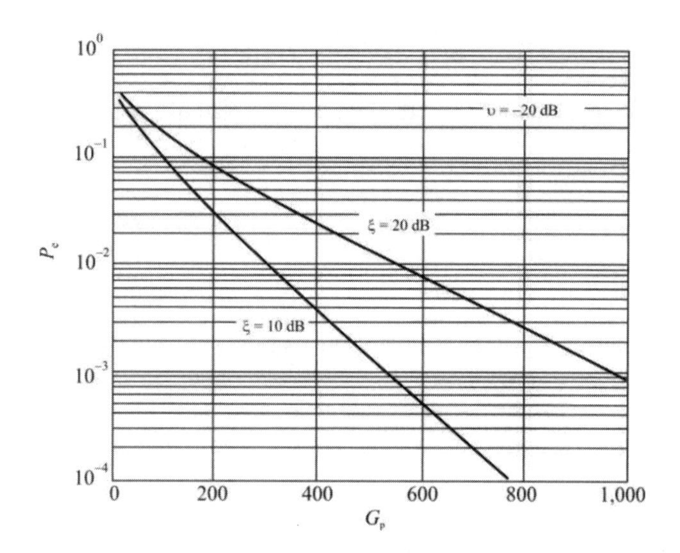

Figure 10.9 BBN jamming performance against QPSK DSSS systems as the processing gain, G_p, is varied for $\upsilon = -20$ dB.

10.4.2.1 Coherent Reception

For coherent reception, based on (3.49), the BER is given by

$$P_e = \frac{1}{2}\text{erfc}\left[\frac{1}{\sqrt{\left(\frac{E_s^2}{4\beta\Lambda}\right)^{-1} + \left(\frac{E_s}{J_0 + N_0}\right)^{-1}}}\right] \tag{10.44}$$

where the jamming noise spectral density is given by J_0 W Hz^{-1}.

10.4.2.2 Noncoherent Reception

Again, based on (3.150), the background noise level is raised by the jammer producing the BER expression

$$P_e = \text{erfc}\left[\frac{1}{\sqrt{\left(\frac{E_s^2}{8\beta\Lambda}\right)^{-1} + 4\left(\frac{E_s}{J_0 + N_0}\right)^{-1} + 2\beta\left(\frac{E_s}{J_0 + N_0}\right)^{-1}}}\right] \tag{10.45}$$

10.5 PBN Jamming of DSSS Systems

In PBN jamming the power from the jammer is spread over a bandwidth W_J that is narrower than the whole spread bandwidth W_{ss}. The fraction of W_{ss} covered by the jammer is given by

$$\gamma = \frac{W_J}{W_{ss}} \tag{10.46}$$

and the spectral density of the jammer is given by

$$\begin{aligned} S_J &= \frac{J}{W_J} \\ &= \frac{J}{W_{ss}} \frac{W_{ss}}{W_J} \\ &= \frac{J_0}{\gamma} \end{aligned} \tag{10.47}$$

where, as before, J_0 is the jammer energy density as if the jammer power were spread over the whole spreading bandwidth W_{ss}. Both the jammer energy as well as the thermal noise affect performance, although frequently it is assumed that the jammer energy is larger than the thermal noise so that the latter can be ignored. That is normally the case in the portion of the spectrum covered by W_J but not outside this region. It typically takes $\upsilon > 10$ dB for effective communication performance and $\xi = 0$ dB implies that the jamming signal is 10 dB higher than the noise level. Therefore, ignoring the noise in the region of the spectrum covered by the jammer is justified.

Just as for other forms of jamming, techniques have been investigated for suppression of the effects of PBN jamming [21, 22].

Figure 10.10 illustrates the spectra for PBN jamming of DSSS signals. The jammer signal could be centered at the center frequency of the signal as shown in Figure 10.10(a) or it could be offset as shown in Figure 10.10(b). In fact, it need not be contiguous as shown in Figure 10.10 for the following analysis to apply.

If $S_J(f)$ denotes the density function of the jamming signal and that signal is Gaussian noiselike, then for BPSK and QPSK the BER is given by [23, 24]

$$P_e = \frac{1}{2}\text{erfc}\left\{\sqrt{\frac{E_s}{N_T}}\right\} \tag{10.48}$$

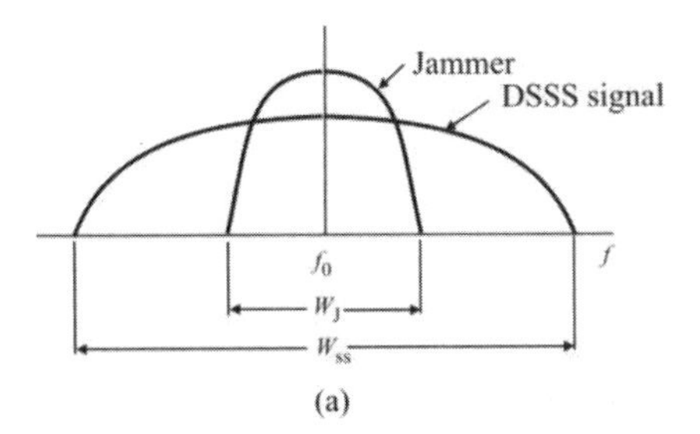

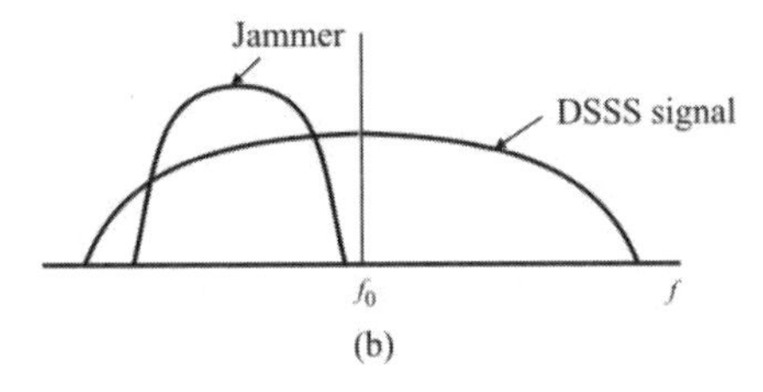

Figure 10.10 PBN jamming of DSSS signals. Shown in (a) is when the jammer is centered on the center frequency of the signal while (b) shows the jammer offset.

where the modified noise component density is given by

$$N_{\mathrm{T}} = N_0 + \frac{2T_{\mathrm{b}}}{G_{\substack{\lim \\ \delta x \to 0}}} \int_{-\infty}^{\infty} S_{\mathrm{J}}(f)\mathrm{sinc}^2[(f - f_0)T_{\mathrm{c}}]df \qquad (10.49)$$

In the ideal case when $S_J(f)$ is flat, centered at f_J with bandwidth W_J then

$$S_{\mathrm{J}}(f) = \frac{J}{W_{\mathrm{ss}}} \qquad (10.50)$$

and this noise component becomes [23]

$$N_{\mathrm{T}} \approx N_0 + \frac{JT_{\mathrm{s}}}{G_{\mathrm{p}}W_{\mathrm{J}}} \int_{f_{\mathrm{J}}-\frac{W_{\mathrm{J}}}{2}}^{f_{\mathrm{J}}+\frac{W_{\mathrm{J}}}{2}} \mathrm{sinc}^2[(f - f_0)T_{\mathrm{c}}]df \qquad (10.51)$$

The BER for uncoded BPSK and QPSK DSSS in PBN jamming is the same and is given by [24]

$$P_e = Q\left(\sqrt{\frac{E_s}{N_0/2 + S_J(f_{IF})}}\right) \tag{10.52}$$

where $S_J(f_{IF})$ is the spectral density of the jammer at the IF frequency of the receiver within the receiver's bandwidth. Assuming the jammer frequency is offset from f_0, for narrow bandwidths where $W_J << W_{ss}$ and the sinc^2 function can be assumed constant over the jammer bandwidth of interest,

$$S_J(f_{IF}) = J_0 \text{sinc}^2\left[(f_0 - f_J)\frac{2}{W_{ss}}\right] \tag{10.53}$$

where f_J is the jammer center frequency. Thus, when $W_J << W_{ss}$ and $f_J = f_0$,

$$S_J(f_{IF}) = J_0 \tag{10.54}$$

and

$$P_e = Q\left\{\sqrt{\frac{2E_s}{N_0 + 2J_0}}\right\} = Q\left\{\sqrt{\frac{2}{\frac{N_0}{E_s} + \frac{2J_0}{E_s}}}\right\} = Q\left\{\sqrt{\frac{2}{\frac{N_0}{E_s}\frac{W_{ss}}{W_{ss}} + \frac{2J/W_{ss}}{RT_s}}}\right\}$$
$$= Q\left\{\sqrt{\frac{2}{\frac{P_N}{R\frac{T_s}{T_c}} + \frac{2J}{R\frac{T_s}{T_c}}}}\right\} = Q\left\{\sqrt{\frac{2G_p}{\frac{1}{\upsilon} + 2\xi}}\right\} \tag{10.55}$$

assuming that $R_s = W_F$ and inserting $T_c = 1/W_{ss}$. This function is plotted in Figure 10.11 versus the processing gain, G_p. Clearly, if the jammer can overcome the processing gain of the DSSS system, significant deleterious effects can occur in the receiver. When $\upsilon = -10$ dB for $\xi = 10$ dB, coding gains less than 13 dB can be tolerated while if $\xi = 20$ dB, coding gains of over 20 dB are still effective.

These results are not very sensitive to the SNR level at these levels of JSR. The DSSS processing gain when using PBN jamming BPSK or QPSK signals is 3 dB less than for BBN jamming. Thus, PBN jamming is the preferred choice against these signal types.

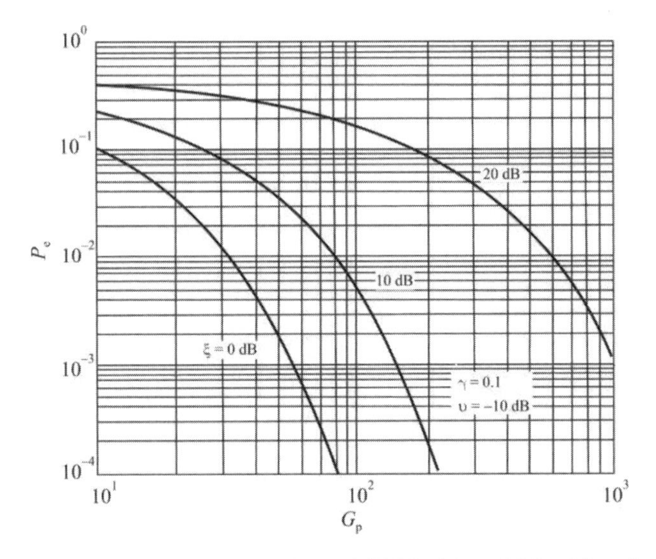

Figure 10.11 Bit error rate for uncoded BPSK and QPSK with partial band jamming versus the processing gain when $\gamma = 0.1$ and $\upsilon = -10$ dB.

10.6 Pulse Jamming of DSSS Systems

A pulsed jammer transmits BBN power for a fraction of the time specified by γ and is off for the rest of the time, the fraction of which is $1 - \gamma$. If a pulsed jammer and nonpulsed BBN jammer have the same average power, then the pulsed jammer will have a larger peak power since it is not always emitting. The bandwidth of the pulse jammer when it is on is assumed to be W_{SS} [25]. As with noise jamming, techniques have been examined to mitigate the effects of pulse jamming on DSSS systems [26].

As with any pulsed signaling, including jamming, the receiver characteristics are important to evaluate the effectiveness of the jamming waveform. The signal must be present long enough for the receiver filters to settle in order for the analysis that follows to apply.

10.6.1 Pulsed Jamming of Uncoded DSSS Systems

The average bit error probability due to noise and the jamming is given by

$$P_e = (1-\gamma)P_{e_1}\left(E_s / N_1\right) + \gamma P_{e_2}\left(E_s / N_2\right) \tag{10.56}$$

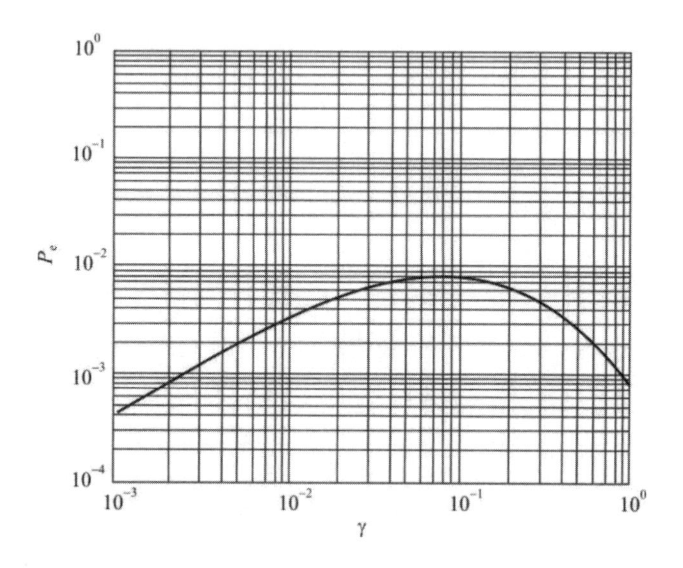

Figure 10.12 Pulsed jamming performance against BPSK DSSS when $\upsilon = -10$ dB, $G_{\text{p,dB}} = 20$ dB, and $\xi = 10$ dB. The variable γ is the fraction of time the pulsed BBN jammer is on.

where P_{e_1} is the BER when the jammer is off and P_{e_2} is the BER when the jammer is on. If $N_{0,1}$ represents the noise density when the jammer is off, and $N_{0,2}$ denotes the noise density when the jammer is on, then $N_1 = N_{0,1}$ and $N_{0,2} = N_0 + J_0$. P_{e_1} is therefore due to noise and P_{e_2} is due to the jammer and noise. The average BER is given by [18]

$$P_e = (1-\gamma)Q\left(\sqrt{2G_p\upsilon}\right) + \gamma Q\left(\sqrt{\frac{2\gamma G_p}{\frac{\gamma}{\upsilon}+\xi}}\right) \tag{10.57}$$

where it is assumed that the jammer is on for at least T_b, the bit time. This expression is valid for coherent BPSK, QPSK, and MSK.

The pulsed jammer performance is shown in Figure 10.12 for $\upsilon = -10$ dB and $G_p = 100$. There is a value of γ that maximizes P_e, denoted by γ^*, as illustrated in Figure 10.12. After a little calculus, from (10.57) [27]

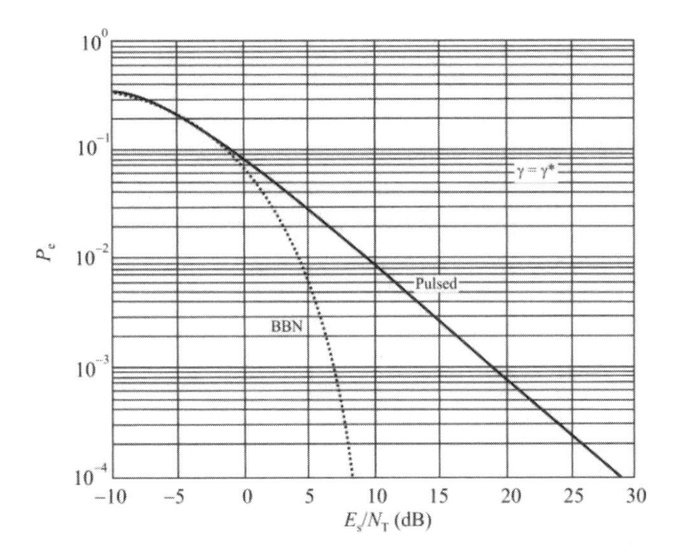

Figure 10.13 BER performance of uncoded BPSK DSSS comparing pulsed jamming at γ^* with BBN jamming.

$$\gamma^* = \begin{cases} \dfrac{0.709}{E_s / N_T}, & E_s / N_T > 0.709 \\ 1, & E_s / N_T \le 0.709 \end{cases} \tag{10.58}$$

with corresponding bit error probability

$$P_e^* = \begin{cases} \dfrac{0.083}{E_s / N_T}, & E_s / N_T > 0.709 \\ Q\left(\sqrt{\dfrac{2E_s}{N_T}}\right), & E_s / N_T \le 0.709 \end{cases} \tag{10.59}$$

This BER is plotted in Figure 10.1, which illustrates the dramatic benefits of pulsed BBN jamming over continuous BBN jamming of uncoded BPSK DSSS systems. The pulse jammer enjoys about a 15 dB advantage over the BBN barrage jammer at a BER of 10^{-4}. At an E_s / N_T of –3 dB and less, the two jamming techniques produce about the same results. Figure 10.13 assumes that the optimum pulse width is used.

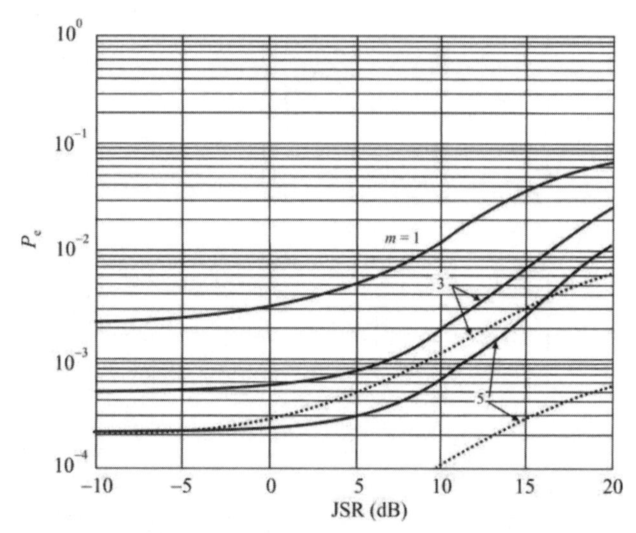

Figure 10.14 Pulsed jamming performance against repeat coded BPSK DSSS with soft decoding; solid lines are without JSI while dotted are with JSI. For $m = 1$, these curves are the same. In this example, $G_p = 100$, approximately that of 3G+ PCS systems, $\upsilon = -10$ dB, and $\gamma = 0.3$.

10.6.2 Pulsed Jamming of Repeat Coded BPSK DSSS Systems

Interleaving the data bits before multiplying by the spreading sequence spreads the data bits out over time, usually separated by enough time so that the resulting spread data bit chips are independent. With no interleaving and associated deinterleaving, the BER performance of repeat-coded BPSK DSSS is the same as uncoded [28] and there is no advantage to coding.

With soft decision decoding and unknown channel state [29]

$$P_e = \sum_{k=\left\lfloor \frac{m+1}{2} \right\rfloor}^{m} \binom{m}{k} \gamma^k (1-\gamma)^{m-k} Q\left(\sqrt{\frac{2mE_s}{kN_T}\gamma}\right) \tag{10.60}$$

where m is the diversity of the code; that is, the number of times the bits are repeated. By noting that $W_J = W_{ss}$, $N_T = N_0 + J_0$, $R = E_b R_s$, $P_N = N_0 W_{ss}$, $G_p = W_{ss}/R_s$, $\upsilon = R/P_N$, $\xi = J/R$, and $J = J_0 W_J$, then (10.60) can be expressed in terms of power ratios as

$$P_e = \sum_{k=\left\lceil \frac{m+1}{2} \right\rceil}^{m} \binom{m}{k} \gamma^k (1-\gamma)^{m-k} Q\left(\sqrt{\frac{2m\gamma G_p}{\frac{k}{\upsilon} + k\xi}}\right) \tag{10.61}$$

which is plotted in Figure 10.14 for $\gamma = 0.3$. The deleterious effects of coding on jamming performance at low JSR are evident in Figure 10.14. We should keep in mind, however, that when $m = 3$, the effective data rate reduces to 1/3 what it was when uncoded, 1/5 when $m = 5$, and so forth. The communicator can punch through the jamming this way but pays a price for it.

Without JSI the soft decision decoder is basing its decisions on bit reliability computed without knowledge of the presence or absence of a jamming signal. Other information then is used, such as estimates of the noise level and the channel fading characteristics.

When the jammer state is known (JSI is available by some means) and with soft decision decoding [30]

$$P_e = \gamma^m Q\left(\sqrt{\frac{2E_s}{N_T}\gamma}\right) \tag{10.62}$$

yielding

$$P_e = \gamma^m Q\left(\sqrt{\frac{2\gamma G_p}{\frac{1}{\upsilon} + \xi}}\right) \tag{10.63}$$

which is also plotted in Figure 10.14.

When the receiver knows the JSI, the pulse jamming performance significantly degrades as illustrated in Figure 10.14. For the parameters in this example, repeat coding effectively defeats the pulse jamming attempts. At $P_e = 10^{-2}$, without JSI, $m = 3$ repeat coding degrades JSR performance by about 7 dB and $m = 5$ about 10 dB. With JSI the degradation is even greater.

Jamming performance variation with γ is illustrated in Figure 10.15. The JSI in this case significantly degrades the jamming performance, even for relatively small amounts of coding.

For repeat codes with m symbols per data bit [31], hard decision decoding, and unknown jammer state the BER is given by

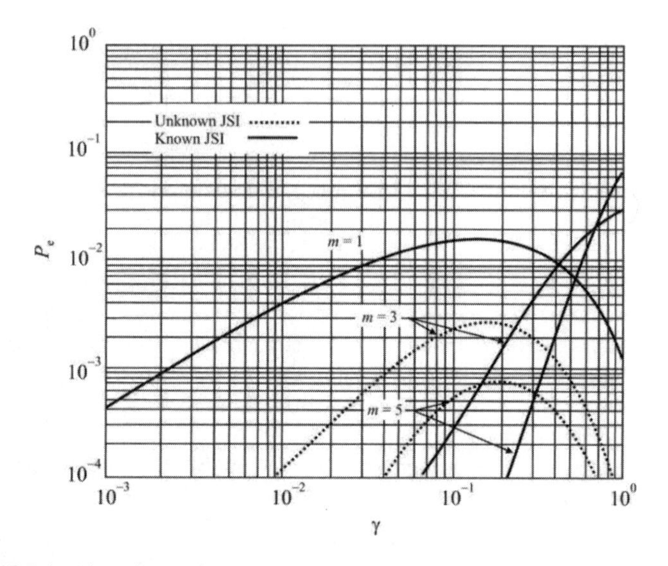

Figure 10.15 Pulsed jamming performance against repeat coded BPSK varying γ with soft decision decoding. Here, $G_p = 100$, $\upsilon = -10$ dB, and $\xi = 10$ dB. When $m = 1$, the JSI and no JSI curves are the same.

$$P_e = \sum_{k=\left\lfloor \frac{m+1}{2} \right\rfloor}^{m} \binom{m}{k} \Lambda^k (1-\Lambda)^{m-k} \tag{10.64}$$

where

$$\Lambda = \gamma Q\left(\sqrt{\frac{2E_s}{mN_T}\gamma}\right) \tag{10.65}$$

Therefore,

$$\Lambda = \gamma Q\left(\sqrt{\frac{2\gamma G_s}{\frac{m}{\upsilon} + m\xi}}\right) \tag{10.66}$$

P_e is plotted in Figures 10.16 for $\gamma = 0.3$, $G_p = 100$, and $\upsilon = -10$ dB, although the results are essentially independent of the SNR at these levels of jamming. The other parameters for this figure are similar to those for 3G+ PCS. Comparing the

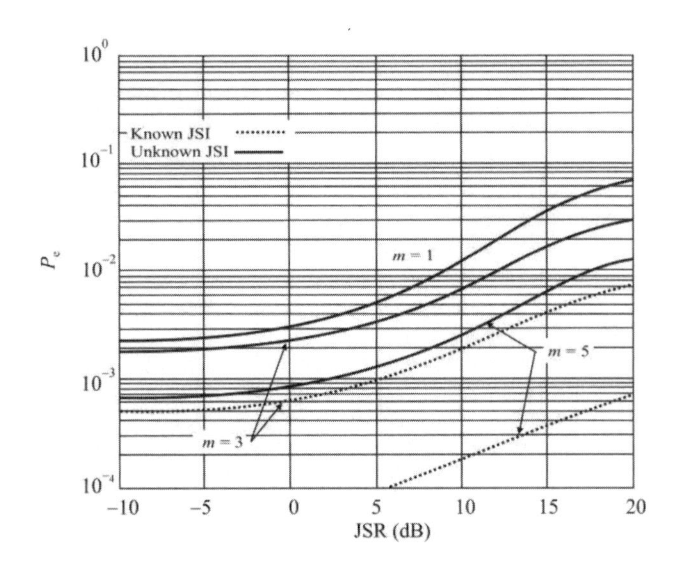

Figure 10.16 Pulsed jamming performance against repeat coded BPSK DSSS for hard decision decoding, when $\gamma = 0.3$ and $G_p = 20$ dB, and $\upsilon = -10$ dB, no JSI.

curves here with those in Figure 10.14, as expected, jamming performance is better against hard decision decoding when JSI is not available.

With JSI and hard decision decoding the BER is given by

$$P_e = \gamma^m \sum_{k=\left\lfloor \frac{m+1}{2} \right\rfloor}^{m} \binom{m}{k} \Lambda^k (1-\Lambda)^{m-k} \tag{10.67}$$

where Λ is given by (10.65) and (10.66) without the leading γ multiplier. In this case the jamming performance decreases somewhat as seen in Figures 10.16 when $\gamma = 0.3$. Again, the variation with γ for hard decision decoding is illustrated in Figure 10.17. Without JSI there is an optimum γ that maximizes P_e; however, repeat coding also decreases jammer performance for hard decision decoding.

A comparison of soft decision and hard decision decoding when JSI is not available is shown in Figure 10.18 for $\gamma = 0.3$, $G_{p,dB} = 20$ dB, and $\upsilon = -10$ dB. Pulsed jammer performance is better against hard decision decoding by about 10 dB at 10^{-2} BER when there is no coding ($m = 1$). With minimal coding ($m = 3$ in this case), P_e never reaches acceptable jamming performance for either decoding method. Likewise, pulsed jamming performance is shown in Figure 10.19 for the same parameters except $\gamma = 0.3$. In this case, as expected, jamming performance

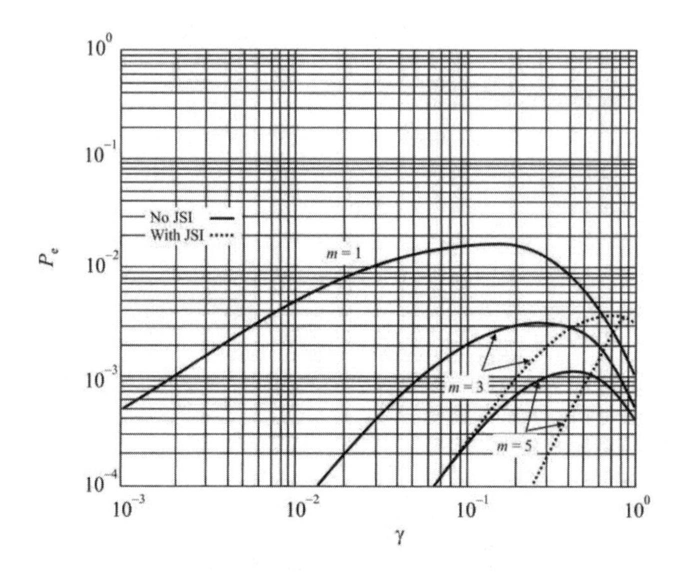

Figure 10.17 Pulsed jamming performance against repeat coded BPSK using hard decoding. $G_p = 100$, $\upsilon = -10$ dB, and $\xi = 10$ dB.

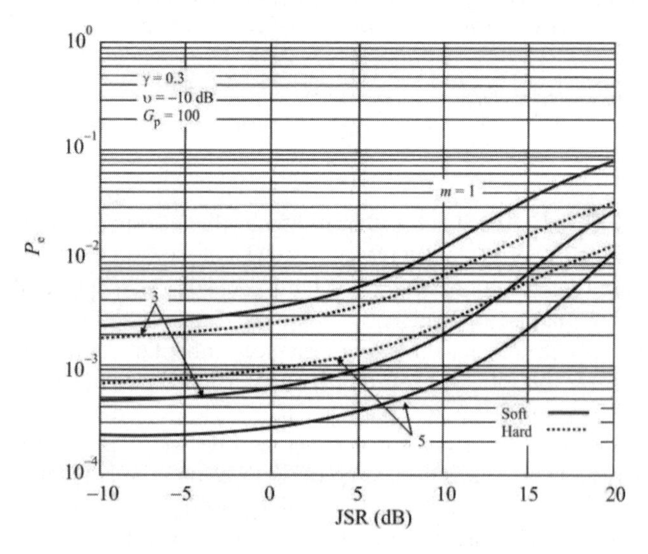

Figure 10.18 Pulsed jammer performance against repeat coded BPSK DSSS when $\gamma = 0.3$, $G_p = 100$, $\upsilon = -10$ dB, and the JSI is unknown. When $m = 1$, the decision method, of course, makes no difference.

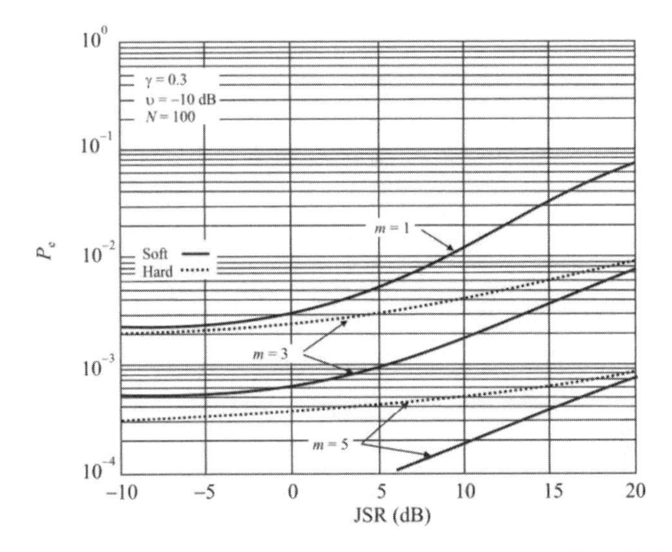

Figure 10.19 Pulsed jammer performance comparison of soft decision decoding with hard decision decoding against BPSK when the JSI is known, $\gamma = 0.3$, $\upsilon = -10$ dB, and $G_p = 100$.

improved but only achieves acceptable jamming performance when the signaling has no coding and high values of JSR.

10.7 Tone Jamming of DSSS Systems

10.7.1 Introduction

Tone jamming employs one of the easiest waveforms to generate and use. As opposed to BBN and PBN jamming discussed previously, tone jamming minimizes friendly fratricide so is less likely to lead to objections to its use. This section follows [5] closely.

Tone jamming can employ one or multiple tones. The spectrum when one tone is used is illustrated in Figure 10.20 while the spectrum in the case of MT jamming is illustrated in Figure 10.21. The jamming tones can be located arbitrarily, but some locations produce better results than others. Several techniques have been developed as countermeasures for tone jamming of DSSS systems [32–42].

10.7.2 Jammer

The total jamming signal at the receiver shown in Figure 10.2 is given by

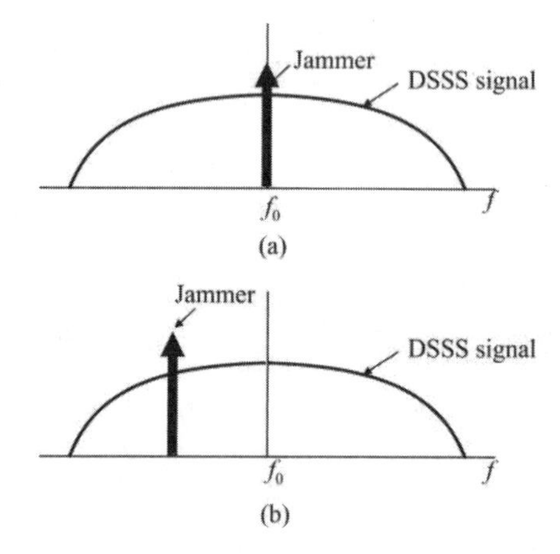

Figure 10.20 Single tone jamming of a DSSS signal: (a) tone at f_0 and (b) tone away from f_0.

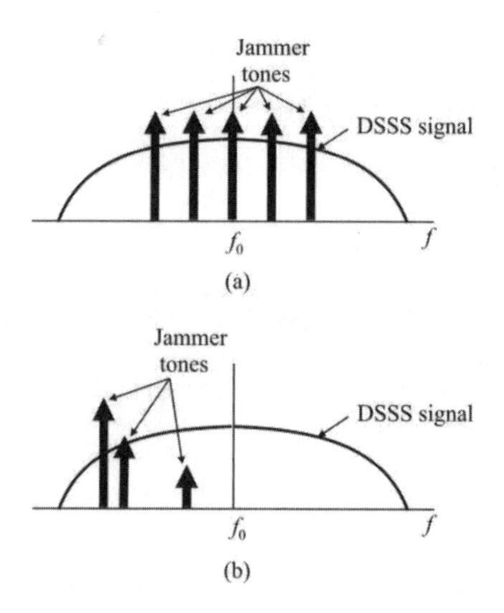

Figure 10.21 MT jamming of DSSS signals: (a) equal amplitude, equally spaced tones centered at f_0 and symmetric about the center and (b) unequal amplitude tones offset from the center frequency and unequally spaced.

$$j(t)=\sum_{n=1}^{N_\mathrm{J}}\sqrt{2J_n}\cos(\omega_n t+\phi_n) \tag{10.68}$$

where J_n is the average power in the nth tone, ω_n is its frequency, ϕ_n is its phase offset from the target signal that we assume to be uniformly distributed on $(0, 2\pi]$, and N_J is the number of jammer tones $(N_\mathrm{J} \ge 1)$. $\cos(\omega_\mathrm{j}t+\phi_\mathrm{j})$ and $\sin(\omega_\mathrm{j}t+\phi_\mathrm{j})$ are the jammer components for the in-phase and quadriphase paths, respectively.

10.7.3 Received Signals

Including the propagation delays between the transmitter and receiver (T_d) and jammer and receiver (T_J), the received signal $r(t)$ is given by

$$r(t)=s_\mathrm{s}(t-T_\mathrm{d})+j(t-T_\mathrm{J})+n(t) \tag{10.69}$$

Since we are assuming the receiver to be a linear system, the principle of superposition holds for its response to these three inputs so we can calculate the correlator input components due to the three received signals separately: data, jammer, and noise.

10.7.3.1 Received Data Component

The component of the received signal due to the data signal is given by

$$\begin{aligned}s_\mathrm{s}(t-T_\mathrm{d})=&\sqrt{\frac{R}{2}}d(t-T_0-T_\mathrm{d})w^\mathrm{I}(t-T_\mathrm{d})c^\mathrm{I}(t-T_0-T_\mathrm{d})\cos(\omega_0 t+\phi_\mathrm{r})\\&+\sqrt{\frac{R}{2}}d(t-T_\mathrm{d})w^\mathrm{Q}(t-T_\mathrm{d})c^\mathrm{Q}(t-T_\mathrm{d})\sin(\omega_0 t+\phi_\mathrm{r})\end{aligned} \tag{10.70}$$

where $\phi_\mathrm{r}=\phi_\mathrm{s}-\omega_0 T_\mathrm{d}$.

As mentioned, we assume that the system is coherent and that the receiver is synchronized to the transmitter so that

- $\phi_\mathrm{r}\approx\hat{\phi}_\mathrm{s}$, which means that phase lock has been achieved.
- The spreading sequences are synchronized [5].

- $\hat{T}_d$ is the receiver's estimate[1] of T_d.

In Figure 10.2, the signal $\tilde{y}_1^I(t)$ is the received data signal passing through the weighting multiplier, the first mixer, which converts the signals to a convenient IF frequency, and a multiplier (which despreads the signal) and bandpass filter. We assume that the sum frequency component is filtered away so that

$$\begin{aligned}\tilde{y}_1^I(t) &= \sqrt{\frac{1}{2}} s_d(t-\hat{T}_d) w^I(t-\hat{T}_d) \times 2\cos[(\omega_0+\omega_{IF})t] c^I(t-T_0-\hat{T}_d) \\ &= \frac{\sqrt{R}}{2} d(t-T_0-\hat{T}_d)[w^I(t-\hat{T}_d)]^2 \cos(\omega_{IF}t-\phi_r) - \frac{\sqrt{R}}{2} d(t-\hat{T}_d) \\ &\quad \times w^I(t-\hat{T}_d) w^Q(t-\hat{T}_d) c^I(t-T_0-\hat{T}_d) c^Q(t-\hat{T}_d)\sin(\omega_{IF}t-\phi_r)\end{aligned} \quad (10.71)$$

The signal for the I-path integrator input due to the data signal component is denoted by $y_1^I(t)$, and is given by

$$\begin{aligned}y_1^I(t) &= \left[x_1^I(t)\times 2\cos(\omega_{IF}t-\phi_r)\right]_{\text{diff freq}} \\ &= \frac{\sqrt{R}}{2} d(t-\hat{T}_d)[w^I(t-\hat{T}_d)]^2\end{aligned} \quad (10.72)$$

Similarly, we can obtain the data signal component at the quadrature phase integrator input as

$$\begin{aligned}y_1^Q(t) &= \left[x_1^Q(t)\times 2\cos(\omega_{IF}t-\phi_r)\right]_{\text{diff freq}} \\ &= \frac{\sqrt{R}}{2} d(t-\hat{T}_d)[w^Q(t-\hat{T}_d)]^2\end{aligned} \quad (10.73)$$

10.7.3.2 Received Jammer Component

In the in-phase path, the despreading mixer output jammer component passing the bandpass filter is

---

[1] Modern DSSS receivers are implemented with rake fingers, one finger for each estimated multipath component. In order to set the parameters for these fingers, estimates of the time delays for these paths are necessary. This information is therefore available anyway for the time delay of any direct path.

$$
\begin{aligned}
x_2^{\mathrm{I}}(t) &= \sqrt{\frac{1}{2}} j(t-\hat{T}_{\mathrm{d}}) w^{\mathrm{I}}(t-\hat{T}_{\mathrm{d}}) \times 2\cos[(\omega_0+\omega_{\mathrm{IF}})t] c^{\mathrm{I}}(t-T_0-\hat{T}_{\mathrm{d}})] \\
&= \sum_{n=1}^{N_{\mathrm{J}}} \sqrt{J_n} w^{\mathrm{I}}(t-\hat{T}_{\mathrm{d}}) c^{\mathrm{I}}(t-T_0-\hat{T}_{\mathrm{d}}) \cos[(\Delta\omega_n+\omega_{\mathrm{IF}})t-\phi_{\mathrm{n}}]
\end{aligned} \tag{10.74}
$$

where $\Delta\omega_n = \omega_0 - \omega_n$ and ϕ_n is the sum of the jammer phase offset of the nth tone and the jammer to receiver propagation delay $\omega_n T_{\mathrm{J}}$. The output of the second mixer due to the jammer component is

$$
y_2^{\mathrm{I}}(t) = \sum_{n=1}^{N_{\mathrm{J}}} \sqrt{J_n} w^{\mathrm{I}}(t-\hat{T}_{\mathrm{d}}) c^{\mathrm{I}}(t-T_0-\hat{T}_{\mathrm{d}}) \cos(\Delta\omega_n t - \phi_n + \phi_{\mathrm{r}}) \tag{10.75}
$$

In a similar fashion, the second mixer output due to the jammer in the Q-path is determined to be

$$
y_2^{\mathrm{Q}}(t) = \sum_{n=1}^{N_{\mathrm{J}}} \sqrt{J_n} w^{\mathrm{Q}}(t-\hat{T}_{\mathrm{d}}) c^{\mathrm{Q}}(t-\hat{T}_{\mathrm{d}}) \sin(\Delta\omega_n t - \phi_n + \phi_{\mathrm{r}}) \tag{10.76}
$$

10.7.3.3 Received Noise Component

We assume that the received noise is band-limited to approximately the signal bandwidth, which would be the case in practice due to filtering in the receiving chain. We assume an AWGN channel so the noise at the output of the bandpass filter is bandlimited Gaussian noise. Since the receiver input noise is $n(t)$, then for the in-phase path, the noise at the output of the bandpass filter is

$$
\begin{aligned}
x'^{\mathrm{I}}(t) &= n(t) w^{\mathrm{I}}(t-\hat{T}_{\mathrm{d}}) c^{\mathrm{I}}(t-T_0-\hat{T}_{\mathrm{d}}) \times 2\cos[(\omega_0+\omega_{\mathrm{IF}})t] \\
&= w^{\mathrm{I}}(t-\hat{T}_{\mathrm{d}}) c^{\mathrm{I}}(t-T_0-\hat{T}_{\mathrm{d}}) [n_1^{\mathrm{I}}(t)\cos(\omega_{\mathrm{IF}}t) - n_2^{\mathrm{I}}(t)\sin(\omega_{\mathrm{IF}}t)]
\end{aligned} \tag{10.77}
$$

where

$$
\begin{aligned}
n_1^{\mathrm{I}}(t) &= 2n(t)\cos(\omega_0 t) \\
n_2^{\mathrm{I}}(t) &= 2n(t)\sin(\omega_0 t)
\end{aligned}
$$

are the noise signal components for the in-phase path. The statistics of $n_1^{\mathrm{I}}(t)$ and $n_2^{\mathrm{I}}(t)$ are independent.

The noise at the input of the correlator in the in-phase path is

$$\begin{aligned} x^{\dagger \mathrm{I}}(t) &= w^{\mathrm{I}}(t-\hat{T}_{\mathrm{d}})c^{\mathrm{I}}(t-T_0-\hat{T}_{\mathrm{d}})[n_1^{\mathrm{I}}(t)\cos(\omega_{\mathrm{IF}}t) \\ &\quad -n_2^{\mathrm{I}}(t)\sin(\omega_{\mathrm{IF}}t)]\times 2\cos(\omega_{\mathrm{IF}}t+\hat{\phi}_{\mathrm{r}})_{\text{diff freq}} \\ &= w^{\mathrm{I}}(t-\hat{T}_{\mathrm{d}})c^{\mathrm{I}}(t-T_0-\hat{T}_{\mathrm{d}})[n_1^{\mathrm{I}}(t)\cos\hat{\phi}_{\mathrm{r}} - n_2^{\mathrm{I}}(t)\sin\hat{\phi}_{\mathrm{r}}] \end{aligned} \quad (10.78)$$

Likewise, the noise at the input of the correlator of the quadrature path is

$$\begin{aligned} x^{\dagger \mathrm{Q}}(t) &= w^{\mathrm{Q}}(t-\hat{T}_{\mathrm{d}})c^{\mathrm{Q}}(t-T_0-\hat{T}_{\mathrm{d}})[n_1^{\mathrm{Q}}(t)\cos(\omega_{\mathrm{IF}}t) \\ &\quad -n_2^{\mathrm{Q}}(t)\sin(\omega_{\mathrm{IF}}t)]\times 2\cos(\omega_{\mathrm{IF}}t+\hat{\phi}_{\mathrm{r}})_{\text{diff freq}} \\ &= w^{\mathrm{Q}}(t-T_{\mathrm{d}})c^{\mathrm{Q}}(t-T_{\mathrm{d}})[n_1^{\mathrm{Q}}(t)\cos\phi_{\mathrm{r}} - n_2^{\mathrm{Q}}(t)\sin\phi_{\mathrm{r}}] \end{aligned} \quad (10.79)$$

10.7.4 Correlator Outputs

The correlator outputs due to the noise at the sampling time are

$$x^{\mathrm{I}}(t) = \frac{A}{T_{\mathrm{s}}}\int_0^{T_{\mathrm{s}}} x^{\dagger \mathrm{I}}(t)dt \quad (10.80)$$

$$x^{\mathrm{Q}}(t) = \frac{A}{T_{\mathrm{s}}}\int_0^{T_{\mathrm{s}}} x^{\dagger \mathrm{Q}}(t)dt \quad (10.81)$$

respectively, when A represents the gain of the integrators shown in Figure 10.2. Since $x^{\dagger \mathrm{I}}(t)$ and $x^{\dagger \mathrm{Q}}(t)$ are Gaussian functions and the correlators are linear, $x^{\mathrm{I}}(t)$ and $x^{\mathrm{Q}}(t)$ are also Gaussian functions. The means of $x^{\mathrm{I}}(t)$ and $x^{\mathrm{Q}}(t)$ are zero and their variances are

$$\begin{aligned} (\sigma^2)^{\mathrm{I}} &= \mathcal{E}\{x^{\mathrm{I}}(t)x^{\mathrm{I}}(t)\} \\ &= \left(\frac{A}{T_{\mathrm{s}}}\right)^2 \int_0^{T_{\mathrm{s}}}\int_0^{T_{\mathrm{s}}} \mathcal{E}\{n^{\mathrm{I}}(v)n^{\mathrm{I}}(u)\}w^{\mathrm{I}}(v)w^{\mathrm{I}}(u)c^{\mathrm{I}}(v)c^{\mathrm{I}}(u)\cos^2\hat{\phi}_{\mathrm{r}}dudv \\ &\quad + \left(\frac{A}{T_{\mathrm{s}}}\right)^2 \int_0^{T_{\mathrm{s}}}\int_0^{T_{\mathrm{s}}} \mathcal{E}\{n^{\mathrm{I}}(v)n^{\mathrm{I}}(u)\}w^{\mathrm{I}}(u)w^{\mathrm{I}}(v)c^{\mathrm{I}}(u)c^{\mathrm{I}}(v)\sin^2\hat{\phi}_{\mathrm{r}}dudv \\ &= \left(\frac{A}{T_{\mathrm{s}}}\right)^2 \int_0^{T_{\mathrm{s}}}\int_0^{T_{\mathrm{s}}} \frac{N_0}{2}\delta(u-v)w^{\mathrm{I}}(u)w^{\mathrm{I}}(v)c^{\mathrm{I}}(u)c^{\mathrm{I}}(v)\cos^2\hat{\phi}_{\mathrm{r}}dudv \end{aligned}$$

$$= \left(\frac{A}{T_s}\right)^2 \int_0^{T_s} \frac{N_0}{2}[w^I(u)]^2[c^I(u)]^2 du$$

$$= \frac{N_0 A^2}{2T_s}[w^I(t)]^2 \qquad (10.82)$$

and

$$(\sigma^2)^Q = \mathcal{E}\{x^Q(t)x^Q(t)\}$$
$$= \left(\frac{A}{T_s}\right)^2 \int_0^{T_s} \frac{N_0}{2}[w^Q(u)]^2[c^Q(u)]^2 du$$
$$= \frac{N_0 A^2}{2T_s}[w^Q(t)]^2 \qquad (10.83)$$

Let $y_3^I(t) = (\sigma^2)^I$, $y_3^Q(t) = (\sigma^2)^Q$, and $y_3(t) = y_3^I(t) + y_3^Q(t)$, for BPSK/BPSK and for QPSK/BPSK, the noise power $y_3(t) = N_0 A^2 / T_s$. In the in-phase path, the correlator input due to the signal and jammer components is $y_i(t) = y_1^I(t) + y_2^I(t)$; in the quadrature path the correlator inputs due to the signal and jammer components are $y_i(t) = y_1^Q(t) + y_2^Q(t)$. The output of both correlators due to the noise component is $y_3(t)$. The results obtained can be used to calculate the performance for each modulation.

10.7.5 Single-Tone Jamming

In the presence of noise alone, or broadband jamming with equivalent spectral density, the performance of coherent spread systems is exactly the same as that of their unspread counterparts. That is, the presence of quadriphase or biphase chip spreading on BPSK data does not alter the fact that the signals are antipodal, and hence the error rate performance is given by

$$P_e = Q\left(\sqrt{\frac{2E_s}{N_0}}\right) \qquad (10.84)$$

where $Q(x)$ is Marcum's Q-function that is the probability that a Gaussian random variable exceeds its mean by x standard deviations, E_b is the received energy per information bit, and N_0 is the single-sided noise spectral density. This statement is also true for QPSK data modulation.

The case of single-tone jamming is considered here and MT jamming will be investigated in Section 10.7.7. The single-tone jamming notion is illustrated in Figure 10.20. For single tone jamming, there are two cases considered here depending on the placement of the jamming tone:

- The jamming signal frequency is the same as the DSSS carrier frequency (close to it but not on it—directly on it attenuates the jammer signal unnecessarily, see Figure 10.5).
- The jamming frequency is offset from the DSSS carrier frequency.

The DSSS signals considered here could have BPSK or QPSK data modulation and the spreading can be biphase or quadriphase. In general, the jamming effectiveness is dependent on the phase of the jamming signal relative to that of the target signal and the resulting probability of a bit error is denoted $\Pr\{e|\theta\}$ where θ is that phase difference. To remove this dependency it is necessary to average over $[0, 2\pi)$ as

$$P_e = \frac{1}{2\pi}\int_0^{2\pi} \Pr\{e|\theta\} d\theta \tag{10.85}$$

A single-tone jammer places a high power CW tone at some place in the frequency spectrum. That place would normally depend on the target and the goal would be to optimize that placement in some sense to the jammer's favor. When the jammer is placed within W_{ss}, it is assumed herein that any effects of wideband filtering at the receiver have no effect on the jammer signal. In other words it is far enough from the filter band edges that this filtering does not attenuate it significantly.

10.7.5.1 Biphase Spreading and BPSK Data Modulation

The BER for biphase spreading and BPSK data modulation with a long code and with a single-tone jamming signal is given by [5]

$$\Pr\{e|\theta\} = Q\left\{\frac{1 - g_1(\theta)\cos\theta}{\left[\left[\frac{1}{2N\upsilon} + \frac{1}{2N}\xi \operatorname{sinc}^2\left(\frac{\Delta\omega T_s}{2N}\right)\left(1+\frac{1}{L}\right)\left(1+g_2(\theta)\right)\right]\right]^{1/2}}\right\} \tag{10.86}$$

where

$$g_1(\theta) = \sqrt{\xi}\frac{1}{L}\text{sinc}\left(\frac{\Delta\omega T_s}{2N}\right)\frac{\text{sinc}\left(\frac{\Delta\omega T_s}{2}\right)}{\text{sinc}\left(\frac{\Delta\omega T_s}{2N}\right)} \tag{10.87}$$

$$g_2(\theta) = \frac{\text{sinc}(\Delta\omega T_s)}{\text{sinc}\left(\frac{\Delta\omega T_s}{N}\right)}\cos(2\theta) - \frac{2N}{L}\frac{\text{sinc}^2\left(\frac{\Delta\omega T_s}{2}\right)}{\text{sinc}^2\left(\frac{\Delta\omega T_s}{2N}\right)}\cos^2\theta \tag{10.88}$$

and $\Delta\omega$ is the frequency offset of the tone.

When the length of the PN code, L, is large compared to N, the number of spreading chips in a data symbol (and therefore equal to the processing gain, G_p), those terms preceded by the $1 / L$ factor can be neglected in (10.86) with little loss of generality, which simplifies the P_e expression to [5]

$$\Pr\{e|\theta\} = Q\left\{\left[\left[\frac{1}{2N\upsilon} + \frac{1}{2N}\xi\text{sinc}^2\left(\frac{\Delta\omega T_b}{2N}\right)\right] \times\left[1 + \frac{\text{sinc}(\Delta\omega T_b)}{\text{sinc}\left(\frac{\Delta\omega T_b}{N}\right)}\cos(2\theta)\right]\right]^{-1/2}\right\} \tag{10.89}$$

When the modulation is BPSK, then the performance is as indicted by (10.89). In the case of $\Delta\omega = 0$, (10.89) is illustrated in Figure 10.22 for $\upsilon = -10$ dB.

When $\Delta\omega \neq 0$ the tone can be placed anywhere within W_{ss}. Here, (10.86) generates the curves illustrated in Figure 10.23. Again at $\xi \approx G_p$, $P_e \approx 10^{-1}$. Figure 10.24 illustrates the effects of varying the frequency offset of the jammer tone for small processing gain conditions. For $G_{p,dB}$ = 13 dB, jamming performance decreases the further away from f_0 the jamming tone is placed. For $G_{p,dB}$ = 20 dB, the jamming performance remains relatively constant irrespective of how far away the tone is placed.

The average probability of error is calculated using (10.85). For coherent target links, assuming a uniform PDF for θ in (10.85) may not be a good approximation. The phase is likely to be correlated from one symbol to the next.

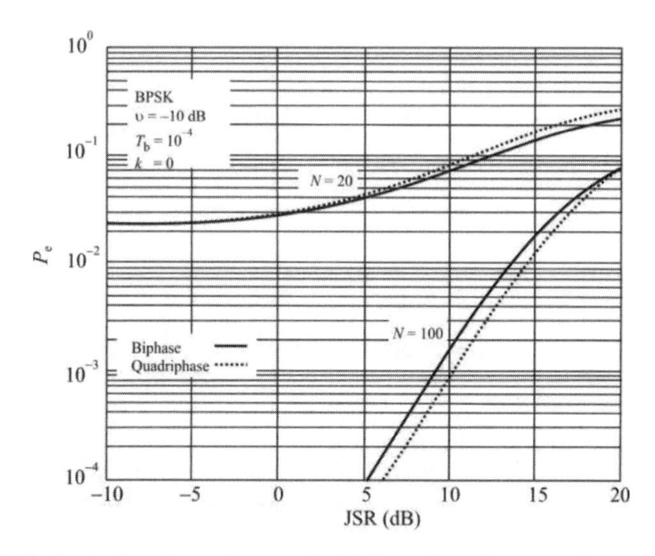

Figure 10.22 Single-tone jamming performance for BPSK and jamming tone at f_0 with long codes. In this case $T_b = 10^{-4}$, and $\upsilon = -10$ dB.

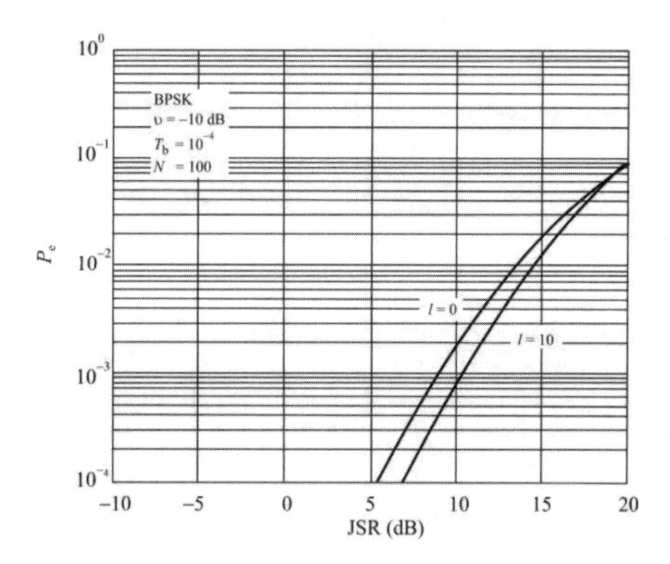

Figure 10.23 Single-tone jamming performance against DSSS BPSK with long codes. The jammer tone is offset from the center frequency by a factor l / T_s. The performance curves are the same for biphase and quadriphase modulation. $\upsilon = -10$ dB, $T_b = 100$ μs, $G_{p,dB} = 20$ dB.

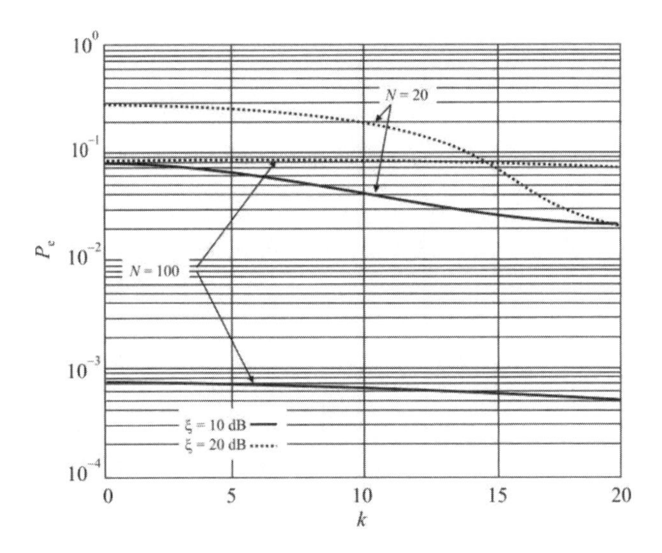

Figure 10.24 Effects of varying the jammer tone frequency offset, where l is the multiple of the data rate the jamming tone is offset from the signal center frequency and long codes. The same curves apply for both biphase and quadriphase spreading with BPSK data modulation. υ = –10 dB and $T_b = 10^{-4}$.

The performance of biphase spreading and BPSK data depends on the relative phase between the jammer and the signal even with $\Delta\omega = 0$ and $L \to \infty$. For example, in (10.86), for $\Delta\omega = 0$ and $L \to \infty$, with $\theta = 90^\circ$ the jamming effect vanishes completely, while $\theta = 0^\circ$ yields the best jammer performance. Figure 10.25 shows about a 2 dB difference between the best jamming performance and that obtained by averaging over θ using (10.85).

For an on-tune jammer $(\Delta\omega \to 0)$, (10.86) becomes

$$\Pr\{e|\theta\} = Q\left\{\frac{1-\sqrt{\xi}\dfrac{1}{L}\cos\theta}{\left[\dfrac{1}{2N\upsilon}+\dfrac{\xi}{N}\left(1+\dfrac{1}{L}\right)\left(1-\dfrac{N}{L}\right)\cos^2\theta\right]^{1/2}}\right\} \tag{10.90}$$

10.7.5.2 Biphase Spreading BPSK Data Modulation: Short Codes

If $L \sim N$, then the Gaussian approximation is not valid even though $N >> 1$ because k_1, the number of "one" chips in a bit interval, is not independent bit by bit. Therefore, k_1 is not an r.v. but a deterministic quantity.

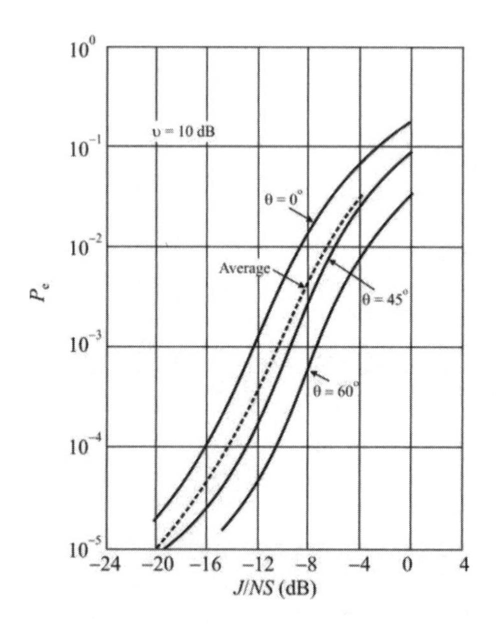

Figure 10.25 Biphase spreading with BPSK with long codes in CW jamming. P_e versus the decoded JSR. (After: [5]. © 1983 IEEE. Reprinted with permission.)

As illustrated in Figure 10.5, on-tune jamming ($\Delta\omega \rightarrow 0$) is not the best place to put a jammer tone due to the severe attenuation of the tone there. It would be better to place the jammer at a frequency $1/T_s$ higher or lower than at the carrier frequency. In that case the contribution of the jammer to the decision is proportional to $(L+1)/L^2 \sim 1/L$ rather than $1/L^2$. In general, if the jammer is off-tuned by a multiple l of the data bit rate, then

$$\Pr\{e|\theta\} = Q\left[\sqrt{\frac{2E_s}{N_0}}\left(1 - \frac{1}{L}\operatorname{sinc}\frac{l\pi}{L}\sqrt{\xi}\cos\theta\right)\right] \tag{10.91}$$

Equation (10.91) can be put in terms of power ratios, yielding

$$\Pr\{e|\theta\} = Q\left[\sqrt{2L\upsilon}\left(1 - \frac{1}{L}\operatorname{sinc}\frac{l\pi}{L}\sqrt{\xi}\cos\theta\right)\right] \tag{10.92}$$

This expression is not very sensitive to l as we would expect, since l is a multiplier of π / N with N typically much larger than l.

Consider the model in Figures 10.1 and 10.2 for the case of biphase spreading/BPSK, and let $w_{\rm I}(t) = \sqrt{2}$, $w_{\rm Q}(t) = 0$, and $T_0 = 0$. By doing this, only the in-phase channel remains. Equation (10.72), representing the data signal, and (10.75), representing the jammer signal, become

$$y_1(t) = \sqrt{R}d(t) \tag{10.93}$$

$$y_2(t) = \sqrt{2J}c^{\rm I}(t)\cos(\Delta_\omega t - \phi_{\rm j} + \phi_{\rm r}) \tag{10.94}$$

We normalize the correlator output by letting $A = \sqrt{1/R}$; then the output data component becomes ±1. The result (due to the signal and jammer components only) is given by

$$\begin{aligned} z_2(t) &= \frac{A}{T_{\rm s}}\int_0^{T_{\rm s}} \sqrt{R}d(t)dt + \frac{A}{T_{\rm s}}\int_0^{T_{\rm s}} \sqrt{2J}c^{\rm I}(t)\cos(\Delta\omega t - \phi_{\rm j} + \phi_{\rm r})dt \\ &= \pm 1 + \frac{1}{T_{\rm s}}\sqrt{2\xi}\int_0^{T_{\rm s}} c^{\rm I}(t)\cos(\Delta\omega t + \phi)dt \end{aligned} \tag{10.95}$$

and let

$$z_2(nT_{\rm s}) = \frac{1}{T_{\rm s}}\int_{(n-1)T_{\rm s}}^{nT_{\rm s}} \sqrt{2\xi}c^{\rm I}(t)\cos(\Delta\omega t + \phi)dt \tag{10.96}$$

where $\phi = \phi_{\rm r} - \phi_{\rm j}$.

The error probability is a function of the statistics of z_2. The mean of z_2 is the ensemble average over all possible spreading code sequences during the integration time which is given by

$$\overline{z}_2(nT_s) = \mathcal{E}\left\{\frac{1}{T_{\rm s}}\int_{(n-1)T_{\rm s}}^{nT_{\rm s}} \sqrt{2\xi}c^{\rm I}(t)\cos(\Delta\omega t + \phi)dt\right\} \tag{10.97}$$

which is the integral z_2, (10.96), over one symbol duration. Since $c^{\rm I}(t)$ takes on values of ±1 with $\Pr\{c(t) = 1\} = \Pr\{c(t) = -1\} = ½$, a zero average results for each coded symbol, that is $\overline{z}_2 = 0$.

A more complicated case is to evaluate the totality of the integral z_2 in (10.96) over each symbol. The integral over each chip is evaluated and the results are summed as

$$\begin{aligned} z_2(nT_s) &= \frac{1}{T_s}\sqrt{2\xi}\sum_{i=0}^{N-1} c_i^{\mathrm{I}} \int_{iT_c}^{(i+1)T_c} \cos(\Delta\omega t + \phi)dt \\ &= \frac{1}{T_s}\sqrt{2\xi}\sum_{i=0}^{N-1} c_i^{\mathrm{I}} \frac{1}{\Delta\omega}\{\sin[\Delta\omega(i+1)T_c + \phi] - \sin(\Delta\omega i T_c + \phi)\} \end{aligned} \tag{10.98}$$

where the spreading code sequence, c_i^{I} takes on values of ±1, and N is the number of spreading waveform chips in a data symbol [$N = G_p$]. Since (10.98) is independent of whether the code sequence repeats within the symbol or not, it is valid for short and long codes. Equation (10.98) can be written as

$$z_2(nT_s) = \frac{T_c}{T_s}\sqrt{2\xi}\frac{\sin(\Delta\omega T_c/2)}{\Delta\omega T_c/2}\sum_{i=0}^{N-1} c_i^{\mathrm{I}} \cos(i\Delta\omega t + \psi) \tag{10.99}$$

where $\psi = \Delta\omega T_c/2 + \phi$. The mean of z_2 is zero, and the variance is given by

$$\sigma_{z_2}^2 = \mathcal{E}\left\{ \begin{aligned} &\left(\frac{T_c}{T_s}\right)^2 2\xi \left[\frac{\sin(\Delta\omega T_c/2)}{\Delta\omega T_c/2}\right]^2 \\ &\times \sum_{j=1}^{N-1} c_j^{\mathrm{I}} \cos(\Delta\omega j T_c + \psi) \sum_{i=1}^{N-1} c_i^{\mathrm{I}} \cos(\Delta\omega i T_c + \psi) \end{aligned} \right\} \tag{10.100}$$

When the spreading codes are independent with equally random binary digits, then $\mathcal{E}\{c_j^{\mathrm{I}} c_i^{\mathrm{I}}\} = 1$, for $j = i$, and $\mathcal{E}\{c_j^{\mathrm{I}} c_i^{\mathrm{I}}\} = 0$, for $j \neq i$; that is

$$\mathcal{E}\{c_j^{\mathrm{I}} c_i^{\mathrm{I}}\} = \delta_{ji} \tag{10.101}$$

Using

$$\sum_{i=0}^{N-1} \cos(iy + x) = \frac{\cos\{x + [(N-1)/2)y]y\}\sin(Ny/2)}{\sin(y/2)} \tag{10.102}$$

(see [43]), (10.100) becomes

$$\sigma_{z_2}^2 = \frac{2\xi}{N^2}\left[\frac{\sin(\Delta\omega T_c/2)}{\Delta\omega T_c/2}\right]^2 \sum_{i=0}^{N-1}\cos^2(i\Delta\omega t+\psi)$$
$$= \frac{\xi}{N}\left[\frac{\sin(\Delta\omega T_c/2)}{\Delta\omega T_c/2}\right]^2 \left\{1+\frac{\cos[2\phi+(N-1)\Delta\omega T_c]\sin(N\Delta\omega T_c)}{N\sin(\Delta\omega T_c)}\right\} \quad (10.103)$$

The variance of the output of the integrator is the sum of the variance due to the noise, (10.82) and (10.83), and the variance due to the spread jammer (10.103). Assuming that a –1 data symbol was transmitted, an error occurs whenever the integrator output is positive [44], that is,

$$P_e = \Pr\{z_2(nT_s)+y_3(t)>1.0\}$$
$$= \int_1^{\infty}\frac{1}{\sqrt{2\pi}\sigma}e^{-\alpha^2/(2\sigma^2)}d\alpha = Q(1/\sigma) \quad (10.104)$$

where $\sigma^2 = \sigma_{z_2}^2 + y_3$ and $Q(1/\sigma)$ is the complementary error function. When $\Delta\omega \neq 0$ and there is a high SNR, the variance y_3 is negligible. Substituting (10.103) into (10.104), the error probability ensues as

$$P_e = Q\left\{\left[\frac{2N}{2\xi\left[\frac{\sin(\Delta\omega T_c/2)}{\Delta\omega T_c/2}\right]^2\left[1+\frac{\cos[2\phi+(N\Delta\omega T_c)]\sin(N\Delta\omega T_c)}{N\sin(\Delta\omega T_c)}\right]}\right]^{1/2}\right\}$$
$$= Q\left\{\left[\frac{2E_b}{2E_j\left[\frac{\sin(\Delta\omega T_c/2)}{\Delta\omega T_c/2}\right]^2\left[1+\frac{\cos\left[2\phi+(N\Delta\omega T_c)\right]\sin(N\Delta\omega T_c)}{N\sin(\Delta\omega T_c)}\right]}\right]^{1/2}\right\} \quad (10.105)$$

Since $N = T_s/T_c$, $E_b = RT_s$, $E_j = JT_b$, $\xi = J/R$, and with $\Delta\omega$ as a parameter, the performance curves are plotted in Figure 10.26. For this figure, $N = 4$. Also, for illustration purposes, it is assumed that $\phi = 0$. When $N = 128$, the results in Figure 10.27 ensue with the same assumptions. It is clear from these two sets of curves that as the spreading code gets larger (and therefore lower data throughput since

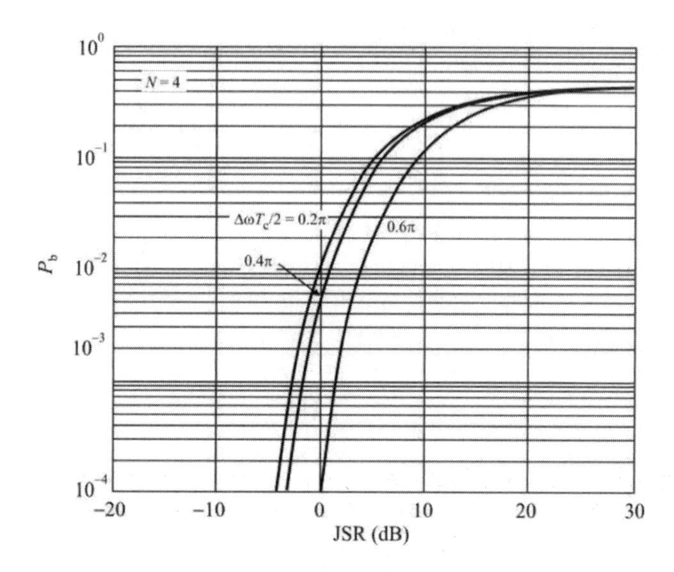

Figure 10.26 BER versus JSR when $N = 4$, SNR is high, and $J >> P_N$. This corresponds to maximum data rate in 3G+ phones since $N = 4$ is the minimum coding rate. This is for minimum coding protection.

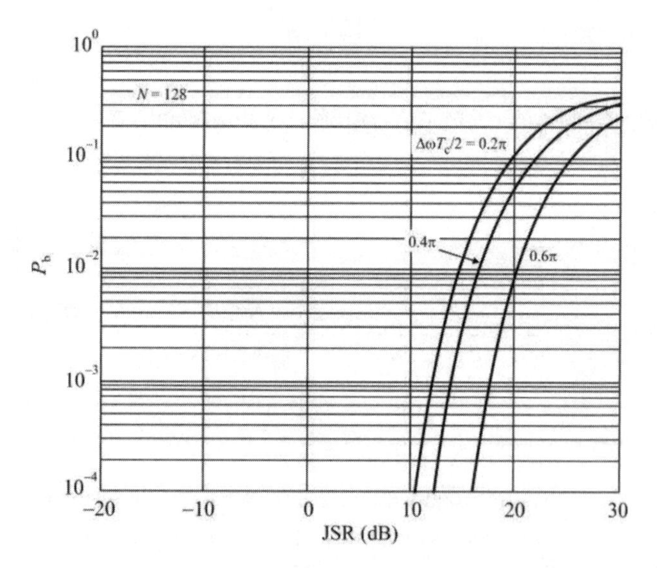

Figure 10.27 BER when $N = 128$. The SNR is high and $J >> P_N$. This corresponds to the minimum throughput in 3G+ PCS, which also corresponds to maximum coding protection.

the transmitted bit rate stays constant), jamming gets more difficult and larger jammer power is required. It is relatively easy to cause a BER of 10^{-2} when $N = 4$ while for $N = 128$, about 15 dB more jamming power is required.

10.7.5.3 Quadriphase Spreading and BPSK Data Modulation

With quadriphase chip spreading and BPSK data modulation, the decision statistic is

$$\begin{aligned} z(\theta) = \pm\sqrt{E} + \frac{\sqrt{J}}{\sqrt{2T_s}} \int_{T_0}^{T_0+T_s} c^{\mathrm{I}}(t)\cos(\Delta\omega t + \theta)dt \\ + \frac{\sqrt{J}}{\sqrt{2T_s}} \int_{T_0}^{T_0+T_s} c^{\mathrm{Q}}(t)\sin(\Delta\omega t + \theta)dt + n \end{aligned} \tag{10.106}$$

Where n is a zero mean Gaussian r.v. with variance $N_0/2$. It represents the response of the receiver to thermal noise. The BER is

$$\Pr\{\mathrm{e}|\theta\} = Q\left\{\frac{1 - g_1(\Delta\omega)(\cos\theta + \sin\theta)}{\sqrt{\frac{1}{2N\upsilon} + \frac{1}{2N}\xi\,\mathrm{sinc}^2\left(\frac{\Delta\omega T_s}{2N}\right)\left[\left(1 + \frac{1}{L}\right)(1 - g_2(\Delta\omega)) - g_3(\Delta\omega)\sin 2\theta\right]}}\right\} \tag{10.107}$$

where

$$g_1(\Delta\omega) = \sqrt{\frac{\xi}{2}}\frac{1}{L}\,\mathrm{sinc}\left(\frac{\Delta\omega T_s}{2N}\right)\frac{\mathrm{sinc}\left(\frac{\Delta\omega T_s}{2}\right)}{\mathrm{sinc}\left(\frac{\Delta\omega T_s}{2N}\right)} \tag{10.108}$$

$$g_2(\Delta\omega) = \frac{N}{L}\frac{\mathrm{sinc}^2\left(\frac{\Delta\omega T_s}{2}\right)}{\mathrm{sinc}^2\left(\frac{\Delta\omega T_s}{2N}\right)} \tag{10.109}$$

$$g_3(\Delta\omega) = \frac{1}{L}\frac{\mathrm{sinc}(\Delta\omega T_s)}{\mathrm{sinc}\left(\frac{\Delta\omega T_s}{N}\right)} + \frac{N}{L^2}\frac{\mathrm{sinc}^2\left(\frac{\Delta\omega T_s}{2}\right)}{\mathrm{sinc}^2\left(\frac{\Delta\omega T_s}{2N}\right)} \tag{10.110}$$

When $L >> N$ the BER becomes [5]

$$P_e = Q\left\{\left[\frac{1}{2N\upsilon} + \frac{1}{2N}\xi\text{sinc}^2\left(\frac{\Delta\omega T_s}{2N}\right)\right]^{-1/2}\right\} \tag{10.111}$$

This function is also illustrated in Figure 10.22. Note that this P_e is now independent of the jammer phase.

At low JSR, for BPSK data modulation, single-tone jamming is more effective on biphase spreading than on quadriphase spreading. Where $\xi \approx G_p$, this reverses and it becomes more effective against quadriphase. In addition, note that at this SNR, and lower values as well, $P_e \approx 10^{-1}$ when $\xi \approx G_p$. At high JSRs ($\xi > 10$ dB or so), the jamming performance is almost independent of the SNR.

The case when $\Delta\omega \neq 0$ is also illustrated in Figure 10.23. The effects of varying the tone frequency offset from the DSSS carrier frequency are the same as for biphase spreading shown in Figure 10.24.

10.7.5.4 Quadriphase Spreading and BPSK Data Modulation: Short Codes

For the quadriphase spreading and BPSK encoded data, the BER for an ontune jammer is given by

$$\Pr\{e|\theta\} = Q\left(\sqrt{\frac{2E}{N_0}}\left[1 - \frac{1}{N}\sqrt{\frac{\xi}{2}}(\cos\theta + \sin\theta)\right]\right) \tag{10.112}$$

which, as above, can be written in terms of power ratios as

$$\Pr\{e|\theta\} = Q\left(\sqrt{2N\upsilon}\left[1 - \frac{1}{N}\sqrt{\frac{\xi}{2}}(\cos\theta + \sin\theta)\right]\right) \tag{10.113}$$

From (10.112) it is clear that if $P(e|\theta)$ is averaged over θ $[0, 2\pi)$ using (10.85), then biphase spreading and quadriphase spreading both provide the same performance for BPSK data encoding. On the other hand, for a coherent jammer ($\theta = 0^\circ$), quadriphase spreading has a 3 dB advantage over the biphase spreading. These results are plotted in Figure 10.28 for N = 4, 32, and 64. If we accept $P_e \geq 10^{-2}$ as our measure of EW success, it is clear from Figure 10.28 that for

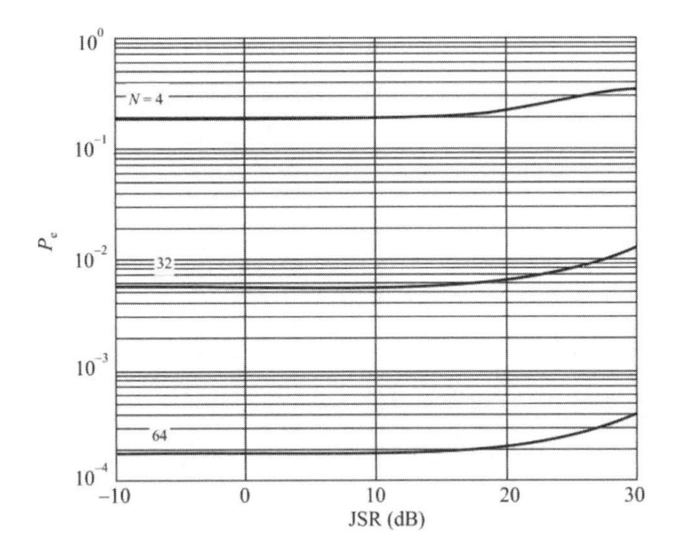

Figure 10.28 BER for biphase and quadriphase spreading and BPSK data encoding. $\upsilon = -10$ dB and $k = 0$.)

$N \leq 32$ the spreading provides little to no AJ protection, although, to be fair, the PCS standards were not developed with AJ performance in mind.

If the jammer frequency offset is an integer multiple l of the bit rate, then (10.112) can be written as

$$\Pr\{e|\theta\} = Q\left(\sqrt{\frac{2E}{N_0}}\left[1 - \frac{1}{N}\sqrt{\xi}\mathrm{sinc}\left(\frac{l\pi}{N}\right)\cos\left(\theta + \frac{\pi}{4}\right)\right]\right) \tag{10.114}$$

which, again, must be averaged over $(0, 2\pi)$ to determine P_e with (10.85).

10.7.5.5 Biphase Spreading and QPSK Data Modulation

For biphase spreading and QPSK data modulation the BER can be written as

$$\Pr\{e|\theta\} = \frac{1}{2}Q\left\{\frac{1 - g_1(\Delta\omega)\cos\theta}{\left[\left[\frac{1}{2N\upsilon} + \frac{1}{2N}\xi\mathrm{sinc}^2\left(\frac{\Delta\omega T_s}{2N}\right)\right]\left[\left(1 + \frac{1}{L}\right)[1 + g_2(\Delta\omega)]\right]\right]^{1/2}}\right\}$$

$$+\frac{1}{2}Q\left\{\frac{1-g_1(\Delta\omega)\sin\theta}{\left[\left[\frac{1}{2N\upsilon}-\frac{1}{2N}\xi\mathrm{sinc}^2\left(\frac{\Delta\omega T_\mathrm{s}}{2N}\right)\right]\left[\left(1+\frac{1}{L}\right)[1-g_3(\Delta\omega)]\right]\right]^{1/2}}\right\} \tag{10.115}$$

where

$$g_1(\Delta\omega)=\sqrt{2\xi}\frac{1}{L}\mathrm{sinc}\left(\frac{\Delta\omega T_\mathrm{s}}{2N}\right)\frac{\mathrm{sinc}(\Delta\omega T_\mathrm{s})}{\mathrm{sinc}\left(\frac{\Delta\omega T_\mathrm{s}}{2N}\right)} \tag{10.116}$$

$$g_2(\Delta\omega)=\frac{\mathrm{sinc}(2\Delta\omega T_\mathrm{s})}{\sin c\left(\frac{\Delta\omega T_\mathrm{s}}{N}\right)}\cos(2\theta)-\frac{4N}{L}\frac{\mathrm{sinc}^2(\Delta\omega T_\mathrm{s})}{\mathrm{sinc}^2\left(\frac{\Delta\omega T_\mathrm{s}}{2N}\right)}\cos^2\theta \tag{10.117}$$

$$g_3(\Delta\omega)=\frac{\mathrm{sinc}(2\Delta\omega T_\mathrm{s})}{\mathrm{sinc}\left(\frac{\Delta\omega T_\mathrm{s}}{N}\right)}\cos(2\theta)+\frac{4N}{L}\frac{\mathrm{sinc}^2(\Delta\omega T_\mathrm{s})}{\mathrm{sinc}^2\left(\frac{\Delta\omega T_\mathrm{s}}{2N}\right)}\sin^2\theta \tag{10.118}$$

With the assumption that $L >> N$ (but not $\Delta\omega = 0$), P_e is given by

$$\begin{aligned}\Pr\{\mathrm{e}|\theta\}=\frac{1}{2}Q&\left\{\left[\frac{1}{2N\upsilon}+\frac{1}{2N}\xi\mathrm{sinc}^2\left(\frac{\Delta\omega T_\mathrm{s}}{2N}\right)\times\left[1+\frac{\mathrm{sinc}(2\Delta\omega T_\mathrm{s})}{\mathrm{sinc}\left(\frac{\Delta\omega}{N}\right)}\cos(2\theta)\right]\right]^{-1/2}\right\}\\&+\frac{1}{2}Q\left\{\left[\frac{1}{2N\upsilon}+\frac{1}{2N}\xi\mathrm{sinc}^2\left(\frac{\Delta\omega T_\mathrm{s}}{2N}\right)\times\left[1-\frac{\mathrm{sinc}(2\Delta\omega T_\mathrm{s})}{\mathrm{sinc}\left(\frac{\Delta_\omega}{N}\right)}\cos(2\theta)\right]\right]^{-1/2}\right\}\end{aligned} \tag{10.119}$$

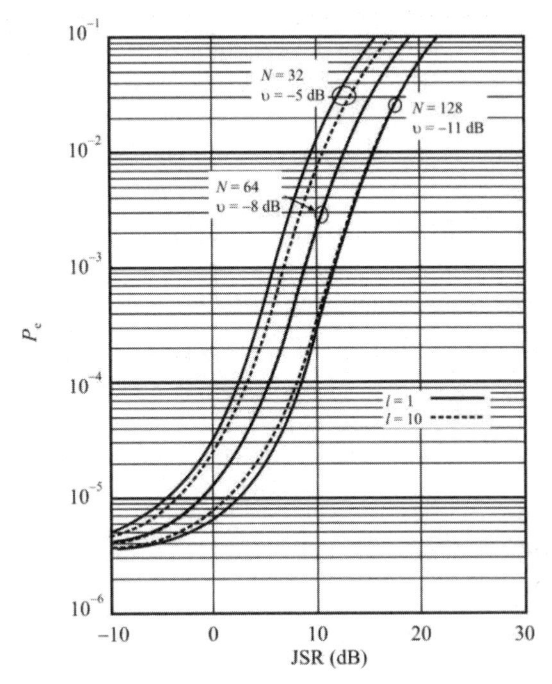

Figure 10.29 BER for biphase spreading and QPSK data modulation, short codes, with jammer tone offset. $\Delta\omega = k/2\pi T_s$, $\upsilon = -10$ dB, and $k = 1$ and 10.

The cases for both $\Delta\omega$ and $\Delta\omega \neq 0$ are illustrated in Figure 10.23. The effects of offsetting the jammer tone from the DSSS center frequency are not significant in this case.

Employing (10.85) with (10.119), the cases for $l = 1$ and 10 are illustrated in Figure 10.29. Since the frequency offset from the carrier (or IF) is given by l/T_s, when $l = 1$ the tone offset was at the first offset position, and when $l = 10$, the jammer tone was at the tenth. In neither case was there a jammer tone at the target center frequency. The effects of offsetting the jammer tone from the DSSS center frequency are not significant.

If the conditions similar to those discussed in (10.90) are satisfied, then for $\Delta_\omega \to 0$ (ontune jammer) and $L >> N$, (10.119) reduces to

$$\Pr\{e|\theta\} = \frac{1}{2}Q\left\{\left[\frac{1}{2N\upsilon} + \frac{\xi}{N}\cos^2\theta\right]^{-1/2}\right\} + \frac{1}{2}Q\left\{\left[\frac{1}{2N\upsilon} + \frac{\xi}{N}\sin^2\theta\right]^{-1/2}\right\} \quad (10.120)$$

Using (10.85), the curves generated by this function are nearly identical to the corresponding curves in Figure 10.22; P_e is slightly larger for higher JSR values (above about 25 dB).

Comparing (10.86) and (10.119), we see that an offtuned jammer is as effective against biphase spread/BPSK data and biphase spread/QPSK data when $L >> N$.

10.7.5.6 Quadriphase Spreading and QPSK Data Modulation: Long Codes

When $L >> N$ the type of spreading for QPSK data modulation makes no difference. Therefore the results are the same as for BPSK, (10.119).

10.7.6 Single-Tone Jamming of Chaotic Systems

The performance of chaotic systems using CSK modulation in the presence of a single jammer tone was derived by Lau et al. [45].

10.7.6.1 Coherent Reception

Under the assumptions discussed in Section 3.13.5, the BER for binary CSK communication systems with a single tone jammer within the passband of the receiver of power P_J is given by

$$P_e = \frac{1}{2}\text{erfc}\left[\frac{1}{\sqrt{\left(\dfrac{E_b^2}{4\beta\Lambda}\right)^{-1} + \upsilon^{-1} + \left(\dfrac{E_b}{2J}\right)^{-1}}}\right] \tag{10.121}$$

This BER is illustrated in Figure 10.30 for representative values of the parameters.

10.7.6.2 Noncoherent Reception

Likewise, for noncoherent reception in binary DCSK communication systems with a jammer tone present in the passband of power P_J, the BER is given by

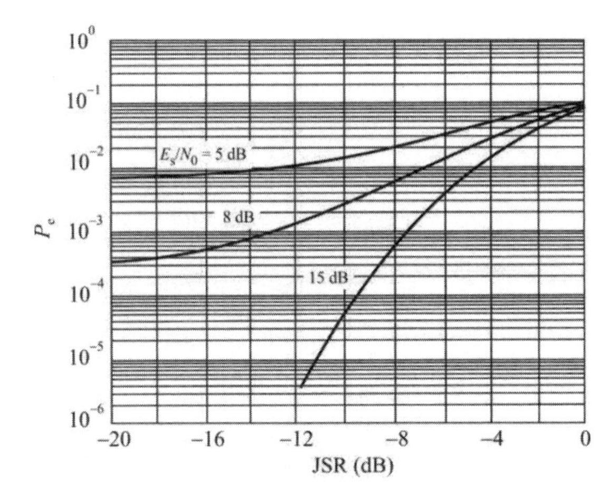

Figure 10.30 BER for Chaotic coherent CSKJ communication system. In this example $2\beta = 200$ and the normalized jamming frequency is 3.12735. (After: [45]. © IEEE, 2002. Reprinted with permission.)

$$P_e = \frac{1}{4}\int_{-1}^{1}\left\{\begin{aligned}&\operatorname{erfc}\left[\frac{\dfrac{E_b}{2}+\beta J\cos(\pi f)-\dfrac{J\sin(\pi f)}{\sin(\pi f/\beta)}w}{\sqrt{2\beta\Lambda+4E_b J\cos^2\left(\dfrac{\pi f}{2}\right)+E_b N_0+2\beta J N_0+\dfrac{\beta N_0^2}{2}}}\right]\\&+\operatorname{erfc}\left[\frac{\dfrac{E_b}{2}-\beta J\cos(\pi f)+\dfrac{J\sin(\pi f)}{\sin(\pi f/\beta)}w}{\sqrt{2\beta\Lambda+4E_b J\sin^2\left(\dfrac{\pi f}{2}\right)+E_b N_0+2\beta J N_0+\dfrac{\beta N_0^2}{2}}}\right]\end{aligned}\right\} \times\frac{1}{\pi\sqrt{1-w^2}}dw \tag{10.122}$$

Here, f is the normalized jammer frequency given by $f = f_J T_b$ where f_J is the actual jammer frequency and T_b is the bit time. This BER is illustrated in Figure 10.31 for representative values. For all values of SNR considered in the example shown in Figure 10.31, P_e rises above 10^{-1} around 0 dB JSR.

10.7.7 Multitone Jamming

If the jammer employs MT, then (10.86), (10.107), and (10.119) can still be used to calculate the BER. For a N_J tone jammer, (10.107) with $L >> N$ (10.107) can be modified to

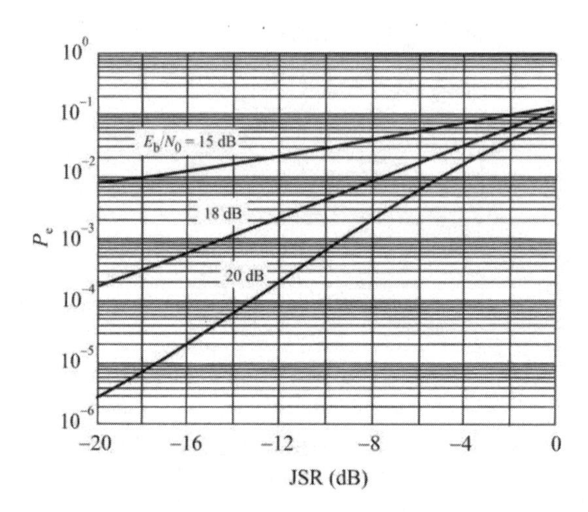

Figure 10.31 BER for binary chaotic noncoherent communication system. For this example 2β = 200 and the normalized jamming frequency is 3.828734. (After: [45]. © IEEE, 2002. Reprinted with permission.)

$$P_e = Q\left\{\left[\frac{1}{2N\upsilon} + \sum_{n=1}^{N_J} \frac{\xi_n}{2N}\left[\frac{\operatorname{sinc}\left(\frac{\Delta\omega_n T_s}{2N}\right)}{\frac{\Delta\omega_n T_s}{2N}}\right]^2\right]^{-1/2}\right\} \tag{10.123}$$

where ξ_n is the JSR in the nth tone, $\Delta\omega_n = \omega_n - \omega_c$ is the frequency offset for the ith tone, and $J = J_1 + J_2 + \ldots + J_k$. Note that J is the total jamming power and assumed to stay constant and equally divided among the tones. We note that we have neglected the contribution due to the jammer phase in this simple extension.

We can conclude from (10.123), when $L >> N$ the best case jamming is a single tone jammer with $\Delta\omega \to 0$ because for a given N and R and assuming $J = J_1 + J_2 + \ldots + J_{N_J}$

$$\sum_{n=1}^{N_J} \frac{\xi_n}{2N} \operatorname{sinc}^2\left(\frac{\Delta\omega T_s}{2N}\right) < \frac{\xi}{2N} \tag{10.124}$$

The same conclusion also holds for biphase spreading and BPSK data and quadriphase spreading and QPSK data encoding. These conclusions are for the case when the terms corresponding to $1/L$ are dropped in (10.107) and (10.119). The results are not the same for short codes where off-tuning by $1/T_s$ produces better results.

When MT jamming tones are used as the jamming waveform against BPSK DSSS signals, results ensue that are the same as partial band jamming discussed previously. This is based on assuming that the output of the DSSS despreader at the receiver is Gaussian noise.

The notion is illustrated in Figure 10.21. Five equal amplitude and equally spaced tones, centered on and symmetric around f_0 are shown in Figure 10.21(a). The more general case of three unequal amplitude and unequally spaced tones asymmetric around f_0 is illustrated in Figure 10.21(b). Milstein, Davidovice, and Schilling examined the effects of multiple tones on DSSS signals [44].

The development in Section 10.7.5 can be extended to the case when there are multiple tones arbitrarily spaced within the DSSS signal. Those extensions are presented in this section. Let J_n denote the power of the nth jammer tone at the receiver, Δf_n denote the frequency difference between the nth jammer tone and the DSSS carrier frequency, and θ_n denote the phase differences. Furthermore suppose there are N_J jammer tones present.

In the case of multiple tones, the jammer power is split into some number of channels, and the power per channel is therefore reduced. We will assume here that the same total power results irrespective of the number of jammed channels. Figure 10.21 illustrates the notion of multiple tone jammer channels.

For $L >> N$ and including the effects of phase offsets, it is necessary to average over $(0, 2\pi)$, for each θ_k. Thus

$$P_e = \left(\frac{1}{2\pi}\right)^{N_J} \underbrace{\int_0^{2\pi}\int_0^{2\pi}\cdots\int_0^{2\pi}}_{N_J \text{ times}} \Pr\{e|\theta_1,\theta_2,\cdots,\theta_{N_J}\}d\theta_1 d\theta_2\cdots d\theta_{N_J} \tag{10.125}$$

where $\Pr\{e|\vec{\theta}\} = \Pr\{e|\theta_1,\theta_2,\cdots,\theta_{N_J}\}$ is given by the expressions for the particular combinations of spreading method and data modulation.

10.7.7.1 Biphase Spreading with BPSK Modulation: Long Codes

For BPSK and biphase spreading the error probability is

$$\Pr\{e|\vec{\theta}\} = Q\left\{\left[\frac{1}{2N\upsilon} + \frac{1}{2N}\sum_{n=1}^{N_J}\left(\xi_n \text{sinc}^2\left(\frac{\Delta\omega_n T_b}{2N}\right) \times\left[1 + \frac{\text{sinc}(\Delta\omega_n T_b)}{\text{sinc}\left(\frac{\Delta\omega_n T_b}{N}\right)}\cos(2\theta_n)\right]\right)\right]^{-1/2}\right\} \quad (10.126)$$

where ξ_n is the JSR of the nth tone, $\Delta\omega_n = \omega_n - \omega_c$ is the frequency offset for the nth tone, $\vec{\theta} = \begin{bmatrix}\theta_1 & \theta_2 & \dots & \theta_{N_J}\end{bmatrix}$ is a vector of jammer phase offsets, and the total jamming power, which for comparison purposes, was assumed to stay constant, is given by $J = J_1 + J_2 + \ldots + J_{N_J}$. P_e can then be found by inserting (10.126) into (10.125). When these results are plotted, for N = 50 and 100 and for n = 1, n = 1, 2, and n = 1, 2, and 3, the results are equivalent. Jamming additional channels provides neither improvement nor degradation in jamming performance. As such, because of implementation complexity, n =1 should be used.

10.7.7.2 Quadriphase Spreading and BPSK Data Modulation: Long Codes

For BPSK and quadriphase spreading, $\Pr\{e|\theta_1,\theta_2,\cdots,\theta_{N_J}\}$ is independent of the phase offsets and is given by

$$P_e = Q\left\{\left[\frac{1}{2N\upsilon} + \frac{1}{2N}\sum_{n=1}^{N_J}\xi_n \text{sinc}^2\left(\frac{\Delta\omega_n T_b}{2N}\right)\right]^{-1/2}\right\} \quad (10.127)$$

10.7.7.3 Both Spreading Methods and QPSK Data Modulation: Long Codes

For both spreading methods and QPSK, the BER is given by

$$\Pr\{e|\theta_1,\theta_2,\cdots,\theta_{N_J}\} =$$

$$\frac{1}{2}Q\left\{\left[\frac{1}{2N\upsilon}+\frac{1}{2N}\sum_{n=1}^{N_\mathrm{J}}\left(\begin{array}{l}\xi_n \mathrm{sinc}^2\left(\dfrac{\Delta\omega_n T_\mathrm{b}}{2N}\right)\\ \qquad\times\left[1+\dfrac{\mathrm{sinc}(2\Delta\omega_n T_\mathrm{b})}{\mathrm{sinc}(\Delta\omega_n)}\cos(2\theta_n)\right]\end{array}\right)\right]^{-1/2}\right\}$$

$$+\frac{1}{2}Q\left\{\left[\frac{1}{2N\upsilon}-\frac{1}{2N}\sum_{n=1}^{N_\mathrm{J}}\left(\begin{array}{l}\xi_k \mathrm{sinc}^2\left(\dfrac{\Delta\omega_n T_\mathrm{b}}{2N}\right)\\ \qquad\times\left[1+\dfrac{\mathrm{sinc}(2\Delta\omega_n T_\mathrm{b})}{\mathrm{sinc}\left(\dfrac{\Delta\omega_{n_n}}{N}\right)}\cos(2\theta_n)\right]\end{array}\right)\right]^{-1/2}\right\} \quad (10.128)$$

These results are illustrated in Figure 10.32 for BPSK data modulation and biphase spreading when $N_\mathrm{J} = 1$ compared with $N_\mathrm{J} = 3$. The MT jamming performance is lower than single tone because the power per tone is 1/3 that for the single tone, although at $P_\mathrm{e} > 10^{-1}$ the differences are insignificant. The effect of averaging over the phase offset is to decrease the effectiveness of the jamming. In this example, at $\xi = 10$ dB, P_e reduces from 0.006 to 0.007 as N_J is increased from 1 to 3.

When the effects of the phase difference between the jammer tones and the target signal are ignored, the BER from (10.86), (10.111), and (10.107) all produce the same result for MT jamming given by

$$P_\mathrm{e} = Q\left\{\left[\frac{1}{2N\upsilon}+\frac{1}{2N}\sum_{n=1}^{N_\mathrm{J}}\xi_n \mathrm{sinc}^2\left(\frac{\Delta\omega_n T_\mathrm{b}}{2N}\right)\right]^{-1/2}\right\} \quad (10.129)$$

where the total jamming power at the receiver is held constant. This function is illustrated in Figure 10.33 for $\upsilon = -10$ dB, $G_\mathrm{p} = 20$ dB, and $T_\mathrm{b} = 100$ μs. For this illustration, the tones are placed at multiples of $1/T_\mathrm{b}$, and the multiples start at zero and go up. Therefore there is always a tone at zero offset, if there is a tone at $2/T_\mathrm{b}$ then there is a tone at zero and $1/T_\mathrm{b}$ as well, and so forth. So with a single tone, P_e does not reach 10^{-1} until the spreading gain is overcome. With multiple tones, however, $P_\mathrm{e} = 10^{-1}$ is reached at lower values of JSR. For example, when $N_\mathrm{J} = 10$, $P_\mathrm{e} = 10^{-1}$ is achieved at $\xi = 10$ dB, an improvement of 10 dB.

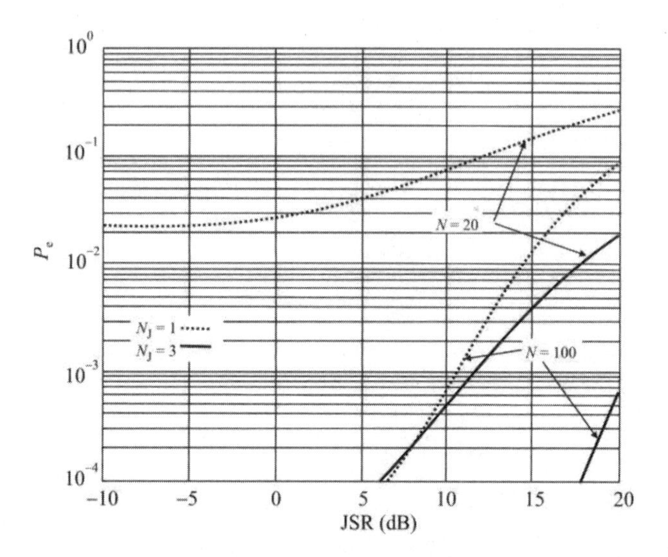

Figure 10.32 Multiple tone jamming performance against BPSK data modulation with biphase spreading for $N_J = 1$ and $N_J = 3$ when phase offset effects are included. When $N_J = 1$, the tone is at the DSSS system center frequency. When $N_J = 3$, the tones are offset by $1/T_b$, $5/T_b$, and $10/T_b$. In this example, $T_b = 100$ μs and $\upsilon = -10$ dB.

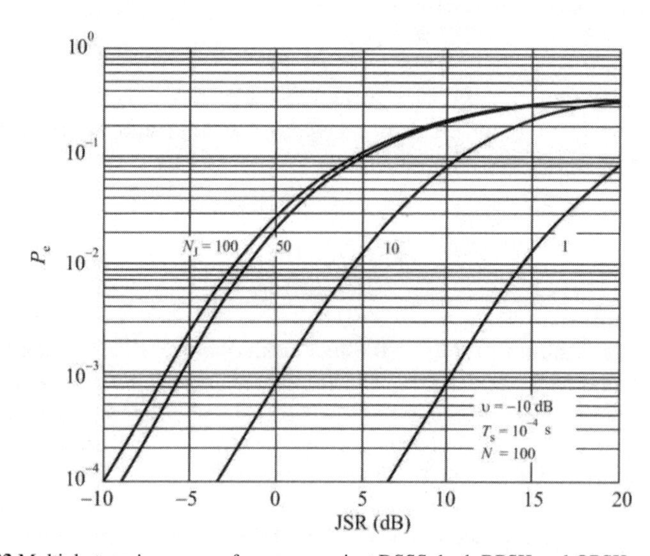

Figure 10.33 Multiple tone jammer performance against DSSS, both BPSK and QPSK as well as both spreading techniques. $\upsilon = -10$ dB, $G_p = 20$ dB, and $T_b = 100$ μs. N_J is the number of jammer tones.

10.7.7.4 Multitone Jamming of Biphase or Quadriphase Spreading and BPSK Data Modulation Long and Short Codes

Chen derived some results for BPSK spreading/BPSK data modulation and QPSK spreading/BPSK data modulation in the presence of multiple jammers [2]. The results for the two cases, as expected, are the same. In addition there was no assumption about PN code length in the development of the BER so this approach works for long as well as short codes. The only requirement is that the noise is Gaussian.

The BER is given by

$$P_{\mathrm{e}} = Q\left\{ (NN_{\mathrm{J}})^{1/2} \left\{ \xi \sum_{n=1}^{N_{\mathrm{J}}} \left[\begin{array}{l} \operatorname{sinc}^2\left(\Delta\omega_n T_{\mathrm{c}}/2\right) \\ \times \left[1 + \dfrac{\cos\left(2\phi + (N\Delta\omega_n T_{\mathrm{c}})\right)\sin(N\Delta\omega_n T_{\mathrm{c}})}{N\sin(\Delta\omega_n T_{\mathrm{c}})} \right] \end{array} \right] \right\}^{-1/2} \right\} \quad (10.130)$$

where $\xi = J/R$, J is the jammer power for each of the N_{J} tones, assumed to be the same, R is the power of the signal, $\phi = \theta_{\mathrm{r}} - \phi_n, \theta_{\mathrm{r}}$ is the phase offset of the receiver, ϕ_n is the phase offset of the nth jammer tone, and P_{N} is the noise power at the receiver (that is, $\upsilon = R/P_{\mathrm{N}}$). This model also excludes a tone at the target frequency (no offset).

Results are illustrated in Figure 10.34 for coherent jammer tones and a phase locked receiver so that $\phi = 0$ when the SNR = –11 dB and N = 128, typical 3G+ PCS values. The noise power, P_{N}, from the correlators shown in Figure 10.2 is given by

$$P_{\mathrm{N}} = \frac{N_0 A^2}{T_{\mathrm{s}}} \quad (10.131)$$

and the results shown in Figure 10.34 assume that the correlator integration gains have been normalized to $A = 1/\sqrt{R}$ so that the correlator output due to the data component is ±1.

10.7.8 Comparison of Various Strategies

Figure 10.35 shows the predicted performance of biphase PN with fixed υ = 10 dB, assuming three cases:

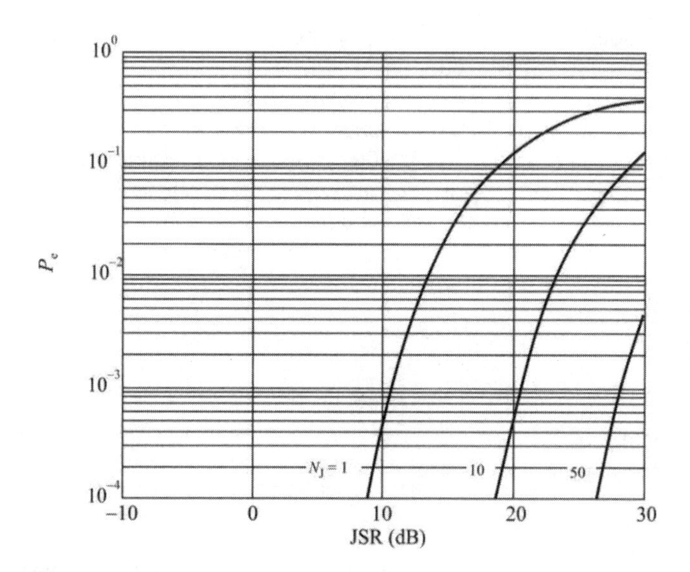

Figure 10.34 BER of BPSK BPSK. NJ is the number of jammer tones.

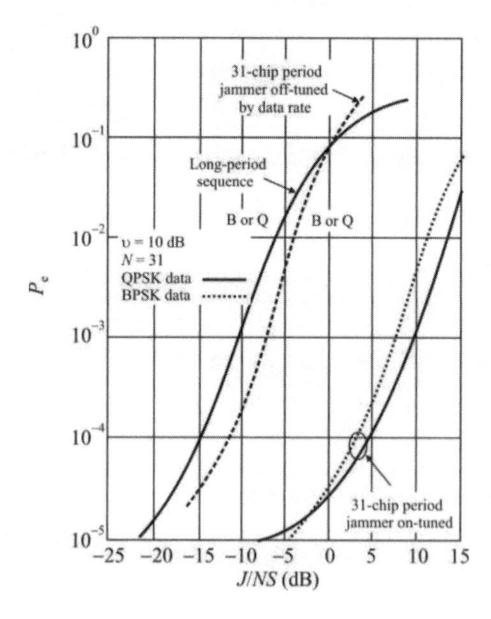

Figure 10.35 BER for biphase spreading versus total JSR. (After: [5]. © 1983 IEEE Reprinted with permission.)

- On-tune jammer, short code;
- Jammer off-tune by $1/T_s$, short code;
- Long (random) code.

In all cases, the number of chips per symbol, N, is 31. These results indicate that if the jammer is offtuned by one symbol rate, the jamming performance against short codes is much better than if the jammer were ontune, and is in fact comparable to performance with the biphase long code. This effect is due primarily to the suppression of the signal when the jammer is ontuned as illustrated in Figure 10.5.

We might draw the conclusion from Figure 10.35 that short codes are preferable to long codes where the jammer is offtuned by the bit rate. This conclusion, however, overlooks the serious vulnerability of systems employing short codes. It is reasonably straightforward for an EW system to discover a short code that is sent repeatedly. The target loses all AJ processing gain against this jamming strategy. Therefore, short codes can be (and are) used in many non-adversarial spread-spectrum systems such as ranging and CDMA systems. Long codes are necessary if vulnerability to EW is a consideration.

10.7.9 Summary

Analytical expressions for the BER performance of a DSSS system have been derived for a tone jammer with an arbitrary frequency offset. From these results it can be shown that BPSK and QPSK data provide the same performance for long period sequences. Quadriphase chip spreading provides 2–3 dB better performance than biphase chip spreading for the long period sequences. These results are generalized for MT jammers. For moderate length sequences (longer than one symbol but not much longer), the BER will depend on a particular spreading sequence and will be different for different spreading sequences because the random effects do not average out.

10.8 Concluding Remarks

The effectiveness of the principal types of jamming on DSSS signals was presented in this chapter along with some illustrative examples using typical parameters. These types include BBN jamming, PBN jamming, pulsed jamming, and tone jamming. The last of these includes both single tone jamming and MT jamming.

Only PSK modulations were discussed. That is because BPSK and QPSK are the most prolific types of modulations for DSSS systems. The latter includes offset QPSK and MSK, both of which exhibit reduced sidelobe levels in the frequency domain, which reduces the interference caused in nearby frequency channels. They do, however, increase the amount of ISI in general (which is a function of timing).

Repeat coding is ineffective against all but pulsed jamming because for jammer signals that are steadily present, the decorrelation process in a DSSS receiver converts the jammer signal to a post-detection signal that has AGWN properties and the noise floor is effectively raised by the detection process. The jammer noise is spread across all frequencies and all time so that there is no place where the communication signal can avoid the jammer signal. Therefore, for fixed signal power, the energy per bit is decreased with repeat coding thereby increasing the jammer effectiveness. For pulsed jamming, there are times when the jammer signal is not present so the possibility exists to receive a data symbol that is degraded by thermal noise only.

The length of the spreading code in the DSSS system affects the ability to jam. Short codes, which repeat themselves after one or a few data bits, are quite vulnerable to attempts to ascertain the underlying code sequence, which, in some cases such as PCS, is not a problem. In situations where the DSSS coding is used for LPI or AJ purposes, however, the vulnerability of short codes cannot be tolerated. This is true for most military applications. Long codes, which repeat themselves after many data bits, are far less vulnerable.

Whereas for PBN, placing the center of the jammed band at the center of the DSSS signal is optimum; exactly the opposite is true for MT jamming. A tone placed at the DSSS center frequency suffers severe attenuation by the decorrelation process.

References

[1] Geraniotis, E. A., "Performance of Noncoherent Direct-Sequence Spread-Spectrum Multiple-Access Communications," *IEEE Journal on Selected Areas in Communications,* Vol. SAC-3, No. 5, September 1985, pp. 687–694.

[2] Chen, G. L., and R. R. Adhami, "Performance of BPSK/BPSK and QPSK/BPSK Spread Spectrum Communication Systems in the Presence of Multiple-Tone Jammer," *Proceedings 1989 IEEE Southeastcon*, 1989, pp. 843–849.

[3] Garber, F. D., and M. B. Pursley, "Performance of Offset Quadriphase Spread Spectrum Multiple Access Communications," *IEEE Transactions on Communications*, Vol. COM-29, March 1982, pp. 305–313.

[4] Lehnert, J. S., and M. B. Pursely, "Multipath Diversity Reception of Spread Spectrum Multiple Access Communications," *IEEE Transactions on Communications*, Vol. COM-35, November 1987, pp. 1189–1199.

[5] Lunayach, R. S., "Performance of a Direct Sequence Spread-Spectrum System with Long Period and Short Period Code Sequences," *IEEE Transactions on Communications*, Vol. COM-31, No. 3, March 1983, pp. 412–419.

[6] Chen, G. L., and R. Adhami, "The Error Probability for the BPSK/BPSK and OQPSK/BPSK Spread Spectrum Communication Systems in the Presence of Single tone Jammer," *Proceedings 21st Southeastern Symposium on System Theory*, 1989, pp. 584–588.

[7] Simon, M. K., J. K. Omura, R. A. Scholtz, and B. K. Levitt, *Spread Spectrum Handbook,* New York: McGraw-Hill, 1994, pp. 414–419.

[8] Torrieri, D. J., *Principles of Secure Communication Systems*, 2nd ed., Norwood, MA: Artech House, 1992, p. 117.

[9] Pursley, M. B., D. V. Sarwate, and W. E. Stark, "Error Probability for Direct-Sequence Spread-Spectrum Multiple-Access Communications—Part I: Upper and Lower Bounds," *IEEE Transactions on Communications,* Vol. COM-30, No. 5, May 1982, pp. 975–984.

[10] Geraniotis, E. A. and M. B. Pursley, "Error Probability for Direct-Sequence Spread-Spectrum Multiple-Access Communications—Part II: Approximations," *IEEE Transactions on Communications,* Vol. COM-30, No. 5, May 1982, pp. 985–995.

[11] Borth, D. E., and M. B. Pursley, "Analysis of Direct-Sequence Spread-Spectrum Multiple-Access Communication Over Ricean Fading Channels," *IEEE Transactions on Communications,* Vol. COM-27, No. 10, October 1979, pp. 1566–1577.

[12] Milstein, L. B., and D. L. Schilling, "Performance of a Spread Spectrum Communication System Operating over a Frequency-Selective Fading Channel in the Presence of Tone Interference," *IEEE Transactions on Communications,* Vol. COM-30, No. 1, January 1982, pp. 240–247.

[13] Yoon, J., and J. F. Doherty, "A Fast Adaptive Receiver for Interference Suppression in DS-CDMA Systems," *Proceedings IEEE MILCOM*, 1991.

[14] Fang, L., and L. B. Milstein, "Performance of Successive Interference Cancellation for a Multicarrier DS/CDMA System," *Proceedings IEEE MILCOM*, 1999.

[15] Yoon, J. and J. F. Doherty, "Adaptive Interference Suppression in CSMA Systems with Large Processing Gains," *Proceedings IEEE MILCOM*, 1999.

[16] Lok, T. M., T. F. Wong, and J. S. Lehnert, "Blind Adaptive Reception for MC-CDMA Systems with Interference Suppression," *Proceedings IEEE MILCOM*, 1999.

[17] Khoshbin-Ghomash, H. and R. F. Ormondroyd, "An Adaptive Neural Network Receiver for CDMA Multi-User Interference Cancellation in Multipath Environments," *Proceedings IEEE MILCOM*, 1999.

[18] Simon, M. K., J. K. Omura, R. A. Scholtz, and B. K. Levitt, *Spread Spectrum Handbook,* New York: McGraw-Hill, 1994, p. 149.

[19] Peterson, R. L., R. E. Ziemer and D. E. Borth, *Introduction to Spread Spectrum Communications,* Upper Saddle River, NJ: Prentice Hall, 1995, p. 329.

[20] Schilling, D. L., L. B. Milstein, R. L. Pickholtz, and R. W. Brown, "Optimization of the Processing Gain of an M-ary Direct Sequence Spread Spectrum Communication System," *IEEE Transactions on Communications*, Vol. COM-28, No. 8, August 1980, pp. 1389–1398.

[21] Sharnain, P., and L. B. Milstein, "Minimum Mean Square (MMSE) Receiver Employing 16-QAM in CDMA Channel with Narrowband Gaussian Interference," *Proceedings IEEE MILCOM*, 1999.

[22] Thomas, J., and E. Geraniotis, "Joint Iterative MMSE Multiuser Detection and Narrowband Jammer Suppression in Coded DS-CDMA Channels," *Proceedings IEEE MILCOM*, 1999.

[23] Torrieri, D. J., *Principles of Secure Communication Systems*, 2nd ed., Norwood, MA: Artech House, 1992, p. 127.

[24] Peterson, R. L., R. E. Ziemer and D. E. Borth, *Introduction to Spread Spectrum Communications,* Upper Saddle River, NJ: Prentice Hall, 1995, p. 339.

[25] Gui, X. and T. S. Ng, "Performance of DS SS System Under On-Off Wideband Jamming," *Electronic Letters*, January 1997.

[26] Liaang, C-P, and W. E. Stark, "Turbo Codes in DS-SS with Adaptive Nonlinear Suppression of Impulsive Interference," *Proceedings IEEE MILCOM*, 1999.

[27] Peterson, R. L., R. E. Ziemer and D. E. Borth, *Introduction to Spread Spectrum Communications,* Upper Saddle River, NJ: Prentice Hall, 1995, p. 351.

[28] Simon, M. K., J. K. Omura, R. A. Scholtz, and B. K. Levitt, *Spread Spectrum Handbook,* New York: McGraw-Hill, 1994, p. 157.

[29] Simon, M. K., J. K. Omura, R. A. Scholtz, and B. K. Levitt, *Spread Spectrum Handbook,* New York: McGraw-Hill, 1994, p. 160.

[30] Simon, M. K., J. K. Omura, R. A. Scholtz, and B. K. Levitt, *Spread Spectrum Handbook,* New York: McGraw-Hill, 1994, p. 165.

[31] Simon, M. K., J. K. Omura, R. A. Scholtz, and B. K. Levitt, *Spread Spectrum Handbook,* New York: McGraw-Hill, 1994, p. 161.

[32] Ketchum, J. W., and J. G. Proakis, "Adaptive Algorithms for Estimating and Suppressing Narrow-Band Interference in PN Spread-Spectrum Systems," *IEEE Transactions on Communications,* Vol. COM-30, No. 5, May 1982, pp. 913–924.

[33] Li, L. M., and L. B. Milstein, "Rejection of Narrow-Band Interference in PN Spread-Spectrum Systems Using Transversal Filters," *IEEE Transactions on Communications,* Vol. COM-30, No. 5, May 1982, pp. 925–929.

[34] Milstein, L. B., and P. K. Das, "An Analysis of a Real-Time Transform Domain Filtering Digital Communication System—Part I: Narrow-Band Interference Rejection," *IEEE Transactions on Communications*, Vol. COM-28, No. 6, June 1980, pp. 816–824.

[35] Milstein, L. B., "Interference Rejection Techniques in Spread Spectrum Communications," *Proceedings of the IEEE,* Vol. 76, No. 6, June 1988, pp. 657–671.

[36] Gao, X., C.-S. Li, and X-R Lai, "Effect of Tracking Error on DS-CDMA Partial Parallel Interference Cancellation," *Proceedings of 2000 IEEE Wireless Communication and Networking Conference*, 2000.

[37] Zhang, Y., and J. Dill, "An Anti-Jamming Algorithm Using Wavelet Packet Modulated Spread Spectrum," *Proceedings IEEE MILCOM*, 1991, http://www.greenhouse.com/society/TacCom/milcom_99_papers.html.

[38] Poor, H. V., and X. Wang, "Code-Aided Adaptive Narrowband Interference Suppression for Direct-Sequence Spread-Spectrum Communications," *Proceedings IEEE MILCOM*, 1997.

[39] Comley, V. E., "C.W. Interference Excision in a DS/SS Communication System Using Spectrally Defined Spreading/Despreading Functions," *IEEE MILCOM Proceedings*, 1998, http://www.greenhouse.com/society/TacCom/milcom_98_papers.html.

[40] Singh, R. "Blind Interference Suppression for DS-CDMA," *Proceedings IEEE MILCOM* 1999.

[41] Scholtz, R. A., "The Spread Spectrum Concept," *IEEE Transactions on Communications*, Vol. COM-25, August 1977, pp. 748–755.

[42] Torrieri, D. J., *Principles of Secure Communication Systems,* 2nd ed., Norwood, MA: Artech House, 1992, pp. 123–124.

[43] Gradshteyn, I. S., and I. W. Ryzhik, *Tables of Integrals Series and Products*, New York: Academic Press, 1965.

[44] Milstein, L. B., S. Davidovice, and D. L. Schilling, "The Effect of Multiple-Tone Interference on a Direct Sequence Spread Spectrum Communication System," *IEEE Transactions on Communications*, Vol. COM-30, March 1982, pp. 346–445.

[45] Lau, F. C. M., M. Ye, C. K. Tse, and S. F. Hau, "Anti-Jam Performance of Chaotic Digital Communication Systems," *IEEE Transactions on Circuits and Systems–I: Fundamental Theory and Applications*, Vol. 49, No. 10, October 2002, pp. 1486–1494.

Chapter 11

Electronic Warfare and Fast Frequency Hopping Systems

11.1 Introduction

FFHSS systems, where there is one or more hop dwells per data bit, were devised as a countermeasure to follower ES and EA systems. If the hopping is fast enough, depending on the physical relationship of the communication transmitter and receiver relative to the placement of the jammer, the signal will have moved onto the next hop frequency before the jamming signal arrives at the receiver, thereby making following jamming ineffective. This chapter presents the performance of follower jamming, BBN jamming, PBN jamming, and MT jamming of FFHSS systems, while Chapter 12 presents jamming performance for SFHSS systems with these same techniques [1–6].

In general, the type of jammer that performs best for SFHSS targets is different from that used against fast frequency hopping targets. Whereas following jamming can be rendered ineffective by hopping faster, follower jamming can be very effective against SFHSS.

In this analysis it is assumed that the jammer knows all of the pertinent target parameters except the hopping code. Thus, the hopping band, the channel information such as bandwidth, center frequencies, the location of the bandpass filters relative to the channel center frequencies, the bit rate, the symbol rate, and any coding that is applied, are all known parameters. Most of the time, several, but not all, of these parameters will be known. What is not known must be measured from the signals received by the associated ES system.

This chapter is structured as follows. We begin by discussing the channel structure for FFHSS signals. That is followed by a brief review of the receiver architecture used for intercept of FFHSS systems and what the implications are for intercepting multiple hops per data bit. We then cover the performance of BBN jamming of these signals. Follower jamming performance is covered next,

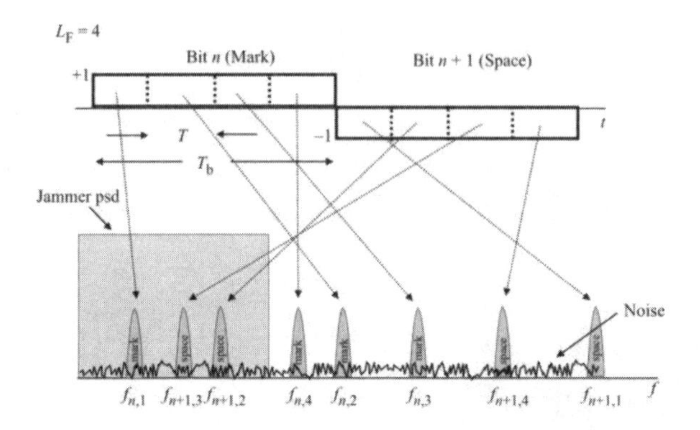

Figure 11.1 Channel structure for fast frequency hopping.

followed by the performance analysis of PBN jamming. Next we discuss tone jamming performance. The chapter concludes with a brief review of the performance of pulsed jamming against FFHSS systems.

11.2 Channel Structure

The channel structure with the jammer present is shown in Figure 11.1. Depending on the jamming strategy, the shaded region would change, of course. Figure 11.1 illustrates partial-band contiguous jamming. In this chapter, a *channel* is defined as the portion of the spectrum containing a single MFSK data tone. This is not the normal definition, where a channel is defined as the portion of the spectrum within which the entire modulated signal is located. The individual channel structure is shown in Figure 11.2 for BFSK. The two tone frequencies representing the mark and space are normally located in the center of the bands symmetrically on either

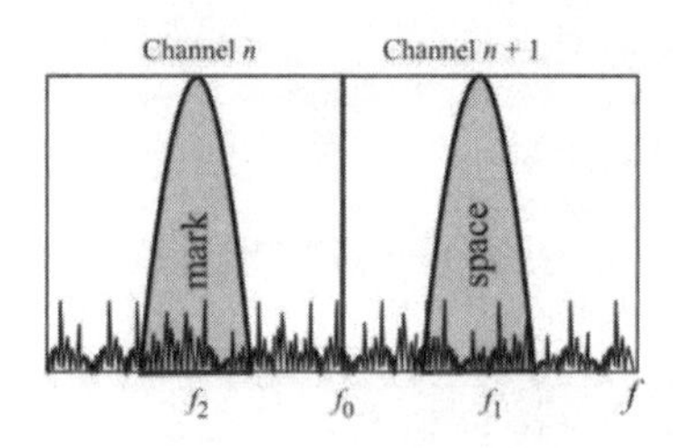

Figure 11.2 Detailed channel structure for a contiguous channel spaced BFSK FFHSS signal.

side of the frequency f_0. We will assume such a channel structure herein, although in practice the two channels need not be adjacent (although they normally are).

Only noncoherent FFHSS systems will be considered herein. This is the most prolific form of FFHSS system. Whereas it is not difficult to build SFHSS systems that are coherent within a dwell, it is difficult to maintain coherency as the synthesizers in the transmitter and receiver change over large frequency spans for either SFHSS or FFHSS systems. It is possible to construct coherent FFHSS systems, however. For those interested in how effective jammers can be against coherent systems, [7–13] should be consulted.

Another area that is not addressed herein are the effects of fading on FFHSS systems. Fading impacts the BER of these systems, however [14–18]. Such fading is normally frequency dependent and as such is somewhat mitigated by FFHSS systems since multiple frequencies are used for each data bit.

There are two practical waveforms to use for follower jamming FFHSS systems: narrowband noise and tones. In either case, the detected tone channel could be jammed, the complementary channel could be jammed, or both could be jammed. As will be shown, when using either narrowband noise or tones above $\xi \approx 0$ dB, the complementary channel must be jammed because placing additional energy only at the detected tone channel enhances the receiver's ability to properly detect the signal.

There are two particular effects that must be considered with FFHSS systems. The first is the effects of multiple dwells per data bit. Simply causing one dwell to be received in error does not ensure that the data bit will be detected in error since there are other dwells for that data bit. The second consideration, not unique to FFHSS but still noteworthy, is the effect of only jamming parts of dwells, either individually or parts of the set corresponding to a data bit. This latter condition can arise in three cases:

- In follower jamming when the signal must first be detected in order to be jammed;
- In partial band noise jamming when only a portion of the spectrum is covered by the jammer at any given time;
- In partial band MT jamming when multiple tones are placed in the spectrum but not every data tone frequency contains a jamming signal.

11.3 Receiver Architecture

For simplicity, this analysis is restricted to BFSK systems. The receiver structure for noncoherent FFHSS BFSK considered here is shown in Figure 11.3. This is a hard decision receiver in that statistics about the detection process are not used in

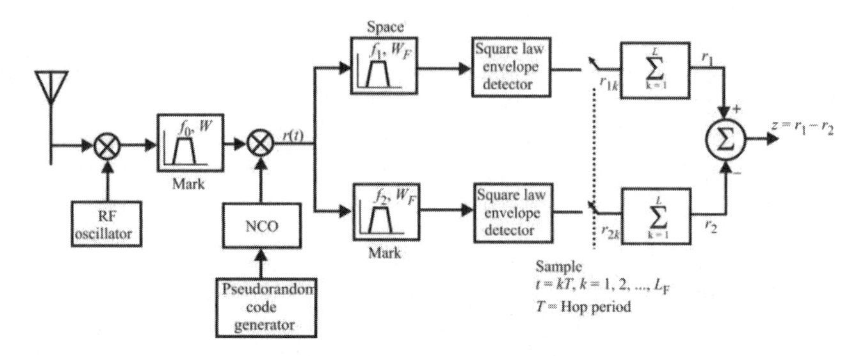

Figure 11.3 FFHSS receiver and noncoherent BFSK detector.

the decision as to which symbol was sent. It is also possible to consider soft decision decoding where the detection statistics are taken into account [19–28]. The incoming signal somewhere in bandwidth W_{ss} is first frequency converted to some convenient center frequency and then filtered by a bandpass filter of width W Hz. These operations have no effect on the analysis. Next, the signal is mixed with a local oscillator that is tuned to the correct center frequency where the signal is located. It is assumed that the local pseudo-random code generator and time epochs are in sync with those at the transmitter. Offset from that frequency are the two tone frequencies f_1 and f_2, the former representing a space and the latter a mark. Passing through both bandpass filters is noise and possibly a jamming signal. Passing through one of the filters is also the signal at that instant. The filter outputs are then detected with a square law device and then sampled forming the sampled signals r_{1k} and r_{2k}, $k = 1, 2, \ldots, L_F$. These statistical variables are summed from one to L_F. The mark sum is subtracted from the space sum forming test statistic z. If z if less than zero a mark is declared and if z is greater than zero a space is declared.

11.4 Multiple Dwells Per Data Bit

For FFHSS, there are L_F hops per data bit and the probability of receiving a data bit in error is determined by receiving over half of the individual dwells in error. Therefore, there will be a data bit detection error if $\lfloor L_F/2 \rfloor + 1$ or more individual dwells are detected in error, where $\lfloor \bullet \rfloor$ denotes the integer part of the argument.

A majority of the dwells need to be detected correctly in order to correctly detect the data bit. Because of the memoryless channel assumption, the probability of correctly detecting a majority of dwells is given by

$$P_{\rm D} = \sum_{k=\left\lfloor \frac{L_{\rm F}}{2} \right\rfloor+1}^{L_{\rm F}} \binom{L_{\rm F}}{k} (1-P_{\rm d})^{L_{\rm F}-k} P_{\rm d}^k \tag{11.1}$$

where $P_{\rm d}$ is the probability of correctly detecting the data bit on the kth dwell. This expression is due to the fact that there are

$$\binom{A}{a} = \frac{A!}{a!(A-a)!} \tag{11.2}$$

ways of taking a items from a set of A things.

As a specific example, suppose $L_{\rm F}$ = 5 hops per data bit. Then a correct detection will occur for any of the following conditions:

d_1, d_2, d_3
d_1, d_2, d_4
d_1, d_2, d_5
d_1, d_3, d_4
d_1, d_3, d_5
d_1, d_4, d_5
d_2, d_3, d_4
d_2, d_3, d_5
d_2, d_4, d_5
d_3, d_4, d_5
d_1, d_2, d_3, d_4
d_1, d_2, d_3, d_5
d_1, d_2, d_4, d_5
d_1, d_3, d_4, d_5
d_2, d_3, d_4, d_5
d_1, d_2, d_3, d_4, d_5

where d_k refers to a correct detection at the kth dwell. Hence, there are 16 ways to correctly detect a data bit: 10 with 3 detections, 5 with 4 detections, and one with 5 detections. Computing (11.1), we find that $P_{\rm D} = 10P_{\rm d}^3 + 5P_{\rm d}^4 + 1P_{\rm d}^5$.

11.5 BBN Jamming of FFHSS Systems

Like DSSS, in BBN jamming against FFHSS systems, the background noise in all channels is raised by the level of the jammer noise power density. The communication performance, and therefore the effectiveness of the jammer, can be determined in the same way as for noise only, with the jammer power added to the noise power.

The analysis of the effectiveness follows that in Section 3.11 with the noise level adjusted to be $N_T = N_0 + J_0$, where J_0 is the amount of noise jamming power per Hertz added by the jammer. This is true for both coded and uncoded signals.

In this case, $\gamma = 1$. The probability of correctly detecting a data bit per dwell is given by

$$P_d = 1 - P_{e_k} \tag{11.3}$$

and the probability of a missed detection on a dwell is

$$P_{e_k} = \frac{1}{2}\exp\left(-\frac{1}{2}\frac{R}{P_t + J}\right) \tag{11.4}$$

so

$$P_D = \sum_{k=\left\lfloor \frac{L_F}{2} \right\rfloor + 1}^{L_F} \binom{L_F}{k} \left[\frac{1}{2}\exp\left(-\frac{1}{2}\frac{R}{P_N + J}\right)\right]^{L_F - k} \left[1 - \frac{1}{2}\exp\left(-\frac{1}{2}\frac{R}{P_N + J}\right)\right]^k \tag{11.5}$$

and the overall BER is given by

$$P_e = 1 - P_D \tag{11.6}$$

This jamming performance for BBN jamming is illustrated in Figures 11.4 and 11.5 for $L_F = 1$ and $L_F = 3$, respectively. In the former, $P_e > 10^{-1}$ for $\xi > -5$ dB when $\upsilon = 20$ dB and $\xi > -7$ dB when $\upsilon = 10$ dB. For lower values of the JSR the difference in SNR becomes more apparent. When $L_F = 3$, the corresponding JSR is $\xi = -3$ dB for $\upsilon = 20$ dB and $\xi = -4$ dB for $\upsilon = 10$ dB, reflecting the jamming degradation caused by the frequency diversity.

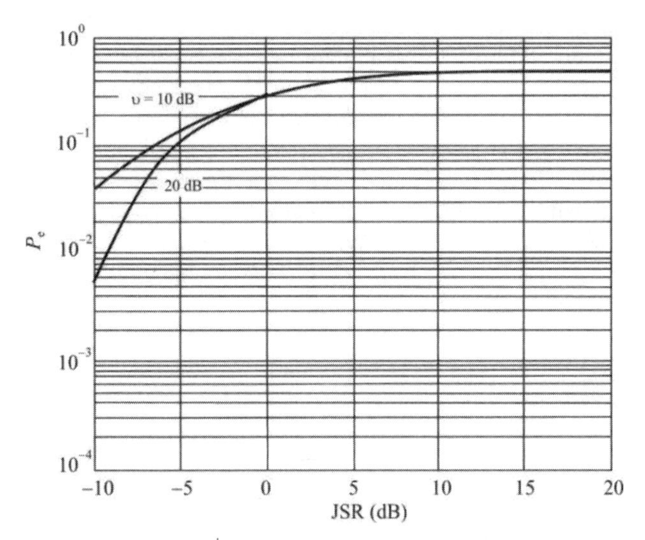

Figure 11.4 BBN jamming performance for noncoherent BFSK when $L_F = 1$.

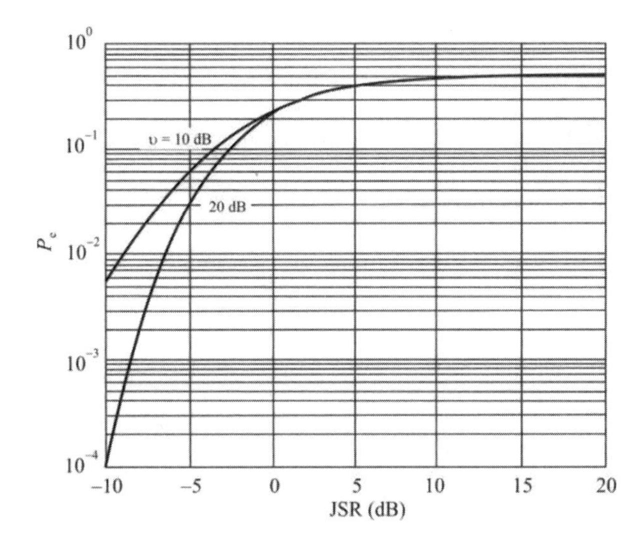

Figure 11.5 BBN jamming performance with BFSK when $L_F = 3$.

Milstein et al. analyzed the effects of BBN jamming on FFHSS systems when these FHSS systems employed representative coding [8]. Their results indicate that for $N_J = 100$ frequency channels out of $N_c = 1{,}000$ covered by the jammer and with $\xi = 10$ dB, a (23, 12) Golay code when $N = 521$ of the 1,000 channels were used for hopping degrades jamming performance by about 3 dB, and a (7, 4) Hamming code when $N = 571$ channels degrades jamming performance by about 2 dB at this same BER. Increasing the JSR to 30 dB totally negated the effects of coding and the BER was maintained above 10^{-1}. When $\xi = 30$ dB the codes were beyond the point where error extension sets in for these codes. Achieving these levels of JSR, however, is difficult when the receiver employs limiters prior to detection.

11.6 Follower Jamming of FFHSS Systems

It is possible to employ a follower jammer against FFHSS systems. The necessary speed to perform the sorting and tracking exacerbates the problem, however. For any scenario, it is possible to determine a hop rate for which follower jamming cannot be effective because the transmitter (and receiver) have moved to the next frequency before the jammer can put energy into the receiver [29–32].

Follower jamming performance equations were presented in Chapter 4. Those results are used in this section to examine the expected performance of jamming techniques against FFHSS targets.

The question naturally arises whether it is better to jam one, two, or three channels. Furthermore, if only one frequency is jammed, which should it be: the detected channel or one of those adjacent to it? If two are jammed, which two: the detected channel along with an adjacent channel or the two adjacent channels? If all three are jammed then it is clear which channels are jammed. The lower the number of channels jammed, the more power can be delivered in the jammed channels while for more jammed channels, the power per channel is less. Figure 11.6 shows a comparison of one channel versus three based on

$$P_{e_k,\text{1channel}} = (1-\gamma)\frac{1}{2}\exp\left(-\frac{1}{2}\upsilon\right) + \gamma\frac{1}{2}\frac{P_N + J}{2P_N + J}\exp\left(-\frac{1}{\frac{2}{\upsilon}+\xi}\right) \tag{11.7}$$

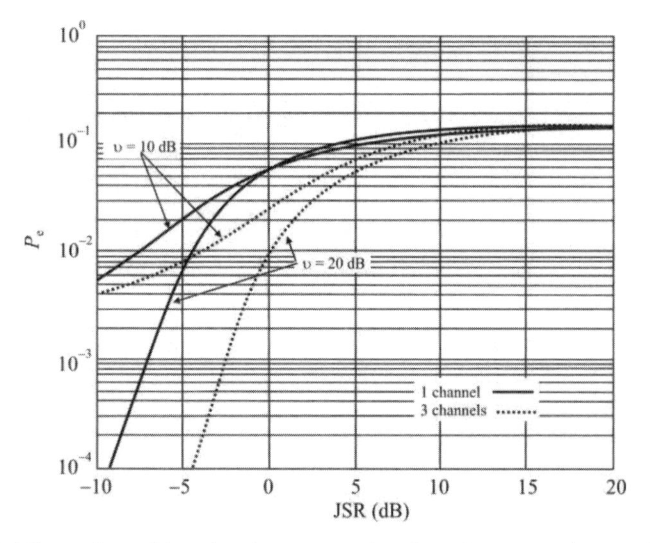

Figure 11.6 Comparison of jamming three consecutive channels versus only one adjacent channel when $\gamma = 0.3$ and $L_F = 1$.

$$P_{e_k,3\text{channel}} = (1-\gamma)\frac{1}{2}\exp\left(-\frac{1}{2}\upsilon\right) + \gamma\frac{1}{2}\exp\left(-\frac{1}{\dfrac{2}{\upsilon}+\dfrac{\xi}{3}}\right) \tag{11.8}$$

$$\begin{aligned} P_{b1} &= 1 - P_{e1,k} \\ P_{b3} &= 1 - P_{e3,k} \end{aligned} \tag{11.9}$$

and $P_{D,1}$ is given by (11.1) with $P_d = P_{d,1}$ while $P_{D,3}$ is given by (11.1) with $P_d = P_{D,3}$. Finally,

$$\begin{aligned} P_{e,1} &= 1 - P_{D,1} \\ P_{e,3} &= 1 - P_{D,3} \end{aligned} \tag{11.10}$$

Recall that γ is the fraction of a dwell that is jammed by the follower jammer. From Figure 11.6 it is clear that in this case jamming one channel is better, even though the probability of hitting the correct channel is less, although for $P_e > 10^{-1}$

Table 11.1 Bit Error Probabilities for NBN/Follower Jamming

Jammer Energy Location	Bit Error Per Dwell (Equation)
Data channel only	(8.39)
Complementary channel only	(8.41)
Two channels	(8.43)
Two adjacent channels	(8.45)
Three channels	(8.47)

the differences are not large. These results are illustrative; however, they are relatively insensitive to variations in the SNR.

11.6.1 Follower/NBN Jamming

The follower/NBN jamming performance is based on (11.1), (11.3), and (11.6) with the probability of a data bit error on each dwell as given in Table 11.1. Figure 11.7 shows comparisons of the jamming performance when υ = 10 dB, $\gamma = 0.1$, and $L_F = 1$. P_{data} refers to jamming only the data channel, P_{comp} refers to jamming a single adjacent channel, which may or may not be the complementary channel, P_2 refers to jamming 2 channels, one of which is the data channel, P_{adj} refers to jamming the 2 channels adjacent to the data channel, and P_3 refers to jamming all three channels. For the values of SNR illustrated and only 10% of the hop jammed, the BER does not reach the goal of 10^{-1} for any jamming method. Jamming a single adjacent channel (the complementary channel half the time) produced the best jamming results, leveling to BER of about 5×10^{-2} at about ξ = 10 dB.

When the fraction of hop jammed is increased to $\gamma = 0.3$, the results in Figure 11.8 ensue. Again, it is illustrated that jamming the data channel alone is not effective at $\xi > 0$ dB, but jamming the complementary channel achieves an acceptable BER at $\xi \approx 4\text{dB}$. The other methods level to $P_e \approx 7 \times 10^{-2}$ at $\xi \approx 10\text{dB}$.The effects of increasing the number of hop dwells per data bit are illustrated next. The results are shown in Figure 11.9 for when the jammed fraction $\gamma = 0.1$ and $L_F = 3$. Within the ranges considered, only the complementary channel jammer produced errors approaching 10^{-2}. Clearly, increasing the diversity is an effective countermeasure to EA.

When the jammed fraction $\gamma = 0.3$, the results improved somewhat as shown in Figure 11.10. The BER never rises to 10^{-1} but levels to about 6×10^{-2} at $\xi \approx 15$ dB for complementary channel jamming. The other jamming approaches barely

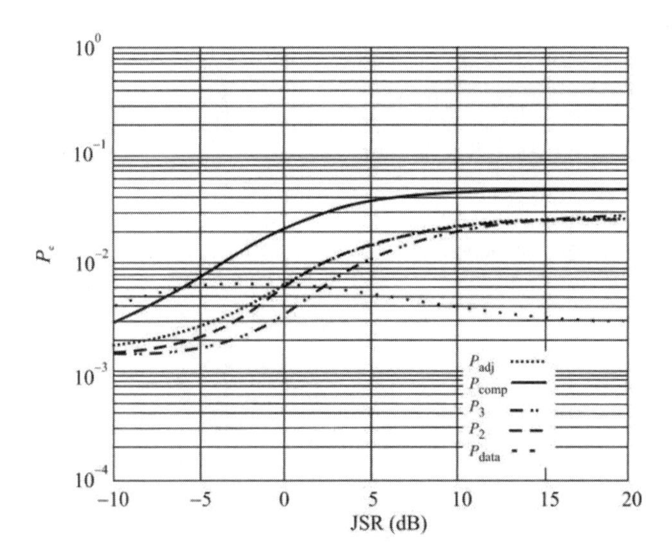

Figure 11.7 Follower jamming performance with NBN jamming when υ = 10 dB, γ = 0.1, and L_F = 1.

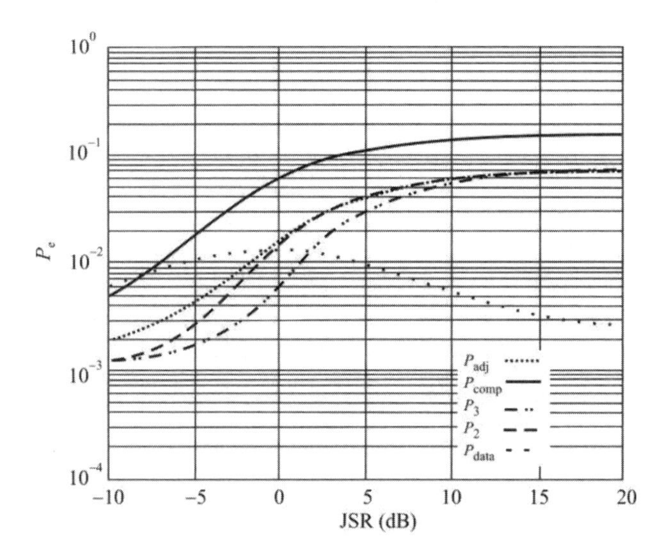

Figure 11.8 Follower jammer NBN performance when υ = 10 dB, γ = 0.3, and L_F = 1 (one hop per bit).

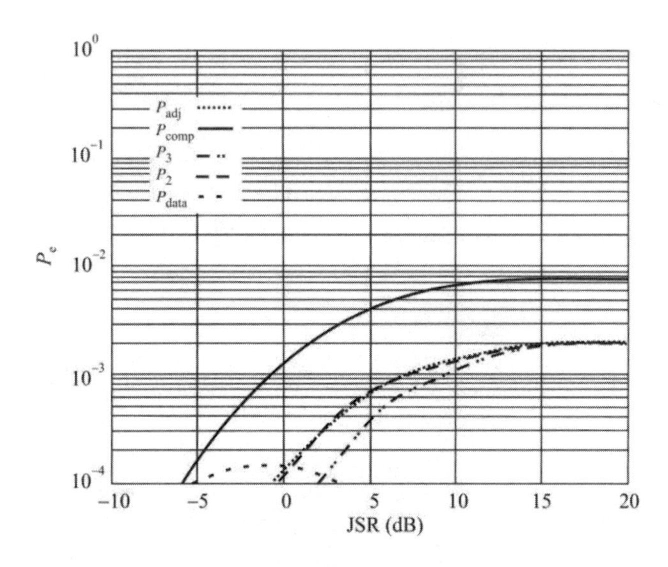

Figure 11.9 Follower jammer performance with NBN when υ = 10 dB, γ = 0.1 and $L_F = 3$.

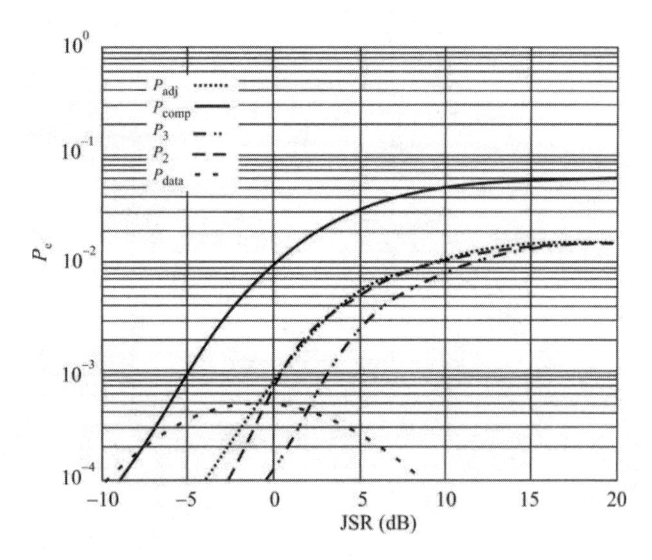

Figure 11.10 Follower jamming performance with NBN when υ = 10 dB, γ = 0.3 and $L_F = 3$.

produce $P_e > 10^{-2}$. For more reasonable level of JSR (around 0 dB), no method procduces $P_e > 10^{-2}$.

When the SNR is increased to 20 dB, for $\gamma = 0.1$ and $L_F = 1$, the calculated performance is as shown in Figure 11.11. Comparing this with Figure 11.7, at lower jamming levels the BER decrease. Above about $\xi \approx 0$ dB, the BER is not very sensitive to the SNR.

Comparing Figure 11.12 with Figure 11.8, the same conclusions can be reached. The jamming performance is not very dependent on the SNR above $\xi \approx 0$ dB or so. Above $\upsilon \approx 5$ dB, the BER rises above 10^{-1} for jamming the complementary channel, and the other methods approach 10^{-1} for large levels of jamming.

Figure 11.13 shows the jamming performance when the signal $\upsilon = 20$ dB, $\gamma = 0.1$, and $L_F = 3$. The larger number of dwells per data bit decreases the jammer performance by almost an order of magnitude, keeping the data throughput above 2×10^{-2}. That performance loss is regained, however, if the dwell fraction is raised to 0.3 as illustrated in Figure 11.14. None of the jamming techniques produces an adequate BER, however.

The energy detector at the receiver makes a judgment as to which tone was sent depending on the amount of energy emerging from the filters. Placing noise energy in the data channel without placing it in the complementary channel still effectively increases the energy in that channel, thereby enhancing the receiver's ability to correctly determine the tone that was sent and, therefore, data channel jamming is ineffective.

Therefore, one conclusion for narrowband noise jamming is that jamming one of the adjacent channels (either one, one of which will be the complementary channel half the time) produces the highest BER in most cases. Also, it is important to jam as much of the hop as possible, especially if the FFHSS system employs diversity. Above a JSR around zero, the BFSK signal performance with jamming present is not very sensitive to the SNR level.

11.6.2 Follower/Tone Jamming

The expressions for the BER with tone jamming are given by (11.1), (11.3), and (11.6) with the error rate per dwell as specified in Table 11.2. Comparisons of follower jamming approaches with tone jamming are shown in Figures 11.15 through 11.23. When $\upsilon = 10$ dB, $\gamma = 0.1$, and $L_F = 1$ dwell per data bit, the results are shown in Figure 11.15. At lower values of JSR, the strategy of jamming both the data channel and one of the complementary channels is best. Even jamming the data channel alone is a reasonable approach. But as the JSR increases past about 1 dB, jamming one of the adjacent channels produces the best jamming

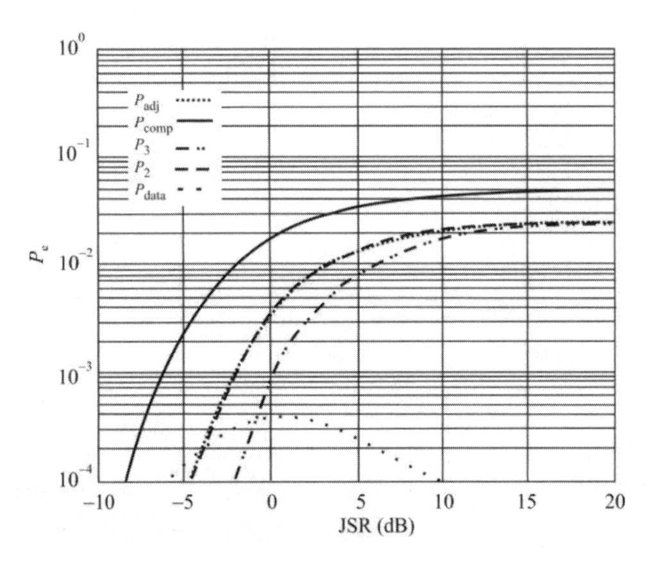

Figure 11.11 Follower NBN performance when $\upsilon = 20$ dB, $\gamma = 0.1$, and $L_F = 1$.

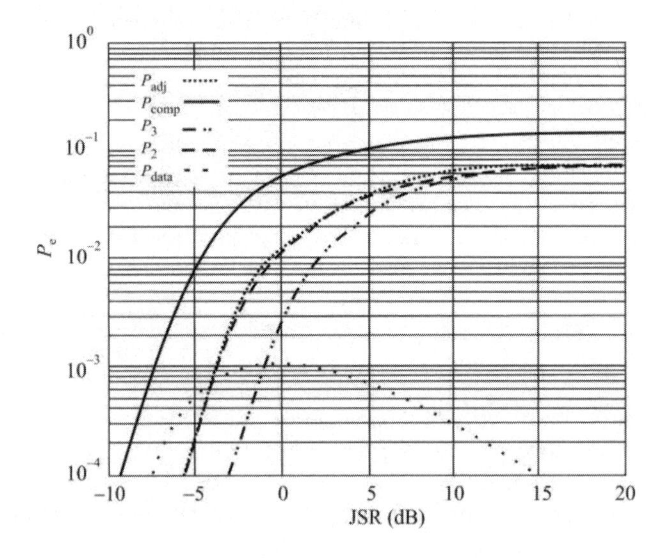

Figure 11.12 Follower performance with NBN when $\upsilon = 20$ dB, $\gamma = 0.3$, and $L_F = 1$.

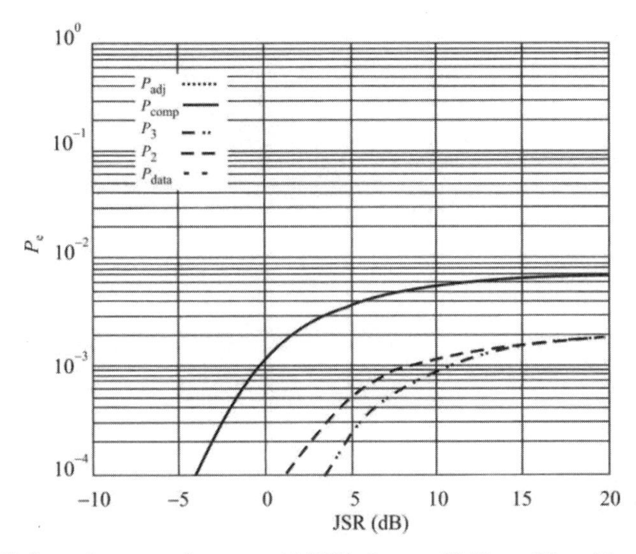

Figure 11.13 Follower jammer performance with NBN when $\upsilon = 20$ dB, $\gamma = 0.1$, and $L_F = 3$.

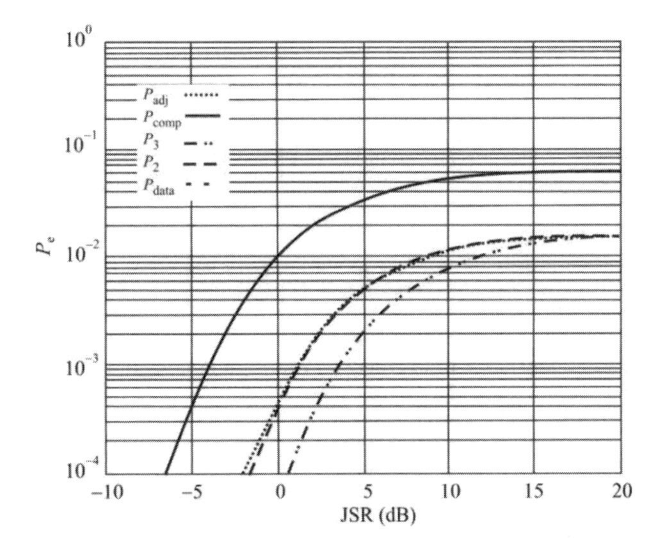

Figure 11.14 Follower NBN performance when $\upsilon = 20$ dB, $\gamma = 0.3$ and $L_F = 3$.

Table 11.2 Bit Error Probabilities for Tone/Follower Jamming

Jammer Tone Location	Bit Error Per Dwell (Equation)
Complementary channel only	(8.2) with P_{e1} per (8.1) and P_{e2} per (8.49)
Data and complementary channels	(8.2) with P_{e1} per (8.1) and P_{e2} per (8.51)
Two adjacent channels	(8.2) with P_{e1} per (8.1) and P_{e2} per (8.53)
Three channels	(8.2) with P_{e1} per (8.1) and P_{e2} per (8.55)

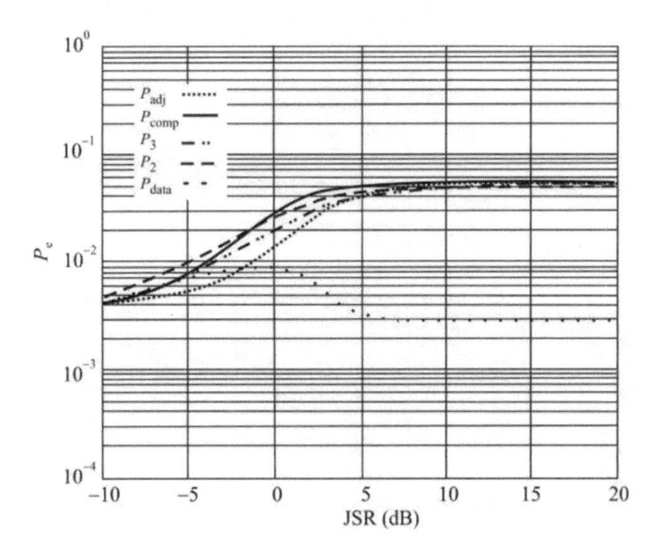

Figure 11.15 Comparison of the jamming strategies for follower jamming with tone jamming waveforms, $\upsilon = 10$ dB, $\gamma = 0.1$, $L_F = 1$.

performance. The effects of jamming just the data channel are pronounced in these results and should be avoided unless it is known that the JSR is weak, under which circumstance that strategy produces reasonable results.

The results for the same SNR and L_F but with the jammed fraction increased to 0.3 are shown in Figure 11.16. Again, adequate performance starts at approximately 1 dB JSR for the best strategies of those considered. For larger JSR, jamming one of the two adjacent channels produces the highest BER. Above $\xi \approx 3$ dB, P_e is higher than 10^{-1} for all strategies except jamming the data channel only. At high JSR, any of the strategies would be adequate for data signals. At low JSR, even jamming the data channel produces positive jamming results.

When the SNR is 10 dB, the jamming fraction is 10%, and the number of hops per data bit is 3, considerable jamming degradation occurs as shown in Figure 11.17. A BER of 1% is about all that any of the strategies could accomplish, with most of them significantly worse than that. This is further evidence that follower jamming against FFHSS systems when $L_F > 1$ is largely unproductive when the fraction of dwell jammed is low. Jamming performance improves considerably when the fraction increases, however, as shown in Figure 11.18. Even at $L_F = 3$, the BER increases beyond 10^{-1} when jamming the adjacent channels at JSR above about 5 dB.

The results for when $\upsilon = 20$ dB, $\gamma = 0.1$, and $L_F = 1$ are shown in Figure 11.19. Even with no diversity the BER never rises to 10^{-1} for any jamming strategy. Above $\xi \approx 0$ to 5 dB, all viable approaches settle to $P_e \approx 4 \times 10^{-2}$. For the same SNR, increasing the spectrum fraction to 0.3 significantly improved the jamming performance as illustrated in Figure 11.20. In that case, around 0 dB, jamming the complementary channel was effective.

As expected, when diversity was added to the signal the jamming performance degrades as shown in Figure 11.21. The BER of 10^{-1} was not achieved with any of the jamming approaches. Much of the jamming performance returned, however, when $\gamma = 0.3$, with the adjacent channel performance rising above 10^{-1} around 3 dB as seen in Figure 11.22.

11.6.3 Summary for Follower Jamming

Comparing Figures 11.7 through 11.14 for NBN jamming with corresponding Figures 11.15 through 11.22, it is evident that, at least theoretically, when jamming one of the two adjacent channels, the performance on NBN and tone jamming are about the same. For the other approaches to follower jamming, using tones is somewhat more effective for like parameters. This effectiveness manifests in the jammer's ability to produce a higher BER with a specific JSR.

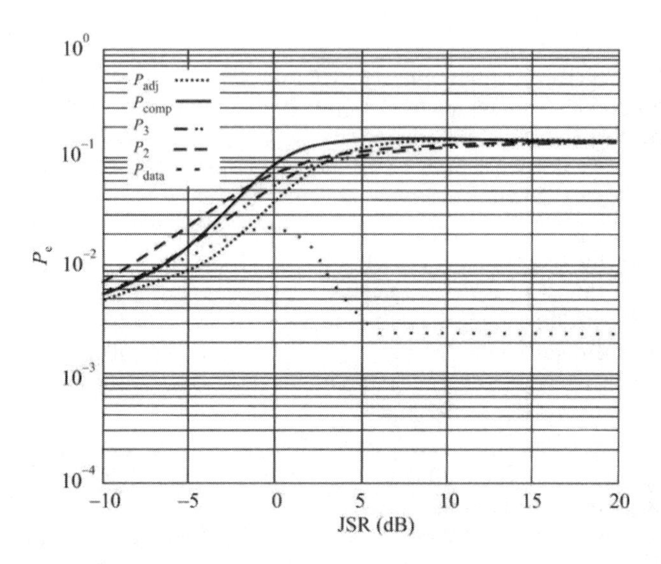

Figure 11.16 Comparison of the jamming strategies for follower jamming with tone jamming waveforms, $\upsilon = 10$ dB, $\gamma = 0.3$, $L_F = 1$.

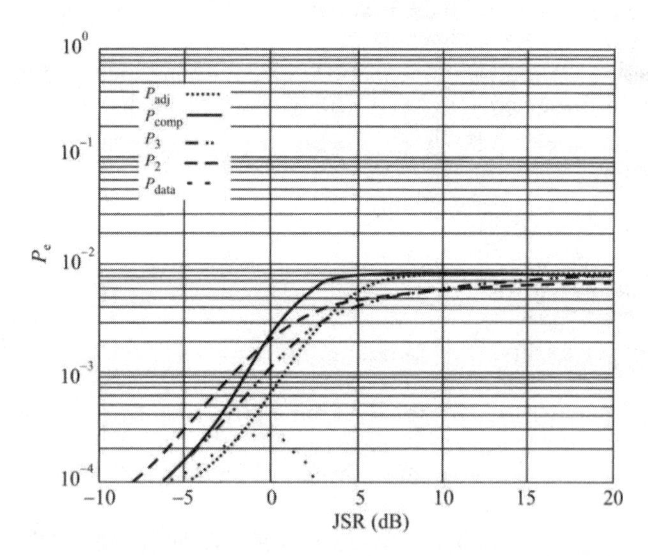

Figure 11.17 Comparison of the jamming strategies for follower jamming with tone jamming waveforms, $\upsilon = 10$ dB, $\gamma = 0.1$, $L_F = 3$.

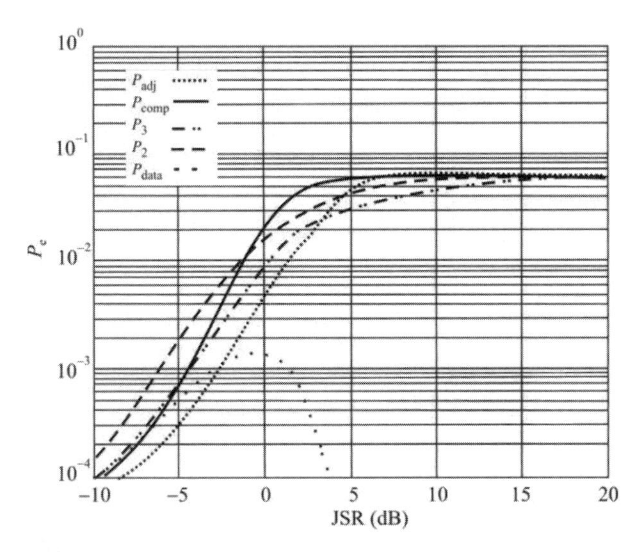

Figure 11.18 Comparison of the jamming strategies for follower jamming with tone jamming waveforms, $\upsilon = 10$ dB, $\gamma = 0.3$, $L_F = 3$.

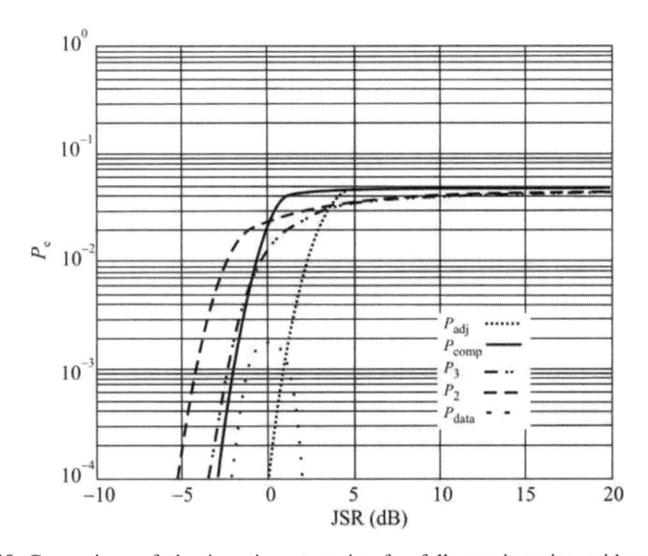

Figure 11.19 Comparison of the jamming strategies for follower jamming with tone jamming waveforms, $\upsilon = 20$ dB, $\gamma = 0.1$, $L_F = 1$.

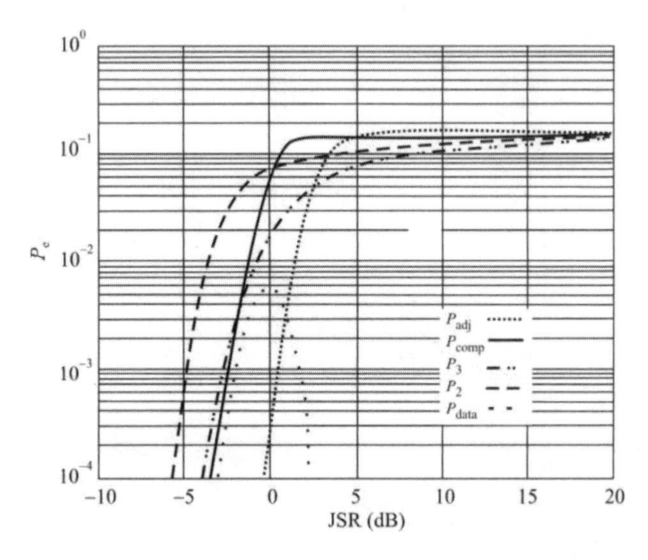

Figure 11.20 Comparison of the jamming strategies for follower jamming with tone jamming waveforms, $\upsilon = 20$ dB, $\gamma = 0.3$, $L_F = 1$.

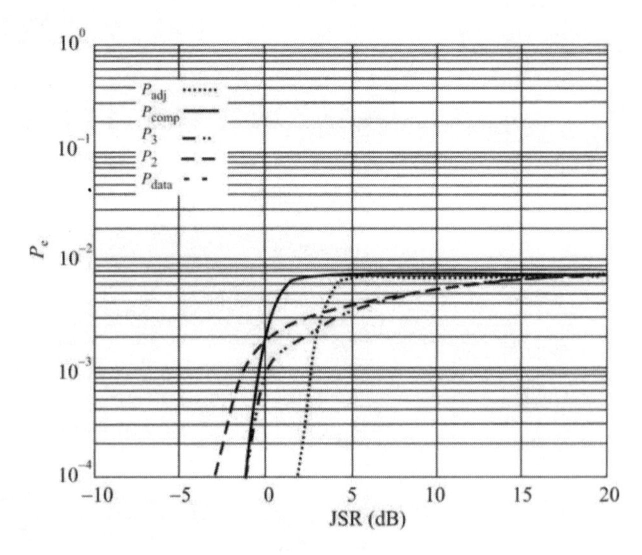

Figure 11.21 Comparison of the jamming strategies for follower jamming with tone jamming waveforms, $\upsilon = 20$ dB, $\gamma = 0.1$, $L_F = 3$.

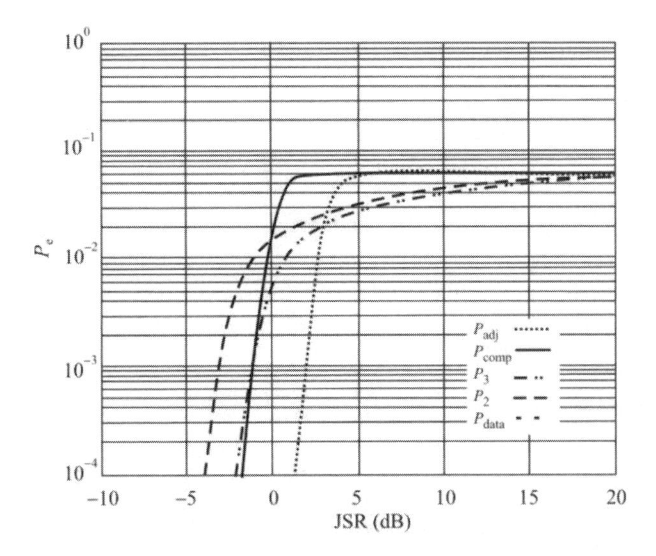

Figure 11.22 Comparison of the jamming strategies for follower jamming with tone jamming waveforms, $\upsilon = 20$ dB, $\gamma = 0.3$, $L_F = 3$.

Around the critical region where $\xi \approx 0$ dB, jamming the complementary channel is the best all-around approach.

For high SNR, the transition from low to high BER occurs more sharply for tone jamming, although the ultimate BER achieved is about the same. This is a disadvantage for the tone jammer because the transition from being in a jam condition to no-jam occurs over a very narrow range of JSR. Since the NBN performance falls off more slowly, the jamming performance degrades more gracefully.

The deleterious effects of increasing the frequency diversity in FFHSS systems can be addressed by increasing the dwell coverage fraction. For a given operational configuration this can only be accomplished by decreasing the processing time since the signal propagation speed is fixed. If the jammer is mobile (as on a UAS for example) it can be moved closer to the path between the transmitter and receiver to gain some speed benefits, however.

11.7 PBN Jamming of FFHSS Systems

PBN jamming signals are also useful for jamming FFHSS systems. A dwell is jammed when it hops into the part of the spectrum containing the jamming waveform, although this does not necessarily cause that dwell to be detected in error. Likewise, when the dwell hops into a part of the spectrum that is not jammed, the thermal noise can still cause a detection error even though there is no jamming signal present. Note that, as discussed in Chapter 4, the portion of the spectrum being jammed with PBN need not be contiguous for these results to apply.

While the jammer will move the bands being jammed around in the spectrum to avoid countermeasures by the communication system, such as precluding hopping to that portion, in general this movement is neither associated with the pattern nor timing of the hopped system. Compared with the FFHSS hop timing, changing the location of the PBN signal is relatively slow.

11.7.1 Uncoded Signals

PBN jamming of uncoded FFHSS signals will be considered here. The partial band jammer occupies the bandwidth shown in Figure 11.1, which is a fraction of the whole bandwidth. The total hop bandwidth is given by W_{ss} while the jammer bandwidth is given by W_J.

The PBN method of jamming a FFHSS system is illustrated in Figure 11.1 for $L_F = 4$ hops per bit. During a single data bit interval, the transmitter and receiver change frequencies $L_F = 4$ times at equal intervals T seconds long. The specific frequencies will depend on whether a mark or space is transmitted. In the next bit interval four (probably) distinct frequencies are chosen.

Lee et al. [33] performed a general analysis of the performance of an FFHSS system with PBN jamming using square-law detectors and noncoherent combining of the decisions per hop. The detector structure analyzed by Lee et al. is shown in Figure 11.3. The following is a simplified analysis of PBN that applies to the same receiver structure and will serve the purpose here. There will be an error detecting a dwell if the SNR and JSR parameters are appropriate and the dwell lands within the portion of the spectrum being jammed. The fraction of the spectrum jammed is given by

$$\gamma = \frac{W_J}{W_{ss}} \tag{11.11}$$

When the portion of the spectrum the hop lands in is not jammed, the BER is given by

$$P_{e_1} = \frac{1}{2}\exp\left(-\frac{1}{2}\upsilon\right) \tag{11.12}$$

In the portion of the spectrum where the jammer lies, the probability of a dwell error is

$$P_{e_2} = \frac{1}{2}\exp\left(-\frac{1}{2}\frac{E_c}{N_0 + J_0'}\right) \tag{11.13}$$

where

$$J_0' = \frac{J_0}{\gamma} \tag{11.14}$$

is an elevated level of jammer noise density. Therefore

$$\begin{aligned} P_{e_2} &= \frac{1}{2}\exp\left(-\frac{1}{2}\frac{E_c}{N_0 + \frac{J_0}{\gamma}}\right) \\ &= \frac{1}{2}\exp\left(-\frac{1}{2}\frac{\gamma E_c}{\gamma N_0 + J_0}\right) \end{aligned} \tag{11.15}$$

Since $E_c = RT_c$, $J = J_0 W_F$, and $P_N = N_0 W_F$,

$$\begin{aligned} P_{e_2} &= \frac{1}{2}\exp\left(-\frac{1}{2}\frac{\gamma RT_c}{\frac{\gamma P_N}{W_F} + \frac{J}{W_F}}\right) \\ &= \frac{1}{2}\exp\left(-\frac{1}{2}\frac{\gamma}{\frac{\gamma}{\upsilon} + \xi}\right) \end{aligned} \tag{11.16}$$

when it is assumed that $T_c W_F = 1$.

At high enough signal levels and low enough jamming power, an optimum value of γ is discernable that maximizes the BER. This can be seen in Figures 11.23 and 11.24 where the BER is plotted versus γ when the $\upsilon = 10$ and 20 dB and $L_F = 4$. The peaks of these curves is the optimum jamming value for γ. Unfortunately, for reasonable values of JSR for the SNRs considered here, that optimum γ is 1, implying BBN jamming.

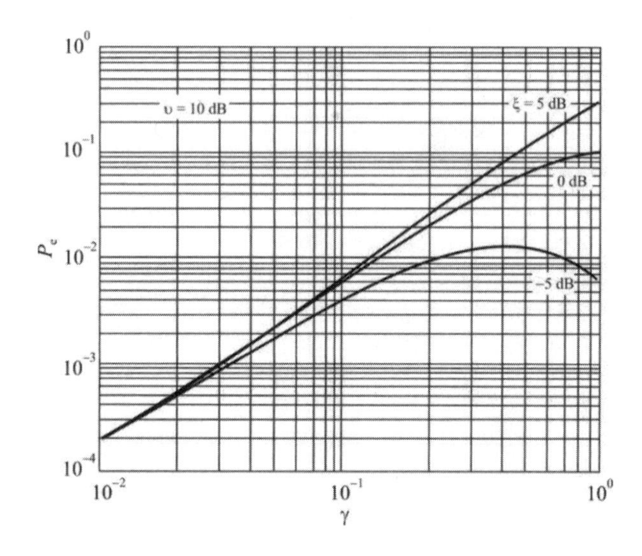

Figure 11.23 PBN jamming performance as a function of the fraction of dwell jammed when υ = 10 dB for $L_F = 3$.

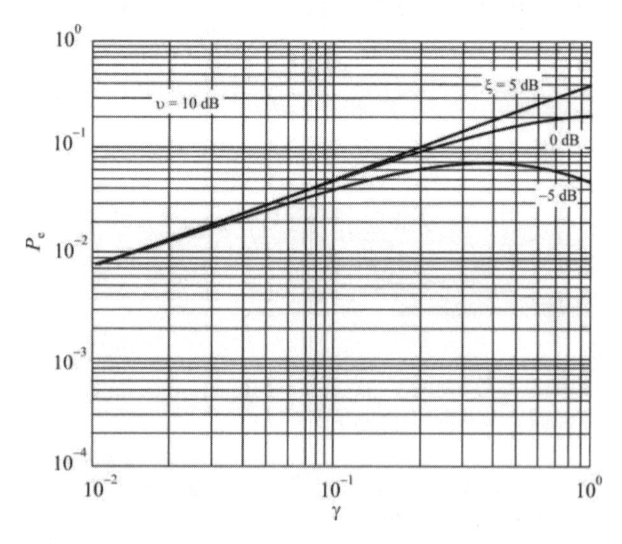

Figure 11.24 PBN jamming as a function of the hop dwell fraction for υ = 10 dB and $L_F = 1$.

In the low VHF band (30–90 MHz, 25 kHz channels), there are approximately 2,400 channels. Jamming 10% of them results in jamming 240 channels while jamming 30% results in jamming 720 channels. In the UHF frequency band (225–400 MHz, 50 kHz channels) there are 3,500 channels. In that case 10% amounts to 350 channels while 30% amounts to 1,050 channels. Noise jamming easily causes friendly fratricide, be it BBN or PBN. Jamming such large portions of the spectrum could only be applied under special circumstances due to this fratricide.

11.7.2 Error Coded Signals

When the FHSS signals employ error coding some degradation in jamming performance will occur. This is illustrated in Figure 11.25 for a few typical error-correcting codes. For ξ = 10 dB, the error correcting codes considered here degrade jamming performance by between 1 and 3 dB at $P_e = 10^{-2}$. When ξ = 30 dB, however, the codes are in the error extension region and $P_e \approx 10^{-1}$ for all the SNRs considered [34–36]. Such a large JSR is difficult to achieve, however, due to the signal limiters that are normally employed prior to the detectors in FHSS systems.

11.7.3 Narrowband Noise Jamming

When the bandwidth of the jammer signal matches that of the channel, jamming can still be effective. The FFHSS signal must hop into that channel, however. The performance is illustrated in Figure 11.26 for this case. Shown there are BER for uncoded as well as coded signals.

Against uncoded signals the BER falls fast with this jamming approach even when ξ = 30 dB because of the low probability of the FFHSS signal encountering the jamming signal.

11.8 Single-Tone Jamming of FFHSS Systems

Single-tone jamming was examined by Milstein et al. [8] against uncoded and coded FHSS systems. For single-tone jamming, example results are shown in Figure 11.27. Such a strategy places all the jammer power at a single frequency. At best, this strategy will jam a single data bit, and that is when $L_F = 1$. For any $L_F > 1$, the data bit is susceptible to only thermal noise at any other frequency. Single-tone jamming, therefore, is not a viable technique and is not used for FFHSS systems.

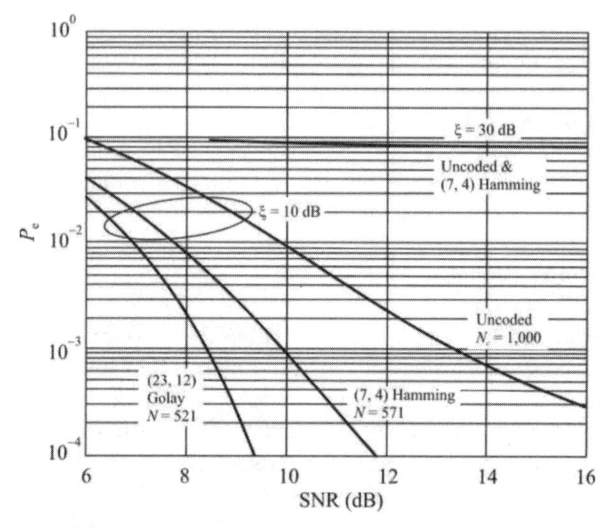

Figure 11.25 PBN jamming performance against some common error correcting codes. In this illustration $\gamma = 0.1$. It is assumed here that $W_F T_c = 1$. (Source: [8]. © IEEE 1980. Reprinted with permission.)

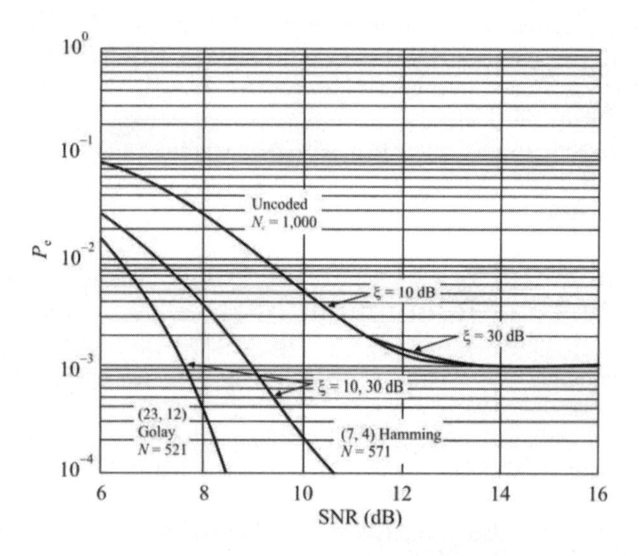

Figure 11.26 Narrowband noise jamming performance. In this illustration the jammer occupies a bandwidth only one channel wide. In addition, N_c = 1,000 channels while the encoded signals shown have the number of channels available as shown. It is assumed here that $W_F T_c = 1$. (Source: [8]. © IEEE 1980. Reprinted with permission.)

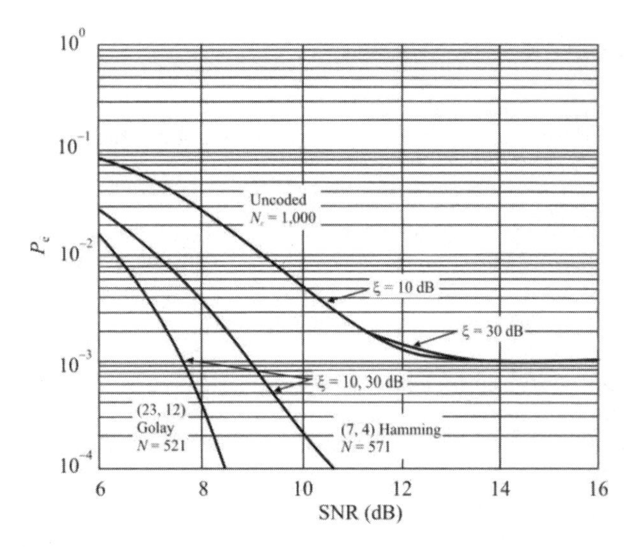

Figure 11.27 Single tone jammer performance against uncoded and coded FHSS systems. $W_F T_c = 1$ is assumed. (Source: [8]. © IEEE 1980. Reprinted with permission.)

11.9 Multiple-Tone Jamming of FFHSS Systems

MT jamming can be effective against FFHSS signals. The jamming tones are placed at N_J frequencies. For simplicity, it is assumed that the possible locations of the hopping tones in the spectrum are known, even though the hop sequence is not. This assumption would typically be satisfied in practice. When the jammer hops to one of these frequencies with a data tone, that tone is actually enhanced by the jammer and detection performance at the receiver is improved. The advantage of MT jamming occurs when the tone is placed so that the complementary channel associated with a hop receives the jammer energy. In that case, the jamming effectiveness is a function of SNR and JSR at the receiver [8, 37].

As in PBN jamming, the multiple tones need not be contiguous. In addition, the jammer would typically change the frequency of the tones periodically to avoid countermeasures by the transmitter and receiver.

For completeness, it should be noted that generating multiple tones with a single high-power amplifier at a jammer is not necessarily the best method of providing jamming signals. It can be shown that the power per tone when doing so decreases approximately as the square of the number of such tones generated [37]. Thus, the efficiency of the jammer is compromised.

11.9.1 Uncoded FFHSS MFSK Signals

FFHSS signals are, by definition, frequency diversity coded since a single data bit is transmitted on several hop dwells. Therefore, the advantages of this coding as discussed in Section 3.13.3.2 and in this chapter are present in all FFHSS signaling schemes. Hence, when coding is discussed for such signals, other forms of coding are implied. This will be discussed in Section 11.9.2.

For MT jamming, the jamming tones are fixed in the spectrum. For uncoded FFHSS signals, a bit error will occur when the dwell lands in a channel if the dwell hits the complementary channel of that data channel. If the dwell hits the data channel, the detection performance of the receiver is enhanced because the jammer tone energy adds to that of the data tone. The data tone and complementary tone need not be contiguous in this case.

Let N_c denote the number of channels in W_{ss}. The fraction of the channels jammed is given by

$$\gamma = \frac{N_J}{N_c} \tag{11.17}$$

The probability of jamming the complementary channel, given that the dwell fell on the channel, is given by [38]

$$P_{e_2} = \frac{1}{2}\left[Q\left(\sqrt{\frac{J}{P_N}}, \sqrt{\upsilon} \right) - \frac{1}{2}\exp\left(\frac{S+J}{2P_N} \right) I_0\left(\frac{\sqrt{SJ}}{P_N} \right) \right] \tag{11.18}$$

The probability of a bit error is therefore given by

$$\begin{aligned}
P_e &= \Pr\{\text{Jamming the complementary channel} \\
&\qquad |\text{Do not land on channel}\} \\
&\qquad \times \Pr\{\text{Do not land on channel}\} \\
&\qquad + \Pr\{\text{Jamming the complementary channel} \\
&\qquad |\text{Land on channel}\} \\
&\qquad \times \Pr\{\text{Land on the channel}\} \\
&= (1-\gamma)\frac{1}{2}\exp\left(-\frac{1}{2}\upsilon \right)
\end{aligned}$$

$$+\gamma\frac{1}{2}\left[Q\left(\sqrt{\frac{J}{P_\text{N}}},\sqrt{\upsilon}\right)-\frac{1}{2}\exp\left(\frac{S+J}{2P_\text{N}}\right)I_0\left(\frac{\sqrt{SJ}}{P_\text{N}}\right)\right] \quad (11.19)$$

11.9.2 Error-Coded FFHSS MFSK Signals

The types of coding applicable to FFHSS signals, other than frequency diversity coding, in general, are the same available for other digital modulations. Convolutional and block coding techniques discussed in Section 3.4 apply here as well.

11.10 Pulsed Jamming of FFHSS Systems

Pulsed jamming is not as effective against FFHSS systems as it is against the DSSS systems discussed in Chapter 10. PBN jamming has similar effects on FFHSS systems that pulsed has on DSSS [39].

11.11 Concluding Remarks

Various techniques for jamming FFHSS signals were presented in this chapter. In addition to BBN, follower jamming using NBN and tones, as well as PBN and MT jamming were discussed. Jamming the synchronization process in FFHSS systems is also possible, and some discussion on the significance of doing this was included.

A comparison of BBN, PBN, and multitone jamming against FFHSS systems is shown in Figure 11.28 for $\gamma = 0.1$ and $L_\text{F} = 1$. For this comparison, it is assumed that the hop rate, R_h, is numerically equal to the noise filter bandwidth in the receiver so that $T_\text{c}W_\text{F} = 1$. It is also assumed that for PBN jamming, the optimal jamming fraction is used by the jammer.

Recall that in the far positive region of SNR for uncoded BFSK and for the optimum fraction in PBN jamming, the BER falls off linearly as the SNR increases, yielding a significant advantage to the jammer. In the region of interest for successful jamming, however, no such advantage exists for PBN jamming. BBN jamming in the region of positive JSRs is the better strategy, ceteris paribus, when $\gamma = 0.1$ and $L_\text{F} = 1$.

When γ is increased to 0.3, both PBN and MT jamming improves as illustrated in Figure 11.29. Of course BBN jamming is not changed. In this case, PBN with a

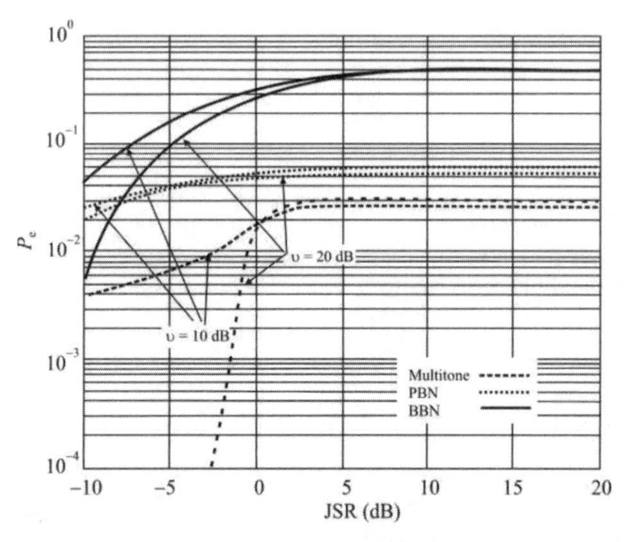

Figure 11.28 Comparison of BBN, PBN, and MT jamming of BFSK FFHSS signals when $\gamma = 0.1$ and $L_F = 1$.

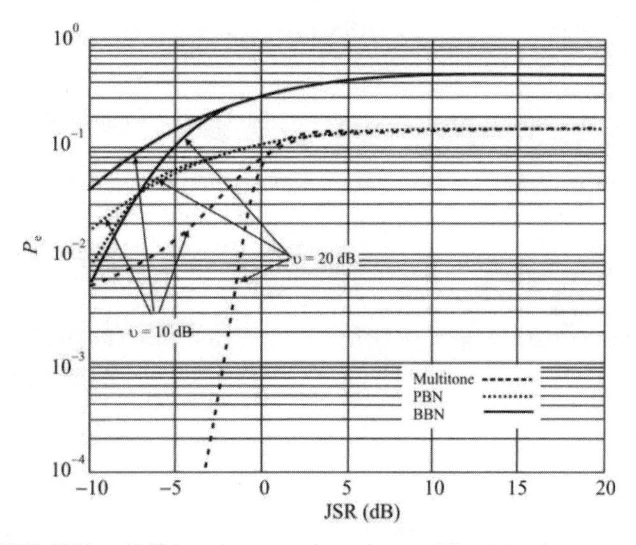

Figure 11.29 BBN, PBN, and MT jamming comparison when $\gamma = 0.3$ and $L_F = 1$.

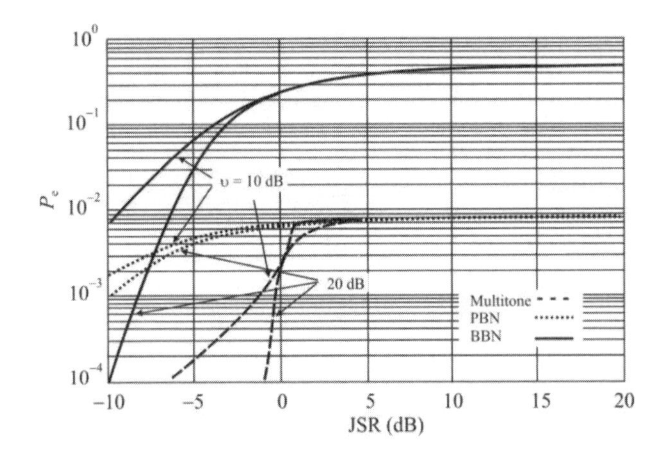

Figure 11.30 BBN, PBN, and MT jamming of FFHSS BFSK when $\gamma = 0.1$ and $L_F = 3$.

JSR greater than about 3 dB provides successful jamming, irrespective of the SNR. MT jamming, while better than when $\gamma = 0.1$, still does not achieve the 10^{-1} BER goal.

On the other hand, when L_F is increased to 3, BBN jamming improves while the other two worsen, as illustrated in Figures 11.30 and 11.31. BBN gets better because there is less signal energy per tone since the available energy is spread over multiple (3 in this case) tones. Neither value of γ is sufficient to provide successful jamming when frequency diversity is included in the signaling scheme.

For the ranges of variables considered here, when the SNR is relatively low and the processing gain marginal, the MT technique rises to a better P_e faster than the other two, but at higher JSRs (above about 10 dB), the other two are better. MT jamming is the better approach for higher SNR and better processing gain. In both cases of MT jamming, the goal of a 10^{-1} BER is almost reached at $\xi \approx 0$ dB.

FFHSS, as an ECCM technique, was devised to thwart ES and EA attempts. If the hopping is fast enough, it can indeed accomplish these tasks. Implementation of FFHSS systems is considerably more difficult than SFHSS systems, which are discussed in the next chapter.

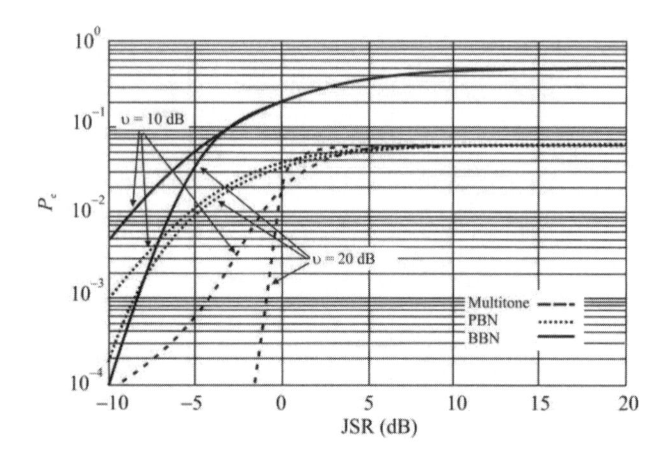

Figure 11.31 BBN, PBN, and MT jamming performance for FFHSS BFSK when $\gamma = 0.3$ and $L_F = 3$.

References

[1] Simon, M. K., J. K. Omura, R. A. Scholtz, and B. K. Levitt, *Spread Spectrum Handbook,* New York: McGraw-Hill, 1994, pp. 462–664.

[2] Peterson, R. L., R. E. Ziemer and D. E. Borth, *Introduction to Spread Spectrum Communications,* Upper Saddle River, NJ: Prentice Hall, 1985, Ch. 6.

[3] Dixon, R. C., *Spread Spectrum Systems*, 2nd ed., New York: John Wiley & Sons, 1984, Section 2.2.

[4] Pettit, R. H., *ECM and ECCM Techniques for Digital Communications,* Belmont, CA: Lifetime Learning Publications, 1982, pp. 90–97.

[5] Nicholson, D. L., *Spread Spectrum Signal Design LPE and AJ Systems,* Rockville, MD: Computer Science Press, 1988, pp. 171–173.

[6] Torrieri, D. J., *Principles of Secure Communication Systems,* 2nd ed., Norwood, MA: Artech House, 1992, Ch. 3.

[7] Simon, M. K., J. K. Omura, R. A. Scholtz, and B.K. Levitt, *Spread Spectrum Handbook,* New York: McGraw-Hill, 1994, Vol. II, Ch 3.

[8] Milstein, L. B., R. L. Pickholtz, and D. L. Schilling, "Optimization of the Processing Gain of an FSK-FH System," *IEEE Transactions on Communications,* Vol. Com-28, No. 7, July 1980, pp. 1062–10711.

[9] Simon, M. K., and A. Polydoros, "Coherent Detection of Frequency-Hopped Quadrature Modulations in the Presence of Jamming—Part I: QPSK and QASK Modulations," *IEEE Transactions on Communications,* Vol. Com-29, No. 11, November 1981, pp. 1644–1660.

[10] Simon, M. K., "Coherent Detection of Frequency-Hopped Quadrature Modulations in the Presence of Jamming—Part II: QPR Class I Modulation," *IEEE Transactions on Communications,* Vol. Com-29, No. 11, November 1981, pp. 1661–1668.

[11] Simon, M. K., G. L. Huth, and A. Polydoros, "Coherent Detection of QASK for Frequency-Hopping Systems—Part I: Performance in the Presence of a Gaussian Noise Environment," *IEEE Transactions on Communications,* Vol. Com-30, No. 1, January 1982, pp. 158–164.

[12] Simon, M. K., "Coherent Detection of QASK for Frequency-Hopping Systems—Part II: Performance in the Presence of Jamming," *IEEE Transactions on Communications,* Vol. Com-30, No. 1, January 1982, pp. 165–172.

[13] Kaleh, G. K., "Performance Comparison of Frequency-Diversity and Frequency-Hopping Spread-Spectrum Systems," *IEEE Transactions on Communications,* Vol. 45, No. 8, August 1997, pp. 910–911.

[14] Milstein, L. B., and D. L. Schilling, "The Effect of Frequency-Selective Fading on a Non-Coherent FH-FSK System Operating with Parial-Band Tone Interference," *IEEE Transactions on Communications,* Vol. Com-30, No. 5, May 1982, pp. 904–911.

[15] Simon, M. K., J. K. Omura, R. A. Scholtz, and B.K. Levitt, *Spread Spectrum Handbook,* New York: McGraw-Hill, 1994, pp. 225–233.

[16] Peterson, R. L., R. E. Ziemer and D. E. Borth, *Introduction to Spread Spectrum Communications,* Upper Saddle River, NJ: Prentice Hall, 1985, Ch. 8.

[17] Teh, K. C., C. Kit, and K. H. Li, "Partial-Band Jamming Rejection of FFHSS/BFSK with Product Combining Receiver over a Rayleigh-Fading Channel," *IEEE Communication Letters,* Vol. 1, No. 3, May 1997, pp. 64–66.

[18] Teh, K. C., A. C. Kot, and K. H. Li, "Multitone Jamming Rejection of FFHSS/BFSK Spread-Spectrum System Over Fading Channels," *IEEE Transactions on Communications,* Vol. 46, No. 8, August 1998, pp. 1050–1057.

[19] Lee, J. S., L. E. Miller, and Y. K. Kim, "Probability of Error Analysis of a BFSK Frequency-Hopping System with Diversity under Partial-Band Jamming Interference—Part II: Performance of Square-Law Nonlinear Combining Soft Decision Receivers," *IEEE Transactions on Communications,* Vol. Com-32, No. 12, December 1984, pp. 1243–1250.

[20] Viswanathan, R. and S. C. Gupta, "Performance Comparison of Likelihood, Hard-Limited, and Linear Combining Receivers for FH-MFSK Mobile Radio—Base–to–Mobile Transmission," *IEEE Transactions on Communications,* Vol. Com-31, No. 5, May 1983, pp. 670–677.

[21] Viswanathan, R. and S. C. Gupta, "Nonparametric Receiver for FH-MFSK Mobile Radio," *IEEE Transactions on Communications,* Vol. Com-33, No. 2, February 1985, pp. 178–184.

[22] Miller, L. E., J. S. Lee, and A. P. Kadrichu, "Probability of Error Analysis of a BFSK Frequency-Hopping System with Diversity under Partial-Band Jamming Interference—Part III: Performance of a Square-Law Self-Normalizing Soft Decision Receiver," *IEEE Transactions on Communications,* Vol. Com-34, No. 7, July 1986, pp. 669–675.

[23] Viswanathan, R. and K. Taghizadeh, "Diversity Combining in FH/BFSK Systems to Combat Partial Band Jamming," *IEEE Transactions on Communications,* Vol. Com-36 No. 9, September 1988, pp. 1062–10611.

[24] Teh, K. C., K. K. Li, and A. C. Kot, "Performance Analysis of an FFHSS/BFSK Linear-Combining Receiver Against Multitone Jamming," *IEEE Communication Letters,* Vol. 2, No. 8, August 1998, pp. 205–207.

[25] Teh, K. C., "Multitone Jamming Rejection of FFHSS/BFSK Self-Normalizing Receiver," *Proceedings IEEE MILCOM*, 1998, http://www.greenhouse.com/society/TacCom/milcom _98_papers.html.

[26] Lee, J. S., L. E. Miller, and R. H. French, "The Analysis of Uncoded Performance for Certain ECCM Receiver Design Strategies for Multihops/Symbol FH/MFSK Waveforms," *IEEE Journal on Selected Areas in Communications*, Vol. SAC-3, No. 5, September 1985, pp. 611–621.

[27] Levitt, B. K., "Strategies for FH/MFSK Signaling with Diversity in Worst-Case Partial-Band Noise," *IEEE Journal on Selected Areas in Communications*, Vol. SAC-3, No. 5, September 1985, pp. 622–626.

[28] Yoon, Y., K. Lee, D. Kim, and K. Kim, "Performance Improvement of a Fast FH-FDMA System by the Clipped-Linear Combining Receiver," *Proceedings IEEE MILCOM*, 2001

[29] Felstead, E. B., "Follower Jammer Considerations for Frequency Hopped Spread Spectrum," *Proceedings IEEE MILCOM*, 1998, http://www.greenhouse.com/society/TacCom/milcom _98_papers.html.

[30] Hassan, A. A., W. E. Stark, and J. E. Hershey, "Frequency-Hopped Spread Spectrum in the Presence of a Follower Partial-Band Jammer," *IEEE Transactions on Communications,* Vol. 41, No. 7, July 1993, pp. 1125–1131.

[31] Hassan, A. A., J. E. Hershey, and J. E. Schroeder, "On a Follower Tone-Jammer Countermeasure Technique," *IEEE Transactions on Communications,* Vol. 43, No. 2/3/4, February/March/April 1995, pp. 754–756.

[32] Torrieri, D. J., "Fundamental Limitations on Repeater Jamming of Frequency-Hopping Communications," *IEEE Journal on Selected Areas in Communications*, Vol. 7, No. 4, May 1989, pp. 569–575.

[33] Lee, J.S., R.H. French and L.E. Miller, "Probability of Error Analysis of a BFSK Frequency-Hopping System with Diversity under Partial-Band Jamming Interference–Part I: Performance of Square-Law Linear Combining Soft Decision Receiver," *IEEE Transactions on Communications*, Vol. Com-32, No. 6, June 1984, pp. 645 –653.

[34] Stark, W. E., "Coding for Frequency-Hopped Spread-Spectrum Communication with Partial-Band Interference—Part I: Capacity and Cutoff Rate," *IEEE Transactions on Communications,* Vol. Com-33, No. 10, October 1985, pp. 1036–1044.

[35] Stark, W. E., "Coding for Frequency-Hopped Spread-Spectrum Communication with Partial-Band Interference—Part II: Coded Performance," *IEEE Transactions on Communications,* Vol. Com-33, No. 10, October 1985, pp. 1045–1057.

[36] Chu, M. J. and W. E. Stark, "Asymptotic Performance of a Coded Communication System with Orthogonal Signaling in Parial Band Jamming," *Proceedings IEEE MILCOM*, 1998, http://www.greenhouse.com/society/TacCom/milcom_98_papers.html.

[37] Simon, M. K., "The Performance of M-ary FH-DPSK in the Presence of Partial-Band Multitone Jamming," *IEEE Transactions on Communications,* Vol. Com-30, No. 5, May 1982, pp. 953–958.

[38] Levitt, B. K., "FH/MFSK Performance in Multitone Jamming," *IEEE Journal on Selected Areas in Communications*, Vol. SAC-3, No. 5, September 1985, pp. 627–643.

[39] Milstein, L. B., S. Davidovici, and S. L. Schilling, "Coding and Modulation Techniques for Frequency-Hopped Spread-Spectrum Communications over a Pulse-Burst Jammed Rayleigh Fading Channel," *IEEE Journal on Selected Areas in Communications*, Vol. SAC-3, No. 5, September 1985, pp. 644–651.

Chapter 12

Electronic Warfare and Slow Frequency Hopping Systems

12.1 Introduction

Slow frequency hopping is when the communication system changes its frequency on a regular basis and there is more than one data bits transmitted per dwell at each frequency. A typical hop rate is around 100 hps, which means that the transmitter dwells at each frequency for somewhat less than 10 ms. It must be less than 10 ms per frequency because it takes a finite amount of time to switch between frequencies and the filters to settle.

In this chapter it is assumed that the FHSS system employs noncoherent, MFSK modulation. The tone frequencies of the target are given by $f_1, f_2, \ldots, f_M$ as shown in Figure 12.1. It is assumed that these tones are placed symmetrically on each side of the carrier frequency f_0. The instantaneous bandwidth of the jammer is given by W_J and the total bandwidth covered is given by W_{ss}. The target frequency need not be known exactly and the epoch (timing) information of the target signal is not known.

As in Chapter 11, the effects of frequency selective channel fading will only be briefly addressed for SFHSS systems [1]. This is not to imply that such fading is not important, but the goal here is to focus on the effects of jamming.

Error control coding is an important countermeasure for jamming SFHSS systems as discussed in Chapter 3. Turbo coding has recently been devised as a method of error control coding for SFHSS systems [2–6]. Other coding has also successfully been applied to SFHSS signals for error control [7]. When results as to the effectiveness of jamming against these signals are available, they are included here.

For the most part, in this chapter it is assumed that the transmitter and receiver have synchronized epochs and hopping codes. Some results are presented, however, for the case of attacking SFHSS code acquisition processes.

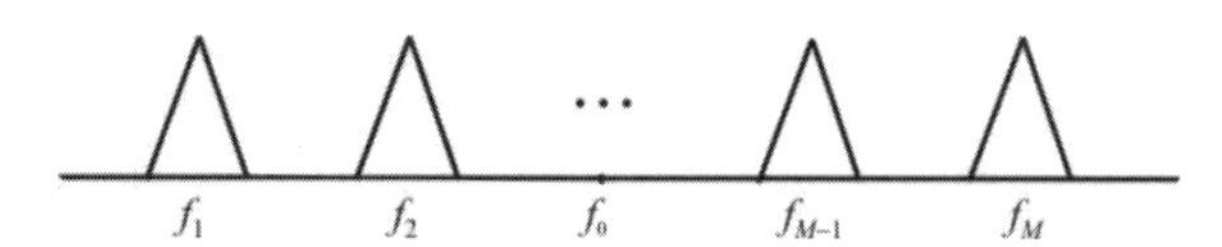

Figure 12.1 MFSK spectrum.

The structure of this chapter is as follows. As in Chapter 11, we begin by analyzing the performance of BBN jamming on SFHSS systems. We follow this with the effects of PBN jamming followed by MT jamming performance. Follower jamming, including both NBN and MT techniques, is addressed next. We conclude the chapter with a brief discussion of jamming error-coded MFSK signals.

12.2 BBN Jamming of SFHSS Systems

Just as for FFHSS jamming where broadband noise raises the noise floor everywhere in the spectrum, the same applies to SFHSS jamming. The analysis in Section 11.5 applies here as well. The biggest difference is that for SFHSS systems, the tones are phase incoherent from one hop to the next but within a hop they may be phase coherent.

12.2.1 Uncoded

The bit error rate for uncoded, incoherent, orthogonal MFSK is given by [8, 9]

$$P_\text{b} = \frac{1}{2(M-1)} e^{-\frac{lE_\text{b}}{2N_\text{T}}} \sum_{q=2}^{M} \binom{M}{q} (-1)^q e^{\frac{lE_\text{b}(2-q)}{2qN_\text{T}}} \tag{12.1}$$

where $N_\text{T} = N_0 + J_0$ is the total noise density due to thermal noise and the jammer, and $l = \log_2 M$. This function is plotted in Figure 12.2 for BFSK ($M = 2$) where it is assumed that the detector filter bandwidth, W_F, is numerically equivalent to the data rate; that is, $W_\text{F} = R_\text{b}$. Figure 12.3 shows the jamming performance for $M = 4$. BBN jamming is more effective against BFSK than QFSK with a JSR that is about 5 dB less at $P_\text{e} = 10^{-1}$.

The communication system processing gain in this case is given by [10]

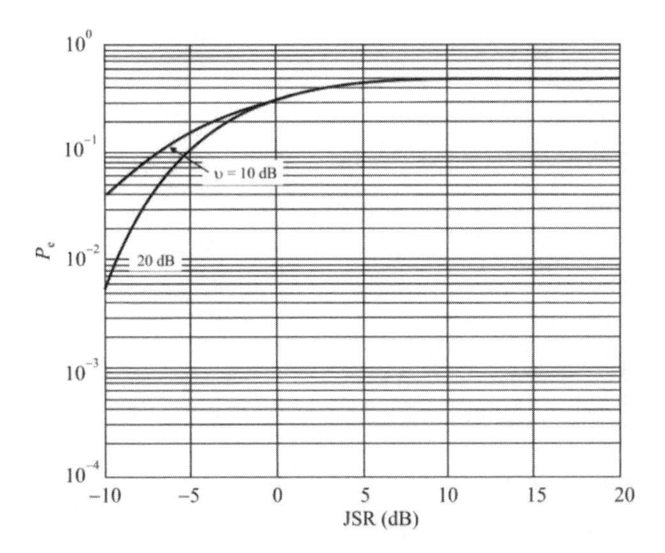

Figure 12.2 BBN jammer performance against uncoded SFHSS noncoherent BFSK FHSS including the effects of noise.

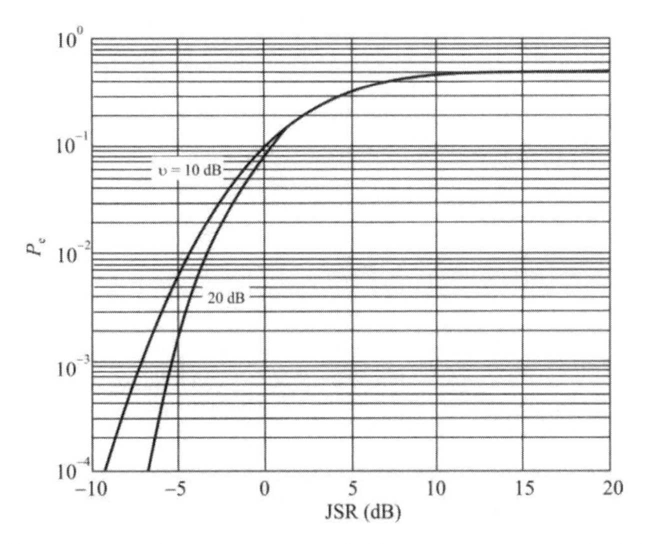

Figure 12.3 BBN jamming performance against uncoded SFHSS noncoherent QFSK FHSS system.

$$N = \frac{W_{ss} \log_2 M}{MR_b} \tag{12.2}$$

The processing gain in dB is given by $G_{p,dB} = 10 \log(N)$. For the example illustrated, $M = 2$, suppose $R_b = 20$ kbps and $W_{ss} = 60$ MHz. Then $G_{p,dB} = 32$ dB.

12.2.2 Error Coded

Frequency diversity coding transmits the SFHSS bit at multiple frequencies, which is the same definition as FFHSS systems. Therefore, jamming performance against diversity coded SFHSS systems have the same performance as uncoded FFHSS systems as presented in Chapter 11.

Other forms of coding can be applied to SFHSS systems, however, and their communication performance is better than diversity coding. The SFHSS signals can be treated quasi-statically to determine the effects of coding. Quasi-static analysis means that within a dwell, the data modulation can be treated as if the signal were not hopping.

12.3 PBN Jamming of SFHSS Systems

It is important to recall that the channels jammed by a PBN jammer need not be contiguous. The jammer could split its energy over several disjoint channels if it is advantageous to do so. This is shown in Figure 11.1(d). Also, the jammer would likely change the portions of the spectrum being jammed periodically to avoid effective FHSS system countermeasures. The fraction of the spectrum covered by the jammer is given by

$$\gamma = \frac{W_J}{W_{ss}} \tag{12.3}$$

For contiguous PBN jamming, a signal at a single center frequency is modulated with a noise waveform that has a frequency bandwidth that is normally considerably wider than the channel bandwidth considered above.

With PBN noise jamming, when the signal hops to a new frequency, that channel may be in the band that is being jammed or not. If it is, then the total noise present will be the sum of the jammer noise and the thermal noise. If that channel is not being jammed then the signal degradation is due to the thermal noise only.

12.3.1 Uncoded SFHSS MFSK Systems

The probability of a symbol error in the presence of a jammer is given by the probability of a symbol error when the jammer is present times the probability of the jammer being present plus the probability of a symbol error when the jammer is absent times the probability of the jammer being absent. Thus,

$$P_s = P_S(\text{error}|N_p)P(N_p) + P_S(\text{error}|N_n)P(N_n) \tag{12.4}$$

Now,

$$\begin{aligned} P(N_p) &= \gamma \\ P(N_n) &= 1-\gamma \end{aligned} \tag{12.5}$$

Using (12.1) and (12.5) in (12.4) yields

$$\begin{aligned} P_s = {} & (1-\gamma)\frac{1}{2(M-1)}e^{-\frac{lE_b}{2N_0}}\sum_{q=2}^{M}\binom{M}{q}(-1)^q e^{\frac{lE_b(2-q)}{2qN_0}} \\ & +\gamma\frac{1}{2(M-1)}e^{-\frac{lE_b}{2N_T}}\sum_{q=2}^{M}\binom{M}{q}(-1)^q e^{\frac{lE_b(2-q)}{2qN_T}} \end{aligned} \tag{12.6}$$

where $N_T = N_0 + J_0$ and $J_0 = J/\gamma$. The SNR can be expressed as

$$\frac{E_b}{N_T} = \frac{E_b}{N_0 + J_0} = \frac{ST_b}{\frac{P_N}{W_F} + \frac{J}{W_F}} = \frac{ST_bW_F}{P_N + J} = \frac{T_bW_F}{\frac{1}{\upsilon} + \xi} \tag{12.7}$$

When $W_F = R_b$, $T_bW_F = 1$ and

$$\frac{E_b}{N_T} = \frac{1}{\frac{1}{\upsilon} + \xi} \tag{12.8}$$

and

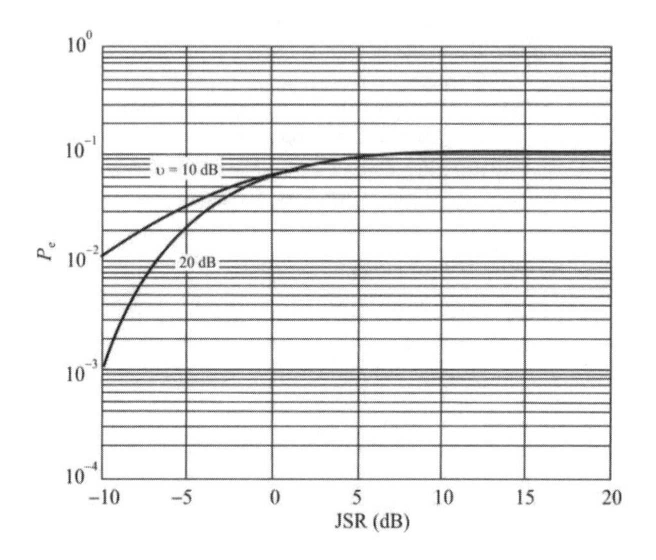

Figure 12.4 Partial band jamming performance against SFHSS BFSK systems when the fraction of the band covered $\gamma = 0.2$.

$$\frac{E_b}{N_0} = \upsilon \tag{12.9}$$

This function is plotted in Figure 12.4 for BFSK when $\gamma = 0.2$.

In the analysis of PBN jamming FHSS systems, it is often assumed that the noise can be ignored compared to the jammer. That assumption is marginal. Most of the time a frequency-hopping receiver is dealing with background noise, not the partial band jammer. Of course, in a crowded environment the biggest problem will be cochannel interference anyway, not a jammer. However, by ignoring the effects of thermal noise on the BER performance it is assured that the jammer will produce *at least* the BER computed, so it represents a worst case for the jammer.

The approximate effects of considering such cochannel interference, with an emphasis on the interference effects on coarse acquisition of FHSS signals, was considered by Putnam [11] as described in Chapter 4. In that analysis, it was assumed that the interference was received at the same signal level as the target signal. A comparison was made for three coarse acquisition schemes: serial search, matched filter, and the combination, two-level acquisition. Typical results are shown in Figure 12.5 for the two-step acquisition. For the parameters considered, $P_{fa} = 10^{-9}$, $M = 5$, M being the number of matched filters used, $K = 10$, $K = T_1/T_c$, where T_1 is the duration of the time an active correlator takes and T_c is the hop

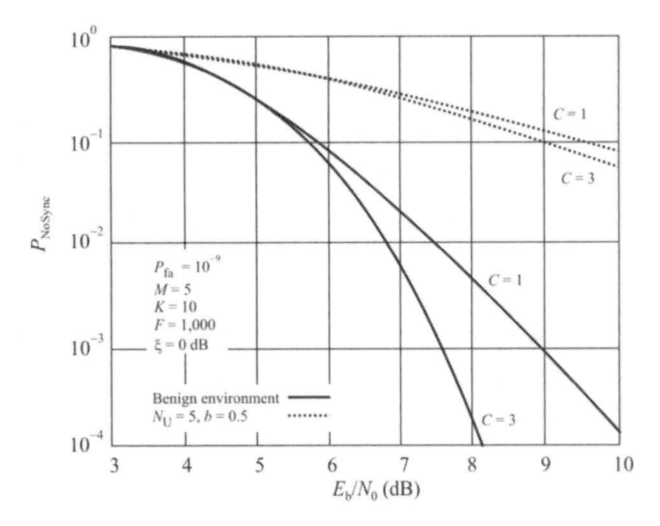

Figure 12.5 Effects of cochannel interference on coarse acquisition with the two-step approach. (Source: [11]. © IEEE 1983. Reprinted with permission.)

interval, $N_c = 1,000$ is the number of hopping channels available, C is the number of active correlators considered, N_U is the number of simultaneous users of the channels, and b is the average fractional power in the scattering components of the signal. For a detailed explanation of these parameters see [11]. For as few as five users, the probability of missed acquisition (P_{NoSync}) increases by over three orders of magnitude. Clearly, interference is a significant limit to coarse acquisition in SFHSS systems.

Assume initially that the signal is BFSK and that the noise can be neglected. In addition, assume that the jammer distributes its total power over the entire bandwidth W_{ss}. Denoting the jammer power by J, the power spectral density of the jammer is J/W_{ss} and $N_T = J/W_{ss}$. Since $E_b = R/R_b$,

$$\frac{E_b}{N_0} = \frac{\dfrac{R}{R_b}}{\dfrac{J}{W_{ss}}} = \frac{R}{J}\frac{W_{ss}}{R_b} \tag{12.10}$$

W_{ss}/R_b is the ratio of the total bandwidth to the data rate and, assuming that the spectral efficiency of the modulation process (= R_b/W_F) is one, is equal to the processing gain of the AJ communication system. Therefore the processing gain of the system improves the bit error performance by this amount. The jammer is placing energy in all of the frequency channels while the communicator is occupying only one at a time. The jammer is therefore wasting $[(N-1)/N]J$ of its power. Only (J/N) of the power is effectively jamming the receiver and the effective jamming power is P_J/W_{ss}. The probability of a bit error is therefore

$$P_e = \frac{1}{2}\exp\left(-\frac{1}{2}\frac{RW_{ss}}{J}\right) = \frac{1}{2}\exp\left(-\frac{1}{2}\frac{1}{\xi / W_{ss}}\right) \tag{12.11}$$

Now consider the case when the jammer distributes its energy over only a portion of the bandwidth given by γ. Denote the probability of bit error due to the jammer by P_J. Neglecting the effects of the noise in the portion of the spectrum unoccupied by the jammer, then

$$E_b = \frac{R}{R_b} \tag{12.12}$$

$$N_0 = \frac{J}{\gamma W_{ss}} \tag{12.13}$$

$$\frac{E_b}{N_0} = \frac{R}{R_b}\frac{\gamma W_{ss}}{J} = \frac{R\gamma N}{J} \tag{12.14}$$

$$P_J = \frac{1}{2}e^{-\frac{1}{2}\frac{R\gamma N}{J}} = \frac{1}{2}e^{-\frac{1}{2}\frac{\gamma}{\xi/N}} \tag{12.15}$$

Since the effects of noise in the portion of the spectrum unoccupied by the jammer are ignored, then the probability of a bit error is zero in this region. Then the overall probability of bit error is given by

$$P_b = \gamma P_J + (1-\gamma)0 = \gamma P_J = \frac{\gamma}{2}e^{-\frac{1}{2}\frac{R\gamma N}{P_J}} = \frac{\gamma}{2}e^{-\frac{1}{2}\frac{\gamma}{\xi/N}} \tag{12.16}$$

By differentiating this expression with respect to γ and setting this equal to zero, the value of γ that maximizes P_e can be found. Doing so yields

$$\gamma^* = \frac{2}{NR / J} \tag{12.17}$$

Table 12.1 Parameters for MFSK

$M = 2^K$	C	X_0	JSR at 10^{-5} (dB)	M-ary Coding Gain
2	0.3679	2	-45.66	0
4	0.2329	1.192	-43.67	1.99
8	0.1954	0.927	-42.91	2.75
16	0.1813	0.798	-42.59	3.07
32	0.1746	0.723	-42.43	3.21

Source: [12].

Note that at $NR/J = 2$, $\gamma^* = 1$, and the partial band becomes the whole band. Therefore, this analysis only applies for $NR/J \geq 2$. When $NR/J < 2$ there is no value of γ less than one that applies, yielding a probability of bit error of

$$P_e = \begin{cases} \dfrac{1}{2} e^{-\frac{1}{2\xi/N}}, & \dfrac{1}{\xi/N} < 2 \\ \dfrac{e^{-1}}{\dfrac{1}{\xi/N}}, & \dfrac{1}{\xi/N} \geq 2 \end{cases} \tag{12.18}$$

In general, this analysis can be extended to where the probability of symbol error for uncoded M-ary orthogonal communications with noncoherent detection where the worst-case SER (worst-case in the sense of the communicator in that the jammer can do no better) is given by [12]

$$P_s = \begin{cases} \dfrac{C}{E_b / J_0}, & E_b / J_0 > X_0 \\ \dfrac{1}{2(M-1)} \displaystyle\sum_{i=2}^{M} (-1)^i \binom{M}{i} e^{-\frac{l(i-1)E_b}{iJ_0}}, & E_b / J_0 \leq X_0 \end{cases} \tag{12.19}$$

where $M = 2^K$ and the optimum fraction $\gamma^* = X_0 N_J / E_b$. The values for the parameters as a function of M are given in Table 12.1.

The expression above for $M = 2$ is plotted in Figure 12.6 for noncoherent BFSK and two values of processing gain. Below $N/\xi = 2$, P_b falls off linearly with ξ as opposed to exponentially when $N/\xi > 2$. Therefore, partial band jamming is considerably more effective at interfering with the communicator than BBN jamming in this region. If the jammer knows, can determine, or can estimate the SNR of the target receiver it can adjust its power and spectrum fraction to achieve the desired level of P_s.

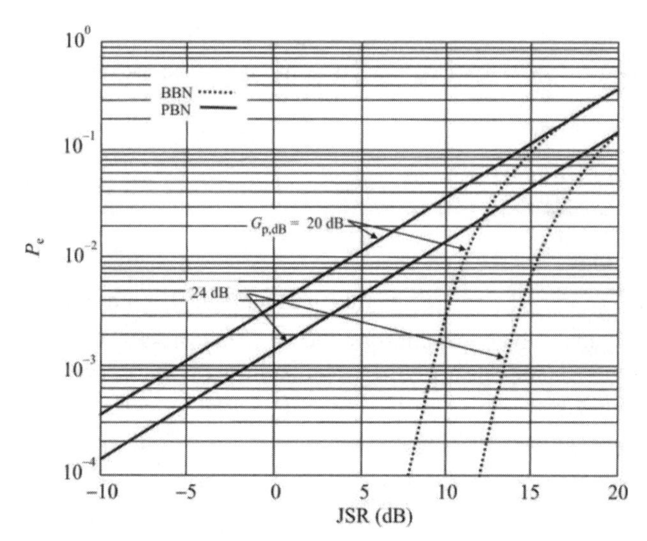

Figure 12.6 Comparison of BBN jamming versus PBN band jamming for representative processing gains (20 dB and 24 dB).

The jammer is prudent to choose the optimum value for γ rather than jam the whole band. The advantage of PBN jamming is evident from this chart. At a 10^{-3} BER, the PBN jammer has about a 15 dB advantage over the BBN jammer. This advantage can be almost completely restored with coding and interleaving as long as the communication system can maintain operation in the low bit error rate region. When a channel BER of 5% or so is established by the jammer then the coding gain almost totally disappears as discussed in Section 3.3.

The dependence of P_e on the processing gain is illustrated in Figure 12.7 where P_e is plotted versus N. The performance advantage of PBN jamming is clearly evident, keeping P_e relatively high even as the processing gain is increased tenfold. It should be noted, however, that JSR decreases as the fraction of the band jammed increases. This is because it is more prudent to put the jamming energy into a smaller segment so that the communicator does not see the jammer as much but when they do meet, the jammer has more concentrated energy so the communicator has a harder time getting the bits through without error.

12.3.2 Error-Coded SFHSS MFSK Systems

Utilizing the jammer state information then, when it is assumed that the jamming noise density is sufficiently larger than the background noise, the bit error probability when only noise is present can be ignored; then if a symbol is not jammed, an error-free decision can be made as to whether the symbol was a mark

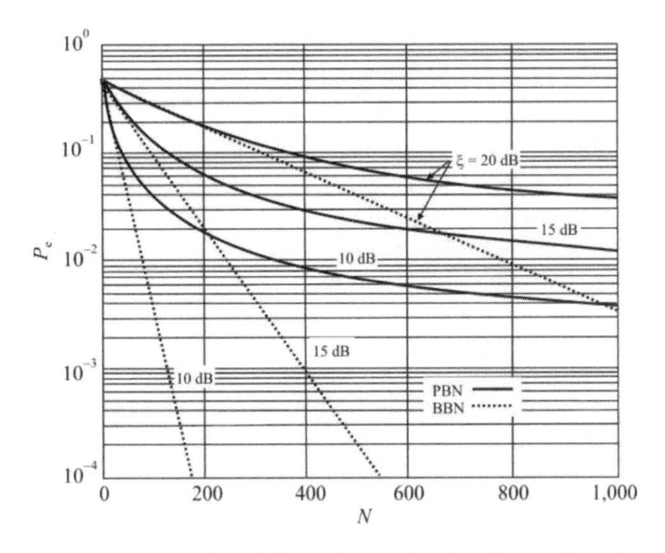

Figure 12.7 PBN jamming compared to BBN jamming as a function of processing gain.

or a space. With diversity, if all the symbols were jammed then the decision as to whether the transmitted symbol was a mark or a space is based on the largest detector output summed over the diversified symbols. This is referred to as soft decision detection [13].

Worst-case PBN jamming performance against diversity coded BFSK is shown in Figure 12.8 [14]. In this chart, m refers to the diversity of the code. The diversity is the number of equal energy subsymbols derived from the data bits. In repetition coding, for example, it is the number of frequencies transmitted per information bit. Also shown on the chart is the optimum value of the diversity, denoted m^*, which is a function of the SNR.

Ma and Poole [12] characterized several codes used for MFSK in partial band noise jamming and compared them to the uncoded case. The results for binary error-correcting codes when the worst-case parameters are used is shown in Figure 12.9. Figure 12.9 assumes hard decision decoding and no JSI availability. Table 12.2 shows numerically the performance of the codes considered under the worst-case situations for a BER of 10^{-5}. At reasonable values of JSR (> –10 dB), the addition of coding to the FHSS system actually increases jamming performance!

Shown in Figure 12.10 is the E_b/J_0 required for BFSK to produce a $P_e = 10^{-5}$ versus the fraction of the spectrum jammed for binary coding without diversity. Again, this data assumes hard decision decoding with no JSI. An optimum value

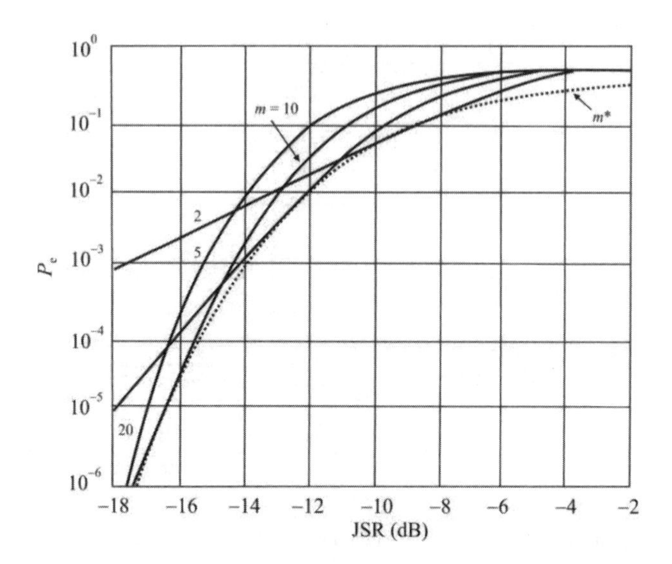

Figure 12.8 Worst-case PBN jamming performance against BFSK with diversity *m* chips/bit when the thermal noise is ignored. Since thermal noise is ignored, $\upsilon = 1/\xi$. (After: [14]. © 1994, McGraw-Hill. Reprinted with permission.)

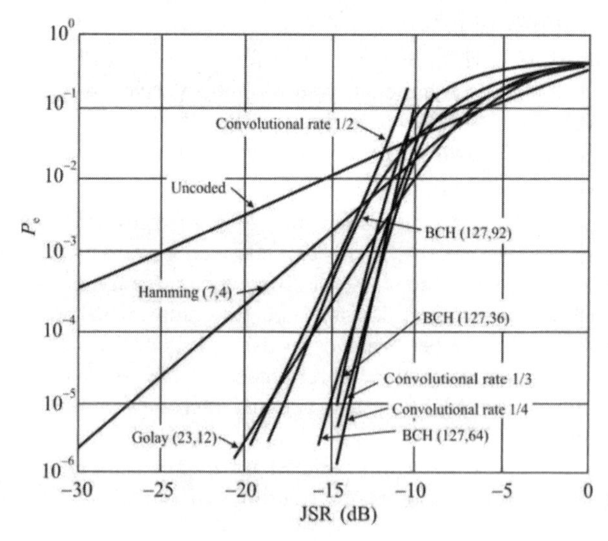

Figure 12.9 Jamming performance against binary error correcting codes in worst-case PBN jamming. (After: [12]. © IEEE 1984. Reprinted with permission.)

Table 12.2 Characteristics of Some Common Codes

Code	Effective Code Rate ($R = lr$)	Spectrum Fraction	JSR at 10^{-5} (dB)	Coding Gain (dB)
Hamming (7,4)	4/7	0.006983	–27.0	12.66
Golay (23,12)	12/23	0.05053	–12.8	26.86
BCH (127,92)	92/127	0.03573	–12.8	26.86
BCH (127,64)	64/127	0.1071	–15.3	30.36
BCH (127,36)	36/127	0.2391	–14.7	30.96
Convolutional (2,1)7	1/2	0.0762	–17.2	28.46
Convolutional (3,1)7	1/3	0.301	–12.0	32.66
Convolutional (4,1)7	1/4	0.43	–12.7	32.96
Convolutional (8,1)7	1/8	0.869	–12.65	33.10

Source: [12].

of γ, γ^*, can again be discerned, which depends on the type of coding employed. At values of γ other than γ^* the SNR required to produce a BER of 10^{-5} is smaller.

Closed-form error probabilities for coded performance are difficult to obtain. Instead bounds are often used to evaluate performance; one of these is the Chernoff bound. This bound for communication performance with coding and diversity utilizing soft decision decoding and assuming JSI is available is given by [12]

$$P_s \le (2^l - 1)\frac{1}{2}\left[\frac{4\gamma}{3}e^{-\frac{\gamma l E_b}{3mJ_0}}\right]^m \tag{12.20}$$

where

m = diversity, redundant transmissions per symbol;
L_S = diversity, redundant transmissions per bit;
$R = K_r$, the effective code rate;
$M = 2^l$.

In this case $M = 2$, so $M = L_S$. Finding the optimum γ yields

$$\gamma^* = \frac{3mJ_0}{lE_b} \tag{12.21}$$

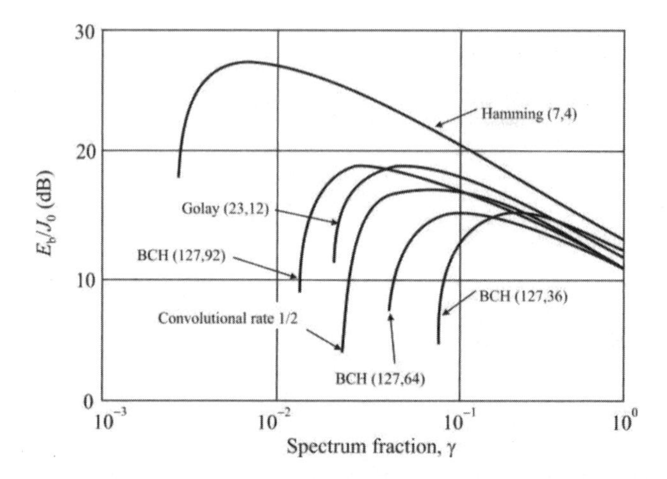

Figure 12.10 The E_b/J_0 at a BER of 10^{-5} for binary coding without diversity. Since thermal noise is ignored, $E_b/J_0 = \upsilon = 1/\xi$. (After: [12]. © IEEE 1984. Reprinted with permission.

This is the *quasi-worst-case spectrum fraction* because it is based on a bound and not an actual equality. Finding the optimum diversity *m* yields

$$m^* = \frac{l}{4}\frac{E_b}{N_0} \tag{12.22}$$

where again, m^* is the *quasi-optimum diversity* for the same reason. Substituting this into the equation for γ^* yields

$$\gamma^* = \frac{3}{4} \tag{12.23}$$

irrespective of the other parameters. This then causes the channel symbol error probability to become

$$P_s^* \le \frac{2^l - 1}{2} e^{-\frac{lE_b}{4J_0}} \tag{12.24}$$

for the MFSK case and since $l = 1$ and $P_e = P_s$ for BFSK,

$$P_e^* \le \frac{1}{2} e^{-\frac{E_b}{4J_0}} \tag{12.25}$$

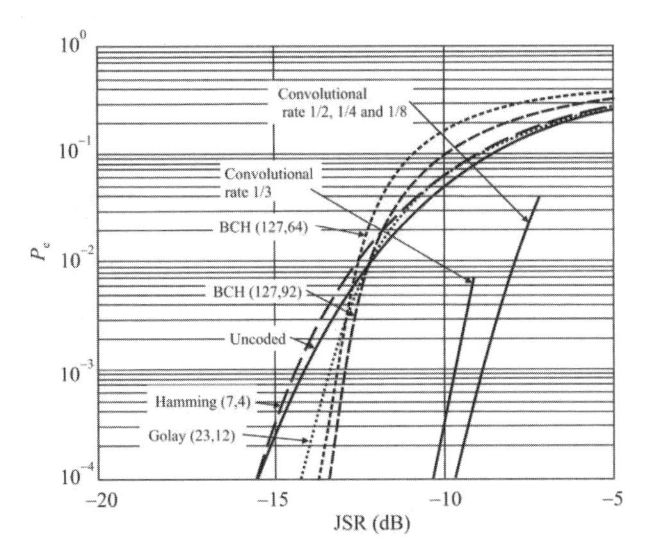

Figure 12.11 PBN jamming performance against binary error correcting codes employing optimum diversity for BFSK ignoring the effects of thermal noise. (After: [12]. © IEEE 1984. Reprinted with permission.)

Thus, there is an optimum spectrum fraction and optimum amount of frequency diversity for worst-case jamming effectives. Figure 12.11 shows this worst-case jamming effectiveness with PBN jamming for binary error correction codes with optimum diversity included [12]. Note that for many of the codes, above $\xi \approx -12$ dB, uncoded transmission outperforms coding. This data applies to FFHSS as well as SFHSS with bit interleaving sufficient to approximate a memoryless channel.

The required JSR to achieve a 10^{-5} BER as a function of the spectrum fraction covered by a PBN jammer against codes utilizing optimum diversity is shown in Figure 12.12. While $P_e > 10^{-5}$ is not adequate for most jamming attacks, it does illustrate that with as little as 1% of the spectrum jammed, the JSR need not be that large to be effective.

Table 12.3 shows the error performance of diversity only MFSK with optimum diversity in the best case jamming environment [12]. Table 12.4 shows the error performance when coding is employed in the worst-case jamming situation [12].

There is an optimum amount of diversity depending on the SNR. The upper bound on the BER for BFSK for the optimum diversity and the worst-case band fraction of $\gamma = 0.75$ in PBN jamming is shown in Figure 12.13 [15]. Comparing

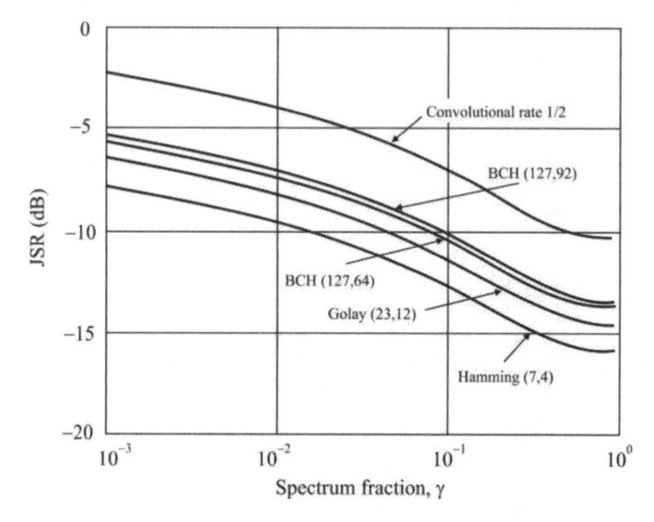

Figure 12.12 Effects of various spectrum fractions covered by a PBN jammer against coded communications employing optimum diversity. (After: [12]. © IEEE 1984. Reprinted with permission.)

this with Figure 12.4 it is clear that including the diversity has restored to the communicator most of the SNR advantage that the PBN jammer had without diversity. In this chart the optimum diversity is given by

$$m^* = \frac{1}{4}\frac{1}{\xi} \tag{12.26}$$

as long as $\xi \leq -6$ dB. For larger values of m than m^*, BER performance degrades because of the effects of noncoherent combining in the receiver.

The comparison shown in Figure 12.14 [16] shows the performance gain of coding (diversity) for noncoherent BFSK compared with BBN jamming without diversity and PBN jamming with no diversity. It again shows that the diversity processing removes most of the jammer gain over no diversity. The approximately 30 dB gained by the jammer due to the linear relationship between JSR and P_b disappears at low values of SNR. At reasonable values of JSR (above about −10 dB), however, PBN jamming with γ^* when the communication system uses diversity m^* the BER increases over the case of no diversity.

In a Rayleigh fading communication channel (not the jammer channel), the energy per bit of the signal at the receiver is described by the density function

Table 12.3 Jammer Effectiveness Against Diversity Only MFSK with Optimum Diversity—Chernoff Bound/Union Bound

M	K	E_b/J_0 at 10^{-5} BER (dB)	Diversity Gain (dB)	Optimum Diversity per bit, L*	Optimum Diversity per Symbol
2	1	12.36	29.30	12.82	-
4	2	12.62	30.05	5.76	12.52
8	3	12.12	30.79	4.06	12.18
16	4	12.10	31.45	3.22	12.88
32	5	12.36	32.06	2.72	12.60
64	6	12.79	32.63	2.38	14.28

Source: [12].

Table 12.4 Jammer Effectiveness Against Optimal Diversity with Worst-Case Jamming

Code	Effective Code Rate ($R = K_r$)	Best Case Jamming Spectrum Fraction, γ*	Optimum Diversity per Encoded Bit, L*	JSR at 10^{-5} BER (dB)	Diversity Gain (dB)	Error Correcting Coding Gain (dB)
Hamming (7,4)	4/7	¾	5.97	−12.21	29.30	0.15
Golay (23,12)	12/23	¾	3.99	−14.85	29.30	1.51
BCH (127,92)	92/127	¾	4.31	−12.77	29.30	2.59
BCH (127,64)	64/127	¾	3.13	−12.95	29.30	2.41
BCH (127,36)	36/127	¾	2.43	−15.35	29.30	1.01
Conv (2,1)7	½	¾	1.53	−12.88	29.30	5.48
Conv (3,1)7	1/3	¾	0.91	−12.40	29.30	5.96
Conv (4,1)7	¼	¾	0.77	−12.88	29.30	5.48
Conv (8,1)7	1/8	¾	0.39	−12.88	29.30	5.48
Hamming (31,26)7	26/31	¾	6.86	−15.15	29.30	1.21
Golay (24,12)7	12/24	¾	4.02	−15.07	29.30	1.29

Source: [12].

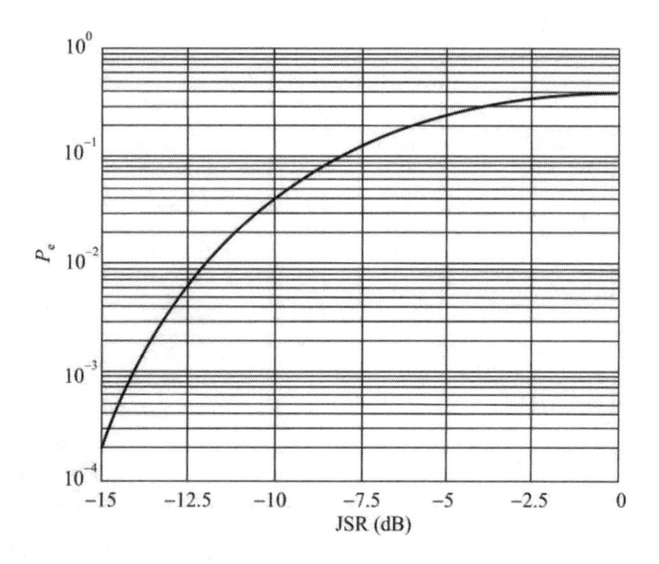

Figure 12.13 Worst-case PBN jamming when the signal has optimum diversity. (After: [15]. © McGraw-Hill 1994. Reprinted with permission.)

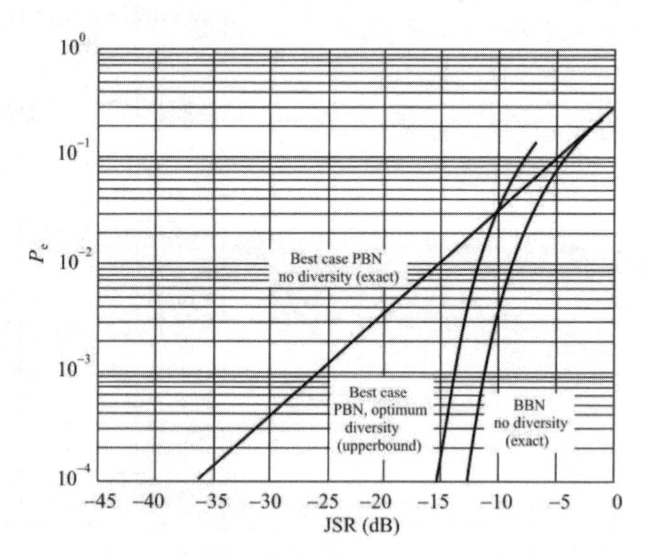

Figure 12.14 Jamming gain degradation when the communicator uses diversity for noncoherent BFSK. This is PBN jamming with no diversity, optimum diversity, and BBN jamming. (After: [16]. © McGraw-Hill 1994. Reprinted with permission.)

$$p(E_S) = \frac{1}{m_b} e^{-\frac{E_S}{m_b}} \tag{12.27}$$

where m_b is the mean value of the energy per bit. In this case, the BER for noncoherent BFSK is given by [17]

$$P_e = \frac{1}{\frac{2}{\gamma} + \frac{m_b}{J}} \tag{12.28}$$

This function is plotted in Figure 12.15. At $\xi = 0$ dB ($m_b/J = 0$ dB), $\gamma \approx 0.2$ is adequate to achieve a BER = 10^{-1}. As the JSR decreases, this fraction becomes larger, requiring about $\gamma \approx 0.5$ at $\xi \approx -9$ dB. For lower JSR values, the 10^{-1} BER goal is not achievable. Such JSR levels are possible if the jammer can get close to the target receiver.

If, in addition, the jammer channel is also subject to Rayleigh fading, then the jammer power density follows a similar probability density function, to wit,

$$p(J) = \frac{1}{m_J} e^{-\frac{J}{m_J}} \tag{12.29}$$

where m_J is the mean value of the jamming power. The mean values m_b and m_J are different because:

- The transmitter ERPs are different.
- The propagating paths are not the same and have different characteristics.

In that case the BER is bounded by

$$P_e \leq \frac{1}{\frac{2}{\gamma} + \frac{m_b}{m_J}} \tag{12.30}$$

The graph for this function is the same as Figure 12.15 with the SNR defined as

$$\upsilon = \frac{m_b}{m_J} \tag{12.31}$$

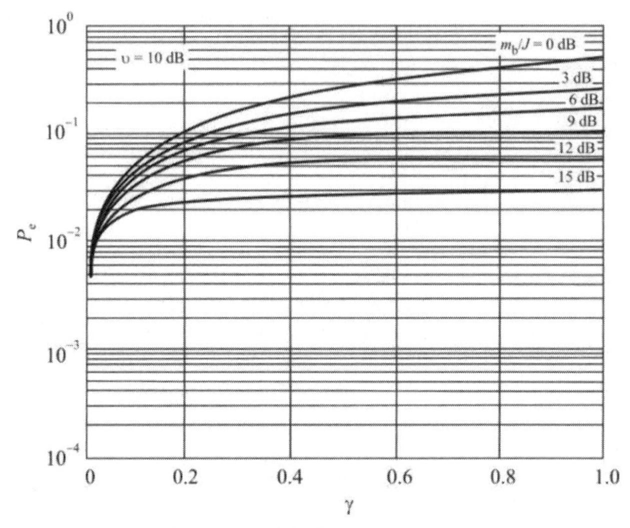

Figure 12.15 Probability of error for a Rayleigh fading communication channel and PBN jamming.

In this case the peak value of the BER occurs at $\gamma = 1$ so the optimum jamming strategy is to jam the full band. However, $P_e > 10^{-1}$ for $\upsilon < 6$ dB at $\gamma \approx 0.2$ to 0.3, so, again, relatively small amounts of jammed spectrum can produce favorable results.

12.4 Multitone Jamming of SFHSS Systems

One technique for jamming FSK signals is to broadcast narrowband signals at the tone frequencies used by the communication system. As described in Chapter 8, such jamming is referred to as multitone jamming. The fraction of the tones jammed, N_J, compared to the totality of channels in W_{ss} is given by γ, the same quantity as for partial-band jamming. That is,

$$\gamma = \frac{N_J}{N_c} \tag{12.32}$$

where

$$N_c = \frac{W_{ss}}{W_c} \tag{12.33}$$

and W_c is the width of one channel.

It is not necessary to hit the tone frequencies exactly. It is sufficient to put the tone anywhere within the pass band of the receiver filters to be effective. It is also possible to emit energy at randomly selected frequencies. This type of jamming is called *independent multitone jamming*.

The number of tones per jammed channel in MT jamming is given by n. Note that all channels are not necessarily jammed, just those that have n tones in them.

When a mark (or space) is sent the following possibilities arise: (1) one of the tones is at the frequency of the mark and one or more are at some other location, or (2) none of the tones is at the mark frequency. For the first case, the power or the signal at the mark frequency coupled with the jammer power at the same frequency would cause the receiver to decide correctly that a mark was sent. In the second case, successful jamming would depend on the relative power levels between the mark signal and one (or more) of the jamming signals. However, when the jammer power is larger than the mark signal only one of the jammer tones is necessary, not multiples. Therefore this strategy wastes the jammer power. Thus a single tone per channel at most is a better strategy, spreading the jammer power into more channels.

This tone, placed at one of the data tone frequencies, will cause an erroneous detection decision $(M-1)/M$ times when the tone sent is evenly distributed over the M tones. The other decision will be correct since the jammer tone is at the tone frequency that was actually sent.

12.4.1 Uncoded SFHSS MFSK Systems

For BFSK there are two possible values for n: $n = 1$ and $n = 2$. For $n = 1$, a symbol error will occur when the jamming tone power is larger than the symbol tone power and the jammer tone frequency is at the complementary tone frequency. The probability of this is $\gamma / 2$. Since these are independent events, the probability of a bit error is given by

$$P_b = \frac{\gamma}{2} P_e \tag{12.34}$$

where P_e is given by [18, 19]

$$P_e = \frac{1}{2} Q\left(\sqrt{\frac{J}{P_N}}, \sqrt{\frac{R}{P_N}} \right) \tag{12.35}$$

when the target signal is given by

$$r(t) = \sqrt{2R} \cos(2\pi f_i t), \; kT_s \le t < (k+1)T_s; \qquad i = 1,2; \; k = 0,1,\ldots \tag{12.36}$$

R is the average power in the signal, T_s is the symbol duration, and the jammer tone at the receiver is given by

$$j(t) = \sqrt{2J} \cos[2\pi f_i t + \phi_i(t)], \; i = 1,2 \tag{12.37}$$

J is the average jammer tone power, P_N is the thermal noise power in the BFSK filter, and, lastly, $Q(x, y)$ is Marcum's Q-function from Appendix A. This function is plotted in Figure 12.16 for υ = 10 dB and 20 dB when γ = 10/2,400 and in Figure 12.17 for γ = 100/2,400. The BER increases until ξ = 1 to 3 dB, after which it remains constant irrespective of γ and the SNR. In neither case is a BER of 10^{-1} achieved.

When $n > 1$ and MFSK in a jammed channel, then at least two of the detectors will have significant energy in them. At least one of them is assured to be in a complementary channel to the correct symbol detector. However, one of them may be in the correct symbol detector or in another complementary channel. If the jammer tone is in the correct detector, then, as explained in Section 6.1.3, the phase relationship between the symbol tone and the jammer tone may make that detector output greater or smaller than the jammed complementary channel depending on the phase relationship of the jammer tone to the data tone. Of course, if the second jammer tone is at another complementary frequency, then a detection error is forced if the jammer tone is more powerful than the data tone, and no detection error occurs otherwise. Thus, even though the channel contains jamming tones and one of these is at the same frequency as the data symbol, the correct symbol may still be detected.

The jamming performance depends on, among other things, the number of tones in each channel. For BFSK, there are two possible tones per channel and with n = 2 MT jamming one or two of these tone frequencies can be jammed, depending on whether the channel is jammed and, of course, the SNR.

The probability of generating a detection error for BFSK is given by

$$P_e = \Pr\{e | \text{land on 2 tones}\} \Pr\{\text{land on 2 tones}\}$$

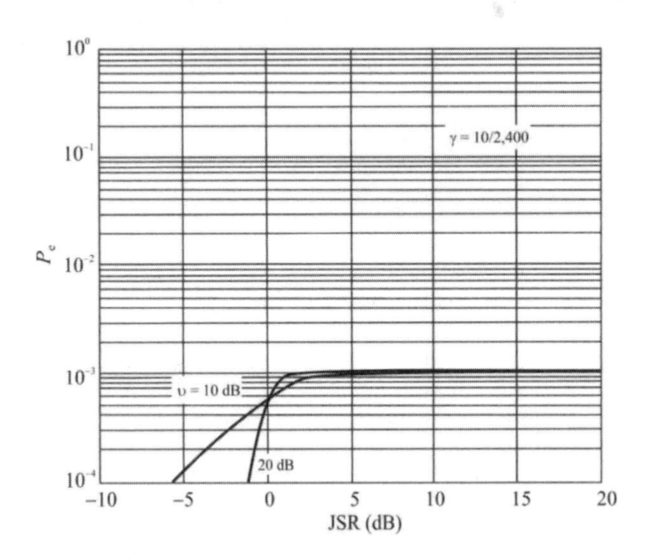

Figure 12.16 Multitone jamming performance for SFHSS uncoded, noncoherent BFSK for $n = 1$ when $\gamma = 10 / 2{,}400$.

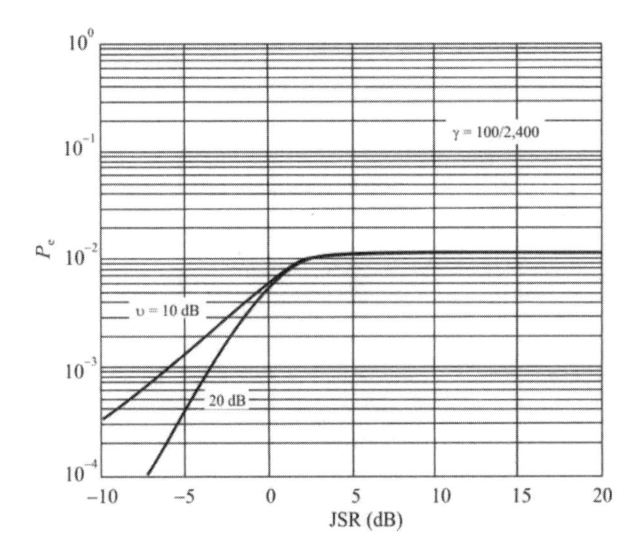

Figure 12.17 Multitone jamming performance against SFHSS uncoded, noncoherent BFSK for $n = 1$ multitone jamming when $\gamma = 100 / 2{,}400$.

$$+\Pr\{\mathrm{e}|\text{land on 1 tone}\}\Pr\{\text{land on 1 tone}\}$$
$$+\Pr\{\mathrm{e}|\text{land on 0 tones}\}\Pr\{\text{land on 0 tones}\} \qquad (12.38)$$

since these are independent events. The second of the probabilities in each line are given by [19]

$$\Pr\{\text{land on 2 tones}\}=\frac{N_\mathrm{J}}{N_\mathrm{c}}\frac{N_\mathrm{J}-1}{N_\mathrm{c}-1} \qquad (12.39)$$

$$\Pr\{\text{land on 1 tone}\}=\frac{N_\mathrm{c}-N_\mathrm{J}}{N_\mathrm{c}}\frac{N_\mathrm{c}-N_\mathrm{J}}{N_\mathrm{c}-1} \qquad (12.40)$$

$$\Pr\{\text{land on 0 tones}\}=\frac{N_\mathrm{c}-N_\mathrm{J}}{N_\mathrm{c}}\frac{N_\mathrm{c}-N_\mathrm{J}-1}{N_\mathrm{c}-1} \qquad (12.41)$$

where

N_c = number of channels in W_ss;
N_J = number of jammer tones (n = one or two per jammed channel).

Denoting the first probability on each line in (12.38) by P_2, P_1, and P_0, respectively, then [20]

$$P_2=\frac{1}{2\pi}\int_0^{2\pi}\left\{Q\left[\sqrt{\frac{J}{P_\mathrm{N}}},\frac{D(x)}{\sqrt{2P_\mathrm{N}}}\right]-\frac{1}{2}\exp\left[-\frac{2J+D^2(x)}{4P_\mathrm{N}}\right]I_0\left[\frac{\sqrt{2J}D(x)}{2P_\mathrm{N}}\right]\right\}dx \qquad (12.42)$$

with

$$D^2(x)=2R+2J+4\sqrt{RJ}\cos(x)$$

$$P_1=\frac{1}{2}Q\left[\sqrt{\frac{J}{P_\mathrm{N}}},\sqrt{\upsilon}\right] \qquad (12.43)$$

and

$$P_0=\frac{1}{2}\exp\left(-\frac{1}{2}\upsilon\right) \qquad (12.44)$$

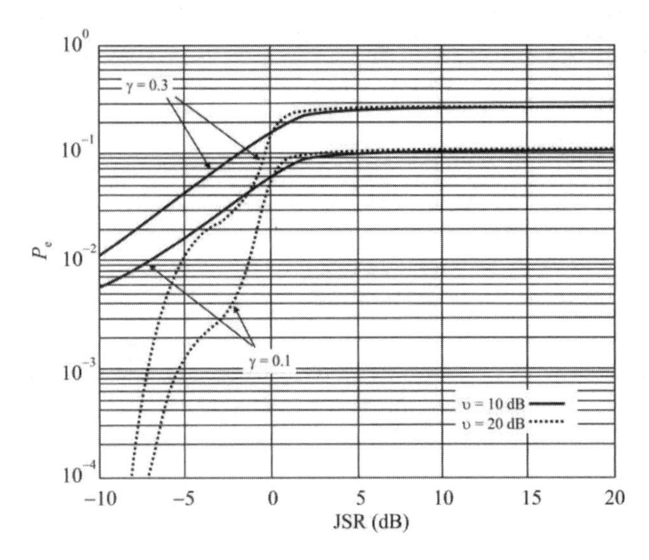

Figure 12.18 Performance of MT jamming against uncoded, noncoherent BFSK SHF targets.

The MT jammer performance is illustrated in Figure 12.18 for two values of SNR and spectrum fractions. The spectrum fractions are given by (12.32) and represent the fraction of the channels with jammer tones in them. In all cases, effective jamming occurs when $\xi > 2$ or 3 dB irrespective of the SNRs considered here.

Peterson [21] also points out that tone jamming is more efficient than PBN jamming with the optimal band fraction γ according to Table 12.5. To glean the significance of this analysis, consider the example from earlier. Suppose the jammer is mounted in a UAS with an ERP of $J_T = 100W$. Suppose the target communication networks need a BER of 10^{-3} or better to work properly (this is one bit in 1,000 or fewer in error) and they employ omnidirectional antennas with 0 dB gain. The chart in Figure 12.19 [22] indicates that in a Rayleigh fading channel, an SNR of 30 dB or more is required, which corresponds to a factor of 1,000. (Most channels in the VHF range exhibit either Rayleigh or Ricean fading.

Table 12.5 Comparison of Tone Jamming to PBN Jamming

M	Efficiency Improvement of Tone Jamming Over Partial Band Jamming (dB)
2	4.3
4	6.3
8	8.3
16	12.5

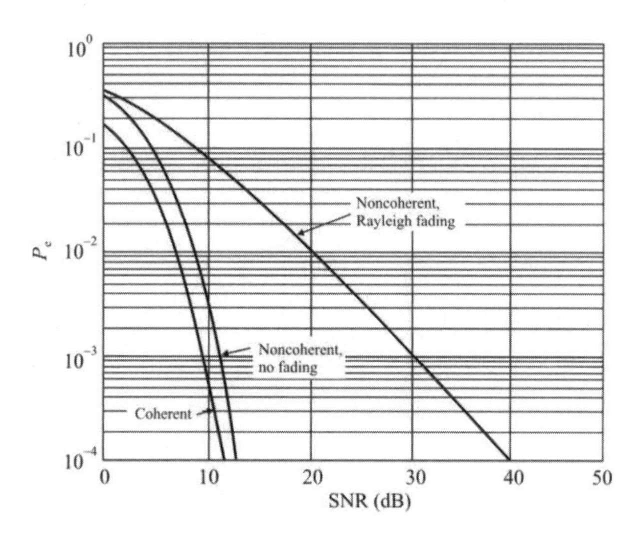

Figure 12.19 FSK BER performance versus SNR in a Rayleigh fading channel. (Source: [22]. © Academic Press 1971. Reprinted with permission.)

Rayleigh fading is due to specular or diffuse signal components only while Ricean fading has a significant direct component as well.) Suppose the total hop bandwidth is 50 MHz at a middle frequency of 60 MHz, a data rate of 25 kbps, and a suburban noise environment. Since $E_b / N_J = R / J\ W_{ss} / R_b$ then the required S/J is $10^3 \times 25 \times 10^3/50 \times 10^6 = 0.5$. This corresponds to a JSR of 2 (3 dB). From Figure 12.20, at any link distance against 1 W or 10 W targets the links are jammed. For targets of 100 W, the jammer must be very close and even then only for the 30 km link distance.

12.4.2 Error-Coded SFHSS MFSK Signals

The performance of MT jamming when the BFSK signal is coded with time diversity m is shown in Figure 12.21 [23]. In the horizontal region every channel has a tone in it so the signal is jammed with probability one. As always, it should be remembered that diversity reduces data throughput. When $m = 5$, the data bit is sent five times so the data throughput is reduced about five times, but a $\xi = -8$ dB causes $P_e > 10^{-1}$ in this case.

As shown above for partial band jamming, the jamming SNR advantage is largely lost because of diversity in the communication signal. This is shown in Figure 12.21. When diversity is not used in the communication system, the jammer using MT jamming has a 25 dB SNR advantage over the receiver at a

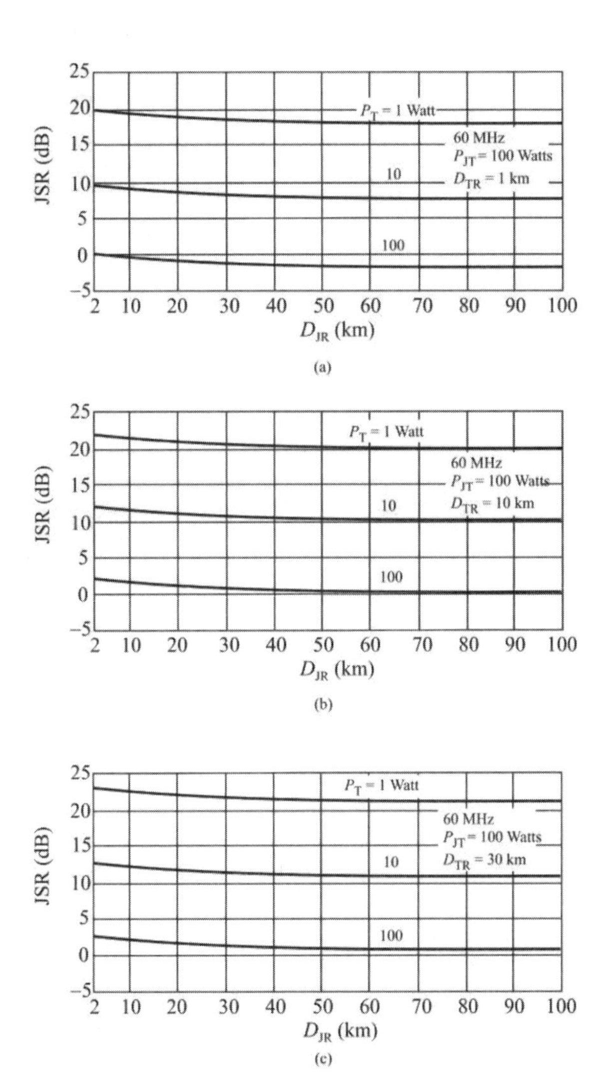

Figure 12.20 JSRs versus jammer to receiver distance for three target link distances: (a) D_{TR} = 1 km, (b) D_{TR} = 10 km, and (c) D_{TR} = 30 km.

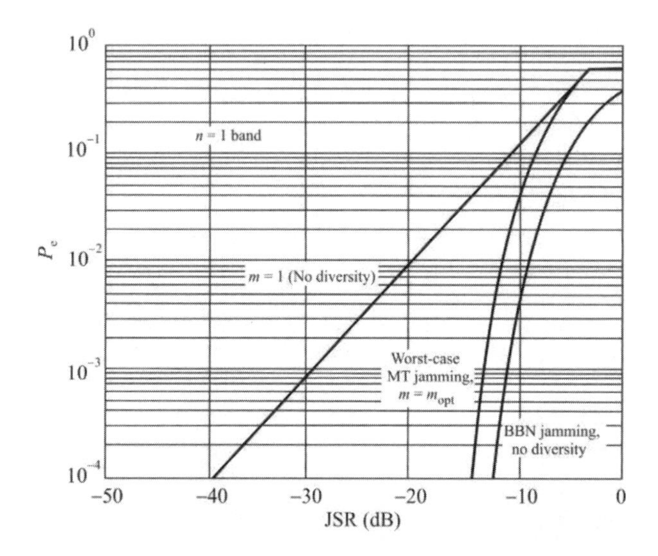

Figure 12.21 MT jamming of a time diversity coded BFSK signal. The time diversity of the code is denoted by m. (After: [23]. © McGraw-Hill 1994. Reprinted with permission.)

BER of 10^{-4}. With the optimum amount of diversity the receiver gets this jammer advantage back to within 1.6 dB of BBN jamming. Again though, adding time diversity reduces data throughput since a single data bit must be sent several times.

Another technique is to send pilot bits whose value and location are known (in time and frequency) and constant. These pilot bits are used to estimate the channel parameters. Such tones (bits) can be used to estimate multipath, for example, which may be necessary anyway for the channel to work correctly.

12.5 Follower Jamming of SFHSS Systems

As for FFHSS systems, a follower jammer attempts to follow an SFHSS signal as it changes frequencies. Measurements on intercepted energy are made and characteristics are measured to try to correlate one hop with the last. Once a determination is made that a hop correlates with the target of interest, a jamming signal is initiated at that frequency. That jamming waveform could be of many types, but the most common are narrowband noise and one or a few tones [24]. Both increase the energy in the filters in the target receiver.

As opposed to FFHSS systems, where consideration of jamming one, two, or all three channels (data tone, one of the other adjacent channels, two adjacent channels only, or all three channels) needs to be addressed, for SFHSS there is no distinction between jamming the data channel and its complementary, since it is not known which to call which. The net effect of jamming only one channel is the same whichever is chosen, which is to reduce the BER by half what it would be otherwise since, on average, half of the data bits are jammed in that bit's complementary channel. Also, for SFHSS, since the complementary channel can be determined, it makes no sense to jam the channel on the other side of whichever tone is called the data channel. Therefore, the follower schemes that make sense for SFHSS are not exactly the same as for FFHSS. The ones that do potentially make sense jam only one channel or two (both data channels).

The jamming performance will depend on which scheme is chosen. What makes follower jamming different for SFHSS systems from FFHSS systems is twofold:

- The complementary channel can be determined by measurement so it is not an unknown.
- Since several data bits are sent per dwell, successfully jamming the channel results in jamming the data bits, whereas for FFHSS, the data bit is also transmitted on one or more additional dwells due to the inherent frequency diversity.

Narrowband in this case refers to a channel width that is approximately half (for BFSK) its normal definition so that only one of the data tones is jammed. In the low VHF range, for example, the bandwidth would be approximately 12.5 kHz while in the HF range it would be approximately 1.5 kHz.

This argument assumes that BFSK was the modulation, but the arguments are easily extended to include MFSK in the logical way. Jamming performance, of course, will depend on how many tones are present. In this section, uncoded BFSK will be assumed.

Therefore the probability of bit errors can be expressed as in Chapter 11, suitably modified for SFHSS systems. Again, let γ denote the fraction of the dwell upon which jamming energy is applied. The probability of a bit error for an individual dwell is given by

$$P_\text{e} = (1-\gamma)\frac{1}{2}\exp\left(-\frac{1}{2}\upsilon\right) + \gamma P_{\text{e}_2} \tag{12.45}$$

where the first term accounts for the nonjammed portion of the dwell. P_{e_2} is determined by the particular jamming technique.

12.5.1 Noise Jamming

For noise jamming, the noise can be one channel wide or can cover both channels. In the latter case, the jamming performance is the same as PBN jamming when the correct hop association is accomplished. In this case, however, γ corresponds to the fraction of the hop jammed as opposed to the fraction of the spectrum covered by the jammer.

12.5.1.1 Single Channel

If only one of the two data channels is jammed with a narrowband noise signal of power P_J, then following Torrieri [25], from (3.89), setting $B_1 = 0$, $B_2 = 0$, $P_{N_1} = P_N + J, P_{N_2} = P_N$, and $A = \sqrt{2R}$, and carrying out the calculus yields

$$P_{e_2} = \frac{1}{2}\exp\left(-\frac{1}{\frac{2}{\upsilon}+\xi}\right) \tag{12.46}$$

12.5.1.2 Both Channels

If noise of power P_J is applied to both data channels then the results are the same as if BBN jamming were applied. The resulting probability of bit error is [18]

$$P_{e_2} = \frac{1}{2}\exp\left(-\frac{1}{2}\frac{1}{\frac{1}{\upsilon}+\xi}\right) \tag{12.47}$$

12.5.2 Tone Jamming

When unmodulated tones are used as the jammer waveforms, it is assumed that the tones are placed within the bandpass of the receiver filters of width W_F. The tones need not be at exactly the same frequency as the data tones to effectively raise the energy level in the complementary channel filter [26].

12.5.2.1 Single Tone

When only a single tone is used as the jamming waveform then the probability of a bit error is given by [18]

$$P_{e_2} = \frac{1}{2} Q\left(\sqrt{\frac{J}{P_N}}, \sqrt{\upsilon} \right) \tag{12.48}$$

12.5.2.2 Two Tones

When both of the channels are jammed with tones then [18]

$$P_{e_2} = \frac{1}{2\pi} \int_0^{2\pi} \left\{ Q\left[\sqrt{\frac{J}{P_N}}, \frac{D(x)}{\sqrt{2P_N}} \right] - \frac{1}{2} \exp\left[-\frac{2J + D^2(x)}{4P_N} \right] I_0 \left[\frac{\sqrt{2J}\, D(x)}{2P_N} \right] \right\} dx \tag{12.49}$$

where

$$D^2(x) = 2R + 2J + 4\sqrt{RJ}\cos(x)$$

12.5.3 Comparison

Comparisons of these four approaches to follower jamming SFHSS systems for representative values of the parameters are shown in Figure 12.22 through 12.25. In Figure 12.22, υ = 10 dB and $\gamma = 0.1$. All of the jamming approaches produce similar results, and the BER never achieves 10^{-1}. When the fraction of the dwell jammed is raised to $\gamma = 0.3$, the results in Figure 12.23 ensue against the same SNR. Again, all techniques produced similar results, achieving a 10^{-1} BER between 0 and 2 dB JSR.

For a 20 dB SNR and $\gamma = 0.1$, the results are shown in Figure 12.24. At higher values of JSR the results are similar to those for ν = 10 dB. At lower JSRs values, however, the results are different. When $\gamma = 0.3$, the performance is shown in Figure 12.25.

These results indicate that the particular method of follower jamming SFHSS targets is not a critical factor. The timing, however, as well as the ability to accurately identify the new hop as belonging to the target of interest *is* critical. In dense target environments it becomes extremely difficult to associate one hop with the last one from the same transmitter.

Performance of the various techniques as the dwell fraction is varied is shown in Figure 12.26 for a few values of JSR. As expected, when JSR is greater than

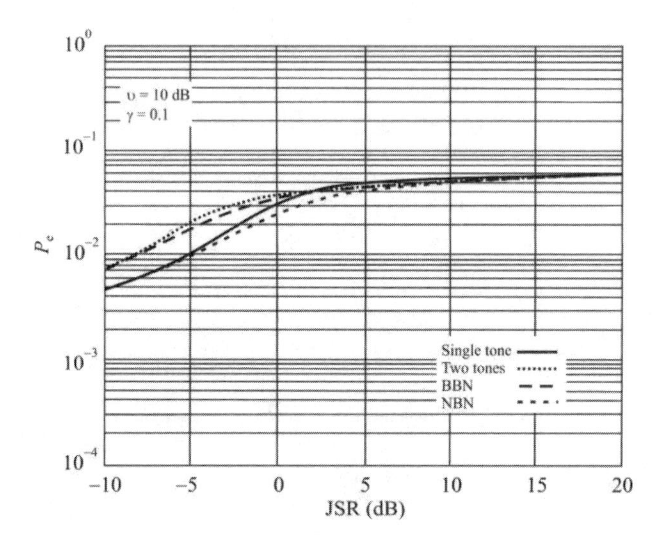

Figure 12.22 Follower jamming performance against SFHSS systems when ν = 10 dB and γ = 0.1.

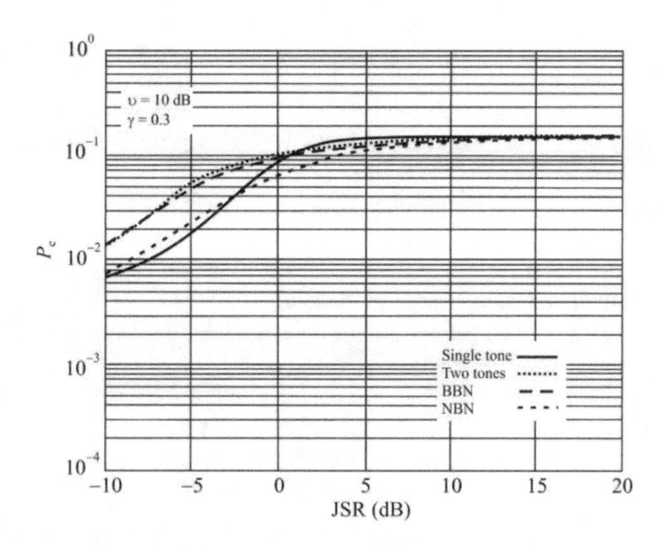

Figure 12.23 Follower jamming performance against SFHSS when υ = 10 dB and γ = 0.3.

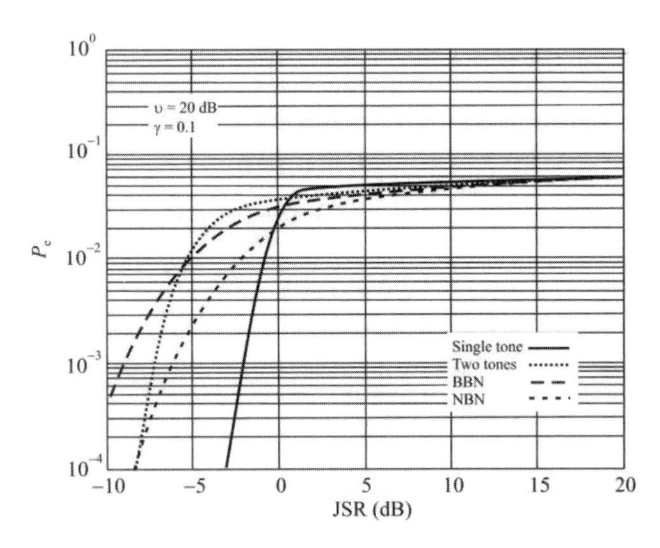

Figure 12.24 Follower jammer performance against SFHSS when υ = 20 dB and γ = 0.1.

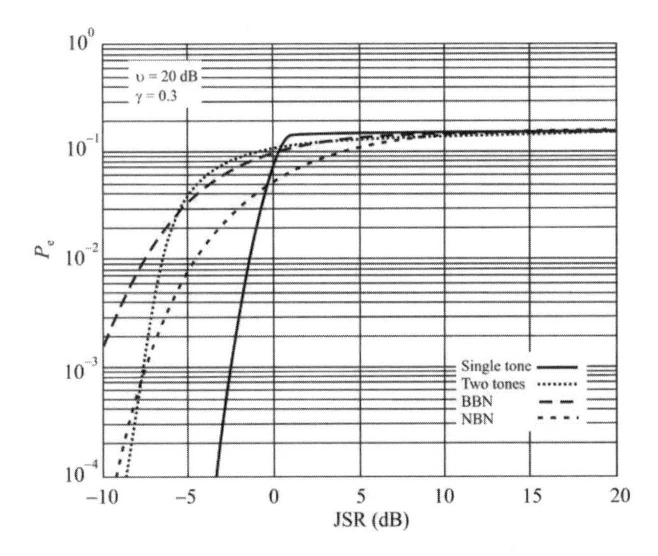

Figure 12.25 Follower jammer performance against SFHSS when υ = 20 dB and γ = 0.3.

0 dB, the performance is not improved much. Also, the performance degrades gracefully as the fraction decreases.

12.6 Error-Coded MFSK Jamming

SFH signals can be treated quasi-statically. That means that within a dwell, where many data bits are sent, the signal can be treated as if it were not hopping. This is particularly useful for coded MFSK signals since there is a wealth of knowledge about such signals. References [27–29] are suggested sources.

12.7 Concluding Remarks

A comparison of BBN jamming, best case PBN jamming, and best case $n = 1$ MT jamming of noncoherent, orthogonal uncoded 2-FSK is shown in Figure 12.27 [30]. Clearly the $n = 1$ multitone is the best for uncoded BFSK signals. Additionally, beyond a fairly low JSR threshold, both worst-case BN and worst-case MT have an inverse linear relationship to the SNR, a significant advantage for the jammer as opposed to spreading the jamming energy over the whole hopping bandwidth. This demonstrates the importance of coding for BFSK communications, which at low P_e removes the jammer advantage. It is possible to "punch through" coding applied to BFSK signals. This requires a channel BER of about 0.04 to 0.12. When this happens the decoded BER is approximately equal to the channel BER that, for many purposes, precludes effective communications. At low E_b / N_0, or, equivalently when $R_b = W_F$, at high enough JSR, adding coding actually decreases performance. For $E_b / N_0 < 6$ dB, all of these coding schemes produce a BER larger than when uncoded. For reasonable values of JSR, coding makes digital signals very vulnerable to PBN jamming. Only at values of JSR less than –5 dB or lower is coding effective.

The required decoded BER depends on the application. Many situations, such as voice communications, can tolerate a significant BER and still be effective. Many others cannot. Thus it is important to know the target set for the jammer for the EA to be effective.

It is possible for BFSK communication systems to add more and more coding as the communication environment gets noisier and noisier. When it does so, it either decreases the information throughput rate while holding the required bandwidth constant or it increases the bandwidth required while holding the information data rate constant. In most realistic military scenarios, the bandwidth is the limiting factor so normally the information data rate is decreased.

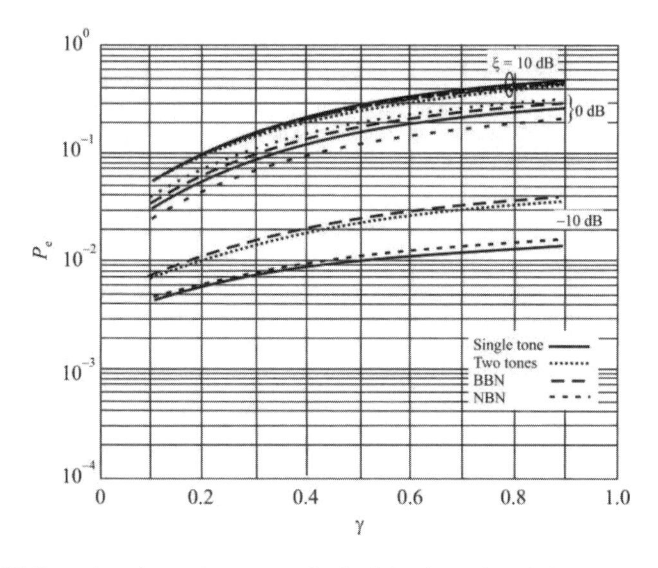

Figure 12.26 Follower jamming performance as the dwell fraction, γ, is varied.

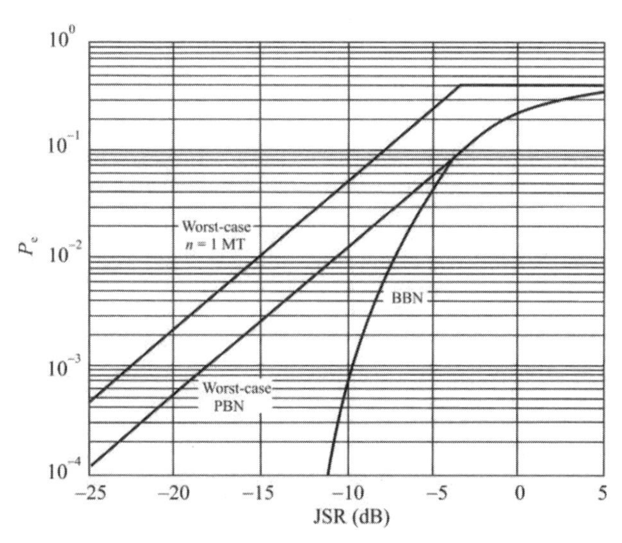

Figure 12.27 Comparison of BBN, worst-case PBN, and n = 1 MT jamming against uncoded, noncoherent orthogonal BFSK. (Source: [30]. © McGraw-Hill 1994. Reprinted with permission.)

As can be noted by observing Figures 12.9, 12.11, and 12.14, it is frequently true that adding coding shortens the region from low BER to high BER. This can have the effect of successfully communicating at one moment to noncommunications the next.

References

[1] Geraniotis, E., and M. B. Pursley, "Error Probabilities for Slow-Frequency-Hopped Spread-Spectrum Multiple-Access Communications over Fading Channels," *IEEE Transactions on Communications,* Vol. Com-30, No. 5, May 1982, pp. 996–1009.

[2] Kang, J. H., and W. E. Stark, "Performance of Turbo-Coded FH-SS with Partial-Band Interference and Rayleigh Fading," *Proceedings IEEE MILCOM*, 1998, http://www.greenhouse.com /society/TacCom/milcom_98_papers.shtml.

[3] Jordan, M. A., "Turbo Code Performance in Partial-Band Jamming," *Proceedings IEEE MILCOM*, 1998, http://www.greenhouse.com/society/TacCom /milcom_98_papers.shtml.

[4] Wang, C. C., "On the Performance of Turbo Codes," *Proceedings IEEE MILCOM*, 1998.

[5] Kang, J. H., and W. E. Stark, "Performance of Turbo-Coded FH-SS with Partial-Band Interference," *Proceedings ISIT*, 1998.

[6] Kang, J. H. and W. E. Stark, "Turbo Codes for Noncoherent FH-SS with Partial Band Interference," *IEEE Transactions on Communications*, Vol. 46, No. 11, November 1998, pp. 1451–1458.

[7] Pouttu, A., and J. Juntti, "Performance of Coded Slow Frequency Hopping M-ary FSK with Clamped Average Energy Metric in Follow-On and Partial Band Jamming," *Proceedings IEEE MILCOM*, 1997.

[8] Arthurs, E., and H. Dym, "On the Optimum Detection of Digital Signals in the Presence of White Gaussian Noise: A Geometric Interpretation and a Study of Three Basic Data Transmission Systems," *IRE Transactions on Communication Systems,* Vol. CS-10, December 1962, pp. 336–372.

[9] Peterson, R. L., R. E. Ziemer, and D. E. Borth, *Introduction to Spread Spectrum Communications,* Upper Saddle River, NJ: Prentice Hall, 1985, p. 333.

[10] Peterson, R. L., R. E. Ziemer, and D. E. Borth, *Introduction to Spread Spectrum Communications,* Upper Saddle River, NJ: Prentice Hall, 1985, p. 334.

[11] Putnam, C. A., S. S. Rappaport, and D. L. Schilling, "A Comparison of Schemes for Coarse Acquisition of Frequency-Hopped Spread-Spectrum Signals," *IEEE Transactions on Communications*, Vol. COM-31, No. 2, February 1983, pp. 183–189.

[12] Ma, H. H., and M. A. Poole, "Error-Correcting Codes Against the Worst-Case Partial Band Jammer," *IEEE Transactions on Communications*, Vol. COM-32, No. 2, February 1984, pp. 124–133.

[13] Tedesso, T. W., and R. C. Robertson, "Performance Analysis of a SFHSS/NCBFSK Communication System with Rate 1/2 Convolutional Coding in the Presence of Partial-Band Noise Jamming," *Proceedings IEEE MILCOM*, 1998.

[14] Simon, M. K., J. K. Omura, R. A. Scholtz, and B. K. Levitt, *Spread Spectrum Handbook,* New York: McGraw-Hill, 1994, p. 502.

[15] Simon, M. K., J. K. Omura, R. A. Scholtz, and B. K. Levitt, *Spread Spectrum Handbook,* New York: McGraw-Hill, 1994, p. 500.

[16] Simon, M. K., J. K. Omura, R. A. Scholtz, and B. K. Levitt, *Spread Spectrum Handbook,* New York: McGraw-Hill, 1994, p. 506.

[17] Nicholson, D. L., *Spread Spectrum Signal Design LPE and AJ Systems,* Rockville, MD: Computer Science Press, 1988, pp. 142–145.

[18] Torrieri, D. J., *Principles of Secure Communication Systems,* 2nd ed., Norwood, MA: Artech House, 1992, p. 19.

[19] Pettit, R. H., *ECM and ECCM Techniques for Digital Communication Systems,* Belmont, CA: Lifetime Learning Publications, 1982, pp. 85–94.

[20] Torrieri, D. J., *Principles of Secure Communication Systems,* 2nd ed., Norwood, MA: Artech House, 1992, pp. 18–20.

[21] Peterson, R. L., R. E. Ziemer, and D. E. Borth, *Introduction to Spread Spectrum Communications,* Upper Saddle River, NJ: Prentice Hall, 1985, p. 370.

[22] Whalen, A. D., *Detection of Signals in Noise,* New York: Academic Press, 1971, p. 212.

[23] Simon, M. K., J. K. Omura, R. A. Scholtz, and B. K. Levitt, *Spread Spectrum Handbook,* New York: McGraw-Hill, 1994, p. 513.

[24] Felstead, E. B., "Follower Jammer Considerations for Frequency Hopped Spread Spectrum," *Proceedings IEEE MILCOM,* 1998, http://www.greenhouse.com/society/TacCom/milcom_98_ papers.shtml.

[25] Torrieri, D. J., *Principles of Secure Communication Systems,* 2nd ed., Norwood, MA: Artech House, 1992, p. 18.

[26] Hassan, A. A., W. E. Stark, and J. E. Hershey, "Error Rate for Optimal Follower Tone-Jamming," *IEEE Transactions on Communications,* Vol. 44, No. 5, May 1996, pp. 546–548.

[27] Simon, M. K., J. K. Omura, R. A. Scholtz, and B. K. Levitt, *Spread Spectrum Handbook,* New York: McGraw-Hill, 1994, pp. 495–598.

[28] Proakis, J. G., *Digital Communications*, 3rd ed., New York: McGraw-Hill, 1995.

[29] Gagliardi, R. M., *Introduction to Communications Engineering*, 2nd ed., New York: John Wiley and Sons, 1988.

[30] Simon, M. K., J. K. Omura, R. A. Scholtz, and B. K. Levitt, *Spread Spectrum Handbook,* New York: McGraw-Hill, 1994, p. 484.

Chapter 13

Electronic Warfare and Ultrawideband Systems

13.1 Introduction

As pointed out in Chapter 3, UWB signals have significant processing gain so we would expect problems trying to jam them with any kind of reasonable power in the jammer. Therefore a different approach than simply overpowering the UWB signals is in order. One method is to try to attack the inherent nonlinearities in the UWB receiver.

Also, at least for the foreseeable future, UWB communication technology, in the United States and Europe at least, will be limited to *personal area networks* (PANs) because of the high potential for interfering with other narrower communication services. This is not necessarily the case for the rest of the world, however. UWB systems have been shown to work effectively over links of several kilometers at microwatt power levels, so usage for more general purpose wireless communications, such as cellular phone systems, is certainly possible. In fact, some estimates of cellular coverage with UWB technology have the number of subscribers per cell in the 1,000 range whereas limits at the time of this writing are in the range of 50–100 per cell.

UWB systems transmit information with very short duration pulses rather than by modulating a carrier to move the information signal elsewhere in frequency. The large bandwidth of a UWB system is determined by the pulse shape and duration. This large system bandwidth relative to the information bandwidth allows UWB systems to operate with a (very) low power spectral density. A low power spectral density implies an inherent covertness of UWB, due to the fact that the UWB signal may be near or below the noise floor of an EW receiver, much the same as DSSS signals can be.

We investigate the detectability of UWB systems with noncooperative EW systems in this chapter. UWB signals are hard to detect because a limited amount of power is distributed across a very wide bandwidth, as discussed in Chapter 3. This is not the primary reason for a communicator to use UWB signaling, however. It is principally a technique for sharing common bandwidth among many users.

Since the details of the modulation process are not known to a noncooperative EW receiver, the interceptor cannot use a matched filter or a correlator and must treat the signal as a random process and base its detection on the presence or absence of energy compared to the background noise. As described in Chapter 2, the module that does this is an energy detector.

This chapter is structured as follows. We begin with an evaluation of the detectability of a UWB signal, and it should surprise no one that the detector of choice is the radiometer. That is followed by an analysis of jamming performance. In particular, the techniques we look at for jamming UWB signals are BBN, MT, tone, pulsed, PBN, and NBN.

13.2 Detecting UWB Signals

We examine the detectability of UWB signals in this section. Two measures of detectability will be described. The first measure computes P_d versus the SNR required to achieve some measure of performance, usually P_fa and P_d. The second measure is the ratio of the distance an intercepting interloper must be to the distance between the target transmitter and receiver to achieve a measure of performance, again usually P_d and P_fa.

13.2.1 Modulations

The UWB system considered here employs time hopping as an LPD/LPI technique. The signal transmitted by the lth user is given by

$$s^{(l)}(t) = \sum_i A^{(l)}_{\lfloor i/N_\mathrm{s} \rfloor} w\left[t - iT_\mathrm{f} - c_i^{(l)} T_\mathrm{c} - \delta d^{(l)}_{\lfloor i/N_\mathrm{s} \rfloor} \right] \tag{13.1}$$

where t is the transmitter clock time. The frame time T_f typically is a hundred to a thousand times the pulse width resulting in a signal with a very low duty cycle. The superscript (l) represents the lth user, $A^{(l)}_{\lfloor\ \rfloor}$ is the amplitude of the ith pulse based on the energy per pulse and the data sequence, N_s is the number of pulses used to represent one data symbol, $w()$ is the pulse shape with normalized energy, $c_i^{(l)}$ is a pseudorandom, repetitive time hopping sequence, T_c is the time hop delay, δ is the PPM time delay parameter, $d^{(l)}_{\lfloor\ \rfloor}$ is a function of the data sequence, and the $\lfloor\ \rfloor$ notation represents the integer portion of the argument. $w(t)$ represents the transmitted Gaussian monocycle or one of the other possibilities discussed in Chapter 3. $n(t)$ is AWGN.

Possible modulation schemes that have been proposed for UWB communications systems are pulse amplitude modulation (PAM), on/off keying (OOK), bipolar signaling, pulse position modulation (PPM), and various combinations of these. An M-ary PPM scheme, bipolar signaling, and an M-ary combination of bipolar signaling and PPM (biorthogonal signaling) are considered here.

For the M-ary PPM, the data is contained in one of M possible time delays, so $d^{(l)}_{\lfloor\ \rfloor} = 0,1,\ldots,M-1$ and $A^{(l)}_{\lfloor\ \rfloor} = \sqrt{E_{\text{p}}}$. For bipolar signaling, the data is contained in the polarity of the transmitted pulses, so $d^{(l)}_{\lfloor\ \rfloor} = 0$ and $A^{(l)}_{\lfloor\ \rfloor} = \pm\sqrt{E_{\text{p}}}$. For biorthogonal signaling, the data is contained in the polarity of the pulse and one of $M/2$ possible time delays, such that $d^{(l)}_{\lfloor\ \rfloor} = 0,1,\ldots,M/2-1$ and $A^{(l)}_{\lfloor\ \rfloor} = \pm\sqrt{E_{\text{p}}}$.

13.2.2 Required SNR Measure of Effectiveness

13.2.2.1 Radiometer Detectors

We will use the radiometer as the architecture to examine the detectability of UWB signals. We discussed radiometers at length in Chapter 2 where, in particular, the fundamental radiometer shown in Figure 13.1 was characterized. This section describes a wideband radiometer system, and introduces the multi-radiometer detector, in the form of a time-channelized radiometer.

As pointed out in Chapter 2, the optimal receiver for detection of a signal in AWGN is the energy detector or the radiometer. It is easily implemented in hardware and can be used to detect SS signals. Detection is accomplished by calculating the energy of the received message and comparing with a predetermined threshold. As illustrated in Figure 13.1, it consists of a bandpass filter of bandwidth W, a squaring device, and an integrator with an integration time T set equal to the observation time interval.

The performance of the radiometer is characterized by the probability of detection P_{d} when the signal is present (along with the noise, of course) and the probability of false alarm P_{fa} when only noise is present. The signal power-to-noise PSD, R/N_0, is usually the parameter specified to achieve the P_{d} and P_{fa} goals. We will use R/N_0 as a measure of the covertness of the communication system under consideration.

When the signal is absent and the input to the radiometer is strictly noise with two-sided PSD $N_0/2$ and the statistics of the output of the radiometer U are

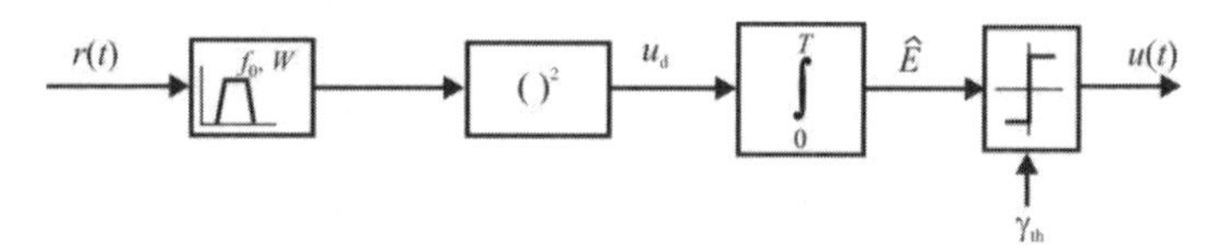

Figure 13.1 Radiometer UWB detector.

normalized, the normalized random variable $Y = 2U / N_0$ has a central chi-square distribution with $v = 2TW$ degrees of freedom where T and W are the observation time interval and the filter bandwidth of the radiometer respectively, as illustrated in Figure 13.1. The PDF of Y is given by [1]

$$p_{\mathrm{n}}(y) = \frac{1}{2^{u/2}\Gamma\left(\frac{u}{2}\right)} y^{(u-2)/2} e^{-y/2}, \qquad y \geq 0 \tag{13.2}$$

When the signal is present at the input to the radiometer with energy E measured over time T, then the random variable Y has a noncentral chi-square distribution with $2TW$ degrees of freedom and noncentral parameter $\lambda = 2E/N_0$. The PDF of Y can then be written as [1]

$$p_{\mathrm{s+n}}(y) = \frac{1}{2}\left(\frac{y}{\lambda}\right)^{(u-2)/4} e^{-(y+\lambda)/2} I_{(u-2)/2}\left(\sqrt{y\lambda}\right), \quad y \geq 0 \tag{13.3}$$

where $I_n(z)$ is the nth order modified Bessel function of the first kind. The P_{fa} and P_{d} values of the radiometer can be determined by integrating (13.2) and (13.3), respectively, as

$$P_{\mathrm{fa}} = \int_{2\gamma_{\mathrm{th}}/N_0}^{\infty} p_{\mathrm{n}}(y)dy \tag{13.4}$$

and

$$P_{\mathrm{d}} = \int_{2\gamma_{\mathrm{th}}/N_0}^{\infty} p_{\mathrm{s+n}}(y)dy \tag{13.5}$$

where γ_{th} is the trigger threshold.

The threshold γ_{th} and the R/N_0 are determined in order to satisfy the performance criteria of the radiometer, given in (13.4) and (13.5). However, closed form solutions for (13.4) and (13.5) are not available so they must be evaluated numerically. For large TW, the chi-square and the noncentral chi-square density functions asymptotically converge to a Gaussian by the central limit theorem.

Probability of Error

Assuming that N_u users are active in the system, the received signal $r(t)$ can be expressed as [1, 2],

$$r(t) = \sum_{l=1}^{N_u} \alpha^{(l)} s^{(l)}[(t - \tau^{(l)}] + n(t) \tag{13.6}$$

where $\alpha^{(l)}$ is the attenuation of the signal over the channel and $\tau^{(l)}$ represents the timing mismatch between the transmitter and the receiver. Under the assumption that the interfering signals have the same power level as that of the target, and the time offsets of interfering pulses arriving at the receiver are independent and uniformly distributed r.v.s over the frame time $[0, T_f)$, the total interference can be modeled as a zero mean Gaussian random process [1, 2]. The probability of error is given as

$$P_e(N_u) = \frac{1}{\sqrt{2\pi}} \int_{\sqrt{\gamma(N_u)}}^{\infty} \exp\left(-\frac{y^2}{2}\right) dy \tag{13.7}$$

where $\gamma(N_u)$ is the radiometer output SNR for N_u users and can be expressed as

$$\gamma(N_u) = \frac{1}{\dfrac{\sigma_n^2}{(\alpha_1 N_s m_p)^2} + \dfrac{1}{N_s}\dfrac{\sigma_a^2}{m_p}\sum_{l=2}^{N_u}\left(\dfrac{\alpha_l}{\alpha_1}\right)^2} \tag{13.8}$$

where m_p is the receiver correlator output for a single received pulse, σ_a^2 is the variance of the interfering users, σ_n^2 is the noise power, and α_l is the attenuation of the signal from the lth user. Without loss of generality we let α_1 be the amplitude for the target signal. The SNR for the single user case can be written as

$$\gamma(N_1) = \frac{(\alpha_1 N_s m_p)^2}{\sigma_n^2} \tag{13.9}$$

The power of the signal at the output of the energy detector bandpass filter, $y(t)$, is $\sigma_y^2 = N_0 W_\mathrm{i}$ in the absence of a signal $(\hat{H}_0)$ of the filter and $\sigma_y^2 + \sigma_s^2 = (N_\mathrm{signal} + N_0)W_\mathrm{i}$ in the presence of the signal $(\hat{H}_1)$, where N_signal is the PSD of the signal and N_0 is the one-sided PSD of noise. The signal at the output of the square law device with no signal present, u_0, is given by

$$u_0 = n^2(t), \qquad \text{noise only} \tag{13.10}$$

and with a signal present is

$$u_1 = [r(t) + n(t)]^2, \text{ signal and noise} \tag{13.11}$$

The difference between the expected value of the output of the square law device, u_d, when no signal is present (u_0) and when a signal is present (u_1) is how the radiometer detects a signal. Assuming zero mean noise this is given by

$$\overline{\Delta u} = \mathcal{E}\{u_1\} - \mathcal{E}\{u_0\} = N_\mathrm{signal} W_\mathrm{i} \tag{13.12}$$

The random fluctuations of the energy output of the square law device masks the energy difference. This is due to the variance of the random processes. The signal to be detected can be expected to cause no *variation* in the output power when measured over the whole bandwidth. The variance depends on the noise PSD and is given by [3]

$$\mathrm{var}\{u_\mathrm{d}\} = \sigma_y^4 = (N_0 W_\mathrm{i})^2 \tag{13.13}$$

The integrator in the radiometer time integrates u_d over the observation window T in order to make the constant component $\overline{\Delta u}$ visible by smoothing the random fluctuations σ_y^4. For reliable detection, the number of statistical independent samples, $n_\mathrm{s} \approx W_\mathrm{i} T$, should be large.

When $n_\mathrm{s} >> 1$ the conditional PDF at the integrator's output is given by Edell's model [3]

$$p(\hat{E} | H_i) = \frac{1}{\sqrt{2\pi n_\mathrm{s} \sigma_\mathrm{n}^2}} \exp\left\{ \frac{(\hat{E} - n_\mathrm{s} \bar{u}_\mathrm{di})^2}{2 n_\mathrm{s} \sigma_\mathrm{n}^4} \right\}, \qquad i = 0,1 \tag{13.14}$$

yielding the probability of a false alarm

$$P_{\text{fa}} = \Pr\{\hat{E} \geq \gamma_{\text{th}} | H_0\} = \int_{\gamma_{\text{th}}}^{\infty} p(\hat{E}|H_0)d\hat{E} = Q\left(\frac{\gamma_{\text{th}} - n_{\text{s}}\bar{u}_{\text{d0}}}{\sqrt{n_{\text{s}}\sigma_{\text{n}}^2}}\right) \quad (13.15)$$

and the probability of detection [3]

$$P_{\text{d}} = \Pr\{\hat{E} \geq \gamma_{\text{th}} | H_1\} = \int_{\gamma_{\text{th}}}^{\infty} p(\hat{E}|H_1)d\hat{E} = Q\left(\frac{\gamma_{\text{th}} - n_{\text{s}}\bar{u}_{\text{d1}}}{\sqrt{n_{\text{s}}\sigma_{\text{n}}^2}}\right) = Q\left[Q^{-1}(P_{\text{fa}}) - q_{\text{i}}\right] \quad (13.16)$$

where q_i is given by

$$q_{\text{i}} = \frac{n_{\text{s}}\Delta\bar{u}_{\text{d}}}{\sqrt{n_{\text{s}}\sigma_{\text{n}}^2}} = \sqrt{W_{\text{i}}T}\frac{R}{WN_0} \quad (13.17)$$

P_{fa} is a system parameter that is specified by the requirements of the application, and it is set in the radiometer[1] and the second term is due to the observation window ($W_{\text{i}}T$) of the radiometer and the UWB PSD. We can thus see that the probability of non-cooperative detection for UWB signals is extremely low as R/WN_0 is very small due to the ultrawide bandwidth used and low received power R. P_{d} is highest when W_{i} is equal to W and (6.13) can then be rewritten as

$$q_{\text{i}} = \sqrt{\frac{T}{W}}\frac{R}{N_0} \quad (13.18)$$

13.2.2.2 Time-Channelized Multichannel Radiometer

The radiometer described above is a wideband radiometer in which a single radiometer observes the entire bandwidth W_{i} over time T. Other radiometer architectures improve the total time bandwidth product and thereby the probability of detection. One possibility is the channelized radiometer exemplified in Figure 13.2. M radiometers are used in parallel and each radiometer can observe either a different portion of the spectrum or a different time window, reducing the noise measured by each radiometer [4].

The time interval for each channel of this system is on the same order as the pulse duration, T_{p}, which yields T_0W approximately equal to 1. For these radiometers, Edell's model is no longer valid, but other models for a wide range of TW are given in [1]. Park's [5] and Dillard's [6] models are derived from Barton's

[1]Setting the value of P_{fa} is referred to as *constant false alarm rate* (CFAR) detection.

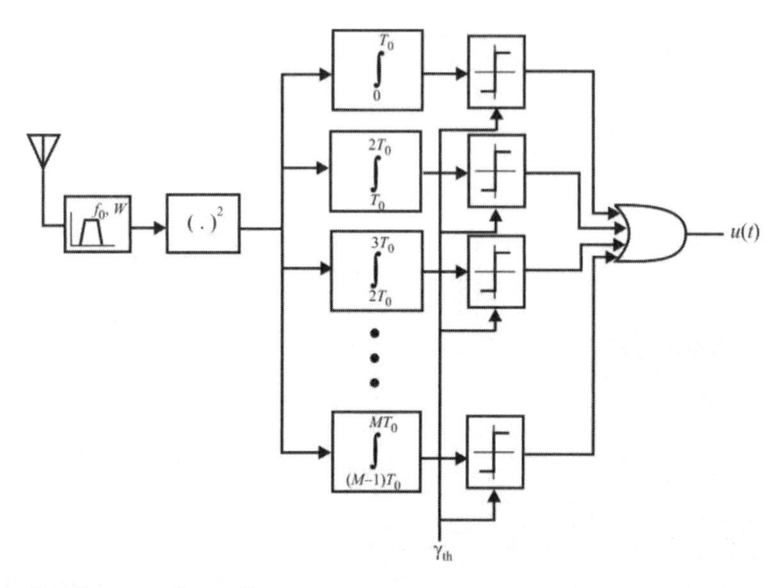

Figure 13.2 Time channelized radiometer.

radar detector loss function [7], which is based on $TW = 1$, so we will use them. Probability of detection for each channel using Park's model can be expressed as

$$P_{d_i} = Q\left[Q^{-1}\left(P_{fa_i}\right) - \sqrt{2Y}\right] \tag{13.19}$$

and probability of detection for each channel using Dillard's model can similarly be expressed as

$$P_{d_i} = Q\left[\sqrt{-2\ln\left(P_{fa_i}\right)} - \sqrt{Y\left(1+\sqrt{1+9.2/Y}\right)}\right] \tag{13.20}$$

where

$$Y = \frac{T_0 R_i^2}{2.3WN_0^2 + R_i N_0} \tag{13.21}$$

and where R_i is the average received power in the ith channel during one time interval, T_0. Here R_l is the power received by radiometer l and P_{fa_i} is the probability of false alarm for the radiometers, assumed to be the same for all.

A time-channelized radiometer gives better detection performance for UWB signals than a frequency channelized configuration. This is due to the fact that UWB systems only transmit a short period of time. The total observation window T is divided among the M radiometers, each observing $T_0 = T / M$ of the window.[2] The optimum M is when T / M is equal to the pulse duration T_p because increasing the time window beyond the pulse duration only increases the noise captured without measuring any additional signal energy.

The detectability of a UWB signal varies depending on what is a priori known about the signal. The ideal detector is a multiradiometer system that is perfectly synchronized with the received pulse and has prior knowledge of pulse width as well as the bandwidth of the SOI. For this case, the bandwidth of the input filter is set to the signal bandwidth and the observation intervals to the signal pulse width (T_0). When less is known about the SOI less sensitivity ensues and the parameters of the multiradiometer system are adjusted according to what is known about the SOI.

Single Signal

The overall performance (i.e., the probability of detection P_d and the probability of false alarm P_{fa}) of the intercept system is a function of the performance (P_{d_i} and the P_{fa_i}) of the individual radiometers. The radiometer analytical models require that only one radiometer with a single observation and integration interval is used for detection and that there is significant signal energy present for the entire duration of the observation interval. Observation intervals of the individual radiometers larger than the pulse width result in the signal being limited to only part of the observation interval because UWB signals are (very) low duty cycle pulses. Furthermore, the Gaussian assumption for the output statistics of the radiometer is invalid because for UWB signals, $TW \sim 1$. For these reasons, the available theoretical models are not applicable, so to determine the detectability we must resort to numerical simulation.

We will assume that the individual radiometers in the multiradiometer system (FBC) are identical; thus the parameters P_{fa_i} and P_{d_i} are the same for each of them. P_{fa} and the P_d for the overall time-channelized radiometer system can be expressed in terms of P_{fa_i} and P_{d_i} as

$$P_d = 1 - \left[(1 - P_{d_i})(1 - P_{fa})^{M-1} \right] \quad (13.22)$$

and

[2] This configuration of radiometers can be accomplished with suitable sample and hold modules and a single radiometer.

$$P_{\text{fa}} = 1 - (1 - P_{\text{fa}_i})^M \tag{13.23}$$

Equations (13.22) and (13.23) are valid for the single user case when there is a single pulse to be detected within a frame and we have M radiometers within the frame.

Given P_{fa} and P_{d} for the multiradiometer system, P_{fa_i} and P_{d_i} can be calculated iteratively using (13.22) and (13.23). The threshold for the individual radiometers is determined by simulation to meet the P_{fa_i} criterion and then, given this threshold value, the R / N_0 is found to meet the required P_{d_i} criterion. We will compare the results so obtained with the required R / N_0 for CDMA by applying the analytical models to the latter since the constraints that render the models inapplicable in UWB do not exist in wideband CDMA.

Since the bandwidth or the pulse width may not be known to the intercept receiver, we first consider the performance of the receiver unaware of the SOI specifications. The multiradiometer system for this receiver is similar in structure to the optimal receiver, but the observation interval, T_0, and the bandwidth of the input filter will likely not be equal to the actual pulse width or the bandwidth of the SOI. We consider the performance when the bandwidth is unknown but the pulse width is known and also the case when the pulse width is unknown but the bandwidth is known. However, for now, synchronization is assumed. We now consider these scenarios in more detail.

Bandwidth Unknown, Pulse Width Known. First we consider the case when the SOI bandwidth is unknown. For this case, the receiver structure is similar to the optimal receiver with the same number of radiometers, M, and the observation interval set equal to the pulse width. If the bandwidth of the input filter of the radiometer is set to be greater than the SOI bandwidth, then the noise power increases. For bandwidths less than the SOI bandwidth the signal power decreases. Both cases reduce the available SNR.

Pulse Width Unknown, Bandwidth Known. We assume in this case that the observation interval of the intercept receiver is different from the actual pulse width of the SOI while we can set the bandwidth to be that of the SOI since it is known. As before, an observation interval that is longer than the pulse width results in excess noise, while an observation interval smaller than the pulse width reduces the signal power. Again both cases reduce the available SNR. Clearly, the number of individual radiometers required to cover a frame is a function of the observation interval, T_0, as is obvious from Figure 13.2. Assuming the observation interval of the receiver is T_{R}, then the total number of radiometers per frame is $M' = T_{\text{f}} / T_{\text{R}} \neq M$ if $T_{\text{R}} \neq T_{\text{p}}$. In this case,

$$P_{\rm d} = 1 - \left[(1 - P_{{\rm d}_i})(1 - P_{\rm fa})^{M'-1}\right] \tag{13.24}$$

and

$$P_{\rm fa} = 1 - (1 - P_{{\rm fa}_i})^{M'} \tag{13.25}$$

We have assumed the received signal to be time synchronized with the observation intervals of the individual radiometers for the cases considered above. For asynchronous signals, there is overlap into adjacent radiometer integration intervals. In the worst-case scenario, exactly half the signal overlaps onto the observation interval of the following radiometer, since the radiometers are aligned in time, thus minimizing signal power in both radiometers. In this case, the signal power input to a particular radiometer is reduced by 3 dB. As a result the required R/N_0 to meet performance criteria increases by 3 dB-Hz. The actual performance of the receiver depends on the amount of signal overlap, and the performance (required SNR) increases as $SNR + 3\kappa$ dB-Hz where $0 < \kappa < 1$. However, since our objective is to obtain an upper bound on signal detectability, the assumption of perfect synchronism is reasonable for our purposes.

Multiple Signals

For L active users in the environment there are L pulses, which are pseudorandomly shifted within a time frame. The multiradiometer receiver structure is still that in Figure 13.2, but because of the presence of more pulses in the system and thus more energy per frame, the performance criterion of the individual radiometers ($P_{{\rm d}_i}$) can be decreased while meeting the system performance criteria. The L pulses may or may not overlap with each other and as a result two cases arise. We assume that all the pulses have equal amplitudes. We also assume that the intercepting receiver has knowledge of the bandwidth and pulse width of the SOI (the most likely case unless there is nothing known about the signal environment).

Nonoverlapping Signals. In this case, we are assuming that the pulses occupy different time slots within a frame. The multiradiometer system for this case is similar to that in the single user case, except that for L users, $P_{\rm d}$ of the overall system is related to the individual radiometer $P_{{\rm d}_i}$, assumed to be equivalent for all radiometer channels, by

$$P_{\rm d} = 1 - \left[(1 - P_{{\rm d}_i})^L (1 - P_{{\rm fa}_i})^{M-L}\right] \tag{13.26}$$

The P_{fa_i} values of the radiometers are insensitive to the number of users since it is a parameter that is obtained assuming no signals are present. Given the performance criteria for the individual radiometers, simulation is used to obtain the R/N_0 required for a specific value of P_{d_i}.

Overlapping Signals. If L users are active in the system and x pulses overlap $(2 \le x \le L)$, then the amplitude increases by a factor of x. The probability of detection for the individual radiometers is given by (assuming complete overlap)

$$P_\text{d} = \begin{cases} 1-\left[(1-P_{\text{d}_i})^\alpha (1-P_{\text{fa}_i})^{M-\alpha}\right], & L \le M \\ 1-\left[(1-P_{\text{d}_i})^\beta (1-P_{\text{fa}_i})^{M-\beta}\right], & L > M, x \ge L/M \\ 1-(1-P_{\text{d}_i})^M, & L > M, x < L/M \end{cases} \tag{13.27}$$

where $\alpha = L - x + 1$ and $\beta = (M/L)(L-x)+1$. Again, complete overlap of pulses is a reasonable assumption since we seek an upper bound on signal detectability.

The detectability is dependent on the number of overlapping pulses, x, which varies between $2 \le x \le L$. Define the average SNR for a given number of users L as

$$SNR_L = \sum_{x=2}^{L} SNR_x p(x) \tag{13.28}$$

where SNR_L is the SNR for L users and SNR_x is the signal-to-noise PDF ratio R/N_0 when L users are active and x pulses overlap. $p(x)$ is the probability that x pulses out of L pulses overlap.

Because $p(x)$ is not normally available in closed form we use numerical simulation to evaluate (13.28). When L users are active each frame has exactly L pulses, which are pseudorandomly placed within the frame. Since a pulse can occupy any time slot in the frame with equal probability, the individual user pulses are uniformly spread within the frame. The performance for a particular frame depends on the *maximum* overlap. The average for a given number of users is calculated by averaging R/N_0 over the number of frames, where the maximum overlap for each frame is the maximum number of users that overlap in a single time slot in the frame. The performance for the CDMA signal is calculated by using the analytical expressions given in (13.19) or (13.20) with allowances made for multiple users resulting in decrease in performance according to

$$SNR_L = SNR_1 - 10\log_{10} L \tag{13.29}$$

Performance

Numerical results for the UWB signals are obtained using Monte Carlo simulation. The users are assumed to be stationary for the duration of the simulation in the peer-to-peer network topology. For simplicity we assume that all the pulses at the multiradiometer detection system have equal amplitudes.

The UWB system uses a Gaussian monocycle of center frequency $f_c = 2$ GHz and pulse width $T_p = 6\times 10^{-10}$ s. The frame time is $T_f = 6\times 10^{-8}$ s resulting in $M = 100$ for the number of time-slots in one frame. The bandwidth of the pulse is $W_p = 4$ GHz.

Ideal Multiradiometer. In the ideal case, the multiradiometer system is assumed to have complete knowledge of the system specifications (i.e., the bandwidth and the pulse width). The performance criteria for the radiometer is assumed to be $P_d = 0.9$ and $P_{fa} = 0.01$. The filter is a fourth-order Butterworth filter with a 4 GHz cutoff. The observation interval of the radiometer is the pulse width, T_p.

The P_{fa_i} and the P_{d_i} values for the individual radiometers are calculated iteratively for the single user case using (13.22) and (13.23). The integrator is reset after each observation interval T_p. The threshold is selected to meet the P_{fa_i} requirements of the individual radiometers. This threshold is then used to calculate the required R/N_0 to meet the P_{d_i} criterion.

The required SNR obtained for UWB in the single user case is compared with that for CDMA in Table 13.1. For CDMA we assume a bandwidth of $W_w = 5$ MHz and a chip rate of 3.7 Mcps. As for the UWB system, performance criteria of $P_d = 0.9$ and $P_{fa} = 0.01$ are assumed for this CDMA system. We used the Park model of the radiometer to compute these values. We see that UWB is much more difficult to detect (by more than 30 dB) than the CDMA system in the single-user case. Note that 100 dB-Hz corresponds to

$$10\log_{10}\gamma = 100 \quad \text{dB-Hz}$$
$$\gamma = 10^{10}$$

Table 13.1 Required SNR for $P_d = 0.9$ and $P_{fa} = 0.01$ for the Ideal Radiometer

Target System	Required SNR (dB-Hz)
UWB	107.78
CDMA	75.12

Source: [8].

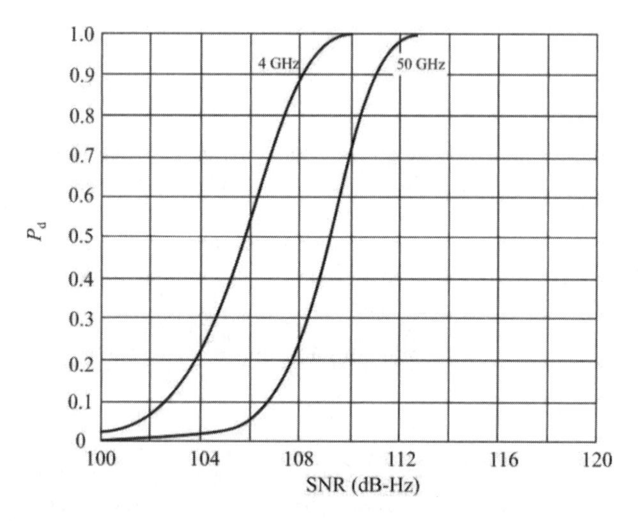

Figure 13.3 Probability of detection versus the normalized SNR for two filter bandwidths, 4 GHz and 50 GHz. (Source: [8]. © IEEE 1991. Reprinted with permission.)

so in a 4 GHz bandwidth

$$SNR = 10\log_{10}\frac{10^{10}}{4\times10^{9}} = 10\log_{10}2.5 = 3.9\,\text{dB}$$

Multiradiometer Performance with Unknown Bandwidth. In this section we evaluate the performance of the multiradiometer detector without prior knowledge of the UWB system bandwidth. The other parameters are the same as in the ideal case. The P_{d_i} and the P_{fa_i} of the radiometer are unchanged from the optimal case because the observation interval is known. Figure 13.3 shows the probability of detection, P_d, as a function of R/N_0 for bandwidths of $W = 4$ GHz and 50 GHz. We can conclude that the multiradiometer performance is not very sensitive to filter bandwidth mismatch.

Multiradiometer Performance with Unknown Pulse Width. The parameters of the suboptimal multi-radiometer system are similar to those in the ideal case, but the P_{fa_i} and the P_{d_i} change according to (13.24) and (13.25) because the number of radiometers within a single time frame vary with the observation interval. Figure 13.4 shows P_d as a function of the R/N_0 for observation periods of the radiometer of 0.6 ns and 60 ns. For the second case, the observation interval is set equal to the frame time; thus a single radiometer occupies the entire frame. This represents the extreme case when maximum noise power is input to the radiometer. We see that

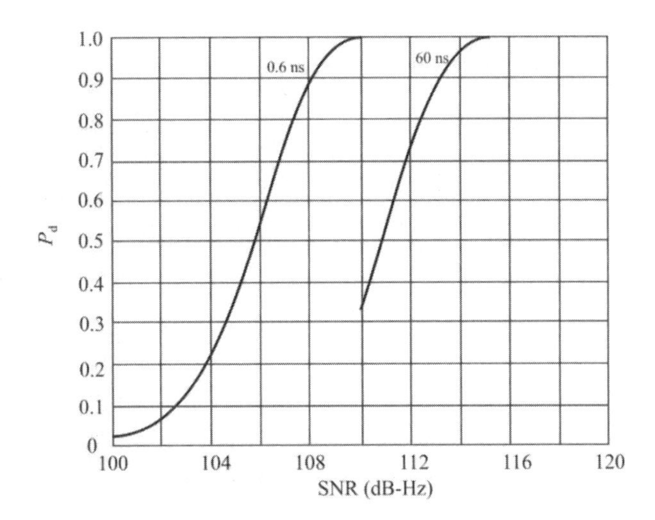

Figure 13.4 Probability of detection as a function of normalized SNR for two observation intervals, 0.6 ns and 60 ns. (Source: [8]. © IEEE 1991. Reprinted with permission.)

multiradiometer receiver performance is relatively insensitive to variations of the observation interval.

Summary

In this section we quantitatively evaluated the UWB detection performance of a multiradiometer intercept receiver. We quantified this performance for a few variations of the radiometer parameters to evaluate how well they match the parameters of the SOI. Results showed that the detector was relatively insensitive to prior knowledge about the bandwidth and pulse width of the UWB signal. In addition it was shown that UWB signals are considerably more difficult to detect than CDMA signals.

13.2.2 Ratio of Distances Measure of Effectiveness

McKinstry and Bueher [9] took a somewhat different approach to evaluate the detectability of UWB systems. We briefly describe their approach and some of their results here.

The pulse shape assumed here is the Gaussian doublet (see Chapter 3) given by

$$w(t)=\sqrt{\frac{16}{3T_\text{p}}}\left[1-16\pi\left(\frac{t}{T_\text{p}}\right)^2\right]\exp\left[-8\pi\left(\frac{t}{T_\text{p}}\right)^2\right] \quad (13.30)$$

where T_p is the pulse width (approximately 99.99% of the energy in $w(t)$ is contained in the interval $-T_\text{p}/2$ to $T_\text{p}/2$).

The pulse repetition time is assumed a constant for this system. As above, the SOI is assumed to be time hopping controlled by a pseudo-random sequence. $d_{[\,]}^{(k)}$ is set to $2T_\text{p}$ to assure orthogonality between possible PPM time shifts.

13.2.2.1 Detection Methods

For radiometers with a large time-bandwidth product, TW, many of the available models discussed in Section 2.7 are very closely approximated by Edell's model. In that case, the probability of detection, P_d, is given by

$$P_\text{d}=Q\left[Q^{-1}(P_\text{fa})-\sqrt{\frac{T}{W}\frac{R}{N_0}}\right], \qquad TW>>1 \quad (13.31)$$

where P_fa is the probability of false alarm in noise alone, R is the average signal power at the radiometer, and N_0 is the noise PSD. This single radiometer model is used in [2] to compare the detectability of a UWB system with more traditional wideband DSSS systems. To quantitatively evaluate performance, [10] determines the probability of detection versus the relative distance between transmitter and interceptor. This assumes a scenario such as that depicted in Figure 13.5. The probability of detection can be found as a function of the ratio d/d_0, the ratio of the distance from the transmitter to interceptor (d) to the distance from the transmitter to the intended receiver (d_0). The interceptor in this situation is attempting to detect the presence of the transmitted signal, not demodulate it. This same metric will be used here to compare the detectability of UWB systems and also to compare different radiometer configurations.

The same time-channelized multiradiometer system as above was used, whose architecture is shown in Figure 13.2. This system consists of M radiometers each of which detects energy over an interval of T_0. This gives an overall observation interval of $T = MT_0$. Considered above is the case when T of the radiometer system is equal to the frame time, T_f, of the UWB system; however, this is not a necessary condition and actually creates some limitations to the performance of the detection system. If T is allowed to be greater than T_f these difficulties are alleviated as will be shown.

As above, the overall desired P_fa can be chosen by setting the probability of false alarm for each channel, P_{fa_i}, to satisfy

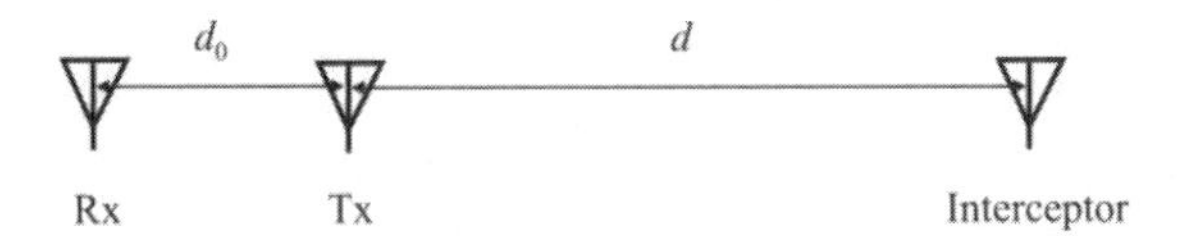

Figure 13.5 Interceptor scenario.

$$P_{\mathrm{fa}} = 1 - \left(1 - P_{\mathrm{fa}_i}\right)^M \tag{13.32}$$

The overall probability of detection of an impulsive UWB signal not known to be synchronized with the radiometer cannot be expressed simply, so we use numerical simulation to determine this relationship.

Simulation of UWB systems employing binary PPM, bipolar modulation, 4-ary PPM, 4-ary biorthogonal modulation, 16-ary PPM, and 16-ary biorthogonal modulation were carried out to evaluate BER performance in AWGN. Simulation of the binary PPM system using a rate 1/4, constraint length 9 convolutional code with Viterbi soft decoding was also performed.

Note that $P_{\mathrm{fa}} \neq P_{\mathrm{fa}_i}$, where P_{fa} is the false alarm for the whole receiver. P_{fa} is the probability that κ or more hop decisions result in a detection when no signal is actually present. We determine that as follows. The probability that none of the M channels has a false alarm is the product of the probabilities of each cell not having a false alarm given by $(1 - P_{\mathrm{fa}_i})^M$. If we assume that the noise in each of the channels is independent of the others, and the channel is AWGN, then the probability of a "1" at the output of the OR gate in the noise-only case will be the probability that at least one of the channels has a false alarm, expressed as

$$P_{\mathrm{o}} = 1 - (1 - P_{\mathrm{fa}_i})^M \tag{13.33}$$

From the binomial expansion theorem, the probability of this occurring exactly n out of the N times will be

$$\binom{N}{n} P_{\mathrm{o}}^n (1 - P_{\mathrm{o}})^{N-n} \tag{13.34}$$

Thus, P_{fa} will be the summation of the probabilities of all possible events exceeding the κ threshold

$$P_{\mathrm{fa}} = \sum_{n=\kappa}^{N} P_{\mathrm{o}}^n (1 - P_{\mathrm{o}})^{N-n} \tag{13.35}$$

We can conclude from this that noncooperative detection of UWB signals has a very low probability of success and the distance required to detect a signal with a reasonable probability of detection is substantially less than the distance at which the target receiver using a matched filter and knowledge of the transmitter can receive the signal.

The above analysis assumes that the receiver knows when the pulses are to occur so it can look for them. In a noncooperative EW system that is not the case. The issue is further compounded when the UWB systems use PN codes to randomize the pulse occurrence time. This PN code is known to the target transmitter and receiver but not to the EW system. The only known answer to this quagmire is to let a wideband radiometer integrate over a reasonably wide bandwidth.

Single Wideband Radiometer Detector

The performance of a single wideband radiometer (represented effectively by Edell's model) with an observation time of 100 μs against a single user employing uncoded UWB SOIs, and a binary PPM system using a rate 1/4, constraint length 9 convolutional code with Viterbi soft decoding is shown in Figure 13.6. This plot shows that higher-order modulations and channel coding can negatively impact detectability.

Channelized Wideband Radiometer Detector

For the channelized radiometer system, the performance of the detector is affected by the parameters chosen. The total interval time for detection decisions was equal to the frame time of the UWB signal above. If the detector's frame is synchronized with the received UWB signal's frame, then when a signal is present, each detector frame will have a channel that overlaps with the UWB pulse transmitted during that frame. But if the detector's frame is not synchronized (exactly half a frame time off, for example) there will be some detector frames that contain no UWB pulses (assuming the pulse in a given frame is randomly distributed due to the time-hopping code). These frames that do not overlap with a pulse will have a very low probability (= P_{fa}) of indicating that a signal is present. This also has a "ceiling" effect on the maximum achievable average probability of detection when the detector is not synchronized. The probability of detection as a function of distance for the uncoded binary PPM signal is plotted in Figure 13.7 for the cases where the detector is perfectly synchronized with the receiver, the detector is exactly half a frame and half a pulse width off from synchronization, and the detector is assumed to have a uniformly random time offset from synchronization. For this comparison the overall observation interval was set equal to the UWB frame duration ($T = T_{\text{f}}$), the observation interval of each channel was set equal to the pulse duration ($T_0 = T_{\text{p}}$), and the overall probability of false alarm was set to

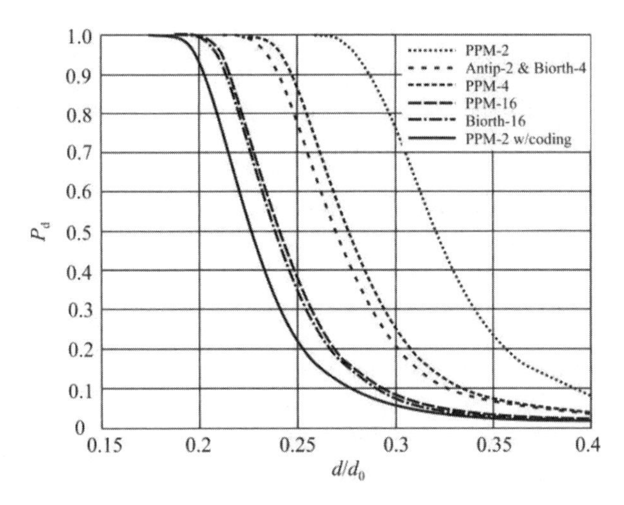

Figure 13.6 Single radiometer UWB SOIs. Covertness of UWB systems employing different modulation schemes. (Detection is by a single radiometer using an observation window of 100 μs, computed using a path loss coefficient of 4, a BER at intended receiver of 10^{-4}, and a P_{fa} of 10^{-2}. The UWB signal has a bandwidth W of 3.25 GHz, a pulse width of 400 ps, and a pulse repetition time of 1 μs). (Source: [9]. © IEEE 2002. Reprinted with permission.)

0.01. All plots giving the results of the channelized radiometer detector are from simulations using Park's model (13.19) for each radiometer channel. The results from the simulations using Dillard's model are nearly identical and thus are not shown.

The probability of detection for the channelized radiometer can be improved by increasing the overall observation interval. A similar effect is seen in the single wideband radiometer when the observation interval is increased [2, 5]. By increasing the overall observation time to multiple frames, the "ceiling" effect seen in Figure 13.7 disappears and the difference due to frame synchronization is greatly minimized.

As can be seen by comparing Figures 13.6 and 13.7, the channelized radiometer is able to detect the UWB signal from a greater distance than the single radiometer (except for very small probability of detection values). It is interesting to note that the channelized radiometer detection curve has a steeper slope than the single radiometer curve. This implies that the distance between almost certain detection and almost certain failure of detection is smaller for the channelized radiometer.

The choice of the observation interval for each channel is also important to the performance of the detector. This choice may be influenced by the hardware difficulties for detecting energy over such short intervals or due to lack of information about the pulse width of the signal to be detected. Figure 13.8

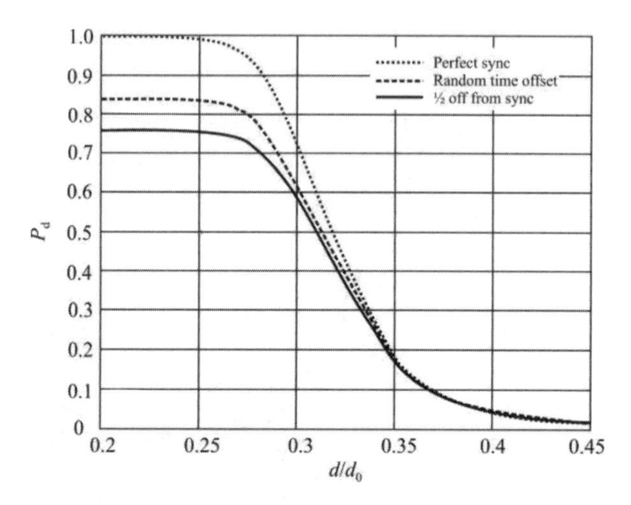

Figure 13.7 Detectability bounds for uncoded binary PPM with a multiradiometer (Source: [9]. © IEEE 2002. Reprinted with permission.)

compares the performance of the channelized detector for different channel observation intervals. As the channel observation interval grows larger than the pulse duration, more noise is captured relative to the energy of the pulse resulting in a reduction of the SNR and the performance of the detector decreases. The performance of the channelized system to detect the binary PPM UWB signal with $T_0 = 25T_p$ (shown in Figure 13.8) yields roughly equivalent performance to the single wideband radiometer. The energy distribution over time of the pulse used determines the optimal value for the channel observation interval for the detector. Simulation revealed this optimum value is slightly less than T_p for detecting the UWB SOI described here.

Summary

The performance of a channelized radiometer detector was evaluated for use in detecting UWB signals and compared to the single wideband radiometer detector. The parameters that impact the performance of such a channelized detector were evaluated and it was shown that it is possible to construct a channelized radiometer detector that improves detectability of UWB signals over a single radiometer if sufficient information about the UWB system is known.

13.3 Jamming UWB Signals

UWB signal jamming can be accomplished with any of the techniques discussed in Chapter 8: barrage, partial band, tone, or multitone. Jamming effectiveness on

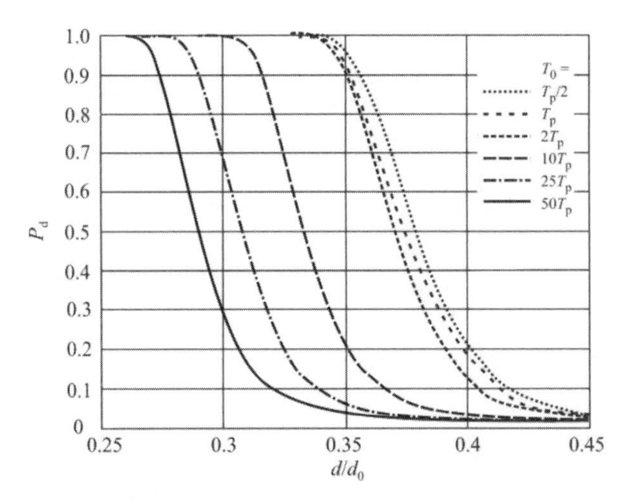

Figure 13.8 Multiradiometer detectability with different observation times for uncoded PPM. (Source: [9]. © IEEE 2002. Reprinted with permission.)

the synchronization and timing modules would depend on the particular system being examined, and thus far there have been very few UWB systems deployed. We examine the effectiveness of the first four techniques mentioned here.

13.3.1 Jamming Effects on UWB Systems

Tone jamming and narrowband interference causes significant problems for UWB systems. This is due to the low transmit power and the large bandwidth that contains multiple jamming tones, when MT jamming is considered. Due to the wide bandwidth of a UWB signal the most likely scenario encountered is tone jamming although the other types of jamming may apply as well.

13.3.2 Processing Gain

The robustness to interference of UWB systems is due to the processing gain. The processing gain in UWB is partly due to the number of pulses per bit (N_S) and partly due to the duty ratio of the signal (T_p / T_f, where T_p is the length of the pulse and T_f is the length of the frame; there are N_S pulses of length T_p each sent in the frame). Reception of a time hopping UWB signal using a matched filter essentially works as a time gate thereby effectively reducing the interference power.

13.3.3 BBN Jamming

Tesi et al. conducted simulation experiments to examine the performance of UWB signals in colored noise [11]. The noise had a total power of 10 dBm, and white noise was filtered with a raised cosine filter to produce the colored noise. Signal power was fixed at 0 dBm. Processing gain was set at 20 dB, which was obtained with 10 dB from the duty cycle factor (T_f / T_p = 10) and 10 dB from coherent integration (L = 10). The SNR was fixed at 8 dB.

Typical results are shown in Figure 13.9. In this case, BPAM was the modulation and time hopping was employed. The impulse had T_p = 1 ns. Similar to DSSS, the processing gain must be overcome before the jamming reached acceptable level.

13.3.4 Tone Jamming

A received signal that is jammed can be represented by

$$r(t) = s(t) + j(t) + n(t) \tag{13.36}$$

where $s(t)$ is the target UWB signal, $j(t)$ is the jamming signal, and $n(t)$ is noise. The jamming PSD is $N_J = J / W_J$ where J is the jamming power. Any nonlinearity or front end saturation of the target receiver are not taken into account.

13.3.5 Multitone Jamming

Hämäläinen et al. documented an analysis of the effects of MT jamming on UWB communication signals [12]. In that analysis, 10 jammer tones were randomly placed within the passband of the UWB system. The total jamming power in all the tones was +15 dBm.

A comparison of the MT jamming performance for BPAM and BPPM is illustrated in Figure 13.10. The goal of 10^{-1} BER was never achieved for either modulation type, which is typical for DSSS signals.

Giorgetti, Chiani, and Win [13] documented the results of an analysis and simulation experiment on the effects of multiple narrowband interferers on UWB systems. The receiver they considered was the common matched filter (MF) receiver illustrated in Figure 13.11. Included in Figure 13.11 are a single target UWB signal at the receiver given by $h_D(t) * s(t)$ where $s(t)$ is the transmitted signal and $h_D(t)$ represents the impulse response of the channel between the transmitter and receiver. Furthermore, N_J jammers are included as $\sum_{j=1}^{N_J} j_j(t)$ where $j_j(t)$ is the jammer signal of the jth jammer given by

$$j_j(t) = \alpha_{J_j} \sqrt{2J_j} \cos\left[2\pi f_j \left(t - \tau_j\right) + \phi_j\right] \tag{13.37}$$

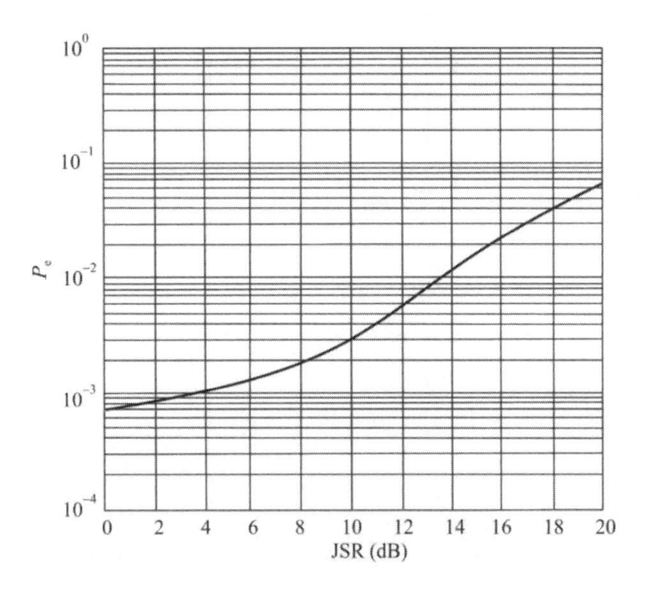

Figure 13.9 BBN jamming performance against a UWB signal where the noise is colored, the modulation is TH-BPAM, and $T_p = 1$ ns. (Source: [11]. © Tesi 2002. Reprinted with permission.)

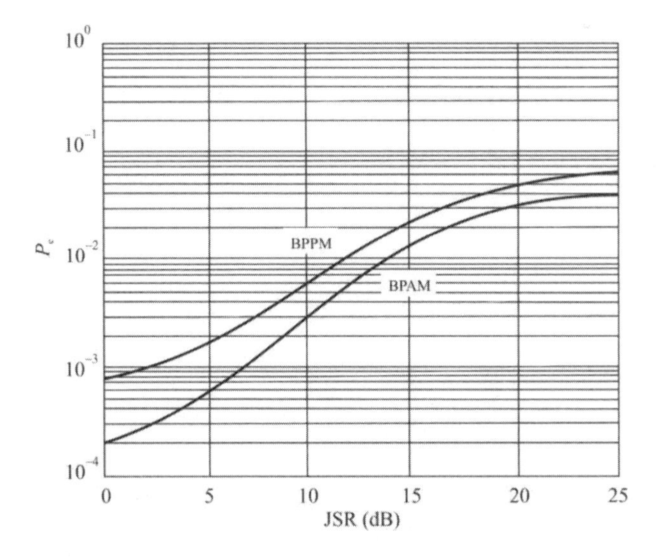

Figure 13.10 MT jamming performance against two UWB modulations. In this case $T_p = 1$ ns and the Gaussian monopulse was the impulse type. $h_D(t)$ is the impulse response of the channel between the transmitter and receiver. (Source: [12]. © IEEE 2002. Reprinted with permission.)

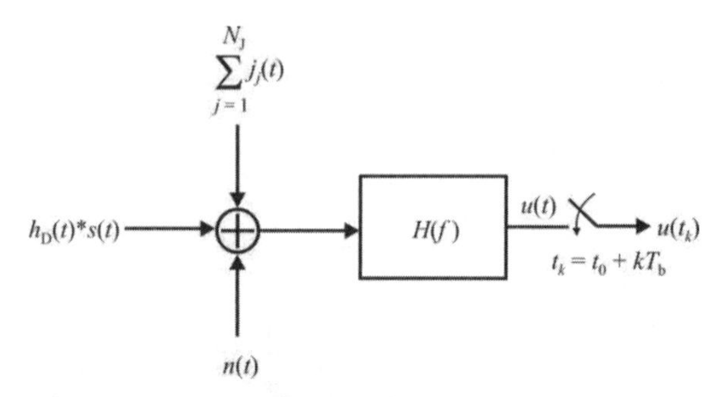

Figure 13.11 UWB receiver and the channel with N_J jammers present. (After: [13]. © IEEE 2005. Reprinted with permission.)

α_{J_j} is the channel gain of the J_jth channel, f_j is the frequency of the jth jammer, ϕ_j is its phase, and J_j is the transmitted power of the jth jammer. $H(f)$ is the transfer function of the MF while $u(t)$ is the output of the MF which is sampled at intervals $t_k = t_0 + kT_b$ yielding $u(t_k)$.

The overall received signal at the target receiver due to the target signal, the N_J jammers, and AWGN $n(t)$ is given by

$$r(t) = \sqrt{E_b}\sum_i r_b(t - iT_b; d_i) + \sum_{j=1}^{N_J}\sqrt{2J_j}r_{J_j}(t) + n(t) \qquad (13.38)$$

The two-sided AWGN PSD is given by $N_0/2$ and E_b represents the average received energy per bit of the target signal. The responses of the channels to the transmitted signal sending data bit d_i, denoted $b(t; d_i)$, and the unit-powered jamming signals $j_j(t)$ are given by

$$r_b(t; d_i) = h_D(t) * b(t; d_i) \qquad (13.39)$$

and

$$r_{J_j}(t) = h_{J_j}(t) * j_j(t) \qquad (13.40)$$

Assume that the jamming signals experience flat fading so that

$$h_{J_j}(t) = \alpha_{J_j}\delta(t - \tau_j), \qquad j = 1, \ldots, N_J \qquad (13.41)$$

where α_{J_j} are the (constant) channel gains, τ_j is the time shift, and $\delta(t)$ is the Dirac delta function. Therefore the received signal due to the jth tone jammer, $r_{\mathrm{J}_j}(t)$, is given by

$$r_{\mathrm{J}_j}(t) = \alpha_{\mathrm{J}_j}\sqrt{2J_j}\cos\left[2\pi f_j\left(t-\tau_j\right)+\phi_j\right] \tag{13.42}$$

Both α_{J_j} and ϕ_j are i.i.d. r.v.s and the latter is assumed to be uniformly distributed over $(0, 2\pi)$.

Now $h_{\mathrm{D}}(t) = \delta(t)$ and $\alpha_{\mathrm{J}_j} = 1, j = 1,\ldots,N_J$ with flat fading and N_{J} tone jammers. Since $r_{\mathrm{b}}(t; d_i) = b(t; d_i)$ the target signal can be written as

$$s_0 = \sqrt{E_{\mathrm{b}}}(1-\rho) \tag{13.43}$$

where ρ is the correlation coefficient between $b(t; 0)$ and $b(t; 1)$ given by

$$\rho = \int_{-\infty}^{\infty} b(t;0)b(t;1)dt \tag{13.44}$$

Let $R = E_{\mathrm{b}} / T_{\mathrm{b}}$ denote the average received power.

13.3.5.1 Fading Channels

When the target channel follows the Nakagami-m distribution with PDF (see Section 2.6.4)

$$p_{\alpha_{\mathrm{D}}}(r) = \frac{2}{\Gamma(m)}\left(\frac{m}{\Omega}\right) r^{2m-1}\exp\left(-\frac{mr^2}{\Omega}\right), \qquad r \geq 0 \tag{13.45}$$

where $\Omega = \mathcal{E}\left\{\alpha_{\mathrm{D}}^2\right\} = 1$ and $m \geq 0.5$ is the fading parameter that is a measure of the fading severity. In the worst case, $m = 0.5$ and the fading PDF is characterized by a one-sided Gaussian distribution, while $m = 1$ corresponds to Rayleigh fading. As $m \to \infty$, the channel is unfading. We assume that the jammer channels experience Rayleigh fading. Then the BER becomes [13]

$$P_{\mathrm{e}} = \frac{\Gamma(m+1/2)}{2\sqrt{\pi}\Gamma(m+1)}\left(\frac{m}{\eta}\right)^m {}_2F_1\left(m, m+\frac{1}{2}; m+1; -\frac{m}{n}\right) \tag{13.46}$$

Where ${}_2F_1\left(m,m+\frac{1}{2};m+1;-\frac{m}{n}\right)$ represents the Gauss hypergeometric function [14] and η is given by

$$\eta=\left[\frac{N_0}{E_b}\frac{2}{1-\rho}+\frac{2}{(1-\rho)^2}\sum_{j=1}^{N_J}\frac{J_j}{R}\frac{|H_0(f_j)|^2}{T_b}\right]^{-1} \quad (13.47)$$

For integer values of m, (13.46) can be written as

$$P_e=\left[\frac{1}{2}-\frac{1}{2}\left(1+\frac{m}{\eta}\right)^{-1/2}\right]^m\sum_{i=0}^{m-1}\binom{m-1+i}{i}\left[\frac{1}{2}+\frac{1}{2}\left(1+\frac{m}{\eta}\right)^{-1/2}\right]^i \quad (13.48)$$

THPPM

Using the bit waveform for PPM

$$b(t;d_i)=b(t-d_i\delta) \quad (13.49)$$

the target signal can be written

$$s(t)=\sqrt{E_b}\sum_i b(t-iN_sT_f-d_i\delta) \quad (13.50)$$

where the unit-energy waveform for each bit is

$$b(t)=\sum_{n=0}^{N_s-1}w(t-nT_f-c_nT_c) \quad (13.51)$$

when N_s is the number of pulses used to transmit a single information bit $d_i\in\{0,1\}$. δ is the pulse position offset, $w(t)$ is the signal pulse with energy $1/N_s$, and E_b is the received bit energy. The pulse repletion time (frame length) T_f and bit duration T_b are related by $T_b = N_sT_f$. Last, $\{c_n\}$ is the time hopping sequence, and T_c is the TH chip width.

The MF transfer function is given by

$$|H_0(f)|=2|W(f)||\sin(\pi f\delta)|\left|\sum_{n=0}^{N_s-1}e^{j2\pi f(nT_f+cnT_c)}\right| \quad (13.52)$$

where $W(f)$ is the Fourier transform of the pulse $w(t)$.

Using the second derivative of the Gaussian monocycle for the received pulse $w(t)$ with energy $1/N_\mathrm{s}$ we get

$$w(t) = \sqrt{\frac{1}{N_\mathrm{s} E_\mathrm{w}}} \left[1 - 4\pi \left(\frac{t}{\tau_\mathrm{w}} \right)^2 \right] e^{-2\pi (t/\tau_\mathrm{w})^2} \tag{13.53}$$

where $E_\mathrm{w} = 3\tau_\mathrm{w}/8$ and τ_w is the pulse duration. Its Fourier transform is given by

$$W(f) = \sqrt{\frac{1}{N_\mathrm{s} E_\mathrm{w}}} \frac{\pi}{\sqrt{2}} \tau_\mathrm{w}^3 f^2 e^{-\frac{\pi}{2} f^2 \tau_\mathrm{w}^2} \tag{13.54}$$

The BER is shown in Figure 13.12 for the case of TH PPM with a single jammer.

In the case when both the target signal and single jammer signal are fading, from (13.46), some results are shown in Figure 13.13.

TH PAM

The bit waveform for TH PAM is given by

$$b(t; d_i) = d_i b(t) \tag{13.55}$$

and the target signal transmitted is given by

$$s(t) = \sqrt{E_\mathrm{b}} \sum_i d_i b(t - i N_\mathrm{s} T_\mathrm{f}) \tag{13.56}$$

In this case the transfer function can be written as

$$|H_0(f)| = 2|W(f)| \left| \sum_{n=0}^{N_\mathrm{s}-1} e^{j2\pi f (n T_\mathrm{f} + c n T_\mathrm{c})} \right| \tag{13.57}$$

13.3.5.2 Frequency-Selective Fading Channels

We now consider the case when a rake receiver is used as the target receiver and when there is a single tone jammer with AWGN.

The frequency-selective multipath fading channel for the target signal is given by

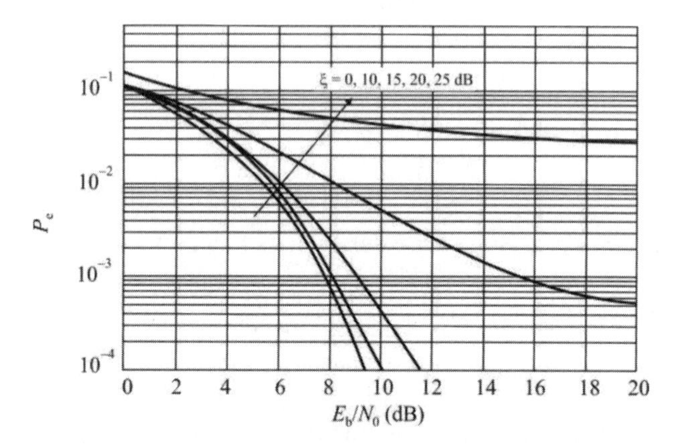

Figure 13.12 Tone jamming performance against unfaded TH PPM when the jammers are faded but the target signal is unfaded. (Source: [13]. © IEEE 2005. Reprinted with permission.)

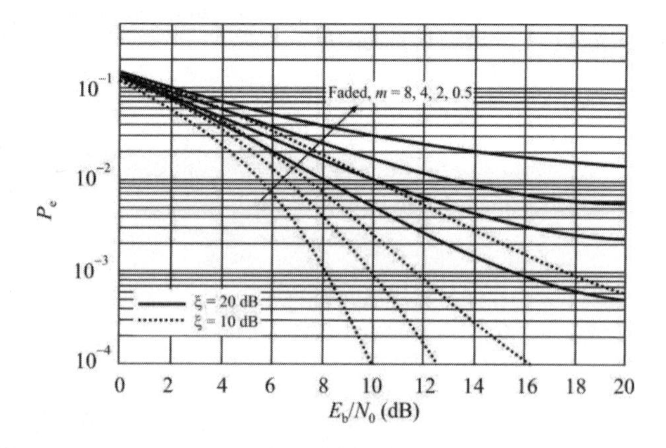

Figure 13.13 Tone jamming performance against TH PPM when both the jammers and target signal are fading. (Source: [13]. © IEEE 2005. Reprinted with permission.)

$$h_D(t) = \sum_{k=1}^{K} h_k \delta(t - t_k) \tag{13.58}$$

where

$$h_k = \alpha_k e^{j\theta_k} \tag{13.59}$$

are statistically independent r.v.s representing the gains of the K paths. In addition $\theta_k \in \{0, \pi\}$ with equal probability represents the phase inversions that can happen due to reflections on the multiple paths. In this case the target signal at the receiver can be written

$$s_0 = \sqrt{E_b}(1-\rho)\alpha_D^2 \tag{13.60}$$

where

$$\alpha_D^2 = \sum_{k=1}^{K} h_k^2 \tag{13.61}$$

The tone jammer contributes $\alpha_J \sqrt{2J} |H(f_J)| \cos\phi$ where α_J is the Rayleigh-distributed amplitude, J is the average jammer power, and ϕ is the jammer phase uniformly distributed over (0, 2π). The noise power at the output of the rake receiver is given by

$$\sigma^2 = N_0(1-\rho)\alpha_D^2 \tag{13.62}$$

The conditional BER, conditioned on the instantaneous channel impulse response, is then

$$P_e = Q(2\tilde{\eta}) \tag{13.63}$$

where

$$\tilde{\eta} = \frac{\alpha_D^2}{\dfrac{N_0}{E_b}\dfrac{2}{1-\rho} + \dfrac{J}{R}\dfrac{2|H_0(f_J)|^2}{T_b(1-\rho)^2}\dfrac{|H_D(f_J)|^2}{\alpha_D^2}} \tag{13.64}$$

and

$$H_{\mathrm{D}}(f)=\sum_{k=1}^{K} h_k e^{-j2\pi f t_k} \tag{13.65}$$

The results when $K = 8$ paths and fingers are illustrated in Figure 13.14.

13.3.6 Pulsed Jamming

The effects of pulsed jamming on DSSS signals was discussed in Chapter 10. Pulsed jamming also affects UWB communication signals in a similar way. Tesi et al. conducted an analysis of the effects of pulsed jamming on UWB signals around 2 GHz [11]. The pulsed signal in that analysis consisted of a truncated sinc pulse with a bandwidth of 60 MHz. The total average power for the jamming pulse was 10 dBm.

Typical results of the analysis are shown in Figure 13.15. As with DSSS, like BBN above, the processing gain was the limiting factor that precluded jamming at acceptable values of JSR.

13.3.7 Partial-Band Noise Jamming

The power scaling parameter ζ in Figure 13.16 represents the ratio between the UWB signal level at the center frequency of the jammer band (f_{J}) to the UWB signal level at the UWB center frequency (f_{c}) and is given by:

$$\zeta=\frac{S_{\mathrm{UWB}}(f_{\mathrm{J}})}{S_{\mathrm{UWB}}(f_{\mathrm{c}})} \tag{13.66}$$

The bit error probability for a UWB BPAM modulated signal in an AWGN channel and tone interference is developed in [15] yielding:

$$P_{\mathrm{e}}=Q\left(\sqrt{\frac{2E_{\mathrm{b}}}{N_0+2\zeta\dfrac{N_{\mathrm{J}}W_{\mathrm{J}}}{W}}}\right) \tag{13.67}$$

Thus, just as in other forms of spread spectrum signaling, the processing gain reduces the PSD of the jammer by spreading the power of the interferer over the UWB bandwidth when despreading the UWB signal. The power scaling factor ζ represents the effectiveness of the jammer. If the jammer is located at a frequency with a low UWB PSD, it has less effect than when the jammer's frequency is close to the maximum UWB PSD.

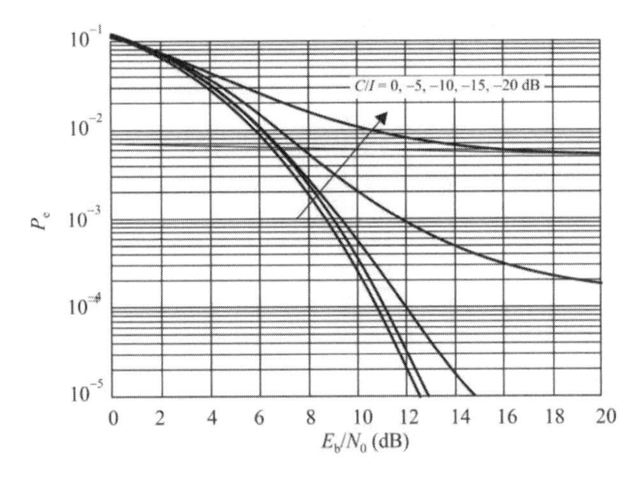

Figure 13.14 Tone jamming performance against TH PPM with a target rake receiver with $K = 8$ fingers. There are 8 paths the target signal transitions and the multipath channels have Nakagami characteristics with $m_k = m_1 \exp[-(k-1)/\gamma], k = 1,\ldots,8, \gamma = 4, m_1 = 3$. (Source: [13]. © IEEE 2005. Reprinted with permission.)

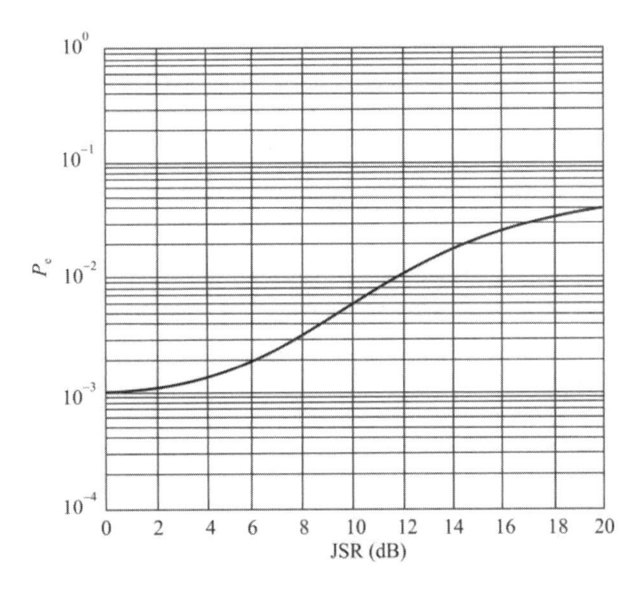

Figure 13.15 Performance of pulsed jamming on UWB communications. In this case the modulation is TH-BPAM, and $T_p = 1$ ns. (Source: [11], © Tesi 2002. Reprinted with permission.)

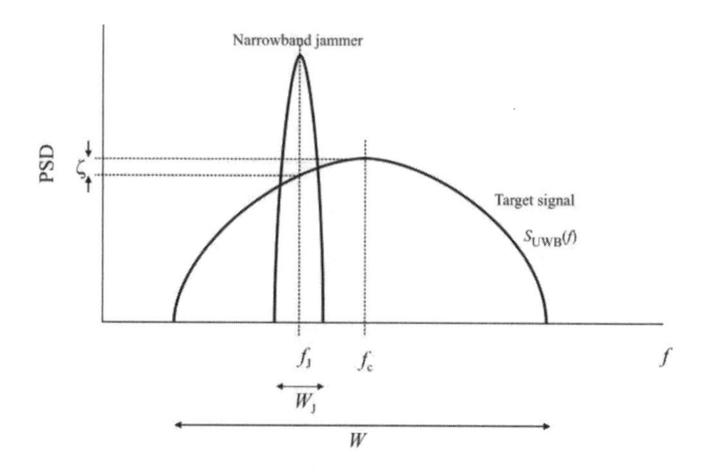

Figure 13.16 PSDs for a UWB signal and jammer. (Source: [15]. © IEEE 2005. Reprinted with permission.)

We assume that the jamming is narrowband when compared to the carrier frequency of the target bandpass signal. In the barrage jamming case, when the entire UWB bandwidth is jammed, the PSD of the jamming signal is $N_J = J/W$, where N_J and J denote one-sided power spectral density of the jammer and jamming power, respectively. With the jammer bandwidth denoted by W_J, the fraction of the partial band jamming PSD that overlaps the target signal PSD is then given by $N_J' = J / W_J$. The symbol error probability P_e for the BPSK and BPAM modulated signal is then given by

$$P_e = Q\left[\sqrt{\frac{E_b}{(N_0 / 2) + \zeta S(f_J)}}\right] \tag{13.68}$$

where E_b and N_0 are bit energy and one-sided noise PSD, respectively, and $S(f_J)$ is the contribution of the jammer energy. When determining the impact of jamming on the target UWB system performance in (4), the weighting is done using the PSD of the target signal over the portion of the spectrum occupied by the jammer, namely,

$$S(f_J) = \frac{N_J'}{W} \int_{f_J - W_J/2}^{f_J + W_J/2} \frac{(2\pi u \sigma)^2}{k^k e^{-k}} e^{\left[-(2\pi u \sigma)^2\right]} du \tag{13.69}$$

where k defines the number of differentiations of the Gaussian pulse (the order).

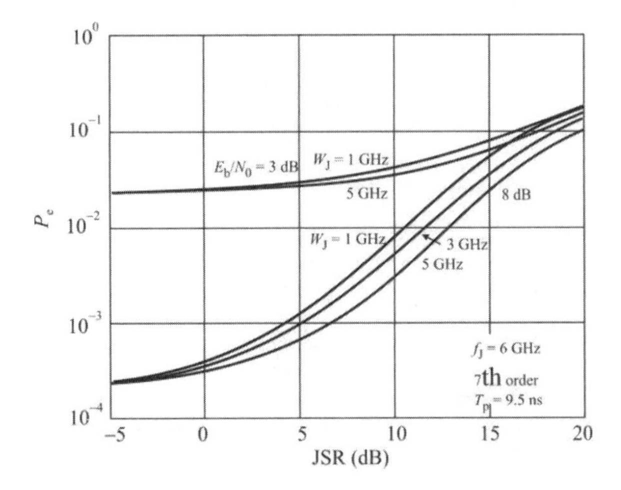

Figure 13.17 Tone jamming performance against UWB signal while varying W_J. The effects of a "narrowband" jammer against UWB signals. Seventh order derivative of the impulse function and $T_p = 500$ ps, $f_J = 6$ GHz. (After: [15]. © IEEE 2005. Reprinted with permission.)

The BER is plotted against different JSR values using the seventh Gaussian pulse derivative (T_p = 0.5 ns) with two E_b/N_0 values—3 dB and 8 dB in Figure 13.17 [15]. Interference is located at 6 GHz having different bandwidths. In Figure 13.18, the corresponding BERs as a function of JSR are given for the first and seventh derivatives of the Gaussian pulse when the center frequency of the interference varies but the bandwidth is fixed to 20 MHz. The impact of the increased desired signal bandwidth on improved sensitivity against the interference can be easily seen from the curves. When the first derivative is used, $W_J/W \approx 4.6\%$ but for the seventh, $W_J/W \approx 0.33\%$ using $W_J = 20$ MHz.

13.3.8 Narrowband Noise Jamming

When $W_J << W$, we can assume that the jammer power is constant over W_J and (13.68) can be simplified into

$$P_e = Q\left[\sqrt{\frac{2}{\frac{R_d N_0}{R} + 2\zeta\xi\frac{R_d}{W}}}\right] \tag{13.70}$$

where ξ is the JSR and R_d and R are data rate and signal power, respectively.

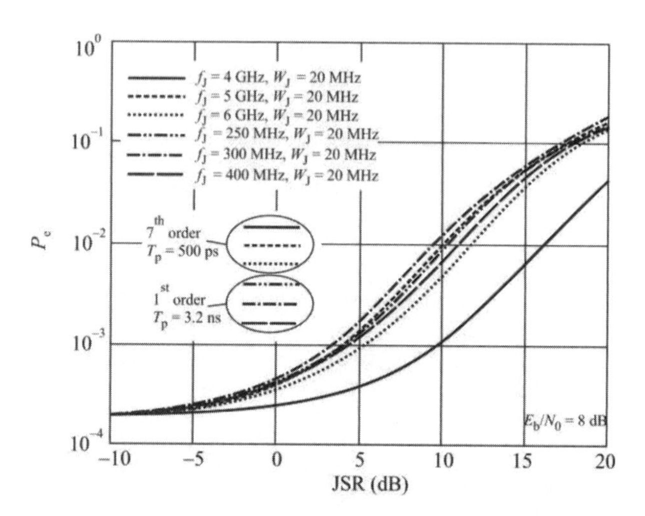

Figure 13.18 UWB BER performance as the jammer frequency is varied. (After: [15]. © IEEE 2005. Reprinted with permission.)

Figure 13.19 is an example of jamming performance against a UWB TH-PPM link for a single tone jammer, with the levels of SNR indicated. For these curves, the pulse time was T_p = 500 ps, the frame time was T_f = 100 ns, the delta offset was δ = 300 ps, there were 2 pulses per symbol, and the jammer was at f_J = 2.412 GHz. Also neither the target link nor the jammer to receiver link experienced fading and the noise was AWGN. This chart is an illustration of the relative immunity of the UWB link to jamming. Comparing this with Figures 13.17 and 13.18 we see that the jamming performance is considerably worse. The reason for this is here we have a single tone, implying zero jamming bandwidth whereas in Figures 13.17 and 13.18, the jammer signal was substantially more broadband and much more of the spectrum was impaired by the jammer.

13.4 Concluding Remarks

In this chapter the detectability and jam capability of an EW system when targeting UWB targets was examined. We determined that UWB signals are difficult to jam because of their processing gain. The processing gain is the product of two factors: T_f/T_p and L, where L represents the number of pulses per symbol. As with DSSS targets, essentially the processing gain must be overcome before the jamming is effective.

UWB signals are detectable with a channelized radiometer; however, doing so is expensive. Essentially a (time) channel is required for each possible pulse time, and, for example, when T_f = 1 μs and T_p = 1 ns, there are 1,000 possible pulse

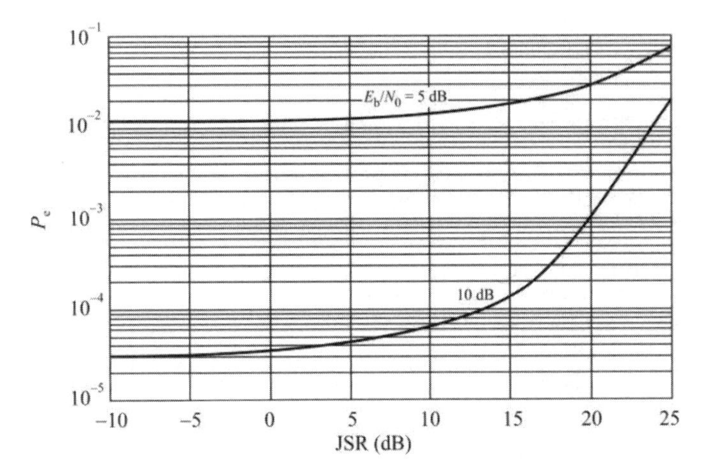

Figure 13.19 BER for TH-PPM vs JSR. Both target link and jammer to receiver link not fading. Single tone jammer. $T_p = 500$ ps, $T_f = 100$ ns, $\delta = 300$ ps, $N_s = 2$ pulses per bit, $f_J = 2.412$ GHz.

locations in each frame. This is further complicated when time hopping using PN codes are used to move the SOI around in the time slots.

References

[1] Mills, R. F., and G. E. Prescott, "A Comparison of Various Radiometer Detection Models," *IEEE Transactions on Aerospace and Electronic Systems*, January 1996, Vol. 32, pp. 467–473.

[2] Weeks, G. D., J. K. Townsend, and J. A. Freebersyer, "Quantifying the Covertness of UWB," *Proceedings of Ultra Wideband Conference*, Washington, D.C., September 1999.

[3] Ipatov, P. V., *Spread Spectrum and CDMA: Principles and Applications*, New York: Wiley, 2005.

[4] Torrieri, D. J., *Principles of Spread-Spectrum Communication Systems*, Norwood, MA: Artech House, 1992.

[5] Park, K. Y. "Performance Evaluation of Energy Detectors," *IEEE Transactions on Aerospace and Electronic Systems*, Vol. AES-14, March 1978, pp. 237–241.

[6] Dillard, R. A. "Detectability of Spread Spectrum Signals," *IEEE Transactions on Aerospace and Electronic Systems*, Vol. AES-15, July 1979, pp. 526–537.

[7] Barton, D. K. "Simple Procedures for Radar Detection Calculations," *IEEE Transactions on Aerospace and Electronic Systems*, Vol. AES-5, September 1969, pp. 837–846.

[8] Bharadwaj, A., and J. K. Townsend, "Evaluation of the Covertness of Time-Hopping Impulse Radio Using a Multi-Radiometer Detection System," *Proceedings IEEE MILCOM*, 2001, Vol. 1, pp. 128–134.

[9] McKinstry, D. R., and R. M. Buehrer, "Issues in the Performance and Covertness of UWB Communications Systems," *MWSCAS-2002, 45th Midwest Symposium on Circuits and Systems*, 2002, Vol. 3.

[10] Weeks, G. D., J. K. Townsend, and J. A. Freebersyer, "A Method and Metric for Quantitatively Defining Low Probability of Detection," *Proceedings IEEE MILCOM*, 1998, Vol. 3, pp. 821–826.

[11] Tesi, R., M. Hämäläinen, J. Iinatti, and V. Hovinen, "On the Influence of Pulsed Jamming and Colored Noise in UWB Transmission," http://www.ee.oulu.fi/~mattih/FWCW2002_final.pdf.

[12] Hämäläinen, M., R. Tesi, J. Iinatti, and V. Hovinen, "On the Performance Comparison of Different UWB Data Modulation Schemes in the Presence of Jamming," *Proceedings Radio and Wireless Conference*, 2002, RAWCON, pp. 83–86.

[13] Giorgetti, A., M. Chiani, and M. Z. Win, "The Effect of Narrowband Interference on Wideband Wireless Communication Systems," *IEEE Transactions on Communications*, Vol. 53, No. 12, December 2005, pp. 2139–2149.

[14] Abramowitz, M., and I. A. Stegun, *Handbook of Mathematical Functions*, New York: Dover, 1964, p. 556.

[15] Hämäläinen, M., and J. Iinatti, "Analysis of Jamming on DS-UWB System," *Proceedings IEEE MILCOM*, 2005, pp. 1–6.

Chapter 14

Electronic Warfare and Hybrid Spread Spectrum Systems

14.1 Introduction

Hybrid spread spectrum signals are made of combinations of the two basic forms of spread spectrum signaling: DSSS and FHSS. Although more complicated to design and build than either DSSS or FHSS systems alone, the combination provides some very attractive features for the communicator. DSSS is an LPD technique that hides the signal, sometimes below the thermal noise level, but in any case, makes the signal look like noise. The FHSS aspect makes the signal difficult to associate one hop with the next with an ES system, therefore making hop dwell association difficult. It also provides for frequency diversity, mitigating the effects of frequency selective fading.

The structure of this chapter is as follows. First, we review what a hybrid SS system is. We then examine jamming performance of coherent hybrid systems, taking a look at BBN, PBN, MT, and NBN follower methodologies. Last we explore noncoherent hybrid SS systems using the same jammer strategies.

14.2 Hybrid SS Systems

Only SFHSS systems are discussed in this chapter. This is because DSSS/FFHSS systems are difficult to build for all but the lowest throughput communication systems. Their application to the mobile, tactical scenario has been limited.

The DSSS signal must be code locked and synchronized in order to retrieve the underlying data. Therefore, either code-coherence must be maintained from hop to hop, or the code must be resynchronized at the new hop frequency. In the latter case, data throughput is reduced. In the former case, if code coherency is lost it also must be reacquired. It is assumed here that for FHSS, noncoherent demodulation is employed. Coherent demodulation of the DSSS signal is considered first then noncoherent DSSS demodulation is examined.

The hybrid processing gain is given by

$$G_p = N_D N_F \tag{14.1}$$

where N_D denotes the DSSS processing gain and the FHSS processing gain is N_F ($= N_c$). These are linear terms, not in decibels. Therefore, substantial overall processing gain is achievable with only moderate individual processing gain.

The appellation SFHSS is somewhat of a misnomer when it comes to hybrid SS systems. It normally means more than one data bit is transmitted on each hop. What it is intended to infer with hybrid systems is that at each hop dwell the signal can be treated as if it were not hopping at all and the associated DSSS signal can be treated as such. This, then, ignores transition times between hops and the accompanying phase acquisition issues, when phase coherence acquisition is required.

As with any hopping system, the signals from multiple sources occasionally collide. If the systems are operating asynchronously, they do not share hop timing (epoch) information. Therefore, when they collide there can be full or partial temporal overlap. It is assumed that even in the asynchronous case, the hop rates are common.

Let N denote the length of the DSSS spreading code. It was discussed in Chapter 10 that for DSSS systems, long spreading codes are desirable for anti-ES reasons. One problem with long codes, however, is that acquiring them at the receiver can take longer than allowed for hybrid systems—code acquisition must be significantly shorter than the hop dwell. Determining the location in the spreading sequence must be done rapidly in such systems so that the majority of the dwell can be used to send data in SFHSS applications. Therefore, short codes are normally used in DSSS/FHSS hybrid systems. Thus, for the majority of this analysis, short DS codes are assumed, so that $N = N_D$. One case is considered later in the chapter where long codes are used.

In this chapter, only uncoded signals are considered.

14.3 Coherent Reception

When signals are coherently detected at a receiver, knowledge of the phase of the incoming signal is required so that the receiver can phase lock to it. When an FHSS signal changes frequency the phase coherency is lost, for coherent demodulation of the DSSS signal the hopping rate must be slow enough so that the phase lock can be regained and meaningful data can be exchanged.

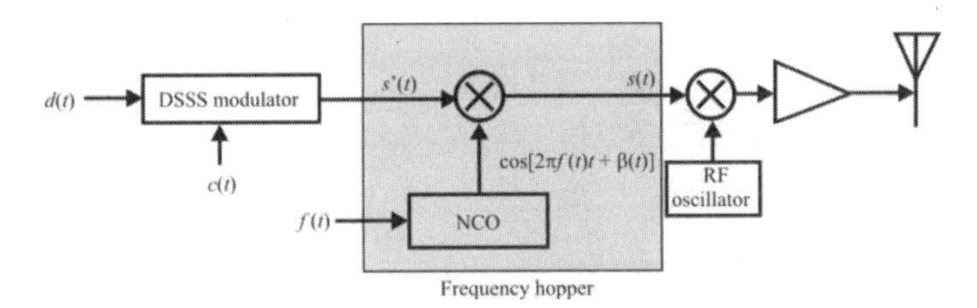

Figure 14.1 Coherent or noncoherent hybrid DSSS/FHSS transmitter. (After: [1]. © IEEE 1985. Reprinted with permission.)

For brevity here, the underlying PN codes are assumed random. Geranoitis [1] considers this case as well as deterministic and Gold coded sequences, the latter of which would be the normal case.

A block diagram of a DSSS/SFHSS transmitter is shown in Figure 14.1. This same diagram applies whether the system is coherent or noncoherent. The coherency occurs in the DSSS modulator before the frequency hopping is applied [1]. The signal from the DSSS modulator for BPSK can be expressed as

$$s'(t) = \sqrt{2S}d(t)a(t)\cos[2\pi f_c t + \phi(t)] \tag{14.2}$$

where S is the average power in the signal, $a(t) = c(t)w(t)$ is the shaped code sequence where $c(t)$ is the code sequence and has values from $\{-1, +1\}$ of duration T_c [$w(t)$ is weighting applied to this sequence to achieve specific signaling characteristics], f_c is the IF, $d(t)$ is the data sequence consisting of bits from $\{-1, +1\}$ of duration T_b, and $\phi(t)$ is the phase imposed by the DSSS modulator. Short DSSS codes are assumed so that there are N code pulses for each data bit and $T_b = NT_c$ and the period of the code sequence is N.

This signal is then hopped by multiplication by $\cos[2\pi f(t)t + \beta(t)]$ where $\beta(t)$ remains constant for the duration of the dwell. The signal from the mixer is

$$s(t) = \frac{\sqrt{2S}}{2}d(t)a(t)\left\{\begin{aligned}&\cos\{2\pi[f_c + f(t)]t + \phi(t) + \beta(t)\}\\ &\qquad + \cos\left[2\pi[f_c - f(t)]t + \phi(t) - \beta(t)\right]\end{aligned}\right\} \tag{14.3}$$

The final RF oscillator puts the resultant signal where desired in the spectrum.

A block diagram of a coherent DSSS/SFHSS receiver is shown in Figure 14.2. The first bandpass filter has a bandwidth equal to the total of the hopped and spread signal. Then the signal is asynchronously dehopped. The dehopped signal is then bandpass filtered with a filter of width W_{ss}, which removes unwanted noise and interference that are not in the pass band of interest, but passes the DSSS

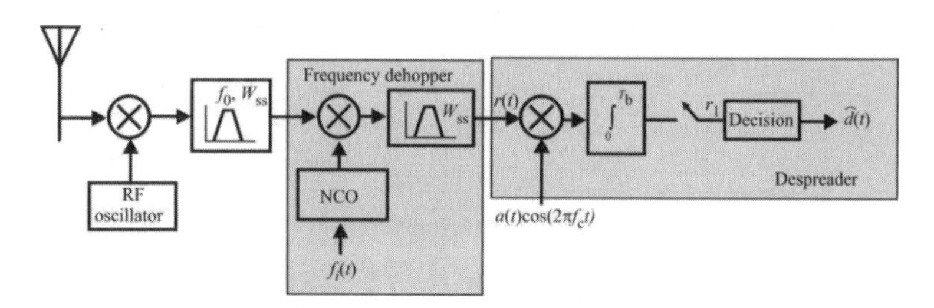

Figure 14.2 DSSS/SFHSS hybrid coherent signal demodulator. (After: [1]. © IEEE 1985. Reprinted with permission.)

signal of bandwidth W_{ss}. The resulting signal is multiplied by a phase-coherent replica of the coded signal at the transmitter, integrated for a bit time, T_b, and a decision made as to which symbol was sent.

The signal received is corrupted by noise in the channel and possibly other interfering signals, which may cause the data bit to be detected in error. For coherent DSSS/FHSS systems, the probability of a symbol error for coherent detection can be approximated as [1]

$$P_\text{e} = \sum_{l=0}^{L-1} \binom{L-1}{l} P^l (1-P)^{L-1-l} P_{\text{e,f}}(l) \tag{14.4}$$

where L is the number of users in the geographical vicinity of each other, and l is the number of full hits. A full hit is total temporal overlap of two or more signals. Partial hits are also possible in asynchronous systems, where only part of a hop dwell collides with another signal. Expression (14.4) is only an approximation because the effects of such partial hits are ignored. $P_{\text{e,f}}(l)$ is the error probability given that l full hits have occurred from the other $L-1$ users. It is assumed that the frequency spectrum is channelized so that there is no overlap of the signals in frequency—all hits are temporal rather than in the frequency domain.

The probability of a full hit is given by

$$P_\text{f} = \left(1 - \frac{1}{N_\text{b}}\right)\frac{1}{N_\text{c}} \tag{14.5}$$

while the probability of a partial hit is given by

$$P_\text{p} = \frac{2}{N_\text{b} N_\text{c}} \tag{14.6}$$

From these, P in (14.4) can be expressed as their sum,

$$P = P_{\mathrm{f}} + P_{\mathrm{p}} \tag{14.7}$$

The last term in (14.4) can be approximated by

$$P_{\mathrm{e,f}}(L) = Q\left[\zeta(L+1)\right] \tag{14.8}$$

where $\zeta(L)$ is calculated in the sequel. As before, let ξ denote the JSR and υ the SNR.

14.3.1 Coherent Asynchronous BPSK DSSS/SFHSS Systems

For BPSK [1],

$$\zeta(L) = \left[\left(\frac{2E_{\mathrm{b}}}{N_0}\right)^{-1} + \frac{L-1}{N}C\right]^{-1/2} \tag{14.9}$$

Assuming $W_{\mathrm{F}} = R_{\mathrm{b}}$, this expression becomes

$$\zeta(L) = \left[(2\upsilon)^{-1} + \frac{L-1}{N}C\right]^{-1/2} \tag{14.10}$$

and where

$$C = \frac{1}{T_{\mathrm{c}}^3}\int_0^{T_{\mathrm{c}}} R_{\mathrm{c}}^2(\tau)d\tau \tag{14.11}$$

when $R_{\mathrm{c}}(\tau)$ is the autocorrelation function of the chipping sequence. Two specific and common examples are [2]

$$C = \begin{cases} 1/3, & \text{rectangular chip waveform} \\ \dfrac{15+2\pi^2}{12\pi^2}, & \text{sine chip waveform} \end{cases}$$

P_{e} is given by (14.4), (14.7), and (14.8).

14.3.2 Coherent Asynchronous QPSK DSSS/SFHSS Systems

The probability of a bit error for coherent asynchronous DSSS/SFHSS systems when the modulation is QPSK is given by (14.4), (14.7), and (14.8) with [1]

$$\zeta(L) = \left[\left(\frac{2E_\text{b}}{N_0} \right)^{-1} + \frac{L-1}{N} C \right]^{-1/2} \tag{14.12}$$

and, again assuming $W_\text{F} = R_\text{b}$,

$$\zeta(L) = \left[(2\upsilon)^{-1} + \frac{L-1}{N} C \right]^{-1/2} \tag{14.13}$$

14.3.3 Coherent Synchronous BPSK DSSS/SFHSS Systems

The probability of a bit error for coherent synchronous DSSS/SFHSS systems when the modulation is BPSK is given by (14.4), (14.7), and (14.8) with [1]

$$\zeta(L) = \left[\left(\frac{2E_\text{b}}{N_0} \right)^{-1} + \frac{L-1}{2N} C \right]^{-1/2} \tag{14.14}$$

Assuming $W_\text{F} = R_\text{b}$,

$$\zeta(L) = \left[(2\upsilon)^{-1} + \frac{L-1}{2N} C \right]^{-1/2} \tag{14.15}$$

14.3.4 Coherent Synchronous QPSK DSSS/SFHSS Systems

The probability of a bit error for coherent synchronous DSSS/SFHSS systems when the modulation is QPSK is given by (14.4), (14.7), and (14.8) with [1]

$$\zeta(L) = \left[\left(\frac{2E_\text{b}}{N_0} \right)^{-1} + \frac{L-1}{N} C \right]^{-1/2} \tag{14.16}$$

and, assuming $W_\text{F} = R_\text{b}$,

$$\zeta(L) = \left[(2\upsilon)^{-1} + \frac{L-1}{N} C \right]^{-1/2} \tag{14.17}$$

14.3.5 BBN Jamming of Coherent DSSS/SFHSS Systems

As mentioned above, DSSS/SFHSS communication systems within range of each other can operate synchronously or asynchronously, depending on the timing information that is available. If all communication nodes have the same timing information, synchronous operation is possible. Otherwise, only asynchronous operation is possible. This issue is exacerbated by the hidden node problem in networks.[1]

14.3.5.1 BBN Jamming of Coherent Asynchronous DSSS/SFHSS Systems

The jamming performance, in general, is different depending on whether the DSSS modulation is BPSK or QPSK. These cases are treated separately.

BPSK

As previously, jamming with BBN jamming of spectral density J_0, raises the noise floor as expressed by N_0 to a higher level

$$N_\mathrm{T} = N_0 + J_0 \tag{14.18}$$

Therefore, the probability of creating a bit error is given by (14.4) with the variables as expressed in (14.5)–(14.8) and

$$\zeta(L) = \left[\left(\frac{2E_\mathrm{b}}{N_\mathrm{T}} \right)^{-1} + \frac{L-1}{N} C \right]^{-1/2}$$

$$= \left[\left(\frac{2E_\mathrm{b}}{N_0 + J_0} \right)^{-1} + \frac{L-1}{N} C \right]^{-1/2}$$

[1] The hidden node problem is when A can hear B and C can hear B but A cannot hear C and vice versa. Timing information is difficult to make common in such cases, especially in mobile networks.

$$= \left[\left(\frac{2E_{\text{b}}}{N_0 + J_0} \frac{W_{\text{ss}}}{W_{\text{ss}}} \right)^{-1} + \frac{L-1}{N} C \right]^{-1/2} \quad (14.19)$$

but $W_{\text{ss}} = N_{\text{F}} W_{\text{F}}$, $N_0 W_{\text{ss}} = P_{\text{N}}$, and $J_0 W_{\text{ss}} = J$. Therefore

$$\zeta(L) = \left[\left(\frac{2N_{\text{F}} W_{\text{F}} E_{\text{b}}}{P_{\text{N}} + J} \right)^{-1} + \frac{L-1}{N} C \right]^{-1/2} \quad (14.20)$$

and when $W_{\text{F}} = R_{\text{b}}$, $W_{\text{F}} E_{\text{b}} = R$, the average power, and

$$\zeta(L) = \left[\left(\frac{2N_{\text{c}} R}{P_{\text{N}} + J} \right)^{-1} + \frac{L-1}{N} C \right]^{-1/2}$$

$$= \left[\left(\frac{2N_{\text{c}}}{\frac{1}{\upsilon} + \xi} \right)^{-1} + \frac{L-1}{N} C \right]^{-1/2} \quad (14.21)$$

QPSK

For QPSK the probability of creating a bit error is given by (14.4) with the variables as expressed in (14.5)–(14.8) and

$$\zeta(L) = \left[\left(\frac{2E_{\text{b}}}{N_{\text{T}}} \right)^{-1} + \frac{2(L-1)}{N} C \right]^{-1/2}$$

$$= \left[\left(\frac{2N_{\text{c}}}{\frac{1}{\upsilon} + \zeta} \right)^{-1} + \frac{2(L-1)}{N} C \right]^{-1/2} \quad (14.22)$$

under the same assumptions as in the last section.

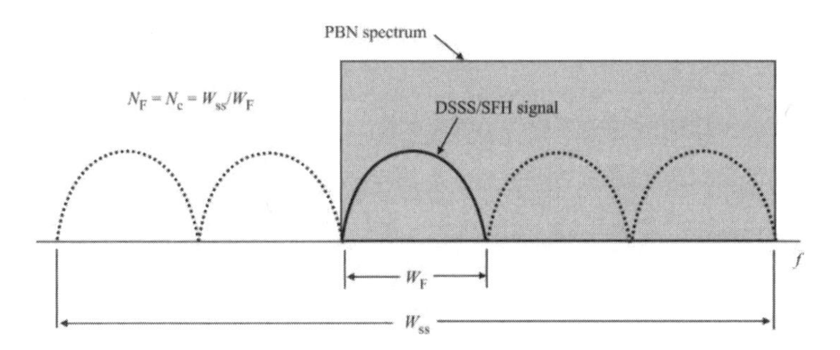

Figure 14.3 Spectra characteristics of PBN jamming of DSSS/FHSS systems.

14.3.5.2 BBN Jamming Coherent Synchronous DSSS/SFHSS Systems

The discussions after (14.4) would lead us to believe that the jamming performance against synchronous and asynchronous systems would be the same. This is because P_e is approximated by assuming that all the hits are synchronous. There is, however, a subtle difference in the way P is calculated. For asynchronous systems, P is given by (14.7). For synchronous systems, $P_p = 0$ and

$$P = P_f \tag{14.23}$$

Other than this difference, P_e is calculated with (14.4) and the variables as expressed in (14.5), (14.6), (14.8), and (14.23).

BPSK

The bit error probability is given by (14.21).

QPSK

The bit error probability is given by (14.22).

14.3.6 PBN Jamming of Coherent DSSS/SFHSS Systems

The notion of PBN jamming of DSSS/FFSS systems is illustrated in Figure 14.3. We assume for convenience and without much loss in generality that the edges of the jamming spectrum are aligned with the edges of two channels and the DSSS spectrum occupies the entirety of one channel of width W_F, with no overlap. The average power in the DSSS signal is then

$$R = E_b W_F \tag{14.24}$$

14.3.6.1 PBN Jamming of Coherent Asynchronous DSSS/SFHSS Systems

Again, the PBN jamming performance depends on the DSSS modulation.

BPSK

When only a portion of the spectrum is jammed then there are channels that contain no jamming energy and channels that do. Those that do contain a higher jammer spectral density given by J_0/γ. Therefore the probability of creating a bit error when the hybrid system lands on a jammed dwell is given by (14.4) with the variables as expressed in (14.5)–(14.8) and

$$\begin{aligned}\zeta(L) &= \left[\left(\frac{2E_b}{N_0 + \dfrac{J_0}{\gamma}}\right)^{-1} + \frac{L-1}{N}C\right]^{-1/2} \\ &= \left[\left(\frac{2\gamma E_b}{\gamma N_0 + J_0}\frac{W_{ss}}{W_{ss}}\right)^{-1} + \frac{L-1}{N}C\right]^{-1/2} \\ &= \left[\left(\frac{2\gamma N_c}{\dfrac{\gamma}{\upsilon} + \xi}\right)^{-1} + \frac{L-1}{N}C\right]^{-1/2}\end{aligned} \tag{14.25}$$

Note that $\upsilon \neq \upsilon_l$, as the former is the SNR at the output of the receiver due to L signals present, while the latter is the SNR due to a single one.

Denoting $P_{e_2} = P_e$ with P_e from (14.4), the overall error rate is

$$P_e = (1-\gamma)\frac{1}{2}\exp\left(-\frac{1}{2}\upsilon\right) + \gamma P_{e_2} \tag{14.26}$$

The performance of jamming coherent asynchronous BPSK systems is illustrated in Figure 14.4. The optimum value of γ is clearly less that 0.1 and the

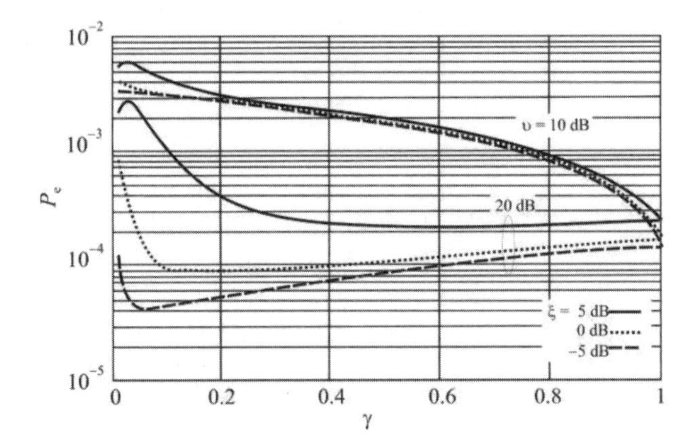

Figure 14.4. PBN jamming performance against coherent asynchronous BPSK systems.

goal of producing $P_e > 10^{-1}$ is not achieved for the values of the parameters considered.

QPSK

As above, the calculation of P_{e_2} for QPSK is calculated with (14.4) and the variables as expressed in (14.5), (14.6), (14.8), and (14.23).

$$\zeta(l) = \left[\left(\frac{2E_b}{N_0 + \dfrac{J_0}{\gamma}}\right)^{-1} + \frac{2(L-1)}{N}C\right]^{-1/2}$$

$$= \left[\left(\frac{2\gamma N_c}{\dfrac{\gamma}{\upsilon} + \xi}\right)^{-1} + \frac{2(L-1)}{N}C\right]^{-1/2} \tag{14.27}$$

The overall probability of a bit error is then given by (14.26).

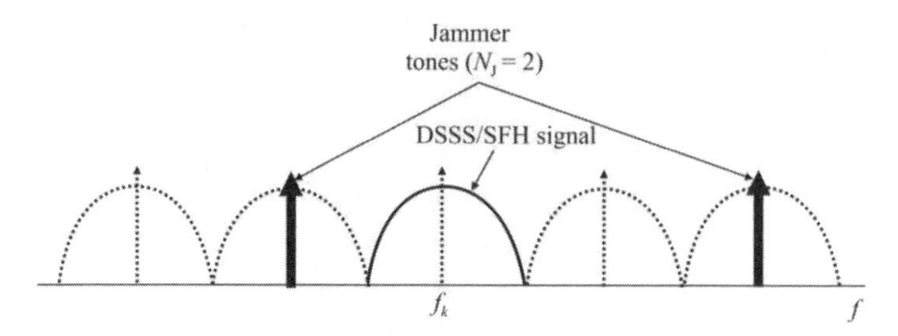

Figure 14.5 Spectra for hybrid DSSS/SFHSS and multitone jammer. Although the jammer is a multitone jammer, within each DSSS channel there is only one tone at any given instant.

14.3.6.2 PBN Jamming of Coherent Synchronous DSSS/SFHSS Systems

The same comments as in Section 14.3.5.2 apply here as well, and the BER is calculated in the same way as in the last section using the appropriate change of variables.

14.3.7 Multitone Jamming of Coherent DSSS/SFHSS Systems

As shown in Chapters 10, 11, and 12, multitone jamming can be effective against both DSSS and FHSS systems. Techniques have been devised to attempt to mitigate the effects of tone jamming against hybrid systems as well. Such techniques consist of filtering out tone interference or jamming tones in the frequency domain by eliminating narrowband regions of the spectrum [3, 4]. In this section the effects of partial hits are ignored; thus, the communication networks are synchronized.

The notion of MT jamming of hybrid DSSS/FHSS systems is illustrated in Figure 14.5 when $N_J = 2$. A tone is placed in several of the DSSS channels and the hybrid signal will either hop into one of the channels with a jammer tone in it, or it will hop into a channel where there is no such tone. Again, the fraction γ of W_{ss} jammed is given by

$$\gamma = \frac{N_J}{N_c} \tag{14.28}$$

Best-case jamming occurs when the jammer knows the hopping center frequency and places a jamming tone near, but not at, that point. Jamming performance degrades when the jamming tone is offset by large distances from f_k, whereas close to the center frequency jamming performance is relatively independent of the tone offset. At zero offset, the despreading process significantly degrades the jamming tone [5] as indicated in Section 10.7.8. Offset from the center frequency by the symbol rate produces the best jamming performance.

Haiou and Naitong [6] performed an analysis of the effects of multitone jammers on coherent QPSK DSSS/SFHSS hybrid systems. The communication signal at the receiver during the *k*th hop dwell is given by

$$r_k(t) = \sqrt{2R}\begin{bmatrix} d_k^{\mathrm{I}}(t)c_k^{\mathrm{I}}(t)\cos\left(\dfrac{\pi t}{T_\mathrm{c}}\right)\cos\left(2\pi f_k t\right) \\ +d_k^{\mathrm{Q}}(t)c_k^{\mathrm{Q}}(t)\sin\left(\dfrac{\pi t}{T_\mathrm{c}}\right)\sin\left(2\pi f_k t\right) \end{bmatrix} \tag{14.29}$$

where R is the average power, $d_k^{\mathrm{I}}(t)$ and $d_k^{\mathrm{Q}}(t)$ are the in-phase and quadrature data sequences, $c_k^{\mathrm{I}}(t)$ and $c_k^{\mathrm{Q}}(t)\, a_{\mathrm{Q}k}(t)$ are the in-phase and quadrature shaped chipping sequences, T_c is the chip time, and f_k is the *k*th hopping frequency.

On the other hand, the total multitone jamming waveform at the receiver is given by

$$j(t) = \sum_{j=1}^{N_J} \sqrt{2J_j}\cos\left(2\pi f_j t + \theta_j\right) \tag{14.30}$$

where J_j is the power in jamming tone *j* and

$$\sum_{j=1}^{N_\mathrm{J}} J_j = J \tag{14.31}$$

f_j is the frequency of the *j*th jamming tone, θ_j is the phase difference between the *j*th jammer tone and the FH carrier, and the number of jamming tones is given by N_J.

The jammer tones after DSSS demodulation by the receiver are given by

$$j_j(t) = \sqrt{2J_j}c(t)\sin\left(2\pi\Delta f_j t - \theta_j\right) \tag{14.32}$$

where $\Delta f_j = f_\mathrm{c} - f_j$ is the frequency difference between the *i*th tone and the DSSS/SFHSS hop center frequency f_c. It is assumed that the frequency offsets $\Delta f_i = \Delta f$ are equal.

The BER performance of this hybrid system is given by

$$P_e = (1-\gamma)\frac{1}{2}\exp\left(-\frac{1}{2}\upsilon\right) + \gamma P_{e_2} \tag{14.33}$$

The second term on the right accounts for the probability of bit error when the hybrid signal hops into a channel that is occupied by a jammer. When it does, the BER is given by P_{e_2} and this happens with a probability given by γ. The first term on the right accounts for the probability of bit error when the signal hops into a channel without a jammer and the error is caused by thermal noise. The channel has just noise and the BER is given by

$$P_e = \frac{1}{2}\exp\left(-\frac{1}{2}\upsilon\right) \tag{14.34}$$

for both BPSK and QPSK when the SNR is given by υ. The probability of this occurring is given by $(1-\gamma)$.

For QPSK, since the BER for both channels is the same, analysis of either channel will yield the same jamming performance.

The expression for P_{e_2} is [6]

$$P_{e_2} = Q\left\{\sqrt{N_\mathrm{D}}\left[\frac{8\xi}{\pi^2}\left[\frac{\cos(\pi\Delta f T_c)}{1-(2\Delta f T_c)^2}\right]^2 + \frac{1}{2\upsilon}\right]^{-1/2}\right\} \tag{14.35}$$

This function is illustrated in Figure 14.6 where $N = N_\mathrm{D} = 7$ and 15,

$$T_c = \frac{T_b}{N} \tag{14.36}$$

$T_b = 50$ μs, $\gamma = 0.1$ and 0.3, $\upsilon = 10$ dB, and $\Delta f = 25$ kHz. These parameters are used for the remaining examples in this chapter.

For the parameters illustrated, MT jamming achieves $P_e > 10^{-1}$ at $\xi = 15$ dB when $\gamma = 0.3$ and $N_\mathrm{D} = 7$ which is about 5 dB higher than BBN and PBN jamming. When $\gamma = 0.1$, inadequate jamming performance ensues for all values of JSR considered. Below about $\xi = 0$ dB, the BER remains relatively constant and independent of γ.

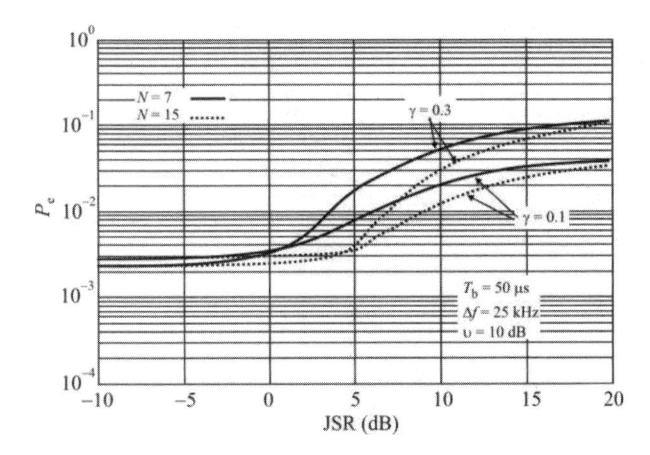

Figure 14.6 MT jammer performance against coherent, synchronous, QPSK, MSK, DSSS/SFHSS systems when υ = 10 dB.

14.3.8 NBN Follower Jamming of Coherent DSSS/SFHSS Systems

Follower jamming can be employed against DSSS/SFHSS systems with any of the jamming waveforms discussed in Chapter 8. The same timing limitations as in Chapter 12 exist for hybrid SS systems, but, because of the difficulty detecting the DS signal rapidly, the problem is exacerbated. Normally it is necessary to process several chips of the DS signal in order to ascertain whether the signal exists in a frequency channel [7].

For SFHSS, there would be, typically, 100 data bits or more during any dwell. Each of these data bits is spread by the DS signal. Since short codes are normally used, it would not be unusual to have on the order of 7–31 chips per data bit in such a system. If $N = 7$ and if the data rate is 10 kbps, then the chip waveform would be 70 kcps yielding a main lobe 3 dB bandwidth of 140 kHz.

Once the target is selected, jamming would be accomplished by attacking the data (DS) channel since the data modulation is PSK. Neglecting the added difficulty of tracking hybrid signals from hop to hop, the jamming performance is the same as for nonhopped DS spread signals discussed in Chapter 10. The tracking can be taken into account in the same way as follower jamming in Chapter 12.

14.3.9 Jamming Coherent DSSS/SFHSS Systems Comparisons

This section presents the performance of BBN and PBN jamming against coherent DSSS/SFHSS systems. Figure 14.7 illustrates a few values of the parameters. Of

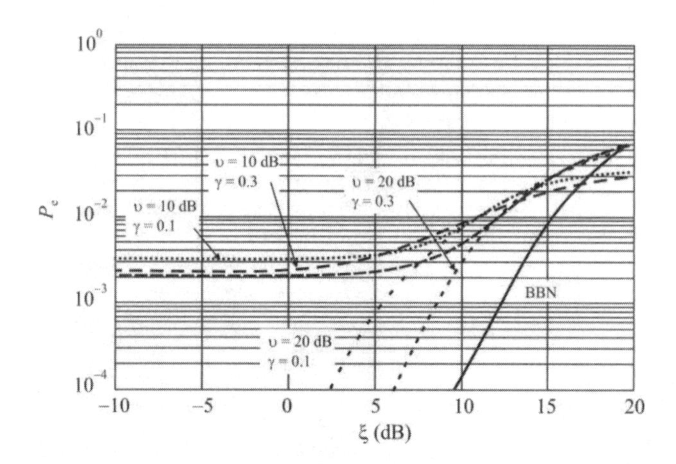

Figure 14.7 PBN and BBN jamming performance against coherent asynchronous BPSK systems. All curves except the one marked BBN are PBN. BBN performance is the same for all parameter values. For this example, $L = 50, N_c = 100$, and $N_b = 10$.

course, the BBN performance does not vary as γ is varied so BBN jamming results in the same curve in Figure 14.7.

For PBN, the goal of $P_e = 10^{-1}$ is not achieved. For $\upsilon = 10$ dB, whether γ is 0.1 or 0.3, the results do not vary significantly. When $\upsilon = 20$ dB, with low values of JSR, $\gamma = 0.1$ actually increases more rapidly as the JSR is increased. For $\xi > 15$ dB, or so, $\gamma = 0.3$ surpasses that of $\gamma = 0.1$, however.

14.4 Noncoherent Reception

For coherent demodulation of a received signal, the local demodulation oscillator must be phase locked to the incoming signal so the phase information can be extracted. In some cases, such phase synchronization is not feasible, for example, when the noise or interference levels are too high. In those cases, noncoherent detection is preferred and can be used to avoid the requirement for the phase lock. In this section the performance of BBN, PBN, and MT jamming against noncoherent DSSS modulation types BFSK, MFSK, and DPSK are examined [8].

For BFSK DSSS modulation, the signal from the DSSS modulator shown in Figure 14.1 that is to be hopped is given by

$$s'(t) = \sqrt{2R}w(t)c(t)\cos\{2\pi[f_c + d(t)\delta]t + \phi(t)\} \tag{14.37}$$

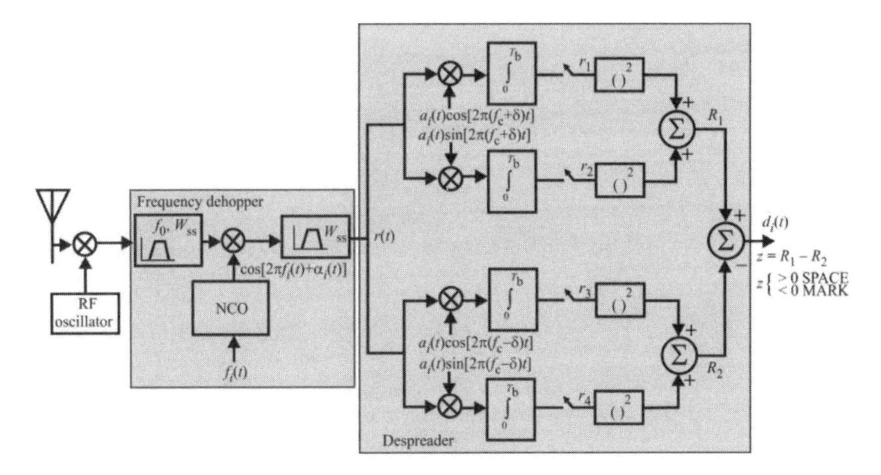

Figure 14.8 Hybrid receiver for noncoherent BFSK modulation.

where R is the average power in the signal; $c(t)$ is the code sequence waveform, which consists of square pulses with values from $\{-1, +1\}$ of duration T_c; $w(t)$ is weighting applied to this sequence to achieve specific signaling characteristics, f_c is the IF; $d(t)$ is the data sequence consisting of bits from $\{-1, +1\}$ and of duration T_b; δ is the frequency offset of the BFSK signal above and below f_c; and $\phi(t)$ is the phase imposed by the DSSS modulator. Short DSSS codes are assumed so that there are N code pulses for each data bit and $T_b = NT_c$ and the period of the code sequence is N.

A receiver for noncoherent BFSK is illustrated in Figure 14.8. The received signal is first noncoherently dehopped after bandpass filtering with bandwidth W_{ss}. The dehopping mixer is then followed by a bandpass filter of bandwidth W_F, that of the DSSS signal. This signal, $r(t)$ is then sent to the DSSS despreader. The upper two integrate and dump channels of the despreader will integrate those chips that are offset from f_c on the additive side while those in the lower two integrate and dump channels in Figure 14.8 will accumulate energy when the chip is offset on the lower side of f_c. After one bit time, T_b, the outputs of the accumulated channels are compared and a bit decision is made.

For MFSK DSSS modulation, (14.37) still applies, but the data bits $d(t)$ are selected from the set $\{1, 2, \ldots, M\}$ and the duration of each symbol is $T_s = T_b \log_2 M$ and the period of the code is $N_{\text{MFSK}} = N \log_2 M$. The receiver for MFSK is the same as that shown in Figure 14.8 with more filter channels added in the normal way.

When the DSSS modulation is DPSK the signal in Figure 14.1 to be hopped, $s'(t)$, is somewhat different. It is given by

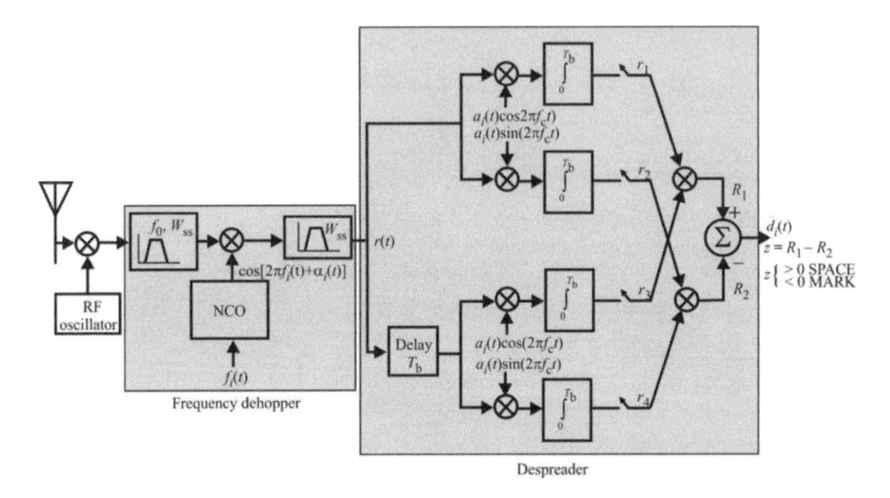

Figure 14.9 Hybrid DSSS/BFSK receiver for noncoherent DPSK waveforms.

$$s'(t) = \sqrt{2R}d(t)w(t)c(t)\cos\left[2\pi f_c t + \phi(t)\right] \tag{14.38}$$

where, in this case, $d(t)$ is a differentially encoded version of the data sequence. Otherwise, the variables are the same. While the transmitter for DPSK is the same as that shown in Figure 14.1, the receiver is different. It is shown in Figure 14.9. After being dehopped by an NCO that is synchronized with the transmitter hopping sequence, the signal is sent to two identical demodulation channels as shown in Figure 14.9, where the lower channel is delayed by one bit time. The in-phase and quadrature-phase channels of the delayed channel are compared to their counterpart in the current symbol to detect the differential phase change. Again, after one bit time a decision is made as to what the accumulated phase change was.

In general, for noncoherent DSSS/FHSS systems, the probability of a symbol error for noncoherent detection can be expressed as [2]

$$P_s = \sum_{l_f=0}^{L-1} \sum_{l_p=0}^{L-1-l_f} P_h(l_f, l_p) P_e(l_f, l_p) \tag{14.39}$$

where L is the number of users in the geographical vicinity of each other, l_f is the number of full hits, and l_p is the number of partial hits. A full hit is total temporal overlap of two or more signals while a partial hit is some, but less than full, temporal overlap. $P_h(l_f, l_p)$ is the probability of l_f full hits and l_p partial hits from the other $L - 1$ users while $P_e(l_f, l_p)$ is the conditional error probability given that l_f full hits and l_p partial hits have occurred. It is assumed that the frequency spectrum is channelized so that there is no overlap of the signals in frequency. All hits are temporal rather than in the frequency domain.

14.4.1 Noncoherent Asynchronous DSSS/SFHSS Systems

In asynchronous hybrid systems, the signals hop independently of one another, changing frequency according to a timing plan that applies to each network independently. Therefore, the signals can interfere with one another either partially or fully. In general, both of these types of collisions can increase the BER, depending on the SNRs involved.

For asynchronous systems, $P_h(l_f, l_p)$ is given by [2]

$$P_h(l_f, l_p) = \binom{L-1}{l_f}\binom{L-1-l_f}{l_p} P_f^{l_f} P_p^{l_p} (1 - P_f - P_p)^{L-1-l_f-l_p} \tag{14.40}$$

where

$$P_f = \left(1 - \frac{1}{N_b}\right)\frac{1}{N_c} \tag{14.41}$$

$$P_p = \frac{2}{N_b N_c} \tag{14.42}$$

where $N_b = T_h/T_b$ = number of data bits per dwell.

14.4.2 Noncoherent Synchronous DSSS/SFHSS Systems

In this case, the term synchronous means that all the radios hop from one frequency to the next at the same time. When the DSSS/FHSS hybrid system is synchronous, there can be only full hits. In that case, l_p makes no sense and (14.39) using (14.40) reduces to

$$P_s = \sum_{l_f=0}^{L-1} \binom{L-1}{l_f} P_f^{l_f} (1 - P_f)^{L-1-l_f} P_e(l_f) \tag{14.43}$$

where

$$P_{\rm f} = \frac{1}{N_{\rm c}} \tag{14.44}$$

14.4.3 BBN Jamming of Noncoherent DSSS/SFHSS Systems

Just as for DSSS and FHSS systems, BBN jamming raises the background noise level as if the noise were that much higher. The jammer signal bandwidth is assumed to be at least as wide as the total bandwidth of the DSSS/FHSS system denoted by $W_{\rm ss}$.

14.4.3.1 Noncoherent Asynchronous DSSS/SFHSS

In this section the BBN jamming performance against BFSK, MFSK, and DPSK modulated noncoherent asynchronous signals will be presented.

BFSK

The conditional probability of error given that $l_{\rm f}$ full hits and $l_{\rm p}$ partial hits have occurred for DSSS/FH hybrid SS system employing BFSK as the DSSS modulation and BBN as the jamming waveform is given by [2]

$$P_{\rm e}(l_{\rm f}, l_{\rm p}) = \frac{1}{2}\exp\left\{-\frac{1}{4}\left[\frac{1}{\dfrac{2E_{\rm b}}{N_0 + J_0}} + \left(l_{\rm f} + \frac{1}{2}l_{\rm p}\right)\frac{C}{N_{\rm D}}\right]^{-1}\right\} \tag{14.45}$$

Considering the BBN jamming density J_0, the power ratios can be determined to be

$$\frac{E_{\rm b}}{N_0 + J_0} = \frac{1}{\dfrac{1}{E_{\rm b}/N_0} + \dfrac{J_0}{E_{\rm b}}} \tag{14.46}$$

but

$$\frac{E_{\rm b}}{N_0} = \frac{R/W_{\rm F}}{P_{\rm N}/W_{\rm ss}} = \frac{R}{P_{\rm N}}\frac{W_{\rm ss}}{W_{\rm F}} = \upsilon N_{\rm c} \tag{14.47}$$

since $W_{ss}/W_F = N_c$, the number of channels when W_F is the instantaneous bandwidth of the DSSS signal. Likewise

$$\frac{J_0}{E_b} = \frac{J/W_{ss}}{R/W_F} = \frac{J}{R}\frac{W_F}{W_{ss}} = \xi\frac{1}{N_c} \tag{14.48}$$

Therefore, (14.45) becomes

$$P_e(l_f, l_p) = \frac{1}{2}\exp\left\{-\frac{1}{4}\left[\frac{\frac{1}{\upsilon}+\xi}{2N_c} + \left(l_f + \frac{1}{2}l_p\right)\frac{C}{2N}\right]^{-1}\right\} \tag{14.49}$$

MFSK

Following similar arguments as used for BFSK, when the DS modulation is MFSK, the probability of error is given by [2]

$$P_e(l_f, l_p) = \sum_{m=1}^{M-1}\binom{M-1}{m}\frac{(-1)^{m+1}}{m+1} \times\exp\left\{-\frac{m}{2(m+1)}\left[\frac{\frac{1}{\upsilon}+\xi}{2N_c\log_2 M} + \left(l_f + \frac{1}{2}l_p\right)\frac{C}{MN\log_2 M}\right]^{-1}\right\} \tag{14.50}$$

DPSK

Likewise, when the DSSS modulation is DPSK, the probability of error is given by

$$P_e(l_f, l_p) = \frac{1}{2}\exp\left\{-\frac{1}{2}\left[\frac{\frac{1}{\upsilon}+\xi}{2N_c} + \left(l_f + \frac{1}{2}l_p\right)\frac{C}{N}\right]^{-1}\right\} \tag{14.51}$$

Figure 14.10 illustrates the BBN jamming performance when an asynchronous DSSS/SFHSS hybrid system uses these modulations for a square chip waveform.

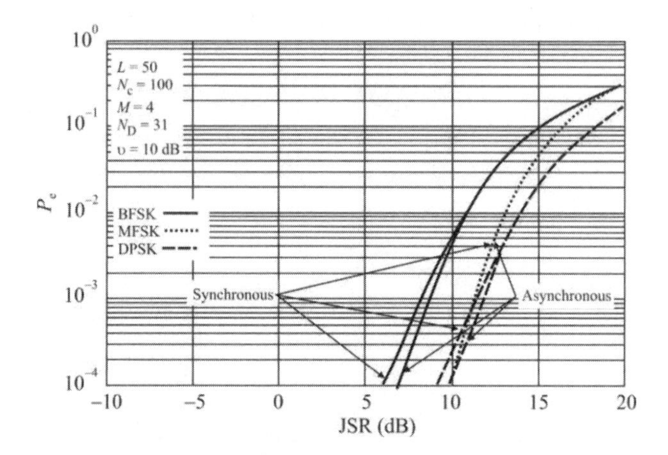

Figure 14.10 BBN jamming performance against noncoherent asynchronous and synchronous DSSS/SFHSS hybrid systems when the various modulations are employed. In this example, $L = 50$, $N_c = 100$, $M = 4$, $N = 31$, $N_b = 10$, and $\upsilon = 10$ dB.

For all of the modulation types it takes a JSR of about 10 dB or more to achieve a BER of 10^{-1}. BBN is most effective against BFSK and least effective against DPSK.

14.4.3.2 Noncoherent Synchronous DSSS/SFHSS

BBN jamming performance against synchronous DSSS/SFHSS systems is presented here. It is assumed that $W_{ss} = N_c W_F$, where the width of each communication channel is W_F, so that the entire spectrum is covered by the jammer.

BFSK

When the DS modulation is BFSK, the probability of error is given by [2]

$$P_e(l_f) = \frac{1}{2}\exp\left\{-\frac{1}{4}\left[\frac{\frac{1}{\upsilon}+\xi}{2N_c} + \frac{l_f}{4N}\right]^{-1}\right\} \tag{14.52}$$

which is the same as (14.45) with $l_p = 0$ and $C = ½$.

MFSK

$$P_\text{e}(l_\text{f}) = \sum_{m=1}^{M-1} \binom{M-1}{m} \frac{(-1)^{m+1}}{m+1} \times \exp\left\{ -\frac{m}{2(m+1)} \left[\frac{\frac{1}{\upsilon}+\xi}{2N_\text{c}\log_2 M} + \frac{l_\text{f}}{2MN\log_2 M} \right]^{-1} \right\} \tag{14.53}$$

which, again, is the same as (14.50) with $l_\text{p} = 0$ and $C = ½$.

DPSK

When the DS modulation is DPSK, the probability of error is given by

$$P_\text{e}(l_\text{f}) = \frac{1}{2}\exp\left\{ -N_\text{D} \left[\frac{\frac{1}{\upsilon}+\xi}{2N_\text{c}} + \frac{l_\text{f}}{2N_\text{D}} \right]^{-1} \right\} \tag{14.54}$$

which, likewise, is the same as (14.51) with $l_\text{p} = 0$ and $C = ½$.

BBN jamming performance against synchronous systems is also shown in Figure 14.10 for these modulation types. For the range of BER of importance here, the jamming performance is virtually the same. What differences there are, occur at low values of JSR and therefore low BER. At these levels, BBN jamming is more effective against synchronous targets. For DPSK, jamming effectiveness is the same for both synchronous and asynchronous targets. Also we can note that, just as for nonhopping DSSS systems discussed in Chapter 10, the DSSS processing gain must be overcome before the jamming is effective.

14.4.4 PBN Jamming of Noncoherent DSSS/SFHSS Systems

As previously, let γ denote the fraction of the frequency spectrum covered by the jammer, this need not be contiguous. Thus,

$$\gamma = \frac{W_\text{J}}{W_\text{ss}} \tag{14.55}$$

Table 14.1 BER Expressions for PBN Jamming

Modulation	**Expression for P_{e_2} (l_f, l_p)**
Asynchronous	
BFSK	(14.49)
MFSK	(14.50)
DBPSK	(14.51)
Synchronous	
BFSK	(14.52)
MFSK	(14.53)
DBPSK	(14.54)

The conditional BER given that l_p and l_f partial and full hits have occurred, respectively, for PBN jamming is given by

$$P_e(l_f, l_p) = (1-\gamma)\frac{1}{2}\exp\left(-\frac{1}{2}\upsilon\right) + \gamma P_{e_2}(l_f, l_p) \tag{14.56}$$

The first term in (14.56) is due to the hop not landing on a jammed channel and no overlap with another emitter so any error is due to noise only, while the second accounts for the case when the communication system hops into a jammed channel. P_{e_2} depends on the type of DS modulation employed and is computed as indicated in this section.

14.4.4.1 Noncoherent DSSS/SFHSS Systems

PBN jamming allows for concentrating a greater amount of power over a smaller portion of the spectrum than BBN and demonstrates the tradeoff between higher power in a smaller portion of the spectrum and broader spectrum coverage, albeit with lower power everywhere.

Table 14.1 gives the expressions for P_{e_2} (k_f, k_p) for asynchronous and synchronous PBN jamming of hybrid systems. As above, in synchronous applications, the epoch information is known among all the communication nodes within range of each other. Therefore, there are no partial hits. PBN jamming performance against asynchronous DSSS/SFHSS hybrid systems is illustrated in Figure 14.11 for typical values of the parameters. The 10^{-1} BER for the FSK modulations is achieved at $\xi \approx 20$ dB, the value of the DSSS processing gain, and is not achieved for DPSK.

Just as for BBN jamming, the differences between asynchronous and synchronous occur at low JSR values. For practical values of JSR where $P_e > 10^{-1}$,

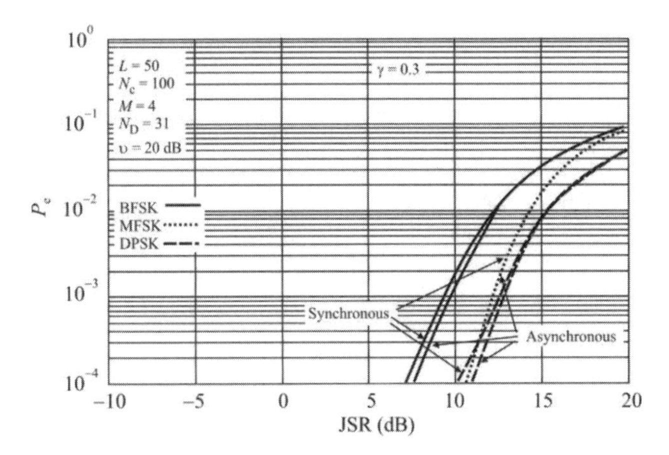

Figure 14.11 PBN jamming performance against noncoherent asynchronous and synchronous DSSS/SFHSS when $\gamma = 0.3$, $\upsilon = 20$ dB, $N_b = 10$, $L = 50$, $N_c = 100$, and $N = 31$.

there is no difference between the two. The performance results are therefore very similar.

14.4.5 Multitone Jamming of Noncoherent DSSS/SFHSS Systems

14.4.5.1 Nonoverlapping Channels

An analysis of MT jamming effects on noncoherent DSSS/SFHSS systems when the DSSS channels were not allowed to overlap, as previously assumed, was conducted by Laxpati [9]. See Figure 14.12(a). Only synchronous systems were considered in that analysis as well. Jamming tone frequency offsets were included, however. With a posteriori detection, the probability of detecting the correct MFSK symbol is given by the probability of not selecting any of the other $M - 1$ incorrect symbols; thus

$$P_c = \int_0^\infty \prod_{\substack{m=1 \\ m \neq i}}^{M} \left\{ 1 - \exp\left[-\frac{u_i^2}{2\sigma_{J_m}^2} \right] \right\} p(u_i) du_i \tag{14.57}$$

assuming that u_i was the sent symbol and where

$$\sigma_{J_m}^2 = \frac{J}{\gamma W_{ss}} E_s \left[1 + \frac{M-1}{N_D} \right] \text{sinc}^2(R_m) \tag{14.58}$$

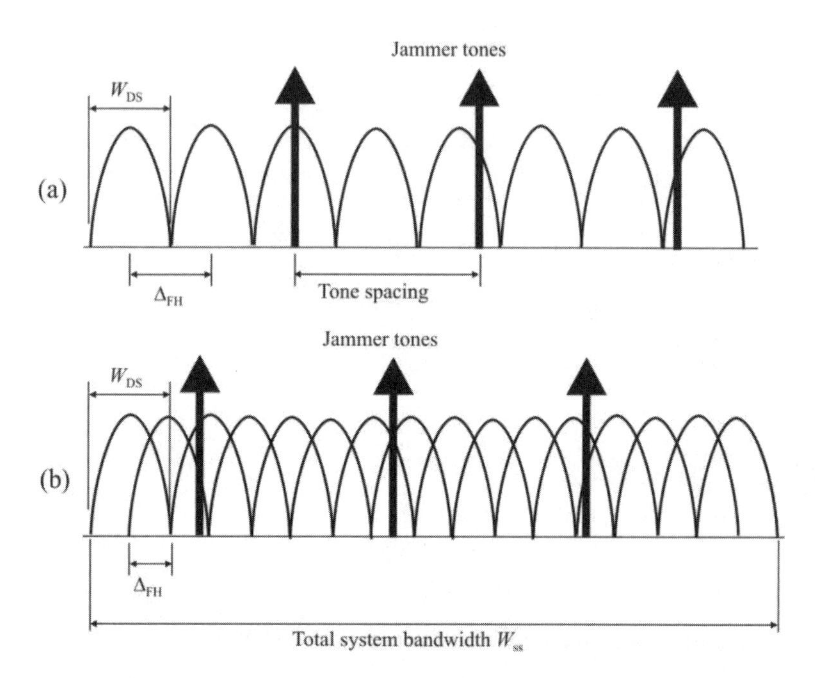

Figure 14.12 Hybrid DSSS/SFHSS spectra with three jammer tones: (a) nonoverlapping DS channels results in three jammed channels, and (b) overlapping channels results in seven jammed channels. (After: [10]. © IEEE 1995. Reprinted with permission.)

is the variance of the jamming waveform with

$$W_{ss} = N_c R_b \tag{14.59}$$

when it is assumed that $W_F = R_b$. Expression (14.57) assumes that for all the incorrect symbols the response of the envelope detector at the receiver follows the Rayleigh distribution. On the other hand, $p(u_i)$ is the probability density function of the Ricean random variable output of the envelope detector at the receiver, u_i, in the correct channel, given by

$$p(u_i) = \frac{u_i}{\sigma^2} \exp\left\{-\frac{u_i^2 + s^2}{2\sigma^2}\right\} I_0\left(\frac{u_i s}{\sigma^2}\right) \tag{14.60}$$

where s is the noncentrality parameter, in this case given by

$$s = E_s \tag{14.61}$$

R_m is a measure of the frequency offset given by

$$R_m = \Delta f\, T_c + \frac{2m - M - 1}{N_D} \tag{14.62}$$

Combining the above and averaging over the M possible symbols yields the probability of correctly detecting the symbol as

$$P_c = \frac{1}{M}\sum_{l=1}^{M}\int_0^\infty \frac{u_l}{\sigma_{J_l}^2}\exp\left(-\frac{u_l^2 + E_s^2}{2\sigma_{J_l}^2}\right) I_0\left(\frac{u_l E_s}{\sigma_{J_l}^2}\right)\prod_{\substack{m=1\\ m\neq l}}^{M}\left[1-\exp\left(-\frac{u_l^2}{2\sigma_{J_m}^2}\right)\right] du_l \tag{14.63}$$

From this expression, the BER can be computed. Let X be a function of the JSR power ratio as

$$X = \frac{1}{\xi}\frac{W_{ss}}{R_b} \tag{14.64}$$

Then by making the change of variables in (14.63) as $y = u_i/\sigma_{Ji}$,

$$P_e = (1-\gamma)\frac{M}{2(M-1)}\exp\left(-\frac{1}{2}\upsilon\right)$$

$$+\gamma\frac{M}{2(M-1)}\left\{1-\frac{1}{M}\sum_{k=1}^{M}\left[\begin{array}{l}\displaystyle\int_0^\infty y\exp\left(-\frac{1}{2}y^2 - \gamma\zeta X G_k\right)\\ \times I_0\left(y\sqrt{2\gamma\zeta X G_k}\right)\\ \displaystyle\times\prod_{\substack{m=1\\ m\neq k}}^{M}\left[1-\exp\left(\frac{1}{2}y^2\frac{G_m}{G_k}\right)\right]dy\end{array}\right]\right\} \tag{14.65}$$

where

$$G_k = \frac{N}{2(N+M-1)\text{sinc}^2(R_k)} \tag{14.66}$$

and

$$\zeta = \log_2(M) \tag{14.67}$$

In this derivation, the fact that

$$E_\text{b} = \frac{E_\text{s}}{\log_2(M)} \tag{14.68}$$

was used.

Expression (14.65) is illustrated in Figure 14.13 for typical parameter values when $\Delta f \approx 0$. There is a 5 dB or so dependence on M, with the lower values being more susceptible to jamming than the larger values. The goal of $P_\text{e} > 10^{-1}$ is never achieved for these values and SNRs. Furthermore, ξ > 10 dB before any significant jamming effects are noted at all.

Figure 14.14 illustrates some of the effects of offsets of the jamming tones. Jammer tones are placed at offsets of $0.5/T_\text{c}$ and $1/T_\text{c}$. The jamming performance is not significantly dependent on the tone offset placement as seen in this example.

14.4.5.2 Overlapping DSSS Channels

Laxpati and Gluck [10] examined the hybrid case with MT jamming when the DSSS channels are allowed to overlap as illustrated in Figure 14.12(b). Again, noncoherent detection of the MFSK signal was assumed. They also considered long codes, irrespective of the DSSS synchronization issues. Their results indicate that the probability of a correct symbol decision in the jth slot is given by

$$P_{\text{c}j} = \frac{1}{M}\sum_{i=1}^{M}\left\{\int_0^\infty \begin{array}{l} y\exp\left(-\dfrac{1}{2}y^2\dfrac{H_{ji}}{\zeta}\right)I_0\left(y\sqrt{2\dfrac{H_{ji}}{\zeta}}\right) \\ \times\displaystyle\prod_{\substack{m=1\\m\neq i}}^{M}\left[1-\exp\left(-\dfrac{y^2H_{jm}}{2H_{ji}}\right)\right]dy \end{array}\right\} \tag{14.69}$$

where

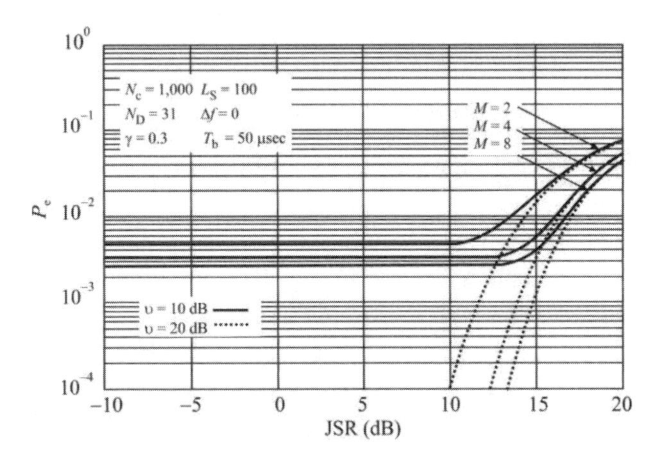

Figure 14.13 MT jamming performance illustration for noncoherent, synchronous BFSK, QFSK, and 8FSK when $\Delta f \approx 0$.

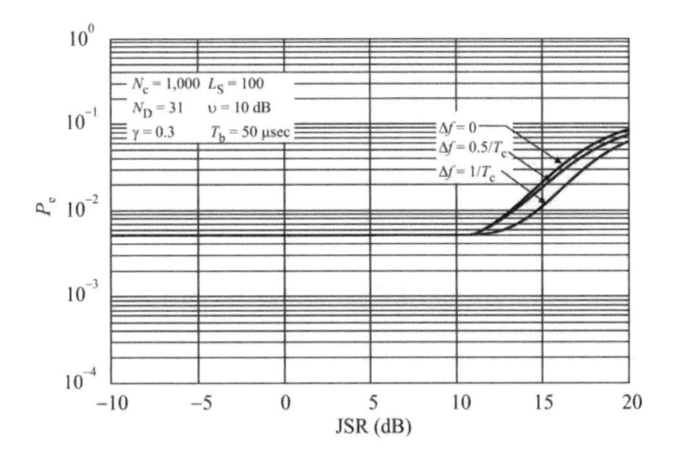

Figure 14.14 Effects of offset jamming tones against noncoherent, synchronous DSSS/SFHSS systems. In this case, the offset is away from $1/T_c$.

$$\zeta = \frac{J / W_{\text{ss}}}{E_{\text{b}}} = \xi \frac{1}{W_{\text{ss}} T_{\text{b}}} = \xi \left(\frac{R_{\text{b}}}{W_{\text{ss}}} \right)$$

is the channel jammer average power density to bit energy ratio, which is the JSR multiplied by the ratio of the data rate to the total system bandwidth. Values of ξ around –10 to 0 dB and $R_{\text{b}}/W_{\text{ss}}$ around –20 dB are typical. Also, using their notation

$$H_{ji} = \frac{\psi(L-1)L_{\text{T}}}{2L_{\text{s}} \sum_{l=1}^{L} \operatorname{sinc}^2(R_{il}) S_{\text{T}}(l+j-1)} \tag{14.70}$$

ψ is the number of information bits per symbol. R_{il} is the offset of the jamming tone in the lth subslot of the jth slot from the ith signal center frequencies and is given by

$$R_{il} = \left(\frac{2i - M - 1}{2M} - \frac{2l - L - 1}{2} - \delta \right) \left(\frac{M}{M(L-1)+1} \right)$$

δ is the factor representing the distance the tones are offset from the channel center frequency, $-1/2 \le \delta \le 1/2$. M is the number of tones or symbols each of which represents ψ information bits. The slot bandwidth after despreading, denoted as W_{DS}, is an integer multiple of the MFSK channel bandwidth and with the slot spacing, denoted as Δ_{FS},

$$W_{\text{DS}} = L_{\text{s}} \Delta_{\text{FH}}$$

N_{J} is the number of jammer tones so that $J_{\text{T}} = J/N_{\text{J}}$ where J is the total jamming power and J_{T} is the jamming power per channel. L_{s} is the number of subslots the communication system has available [8 in Figure 14.12(a)].

$S_{\text{T}}(i)$ is a variable that describes whether a subslot contains a jammer tone and is defined as

$$S_{\text{T}}(i) = \begin{cases} 1, & \text{tone present in subslot } i \\ 0, & \text{no tone present in subslot } i \end{cases}$$

with $i = l + j - 1$ in (14.70).

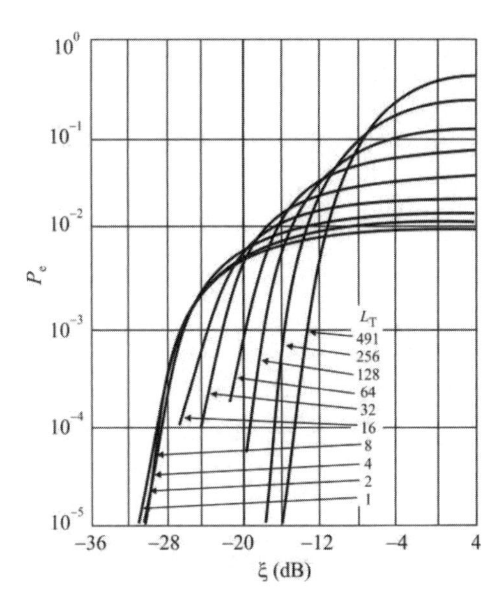

Figure 14.15 Example of results when considering overlapping DS channels in hybrid DSSS/SFHSS systems with noncoherent BFSK. In this case there are 491 channels. $K = 10$, $\delta = 0$. The jamming tones have fixed, minimal spacing. (After: [10]. © IEEE 1995. Reprinted with permission.)

An example of the results of this analysis is illustrated in Figure 14.15. For this example there are 491 channels available to the communication system. As for the nonhopping DSSS systems discussed in Chapter 10, the jamming is relatively ineffective until the DSSS processing gain is overcome.

14.4.6 Jamming DSSS/SFHSS Performance Comparisons

In this section the three types of jamming waveforms are compared against one another for a given modulation type and jamming performance against synchronized timing is compared with that for asynchronous timing.

Figure 14.16 illustrates BBN and PBN jamming performance for asynchronous and synchronous epochs and BFSK modulation. Several observations can be concluded from Figure 14.16:

- BBN and PBN jamming of asynchronous targets produces the same results.
- $P_e > 10^{-1}$ for $\xi > 15$ dB for both values of N.

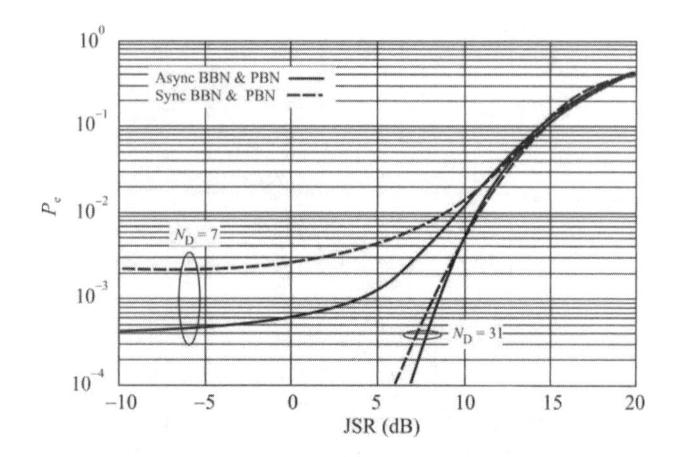

Figure 14.16 BBN and PBN jamming performance against noncoherent BFSK. $L = 50$, $\upsilon = 10$ dB, $\gamma = 10$, $N_c = 100$, $T_b = 50$ μs.

- As expected, DSSS signals with shorter spreading sequence length are more susceptible to jamming.
- For JSR values of interest, synchronous and asynchronous timing react to the same type of jamming in the same way.

If the goal is to produce a BER greater than 10^{-1}, then the selection of PBN or BBN is arbitrary and can be made for other reasons, such as fratricide avoidance.

Figure 14.17 shows a comparison of the three jamming techniques against asynchronous BFSK and QFSK. From this example we can see that BBN jamming at $\xi > 5$ dB or so outperforms the other two techniques. At $\xi < 5$ dB, PBN jamming produces the most errors. MT jamming is the worst performer for all conditions and produces the same results against both BFSK and QFSK with these parameters.

14.5 Concluding Remarks

Jamming performance using standard jamming waveforms against hybrid SS systems was explored in this chapter. The three jamming waveforms considered were BBN, PBN, and multitone. Targets employing BFSK, MFSK, and DPSK modulations were evaluated as well for SFHSS formats.

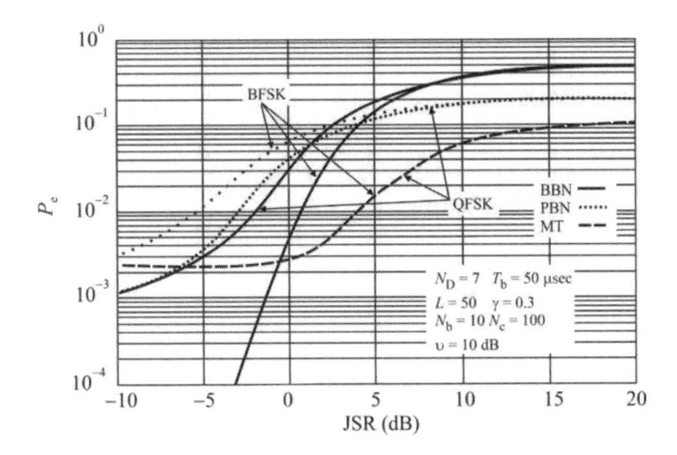

Figure 14.17 Comparison of jamming performance against noncoherent, asynchronous SFHSS BFSK and QFSK when $\upsilon = 10$ dB, $T_b = 50$ μsec, $\Delta f = 25$ kHz, $\gamma = 0.3$, $N = 7$, $N_b = 10$, $N_c = 100$, $L = 50$.

For the examples considered, with parameters nominal for the targets expected, it can be concluded that PBN jamming performs best at low JSR values while BBN jamming is best at higher values.

References

[1] Geraniotis, E. A., "Coherent Hybrid DS-SFHSS Spread-Spectrum Multiple-Access Communications," *IEEE Journal on Selected Areas in Communications,* Vol. SAC-3, No. 5, September 1985, pp. 695–705.

[2] Geraniotis, E. A., "Noncoherent Hybrid DS-SFHSS Spread-Spectrum Multiple Access Communications," *IEEE Transactions on Communications,* Vol. Com-34, No. 9, September 1986, pp. 862–872.

[3] Pouttu, A., J. Juntti, T. Kumpumals, and M. Raustia, "Bit Pattern Matched Phase Interference Suppression in a Hybrid DSSS/FH-System," *Proceedings IEEE MILCOM,* 1999. http://www.greenhouse.com/society/TacCom/milcom_99_papers.html.

[4] Henttu, P., H. Saarnisaari, and S. Aromaa, "Interference Suppression in DSSS/FH System Using Modified Two Sided Adaptive Filter," *Proceedings IEEE MILCOM,* 2001.

[5] Schilling, D. L., L. B. Milstein, R. L. Pickoltz, and R. W. Brown, "Optimization of the Processing Gain of an M-ary Direct Sequence Spread Spectrum Communication System," *IEEE Transactions on Communications,* Vol. Com-28, No. 8, August 1980, pp. 1389–1398.

[6] Haiou, Z., and Z. Naitong, "Performance Analysis of Hybrid DS-SFH/MSK Spread-Spectrum System under Multitone Jamming," *Proceedings IEEE MILCOM,* 1999.

[7] Riddle, L. P., "Performance of a Hybrid Spread Spectrum System Against Follower Jamming," *Proceedings IEEE MILCOM,* 1990, pp. 16.6.1–16.6-5.

[8] Wu, T.-C., C.-C. Chao, and K.-C, Chen, "Capacity of Synchronous Coded DS SFHSS and FFHSS Spread-Spectrum Multiple-Access for Wireless Local Communications," *IEEE Transactions on Communications*, Vol. 45, No. 2, February 1997.

[9] Laxpati, M. A., "Multitone Jamming of Hybrid SFHSS/DS-MFSK Communication Systems," *Electronic Letters,* Vol. 26, No. 5, March 1, 1990, pp. 323–325.

[10] Laxpati, M. A., and J. Gluck, "Optimization of a Hybrid SHF/DS MFSK Link in the Presence of Worst Case Multitone Jamming," *IEEE Transactions on Communications*, Vol. 43, No. 6, June 1995, pp. 2118–2126.

Chapter 15

Characteristics of Urban Terrain

15.1 Introduction

The military cognoscenti have always viewed cities as centers of gravity—something to be either protected or taken away during war. The population centers, transportation hubs, seats of government, sources of wealth, centers for industry, information networks, and key nodes of communication within a nation are normally located in the cities. Forecasts based on the worldwide migration trend from agrarian to industrialized societies predict that 85 percent of the world's population will reside in urbanized areas by 2025. As the world trend toward urbanization increases, the military significance of cities likewise must go up.

Urban areas are today (in many cases), and will be in the future, significant sources of conflict. Cities are where radical ideas ferment and are freely propagated, dissenters and discontents find allies, mixtures of people cause ethnic friction (frequently based on religious beliefs), and disgruntled groups receive media attention. Adversaries may focus on the capture of radio and television stations in an attempt to influence public opinion and attain their political goals. The United States is likely to be forced to neutralize or stabilize some extremely volatile political situations, or provide assistance to allies in need of support, by deploying forces into urban environments [1–4].

For the foreseeable future, western forces will be facing urban gorillas/terrorists as the primary adversary. One of the fundamental tenents of such forces is to take advantage of asymmetric techniques such as manipulating the media and therefore sentiments back home. This is best done in places where there is a considerable population concentration: urban areas [5].

Conducting EW operations in an urban setting brings with it a myriad of problems that don't exist in an open, rural environment (at least not as severely). In this chapter we introduce some of the characteristics of urban terrain, laying the groundwork for discussions of the EW techniques in cities in Chapters 16 and 17.

Signal propagation in cities, especially in areas where there are high-rise buildings, is significantly different from rural terrain. Few of the signals are received over a direct path. To quantify this, models have been developed to primarily estimate signal strength at a receiver. Such models are discussed in Chapter 16. Likewise the noise sources likely to be encountered in urban areas are fundamentally different from rural settings. Where at a quiet receiving site in rural environments a sensitive EW system must deal with principally thermal noise sources, in an urban environment the predominant noise is likely to emerge from man-made sources, which is decidedly non-Gaussian in nature. We focused on understanding these sources of noise in Chapter 2. Finally, in Chapter 17 we consider EW operations in urban terrain and specifically focus on the particular problems such environments pose.

15.2 Military Operations in Urbanized Terrain

Military operations in urban terrain (MOUT) is defined as all military actions planned and conducted on a topographical complex and its adjacent terrain where man-made construction is the dominant feature. It includes combat in cities, which is that portion of MOUT involving house-to-house and street-by-street fighting in towns and cities [6]. MOUT affects the tactical options available to a commander. A built-up area is a concentration of structures, facilities, and populations, such as villages, cities, and towns, which form the economic and cultural focus for the surrounding area.

15.3 Cities

Cities are centers of populations, finance, politics, transportation, communication, industry, and culture. They generally have large population concentrations ranging from tens of thousands to millions of people. Because of their psychological, political, or logistical value, control of cities have often been the scenes of pitched battles.

Operations in built-up areas are normally conducted to capitalize on the operational or tactical significance of a particular city. The side that controls a major city usually has a psychological advantage, which can be enough to significantly affect the outcome of a countrywide conflict. In developing nations, control of only a few cities is often the key to the control of national resources.

The abundance of guerrilla and terrorist operations conducted in built-up areas (e.g., Santo Domingo, Caracas, Belfast, Managua, Beirut, Bagdad, Sana'a) demonstrates the importance many insurgent groups place on urban warfare.

Many cities have expanded dramatically in the past several years, thereby losing their well-defined boundaries as they extended into the countryside. Transportation systems (highways, canals, and railroads) have been built to connect population centers while industries have grown along those connectors, creating "strip areas." Rural areas are connected to the towns by a network of secondary roads.

15.3.1 Multiple Avenues of Approach

Urban terrain is a unique battlespace that provides both the attacker and defender with numerous and varied avenues of approach and fields of fire. The urban battlespace is divided into four basic levels: *building, street, subterranean,* and *air*. Operations can be conducted from above ground, on ground level, inside buildings, or below the ground, typically with fighting on all levels simultaneously.

15.3.1.1 Building Level

Buildings provide cover and concealment; limit or increase fields of observation and fire; and canalize, restrict, or block movement of forces, especially mechanized forces. They provide optimum perches for snipers and antiair weapons. Buildings also provide antitank weapons optimum positioning to allow engagement from above, exploiting an inherent weakness found in most armored vehicles—their top. As we will see, buildings, especially tall ones, significantly increase the space for war fighters to occupy.

15.3.1.2 Street Level

While streets provide the means for rapid advance or withdrawal, forces moving along streets are often canalized by buildings and have little space for off-road maneuver. Because they are more difficult to bypass, obstacles on streets in urbanized areas are usually more effective than those on roads in open terrain.

15.3.1.3 Subterranean Level

Subterranean systems are easily overlooked but can be important to the outcome of operations. These areas include subways, sewers, cellars, and utility systems (see Figure 15.1). The city of Los Angeles alone has more than 200 miles of storm sewers located under the city streets. Both attacker and defender can use subterranean avenues to maneuver to the rear or the flanks of an enemy. These avenues also facilitate the conduct of ambushes, counterattacks, and infiltrations.

Figure 15.1 Subterranean systems. Urban terrain has several opportunities for subterranean systems such as the subway depicted here. Such subterranean systems extend the three dimensional battlespace to a full sphere, not just the hemisphere above ground.

15.3.1.4 Air Level

The air provides another avenue of approach in urban areas. Aviation assets can be used for high speed insertion or extraction of troops, supplies, and equipment. While aviation assets are not affected by obstacles on the streets, they are affected by tall buildings, light towers, signs, power lines, and other aerial obstructions. They are also vulnerable to the man-portable surface-to-air missile threat, crew served weapons, and small arms fire. Helicopters and UAS provide the necessary maneuverability, while fixed wing aircraft would typically be relegated to transport and broad area activities, such as intelligence collection. Significant portions of a city can be affected by EW operations from fixed wing aircraft.

15.3.2 Categories of Built-Up Areas

Built-up areas are sometimes, and somewhat arbitrarily, classified as:

- Villages (populations of 3,000 or less);
- Strip areas (industrialized zones built along roads connecting towns or cities);
- Towns or small cities (populations from 3,000 to 100,000 and not part of a major urban complex);

- Large cities with associated urban sprawl (populations in the millions, covering hundreds of square kilometers).

Military operations that may work well in one of these settings may not work at all in the others. We have learned that the traditional tactics applied to large forces moving across rural terrain, in many cases do not work in urban settings.

15.3.3 Characteristics of Urban Areas

A typical urban environment consists of combinations of one or more city cores, commercial ribbons, core peripheries, residential sprawl, outlying industrial areas, and outlying high-rise areas. Each of the urban area's regions has distinctive characteristics that may weigh heavily and differently in planning for MOUT.

Many cities consist of a core containing high-rise buildings surrounded by a much larger area having buildings of relatively uniform height spread over regions comprising many square blocks, except for isolated clusters of high-rise buildings. In this surrounding urban area the buildings lining one side of a street are adjacent to each other or have passageways between them that are narrower than the width of the buildings. The street grid organizes the buildings into rows that are nearly parallel.

15.3.3.1 City Core

In many cities, the city core has undergone more recent development than the core periphery. As a result, the two regions are often quite different—both in terms of architecture as well as building materials. Typical city cores are made up of high-rise buildings made of concrete and steel which vary in height. Modern urban planning for built-up areas allows for more open spaces between buildings than in old city cores or in core peripheries (see Figure 15.2).

15.3.3.2 Commercial Ribbons

Commercial ribbons are composed of rows of stores, shops, and restaurants that are built along both sides of major streets through built-up areas. Typically, such streets are 25 meters wide or more. The buildings in the outer areas are uniformly two to three stories tall—about one story taller than the dwellings on the streets behind them (see Figure 15.3).

Figure 15.2 Typical urban core.

Figure 15.3 Commercial ribbons.

Figure 15.4 Core periphery.

15.3.3.3 Core Periphery

The core periphery generally consists of streets 12–20 meters wide with continuous fronts of brick or concrete buildings. The building heights are fairly uniform—2 or 3 stories in small towns, 5 to 10 or more stories in large cities (Figure 15.4).

15.3.3.4 Residential Sprawl

Residential sprawl areas consist mainly of low houses or apartments that are one to three stories tall. The area is primarily composed of detached dwellings that are usually arranged in irregular patterns along streets, with many smaller open areas between structures (Figure 15.5).

15.3.3.5 Outlying Industrial Areas

These areas generally consist of clusters of industrial buildings varying from one to five stories in height. Buildings generally vary dramatically in size and composition to match the needs of the particular businesses they house. Industrial parks found in many larger cities are good examples of this category (Figure 15.6).

Figure 15.5 Residential sprawl.

Figure 15.6 Outlying industrial areas.

Figure 15.7 Outlying high-rise area.

15.3.3.6 Outlying High-Rise Areas

These areas are similar in composition to city core areas, but may be composed of clusters of more modern multistory high-rise buildings in outlying parts of the city. Building height and size may vary dramatically (Figure 15.7). Generally, there is more open space between buildings located in the outlying high-rise areas than is found within the city core area.

15.4 Characteristics of Urban Warfare

We cover some of the relevant characteristics of warfare in an urban setting in this section. We will focus on those most relevant to the EW issues of operation in such terrain.

Urban warfare is warfare of vertical envelopment, rather than horizontal, the way most developed militaries have been trained to fight. It requires tactical thinking in three dimensions. Units must maintain "spherical" security at all times. The three-dimensional nature of the urban battlespace requires all-around security. Subterranean pathways through cities can allow forces to relatively easily move past and/or around friendly units, facilitating attacks from surprise directions.

15.4.1 Asymmetric Tactics

While it can be said that as of this writing there are no significant peer adversaries of the United States, what adversaries there are will likely choose asymmetric tactics, attacking the places where the United States is weakest (homeland sympathies, for example). Another example is propaganda via the Internet.

15.4.2 Difficult Terrain

An adversary can hide in buildings and fight through windows. Movement from one building to the next can frequently be accomplished without detection and sometimes such movement can be accomplished without going outside.

Signals typically don't take direct paths in a city. (We will talk much more extensively about signal propagation in the next chapter.) At any given receive site, whether it's the targets' receiver or the receiver of an EW system, the signals intercepted are comprised of several replicas of the target signal, each delayed by a different amount due to the different paths taken from the transmitter to the receiver. Signals are reflected off buildings and diffracted around edges of buildings. To complicate the matter further, moving objects cause the reflections on these paths to be time varying and significant fading in and out of the received signal occurs. This fading can be on the order of 30 dB or more within the distance of a wavelength (tens of centimeters).

15.4.3 Identification of Friend from Foe

When the adversary acts, dresses, looks, and talks like the indigenous population, it can be difficult identifying one from the other. The warfighters can readily blend into the crowd with no one the wiser. With widespread proliferation of PCS devices, including cellular phones and personal digital assistants, an adversarial war fighter can be executing C2 while walking down the street, mixed in with non-combatants.

15.4.4 Underground Enemy

As mentioned, the basements of buildings as well as other subterranean features, such as subways, causes the battlespace environment to occupy more than just the hemisphere defined as above the ground. Significantly sized forces can move undetected through subway tunnels to cross from one side of a city to the other.

Communications is particularly difficult in subterranean settings, however. While signals can propagate down subway tunnels, they do not propagate through the ground at all well. This gets progressively worse as the frequency increases.

This poses issues with C2 of adversarial (and friendly) forces, but also creates intercept and jamming issues by friendly forces.

Urban terrain is a complex and challenging environment. It possesses all of the characteristics of the natural landscape, coupled with manmade construction, resulting in an incredibly complicated and fluid environment that influences the conduct of military operations in unique ways.

At street level the signals emanating from an elevated fixed transmitting antenna are shadowed by the buildings, which gets more severe as the frequency increases. Except along occasional streets aligned with the transmitter, or at very close ranges, the transmitting antenna is not visible from street level. Thus propagation must take place through or between buildings, reflected off one or more, or over the rooftops with signals diffracted at the roofs down to street level.

Signal propagation through buildings is accompanied by loss due to reflection, attenuation, and scattering by exterior and interior walls. While the energy penetrating the row of buildings immediately in front of the MT may be significant, the majority of the propagation path cannot include passing through buildings. When passageways do exist between buildings, they are seldom aligned from row to row and aligned with the transmitting source. As a result, the majority of the paths are either over the rooftops or consist of one or more reflections off the sides of buildings. The field reaching street level results from diffraction of the fields incident on the rooftops in the vicinity of the mobile antenna. While the process by which the energy at the rooftops or reflected from a wall reach the mobile antenna may be quite complicated, the process is still expected to take place in the immediate vicinity of the MT.

15.4.5 Armor

Tanks and armored personnel carriers cannot conveniently operate in cities without extensive dismounted infantry support. Rubble from fallen buildings can get in the way of effective movement. Recovering damaged armored vehicles is more difficult in urban areas for the same reason. Tracked vehicles are preferable to wheeled vehicles in situations where there is likely to be large amounts of rubble in the streets. Otherwise wheeled armored vehicles are faster and preferable.

Tanks and infantry working as an integrated team may prove especially effective. The tank improves success in meeting engagements, countering ambushes, and in maintaining momentum during fire and maneuver.

15.4.6 Fratricide

Fratricide in all forms is a serious problem when fighting in cities because it is harder to identify friend from foe. By definition, the noncombatant civilians make up a substantial portion of the populace, probably the majority of it.

15.4.7 Civilian Structures

Major civilian structures in cities (e.g., hospitals, churches, banks, embassies) are located where they can be tactically useful, they can command key intersections, and/or are built of especially solid construction. Therefore, such facilities are particularly useful to urban defenders.

15.4.8 Artillery

Indirect fire artillery and antiaircraft artillery, as well as direct fire weapons, such as the tubes on tanks, can be useful in urban combat, provided there is little concern about collateral damage. Unfortunately, "collateral damage" is a popular influential topic in PSYOPS, especially if noncombatants, and in particular, children, are involved. Newer GPS-guided artillery fire has proven to be a very effective capability in urban situations; the downside is the expense to acquire these shells.

15.4.9 Electronic Warfare

EW, on the other hand, while technically being an indirect fire weapon, does not directly cause collateral damage—permanent collateral damage anyway. Its effects are only felt directly while the jamming is taking place. The jamming, however, is usually intended to cause a target or targets to act differently in some way than they would if the jamming were not applied; so there is (hopefully) a lasting effect. It is not likely to impact on nontargeted entities, however.

15.4.10 HUMINT

Intelligence, especially from human sources, is absolutely critical to successful urban operations. This form of intelligence is referred to as HUMINT, and goes beyond simply talking to the indigenous population to gain knowledge. Everybody in the battlespace is a potential source of useful HUMINT, including soldiers on patrol and local children.

15.4.11 Ambushes

Hit-and-run ambushes by small groups are a favorite tactic of urban paramilitary forces. The local insurgents are likely to know the terrain better than coalition forces and can take advantage of that.

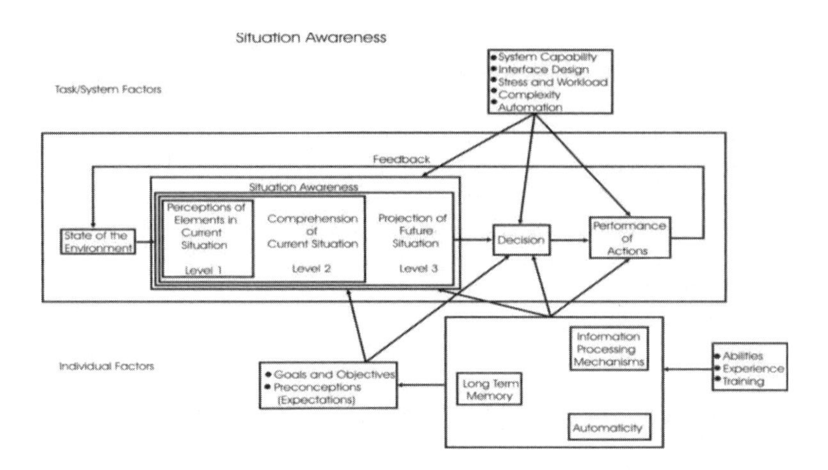

Figure 15.8 Situation awareness model. (Source: [7]. © 1995.)

15.4.12 Situational Awareness

The restricted and compartmented nature of the urban battlespace makes it difficult to gain and maintain situational awareness. Even small unit leaders must learn to quickly recognize and utilize key terrain in the urban environment. They must learn how to determine which buildings or facilities will give them a tactical advantage. This includes such things as tall buildings for observation, command and control, and fields of fire.

Situation assessment can be defined as the process of evaluating the current state of affairs within the decision maker's *area of interest* (AOI). It includes the three levels or stages as depicted in Figure 15.8, which is a model of the situation assessment process as applied to dynamic decision making proposed by Endsley in 1995 [7]. The first step to being aware in a situation is to perceive the environment—to make observations. The second step is to understand what is being observed (level 2). The last step is to project a short time into the future what is likely to happen. Based on this projection, a decision is made as to what action is appropriate and then that action is taken. The action likely changes the environment in some way, providing feedback to what is being observed. Then the process is repeated.

Factors affecting this SA process are comprised of task or system factors that are outside of the person's control, and individual factors. The elements of the former are the system capacity, the interface design through which the person interacts with the system, the stress and workload of the person making the assessment, the complexity of the system or task, and the amount of inherent automation. The elements of the latter include the goals and objectives of the human in the situation and that person's preconceptions or expectations. Contributing to the human factor implications are the long-term memory of the person, that person's information processing ability, and the degree of automatic response that the person possesses (automaticity). These, in turn, are influenced by the person's abilities, experience, and training.

Although not specifically included in Figure 15.8, significant communication is implied in order to develop awareness of situations. Only in the simplest of cases is the flow shown in Figure 15.8 performed by a single person at a single location. With the complexity of combat, many people are involved in the process and they must communicate with one another.

15.5 Typical Tactics of Urban Guerillas

The urban insurgent works alone or in small cells, trying to avoid detection or identification. Being, for the most part, indistinguishable from the local populace, gathering in large groups would be an indication of something unusual happening. Somewhat to the contrary, urban insurgents also create incidents or amass crowds to lure the opposition force into a trap.

Local industry and public services are disturbed by strikes and sabotage. Suicide bombers are frequently a cultural norm that attracts otherwise normal citizens, especially in Muslim-dominated countries. The bombers are convinced that to kill in such a way is a glorious and heavenly thing to do. Widespread disturbances that are designed to stress the resources of the opposing force are organized and executed.

Opportunities for hostile propaganda arise by provoking the opposing force into overreacting. Such reactions can be viewed as overkill if not outright barbarous, especially when the civilian populace is involved in some way. It is perhaps worst, at least to western minds, when civilian children are included. In those cases where there are various factions in a country, especially when they are based on religion, such factions can be provoked into interfactional strife.

Sniping has long been a mainstay tactic of military forces in a conflict. Urban terrain provides ample opportunities for applying these tactics. Witness the slaying of President Kennedy in Dallas in 1963 by Lee Harvey Oswald. There wasn't even a conflict occurring in the United States at the time. Such sniping opportunities occur at roadblocks and outposts, as well as many other places (stop lights).

Lowering public morale by planting explosive devices, either against specific targets or at random, is an effective technique. It causes confusion and destruction, resulting in the sense of no safe place to be.

15.6 Psychological Implications and Operations in Asymmetric Warfare

In the post-Cold-War era, leaders of potential enemies of the United States and other western cultures have discovered that they can win the "hearts and minds" of the world's people through the selected use of real information, disinformation, manipulation of the press, propaganda, and other psychological warfare operations (PsyOps) methods.

It is believed by some analysts that such PsyOps and propaganda programs will continue to have an increasingly more influential impact on future conflicts, and that our military and political leadership should seriously consider expanding efforts by U.S. psychological warfare operations and units to counter these developments.

15.7 Concluding Remarks

We have provided an overview of several of the characteristics of EW in urban terrain. The terrain itself was first discussed, followed by some of the peculiarities such terrain imposes on conducting EW operations.

Modern warfare is being and will be conducted largely in urban settings. This is the nature of the state of affairs in the early twenty-first century. Largely because by far the most likely enemy western, developed militaries will encounter are guerillas aimed at asymmetric warfare and the best terrain for those kinds of endeavors is the urban setting. This is where their actions can receive the most attentions and therefore have the most asymmetric effect.

Some types of modern war fighting targeted at urban settings need not even occur in that setting. The Internet is an example of this; so is broadcast and satellite TV. Significant military actions can be initiated well away from where the effects are intended.

References

[1] Scales, R. H., *Certain Victory: The US Army in the Gulf War*, U.S. Army Command and General Staff College Press, 1994.

[2] Toffler, A., and H. Toffler, *War and Anti-War: Survival at the Dawn of the 21st Century*, New York: Little, Brown, and Company, 1993.

[3] Weinberger, C., and P. Schweizer, *The Next War*, Washington, D.C.: Regnery Publishing, 1996.

[4] Dunnigan, J. F., *Digital Soldiers*, New York: St. Martin's Press, 1996.

[5] Potts, D., (ed.), *The Big Issue: Command and Combat in the Information Age*, Washington, D.C.: CCRP, 2005, Chapters 16 and 17.

[6] MCWP 3-315.3 *Military Operations in Urbanized Terrain*, U.S. Marine Corps, 26 April 1998.

[7] Endsley, M. R., "Toward a Theory of Situation Awareness in Dynamic Systems," *Human Factors: the Journal of the Human Factors and Ergonomics Society*, Vol. 37, No. 1, 1995, pp. 32–64.

Chapter 16

Signal Propagation in Urban Settings

16.1 Introduction

We covered signal propagation characteristics in general in Chapter 2. This chapter is devoted to analyzing signal propagation in urban settings. While wars have been fought in city environments for as long as wars have been fought, the urban setting has become significantly more important in recent years. Such settings have some peculiar characteristics for EW operations and we will discuss some of them here.

This chapter is structured as follows. First we review the properties of urban signal propagation, particularly with many buildings around. We then cover propagation characteristics for microcellular regions followed by an examination of the effects of vegetation. We then have a dialog about the prevailing propagation models specifically designed for urban environments. The chapter concludes with an exploration of the characteristics of indoor signal propagation.

16.2 General Characteristics of Urban Signal Propagation

The vast majority of urban propagation prediction research has been for the purpose of installing cellular and PCS wireless telephone systems. Therefore the characteristics of these systems drive the parameters chosen (such as frequency range considered). The modern cellular and PCS telephone systems worldwide have primarily two frequency bands: one in the 800–900 MHz range and one in the 1,800–1,900 MHz band. Some of the newer network systems (Wi-Fi and WiMAX) include the ISM frequency bands around 2.4 GHz and 5.8 GHz as well. Therefore the frequencies considered in the urban propagating analyses predominantly include these bands.

The most prevalent analytical model for analysis of signal propagation in the VHF and higher ranges in suburban and rural environments is the two ray model. Egli also developed a model for propagation above 40 MHz [1]. That model is based on empirical data collected [1]. Longley and Rice also developed a VHF+ propagation model for use above 20 MHz based on empirical data [2].

The wavelengths at frequencies below about 100 MHz are large compared with the typical structure size in urban environments. Therefore neither the Egli nor the Longley-Rice model would apply. The two ray model may be applied to frequencies down to 20 MHz or so, but the validity of that model to urban environments has not been proven. Furthermore, there is little motivation for developing such a model since the predominant wireless communications in urban environments, as mentioned, are in the UHF and above range.

Signals in the microwave band are common in urban areas, but those links are always point-to-point line of sight. Microwave signals do not propagate through buildings very well. Furthermore, such signals are normally used to exchange PSTN or similar signals, so the links can be carefully engineered and planned; they are not mobile. Two-ray modeling of the propagation paths can be applied in these cases.

16.3 Urban Signal Propagation

16.3.1 Introduction

The understanding of mobile radio channel characteristics is essential for the analysis and design of EW systems that include third, fourth, and beyond generation wireless communication systems. Accurately modeling the radio channel, especially in urban environments, is very difficult to accomplish with any degree of fidelity. In the urban setting, the locations of the BS towers is likely to be well known, and EW planning to deal with these fixed sites is eminently possible. The MSs, on the other hand, are a different story. MSs can be anywhere within the communication system coverage area, which would likely contain many cells, especially in medium to large urban areas. Furthermore, it is likely that the MSs are the more interesting target, especially for geolocation and EA.

Historically, by far the most used techniques for understanding propagation in urban terrain are the empirical models developed by Okumura [3], Hata [4], and COST-231[5] propagation models for elevated antennas. In the 1990s some analytical approaches to analysis of urban signal propagation emerged and are used along with the empirical models. The lower limit on the frequencies of these approaches are generally around 100 MHz or so while the upper limit is about 3 GHz. They are therefore useful for analysis of other types of communication systems as well.

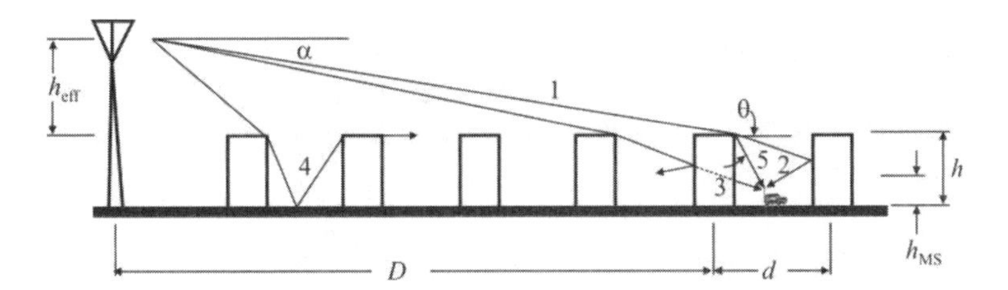

Figure 16.1 Various ray paths for UHF propagation in the presence of buildings. This architecture, where the base antenna is above the roofs of the building does not apply to the urban core. In the urban core, the base antenna is typically lower than the rooftops. (Source: [6]. © IEEE 1988. Reprinted with permission.)

Because the rows of buildings have the form of rectangular obstacles lying on the ground, as seen in the cross section in Figure 16.1, propagation over the rooftops involves diffraction past a series of parallel blocks with dimensions that are large compared to wavelength. At each block a portion of the field will diffract toward the ground. These fields can rejoin those above the buildings only after a series of multiple reflections and diffractions as suggested by the rays labeled 4 in Figure 16.1.

Most of the diffractions are through large angles and/or the fields must be reflected two or more times between the buildings. Each diffraction and/or reflection induces losses in the energy of these waves; therefore, these fields will have small amplitude and can be neglected. Note that while fields reflected between two rows of buildings contribute to the multipath interference between those two rows, they do not contribute to multipath interference between any other two rows.

16.3.2 Properties of Urban Signal Propagation

16.3.2.1 Transmissions

In urban networks, reflections off of objects such as buildings and the ground and transmission into buildings through doors or windows as well as walls play a primary role in the received signal strength. Diffraction and scattering are also important, but from experiments it has been shown that they play a secondary role in many cases. When the object that the signal is hitting is much larger than the wavelength, then the behavior of the signal can be modeled as a ray that is partially reflected off of the object and partially transmitted through the object, the magnitude of both depending on the relative dielectric constant of the material given by

$$\varepsilon_r = \varepsilon_r' + j\varepsilon_r'' \tag{16.1}$$

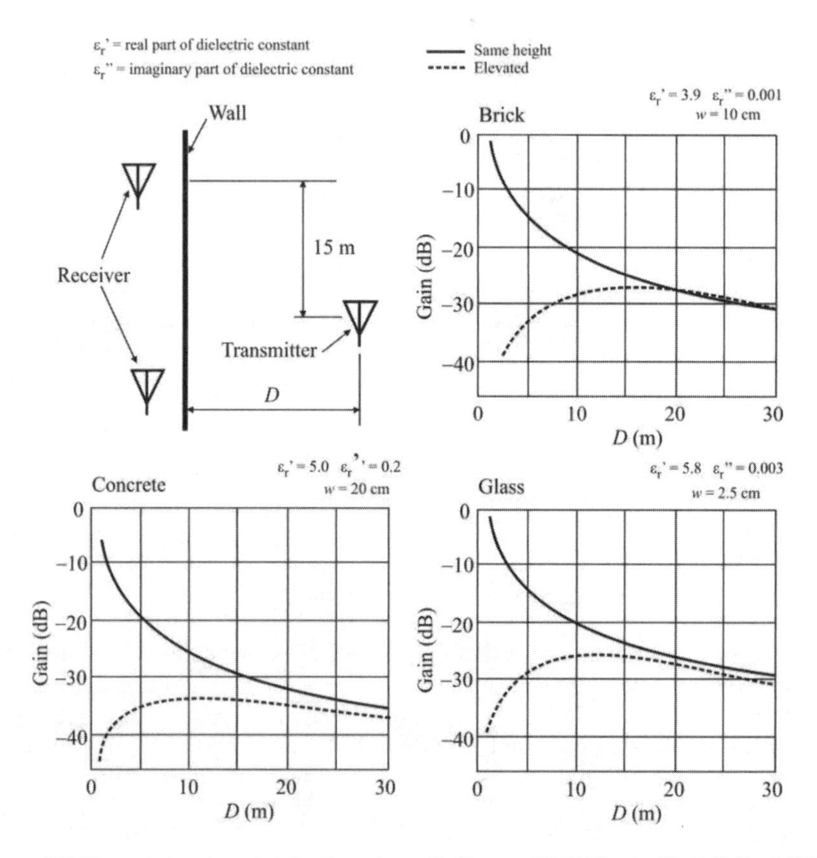

Figure 16.2 Transmission characteristics through a wall. (Source: [7]. © Elsevier North Holland 2007. Reprinted with permission.)

as well as the angle of incidence hitting the wall. In addition, some of the signal energy is partially absorbed by the object.

Consider the received signal strength on one side of a wall with a transmitter on the other side as shown in Figure 16.2. We consider the received signal strength in two locations, specifically, one at the same height as the transmitter and one elevated from the transmitter by 15 m. The elevated position causes the signal to approach the wall from a nonperpendicular angle. At these distances, the signal strength decays according to free-space propagation (r^{-2}) and then intersects the wall and suffers insertion loss. The frequency considered is 2.4 GHz, as in 802.11b/g. Three types of wall materials are examined as shown in Figure 16.2.

First, as expected, the channel gain increases as the distance between the transmitter and receiver decreases when the receiver and transmitter are at the same height. This agrees with our intuition from free-space propagation. We can see the impact of the material by noting the variation in the received signal strength for the different materials. For reference purposes, consider that without the wall, when the transmitter is about 5 m from the wall and a receiver that is at the same height as the transmitter, the channel gain is –14 dB. Thus, when the angle of incidence is 0° (directly horizontal in this case), the wall reduces the channel gain by 0 dB (for glass) to 6 dB (for concrete). The impact of the angle of incidence can be seen when the transmitter and receiver are not at the same height. As we can see, this results in the signal strength first increasing and then decreasing as the distance between the transmitter and receiver increases; this is in opposition to intuition and is caused by the angle of incidence increasing (as measured from the horizontal). The characteristic causing this counterintuitive behavior is that when the ray hits the wall at a grazing angle, only a small amount of signal power is transmitted through the wall. As we see, the point where the signal strength changes from increasing to decreasing depends on the material, and the thickness of that material. It also depends on the height of the receiver.

Figure 16.2 shows variation of the channel gain due to the different building materials. While this simple experiment proves the point that different materials in walls have different effects on signals traversing through them, it is difficult to exactly analytically quantify this behavior since walls, especially exterior walls, are typically constructed with nonhomogeneous material. A second problem is that it is not realistic to know the construction material for each building being considered.

16.3.2.2 Reflections off a Wall

Reflections off a wall are the complementary effect to transmission through it. However, a significant amount of energy can be absorbed by the wall depending on the materials from which it is made and the angle of incidence as we observed in Figure 16.2. Therefore the power transmitted through a wall and the power reflected off of it do not necessarily sum to the power of the incident signal.

Consider the urban canyon illustrated in Figure 16.3. In addition to the direct path between the transmitter and receiver, the signal is repeatedly reflected off of walls and, in a sense, focused down the street. A similar effect can also arise in hallways and tunnels. The result is that the received signal strength may be stronger when propagating down a street than it would be with the same transmitter-receiver distance in free-space. The phase of the signal is affected by the length of the path taken, and in some cases, the signal at the receiver can be completely cancelled when the reflected path length is such that the reflected signal arrives 180° out of phase with the direct path.

Figure 16.3 shows the channel gain down an urban canyon for the same set of materials considered above as well as for the free-space approximation. Note that free-space analysis (d^2) predicts substantially smaller received signal strength than the simulations indicates. For example, when the receiver-transmitter distance is

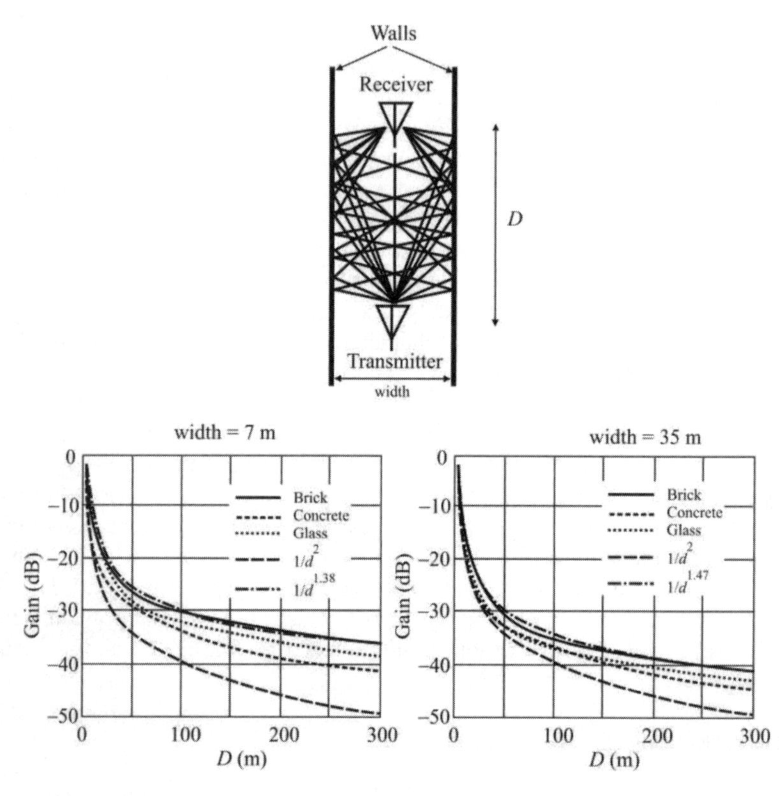

Figure 16.3 Propagation characteristics down an urban canyon. (Source: [7]. © Elsevier North Holland 2007. Reprinted with permission.)

300 m, the difference between the free-space and the model that accounts for reflection ranges from 13 dB to 5 dB, depending on the material and the distance between the walls.

We have also included approximations of the channel gain given by $d^{-\alpha}$, where, for a canyon width of 7 m, $\alpha = 1.38$ and a canyon width of 35 m, $\alpha = 1.47$. These approximations are very close to the characteristics for the brick wall case. In suburban and rural settings it is often assumed that $\alpha \in \{2, 4\}$. However, as this example demonstrates, it is possible that for some paths we have $\alpha < 2$. That is, while buildings may block communication, they may also enhance communication. It is also possible to have $\alpha > 4$ in some settings, for example inside building propagation.

The smoothness of the walls also affects the reflection characteristics. This is due to the wide variation in the reflection coefficient as the angle of incidence varies. Depending on the material and the width of the material, there may be some angles of incidence where no signal is reflected at all. Such angles are called

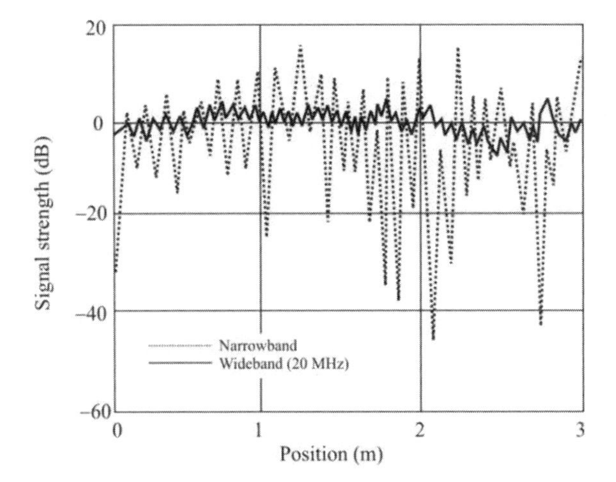

Figure 16.4 Characteristics of rapid fading for narrowband and wideband sources. (Source: [7]. © Elsevier North Holland 2007. Reprinted with permission.)

Brewster angles. In some settings, the wide variation of the reflected signals strength as a function of incident angle results in large fluctuations in the total received signal strength for small movements of the receiver or transmitter antenna.

16.3.2.3 Fading

Wireless signals can experience small-scale fading or multipath fading, which are especially pronounced in urban settings. Such fading results from the constructive and destructive interference of signals at the receiver that follow different paths from the transmitter to the receiver as mentioned above. As a result, because the wavelength is small, a small displacement of the transmitter or the receiver, or for that matter, any mobile object in the vicinity, will cause a change in the interference and hence a change in the received signal strength.

Fast-fading is typically not relevant in wide bandwidth communications such as those often used for data communications (e.g., 802.11, 802.16), or the 3G+ PCS standards, however. This is because the received signal strength is essentially averaged over the bandwidth. Figure 16.4 illustrates a comparison of the fluctuations caused by fading in narrowband versus wideband systems. At some locations, the narrowband signal strength is quite low, even 40 dB less than the mean signal strength of 0 dB. A narrow bandwidth communication will experience such degradations in signal strength. However, when averaged over a sufficiently wide bandwidth, the average signal does not experience such severe degradation. The narrow bandwidth signal experiences rapid and large fluctuations, while the wide bandwidth case experiences much smaller variations.

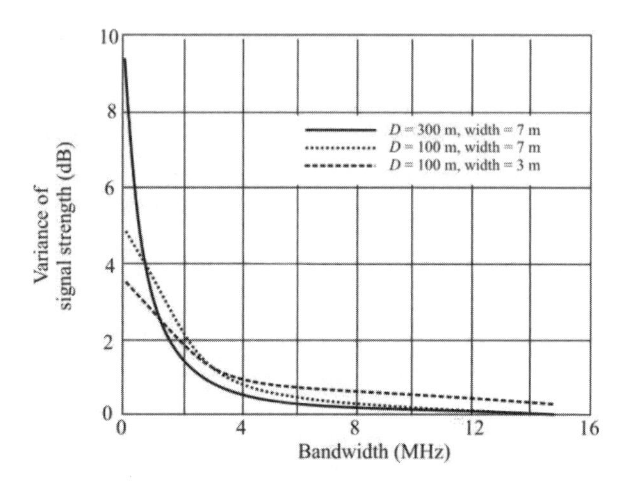

Figure 16.5 Variance of the signal strength for rapid fading. (Source: [7]. © Elsevier North Holland 2007. Reprinted with permission.)

In general, the wider the bandwidth, the less susceptible to fast fading. However, variation in signal strength also depends on the environment. Figure 16.5 shows the variance of the received signal strength over a circle with a one meter radius as a function of the bandwidth. Each curve is for a slightly different environment, but in all cases it is for propagating down the urban canyon as shown in Figure 16.3. In general, the variance decreases as the bandwidth increases; however, the variance of the narrow bandwidth case and the rate that the variation decreases with the bandwidth is environment dependent. In all cases, the variance is quite small when the bandwidth exceeds 10 MHz, while 802.11b has a bandwidth of 22 MHz. Keep in mind, however, that there are other factors besides multipath reflections that could cause large changes in signal strength over small changes in position, that is changes in antenna orientation and passing vehicles.

16.3.2.4 Diffraction

As mentioned, reflections play the most significant role in signal propagation in urban environments. Transmission through walls, windows, and so forth also play a major role. Even though diffraction plays a less important role in many cases, in other cases it can be a major contributor. When a signal is diffracted it "bends" around a corner due to geometrical optical phenomenology to include the edges of buildings as well as the roof lines. The diffracted signal, however, is normally significantly weaker than the original signal; the sharper the bend, the weaker the signal.

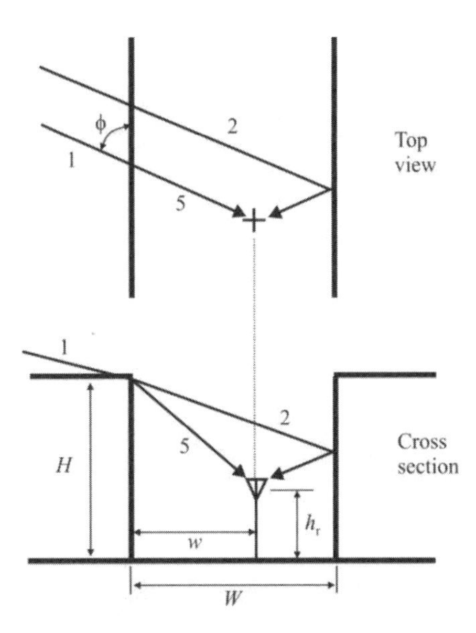

Figure 16.6 RF signal paths close to the MS. (Source: [8]. © IEEE 1984. Reprinted with permission.)

Ikegami et al. documented an analysis coupled with experimental verification of the effects of propagation in urban environments [8]. They posited that there are two dominant rays impinging on an MS on the streets in an urban setting. These two rays consist of one that is diffracted by the last rooftop on the path between the BS and the MS and another that is reflected off the building behind the MS (see Figure 16.6). There could be other paths, but they assumed that all other paths could be ignored compared with the two dominant ones shown.

The expression for the mean field strength they analytically determined based on the two ray assumption is

$$\begin{aligned}\bar{E} = E_0 + 5.8 + 10\log_{10}\left(1+\frac{3}{L_r^2}\right) + 10\log_{10} W \\ -20\log_{10}(H-h_r) - 10\log_{10}(\sin\phi) - 10\log_{10} f \text{ dB}\end{aligned} \tag{16.2}$$

where E_0 is the free space field strength, W, H, ϕ, and h_r are defined in Figure 16.6 expressed in m, degrees, and f is in MHz. L_r is the reflection loss given by the ratio of the amplitudes of the direct and reflected waves.

It is important to note that in the experimental verification of the analytical results, both of these waves were diffracted as shown in Figure 16.6. Therefore the reflected wave would experience more loss than the direct wave. However, the

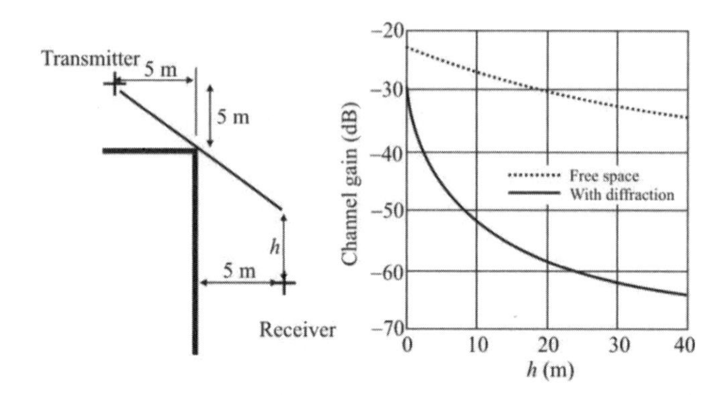

Figure 16.7 Diffraction at a corner of a building, top view. As h increases there are two effects: (1) h is increasing thereby increasing the link distance and (2) the diffraction angle is increasing. (Source: [7]. © Elsevier North Holland 2007. Reprinted with permission.)

angle for path 1 is larger than for path 2 so the strength of the diffracted signal is lower. In addition, the frequencies examined were 209.75, 395.425, and 589.75 MHz as opposed to the 802.11 frequency considered previously of 2.4 GHz.

Consider the example in Figure 16.7, which shows the diffraction around a corner. The diffraction losses are shown in the graph compared with free space losses. As h increases, the receiver antenna moves further away from the corner and the losses increase. Figure 16.8 shows diffraction over a building where two diffractions are required. On both ends, as h increases, the signal must make a sharper bend (i.e., the diffraction angle increases) and, hence, more loss is incurred. It is important to notice how quickly the signal strength decreases as h increases. From Figure 16.8, we can see that diffracting over a building that is only 5 m higher than the transmitter and receiver results in a channel gain that is too small for typical 802.11 communications.

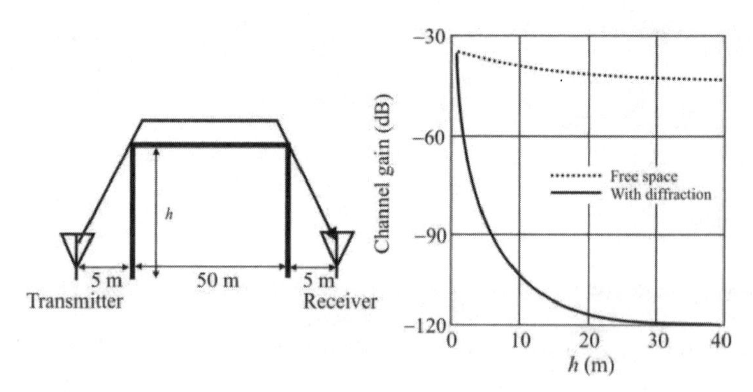

Figure 16.8 Diffraction at the top of buildings, side view (Source: [7]. © Elsevier North Holland 2007. Reprinted with permission.)

While diffracting around two or more corners leads to losses too large for 802.11, it might be able to communicate effectively around a single corner. Consequently, in an urban area, especially in the urban core, a significant portion of the coverage area of 802.11 is due to diffraction.

16.3.2.5 Scattering

Scattering refers to the impact of smaller objects on the propagation (e.g., lampposts, trees, vehicles, people, and office furniture). Scattering also accounts for the unevenness of building walls (e.g., windows, doors, or façades). Scattering is particularly prevalent when the dimensions of the object off of which the signal reflects are on the order of a wavelength (12.5 cm at 2.4 GHz). Without detailed knowledge of the location and dimension of small objects and without details of building walls, it is difficult to accurately analyze the effects of these types of scatterers. The effects of scattering are normally handled by considering the contribution of scatterers stochastically.

Scattering frequently occurs when a wireless signal propagates through vegetation (we discuss vegetation effects later). Indeed, if the vegetation is large and dense, the scattered signal dominates over the direct, nonscattered signal. Even when the vegetation is thin, as it usually is in urban areas, the vegetation can cause loss. At 2.4 GHz, vegetation causes a loss of approximately 0.2 dB per meter of vegetation. Diffraction over vegetation is also common.

16.3.2.6 Time-Varying Channel Gain

While the variability of channel gain is greatly affected by the movement of the transmitter or receiver, the channel gain can also change when the transmitter and receiver are fixed, but objects in the environment move such as passers-by, vehicles, and wind-blown vegetation and street signs. Figure 16.9 shows a typical gain variation as a function of time for a receiver and transmitter as observed along a sidewalk in a large city.

16.3.2.7 Delay Spread

As mentioned earlier, the wireless signal may follow several different paths from the transmitter to the receiver. While one impact of these multiple paths is that the signal may experience multipath fading, another result is that the multiple copies of the signal will be received at different points of time. Essentially, these multiple

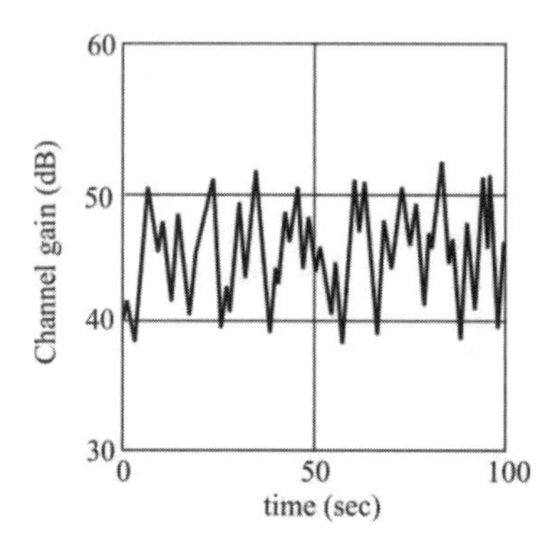

Figure 16.9 Time-varying channel gain. This illustrates that that the gain of a realistic channel is statistically nonstationary.

copies interfere with each other, causing destructive and constructive interference, as mentioned. When this self-interference is considered as noise, it is clear that the effective SNR is decreased by the presence of these delayed copies. When the receiver either at the BS or the MS employs rake receivers, these reflected signals actually increase the signal level so the traditional calculation of SNR considers delayed copies of the transmission as useful signal power. If the rake time delays are not properly estimated or if the rake fingers are not there at all, the traditional SNR is not a good indicator of BER performance.

A channel where multiple copies of the signal arrive at different times is said to have *delay spread*. Two common measures of delay spread are the mean delay spread and the RMS delay spread (see [9] for definitions). With an RMS delay spread of above 60 ns, which is quite common in outdoor urban settings (signals travel at about the speed of light, which is approximately 1 ns per foot, so 60 ns corresponds to a propagation distance of approximately 60 ft), the impact of the delay can be quite severe. It has been found by measurements that indoors the RMS delay spread is typically less than 50 ns due predominantly to the short distances signals travel when reflected off a wall, but outdoors, it can exceed 500 ns. Furthermore, the delay spread typically increases with the distance between the transmitter and receiver. Therefore, rooftop configurations, which are able to propagate considerable distances due to the lack of obstructions at higher heights, are likely to experience large delay spread values.

16.3.2.8 Propagation into and out of Buildings

Measurements at frequencies of 35 MHz to 1,500 MHz have shown that building exterior walls have an attenuation of between 15 and 25 dB. These levels have a log normal distribution with a standard deviation of 8–14 dB. Measured loss through the sides of buildings yield a loss of 3–7 dB through a double pane window and 20–30 dB through a concrete wall.

16.4 Path Loss Predictions for Large Systems

Macrocellular networks make use of high BS antennas to achieve coverage out to distances from the BS that are in the range from 1 to 20 km. For most cities, such distances lie outside the high-rise core, in regions where the buildings are of a more uniform height, and the propagation takes place past many rows of buildings (see Chapter 15). If we restrict the prediction to that of range dependence, we can use average descriptors of the buildings to get reasonable results.

16.4.1 Path Loss

Because the BS antenna is located well above the average building height in such a case, propagation must take place through the buildings, between them, or over the rooftops with the fields diffracted at the roof edges down to the street level. Propagation must primarily take place over the rooftops because signal propagating through the buildings are highly attenuated by losses due to reflection, diffraction, and scattering by exterior and interior walls and the signals cannot readily propagate through the gaps (see Figure 16.1 for notional propagation paths).

To estimate the mean path loss in a macrocellular environment it is assumed that the rows of buildings are replaced by rectangular cylinders lying on the ground. All rows are assumed to have the same height, and each row of buildings is separated by the same distance d as shown in Figure 16.1. The distance d is the average of the separation from back-to-back across the yards and front-to-front across the streets. The mean sector average path loss L in decibels is given by the difference of the sector average received power to the radiated power for isotropic antennas both in decibels, and is given by [10]

$$L_{\mathrm{dB}} = L_{\mathrm{fs}} + L_{\mathrm{rts}} + L_{\mathrm{msd}} \tag{16.3}$$

Here, L_{fs} is the free space path loss given by the Friis equation

$$L_{\mathrm{fs}} = \left(\frac{\lambda}{4\pi D} \right)^2 \tag{16.4}$$

where λ is the wavelength and D is the range, both in comparable units. Expressed in dB this is

$$L_{\text{fs,dB}} = 32.4 + 20\log_{10} f_{\text{MHz}} + 20\log_{10} D_{\text{km}} \tag{16.5}$$

The second term L_{rts} of (16.3) is referred to as the *rooftop-to-street diffraction and scatter loss* component. The third term L_{msd} of (16.3) is the *multiple-screen-diffraction loss* from the BS transmitter antenna to the rooftop of the last building before the mobile receiver.

The rooftop-to-street diffraction and scatter loss L_{rts} was first postulated by Parson and then by Ikegami et al. [8]. Ikegami posited that many rays reach the MS from many different directions after encountering diffractions, reflections, and scattering effects around the receiver. However, close to the MS, propagation may be represented by two main rays. The first of these is diffracted from the rooftop of the building nearest the MS in the direction of the base station, while the second is reflected from the face of the building across the street (paths 1-2 and 1-5 in Figure 16.1). Further, the two main rays are roughly equal. Thus, the rooftop-to-street diffraction and scatter loss, L_{rts}, can be written as

$$L_{\text{rts,dB}} = -10\log_{10}\left\{\frac{2}{\cos\phi}\left[\frac{1}{r_{\text{m}}}\delta^2(\theta)\right]\right\} \tag{16.6}$$

where $\delta(\theta)$ is the diffraction coefficient θ and r_{m} the distance from the edge of the last building down to the mobile are defined in Figure 16.1.

The diffraction coefficient in (16.6) is given by

$$\delta(\theta) = \frac{1}{\sqrt{\pi\beta\cos\phi}}\frac{1}{\theta} \tag{16.7}$$

and is a function on the rooftop edge conditions where the diffraction occurs, which is not typically known. However, for small diffraction angles, θ, the diffraction coefficient is not very sensitive to the boundary conditions. Angle ϕ is defined as in Figure 16.10 and β is the wave number, $\beta = 2\pi/\lambda$. Clearly, when $\phi \sim 90^{\circ}$, propagation is down an urban canyon and the denominator in (16.7) goes to zero. In that case there is no diffraction involved and neither (16.7) nor (16.6) applies.

Signals undergo multiple diffractions as they traverse from the BS to the MS. The field incident on the top of each row of buildings is backward diffracted as well as forward diffracted. Fields that are backward diffracted do not contribute to the signal level at the MS and as a result are neglected. As a result, propagation is one of multiple forward diffraction and is represented by the multiple-screen-diffraction loss

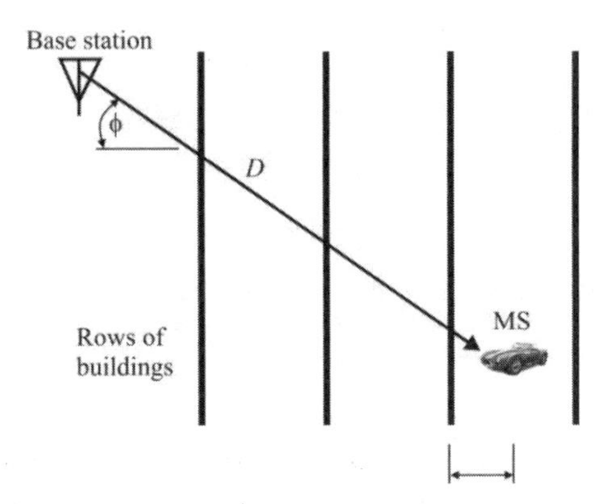

Figure 16.10 Top view of the BS to MS path.

$$L_{\text{msd,dB}} = -10\log_{10}\left[Q_M^2(g_\text{p})\right] \quad (16.8)$$

where $Q_M(g_\text{p})$ is the reduction factor of the incident field from the BS to the last row of buildings just before the MS. The amplitude $Q(g_\text{p})$ of the field at the rooftops due to an incident plane wave of unit amplitude is [11]

$$Q(g_\text{p}) = 3.502g_\text{p} - 3.327g_\text{p}^2 + 0.962g_\text{p}^3 \quad (16.9)$$

that is accurate to within 0.5 dB over the range $0.01 < g_\text{p} < 1$. g_p is a dimensionless parameter given by

$$g_\text{p} = \alpha\sqrt{d\frac{\cos\phi}{\lambda_\text{m}}} \quad (16.10)$$

where α is in radians, d is the distance separation between absorbing screens, and ϕ is the angle that the ray makes with the phase of the last building as illustrated in Figure 16.1.

16.4.2 Vegetation Effects on Path Loss

Torrico, Bertoni, and Lang [12] investigated the effects of tree vegetation on the propagation of UHF signals in urban environments. We briefly discuss their results in this section, indicating that vegetation should not be ignored in the UHF frequency range.

In suburban environments, it is common to see rows of trees planted along the streets, often one in front of each house, and with the trees taller than the houses.

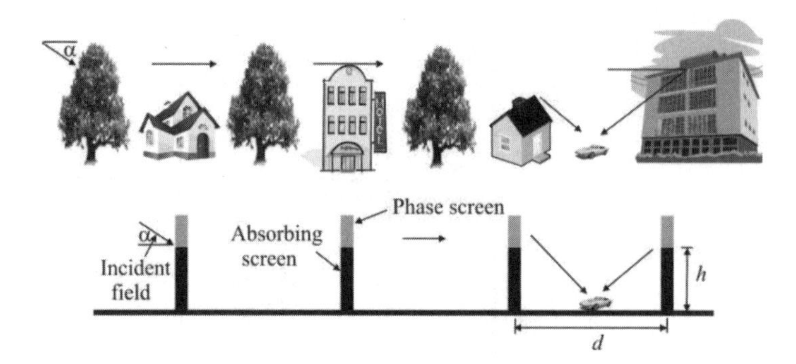

Figure 16.11 Vegetation and how it can be modeled as phase and absorbing screens.

One possible arrangement of rows of trees next to rows of houses is depicted in Figure 16.11. Each row of houses or buildings is represented by an absorbing screen, while the canopy of the adjacent row of trees is represented by a partially absorbing phase screen. With this model, the field passing any row is equal to the field incident on the row multiplied by the transmission coefficient past the tree canopy. The phase and amplitude of the transmission coefficient are found from the width and physical properties of the canopy. Numerical evaluation of the physical optics integral was then used to find the field incident on the next screen.

The attenuation and phase of the partially absorbing phase screen are found by taking the tree canopy to have an elliptical cross-section, with the semi-minor axis equal to a and the semi-major axis equal to b (see Figure 16.12). The center of the canopy is assumed to be at the same height as the buildings, so that the canopy extends a distance b above the buildings. Figure 16.13 shows the attenuation over and above the free space loss. The calculations assume d = 50 m, f = 900 MHz, $\theta = 90^\circ, 89.5^\circ$ (corresponding to $\alpha = 0, 0.5^\circ$), and tree width $2a$ = 4, 8 m for trees that are b = 4 m higher than the buildings. It is seen from the results that for $\alpha = 0.5^\circ$, the field amplitude at the top of the buildings, with and without trees, increases to a settled value for N greater than about 15.

However, for $\alpha = 0$, the field amplitude continues to increase with increasing N. This behavior is similar to that found previously for buildings alone. After ten rows the wider trees are seen to have 4–5 dB more path loss than the buildings by themselves.

16.4.3 Antenna Height Gain

As noted previously, in macrocellular systems the BS antenna is typically well above the average building height. However, the glancing angle α shown in Figure 16.1 is still small—typically less than 1.5° over most of a cell (if h_e = 10 m, then the distance corresponding to 1.5° is 380 m, fairly close to the BS antenna in a macrocellular system). Therefore L_{msd} in (16.3) is not very sensitive to the

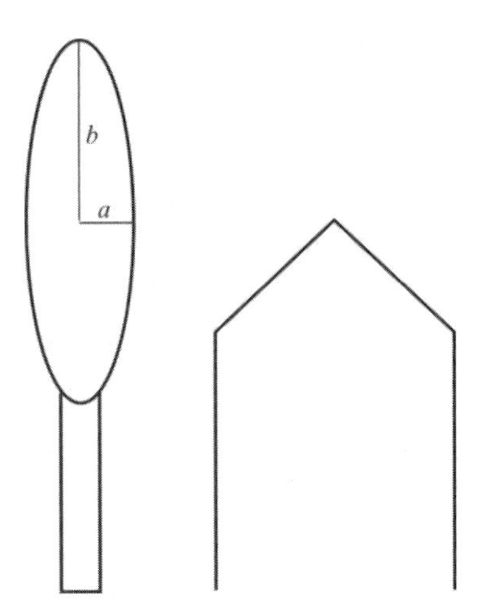

Figure 16.12 Tree cross section.

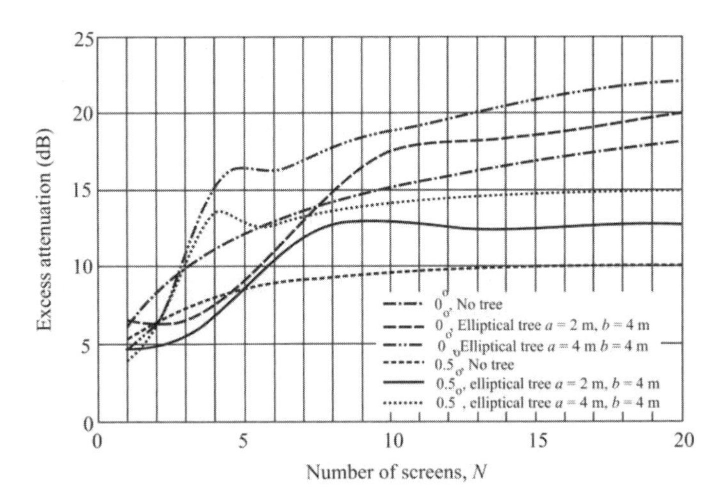

Figure 16.13 Excess attenuation cause by vegetation. (After: [12]. © Torrico. Reprinted with permission.)

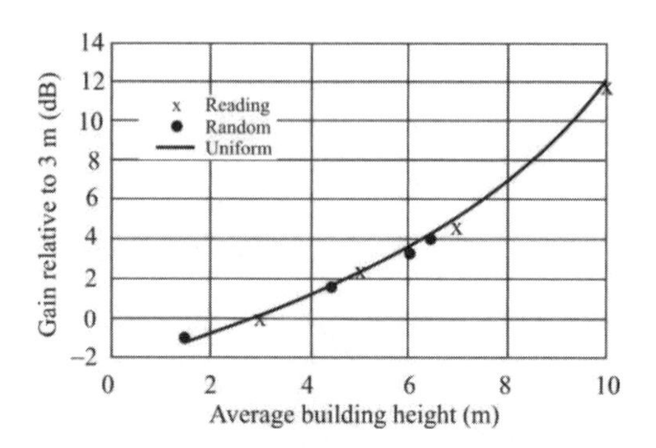

Figure 16.14 Measured and computed height gain. In Reading, United Kingdom, f = 191.25 MHz. (Source: [13]. © IEEE 1992. Reprinted with permission.)

irregularities in the row spacing between buildings/houses, or the lack of parallelism between the rows. Thus, we can use the average distance between buildings/houses, back-to-back across the yard or front-to-front across the street for the rectangular cylinders that are representing the row of houses. Also because of the low glancing angle, α, we can represent the rectangular cylinders as thin absorbing screens.

A typical height gain characteristic is shown in Figure 16.14. In this case increasing the antenna height by a factor of 10 raises the amount of signal received by about 10 dB.

16.4.4 Path Loss Predictions in the High-Rise Urban Core

When one or both ends of the radio link are located in the high rise core, diffraction propagation will take place around the sides of tall buildings, as well as over the tops of lower buildings (see Figure 16.15), along with reflections (the main propagation mechanism) and scattering. If all of the buildings are much taller than the BS antenna, propagation takes place around the buildings through the "urban canyons" formed by the buildings (see Figure 16.16). In large cities, such as parts of Manhattan and Los Angeles, the tall buildings completely fill the blocks, which, for the most part, are arranged in a rectangular grid. For these regions it is possible to characterize the path loss without reference to the shapes of individual buildings because the individual effects tend to average out. However, in most urban cores, the buildings have a more random appearance when viewed over an area of 1 km^2 or greater. Some of the buildings will be low, plazas and other open spaces break up the canyon walls, and the buildings, shape and orientation may not conform to a rectangular grid. In these environments accurate predictions of the path loss over some site require that the actual

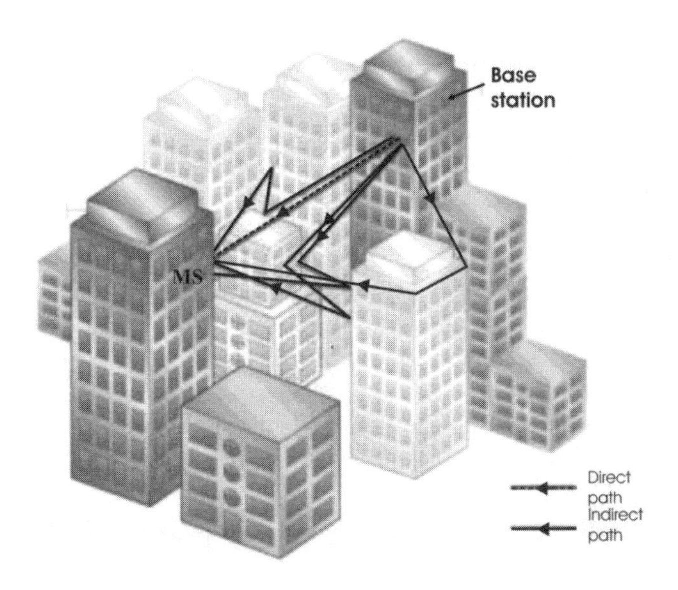

Figure 16.15 High rise urban core signal propagation paths.

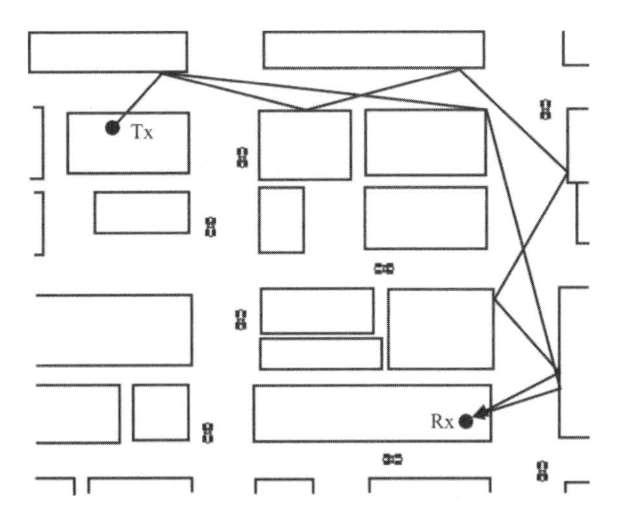

Figure 16.16 Urban canyons viewed from above. These paths are notional. There are too many diffractions and reflections to make the signal at the receiver strong enough with these paths.

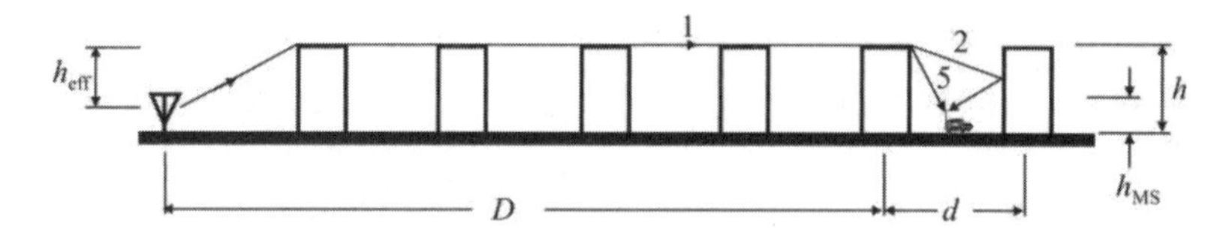

Figuer 16.17 Signal diffraction and scattering when the BS antenna is below the average building height. This is typical of an urban core.

building shapes be accounted for. Such site-specific predictions usually use optical ray tracing methods.

In the urban core, the BS antenna may very well be below the height of the buildings as depicted in Figure 16.17. Diffraction will still occur at the corners of rooftops; however, it will also occur around the vertical edges of buildings causing propagation down the urban canyons (see Figure 16.16). Reflections, however, are still the predominant propagation mechanism in the urban core.

16.5 Path Loss Prediction for Microcellular Systems

Microcellular systems make use of BS antennas at about the height of the buildings or on lampposts to cover cells of radius 1 km or less. For such small cells, the street grid is more likely to be rectangular, and streets are a more significant fraction of the total area in a cell. Averages of the heights and spacing of buildings and vegetation can no longer be applied. In addition, the proximity of buildings to the BS must be taken into account. In low rise areas, the BS antenna may be slightly below the average building heights. In this case the proximity of the buildings may be accounted for through orientation, with respect to the street grid, of the propagation paths over the buildings.

In microcellular systems the fading characteristics close to the BS vary considerably as illustrated in Figure 16.18. This is due primarily to the many reflective paths available when the MS is close to the BS, particularly the ground if nothing else. Moving the MS by a wavelength (cm) can completely cancel the signal at the MS.

16.5.1 Line-of-Sight Propagation Along Streets

On LOS paths in microcellular environments, measurements have shown that a simple two-ray model consisting of the direct and the ground-reflected ray commonly used in suburban and rural propagation studies are sufficient to predict the spatial average propagation loss. Measurements made at 900 MHz with vertically polarized antennas are in close agreement with the two-ray theory [12]. When the signal is plotted versus D on a logarithmic scale it is seen that distinctly different behaviors are obtained before and after a break point. Before the break

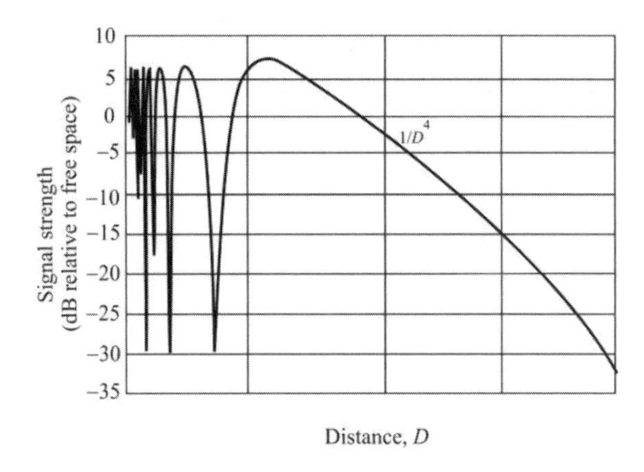

Figure 16.18 Fading characteristics in a microcellular setting.

point, the radio signal oscillates severely due to destructive and constructive combination of the two rays at the receiver, while after the break point, it decreases more rapidly with distance with a typical loss exponent of 4 (i.e., loss $\propto D^{-4}$) as illustrated in Figure 16.18. A rule of thumb for estimating the break point is [14]

$$D_B = \frac{4h_T h_R}{\lambda} \tag{16.11}$$

16.5.2 Propagation over Buildings for Low Antennas

A theoretical approach was developed to predict the path loss for propagation over buildings for low antennas in a microcellular environment where the antennas are above, at, or a little below the average building heights. Evaluation of the two dimensional diffraction process in the plane wave perpendicular to the rows of buildings for low BS antennas required the study of fields radiated by a source that was localized in the vertical plane, rather than an incident plane wave. Such a study was carried out for the special case when the horizontal separation between the BS and first row of buildings is d and is further discussed in [15] for the case when propagation is oblique to the street grid.

The average path loss L in decibels is computed just as it was for the macrocellular case by using (16.3). The only difference lies in how the multiple screen diffraction term L_{msd} of (16.8) is evaluated. The multiple screen diffraction due to propagation past M rows of buildings is computed based on the field reduction factor $Q_M(g_c)$. The dimensionless parameter g_c is given by

$$g_{\mathrm{c}} = h_{\mathrm{e,m}} \sqrt{\frac{\cos\phi}{\lambda_{\mathrm{m}} d_{\mathrm{m}}}} \tag{16.12}$$

where $h_{\mathrm{e,m}}$ is the difference between the BS antenna height $h_{\mathrm{b,m}}$ and the average building height.

16.6 MS to Base Propagation

The astute reader will by this point recognize that we have been focused strictly on the path from the BS antenna to an MS. There are reasons for this. First of all, the propagation paths are bidirectional, at least to first order. We would expect that a path from an MS to the BS would exhibit similar characteristics as the path in the other direction.

The second reason is that cellular and PCS systems utilize a technique of power control in the MSs. A BS controls the power emitted by an MS by adjusting the power dynamically at least several times a second (the exact number depends on the particular system being considered). If there is too much attenuation being experienced by the signal entering the base, the BS simply increases the MSs transmit power. Therefore the propagation loss is not such a major factor on the uplink.

16.7 Propagation Models

16.7.1 Introduction

We should note that monopole antennas rely on an image that theoretically (and practically) extends into the Earth. For an elevated monopole antenna, that image is "elevated" the same amount into the Earth. As such, the image of cellular BS antennas that are elevated on masts or are placed on top of multifloor buildings is a considerable distance into the ground. Therefore, in most cases, the MS is in the near field of the base station's antenna. Therefore EM waves have a reactive component as does the base antenna impedance [16]. Part of the problem is the height of the base antennas. The entire height contributes to the antenna radiation pattern. The models that are generally used for urban propagation analysis do not explicitly take this reactive component into account. They are, however, all based on measurements, so this reactive component is taken into account. Such considerations complicate an accurate mathematical model.

16.7.2 Hata-Okumura Propagation Model

16.7.2.1 Introduction

The Hata-Okumura is the most widely used model in radio frequency propagation for predicting the behavior of cellular transmissions in built up areas. We summarize the Hata-Okumura model in this section.

The Hata propagation model, based on field measurements reported by Okumura [3], is well established and widely used for urban propagation analyses in the UHF range up to 1.5 GHz. An extension of the model toward higher frequencies is found in COST-231 report, which is useable up to 3 GHz.

The Hata-Okumura computation model was first described by Y. Okumura in 1968 [3]. M. Hata simplified the model in 1980 [4], which led to the restriction that only regions with less than a 20 km distance to the transmitter can be predicted.

As only four parameters are required, the computation time is very short. This is an advantage of this model. However, the model neglects the terrain profile between transmitter and receiver (i.e., hills or other obstacles between the transmitter and the receiver are not considered). Also phenomena like reflection and shadowing are not explicitly included.

16.7.2.2 Parameters

The four parameters with their limits are:

- Frequency f (150 ... 1,500 MHz);
- Distance between transmitter and receiver D (1 ... 20 km);
- Antenna height of the transmitter h_T (30 ... 200 m);
- Antenna height of the receiver h_R (1 ... 10 m).

16.7.2.3 Hata Model

In the late 1960s, Yoshihisa Okumura conducted an extensive set of propagation measurements in the Tokyo metropolitan area. From these measurements he developed a set of curves giving the median attenuation relative to free space in an urban area. The computation model was first published by Okumura in 1968 [3]. Some of the Okumura data are shown in Figures 16.19–16.21.

The 1980 model that Hata published as an extension to Okumura's model was suitable for computer applications [4]. The model computes the path loss as a function of transmit and receive antenna heights, path distance, radio frequency, and the type of clutter (urban, suburban, or rural). First, a basic equation is computed that is the loss in urban areas. Correction factors (given next) are applied to this basic equation for suburban and rural areas. Common standard

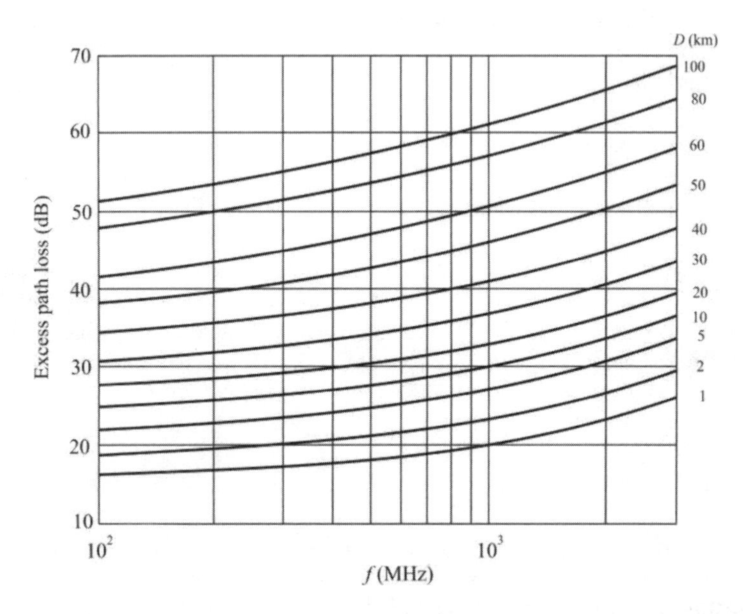

Figure 16.19 Median excess path loss in an urban setting. h_b = 200 m, h_t = 3 m. Okumura data.

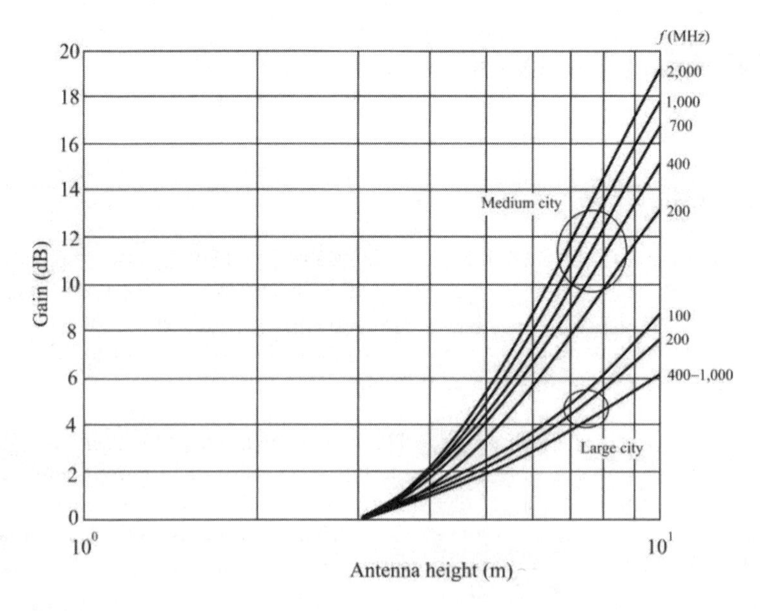

Figure 16.20 MS gain due to height. Decibels relative to 3 m. h_b = 200 m.

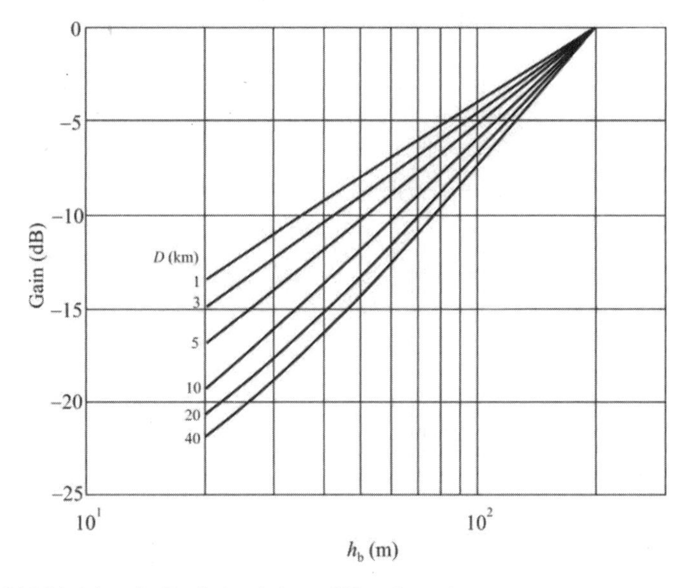

Figure 16.21 BS height gain. Decibels relative to 200 m. h_{MT} = 3 m.

deviations between Okumura-Hata predictions and measured path loss are 10–14 dB.

The model described here is applicable outside of the urban core of high-rise buildings. This is to say that the buildings are assumed to be relatively low rise and more or less uniform in height.

This model is suited for both point-to-point and broadcast transmissions. It assumes that the base antenna is always above the height of any intervening buildings (see Figure 16.1).

There are some legitimate urban targets of concern below the lower valid frequency of 150 MHz. Many contemporary potential adversaries that operate in urban terrain use VHF radios, for example. CB radios at just under 30 MHz are also popular. The characteristics exhibited by this model at 150 MHz provide at least a starting point for analysis.

The Hata model for urban areas as extended by COST 231 (the units of the variables are indicated in the subscripts) is

$$\begin{aligned} L_{\text{U,dB}} &= 69.55 + 26.16\log_{10} f_{\text{MHz}} - 13.82\log_{10} h_{\text{b,m}} \\ &+ (44.9 - 6.55\log_{10} h_{\text{b,m}})\log_{10} D_{\text{km}} + a_{\text{x,dB}}(h_{\text{MS}}) \end{aligned} \tag{16.13}$$

The city size correction factor, $a_{\text{x,dB}}$, is:

For a small or medium sized city,

$$a_{\text{x,dB}}(h_{\text{MT}}) \triangleq a_{\text{m}}(h_{\text{MS}}) = -0.8 + (0.7 - 1.1\log_{10} f_{\text{MHz}})h_{\text{MS,m}} + 1.56\log_{10} f_{\text{MHz}} \quad (16.14)$$

For large cities,

$$a_{\text{x,dB}}(h_{\text{MT}}) \triangleq \begin{cases} a_{2,\text{dB}}(h_{\text{MS}}) = 1.1 - 8.29[\log_{10}(1.54h_{\text{MS,m}})]^2, & 150 \le f \le 200\,\text{MHz} \\ a_{4,\text{dB}}(h_{\text{MS}}) = 4.97 - 3.2[\log_{10}(11.75h_{\text{MS,m}})]^2, & 200 < f \le 1{,}500\,\text{MHz} \end{cases} \quad (16.15)$$

where,

L_U = Path loss in urban areas, dB;
$h_{b,m}$ = Height of BS antenna, m;
$h_{MS,m}$ = Height of MS antenna, m;
f_{MHz} = Frequency of transmission, MHz;
$a_{x,dB}$ = Antenna height correction factor (also referred to as city size correction factor since the larger the city, the higher the base antenna);
D_{km} = Distance between the base and mobile stations, km.

The term "small city" means a city where the mobile antenna height is not more than 10 meters (i.e., $1 \le h_{\text{MS,m}} \le 10\,\text{m}$). Thus the MS is always at ground level, in compliance with the requirement to be outside the urban core.

As the height of the transmitter and the receiver is measured relative to the ground, an effective antenna height h_{eff} is additionally used and added to the antenna height of the transmitter; see Figure 16.1. This also improves the accuracy of the prediction.

Suburban/Rural Areas

This version of the Hata model is applicable to the transmissions just out of the cities and in rural areas where man-made structures are there but not so high and dense as in the cities. To be more precise, this model is suitable where buildings exist, but the MS does not have a significant variation of its height. This model is suited for both point-to-point and broadcast transmissions.

The Model. We define

Table 16.1 Parameters for the Modified Hata Model

Parameter	Definition	Range of Validity
L_{mb}	Modified Hata propagation median value	-
h_b	Base station antenna height (m)	30–300
h_{MS}	MT antenna height (m)	1–10
U	0 = small/medium; 1 = large city	0–1
U_r	0 = open area; 0.5 = suburban; 1 = urban area	0–1
B_l	Percentage of land occucpied by buildings (B_l = 15.849 nominally)	1–100
D	Range (km, not beyond line of sight)	1–100
f	Frequency (MHz)	100–3,000

Source: Siwiak [17].

$$L_{\text{ps,dB}} = -2\left(\log_{10}\frac{f_{\text{MHz}}}{28}\right)^2 - 5.4 \tag{16.16}$$

The equation for the mean propagation loss in suburban and rural areas is

$$L_{\text{SU,dB}} = L_{\text{U,dB}} + L_{\text{ps,dB}} \tag{16.17}$$

where $L_{\text{SU,dB}}$ is the path loss in suburban areas in decibels (other parameters are as defined above).

Open Areas

Some modifications to the equations for open environment are necessary. As above, we define

$$L_{\text{po,dB}} = -4.78(\log_{10} f_{\text{MHz}})^2 + 18.33\log_{10} f_{\text{MHz}} - 40.94 \tag{16.18}$$

Then the excess path loss for an open area is given by

$$L_{\text{O,dB}} = L_{\text{U,dB}} + L_{\text{po,dB}} \tag{16.19}$$

16.7.3 Modified Hata-Okumura Model

Siwiak and Bahreini documented some modifications for the Hata formulas to improve accuracy relative to the Okumura curves [17]. Using the variables in Table 16.1, the accuracy of the Hata formulas can be enhanced over the entire range of validity of the Okumura curves. Define

$$F_1 = \frac{300^4}{f^4 + 300^4} \tag{16.20}$$

$$F_2 = \frac{f^4}{f^4 + 300^4} \tag{16.21}$$

The curvature of the Earth's surface while retaining line-of-sight propagation paths is accounted for with the correction factor

$$S_{\text{ks}} = \left(27 + \frac{f_{\text{MHz}}}{230}\right)\log_{10}\left[\frac{17(h_{\text{b}}+20)}{17(h_{\text{b}}+20)+D^2}\right] + 1.3 - \frac{|f_{\text{MHz}} - 55|}{750} \tag{16.22}$$

The term S_{ks} improves accuracy with respect to the Okumura curves for longer distances.

A linear transition between urban and suburban regions is accounted for with the urbanization parameter U_{r} given by

$$S_{\text{o}} = (1-U_{\text{r}})\left[(1-2U_{\text{r}})L_{\text{po}} + 4U_{\text{r}}L_{\text{ps}}\right] \tag{16.23}$$

Combining the height correction functions (16.14) and (16.15) with frequency transition functions (16.20) and (16.21) and a small/large city parameter U, an overall height correction a_x can be written as

$$a_{\text{x,dB}}(h_{\text{MS}}) = (1-U)a_{\text{m}}(h_{\text{MS,m}}) + U\left[a_2(h_{\text{MS,m}})F_1 + a_4(h_{\text{MS,m}})F_2\right] \tag{16.24}$$

One additional term accounting for the percentage of buildings on the land in the immediate grid under consideration is

$$B_{\text{o}} = 25\log_{10} B_1 - 30 \tag{16.25}$$

Using (16.24) in (16.13) and combining that with (16.22), (16.23), and (16.25) the modified Hata formula can be written as

$$L_{\text{mH}} = L_{\text{U,dB}} + S_{\text{o}} - S_{\text{ks}} + B_{\text{o}} \text{ dB} \tag{16.26}$$

The excess path loss is then given by

$$L_{\text{ex}} = L_{\text{mH}} - L_{\text{fs}} \text{ dB} \tag{16.27}$$

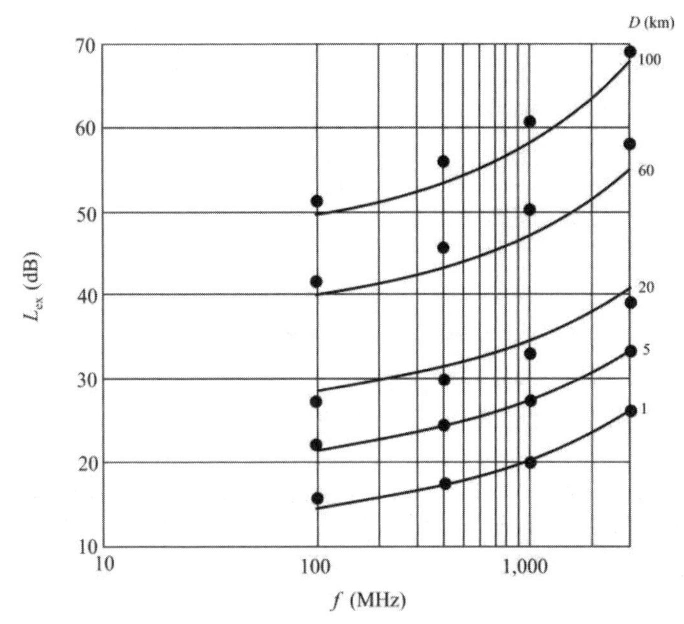

Figure 16.22 Comparison of the modified Hata model with Okumura data. Urban large city, h_b = 200 m, h_{MT} = 3 m This is the additional loss over and above the free space loss given by (16.5). Dark dots are Okumura data points. (Source: [17]. © Artech House 2007. Reprinted with permission.)

Expression (16.27) is compared with Okumura's curves in Figure 16.22, which shows the excess loss over and above that for free space given in (16.5). The modified Hata model is seen to be within about 3 dB of the Okumura data points over the frequency range 100 to 3,000 MHz. Equation (16.26) applies for distances from 1 to 100 km, MS antenna heights of 1–10 m, and fixed-site antenna heights between 30 m and 1,000 m.

16.7.4 Walfisch and Bertoni Model

Walfisch and Bertoni put the urban propagation characteristics on a theoretical basis rather than empirical as the Hata-Okumura model is based [6]. They proposed a method based on numerical integration of the Kirchoff-Huygens integral [18] in optics to determine the fields diffracted past half-screens as illustrated in Figure 16.23. The field in the aperture of the first (n = 0) half-screen is used to compute the field in the aperture of the second screen, and so on. The incident plane wave was assumed to have unit amplitude magnetic intensity $\vec{H}$, polarized along the z-axis in Figure 16.23, and has implied time dependence $\exp(j\omega t)$. The parameters are defined in Figure 16.23 and the UHF frequency range (300 MHz to 3 GHz) is considered.

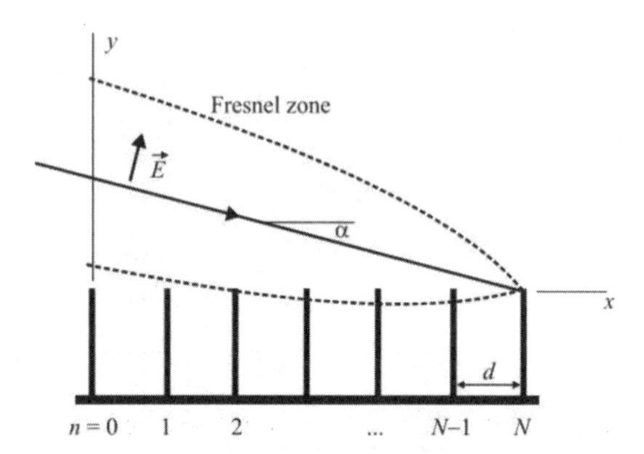

Figure 16.23 First Fresnel zone for the BS signal from an antenna that is taller than the average building height. (Source: [6]. © IEEE 1988. Reprinted with permission)

The free space path loss experienced by RF signals, L_{fs}, is given by the Friis equations (16.4) and (16.5). The excess path loss is given by

$$L_{ex,dB} = 57.1 + \log_{10} f_{MHz} + 18\log_{10} D_{km} - 18\log_{10} h_{b,m} - 18\log_{10}\left(1 - \frac{D_{km}^2}{17h_{b,m}}\right) + A_{dB} \tag{16.28}$$

The influence of the buildings is given by A_{dB} as

$$A_{dB} = 5\log_{10}\left[\left(\frac{d_m}{2}\right)^2 + (h_m - h_{MS,m})^2\right] - 9\log_{10} d_m + 20\log_{10}\left\{\tan^{-1}\left[\frac{2(h_m - h_{MS,m})}{d_m}\right]\right\} \tag{16.29}$$

The total path loss is given by the sum of the free space path loss (16.5) and the excess path loss (16.28)

$$L_{p,dB} = L_{fs,dB} + L_{ex,dB} \tag{16.30}$$

Comparison of these theoretical results with measurements made by Ott and Plitkins [19] in Philadelphia is given in Figure 16.24. A comparison of these theoretical results with the Okumura data is shown in Figure 16.25.

Reflections off the streets back up to the rooftops where it is diffracted toward the MS, as suggested by path 4 in Figure 16.1 also occur.

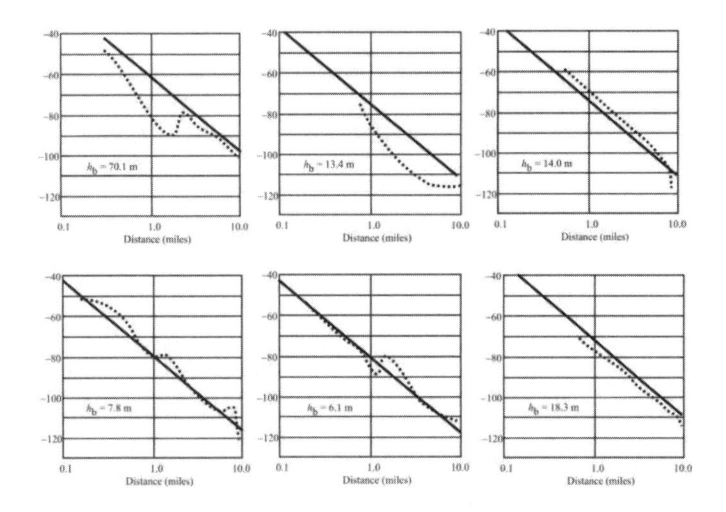

Figure 16.24 Comparison of sector-averaged signal strength for various transmitter antenna heights with theoretical predictions based on the Walfisch model (solid lines). Signal level in dBm, range in miles. (After: [6]. © IEEE 1988. Reprinted with permission)

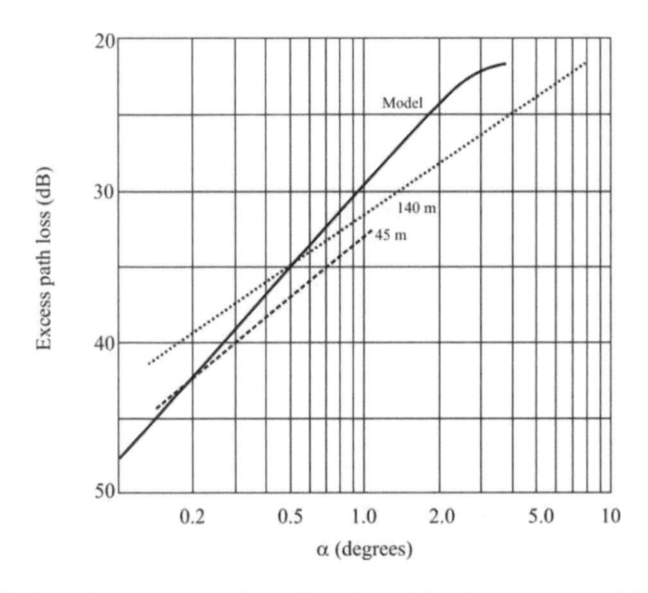

Figure 16.25 Comparison of excess path loss from the model with measurements of Okumura et al. plotted as a function of α at f = 922 MHz and BS heights of 45 and 140 m. (Source: [6]. © IEEE 1988. Reprinted with permission.)

A portion of the field diffracted by each row of buildings reaches street level where it can be detected by the MS. Diffraction down to the MS from the rooftops of neighboring buildings, paths 2 and 5, has previously been proposed as the final stage in the propagation process. In addition there is propagation through buildings as indicated by path 3 that reach the MS at ground level.

With all these possible paths from the BS to the MS, it is still believed that the majority of the energy from the transmitting BS arrives at the MS from diffractions and reflections from the structures in the close vicinity of the MS. As mentioned, it is widely believed that there are two dominant waves impinging on an MS at ground level in urban settings: the one diffracted directly down to the MS from the last rooftop prior to the MS's location and one reflected off the building behind the MS. All other wavefronts are inconsequential relative to these two.

16.7.5 Path Loss in Street Microcells—Two-Slope Model

For ranges less than 500 m and antenna height less than 20 m, some empirical measurements have shown that the received signal strength for LOS propagation along city streets can be accurately described by the two-slope model [20]

$$P = 10\log 10\left[\frac{A}{D^a(1+D/g)^b}\right] \qquad \text{dBm} \tag{16.31}$$

where

D: distance in m.
A: constant.
a and b: parameters that reflect path losses ranging from free space to higher. For a distance close into the BS, free space propagation will prevail so that $a = 2$. At large distances, an inverse-fourth to inverse-eighth power law is experienced so that b ranges from 2 to 6. Note that a and b can have negative values.
g: break point parameter, ranges 150~300 m. Breakpoint occurs where the Fresnel zone between the two antennas just touches the ground assuming a flat surface. This distance is

$$g = \frac{1}{\lambda_c}\sqrt{(\Sigma^2-\Delta^2)^2 - 2(\Sigma^2+\Delta^2)\left(\frac{\lambda_c}{2}\right)^2 + \left(\frac{\lambda_c}{2}\right)^4} \tag{16.32}$$

where

$$\Sigma = h_b + h_{MS}$$

Table 16.2 Break Points for Microcellular

Base Antenna Height, h_b (m)	*a*	*b*	A	Break Point *g*(m)
5	2.30	−0.28	94.5	148.6
9	1.48	0.54	79.8	151.8
15	0.40	2.10	55.5	143.9
19	−0.96	4.72	316.4	158.3

Source: [21].

$\Delta = h_b - h_{MS}$

For high frequencies this distance can be approximated as

$$g = 4\frac{h_b h_{MS}}{\lambda_c} \tag{16.33}$$

It depends on frequency, with the breakpoint at 1.9 GHz being about twice that for 900 MHz. The model parameters obtained by Harley are listed in Table 16.2.

16.8 Indoor Propagation

It is often said that an army should avoid urban terrain due to its difficulties—especially indoors. Yet a good deal of the time modern armies conduct operations in urban terrain; it has become unavoidable with the types of adversaries currently engaged and anticipated. In particular, large cities are becoming more of the twenty-first century battleground than open spaces. As such it is prudent to examine the propagation conditions that exist within buildings as well as into and out of buildings, to the extent that it is different from regular outdoor propagation.

The European COST 231 research project developed a model for indoor office propagation. The excess attenuation (in dB) grows linearly with the number of walls traversed while the effect of floors is nonlinear. The model for path loss for 800–1,900 MHz is [22]

$$L_{\text{in,dB}} = L_{\text{fs,dB}} + 37 + 3.4k_{w1} + 6.9k_{w2} + 18.3N^{\frac{N+2}{N+1} - 0.46} \tag{16.34}$$

with

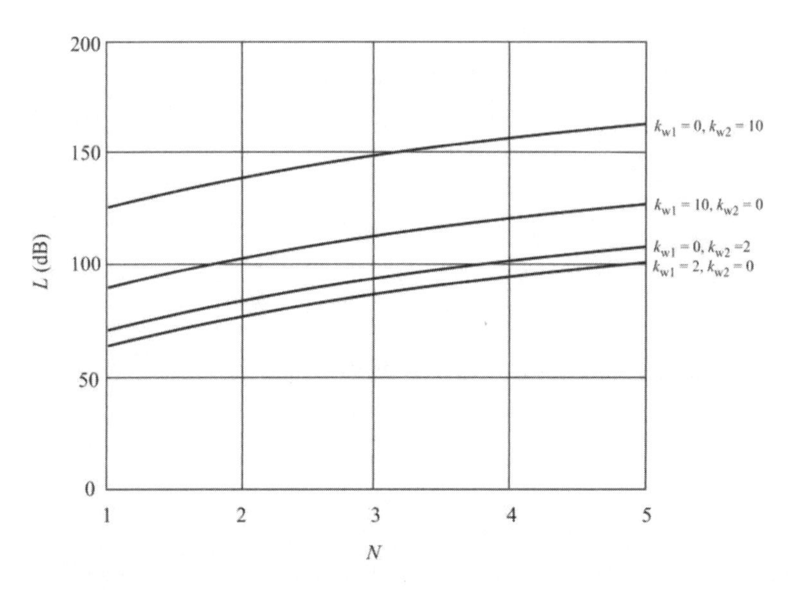

Figure 16.26 Indoor propagation excess path loss (over and above free space path loss). f = 800–1,900 MHz.

$L_{\text{in,dB}}$: the attenuation, dB;
$L_{\text{fs,dB}}$: the free space loss, dB, given by the Friis equation (16.5);
N : the number of traversed floors (reinforced concrete, but not thicker than 30 cm);
k_{w1} : the number of light internal walls (e.g., plaster board), windows, and so forth;
k_{w2} : the number of concrete or brick internal walls.

The sum of terms in (16.34) on the right side excluding the first, called the excess loss, L, is plotted in Figure 16.26 for a few values of the parameters. We see that for $N > 1$ or 2, and even minimal walls, the signals suffer considerable attenuation inside buildings going from floor to floor.

16.9 Concluding Remarks

We discussed the major characteristics of EM wave propagation in urban settings in this chapter. Modeling such propagation is considerably more difficult than in open environments where there are not as many obstructions.

It can be concluded, as indicated in the discussion, that the most significant properties affecting UHF signals in cities are the reflections off buildings and the propagation into and out of buildings, including the absorption properties of the materials in walls as well as its relative dielectric constant. Diffraction and scattering are important parameters as well, but less so than the first two. As indicated, delay spread causing ISI is a consideration in urban signaling but with the rake receiver and related technologies this should not play a significant role in the future.

The models discussed are the most popular for modeling urban propagation. The Hata formulas are based on the measurements of Okumura in Tokyo and are believed to be the most accurate and representative characteristics available. The Walfisch model is not based on collected data but is analytically derived from optical principles; it also is thought to be the most accurate analytical model available.

References

[1] Egli, J. J., "Radio Propagation Above 40 MHz over Regular Terrain," *Proceedings of the IRE*, Vol. 45, No, 10, 1957, pp. 1383–1391.

[2] Longley, A. G., and P. L. Rice, "Prediction of Tropospheric Radio Transmission Loss Over Irregular Terrain, C Computer Method—1968," ESSA Technical Report ERL 79–67, NTIS Access No. 676–874.

[3] Okumura, Y., E. Ohmori, T. Kawano, and K. Fukuda, "Field Strength and Its Variability in VHF and UHF Land-Mobile Radio Service," *Record of the Electronic Communication Laboratory*, Vol. 16, 1968, pp. 825–873.

[4] Hata, M., "Empirical Formula for Propagation Loss in Land Mobile Radio Services," *IEEE Transactions on Vehicular Technology*, Vol. VT-29, 1980, pp. 317–325.

[5] Low, K., "A Comparison of CW-measurements Performed in Darmstadt with the COST-231-Walfisch-Ikegami Model," Rep. COST 231 TD (91) 74, Darmstadt, Germany, 1991.

[6] Walfisch, J., and L. Bertoni, "A Theoretical Model of UHF Propagation in Urban Environments," *IEEE Transactions on Antennas and Propagation,* Vol. 36, No. 12, December 1988, pp. 1788–1791.

[7] Sridhara, V., and S. Bohacek, "Realistic Propagation Simulation of Urban Mesh Networks," *Computer Networks*, Vol. 51, Issue 12, August 2007.

[8] Ikegami, F., S. Yoshida, T. Takeuchi, and M. Umehira, "Propagation Factors Controlling Mean Field Strength on Urban Streets," *IEEE Transactions on Antennas and Propagation*, Vol. 32, 1984, pp. 822–829.

[9] Peterson, R. L., R. E. Ziemer, and D. E. Borth, *Introduction to Spread Spectrum Communications*, Upper Saddle River, NJ: Prentice Hall, 1995, Ch. 8.

[10] Bertoni, H. L., and L. R. Maciel, "Theoretical Prediction of Slow Fading Statistics in Urban Environments," *Proceedings IEEE ICUPC Conference,* 1992, pp. 1–4.

[11] Maciel, L. R., H. L. Bertoni, and H. H. Xia, "Unified Approach to Prediction of Propagation Over Buildings for All Ranges of BS Antenna Height," *IEEE Transactions on Vehicular Technology*, Vol. 42, No. 1, February 1993, pp. 41–45.

[12] Torrico, S. A., H. L. Bertoni, and R. H. Lang, "Modeling Tree Effects on Path Loss in a Residential Environment," *IEEE Transactions on Antennas and Propagation*, Vol. 46, No. 6, June 1998, pp. 872–880.

[13] Bertoni, H. L., and L. Maciel, L. "Theoretical Prediction of Slow Fading Statistics in Urban Environments," *Proceedings IEEE International Conference on Universal Personal Communications* 1992, pp. 1–4.

[14] Poisel, R. A., *Introduction to Communication Electronic Warfare Systems*, 2nd ed., Norwood, MA: Artech House, 2008, p. 73.

[15] Godara, L. C., *Handbook of Antennas in Wireless Communications*, Boca Raton, FL: CRC Press LLC, 2000, Ch. 3.

[16] Jordan, E. C., and K. G. Balmain, *Electromagnetic Waves and Radiating Systems*, 2nd ed., Englewood Cliffs, NJ: Prentice-Hall, 1968.

[17] Siwiak, K., and Y. Bahreini, *Radiowave Propagation and Antennas for Personal Communications,* 3rd ed., Norwood, MA: Artech House, 2007, pp. 208–212.

[18] Hecht, E., *Optics*, 4th ed., Reading, MA: Addison Wesley, 2002.

[19] Ott, G. D., and A. Plitkins, "Urban Path-Loss Characteristics at 820 MHz," *IEEE Transactions on Vehicular Technology,* Vol. VT-27, 1978, pp. 189–197.

[20] Harley, P., "Short Distance Attenuation Measurements at 900 MHz and 1.8 GHz Using Low Antenna Heights for Microcells," *IEEE Journal on Selected Areas in Communications*, Vol. 7, Jan. 1989, pp. 5–11.

[21] Harley, P., "Short Distance Attenuation Measurements at 900 MHz and 1.8 GHz Using Low Antenna Heights for Microcells, *IEEE Journal on Selected Areas in Communications*, Vol. **7,** No 1, January 1989, pp. 5–11.

[22] Davis, J. L., "Indoor Wireless RF Channels," http://www.wireless.per.nl/reference/chaptr03/indoor.htm, accessed August 19, 2010.

Chapter 17

Urban Electronic Warfare

17.1 Introduction

We have discussed the general characteristics of urban signal propagation in Chapter 16, and the predominant sources of noise generated in an urban environment in Chapter 2. Thus having set the properties of communications expected in urban settings, in this chapter we focus our attention on some of the specific issues associated with conducting EW operations targeting these communications in urban settings.

EW in urban settings with the few exceptions discussed in this chapter is just like EW in rural settings. The goal of EA to deny the exchange of information between a target transmitter and receiver by injecting noise or some other signal into the target receiver that overwhelms the target signal and disallows the exchange of information. The primary goal of ES is to provide targeting level data to the EW systems so that the correct links can be targeted. A secondary goal of ES is to garner combat information for situation assessment and subsequent decision making (to the extent that this is different from the primary goal).

This chapter is structured as follows. We first examine some extracts from Army FM 03-03, Urban Operations, which provide some insight as to what the U.S. Army believes is the principal use of EW for urban warfare. That is followed by a discussion of the (more or less) unique challenges in urban EW. We conclude the chapter with a description and conclusions derived from a computer operational simulation of some EW architectures and their tradeoffs for urban EW.

17.2 Electronic Warfare

Army FM 3-06, Urban Operations, describes the U.S. Army's doctrine on urban operations [1]. Included in that doctrine is consideration of EW. EW is defined in paragraph 5-70 as:

> Electronic warfare (EW) includes all actions that use electromagnetic or directed energy weapons to control the electromagnetic spectrum or to attack a threat. Conducting EW in urban areas seeks to achieve much the same results as in other environments. A major consideration in urban areas is collateral effects on portions of the urban infrastructure that rely on the electromagnetic spectrum for service. Thus, precision is a major factor in planning for EW operations. For example, EW attacking a threat's television broadcasts avoids affecting the television broadcasts of neutral or friendly television. Likewise, EW attacking military communications in a large urban area avoids adversely affecting the area's police and other emergency service communications. Urban offensive and defensive operations will have the least restrictions on EW operations while urban stability may have significant constraints on using EW capabilities.

FM 03-06 considers *computer network operation* (CNO) as being particularly important in urban warfare. In paragraphs 5-71–5-74, CNO is presented as follows:

> **Computer Network Operations**
>
> CNO include *computer network attack* (CNA), *computer network defense* (CND), and *computer network exploitation* (CNE). CNO are not applicable to units at corps and below. Echelons above corps (EAC) units will conduct CNA and CNE. If tactical units require either of these network supports, they will request it of EAC units.
>
> ***Computer Network Attack***
>
> Considerations regarding the execution of CNA in urban operations (UOs) are similar to those of EW: CNAs that do not discriminate can disrupt vital civilian systems. However, possible adverse effects on the civilian infrastructure can be much larger—potentially on a global scale. In the short term, CNAs may serve to enhance immediate combat operations but have a debilitating effect on the efficiency of follow-on urban stability operations. Because of these far-reaching effects, tactical units do not execute CNA. CNA is requested of EAC

units. EAC units will receive all requests from lower echelons, carefully consider second- and third-order effects of CNA, and work to ensure its precise application.

Computer Network Defense

In urban operations, CND will require extreme measures to protect and defend the computers and networks from disruption, denial, degradation, or destruction. The nature of the urban environment and configuration of computer networks provides the threat with many opportunities to interdict *local area networks* (LANs) unless monitored by military forces. LANs controlled by military forces are normally more secure than the civilian infrastructure. Commanders should prepare for opportunities by the threat to insert misinformation.

Computer Network Exploitation

CNE consists of enabling operations and intelligence collection to gather data from target or adversary automated INFOSYS or networks. Tactical units do not have the capability for CNE. CNE contributes to intelligence collection at EAC. In urban operations (UO), CNE will be centrally controlled.

17.3 Electronic Isolation

FM 3-06 explains the main purpose of EW in urban environments is to provide electronic isolation [2]:

> Electronic isolation is achieved through offensive information operations (IO). Electronic warfare (particularly two of its components: electronic warfare support and electronic attack) and computer network attack are critical to electronic isolation ... At the operational level, offensive IO aims to quickly and effectively control the information flow into and out of an urban area. This isolation separates the threat's command and control (C2) system in the urban area from its operational and strategic leadership outside the urban area. Offensive IO also focuses on preventing the threat from communicating with civilians through television, radio, telephone, and computer systems. At the

> tactical level, IO aim to isolate the threat's combat capability from its C2 and leadership within the urban area, thus preventing unity of effort within that area. Defensive IO can prevent isolation of friendly forces defending in an urban area.

There are three types of isolation defined in the same document as [3]

> Isolation of an urban environment is often the most critical component of shaping operations. Commanders who's [sic] AO includes operationally significant urban areas often conduct many shaping operations to isolate, or prevent isolation of, those areas from other parts of the AO. Likewise, commanders operating in the urban area focus on isolating decisive points and objectives in the urban area or averting isolation of points that are critical to maintaining their own freedom of action. Isolation is usually the key shaping action that affects UO [*urban operations*]. It applies across full spectrum operations. Most successful UO have effectively isolated the urban area. Failure to do so often contributed to difficult or failed UO. In fact, the relationship between successful isolation and successful UO is so great that the threat often opposes external isolation actions more strongly than operations executed in the urban area (or critical areas within). In some situations, the success of isolation efforts has been decisive. This occurs when the isolation of the urban area compels a defending enemy to withdraw or to surrender before beginning or completing decisive operations. In UO that are opposed, Army forces attempt to isolate the threat three ways: physically, electronically, and, as a resultant combination of these first two, psychologically [see Figure 17.1].

There are many different types of communication systems in most, if not all, major urban areas. Isolating an adversarial group from communicating with sister forces means cutting off all these lines of communication. CNO will certainly be a factor in order to accomplish this—if nothing more than taking a server out by flooding. Mobile, wireless connectivity through Wi-Fi or WiMax must be denied or significantly degraded in addition to the normal methodologies thought of when considering EW. The ability to exchange data and information over cellular phones and PCS systems implies the necessity of EW against wireless phone systems.

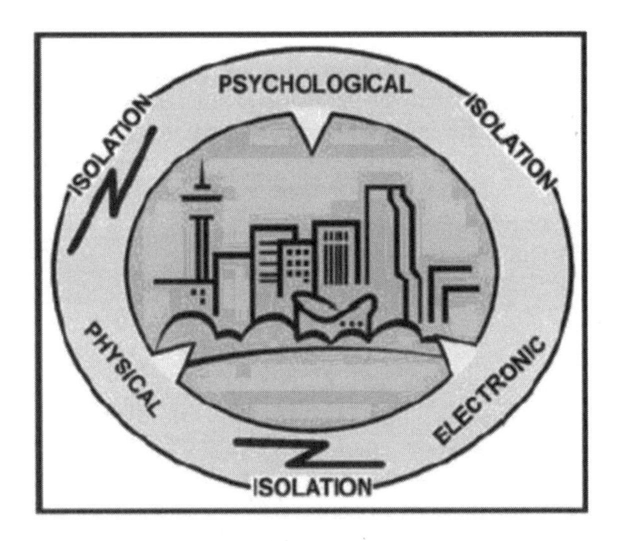

Figure 17.1 Three types of isolation. (Source: [1].)

17.4 Networked Communications

Just as many of the world's developed militaries are making use of the latest in communication technology to effectively use networking as a force multiplier, opposing urban forces have also learned to make use of the benefits of network centric/enabled communications, using modern technologies such as wireless phones, cellular networks, and the Internet. The significant advantage is that their communications merge with normal commercial traffic as the guerillas try to evade EW monitoring [4].

17.5 Improvised Explosive Devices Countermeasures

Remote activation of *improvised explosive devices* (IEDs) is another capability of commercial communications used by urban guerillas. Wireless phones, remote control equipment, and infrared remote controls are some of the activation devices that have been used by insurgents in the Middle East for the past few decades. Military forces are countering this threat with a wide range of countermeasures, constantly improved to attempt to stay ahead of evolving threats. Consequently, unlike the more common IW operations, where jammers are employed by EW units, *urban combat* (UC) requires the use of IED jammers by every combat vehicle, EOD team, and even by dismounted patrols. This trend has led to the development of reliable, effective, "switch on and forget" jammers that can effectively combat remote triggering of IEDs without degradation of the vehicle's or individual's communications system.

This capability does not come without problems, however. For example, it would be desirable to not have the IED jammer on all the time to allow legitimate cellular traffic to transpire. Rather, a *signal initiated jamming* (SIJ) capability would be desired. Unfortunately, once one of these jammers "triggers" on a suspected signal, it could trigger other jammers instead, and so on. Before long, all or a large number of the jammers are triggered.

17.6 Challenges of Urban EW

In addition to, or in lieu of, traditional military tactical push-to-talk radios, typically in the VHF frequency range, an urban RF environment consists of a plethora of "nonstandard" radio systems. The RF environment in an urban setting is expanding at an explosive rate in major cities around the world due primarily to PCS and cellular phone systems as well as mobile Internet services these phones provide.

Even though the tenents of EW are basically the same in an urban environment as they are anyplace else, there are some particular issues with urban terrain that should be discussed. We will cover some of them in this section.

17.6.1 Multipath

In an urban battlespace, signal propagation occurs primarily by the signals taking many paths to get from the transmitter to the receiver. In many cases this is caused by diffraction at the top edge of the building closest to the MT in the direction to the transmitter and reflection off the structure on the other side of the road from that building. This causes two signals to arrive at the MT; however, there is a high likelihood that there are many more. They are just weaker than the two mentioned.

Moving structures such as vehicles or traffic signs blowing in the wind cause these multiple paths to have a nonstationary character. This causes time-variable fading whether the target or EW system is moving or not. The difficulty this presents to an EW system is that most of the existing engineering and physical theory of signal detection, processing, and throughput assumes stationary noise. If this isn't the case then the EW system will likely have degraded performance.

17.6.2 Lack of Direct Wave Signal Component

Except in rare circumstances, there is no direct signal path from the target transmitter to the intercept receiver or from the EA transmitter to the target receiver. As discussed in Chapter 16, almost all signals are reflected or diffracted in an urban environment, especially in the urban core. That makes accurate geopositioning of the transmitter essentially impossible.

It also imposes a different PDF for the amplitude of the intercepted signal. With a direct wave component, the PDF of the amplitude of the intercepted EM wave takes on a Ricean fading distribution whereas for a signal with no direct wave component the amplitude PDF has a Rayleigh fading characteristic. The important significance of this is that the received signal, being composed of the sum of several multipath components, will suffer significant time-varying fading. This makes reliable interception problematic. It also makes it difficult to reliably deliver the jamming waveform to the receiver.

17.6.3 3D Battlespace

Signals can emerge from any direction, including down from the top. Many EW systems utilize dipole or monopole antennas that have theoretical nulls overhead or directly underneath. In theory, signals arriving from these directions will not be detected. In reality, some of the signal will arrive not directly from these directions, but they could still be significantly suppressed.

FM 03-06

> Urban areas present an extraordinary blend of horizontal, vertical, interior, exterior, and subterranean forms superimposed on the natural relief, drainage, and vegetation. An urban area may appear dwarfed on a map by the surrounding countryside. In fact, the size and extent of the urban area of operations is many times that of a similarly sized portion of undeveloped natural terrain. A multi-storied building may take up the same surface area as a small field, but each story or floor contains approximately an equal area as the ground upon which it sits. In effect, a ten-story building can have eleven times more defensible area than "bare" ground—ten floors and the roof. It is the sheer volume and density created by this urban geometry that makes UO resource intensive in time, manpower, and materiel.

This means that 11 times the firepower can occupy the same patch of ground.

17.6.4 Cover and Concealment

In an urban setting there is a plethora of cover and concealment, especially for dismounted infantry. Rooftops and windows provide good firing positions. Especially when elevated (e.g., rooftops), extended communication paths can be established where an EW system at ground level would not be within the line of sight of the receiver.

Movement in urban settings can readily be concealed as forces move from one structure to another. When so doing, they may be exposed only for a brief time before reaching their destination. As mentioned, subterranean pathways may form the ultimate in concealment, letting movements be completely covered.

17.6.5 Noncombatant Population

The noncombatants in an urban setting are likely to be the largest segment present and often determining friend from foe is difficult at best. With the proliferation of commercial communication technology, at the time of this writing, half of the world's population possess and use cellular phone technology.

17.6.6 Cochannel

The increasing use of cellular and personal communication networks in urban settings produces increasing levels of cochannel interference. This is because these systems reuse frequencies in different cells. This frequency reuse causes signals to interfere with one another at an EW system, making it difficult to separate one signal from another. Elevated EW systems have better range, but with frequency reuse an elevated platform sees multiple cells with the same frequencies in use. Elevated stand-off platforms have this problem more than stand-in EW systems or ground-based EW systems.

17.6.7 Elevated Antennas

Antennas are easier to elevate in an urban area than in a rural areas due to tall buildings. This increases an EW system's range because there are fewer opportunities for signal blockage.

This however, could work to a disadvantage. It was already mentioned that the cellular frequencies are reused—usually in cells a few cells distant. Elevating the EW antenna makes it easier for these distant cells to interfere with the cell of interest.

This problem is not unique to cellular/PCS systems. Several (all) of the signals expected in urban environments have limited frequencies over which they can operate (except for those in the ISM band, but even there the communicators must remain within the band—they just don't require a license). All of those systems reuse the available frequencies so they all can potentially cause interference, both to the EW intercept system as well as each other.

17.6.8 Nontraditional Emitters

There are many nontraditional sources of RF signals in urban settings. These include, but are not limited to:

- RC toys;
- Walkie-talkies, both professional and toys;
- Garage door openers;
- CB radios;
- Paging systems;
- Ham radios;
- Licensed and unlicensed private radio systems;
- Licensed business radio systems (e.g. taxis);
- Remote-controlled devices;
    - Lamps;
    - Fans;
    - TVs;
    - Stereos.

Most of these RF sources simply raise the noise floor, thereby lowering the SNR at the intercept receivers. In general, these signals must limit their power and many can operate only in the ISM bands (see Chapter 2). Even though a signal is delegated to the ISM bands does not mean that there is no spillover to other regions of the frequency band, however. With enough such signals in the RF environment there is a point where the SNR gets too bad for effective EW and, in particular, ES operations.

17.6.9 Encryption

Most nonmilitary communications are not encrypted, although there are exceptions. Therefore detection of encrypted communications immediately indicates a probable SOI.

17.6.10 EW Fratricide

Recall that SIJ is when the jammer's intercept receiver detects an SOI and subsequently initiates jamming on that target. This same situation can occur when the intercept receiver of a second jammer detects the jamming signal of the first EW system, and initiates jamming. This scenario could escalate until, in the limit, all jammers in a region are trying to jam each other. This is called EW fratricide.

17.6.11 Power and HVAC

In rural settings, the military typically must bring its power sources with it in the form of fossil-fueled generators. On the other hand, power is readily available in urban settings as is HVAC. In most (all?) large cities, power from the power grid is plentiful and readily available (unless removed).

17.6.12 Satellite Navigation Denied Areas

Satellite navigation devices do not work well in covered areas. Thus inside buildings and underground, navigation must be provided by more traditional means such as maps. Maps may very well be available for such man-made structures such as hotels and subway systems. They would not generally be available for such infrastructure as sewer tunnels. That would provide an advantage for indigenous urban guerillas more familiar with the environment.

Denial of satellite navigation in outdoor environments would have to be provided by EW resources.

17.6.13 Urban Impulsive Noise

We examined urban RF IN at length in Chapter 2. Due to its ubiquity in urban environments, however, it is worth mentioning again. It is perhaps the largest challenge to overcome when conducting EW operations in urban terrain.

17.6.14 Gathering Intelligence

FM 03-06 explains that intelligence collection in urban warfare is different from that for open, more traditional combat operations:

> A complex intelligence environment requiring lower-echelon units to collect and forward essential information to higher echelons for rapid synthesis into timely and useable intelligence for all levels of command. Understanding the multifaceted urban environment necessitates a bottom-fed approach to developing intelligence (instead of a top-fed approach more common and efficient for open terrain and conventional threats). It also emphasizes the need for intelligence reach and a truly collaborative approach to the development and sharing of intelligence.

17.6.15 Friendly C2 Communications

Just as the terrain in an urban setting can affect the ability of an intercept receiver to receive a target signal with a good SNR, those same terrain effects have an impact on friendly forces to communicate with one another, and for the same reasons. VHF and UHF communications in an urban core can be virtually non-existent. That is why there are so many cellular towers in such settings.

17.7 Urban Electronic Warfare Operational Simulation

17.7.1 Introduction

An operational computer simulation experiment of urban EW was conducted to examine a few EW architectures to support networked ground-component forces. The three specific cases examined were:

- A set of distributed thin jammers;
- One and two thick jammers;
- One UAS mounted thin jammer.

Thin and thick jammers get their name from computer networking jargon. Thin jammers have little indigenous capability for ES. They are basically relegated to signal detection, thus identifying energy with frequency. A thick jammer, on the other hand, has sufficient ES capability (usually associated with the presence of an EW operator) to make independent decisions as to the importance of a detected target.

In addition to these, simulations were executed for the baseline where there was no jamming and when jamming was complete—no communications were allowed to transpire. These two cases form the worst case and best case limits of jamming effectiveness.

ES was not modeled for the study principally because the model used does not include the capability for ES. Thus it was assumed that the locations of the targets were known and the closest jammer to a target was tasked with jamming that target. This in effect removed the ES portion of the problem and concentrated the results strictly on the EA aspect. ES was modeled in a separate effort [5] (see also Chapter 4).

17.7.2 EW Methodology

The methodology used to model the effects of jamming are summarized in the following equations:

$$JamPower(\text{dB}) = JamEIRP - JamLoss + RcvGain \tag{17.1}$$

$$RcvPower(\text{dB}) = XmitPwr - XmitLoss + RcvGain \tag{17.2}$$

$$SJR(\text{dB}) = RcvPower - JamPower \tag{17.3}$$

Using the following logic determined whether the jamming was successful or not:

```
IF SJR(dB) < THRESHOLD -> COMMO JAMMED
ELSE -> COMMO SUCCESSFUL
```

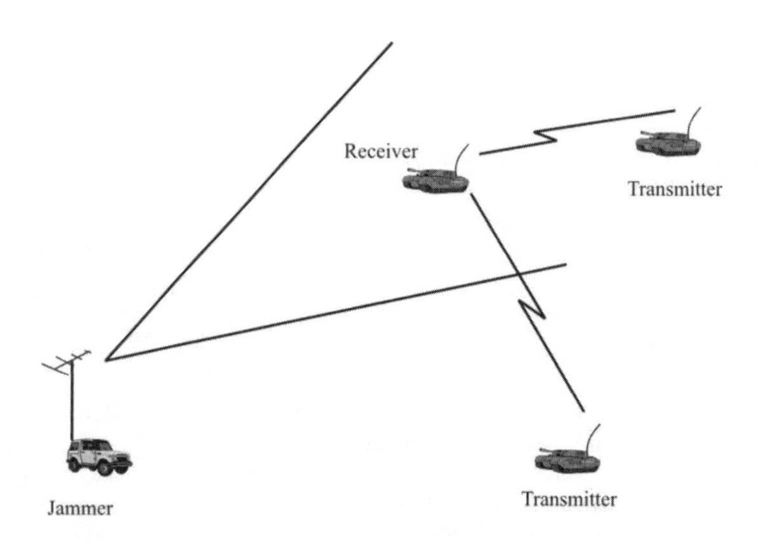

Figure 17.2 Jammer model used in the simulation.

The attenuation losses in (17.1)–(17.3) are calculated using the Hata-Okumura models. Further logic to ascertain jamming success was based on Figure 17.2. If the target receiver was within the jammer's antenna pattern, as illustrated by the lines emanating from the jammer in Figure 17.2, (17.1)–(17.3) are calculated to determine if the message was received.

17.7.3 Modeling EW Limitations

The EW capability had some limitations. In particular:

- Only EA is modeled; ES and EP were not implemented. Therefore modeling accurate follower jamming is not possible. This results in the lack of ability to explicitly play frequency hopping and the effects of propagation effects due to frequency, multipath fading, and so forth were not examined.
- The jammer only operates on one frequency, not a bandwidth of frequencies.

17.7.4 Key Assumptions

The principal assumptions made in the modeling process were:

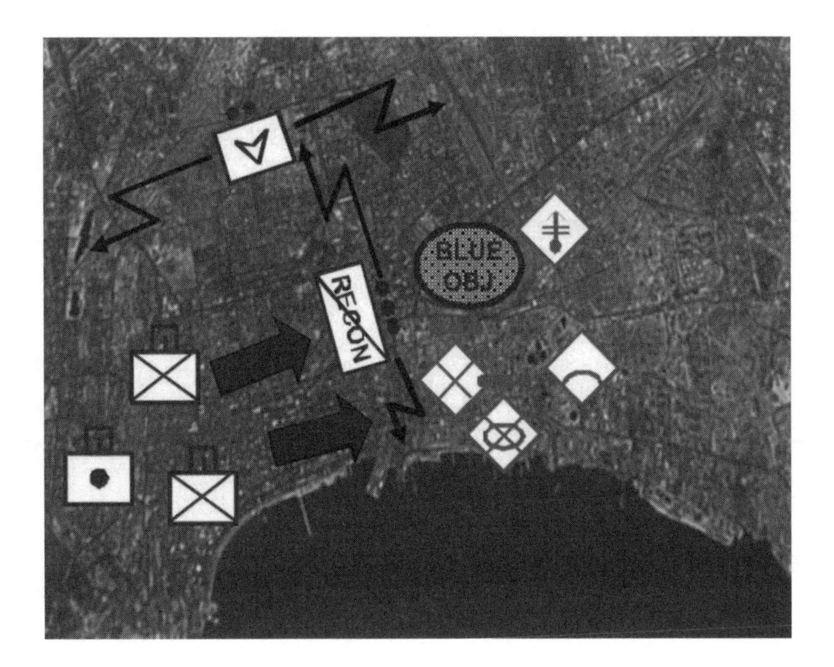

Figure 17.3 Urban scenario.

- Blue maneuver and communications were precoordinated, minimizing negative effects of jamming fratricide.
- Jammer is reactive; it only activates when it receives threat communications and turns off when communications have ceased.
- Red was unaware of blue jamming resulting in red tactics not changing due to jamming.

17.7.5 Urban Scenario

17.7.5.1 Urban Scenario Overview

The urban scenario included portions of a U.S. Army *Future Combat System* (FCS) Combined Armored Battalion CAB(–) with a supporting *Non-Line of Sight* (NLOS) Battalion Bn(–) conducting operations in urban terrain in support of a *Joint Task Force* (JTF) conducting operations to restore a legitimate government (see Figure 17.3.)

In the scenario the urban terrain is restrictive and complex with minimal LOS. Both mounted and dismounted blue assaults were conducted. The ground movements were augmented with joint air support available for precision strikes.

Artillery/NLOS-LS assets provided on-call support. Threat CO(+) consisting of mixed conventional and unconventional forces attempts to defend the government seat of power using adaptive/evasive urban tactics and the red forces reacted dynamically to blue's advancement. They made good use of coverage provided by urban terrain. There was extensive IED placement. However they had limited ADA and mortar support and no artillery or air support. Red forces used buildings to their advantage by using "pop-up" tactics from windows and roofs.

Red forces relied on barriers and IEDs to slow/prevent blue maneuvers causing blue to stop at barriers and remove the obstacles, making them vulnerable. There were two types of IED's in play:

- Typical roadside IED, aimed at destroying passing vehicles (5 of 7 IEDs were of this type);
- IEDs/bombs placed deliberately near barriers where blue infantry would be vulnerable. These were aimed at inflicting damage to blue infantry forces. (Two IEDs near barriers were of this type.)

17.7.6 Role of Electronic Warfare in the Urban Scenario

EW was used in the scenario for three primary purposes:

- Counter red IED remote triggering (7 in scenario);
- Prevent red forces from calling for support from southern units;
- General disruption of red C2 and coordination/synchronization.

17.7.7 Cases Examined

17.7.7.1 Thick Jammer Architecture

The thick jammer used in the cases examined consists of EW (ES and EA) equipment mounted on a small vehicle, such as the HMMWV shown in Figure 17.4. Such a configuration could house up to two operators, but only one jammer was used in the modeling. The transmitter emitted 27 dBW (500 W) and the omnidirectional antenna was mounted atop a 5 m mast. It was assumed that jamming was the sole mission of the vehicle. There was one jammer per CAB(–) included. The jamming vehicle travels behind main combat force and it must stop to provide jamming support.

17.7.7.2 Thin Jammer Architecture

Two versions of the thin jammer were simulated: (1) a jammer mounted in all of the FCS RSTA vehicles in the CAB(–) for which there were five in the simulation

Figure 17.4 Notional thick jammer configuration.

(see Figure 17.5), and (2) a jammer mounted in the Fire Scout UAS (see Figure 17.6). In both cases the thin jammer emitted 20 dBW (100 W).

In the ground configuration the omnidirectional antenna was mounted atop a 3m mast. In addition, jamming was not the sole, or even the principal, mission of the vehicles. The vehicles maneuvered ahead of the main combat force and the jammers were active while maneuvering.

There was one thin UAS jammer in the CAB(–). The UAS maneuvered in a continual racetrack above the city. Because of the flight pattern of the UAS, jamming was the sole mission of UAS (the UAS was not retasked to provide coverage of other space than that required for the jamming racetrack. The jammer was active while maneuvering.

Figure 17.5 Notional RSTA configuration.

Figure 17.6 Helicopter UAS thin jammer configuration.

17.7.8 Base Case Results

17.7.8.1 Loss-Exchange Ratio

The base case LER performance is shown in Figure 17.7. Overall blue force performance is increased with effective jamming present. The ground thin case was close to providing the same benefit as the unrealistic "optimal" case. The reasons for this are:

- The jamming platforms remain near blue forces, protecting them from immediate effects of IEDs.
- The platforms were able to jam while moving, preventing any lapse in protection.

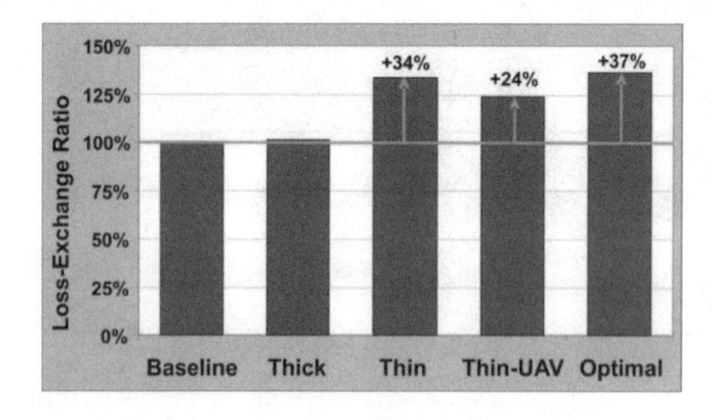

Figure 17.7 Loss exchange ratio (LER) for the base case.

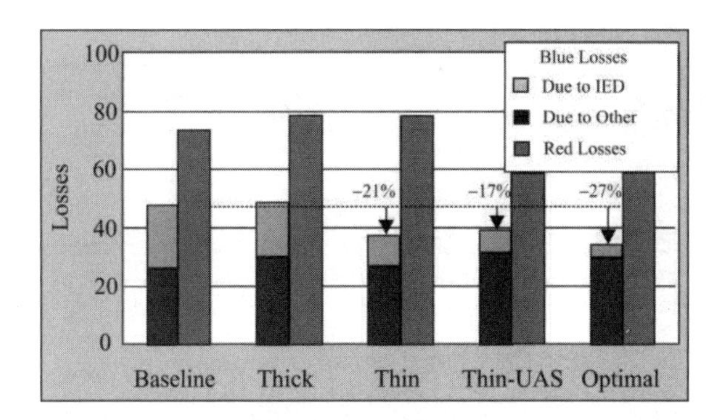

Figure 17.8 Losses base case.

The UAS case does not perform as well due to tall buildings affecting jammer coverage

17.7.8.2 Losses

The red losses remained relatively constant as indicated in Figure 17.8. Jamming was able to significantly reduce IED detonations, decreasing blue losses due to IEDs:

- 92% of IEDs detonated in baseline;
- 56% of IEDs detonated in thin case;
- 69% of IEDs detonated in UAS case;
- 23% of IEDs detonated in optimal case.

In contrast, the thick jamming architecture did not perform well. This is primarily due to the fact that the small blue force had only one thick jammer in the scenario. Furthermore there were lapses in coverage when the jammer must maneuver to its next position.

In complex urban environments, the thin architecture provides the best jamming coverage due to its proximity to blue forces and wide distribution across the battlefield.

17.7.8.3 Jammer Effectiveness over Time

Thick Jammer Performance

The performance versus time for the thick jammer in the urban scenario is shown in Figure 17.9. The thick jammer is effective when it is on but needs to shut-off while maneuvering, which allows the threat to eventually succeed with communications.

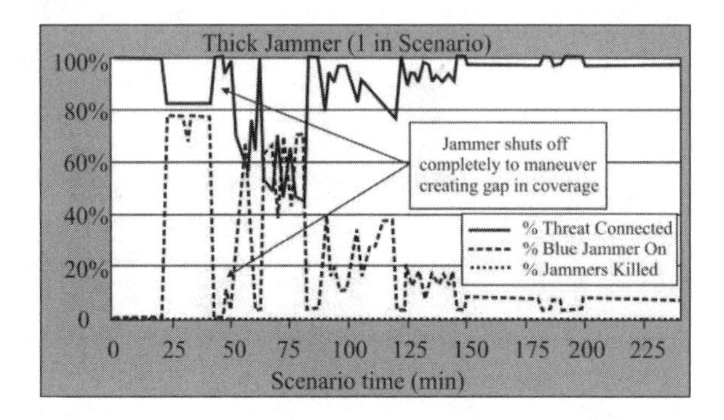

Figure 17.9 Thick jammer performance versus time. The solid line is the percentage of the threat forces that are connected, the dashed line is when the jammer is on, and the dotted line corresponds to the percentage of jammers lost.

Thin Jammer Performance

The ground thin jammer performance over time is shown in Figure 17.10. The jamming platforms are relatively close to the threat so the weaker power of the thin jammer is adequate. The relatively wide distribution of jammers also provides more widespread disruption of threat communication than a single source. In addition, these jammers could remain operational while maneuvering as opposed to the thick jammer assumption that it could not.

UAS Thin Jammer Performance

These results are shown in Figure 17.11. Due to the complex terrain, consisting mostly of (relatively) tall buildings, the UAS performance was not very good—

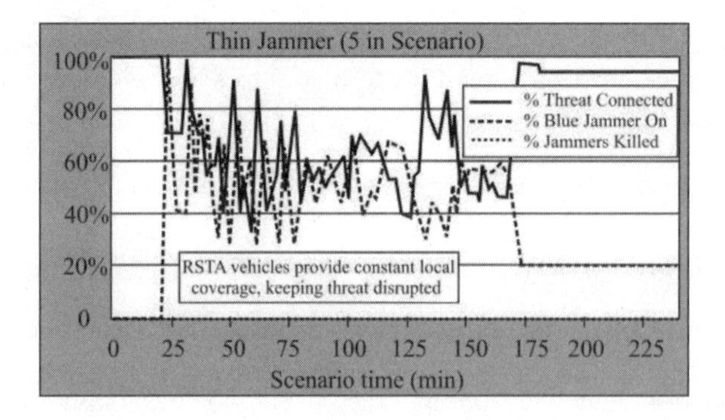

Figure 17.10 Thin jammer performance versus time.

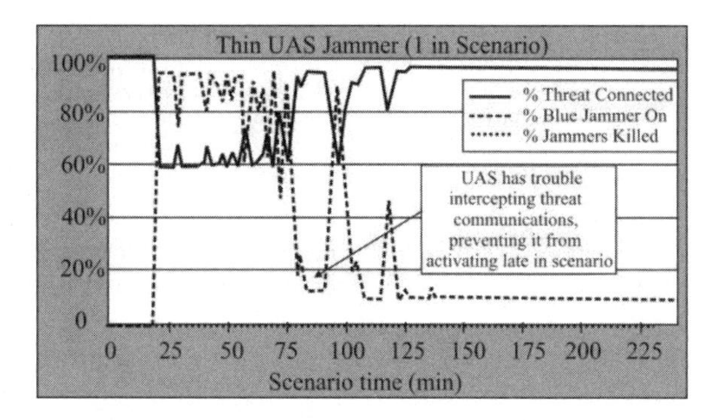

Figure 17.11 Thin UAS jammer performance over time.

especially in the latter stages of the engagement. Throughout the scenario the buildings prevented adequate signal strength from reaching the receiver in the UAS, preventing signal detection, and therefore missing the jamming opportunity.

17.7.9 Conclusions

The scenario represents a small-scale engagement in an urban environment with few players and a heavy IED presence. Red threat communications travel a relatively small distance. The maximum range of a jamming system is not as crucial as urban penetration. The amount of threat C2 communication/coordination is much less than in a large-scale scenario.

A jammer must be in a position to intercept the threat if "reactive" jamming (as opposed to barrage jamming) is used. The jamming platforms must be widely distributed and maneuver with forces to provide constant coverage.

The majority of the key threat players (IEDs, units calling for reinforcement, and so forth) positioned close to the advancing blue force. If jammers are sufficiently distributed among the blue force, the jamming systems only need to influence small local areas, which means that low power is required.

The urban environment is very complex with many parts of the advancing blue force being separated from one another. In the complex urban environment, the ground-based thin architecture is able to provide the best jamming coverage due to its proximity to protected forces and wide distribution throughout the battlespace.

The thick ground-based architecture is ineffective if only one jammer can be allocated to CAB due to it needing to maneuver often to remain with the main force, which requires jamming to be shutoff. Adding another heavy jammer and carefully coordinating maneuvers between the two makes the thick based architecture a viable option.

Thin ground-based architecture performs well in an urban scenario due to it being widely distributed and local to the maneuvering forces, as well as its ability to continue to operate while maneuvering. There was a slight improvement in performance gained when thin jammers added to C2 vehicles as well as the RSTA vehicles due to the RSTA vehicles maneuvering ahead of main force, leaving some portions unprotected.

The thin UAS-based architecture performed adequately but had difficulty with the complex urban terrain. The UAS did not intercept many of the threat communications due to the scarcity of threat communications and blockage from large buildings, resulting in the UAS jammer being inactive for a large portion of the scenario.

In complex urban environments, the ground-based thin architecture is able to provide the best jamming coverage due to its proximity to blue forces and wide distribution across the battlespace. The thin jammer architecture is better suited to the complex terrain due to the large number of jammers present and the adequacy of their range.

17.8 Concluding Remarks

We discussed EW operations in urban terrain in this chapter. In the first part we provided some quotes from current U.S. Army doctrine on EW to glean an overall picture of how modern EW professionals view EW operations in urban terrain. After that, several characteristics of urban EW operations that are different from those in rural situations were presented. Those included indicate that such operations are (or can be) considerably different from rural terrain, even though the fundamental purpose and methods of EW are basically the same.

It is important to note that most of the problems created by the urban terrain are due to the noncooperative nature of EW operations as opposed to the cooperative behavior of the target communication systems. This is a characteristic in common with EW in rural areas. As example of this is the possibility of using RAKE receivers in the target communication receivers. These can be used because of the knowledge of the PN codes in use in the cooperating target systems whereas the EW receiving system would not likely know that code.

In the second part of the chapter we presented a summary of a computer simulation of EW effects in an urban scenario. Three cases were included: (1) a single heavy jammer; (2) several ground-based thin jammers mounted on host vehicles whose primary purpose is not hamming; and (3) a single thin jammer flown on a UAS. The simulation indicated that the distributed ground-based thin jammer architecture produced the best results.

Future warfare will depend extensively on the ability to communicate. Creating dominant battlespace knowledge, where friendly forces know more about an adversary than that adversary knows about friendly forces will only be possible if communications is facilitated.

Analysis results of three architectures based on simulations were presented. It was shown that the best jammer configuration for the urban scenario is the multiple ground-based thin jammers. The primary reasons for this are two-fold:

- The jammers were distributed and therefore could detect more targets and effectively apply jamming power; and
- The jammers could operate on-the-move—they need not stop to operate.

References

[1] FM 3-06 (Army Field Manual 3-06), URBAN OPERATIONS, HQ Department of the Army, 26 October 2006.

[2] Army FM 3-06, Section 6-13. Isolation, https://rdl.train.army.mil/soldierPortal/atia/adlsc/view/public/11645-1/fm/3-06/chap6.htm.

[3] Army FM 3-06, Section 6.11.

[4] *Information Warfare in Urban Combat*, Defense Update, International Online Defense Magazine, Vol. 1, 2006, http://www.defense-update.com/features/du-1-06/urban-c4i-3.htm.

[5] Poisel, R. A., *Introduction to Communication Electronic Warfare Systems*, 2nd ed., Norwood, MA: Artech House, 2008, Chapters 10 and 11.

Chapter 18

Robust Blind Detection and Geolocation of CDMA Signals in an Urban Environment

18.1 Introduction

Spread spectrum signals have been used for secure communications for several decades [1]. Currently, they are also widely used outside the military domain, especially in CDMA systems [2, 3]. Due to their low probability of interception, these signals increase the difficulty of spectrum surveillance, such as might be required by the FCC or law enforcement agencies. The presence of a CDMA transmission is very difficult to detect. In this chapter, we describe an approach to the blind detection and geolocation of CDMA signals far below the noise level. It is based on analysis of the variations of second order statistics estimators of the correlator output with and without a signal present. First a threshold is determined below which the fluctuations in the variance with noise only present remains. The variations in the second-order statistics of signals received at an intercept site are then determined. If there is a CDMA signal hidden within the noise, and the signal employs a short spreading code;[1] these variations rise above the threshold at multiples of the symbol time (1/rate).

The organization of this chapter is as follows. We begin with a presentation of the problem environment we are trying to solve. Next we describe the mathematical model we will use. That is followed by a brief review of the technique used in antenna arrays called *beamforming*, and in particular, extensions to the standard beamformer model to improve its resolution. Next, the actual technique for detection and geolocation of CDMA signals is described. A particular problem in the process we describe is what we call spatial cochannel interference, and we discuss that as well, bounding its effects. Finally we describe

---

[1] Recall that a short spreading code is a code that is repeated one or more times in every symbol.

how the DSSS detection algorithm described in Chapter 10 can be used to assist in the solution to the problems here.

18.2 CDMA Signals

When a DSSS signal is hidden in the noise, the actual fluctuations go above the noise-only threshold. Not only does the method provide a detection of a DSSS signal hidden in noise, but it also provides a precise estimation of the duration of the pseudo-random sequence used by the transmitter.

The performance of the method increases with the number of windows (which itself increases with the duration of the signal used for computing statistics), and also with the length of the pseudo-random sequence used by the transmitter. For example, for pseudo-random sequences of length 127, and 400 analysis windows, the predetection limit is approximately –12 dB, and the computation time is only a few seconds on a PC.

The problem to be solved is to detect and locate CDMA target emitters in an urban environment. The approach proposed is to first divide the *area of interest* (AOI) into appropriately sized cells as illustrated in Figure 18.1. Beams are then formed at two or more intercept sites, and these beams are steered toward each of these cells in turn, the current cell being examined called the *cell of interest* (COI). (Although squares are shown, palpably the shapes of the AOI and COI can be anything.) With CDMA, the AOI will contain many potential targets and they all can occupy the same frequency spectrum.[2] The beamforming process eliminates the extra CDMA targets, which are now interferers, for all directions except the direction of the COI. Using an algorithmic process to detect DSSS signals, if there is a target within the COI, the detection process will determine this, even if the CDMA signal is well below the noise floor. Last, if the target set is using short spreading codes, as explained in Chapter 9, it is possible to blindly determine what the spreading code is. This process is continued until all the cells in the AOI are examined, and then the process is started over. Note that the cells need not be equal.

CDMA signals share the characteristic that they overlap in the same frequency band. Each target signal uses a unique code to decorrelate all of the signals in a frequency channel. The decorrelation process collapses the signal that was transmitted with that unique code to a narrowband, baseband signal that can then be demodulated. The other CDMA signals with different codes do not properly decorrelate, and remain noiselike after the decorrelator.

---

[2] Every cell containing every available channel is not a likely scenario for economic reasons. It is more likely that a cell would contain two or three channels. In the case of IS 95, for example, each channel is 1.25 MHz wide.

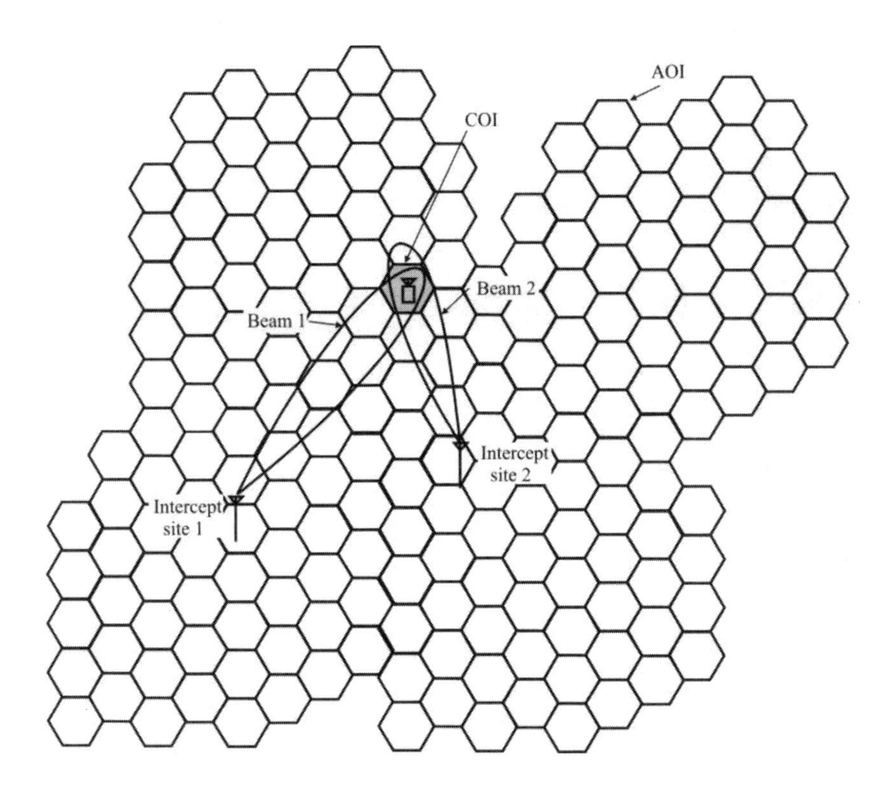

Figure 18.1 An area of interest (AOI) divided into cells of interest (COI).

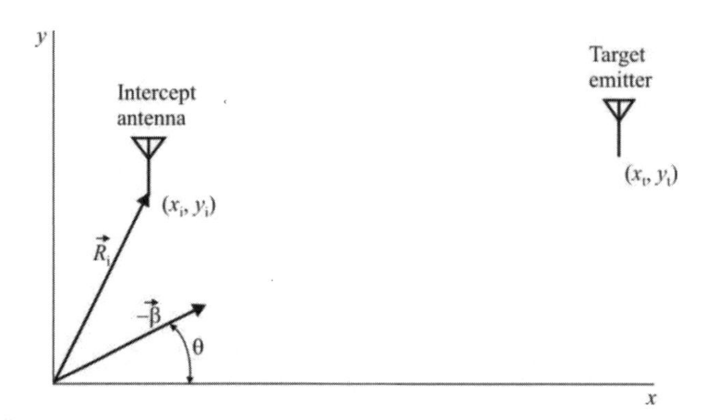

Figure 18.2 Two-dimensional array geometry.

Conventional direction-finding techniques, such as phase interferometry, cannot properly process frequency channels with cochannel interference, which is essentially what a set of CDMA signals is to such techniques.

The thrust of this method of detection and geolocation is to employ spatial filters to separate the target nodes, rather than time filtering or frequency filtering. There are many techniques that can be applied to the problem of spatial filtering. We will discuss a few.

18.3 Parametric Data Model

In this section we describe the model used in the remainder of this chapter.

18.3.1 Sensor Array

A sensor is represented as a point receiver at given spatial coordinates. In the 2-D case and as shown in Figure 18.2, we have $\vec{R}_l = (x_l \quad y_l)^{\mathrm{T}}$. The field measured at sensor l and due to a source at azimuthal DOA θ is given by the solution to the wave equation [4] and is

$$E(\vec{R}_l, t) = s(t)e^{j\omega t - \beta(x_l \cos\theta + y_l \sin\theta)} \tag{18.1}$$

where $\vec{\beta} = \vec{\alpha}\omega$ is the *wave vector*. The magnitude of $\vec{\beta}$, $|\vec{\beta}| = \beta = \omega / c = 2\pi / \lambda$ is called the *wave number*.

If a flat frequency response, say, $g_l(\theta)$, is assumed for the sensor l over the signal bandwidth, its measured output will be proportional to the field at $\vec{R}_l$. Dropping the carrier term $e^{j\omega t}$ for convenience (in practice, the signal is usually downconverted to a low IF or baseband before sampling), the output is modeled by

$$r_l(t) = g_l(\theta)e^{-j\beta(x_l \cos\theta + y_l \sin\theta)} = a_l(\theta)E(\vec{R}_l, t) \tag{18.2}$$

We see that (18.2) requires that the array aperture (i.e. the physical size measured in wavelengths) be much less than the inverse relative bandwidth (f/W). In the array processing literature, this is referred to as the *narrowband assumption*. For an L-element antenna array of arbitrary geometry, the array output vector is obtained as

$$\vec{r}(t) = \vec{a}(\theta)E(\vec{R}_l, t) \tag{18.3}$$

A single signal at the DOA θ thus results in a scalar multiple of the *steering vector*

$$\vec{a}(\theta) = \begin{bmatrix} a_1(\theta) & a_2(\theta) & \cdots & a_L(\theta) \end{bmatrix}^{\mathrm{T}} \tag{18.4}$$

as the array output. Common array geometries are depicted in Figures 18.3 and 18.4. For the *uniform linear array* (ULA) we have $\vec{r}_l = \begin{bmatrix} (l-1)d & 0 \end{bmatrix}^{\mathrm{T}}$, $l = 1, 2, \ldots, L$, and assuming that all elements have the same directivity $g_1(\theta) = g_2(\theta) = \ldots = g_L(\theta) = g(\theta)$, the ULA steering vector takes the form

$$\vec{a}_{\mathrm{ULA}}(\theta) = g(\theta)\begin{bmatrix} 1 & e^{j\beta d\cos\theta} & \ldots & e^{j(L-1)\beta d\cos\theta} \end{bmatrix}^{\mathrm{T}} \tag{18.5}$$

where d denotes the interelement distance. The radius vectors of the *uniform circular array* (UCA) have the form

$$\vec{R}_l = R\begin{bmatrix} \cos[2\pi(l-1)/L] & \sin[2\pi(l-1)/L] \end{bmatrix}^{\mathrm{T}} \tag{18.6}$$

from which the form of the UCA steering vector can easily be derived.

As previously alluded to, a signal source can be associated with a number of characteristic parameters. For the sake of clarity and ease of presentation, refer to Figures 18.3 and 18.4; we assume that θ is a real-valued scalar referred to as the

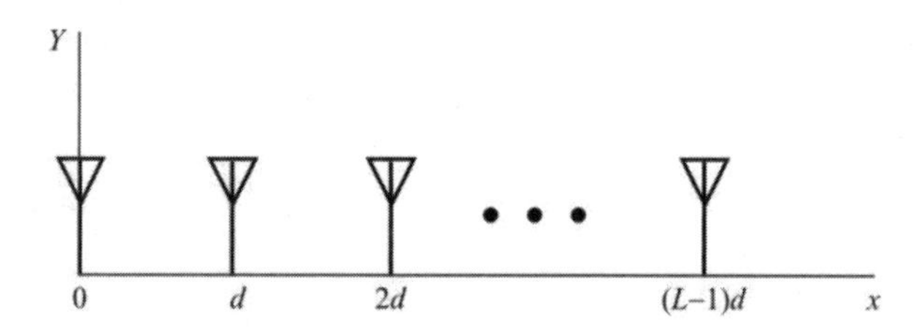

Figure 18.3 Uniform linear array geometry.

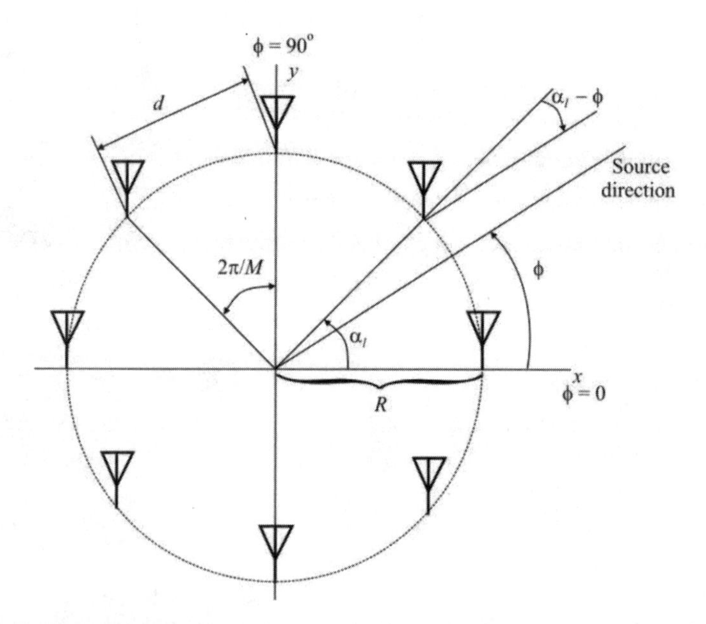

Figure 18.4 Uniform circular array geometry.

direction of arrival (DOA). The extension to the multiple parameters per source case is straightforward.

The superposition principle is applicable assuming a linear receiving system. If M signals, denoted here by $s_m(t)$, impinge on an L-dimensional array from distinct DOAs $\theta_1, ..., \theta_M$, the output vector takes the form

$$\vec{r}(t) = \sum_{m=1}^{M} \vec{a}(\theta_m) s_m(t) \tag{18.7}$$

The output equation can be put in a more compact form by defining a steering matrix and a vector of signal waveforms as

$$\mathbf{A}(\theta) = \left[\vec{a}(\theta_1) \quad \cdots \quad \vec{a}(\theta_m)\right] \qquad (L \times M) \tag{18.8}$$

$$\vec{s}(t) = \left[s_1(t) \quad \cdots \quad s_m(t)\right]^{\mathrm{T}} \tag{18.9}$$

In the presence of an additive noise $\vec{n}(t)$ we now get the model commonly used in array processing

$$\vec{r}(t) = \mathbf{A}(\theta)\vec{s}(t) + \vec{n}(t) \tag{18.10}$$

The methods to be presented all require that $M < L$, which is therefore assumed throughout this chapter. It is interesting to note that in the noiseless case, the array output is then confined to an M-dimensional subspace of the complex L-space, which is spanned by the steering vectors. This is the signal subspace, and this observation forms the basis of subspace-based methods, of which we will present one.

The sensor outputs are appropriately preprocessed and sampled at arbitrary time instances, labeled $t = 1, 2, ..., N$ for simplicity. Clearly, the process $\vec{r}(t)$ can be viewed as a multichannel random process, whose characteristics can be well understood from its first and second order statistics determined by the underlying signals and noise. The preprocessing of the signal is often done in such a way that $\vec{r}(t)$ can be regarded as temporally white.

18.3.2 Spatial Covariance Matrix

The signal parameters that are of interest in this chapter are spatial in nature, and thus require the crosscovariance information among the various sensors, that is the *spatial covariance matrix*

$$\mathbf{R} = \mathcal{E}\{\vec{r}(t)\vec{r}(t)^{\mathrm{H}}\} = \mathbf{A}\mathcal{E}\{\vec{s}(t)\vec{s}(t)^{\mathrm{H}}\}\mathbf{A}^{\mathrm{H}} + \mathcal{E}\{\vec{n}(t)\vec{n}(t)^{\mathrm{H}}\} \tag{18.11}$$

The source covariance matrix is given by

$$\mathcal{E}\{\vec{s}(t)\vec{s}(t)^{\mathrm{H}}\} = \mathbf{S} \tag{18.12}$$

And the noise covariance matrix is

$$\mathcal{E}\{\vec{n}(t)\vec{n}(t)^{\mathrm{H}}\} = \sigma_{\mathrm{n}}^2\mathbf{I} \tag{18.13}$$

The latter covariance structure is a reflection of the noise having a common variance σ_{n}^2 at all sensors and being uncorrelated among all sensors. Such noise is usually termed *spatially white,* and is a reasonable model for receiver noise. However, other MMN sources need not, and probably won't if present, result in spatial whiteness, in which case the noise may need to be prewhitened. More specifically, if the noise covariance matrix is **Q**, the sensor outputs are multiplied by $\mathbf{Q}^{-1/2}$ prior to further processing.[3] The source covariance matrix, **S,** is often assumed to be nonsingular (a rank-deficient **S**, as in the case of coherent signals, is discussed later) or near-singular for highly correlated signals.

The spectral factorization of **R** is given

$$\mathbf{R} = \mathbf{ASA}^{\mathrm{H}} + \sigma_{\mathrm{n}}^2\mathbf{I} = \mathbf{U\Sigma U}^{\mathrm{H}} \tag{18.14}$$

$\mathbf{\Sigma} = \mathrm{diag}\begin{bmatrix}\lambda_1 & \cdots & \lambda_L\end{bmatrix}$ a diagonal matrix of real eigenvalues ordered such that $\lambda_1 \geq \lambda_2 \geq \ldots \geq \lambda_L > 0$. Observe that any vector orthogonal to **A** is an eigenvector of **R** with the eigenvalue σ_{n}^2; there are $L - M$ linearly independent vectors and they make up the noise subspace of **R**. The remaining eigenvalues are all larger than σ_{n}^2, and the corresponding eigenvectors span the signal subspace. Thus we can partition the eigenvalue/eigenvector pairs into signal eigenvectors (corresponding to eigenvalues $\lambda_1 \geq \cdots \geq \lambda_M > \sigma_{\mathrm{n}}^2$), and noise eigenvectors (corresponding to eigenvalues $\lambda_{M+1} = \cdots = \lambda_L = \sigma_{\mathrm{n}}^2$) and (18.14) can be factored accordingly

$$\mathbf{R} = \begin{bmatrix}\mathbf{U}_{\mathrm{s}} & \mathbf{U}_{\mathrm{n}}\end{bmatrix}\begin{bmatrix}\mathbf{\Sigma}_{\mathrm{s}} & \mathbf{0} \\ \mathbf{0} & \mathbf{\Sigma}_{\mathrm{n}}\end{bmatrix}\begin{bmatrix}\mathbf{U}_{\mathrm{s}}^{\mathrm{H}} \\ \mathbf{U}_{\mathrm{n}}^{\mathrm{H}}\end{bmatrix} \tag{18.15}$$

---

[3] $\mathbf{Q}^{-1/2}$ denotes a Hermitian square-root factor of $\mathbf{Q}^{-1}$.

where $\mathbf{\Sigma}_n = \sigma_n^2\mathbf{I}$. This is the so-called EVD of **R**. Since all noise eigenvectors are orthogonal to **A**, the columns of $\mathbf{U}_s$ must span the range space of **A** whereas those of $\mathbf{U}_n$ span its orthogonal complement (the null space of $\mathbf{A}^H$).

18.3.3 Sample Covariance Matrix

The problem of central interest herein is that of estimating the DOAs of target signals impinging on a receiving array, when given a finite data set $\{\vec{x}(t)\}$ observed over t = 1, 2, ..., N. We will primarily focus on reviewing a few techniques based on second-order statistics. All of the earlier formulation assumed the existence of exact quantities (i.e. infinite observation time). It is clear that in practice only sample estimates, which we denote by a hat (i.e., $\hat{\cdot}$) are available. We will estimate **R** with the corresponding *sample covariance matrix*

$$\hat{\mathbf{R}} = \frac{1}{N}\sum_{t=1}^{N} \vec{r}(t)\vec{r}(t)^H \tag{18.16}$$

when there are N frames of data collected. The spectral representation similar to that of **R** is defined as

$$\hat{\mathbf{R}} = \begin{bmatrix} \hat{\mathbf{U}}_s & \hat{\mathbf{U}}_n \end{bmatrix} \begin{bmatrix} \hat{\mathbf{\Sigma}}_s & \mathbf{0} \\ \mathbf{0} & \hat{\mathbf{\Sigma}}_n \end{bmatrix} \begin{bmatrix} \hat{\mathbf{U}}_s^H \\ \hat{\mathbf{U}}_n^H \end{bmatrix} \tag{18.17}$$

Throughout this chapter we assme that we know the number of signals, M, impinging on the antenna array. This is generally a fairly safe assumption since there are good and consistent techniques for estimating M in the event that such information is not available.

18.4 Beamforming

Many approaches have been developed to "steer" the beam of an antenna array in particular directions. This steering is usually implemented to avoid interference from emitters that are not of interest to the receiving system. We will discuss a few of the techniques here, but this is certainly not an exhaustive discussion.

18.4.1 Conventional Delay and Multiply Beamformer

Perhaps the first was the straightforward, standard delay and multiply beamforming (called the conventional or standard beamformer) illustrated in

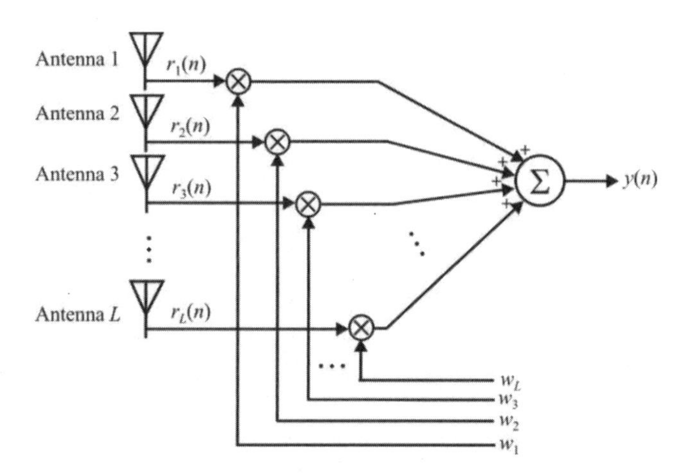

Figure 18.5 Delay and multiply beamformer. The weights, w_i, can be either preset or adaptively determined.

Figure 18.5. In this technique the signals from an array of antennas are multiplied with complex weights. Because the weights are complex, such multiplication modifies both the amplitude as well as the phase of the received signal. The most basic delay and multiply beamformer actually only varies the phase while keeping the amplitude common. The weights can be fixed, as in the case of a fixed beamformer, where the direction of the beam to be steered is known, or they can be changed to steer the beam in some desired direction. For this beamformer, however, the weights are not adaptive. For an array of L antennas, this beamformer will place nulls in the direction of $L - 1$ interfering signals.

Let

M = number of interfering signals impinging on the array;
L = number of antenna elements in the array;
P = number of calibrated angles.

When the signal $s(t)$ is impinging on the array from direction θ_0, the response of the array is given by the product of the signal and the array response to the signal, namely,

$$\vec{r}(t) = \vec{a}(\theta_0)s(t) + \vec{n}(t) \tag{18.18}$$

$\vec{a}(\theta_0)$ [also denoted by $\vec{a}_0(\theta)$] is known as the steering vector corresponding to θ_0, and $\mathbf{A} = [\vec{a}_0, \vec{a}_1, \cdots, \vec{a}_{P-1}]$, the columns of which are the steering vectors corresponding to $\{\theta_i\}_{i=0}^{P-1}$, is called the *array manifold*. The array manifold, therefore, is a calibration matrix for the array. Herein, without loss of generality, we will assume that θ_0 corresponds to the *signal of interest* (SOI) and $\{\theta_i\}_{i=1}^{M}$ corresponds to interfering signals.

The output of the complex multipliers, one for each antenna, are summed to form $y(t)$, to wit,

$$y(t) = \sum_{i=1}^{L} w_i^{\mathrm{H}} r(t) \tag{18.19}$$

The average power associated with $y(t)$ is therefore given by

$$P(\vec{w}) = \frac{1}{N}\sum_{i=1}^{N} |y(t)|^2 = \frac{1}{N}\sum_{i=1}^{N} \vec{w}^{\mathrm{H}} \vec{r}(t)\vec{r}^{\mathrm{H}}(t)\vec{w} = \vec{w}^{\mathrm{H}} \hat{\mathbf{R}} \vec{w} \tag{18.20}$$

where the covariance matrix is

$$\mathbf{R} = \mathbf{R}_{\mathrm{s}} + \mathbf{R}_{\mathrm{i+n}} \tag{18.21}$$

and where $\mathbf{R}_{\mathrm{s}}$ corresponds to the signal covariance matrix and $\mathbf{R}_{\mathrm{i+n}}$ corresponds to the covariance matrix of the interference and noise, is estimated with (18.16).

Using LaGrange multipliers, it can be shown that the optimum weights for the conventional beamformer are given by

$$\vec{w}_{\mathrm{CBF}} = \frac{\vec{a}(\theta_0)}{\sqrt{\vec{a}^{\mathrm{H}}(\theta_0)\vec{a}(\theta_0)}} \tag{18.22}$$

and the spatial spectrum is given by

$$P_{\mathrm{CBF}}(\theta_0) = \frac{\vec{a}^{\mathrm{H}}(\theta_0)\hat{\mathbf{R}}\vec{a}(\theta_0)}{\vec{a}^{\mathrm{H}}(\theta_0)\vec{a}(\theta_0)} \tag{18.23}$$

Unfortunately the conventional beamformer suffers from the problem of low resolution. Figure 18.6 shows the effects, where two signals are arriving with a separation of 10° when the resolution of the array is 12°. The signals cannot be discerned.

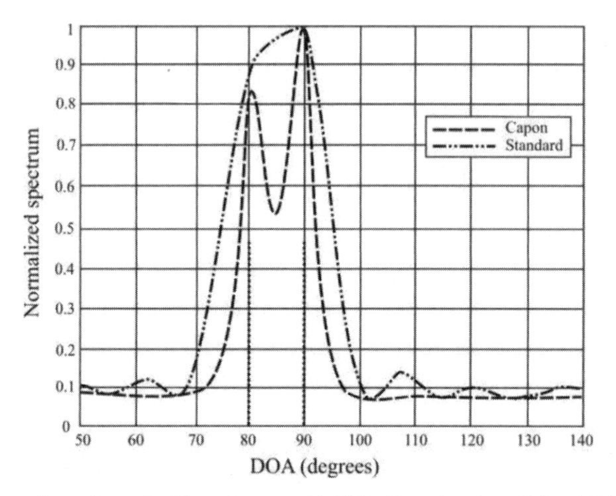

Figure 18.6 Comparison of standard beamformer with SCB. Two signals at 80° and 90°.

18.4.2 Standard Capon Beamformer

The *standard Capon beamformer* (SCB) is an adaptation of the delay and multiply beamformer. A block diagram of a general adaptive beamformer is shown in Figure 18.7. This adaptive beamformer as shown requires knowing the signal of interest (or an estimate of it, which is fairly common in EW applications to include the one discussed in this chapter), designated by $d(k)$. The output of the beamformer in the kth time interval, $y(k)$ can be expressed as

$$y(k) = \vec{w}^{\mathrm{H}}(k)\vec{r}(k) \tag{18.24}$$

The goal of the adaptation processor is to find the optimum weight vector $\vec{w}$ for some optimality criteria. For example, in the case of the Weiner filter, the block diagram that is the same as the one shown in Figure 18.7, the optimality criterion is the minimization of the MSE in the error signal, $e(k)$, given by

$$e(k) = y(k) - d(k) \tag{18.25}$$

In the case of the SCB, the optimality criterion is specified as minimizing the total output power while maintaining a finite, nonzero (unity) gain in the direction of the SOI:

$$\min_{\vec{w}} \vec{w}^{\mathrm{H}}\mathbf{R}\vec{w} \qquad \text{subject to} \qquad \vec{w}^{\mathrm{H}}\vec{a}(\theta_0) = 1 \tag{18.26}$$

when the covariance matrix $\mathbf{R}$ is expressed as

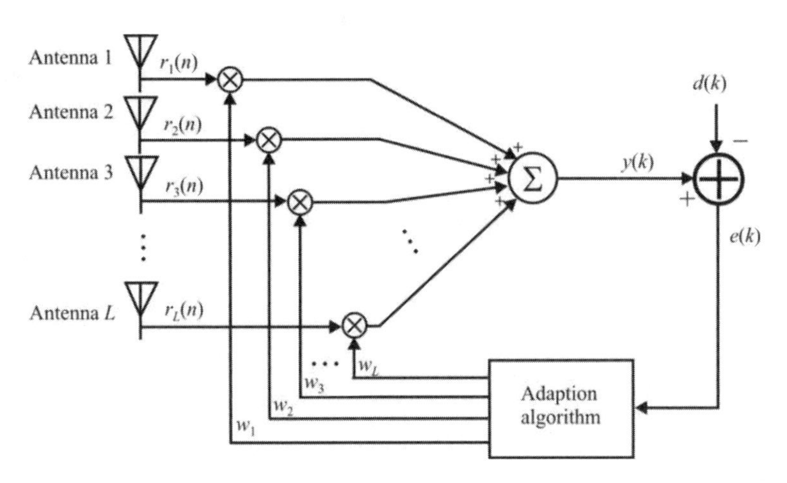

Figure 18.7 Generic adaptive beamformer. The expected signal $d(k)$ must be known or estimated.

$$\mathbf{R} = \sigma_0^2 \vec{a}(\theta_0)\vec{a}(\theta_0)^{\mathrm{H}} + \sum_{m=1}^{M} \sigma_m^2 \vec{a}(\theta_m)\vec{a}(\theta_m)^{\mathrm{H}} + \mathbf{Q} = \mathbf{R}_{\mathrm{s}} + \mathbf{R}_{\mathrm{i+n}} \tag{18.27}$$

The set $\left\{\sigma_0^2, \{\sigma_m^2\}_{m=1}^{M}\right\}$ are the variances (powers) of the SOI, whose power is σ_0^2 and angle of arrival is θ_0, and the M interferers whose powers are given by σ_m^2 and whose angles of arrival are θ_m. In (18.26) and (18.27), $\vec{a}(\theta_m), m = 0,1,\ldots,M$, are the array steering vectors. The total output power is given by $\vec{w}^{\mathrm{H}}\hat{\mathbf{R}}\vec{w}$. $\mathbf{Q}$ is the noise contribution given by

$$\mathbf{Q} = \sigma_{\mathrm{n}}^2 \mathbf{I} \tag{18.28}$$

so that

$$\mathbf{R}_{\mathrm{s}} = \sigma_0^2 \vec{a}(\theta_0)\vec{a}(\theta_0)^{\mathrm{H}} \tag{18.29}$$

and

$$\mathbf{R}_{\mathrm{i+n}} = \sum_{m=1}^{M} \sigma_m^2 \vec{a}(\theta_m)\vec{a}(\theta_m)^{\mathrm{H}} + \mathbf{Q} \tag{18.30}$$

The weight vector can be found by maximizing the array output signal-to-interference-plus-noise ratio (SINR)

$$\text{SINR} = \sigma_s^2 \frac{\left|\vec{w}^{\text{H}}\vec{a}\right|^2}{\vec{w}^{\text{H}}\mathbf{R}_{\text{i+n}}\vec{w}} \tag{18.31}$$

where

$$\mathbf{R}_{\text{i+n}} = \mathcal{E}\left\{[\vec{i}(k)+\vec{n}(k)][\vec{i}(k)+\vec{n}(k)]^{\text{H}}\right\} \tag{18.32}$$

is the $M \times M$ interference-plus-noise covariance matrix, and $\sigma_s^2 = \sigma_0^2$ is the signal power.

From (18.31) it is clear that to maximize the output SINR, the response to the interference and noise, given by $\vec{w}^{\text{H}}\mathbf{R}_{\text{i+n}}\vec{w}$, should be minimized. The solution for the weight vector is found by maintaining a distortionless response toward the desired signal while minimizing the output interference-plus-noise power. Hence, the maximization of (18.31) is equivalent to (18.26).

From (18.26), the following well-known solution can be found for the optimal weight vector:

$$\vec{w}_{\text{opt}} = \alpha\mathbf{R}_{\text{i+n}}^{-1}\vec{a} \tag{18.33}$$

where $\alpha = (\vec{a}^{\text{H}}\mathbf{R}^{-1}\vec{a})^{-1}$ is a normalization constant that does not affect the output SINR (18.31). The solution to (18.33) is commonly referred to as the *minimum variance distortionless response* (MVDR) beamformer, the *sample matrix inversion* (SMI) beamformer, or the *standard Capon beamformer* (SCB).

The resolving power of the SCB is illustrated in Figure 18.6, in comparison to the conventional delay and multiply beamformer. In this case, the two signals are clearly resolved.

The SCB attempts to steer a beam in the direction of the SOI with an array gain of g = 1, while simultaneously steering nulls in the directions of the M interferers.

Generally the covariance matrix, $\mathbf{R}$, is not available and must be estimated. The most common way to do this is with the observed data sequences themselves as

$$\hat{\mathbf{R}} = \frac{1}{N}\sum_{k=1}^{N}\vec{y}(k)\vec{y}^{\text{H}}(k) \tag{18.34}$$

when the observed data vectors are represented by $\vec{y}(k)$. Errors occur, sometimes significant errors, if the SOI is present in the data when this estimate for $\mathbf{R}$ is

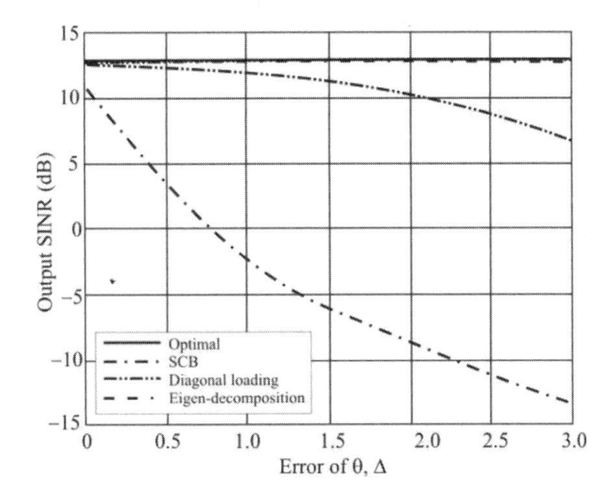

Figure 18.8 Output SNIR comparison of beamforming methods. (υ = 0 dB, INR = 20 dB, K = 1 at –30°, L = 10, N = 50.) (Source: [5]. © EMW Publishing. Reprinted with permission.)

computed. Also, when N is small compared with the number of antenna elements, the estimate for **R** can be in error as well.

18.4.3 Robust Capon Beamformer

There are many other possible beamformer algorithms, including several versions of the *robust Capon beamformer* (RCB). It is known, for example, that the biggest problem with the SCB technique is its supersensitivity to mismatches in the estimated *angle of arrival* (AOA), θ_0, and the closest calibrated $\vec{a}(\theta)$. Differences of only a few degrees can drastically impact its performance as the SOI in this case is treated as interference. Various approaches have been devised, however, to deal with this supersensitivity problem; we will discuss a few of them here.

18.4.3.1 Diagonal Loading

The technique called *diagonal loading* (DL) adds weight to the diagonal elements of the covariance matrix (18.27) producing

$$\min_{\vec{w}} \vec{w}^{\mathrm{H}}(\mathbf{R}+\zeta\mathbf{I})\vec{w} \qquad \text{subject to} \qquad \vec{w}^{\mathrm{H}}\vec{a}(\theta_0)=1 \qquad (18.35)$$

DL can significantly reduce this sensitivity. See Figure 18.8, for example, which shows a few of the applicable techniques. This chart shows the SINR as a function of the steering vector error given by Δ, that is, $\vec{a}(\theta)=\vec{a}(\theta+\Delta)$. The SINR for the

SCB falls off dramatically as the error increases, while the RCB with DL is fairly robust.

The optimum weights for maximizing the array output SNIR are given by a simple adaptation of (18.33) as

$$\vec{w}_{\text{opt}} = \frac{(\hat{\mathbf{R}} + \zeta \mathbf{I})^{-1} \vec{a}}{\vec{a}^{\text{H}} (\hat{\mathbf{R}} + \zeta \mathbf{I})^{-1} \vec{a}} \tag{18.36}$$

The most significant issue with DL is selection of the loading factor ζ. Early approaches for using the DL technique for beamforming used values of ζ that were more or less ad hoc ($\zeta = 10\sigma_{\text{n}}^2$ would be typical, where σ_{n}^2 is the noise power). Only recently have approaches been found for associating the value of ζ with performance requirements.

When the loading parameter, ζ, is fixed, it is referred to as a *fixed diagonal load* (FDL). We describe next a method where the loading parameter is free to vary. It is called a *general linear combination* (GLC)–based robust beamformer.

18.4.4 General Linear Combination–Based Robust Capon Beamforming

As mentioned, the sample covariance matrix $\hat{\mathbf{R}}$ can be a poor estimate of the true covariance matrix $\mathbf{R}$ when N is small, relative to the array dimension M. An improved estimate of R is provided by the GLC-based covariance matrix estimator [6]. This estimator uses the sample covariance matrix $\hat{\mathbf{R}}$ and the identity matrix $\mathbf{I}$ to obtain a more accurate estimate of $\mathbf{R}$ than $\hat{\mathbf{R}}$, and is given by

$$\tilde{\mathbf{R}} = \alpha \mathbf{I} + \beta \hat{\mathbf{R}} \tag{18.37}$$

where $\tilde{\mathbf{R}}$, which should be positive semidefinite ($\tilde{\mathbf{R}} \geq 0$), is the enhanced estimate of $\mathbf{R}$. α and β in (18.37) are chosen by minimizing (an estimate of) the MSE of $\tilde{\mathbf{R}}$, where $\text{MSE}\{\tilde{\mathbf{R}}\} = \mathcal{E}\{\|\tilde{\mathbf{R}} - \mathbf{R}\|\}$. α and β can be constrained such that $\alpha \geq 0$ and $\beta \geq 0$ so that $\tilde{\mathbf{R}} \geq 0$ can be guaranteed.

The values of α and β that minimize an estimate of the MSE of $\tilde{\mathbf{R}}$ are computed in [6] with the results being that

$$\hat{\alpha}_0 = \min \left[\hat{v} \frac{\hat{\rho}}{\left\| \hat{\mathbf{R}} - \hat{v} \mathbf{I} \right\|^2}, \hat{v} \right] \tag{18.38}$$

and

$$\hat{\beta}_0 = 1 - \frac{\hat{\alpha}_0}{\hat{\nu}} \tag{18.39}$$

where

$$\hat{\rho} = \frac{1}{N^2} \sum_{k=1}^{N} \left\| \vec{y}(k) \right\|^4 - \frac{1}{N} \left\| \hat{\mathbf{R}} \right\|^2 \tag{18.40}$$

and

$$\hat{\nu} = \frac{\mathrm{tr}\{\hat{\mathbf{R}}\}}{M} \tag{18.41}$$

with the observed data snapshots given by $\{\vec{y}(k)\}_{k=1}^{N}$.

18.4.4.1 The Beamformer

Based on (18.37)–(18.41) we have a diagonally-loaded enhanced estimate of the covariance matrix

$$\tilde{\mathbf{R}}_{\mathrm{GLC}} = \hat{\alpha}_0 \mathbf{I} + \hat{\beta}_0 \hat{\mathbf{R}} \tag{18.42}$$

Using the enhanced estimate $\tilde{\mathbf{R}}_{\mathrm{GLC}}$ in lieu of $\hat{\mathbf{R}}$ in the SCB formulation yields the following GLC-based robust adaptive beamformer

$$\vec{w}_{\mathrm{GLC}} = \frac{\tilde{\mathbf{R}}_{\mathrm{GLC}}^{-1} \vec{a}}{\vec{a}^{\mathrm{H}} \tilde{\mathbf{R}}_{\mathrm{GLC}}^{-1} \vec{a}} \tag{18.43}$$

Assuming that $\hat{\beta}_0 \neq 0$, the estimate of (18.43) can be written as

$$\tilde{\vec{w}}_{\mathrm{GLC}} = \frac{\left[\frac{\hat{\alpha}_0}{\hat{\beta}_0} \mathbf{I} + \hat{\mathbf{R}} \right]^{-1} \vec{a}}{\vec{a}^{\mathrm{H}} \left[\frac{\hat{\alpha}_0}{\hat{\beta}_0} \mathbf{I} + \hat{\mathbf{R}} \right]^{-1} \vec{a}} \tag{18.44}$$

The output SINR of the resulting beamformer is given by

$$\text{SINR} = \frac{\sigma_0^2 \left| \vec{w}_{\text{GLC}}^{\text{H}} \vec{a} \right|^2}{\vec{w}_{\text{GLC}}^{\text{H}} \mathbf{R}_{\text{n+i}} \vec{w}_{\text{GLC}}} \tag{18.45}$$

and the SOI power estimate is

$$\hat{\sigma}_0^2 = \vec{w}_{\text{GLC}}^{\text{H}} \tilde{\mathbf{R}}_{\text{GLC}} \vec{w}_{\text{GLC}} = \frac{1}{\vec{a}^{\text{H}} \tilde{\mathbf{R}}_{\text{GLC}} \vec{a}} \tag{18.46}$$

Example [6]: We examine a performance comparison, as determined by the output SINR, as N increases for the SCB, FDL, and GLC. We also include the case when the true $\mathbf{R}$ is used (denoted by SINR). For the FDL we set the fixed DL level to equal the noise power, which is assumed known. We assume a ULA with $M = 10$ sensors and with $\lambda/2$ interelement spacing. The far-field narrowband source waveforms and the additive noise are assumed to be temporally white, circularly symmetric, complex Gaussian random processes with zero-mean and variances according to the SNRs. The noise is further assumed to be spatially white, and its covariance matrix is the identity matrix $\mathbf{I}$. Three sources with SNRs 10 dB, 20 dB, and 20 dB are assumed to be present at 0°, 20°, and 60°, respectively. We assume that there are no errors in the array steering vector. The SOI is the first (10 dB) signal and the other two sources are interferers. For each scenario 1,000 Monte Carlo trials were performed.

Figure 18.9 shows the mean of the output SINRs versus the number of snapshots N. As shown, SCB converges to the optimal SINR value very slowly, since $\hat{\mathbf{R}}$ contains the SOI and the SOI power is not small. GLC converges to the optimal value faster than all the other algorithms. FDL gives the second best performance.

18.4.4.2 Summary

In this section we have described a GLC-based robust adaptive beamformer, where the sample covariance matrix $\hat{\mathbf{R}}$, used in the SCB formulation, is replaced by an enhanced covariance matrix estimate $\tilde{\mathbf{R}}$ given by (18.37). GLC is a robust DL approach where the loading elements need not be fixed or the same. We can efficiently obtain the enhanced covariance matrix estimate completely automatically (i.e., without specifying any user parameter). Numerical examples demonstrated the performance of GLC compared with that of the SCB and FDL.

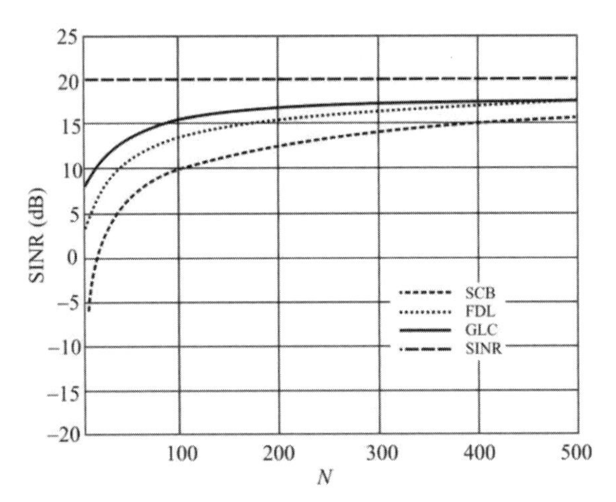

Figure 18.9 Beamformer output SINR versus snapshot number N in absence of array steering vector errors. (Source: [6]. © IEEE 2010. Reprinted with permission.)

18.4.5 Eigendecomposition Method

The most robust RCB method illustrated in Figure 18.8 is based on eigendecomposition of the covariance matrix. As before, there are two (mathematical) subspaces associated with **R**: the signal subspace and the noise subspace. The basis vectors of the signal subspace are orthogonal to the noise subspace and the basis vectors of the noise subspace are orthogonal to the signal subspace. The covariance matrix can be expressed as

$$\mathbf{R} = \mathbf{U}_s \boldsymbol{\Sigma}_s \mathbf{U}_s^H + \mathbf{U}_n \boldsymbol{\Sigma}_n \mathbf{U}_n^H \tag{18.47}$$

where $\mathbf{U}_s$ is a matrix representing the signal subspace and contains the steering vectors, $\vec{a}(\theta)$, corresponding to the AOAs of the signals; $\boldsymbol{\Sigma}_s$ is a diagonal matrix containing the eigenvalues of the signal subspace; $\mathbf{U}_n$ is a matrix that represents the noise subspace and contains the remaining steering vectors, at angles where no signals are arriving; and $\boldsymbol{\Sigma}_n$ is a diagonal matrix containing the eigenvalues of the noise subspace.

First we estimate the value of the covariance function. Then the eigendecomposition of $\hat{\mathbf{R}}$, given by (18.47), is found (there are standard software packages available, for example in MATLAB and MathCad, for this function). The angular spectrum, given by

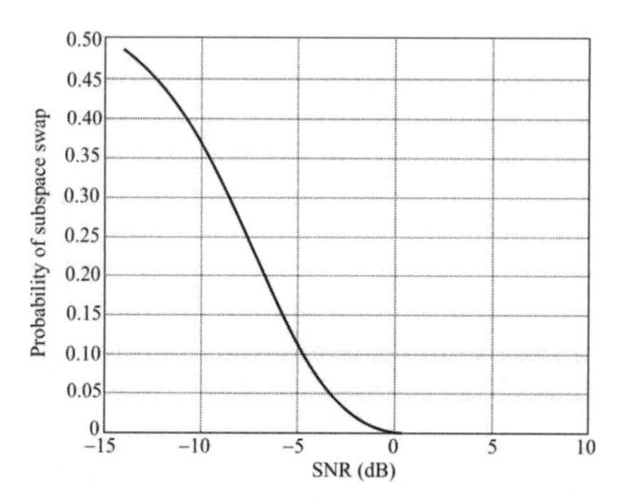

Figure 18.10 Probability of subspace swap. (Source: [6]. © IEEE 1995. Reprinted with permission.)

$$P(\theta) = \frac{1}{\left|\vec{a}(\theta)^{\mathrm{H}} \mathbf{U}_{\mathrm{n}}\right|^2} \tag{18.48}$$

is then calculated. Since the $\vec{a}(\theta)$ are orthogonal to $\mathbf{U}_{\mathrm{n}}$ (the noise subspace steering vectors) at the values of θ corresponding to the angels of arrival of the signals, $\{\theta_k\}$, then $\left|\vec{a}(\theta_k)^{\mathrm{H}} \mathbf{U}_{\mathrm{n}}\right| \rightarrow 0$ and $P(\theta_k) \rightarrow \infty$, or at least peaks at these values of θ. The example results of this method are also shown in Figure 18.8. There is essentially no degradation of SINR as the error is increased, at least for the parameters corresponding to this example.

A drawback of the eigendecomposition method is the requirement to frequently compute the eigendecomposition of $\hat{\mathbf{R}}$. Such an operation requires $O(L^3)$ floating point computations. Methods have been devised, however, to avoid these computations.

It has been observed that RCB eigendecomposition techniques for spatial filtering tend to perform poorly at low SNRs. This is due primarily to the probability of a *subspace swap*, that is, the switching of the noise subspace with the signal subspace, as the value of the signal eigenvalues approach the value of the noise eigenvalues. An analysis was performed by Thomas, Scharf, and Tufts [7] on the susceptibility of the approach to such swapping. Some of those results are shown in Figure 18.10. As can be seen, good performance is maintained down to around 0 dB SNR before significant problems occur.

G_{p} for WCDMA is about 22 dB, as is the processing gain for the other 3G+ standards. This is the maximum processing gain. In the worst case, if the postdetect SNR required is 10 dB (a reasonable goal) then the predecorrelation SNR that is required is –12 dB, which puts the probability of subspace swap at about 0.4. If the spreading code is unknown at the intercept receiver, then about 40% of the time there will be a subspace swap. With the *orthogonal variable spreading factor* (OVSF) coding in the standards, G_p can be as low as 6 dB (4 chips per symbol). With a coding gain of 6 dB, the predecorrelation SNR required is 4 dB, and the probability of subspace swap is low (essentially zero). If the code can be determined, as discussed in Chapter 9, then the postdecorrelation SNR is about 10 dB, and essentially no swapping will occur in either case.

18.5 CDMA Detection and Geolocation[4]

The principal issue with detection and geolocation of CDMA signals is their frequency sharing. CDMA targets (as opposed to DSSS, which is the same technology but different application) are designed to use the same frequency channel among many target radios. As mentioned, traditional DF techniques will not work in such cochannel situations. The beamforming techniques mentioned above (and others) are capable of processing cochannel signals, typically placing nulls in the direction of interferers and a finite gain in the direction of the SOI.

The approach considered here for CDMA signal detection and geolocation relies on the beamformer capability of separating signals in the spatial domain rather than the time or frequency domains. Multiple sensor arrays can be placed in the vicinity of the target AOI. The AOI is divided into cells of size determined by the resolution and accuracy of the sensor array. A 1 km x 1 km region (10^6 m^2), for example could be divided into 10×10 m cells (10^2 m^2), resulting in 10^4 cells. The sensors point beams to the cells (probably sequentially, but that is not necessary). After the nulls are established at each sensor, SOIs appearing at the look direction are determined and if any are detected, signal detection is declared. When two or more sensors detect the presence of the SOI emanating from the COI then a target location is declared at the COI. If 1 ms is required for the processing at each cell, then the entire AOI would be examined in 10 seconds.

Each cell is examined in this way. This processing will result in a map of the AOI indicating the locations (cells) of all of the resolvable SOIs.

One significant drawback of the SCB is the requirement to have a replica of the SOI with which the incoming signal can be compared. In addition, in theory the SOI must contain a training sequence for optimal weight determination. For the problem space being considered here, neither of these conditions applies. The lack of a training sequence can be handled by using (18.16) as an estimate of the autocorrelation matrix, which, in our case, contains noise as well as the SOI. Including the SOI degrades the optimality of the weights so obtained.

---

[4] Patent pending.

The requirement to have a replica of the SOI is not so easily solved. A question to be determined is how exact must this match be in order to compute acceptable weights. For our problem space, we know a considerable amount about the SOI but we do not know it exactly. We know its modulation, for example (BPSK), as well as its chip and data rates. What we don't know is the baseband data and the spreading code.

18.5.1 Spatial Cochannel Interference

As mentioned, CDMA targets are difficult to separate based on frequency because, by design, they overlap 100% in frequency. As depicted in Figure 18.11, SDMA does not totally eliminate cochannel problems, however. As shown, any target transmitter appearing within the beam will provide a certain amount of energy to the intercept site, and this energy will appear as though it is emanating from the target COI.

To take an example of the possibility of occurrence of such cochannel interference, we pose an example. Suppose the AOI is 1 km × 1 km, and the cells are 10 m × 10 m. There are 10^4 such cells in the AOI. If there is a maximum of 60 transmitters in the AOI , then assuming a uniform distribution throughout the AOI, the probability of any one cell containing a target of any variety is 6×10^{-3}, or 0.6%. The worst case for the interceptor is if it is located at one corner of the AOI and the COI is located at the diagonally opposite corner. In that case there are about 140 cells between the interceptor and the COI. The probability of having exactly n cells occupied is given by the binomial distribution

$$P(n) = \frac{N!}{n!(N-n)!} p^n (1-p)^{N-n} \tag{18.49}$$

where N is the total number of possibilities (in our case $N = 140$), and p is the a priori probability of a cell containing a target (in our case $p = 6 \times 10^{-3}$). This distribution is plotted in Figure 18.12 using these numbers. The probability of having no cells occupied is reasonably good, $P \sim 0.4$. The probability of having one, which could be the COI but we have no way of knowing that, is about 0.37.

The first possibility of having undetected spatial cochannel is when $n = 2$, and that probability is about 0.18. The remainder probabilities are as indicated in Figure 18.12. After about $n = 4$ or so, the probabilities are basically nil. Remember that these are worst-case probabilities. In actuality, in most cases they would be less.

This argument presupposes that the beams are only wide enough to cover one cell at range. This equates to a beam that is 0.6° wide for this example. The reasonably achievable beamwidth is maybe 5°. If cells are 37 m × 37 m, then the

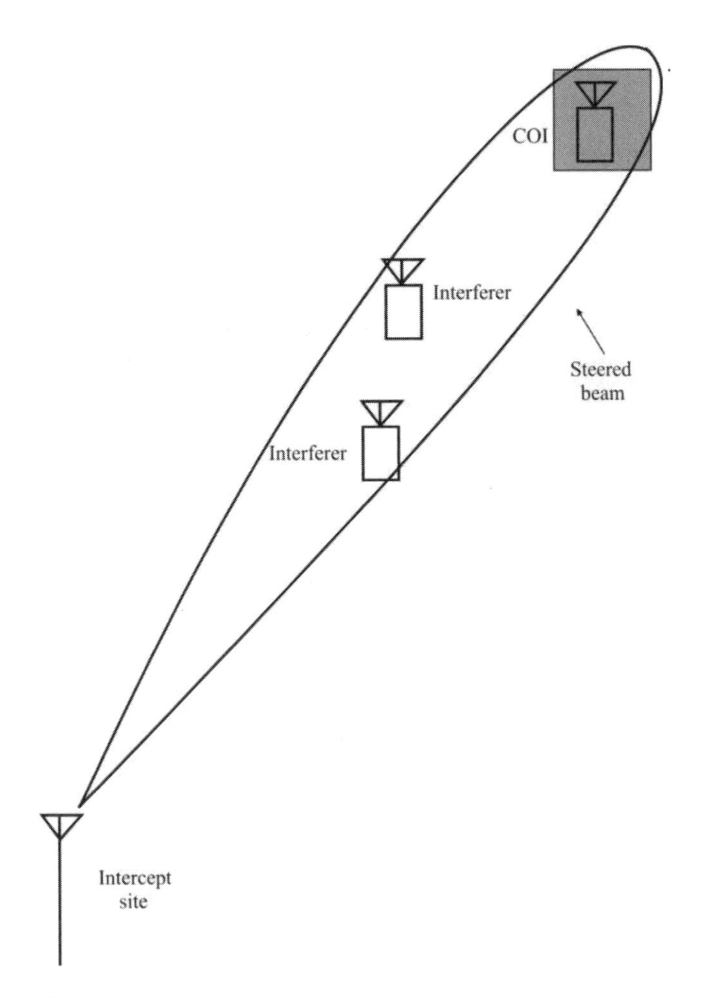

Figure 18.11 Spatial cochannel interference.

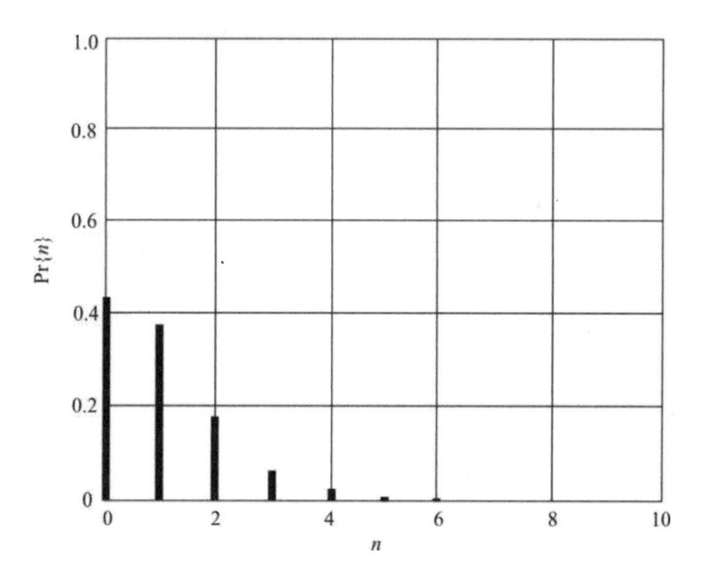

Figure 18.12 Binomial distribution when $N = 140, p = 6 \times 10^{-3}$.

required beamwidth is about 2° and the number of total cells is 841, 29 cells on a side, $N = 41$. These results are shown in Figure 18.13.

There are ways to deal with this spatial cochannel interference; we will discuss a few here.

- Detect the presence of the cochannel interference and subsequently ignore the detection event. The uplink data streams are, in all likelihood, not synchronous. As such, at the intercept site, which, by practical assumption cannot be synchronized with the transmitter, the detected bit stream timing will be askew and variable, depending on the particular geometry of the target and the interferers as well as their signal strength. Such processing could be used to identify multiple transmitters in the beamwidth, and the subsequent detection indication ignored. The probabilities of having cochannel interference are fairly low in the first place. Ignoring a detection and moving on to the next cell would not be required very often.

- Demodulate the detected signals at each intercept site. If clean demodulation of BPSK is possible, the probability is high that it is a good signal without cochannel interference.

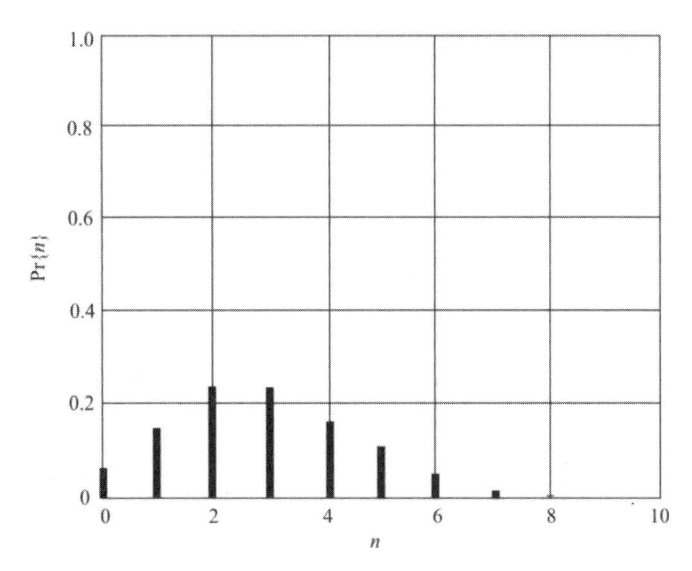

Figure 18.13 Binomial distribution when $N = 41, p = 0.07$.

- Demodulate the detected signals at each intercept site and exchange them among the sites or send them all to a central site. Such demodulation needs to be synchronized so that approximately the same bits are demodulated. At the central site, or at one or more of the intercept sites, if the signals correlate sufficiently then chances are very high that the intercept sites saw the same signal. In that case, the cochannel interference doesn't matter; it's as if it weren't there, and accurate detection can be declared.

The possible cochannel conditions that can occur are shown in Figures 18.14 and 18.15 (of course, there can be more than one target in each beam causing the interference). As these figures point out, there is only one condition that can occur that produces an incorrect decision. That is when there is cochannel interference in both beams but there is no legitimate target in the COI.

Note that this single condition in Figures 18.14 and 18.15 where a detection error occurs is potentially undetectable without further processing. The problem is when there is at least a single target in each beam and none in the COI [Figure18.15(c) with only one false target in the left beam]. In this case, examining the bit stream at each intercept site for timing anomalies will reveal none at either site. Therefore each intercept site will think that there is a target in the COI, although they are each looking at different targets, both incorrect. The

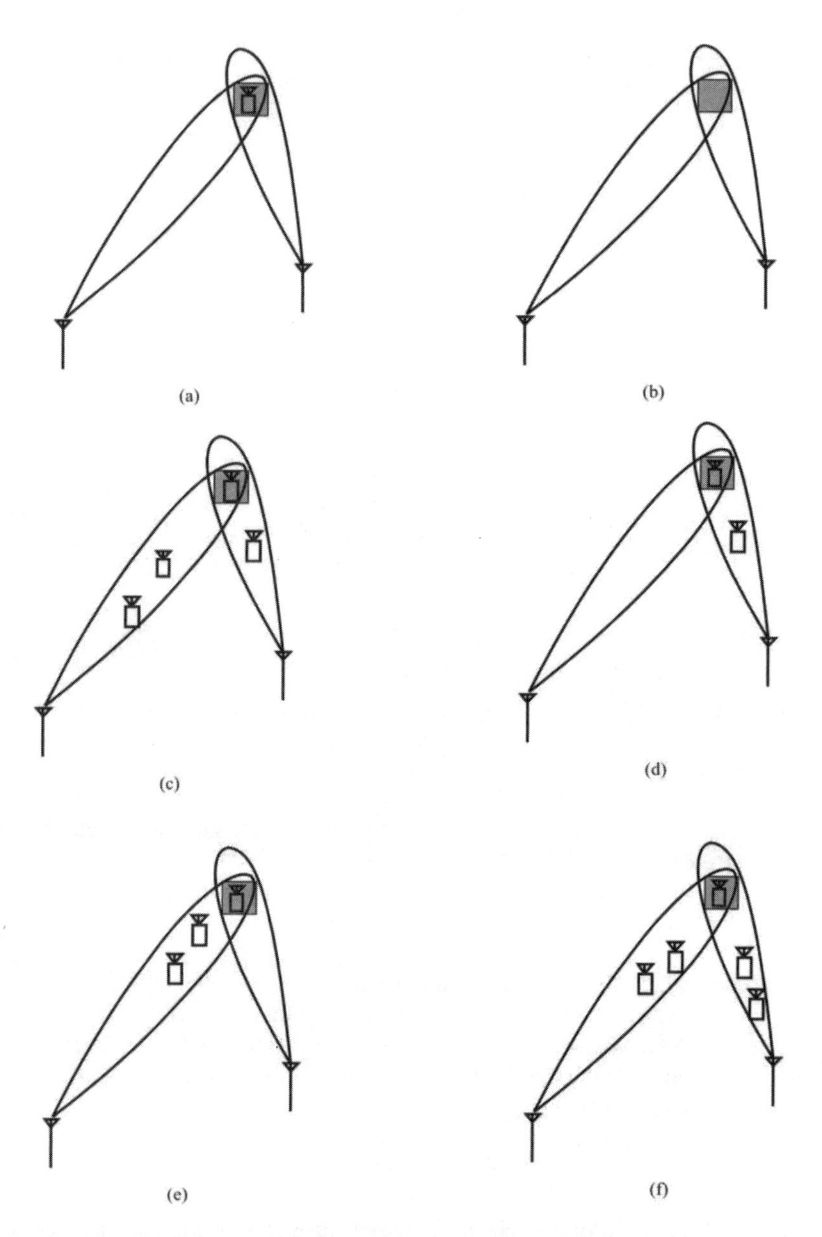

Figure 18.14 Spatial cochannel interference conditions. (a) desired state: target in the COI, no cochannel, correct detection; (b) desired state: no target in COI and no cochannel, correct decision: no target; (c) target in COI, cochannel in both beams, correct detection; (d) target in COI, cochannel in right beam, correct detection decision; (e) target in COI, cochannel in left beam, correct detection decision; and (f) target in COI, cochannel in both beams, correct detection decision.

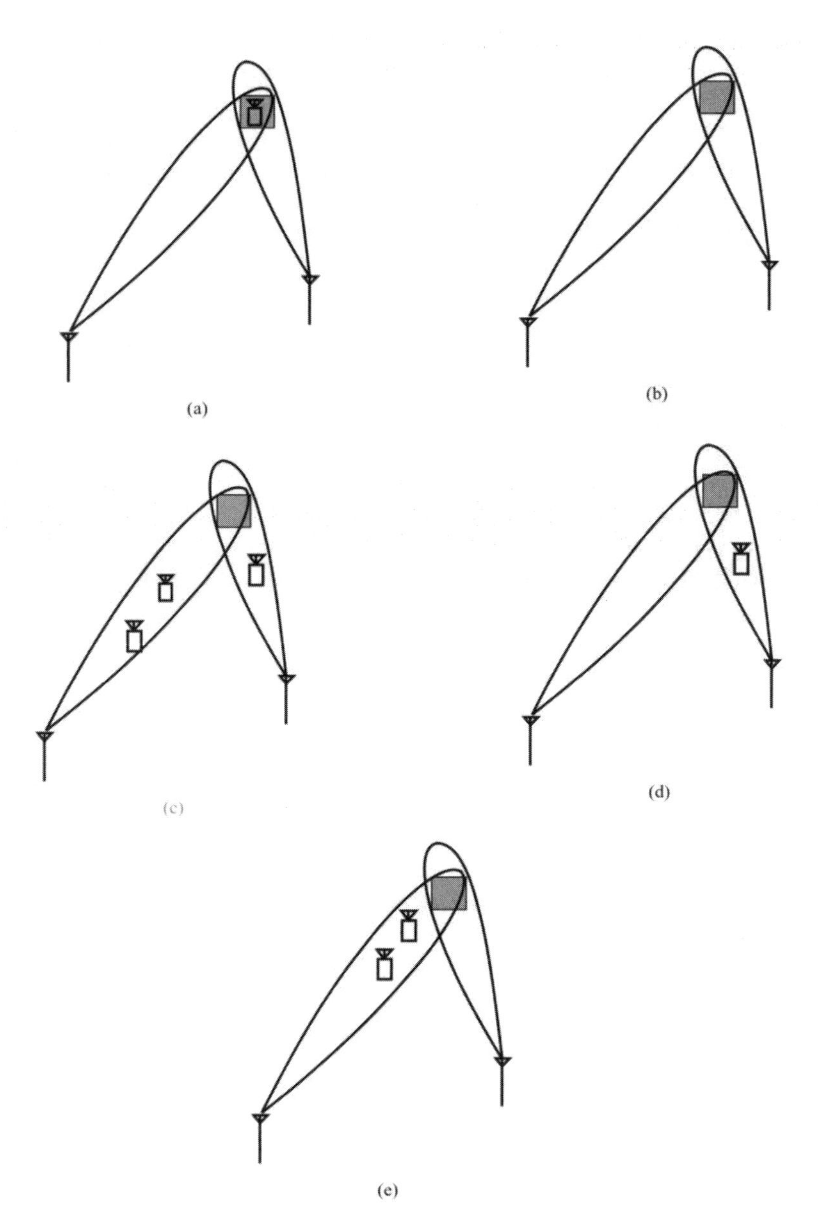

Figure 18.15 Spatial cochannel interference conditions. (a) and (b) same as Figure 18.8 leading to correct decisions; (c) no target in COI, cochannel in both beams, *false alarm*; (d) no target in COI, cochannel in right beam, correct detection decision; and (e) no target in COI, cochannel in left beam. correct detection decision.

only way to detect this problem is to compare the two bit streams from the two intercept sites against each other. There is a very high probability these bit streams will not match, even though the timing of the data collection can be synchronized very tightly. That will indicate that they are looking at two different targets. To accomplish this, however, the detected bit streams must be exchanged or both sent to a common site where the streams could be compared. This would require a datalink of some kind between the intercept sites.

18.6 Blind Identification of CDMA Signals

There have been many blind techniques for detection of DSSS signals proposed. We will discuss one here that has the potential for robust blind detection of CDMA signals, assuming that the signals employ short spreading codes. This technique was developed by Burel and others and is based on the differences between the second order statistics of a signal in noise and that of noise alone.

A procedure was described in Section 4.4.2.5 using the fluctuations of the data for detection of DSSS signals. That procedure is based on the differences between the statistical characteristics of true noise versus pseudo-noise. That procedure can be used here to detect the presence of a DSSS signal in the COI at each of the intercept systems.

18.7 Concluding Remarks

In this chapter we presented a methodology for detecting and geolocating CDMA signals in an urban environment. The idea is to divide the AOI into smaller cells, and search these smaller cells one at a time from two or more intercept sites. Using beamforming technology at the intercept sites, most of the interfering signals not in the COI, but in the channel where the SOI are contained, are eliminated. In most cases, signals can be geolocated successfully with two or more intercept systems. In the few cases where a false detection can occur, we described some ways to ameliorate this problem.

References

[1] Scholtz, R. A., "The Origins of Spread Spectrum Communications," *IEEE Transactions on Communications*, Vol. 30, No. 5, May 1982.

[2] Kim, K. S., et al., "Analysis of Quasi-ML Multiuser Detection of DS/CDMA Systems in Asynchronous Channels," *IEEE Transactions on Communications*, Vol. 47, No. 12, December 1999.

[3] Magill, D. T., F. D. Natali, and G. P. Edwards, "Spread-Spectrum Technology for Commercial Applications," *Proceedings of the IEEE,* Vol. 82, No. 4, April 1994, pp. 572–584.

[4] Jordan, E. C., and K. G. Balmain, *Electromagnetic Waves and Radiating Systems*, Upper Saddle River, NJ: Prentice-Hall, 1968.

[5] Liao, G. S., H. Q. Liu, and J. Li, A Subspace-based Robust Adaptive Capon Beamformer," *Progress in Electromagnetic Research Symposium 2006*, Cambridge, MA, March 2006, pp. 373–375.

[6] Du, L., J. Li, and P. Stoica, "Fully Automatic Computation of Diagonal Loading Levels for Robust Adaptive Beamforming," *IEEE Transactions on Aeraspace and Electronic Systems,* Vol. 46, No. 1, January 2010, pp. 449–458.

[7] Thomas, J. K., L. L. Scharf, and D. W. Tufts, "The Probability of a Subspace Swap in the SVD," *IEEE Transactions on Signal Processing,* Vol. 43, No. 3, March 1995, pp. 730–736.

Appendix A

Q-Function

The Q-function and Marcum's Q-function occur in many places when analyzing signal detection problems in communication theory. This appendix presents a brief discussion of these functions and a method of computing them numerically.

A.1 Q-Function

The *error function* erf(x) is defined as

$$\text{erf}(x) = \frac{2}{\sqrt{\pi}} \int_{-\infty}^{x} e^{-y^2} dy \tag{A.1}$$

while the *complementary error function* is defined as

$$\text{erfc}(x) = \frac{2}{\sqrt{\pi}} \int_{x}^{\infty} e^{-y^2} dy = 1 - \text{erf}(x) \tag{A.2}$$

The function erf(x) represents the area under the zero mean, unit variance Gaussian density from minus infinity to x. On the other hand, erfc(x) represents the area under the "tail" in the zero mean, unit variance Gaussian density, or the area under the curve from x to plus infinity.

Craig derived a simpler form for the erfc function that involves only a finite integral [1]. The advantage of this is that in numerical evaluation of integrals, lower precision is required. Specifically, the form is given by

$$\text{erfc}(x) = \frac{2}{\pi} \int_{0}^{\pi/2} \exp\left[-\frac{x^2}{\sin^2(\phi)} \right] d\phi \tag{A.3}$$

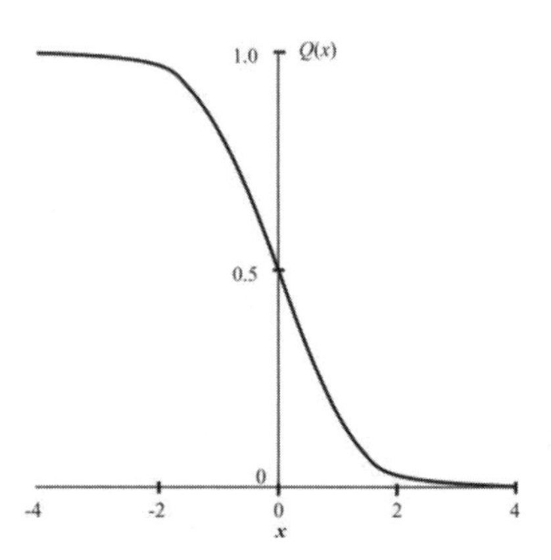

Figure A.1 Q-function

Equation (A.3) is valid for non-negative x, which, of course, is not always the case. If $x < 0$ then we can use

$$\operatorname{erfc}(-x) = 2 - \operatorname{erfc}(x) \tag{A.4}$$

which is valid for nonnegative x.

The Gaussian Q-function is given in terms of the complementary error function as

$$Q(x) = \frac{1}{2}\operatorname{erfc}\left(\frac{x}{\sqrt{2}}\right) \tag{A.5}$$

This function is plotted in Figure A.1. For $x = 0$, the area under both tails of the Gaussian density are equal and therefore are exactly one half. The maximum value of $Q(x)$ is one for large negative values of x, while the minimum is zero for large positive values of x.

Craig also provides us with an alternate expression for the Gaussian Q-function that, like the complementary error function, is given by an integral with finite limits thereby requiring less precision than that provided by (A.5). It is given by

$$Q(x) = \frac{1}{\pi}\int_0^{\pi/2} \exp\left[-\frac{x^2}{2\sin^2\phi}\right] d\phi \tag{A.6}$$

As above, (A.6) is valid for nonnegative x. If x is negative then we can use

$$Q(-x) = 1 - Q(x) \tag{A.7}$$

which is valid for nonnegative x.

A.2 Marcum's *Q*-Function

Marcum's Q-function and the zeroth order Bessel function of the first kind are frequently encountered in the analysis of communication systems as well as in the design and analysis of electronic warfare systems, where, in particular, signal detection problems are analyzed. Marcum's Q-function is an integral of the tail of a probability density function as discussed next. The Bessel functions satisfy the following for a real variable x [2]

$$x^2 \frac{d^2 y}{dx^2} + x\frac{dy}{dx} + (x^2 - n^2)y = 0 \tag{A.8}$$

The function frequently encountered in communication system analysis is where $n = 0$.

Let

$$q(a,x) = xe^{-\left(\frac{x^2+\alpha^2}{2}\right)} I_0(ax) \tag{A.9}$$

where $I_0(y)$ is the zero-order Bessel function of the first kind. Then Marcum's Q-function is defined as

$$Q(a,b) = \int_b^\infty q(a,x)dx \tag{A.10}$$

This function is plotted for several values of a in Figure A.2.

A.2.1 Modified Bessel Function of the First Kind and Zeroth Order

The modified Bessel function of the first kind and zeroth order is given by

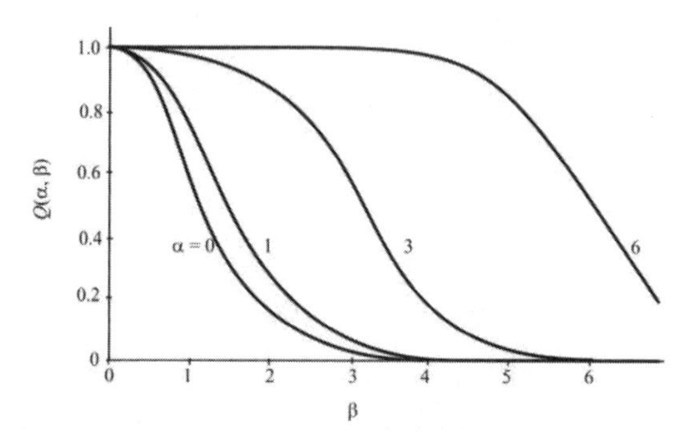

Figure A.2 Marcum Q-function.

$$I_0(x) = \frac{1}{2\pi} \int_{-\pi}^{\pi} e^{x\cos\theta} d\theta \tag{A.11}$$

A plot of this function for small values of x is shown in Figure A.3.

Marcum's Q-function has no closed form solution and so must be determined by numerical means. It is also an improper integral in that one of the limits extends to infinity. In some cases this can cause numerical instability in its computation. Therefore means are sought for more reliable ways to compute it. One such approach was presented by Homier [3]. The expression for this function depends on the relationship between a and b, and is given by:

If $b - a > \varepsilon$,

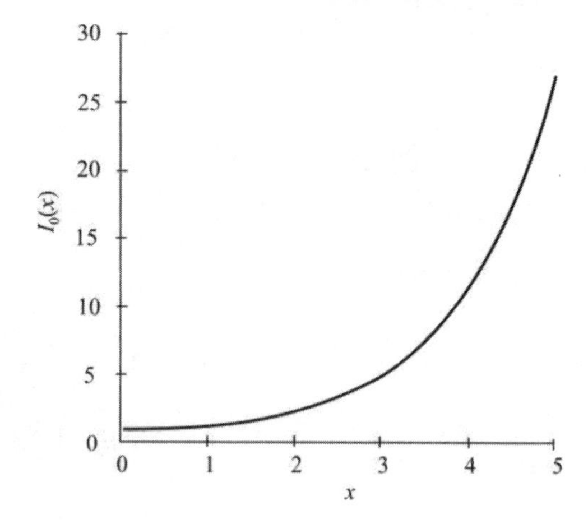

Figure A.3 Modified Bessel function of the first kind and zeroth order.

$$Q(a,b) = \exp\left(-\frac{a^2+b^2}{2}\right)\left[\frac{1-\left(\frac{a}{b}\right)^2}{4\pi}\right] \times \int_0^{2\pi} \frac{\exp(ab\cos\theta)}{1+\left(\frac{a}{b}\right)^2 - 2\left(\frac{a}{b}\right)\cos\theta} d\theta + \frac{1}{2} I_0(ab) \tag{A.12}$$

if $a - b > \varepsilon$

$$Q(a,b) = 1 - \exp\left(-\frac{a^2+b^2}{2}\right)\left[\frac{1-\left(\frac{b}{a}\right)^2}{4\pi}\right] \times \int_0^{2\pi} \frac{\exp(ab\cos\theta)}{1+\left(\frac{b}{a}\right)^2 - 2\left(\frac{b}{a}\right)\cos\theta} d\theta - \frac{1}{2} I_0(ab) \tag{A.13}$$

and if $|a-b| < \varepsilon$

$$Q(a,b) = \frac{1}{2} + \frac{1}{2}\exp\left(-a^2\right) I_0(a^2) \tag{A.14}$$

In these expressions, ε is a parameter that depends on how close a and b are to being equal. When a and b are equal, numerical problems occur, so that is treated as a special case. A typical value for ε is 10^{-3}.

A.3 Generalized *Q*-Function

The generalized Marcum *Q*-function is given by

$$Q_M(a,b) = \int_b^\infty x\left(\frac{x}{a}\right)^{M-1} \exp\left(-\frac{x^2+a^2}{2}\right) I_{M-1}(ax)dx \tag{A.15}$$

where I_{M-1} is the modified Bessel function of the first kind and order $M - 1$. This can be computed as [4]

$$Q_M(a,b) = Q(a,b) + \exp\left(-\frac{a^2 + b^2}{2}\right)\sum_{k=1}^{M-1}\left(\frac{b}{a}\right)^k I_k(ab) \qquad \text{(A.16)}$$

References

[1] Craig, J. W., "A New, Simple and Exact Result for Calculating the Probability of Error for Two-Dimensional Signal Constellations," *Proceedings IEEE MILCOM*, 1991, pp. 25.5.1–25.5.5.

[2] *CRC Standard Mathematical Tables*, 14th ed., The Chemical Rubber Company, 1965, p. 350.

[3] Homier, E., "Numerically Stable Computation of Marcum's Q Function," http://www-scf.usc.edu/~homier/marcum_q.pdf.

[4] Simon, M. K., S. M. Hinedi, and W. C. Lindsey, *Digital Communication Techniques: Signal Design and Detection,* Englewood Cliffs, NJ: Prentice Hall, 1995, Appendix 5A.

Appendix B

Simulated Networks

B.1 Introduction

The specific nets used in the simulations described in the text are shown in this appendix. In those cases where there were only these 12 nets, no other nets were simulated. In those cases when more than 12 nets were simulated, the additional nets were added to the region shown at random, but at tactically significant ranges. All of the numbers in these charts are in kilometers. For readers unfamiliar with the military symbology in the figures, the symbols are defined in Figure B.1.

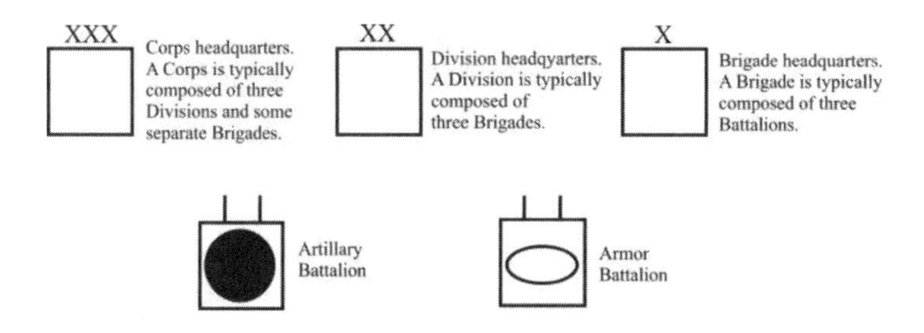

Figure B.1 Symbology used in the network diagrams.

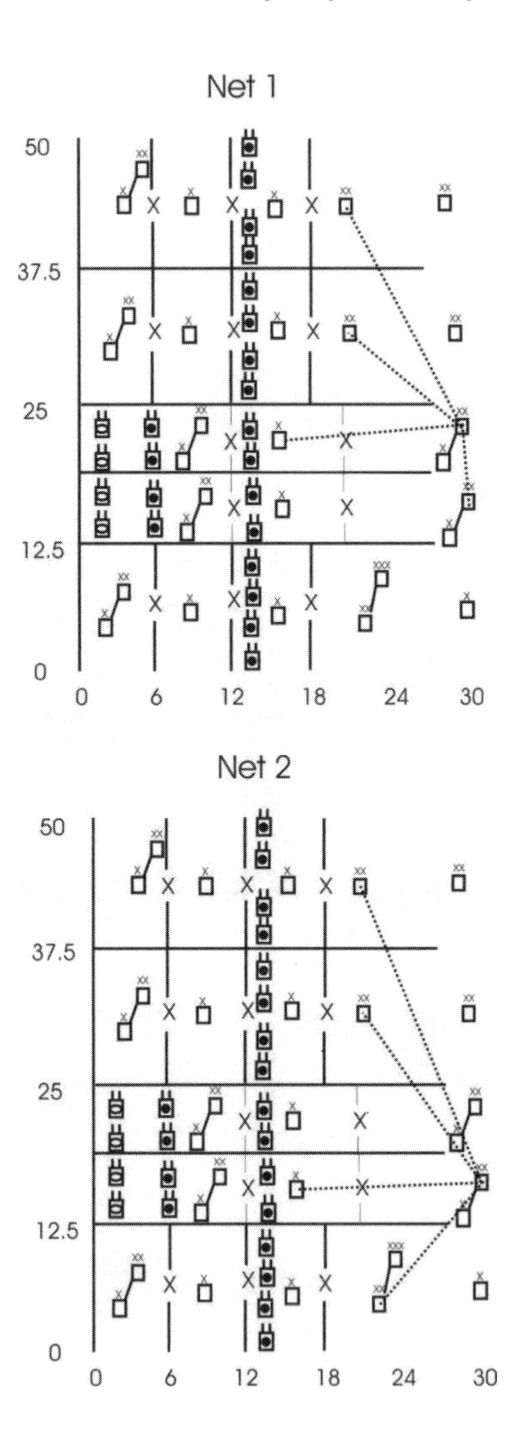

Net 1
50
37.5
25
12.5
0
0
6
12
18
24
30
Net 2
50
37.5
25
12.5
0
0
6
12
18
24
30

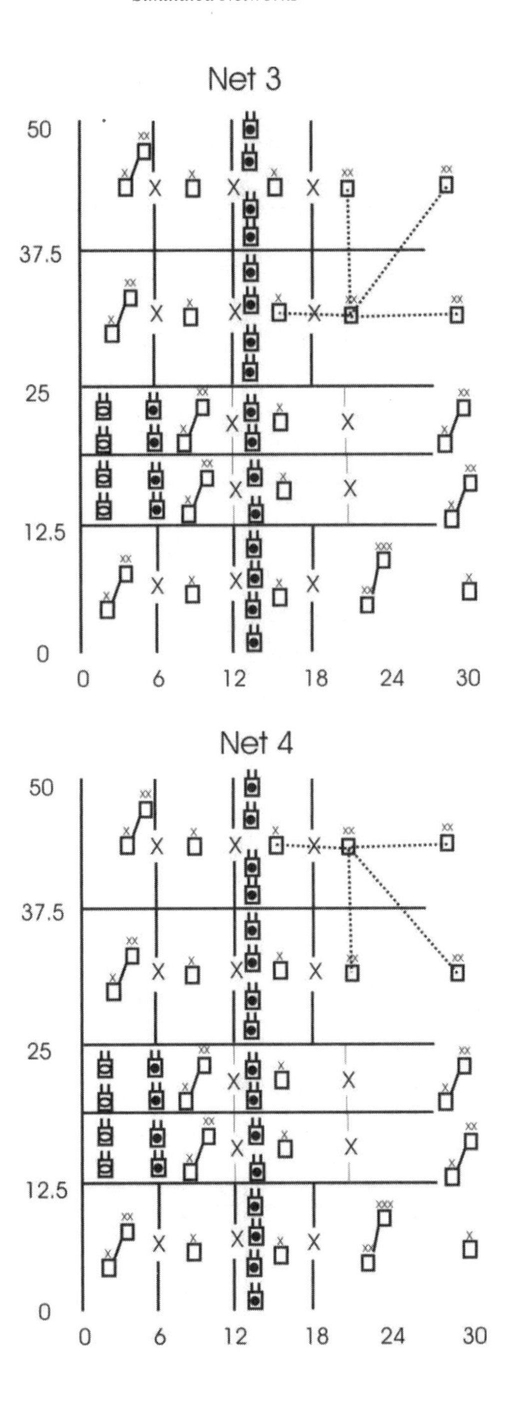
Net 3
50
37.5
25
12.5
0
0
6
12
18
24
30
Net 4
50
37.5
25
12.5
0
0
6
12
18
24
30

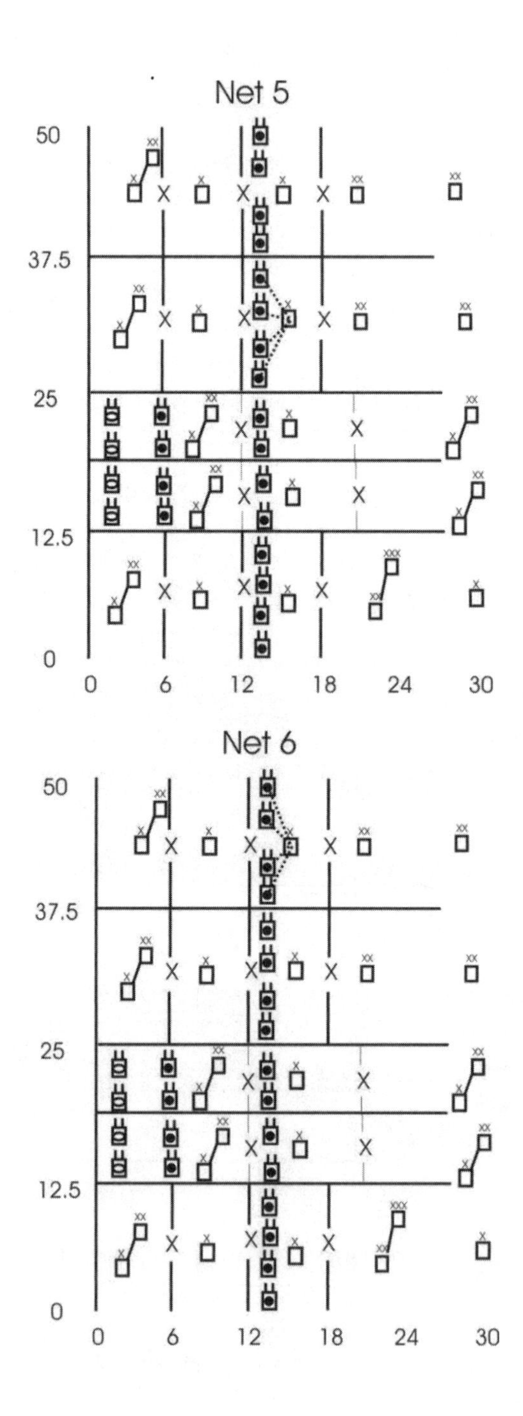
Net 5
50
37.5
25
12.5
0
0
6
12
18
24
30
Net 6
50
37.5
25
12.5
0
0
6
12
18
24
30

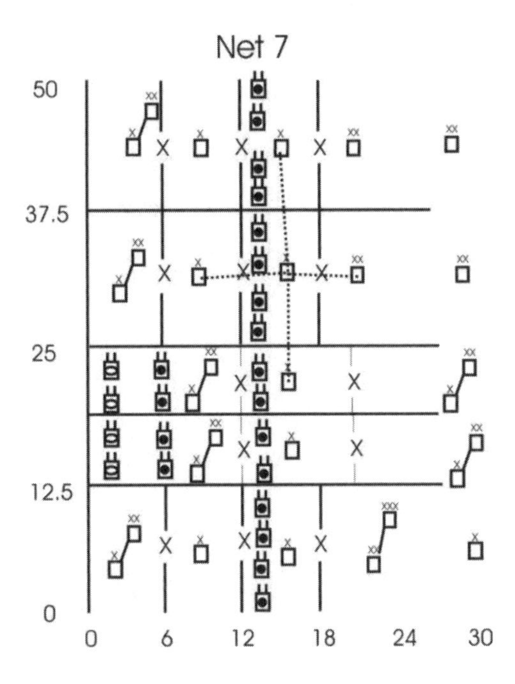
Net 7
50
37.5
25
12.5
0
0
6
12
18
24
30

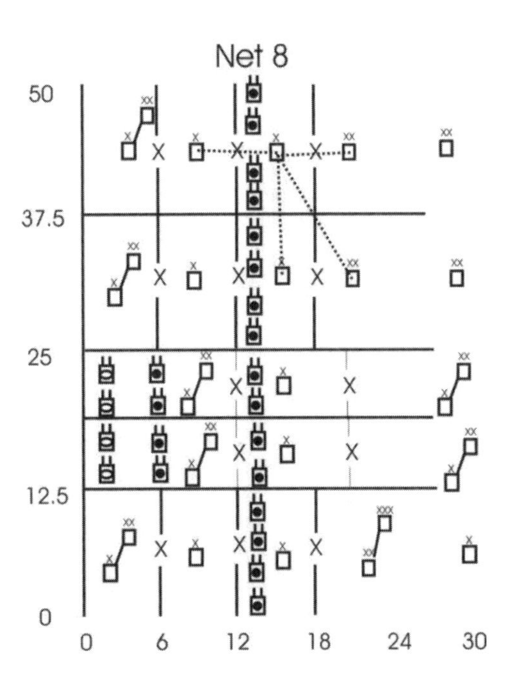
Net 8
50
37.5
25
12.5
0
0
6
12
18
24
30

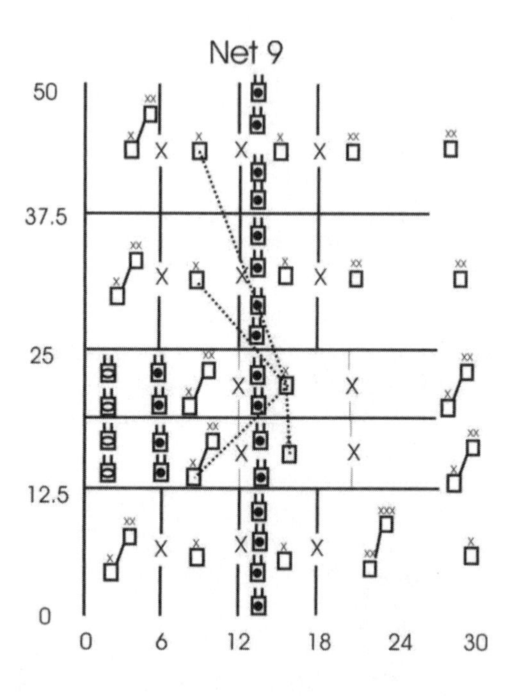
Net 9
50
37.5
25
12.5
0
0
6
12
18
24
30

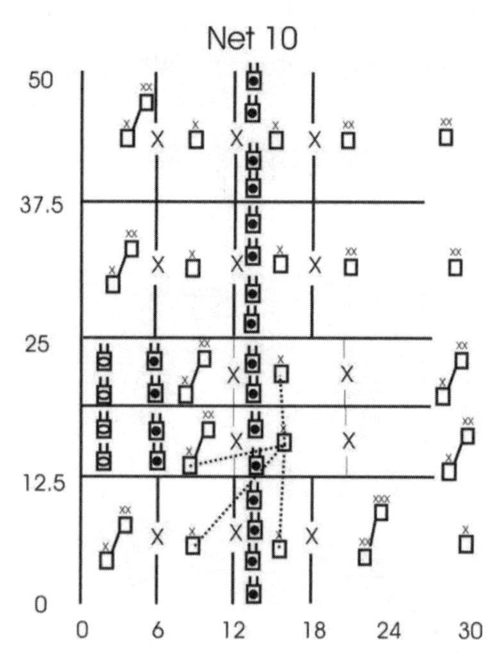
Net 10
50
37.5
25
12.5
0
0
6
12
18
24
30

Net 11
50
37.5
25
12.5
0
0
6
12
18
24
30
Net 12
50
37.5
25
12.5
0
0
6
12
18
24
30

List of Acronyms

AJ	antijam
AM	amplitude modulation
AR	autorregressive
ARQ	automatic repeat request
ASK	amplitude shift key
AWGN	additive white Gaussian noise
BBN	broadband noise
BDPSK	binary differential phase shift key
BER	bit error rate
BFSK	binary frequency shift key
BOOK	binary on-off key
BPAM	binary pulse amplitude modulation
BPSK	binary phase shift key
C2	command and control
CDMA	code division multiple access
CNR	combat net radio
CSI	channel state information
CSK	chaotic shift key
°C	degrees Centigrade
dB	decibel
DCSK	differential chaotic shift key
DE	directed energy
DF	direction finding
DLL	delay lock loop
DNA	Defense Nuclear Agency
DPSK	differential phase shift key
DS	direct sequence
DSSS	direct sequence spread spectrum
EA	electronic attack
ECCM	electronic counter-countermeasures
ECM	electronic countermeasures
EM	electromagnetic
EMCON	emission control
EP	electronic protect
ERP	effective radiated power
ES	electronic support
ESM	electronic support measures
EW	electronic warfare

FBC	filter bank combiner
FCC	Federal Communication Commission
FEC	forward error correction
FFH	fast frequency hopping
FFT	fast Fourier transform
FH	frequency hopping
FHSS	frequency hopped spread spectrum
FLOT	forward line of own troops
FM	frequency modulation
FSK	frequency shift key
G	gain
GHz	gigahertz (10^9 hertz)
GPS	global positioning system
GSM	ground system mobile
JSI	jammer state information
JSR	jam-to-signal ratio
HF	high frequency
IBW	instantaneous bandwidth
IEEE	Institute of Electrical and Electronic Engineers
IF	intermediate frequency
IRE	Institute of Radio Engineers
ISI	intersymbol interference
ISM	instrumentation, scientific, and medical
ITS	Institute for Telecommunications Science
JSR	jam-to-signal ratio
kHz	kilohertz (10^3 hertz)
km	kilometers
kW	kilowatts (10^3 watts)
LAN	local area network
LFSR	linear feedback shift register
LOB	line of bearing
LPD	low probability of detection
LPE	low probability of exploitation
LPI	low probability of intercept
LRS	linear recursive sequence
luf	lowest usable frequency
m	meter
MDS	minimum detectable signal
MFSK	multiple frequency shift key
MHz	megahertz (10^6 hertz)
μsec	microsecond (10^{-6} second)

MPSK	multiple phase shift key
msec	millisecond (10^{-3} second)
MSK	minimum shift key
MT	multitone
muf	maximum usable frequency
NAK	negative acknowledgement
NBN	narrowband noise
NCS	network control station
NTIA	National Telecommunications and Information Administration
nsec	nanosecond (10^{-9} second)
OOK	on-off key
OQPSK	offset quaternary shift key
PAM	pulse amplitude modulation
PB FBC	partial band filter bank combiner
PBN	partial band noise
PCS	personal communication system
pdf	probability density function
PG	processing gain
PN	pseudonoise
PPM	pulse position modulation
PSD	power spectral density
PSK	phase shift key
PSTN	public switched telephone network
PTT	push-to-talk
QPSK	quaternary phase shift key
RF	radio frequency
ROC	receiver operating characteristic
SAW	surface acoustic wave
SER	symbol error rate
SFH	slow frequency hopping
SIR	signal-to-interference ratio
SNR	signal-to-noise ratio
SQPSK	staggered quaternary phase shift key
SS	spread spectrum
TH	time hopping
TID	traveling ionospheric disturbance
UAV	unattended aerial vehicle
UHF	ultra high frequency
UWB	ultrawideband
VCO	voltage controlled oscillator

VHF	very high frequency
VSAT	very small aperture terminal
W	watts
WCDMA	wideband code division multiple access

About the Author

Richard A. Poisel received a B.S. in electrical engineering from the Milwaukee School of Engineering in 1969 and an M.S. in the same discipline from Purdue University in 1971. He spent three years in the military service from 1971 to 1973. After his service he attended the University of Wisconsin, where he received a Ph.D. in electrical and computer engineering in 1977. From 1977 to 2004 he was with the same government organization, which has had several different names and is currently known as the U.S. Army Research, Development, and Engineering Command, Intelligence and Information Warfare Laboratory. During the 1993–1994 academic year, Dr. Poisel attended the MIT Sloan School of Management as a Sloan Fellow, receiving an M.B.A. Initially a research engineer, Dr. Poisel eventually rose to the role of the director of the laboratory on an acting basis from 1997 to 1999. He was appointed chief scientist in 1999 and was relocated to the Army's Intelligence Center at Ft. Huachuca, Arizona, where he served as a technical advisor to the command group. He is currently employed by Raytheon Missile Systems as a senior engineering fellow.

Index

The Artech House Intelligence and Information Operations Series

Concepts, Models, and Tools for Information Fusion, Éloi Bossé, Jean Roy, and Steve Wark

Electronic Intelligence: The Analysis of Radar Signals, Second Edition, Richard G. Wiley

Electronic Warfare for the Digitized Battlefield, Michael R. Frater and Michael Ryan

Electronic Warfare in the Information Age, D. Curtis Schleher

Electronic Warfare Target Location Methods, Second Edition, Richard A. Poisel

EW 101: A First Course in Electronic Warfare, David Adamy

Homeland Security Technology Challenges: From Sensing and Encrypting to Mining and Modeling, Giorgio Franceschetti and Marina Grossi, editors

Information Warfare and Organizational Decision-Making, Alexander Kott, editor

Information Warfare Principles and Operations, Edward Waltz

Introduction to Communication Electronic Warfare Systems, Second Edition, Richard A. Poisel

Knowledge Management in the Intelligence Enterprise, Edward Waltz

Mathematical Techniques in Multisensor Data Fusion, Second Edition, David L. Hall and Sonya A. H. McMullen

Modern Communications Jamming Principles and Techniques, Second Edition, Richard A. Poisel

Modern Communications Receiver Design and Technology, Cornell Drentea

Principles of Data Fusion Automation, Richard T. Antony

Strategem: Deception and Surprise in War, Barton Whaley

Statistical Multisource-Multitarget Information Fusion
Ronald P. S. Mahler

Tactical Communications for the Digitized Battlefield, Michael Ryan and Michael R. Frater

Target Acquisition in Communication Electronic Warfare Systems, Richard A. Poisel